Lecture Notes in Computer Science 709
Edited by G. Goos and J. Hartmanis

Frank Dehne   Jörg-Rüdiger Sack
Nicola Santoro   Sue Whitesides (Eds.)

Algorithms and Data Structures

Third Workshop, WADS '93
Montréal, Canada, August 11-13, 1993
Proceedings

Springer-Verlag
Berlin Heidelberg New York
London Paris Tokyo
Hong Kong Barcelona
Budapest

Volume Editors

Frank Dehne
Jörg-Rüdiger Sack
Nicola Santoro
School of Computer Science, Carleton University
1125 Colonel By Drive, Ottawa, Canada K1S 5B6

Sue Whitesides
School of Computer Science, McGill University
3480 University Street, Montréal PQ, Canada H3A 2A7

CR Subject Classification (1991): F.1-2, E.1, G.2, I.3.5, H.3.3

ISBN 3-540-57155-8 Springer-Verlag Berlin Heidelberg New York
ISBN 0-387-57155-8 Springer-Verlag New York Berlin Heidelberg

Printed in Germany

Typesetting: Camera-ready by author
Printing and binding: Druckhaus Beltz, Hemsbach/Bergstr.
45/3140-543210 - Printed on acid-free paper

PREFACE

The papers in this volume were presented at the Third Workshop on Algorithms and Data Structures (WADS '93). The workshop took place August 11 - 13, 1993, in Montréal, Canada. The workshop alternates with the Scandinavian Workshop on Algorithm Theory (SWAT), continuing the tradition of SWAT '88, WADS '89, SWAT '90, WADS '91, and SWAT '92.

In response to the program committee's call for papers, 165 papers were submitted. From these submissions, the program committee selected 52 for presentation at the workshop. Each paper was evaluated by at least three program committee members, many of whom called upon additional reviewers. In addition to selecting the papers for presentation, the program committee invited the following people to give plenary lectures at the workshop: Mikhail Atallah, Allan Borodin, Richard Cole, Richard Karp, Robert Tarjan, and Andrew Yao.

On behalf of the program committee, we would like to express our appreciation to the six plenary lecturers who accepted our invitation to speak, to all the authors who submitted papers to WADS '93, and to Rosemary Carter of Carleton University for her technical assistance to the program committee. Finally, we would like to express our gratitude to all the people who reviewed papers at the request of program committee members.

August 1993

Frank Dehne
Jörg-Rüdiger Sack
Nicola Santoro
Sue Whitesides

Conference Chair:

S. Whitesides (McGill U.)

Program Committee Chairs:

F. Dehne, J.-R. Sack, and N. Santoro
(Carleton U.)

Program Committee:

M.D. Atkinson (St. Andrew's U.)
H. Attiya (The Technion)
G. Ausiello (U. of Rome)
P. Flajolet (INRIA, Les Chesnay)
Z. Galil (Columbia U.)
S. Hambrusch (Purdue U.)
D. Kirkpatrick (UBC)
M. Klawe (UBC)
R. Kosaraju (Johns Hopkins U.)
J. van Leeuwen (U. of Utrecht)
F. Lombardi (Texas A&M U.)
F. Luccio (U. of Pisa)
J. Matoušek (Charles U.)
I. Munro (U. of Waterloo)
O. Nurmi (U. of Helsinki)
L. Pagli (U. of Pisa)
J. Reif (Duke U.)
R. Seidel (U. of California, Berkeley)
R. Tamassia (Brown U.)
É. Tardos (Cornell U.)
J. Urrutia (U. of Ottawa)
J. Vitter (Duke U.)
D. Wagner (TU Berlin)
S. Whitesides (McGill U.)
P. Widmayer (ETH Zürich)
F. Yao (Xerox PARC)

Invited Speakers:

M. J. Atallah
A. Borodin
R. Cole
R. M. Karp
R. E. Tarjan
A. C. Yao

Sponsored by:

NSERC, Carleton University, and McGill University

ADDITIONAL REFEREES

P. K. Agarwal
E. M. Arkin
D. Avis
A. Bertossi
P. Callahan
R. Canetti
P. Charraresi
J. Chen
Yi.-J. Chiang
R. F. Cohen
L. Devroye
D. Dobkin
H. Edelsbrunner
G. Even
S. Fei
R. Fessler
S. Fogel
G. Frederickson
G. Gambosi
A. Garg
D. Geiger
O. Gerstel
J. Gilbert
M. Goodrich
G. Grahne
R. Grossi
N. V. Hai
X. He
F. Huber
E. Ihler
H. Imai
G. Kant
M. van Kreveld
D. T. Lee
W. Lenhart
S. Leonardi
G. Liotta
E. Lodi
K. Lyons
L. Malmi
L. Margara
R. H. Möhring
C. Montangero
A. Monti
S. Näher
E. Nuutila
P. Orponen
E. Otoo
M.C. Pinotti
K. Pollari-Malmi
R. Ravi
H. Ripphausen-Lipa
K. Romanik
T. Roos
M. Schäffter
P. Scheffler
A. Schuster
Y. Shen
M. Smid
S. Subramanian
S. Suri
S. Tate
I. G. Tollis
J. Vilo
F. Wagner
K. Weihe
M. Wloka

TABLE OF CONTENTS

Computing the All-Pairs Longest Chains in the Plane*

Mikhail J. Atallah
Dept. of Computer Science
Purdue University
West Lafayette, IN 47907.
E-mail: mja@cs.purdue.edu.

Danny Z. Chen
Department of Computer Science and Engineering
University of Notre Dame
Notre Dame, IN 46556.
E-mail: chen@cse.nd.edu.

Abstract

Many problems on sequences and on circular-arc graphs involve the computation of longest chains between points in the plane. Given a set S of n points in the plane, we consider the problem of computing the matrix of longest chain lengths between all pairs of points in S, and the matrix of "parent" pointers that describes the n longest chain trees. We present a simple sequential algorithm for computing these matrices. Our algorithm runs in $O(n^2)$ time, and hence is optimal. We also present a rather involved parallel algorithm that computes these matrices in $O(\log^2 n)$ time using $O(n^2/\log n)$ processors in the CREW PRAM model. These matrices enables us to report, in $O(1)$ time, the length of a longest chain between any two points in S by using one processor, and the actual chain by using k processors, where k is the number of points of S on that chain. The space complexity of the algorithms is $O(n^2)$.

1 Introduction

Problems that involve longest increasing subsequences of a given sequence of numbers have attracted a lot of attention in the past. Probably the most studied version is that of the *longest increasing subsequence* (LIS), for which many $O(n \log n)$ time algorithms are known (e.g., [10, 11, 13], and many others). There is also a well-know connection between increasing subsequences and problems on certain specialized classes of graphs such as permutation graphs, circle and circular-arc graphs, and interval graphs (see, e.g., [12]-[19]). This paper considers the all-pairs version of the problem, whose formulation we state precisely next. We have chosen to formulate it as a problem on

*This research was supported by the Leonardo Fibonacci Institute in Trento, Italy, and by the National Science Foundation under Grant CCR-9202807. Part of this research was done while the first author was visiting LIPN, Paris, France.

points in the plane because our solution techniques are drawn from computational geometry; in terms of sequences, the y coordinate of a planar point corresponds to the value of an entry in the sequence, and the position at which that entry occurs in the sequence is determined by the x coordinate.

A point p is said to *dominate* another point q iff $X(p) \geq X(q)$ and $Y(p) \geq Y(q)$, where $X(p)$ and $Y(p)$ respectively denote the x and y coordinates of point p. Let S be a collection of n points in the plane, and let $\sigma = (p_1, \ldots, p_k)$ be a sequence of points such that each p_i is in S. The sequence σ is *increasing* iff p_i dominates p_{i-1} for all $1 < i \leq k$; such a sequence is called a *chain*, and we say that it *begins* at p_1, that it *ends* at p_k, and that its *length* is k (if the points are weighted then the length of σ is the sum of the weights of its points). The chain σ is *longest* if no other chain starting at p_1 and ending at p_k has greater length than σ.

The problem we consider is that of computing the $n \times n$ matrix D of the lengths of longest chains between pairs of points in S; that is, $D(p,q)$ is equal to the length of a longest p-to-q chain. By convention, for $p \neq q$, if q does not dominate p then $D(p,q) = -\infty$. We also compute an $n \times n$ matrix P (shorthand for "parent") such that $P(p,q)$ is the successor of p in some p-to-q longest chain.

We give a simple $O(n^2)$ time sequential algorithm in the unweighted case. Clearly, knowing P allows one processor to trace a longest p-to-q chain in time proportional to its length.

In parallel, we solve the weighted version of the problem in $O((\log n)^2)$ time using $O(n^2/\log n)$ processors in the CREW PRAM model. We also show that a longest p-to-q chain can be obtained in $O(1)$ time by using k CREW PRAM processors, where k is the number of points of S on that chain. The parallel algorithm bears very little resemblance to the sequential one, which seems hard to "parallelize". It also solves a more general (weighted) version of the problem.

An $O(n^2 \log n)$ time sequential algorithm for this problem is quite trivial to obtain, and to the best of our knowledge this was the best previously known bound for this all-pairs version of the problem. There is a published $O(n^2)$ time algorithm [2] for a special case of this problem: that for chains that start in S and end on a set of points that lie on a vertical line V, where V is to the right of S. In parallel, bounds similar to ours were only known for the special case of the layers of maxima problem, which can be viewed as the version of our problem where the chains of interest begin in S but must end at the point $(+\infty, +\infty)$ [1]. It is actually quite easy to use the methods of [1, 3] to solve the version of the problem where the chains of interest begin in S but must end on a set of points on a vertical line V that is to the right of S.

We now briefly discuss how our approach differs from the one for the above-mentioned special version of the problem, in which all chains start in S and end on a set of points on a vertical line V that is to the right of S. That special version of the problem is substantially easier, both sequentially and in parallel, because for a fixed $p \in S$, the collection of longest chains that begin at p and end on V have the following monotonicity property: Two such longest chains that end at (respectively) q' and q'', $Y(q') < Y(q'')$, can always be chosen such that nowhere is the chain to q'

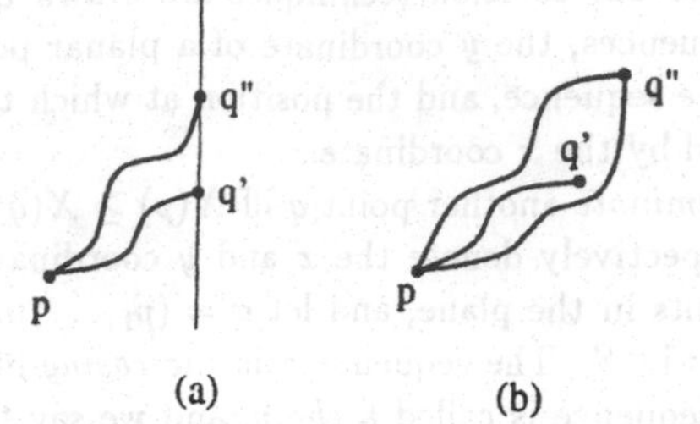

Figure 1: Illustrating (a) how monotonicity holds for some chains, and (b) how it fails to hold for others.

higher (geometrically) than the chain to q'' (intuitively, if it is higher then there is a crossing between the two chains and we can "uncross" them). Figure 1(a) illustrates this. Such a monotonicity property is lacking in the general version of the problem considered here: If q' and q'' do not lie on the same vertical line (see Figure 1(b)) then monotonicity need not hold, in the sense that either one of the two p-to-q'' chains shown could be a unique longest chain to q'', so that such a chain to q'' might go either "above" or "below" a longest p-to-q' chain.

We are unable to obtain an $O(n^2)$ time sequential solution to the weighted version of this problem, and we leave this as an open problem. Our parallel bounds, on the other hand, hold for the weighted version of the problem.

The rest of the paper is organized as follows. Section 2 deals with the sequential algorithm, which is fairly simple. Section 3 gives the parallel algorithm. We have chosen to give the basic terminology and definitions separately for each of the parallel and sequential algorithms, since they have little in common (this way the reader interested in one of the two will not be forced to read material unrelated to her interest). Section 4 concludes by posing some open problems.

2 The Sequential Algorithm

This section gives the $O(n^2)$ time sequential algorithm.

2.1 Preliminaries

The input consists of set S of n points in the plane. For a point $p \in S$, we use $DOM(p)$ to denote the subset of points in S that are dominated by p. A point p of S is a *maximum* in S iff no other point of S dominates p. We use $MAX(S)$ to denote

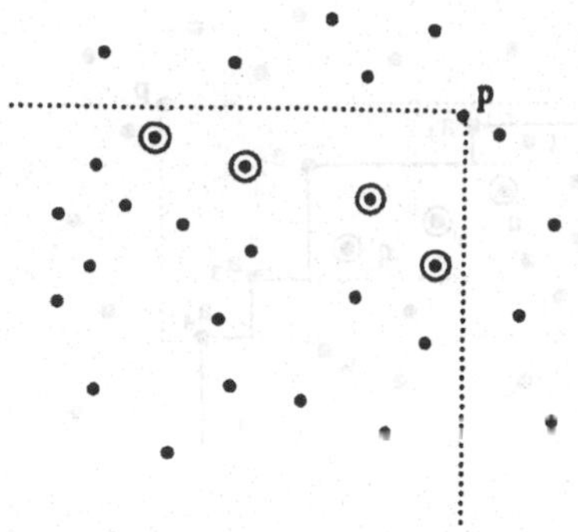

Figure 2: The points of $MD(p)$ are circled.

the set of maxima of S, listed by increasing x coordinates (and hence by decreasing y coordinates). We abbreviate $MAX(DOM(p))$ as $MD(p)$. Figure 2 illustrates the definition of $MD(p)$.

For a point $p \in S$, imagine partitioning $DOM(p) \cup \{p\}$ into k subsets, where $k = \max\{D(q,p) | q \in DOM(p) \cup \{p\}\}$, such that the points in each subset all have longest chains to p of the same length. The subset of $DOM(p) \cup \{p\}$ whose points have a distance j to p is called the j-th *domination layer* of p, denoted by $Layer_j(p)$. For example, $Layer_1(p) = \{p\}$, $Layer_2(p) = MD(p)$, and so on. In general, for each j, $Layer_j(p) = MAX(DOM(p) \cup \{p\} - \cup_{i<j} Layer_i(p))$. We assume that each layer of p is sorted by increasing x-coordinates (hence by decreasing y-coordinates).

It should be clear that, if we were able to compute the domination layers of each $p \in S$, then we would effectively have computed the desired D matrix. Our sequential algorithm will therefore mainly concern itself with the computation of these domination layers and of the P matrix. (The parallel algorithm deals with the weighted version and will use a different approach — in fact most of the definitions given above will not be used in the parallel algorithm.)

2.2 The Algorithm

Below is a high-level description of the sequential algorithm. We are assuming that no two points in S have the same x (resp., y) coordinate, i.e., that if $p, q \in S$ and $p \neq q$ then $X(p) \neq X(q)$ and $Y(p) \neq Y(q)$ (the algorithm can easily be modified for the general case). By convention, walking *forward* (resp., *backward*) along an $MD(p)$ means moving along it by increasing (resp., decreasing) x-coordinates.

Step 1. We first compute $MD(p)$ for every $p \in S$. These $MD(p)$'s can all be easily computed in $O(n^2)$ time as follows. We sort the points by their x coordinates, and then for each $p \in S$ we do the following. From the sorted list we obtain $DOM(p)$, in $O(n)$ time. Then we obtain the maximal elements of $DOM(p)$, also in $O(n)$ time (this

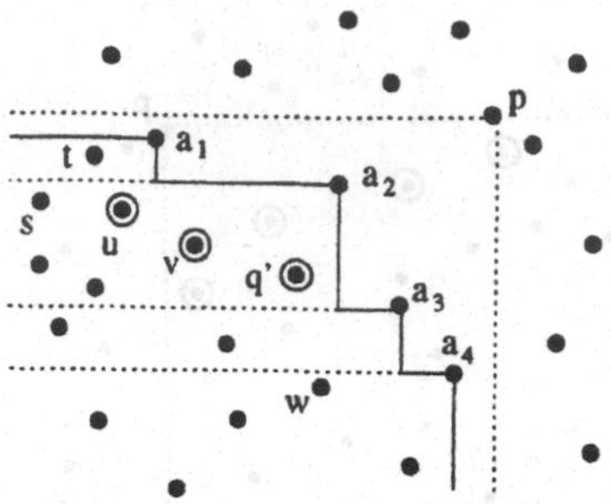

Figure 3: The points of $Interval_{j+1}(p, a_2)$ are shown circled.

is possible since $DOM(p)$ is available sorted). These maximal elements of $DOM(p)$ are, by definition, $MD(p)$.

Step 2. We compute, for each pair $p, q \in S$, the position of $Y(p)$ in the list $Y(MD(q))$, which is the list obtained from $MD(q)$ by replacing every point by its y-coordinate. This is easy to do in $O(n)$ time for a particular $Y(MD(q))$ and all $p \in S$, by merging $Y(MD(q))$ with the sorted list of the y coordinates of the points in S. Doing this once for each $q \in S$ takes a total of $O(n^2)$ time.

Step 3. For each $p \in S$, we obtain the domination layers of p and the column that corresponds to p in the P matrix. We do this in $O(n)$ time for each p, as follows. Clearly, we already have $Layer_1(p)$ ($= \{p\}$) and $Layer_2(p)$ ($= MD(p)$). We obtain $Layer_{j+1}(p)$ from $Layer_j(p)$ in $O(|Layer_j(p)| + |Layer_{j+1}(p)|)$ time, as follows. Let $Layer_j(p) = (a_1, a_2, \ldots, a_k)$, where $X(a_1) < X(a_2) < \cdots < X(a_k)$. We shall walk along the $Layer_j(p)$ list, creating the $Layer_{j+1}(p)$ list as we go along, in left to right order. When we reach a_i while scanning $Layer_j(p)$, we compute the portion of $Layer_{j+1}(p)$ that is in $MD(a_i)$ but not in $DOM(a_{i+1})$; we call this portion $Interval_{j+1}(p, a_i)$ (it forms a contiguous interval of $MD(a_i)$). Figure 3 illustrates the definition of $Interval_{j+1}(p, a_i)$. Note how, in that figure, point w is in $Layer_{j+1}(p) \cap MD(a_2)$ but not in $Interval_{j+1}(p, a_2)$. We shall compute $Interval_{j+1}(p, a_1)$, $Interval_{j+1}(p, a_2)$, $\ldots$, $Interval_{j+1}(p, a_k)$, in that order. While doing this, we maintain a variable called *cutoff* whose significance is that, when we finish processing a_i, *cutoff* contains the rightmost point in $\cup_{1 \le \ell \le i} Interval_{j+1}(p, a_\ell)$; intuitively, *cutoff* is the "dominant" point among those in $\cup_{1 \le \ell \le i} Interval_{j+1}(p, a_\ell)$ as far as the (yet to be computed) lists $Interval_{j+1}(p, a_{i+1}), \ldots, Interval_{j+1}(p, a_k)$ are concerned. In Figure 3, after $Interval_{j+1}(p, a_1)$ is computed, *cutoff* is point t, and after $Interval_{j+1}(p, a_2)$ is computed *cutoff* is point q'.

To determine $Interval_{j+1}(p, a_1)$, we simply start at the beginning of $MD(a_1)$ and walk forward along $MD(a_1)$ until we first reach a point $q \in MD(a_1)$ for which

$Y(q) < Y(a_2)$ (we do not count q as being part of our "walk" along $MD(a_1)$). The (possibly empty) portion of $MD(a_1)$ so traced is obviously equal to $Interval_{j+1}(p, a_1)$. If $Interval_{j+1}(p, a_1)$ is not empty then we set *cutoff* equal to the predecessor of q in $MD(a_1)$, otherwise it remains undefined. For the example shown in Figure 3, $q = u$ and $cutoff = t$. We then proceed to process a_2.

If *cutoff* is undefined (i.e., if $Interval_{j+1}(p, a_1)$ turned out to be empty) then we process a_2 exactly as explained above for a_1. Otherwise we process it as follows. Recall that we already know, from Step 2, the outcome of a hypothetical binary search for $Y(a_3)$ in $Y(MD(a_2))$: Let q' be the predecessor of $Y(a_3)$ in $Y(MD(a_2))$, that is, the lowest point of $MD(a_2)$ whose y-coordinate is larger than $Y(a_3)$. If no such point q' exists on $MD(a_2)$ then surely $Interval_{j+1}(p, a_2)$ is empty and we move on to processing a_3 (leaving *cutoff* unchanged). So suppose that q' exists. If $X(q') < X(cutoff)$ then $Interval_{j+1}(p, a_2)$ is empty and we move on to processing a_3 (leaving *cutoff* unchanged). If $X(q') > X(cutoff)$ then we start at q' and walk backward along $MD(a_2)$ until we reach a point whose x-coordinate is less than $X(cutoff)$; the portion of $MD(a_2)$ so traced is $Interval_{j+1}(p, a_2)$. In Figure 3, the portion so traced is (in that order) q', v, u (point s is not traced because $X(s) < X(t)$). In that case we also update *cutoff* by setting it equal to q' before we proceed to process a_3.

We process $a_3, a_4, \ldots, a_k$ in that order, exactly as explained above except that, when processing a_{i+1}, a_i plays the role of a_1, a_{i+1} plays the role of a_2, and a_{i+2} plays the role of a_3.

Once we have obtained $Layer_{j+1}(p)$ from $Layer_j(p)$, we must compute $P(w, p)$ for every $w \in Layer_{j+1}(p)$ (clearly, such a $P(w, p)$ is in $Layer_j(p)$). This is easily done for all $w \in Layer_{j+1}(p)$ in $O(|Layer_j(p)| + |Layer_{j+1}(p)|)$ time, by merging the two lists $Layer_{j+1}(p)$ and $Layer_j(p)$.

This completes the description of the sequential algorithm. We now turn our attention to the parallel algorithm.

3 The Parallel Algorithm

This section gives the $O((\log n)^2)$ time, $O(n^2/\log n)$ processor algorithm for the weighted version of the problem.

3.1 Preliminaries

The parallel model used is the CREW PRAM, which is the synchronous shared-memory model where concurrent reads are allowed, but no two processors can simultaneously attempt to write in the same memory location (even when they are trying to write the same thing). In what follows, we shall focus on showing that the claimed time complexity can be achieved with an $O(n^2 \log n)$ amount of *work* (= number of operations). This will imply the $O(n^2/\log n)$ processor bound, by Brent's theorem [8]:

Theorem 1 (Brent) *Any synchronous parallel algorithm taking time* T *that consists of a total of* W *operations can be simulated by* P *processors in time* $O((\mathsf{W}/\mathsf{P})+\mathsf{T})$.

Remark: There are actually two qualifications to the above Brent's theorem before one can apply it to a PRAM: (i) at the beginning of the i-th parallel step, we must be able to compute the amount of work W_i done by that step, in time $O(\mathsf{W}_i/\mathsf{P})$ and with P processors, and (ii) we must know how to assign each processor to its task. Both qualifications (i) and (ii) to the theorem will be easily satisfied in our algorithms, therefore the main difficulty will be how to achieve W operations in time T.

Here as in [4], an important method we use involves multiplying special kinds of matrices. Although the situation depicted in Figure 1(b) implies that the structure that gives rise to such matrices is not always available, the fact that we can deal with the situation in Figure 1(a) will be useful. (This will all be made precise later; for now we are only giving an overview.) All matrix multiplications in the algorithm are in the $(\max, +)$ closed semi-ring, i.e., $(M' * M'')(i,j) = \max_k \{M'(i,k) + M''(k,j)\}$. A matrix M is said to be *Monge* [1] iff for any two successive rows $i, i+1$ and columns $j, j+1$, we have $M(i,j) + M(i+1,j+1) \leq M(i,j+1) + M(i+1,j)$. For two point sets A and B in the plane, let matrix M_{AB} contain the lengths of the longest chains that start in A and end in B (by convention, these chains are allowed to go through any points of S on their way from A to B). Now, consider two point sets X and Y, each totally ordered in some way (so we can talk about the predecessor and successor of a point in X or in Y), and such that the rows (resp., columns) of the matrix M_{XY} are as in the ordering for X (resp., Y). Matrix M_{XY} is *Monge* iff for any two successive points p, p' in X and two successive points q, q' in Y, we have $M_{XY}(p,q) + M_{XY}(p',q') \leq M_{XY}(p,q') + M_{XY}(p',q)$. The next lemma characterizes the Monge matrices of chain lengths used in the algorithm.

Lemma 1 *Let* V' *(resp.,* V''*) be a vertical line that contains a set* X *(resp.,* Y*) of points ordered by increasing (resp., decreasing) y-coordinates along* V' *(resp.,* V''*). (Assume that* V' *is to the left of* V''*.) Then the matrix* M_{XY} *of chain lengths between* X *and* Y *is Monge.*

Proof. Obvious. □

The following well-known lemma [3, 1] will be used.

Lemma 2 *Assume that matrices* M_{XY} *and* M_{YZ} *are Monge, with* $|X| = c_1|Y| \leq c_2|Z|$ *for some positive constants* c_1 *and* c_2. *Then* $M_{XY} * M_{YZ}$ *can be computed in* $O(\log|Y|)$ *time and* $O(|X||Z|)$ *work in the CREW PRAM model.*

Remark: Since $*$ is a $(\max, +)$ matrix multiplication, $M_{XY} * M_{YZ}$ need not be Monge.

Lemmas 1 and 2 imply the following.

Lemma 3 *Let* V *(resp.,* V', V''*) be a vertical line that contains a set* X *(resp.,* Y, Z*) of points ordered by increasing y-coordinates along* V *(resp.,* V', V''*). Assume that* $X(V) < X(V') < X(V'')$, *and that* $|X| = c_1|Y| \leq c_2|Z|$ *for some positive constants*

c_1 and c_2. Suppose that, for every increasing chain C from $p \in X$ to $q \in Z$, there is a p-to-q chain C' that is at least as long as C and goes through some $w \in Y$. Then given the matrices M_{XY} and M_{YZ}, the matrix M_{XZ} can be computed in $O(\log |Y|)$ time and $O(|X||Z|)$ work in the CREW PRAM model.

Proof. Let X' (resp., Y', Z') be the same as X (resp., Y, Z) but sorted by decreasing y coordinates. By Lemma 1, $M_{XY'}$ and $M_{Y'Z}$ are both Monge. By Lemma 2, the matrix $M_{XY'} * M_{Y'Z}$ can be computed in $O(\log |Y|)$ time and $O(|X||Z|)$ work. Now, since by hypothesis all the X-to-Z chains can be modified to go through Y without any decrease in their lengths, it follows that the matrix $M_{XY'} * M_{Y'Z}$ is the desired matrix M_{XZ}. □

3.2 The Algorithm for Chain Lengths

The algorithm given in this subsection concerns itself with the computation of chain lengths only, not of the P matrix that describes the n longest chain trees. Including the computation of P here would have cluttered the exposition. The next subsection will deal with the computation of P. In addition, it is not immediately clear that the availability of P makes possible the reporting of a k-point chain in $O(k)$ work and constant time. This too is postponed until the next subsection.

Let $S = \{p_1, \ldots, p_n\}$ where $X(p_1) < \cdots < X(p_n)$. There is a weight associated with each p_i. Let $V_0, V_1, \ldots, V_n$ be vertical lines such that $X(V_0) < X(p_1)$, $X(p_n) < X(V_n)$, and $X(p_i) < X(V_i) < X(p_{i+1})$ for all $i \in \{1, \ldots, n-1\}$.

Let T be a complete n-leaf binary tree. For each leaf v of T, if v is the i-th leftmost leaf in T, then associate with v the region I_v of the plane that is between V_{i-1} and V_i. For each internal node v of T, associate with v the region I_v consisting of the union of the regions of its children. That is, if v has children u and w, then $I_v = I_u \cup I_w$.

Let v be any node of T. Suppose that the left (resp., right) boundary of I_v is V_i (resp., V_j), and let $S_v = S \cap I_v$, that is, S_v is the subset of the input points that lie in I_v. Observe that if v is at a height of h in T then $j - i = 2^h = |S_v|$ (the height of v is the height of its subtree in T, with leaves being at a height of zero). Let L_v (resp., R_v) be the set of points on V_i (resp., V_j) that are the horizontal projections of S_v on V_i (resp., V_j). The points of L_v and R_v are, of course, disjoint from the input set S, and we assign to each of them a weight of zero. Observe that

$$\sum_{v \in T} |L_v| = \sum_{v \in T} |R_v| = O(n \log n),$$

because for each level of T a $p_i \in S$ appears in exactly one S_v of that level, and hence creates at most two extra points, one in L_v and one in R_v (recall that a level of T is the set of nodes in T that have same distance to the root, so that the root is at level zero, its two children at level 1, etc).

There are two phases for the algorithm: Phase 1 is relatively straightforward, while Phase 2 is the key that made our solution possible.

3.2.1 Phase 1

This phase consists of computing, starting at the leaves and going upward in T, one level at a time, the $M_{L_vR_v}$ matrices, which contain the lengths of all the L_v-to-R_v longest chains (chains that begin on L_v and end on R_v, of course possibly going through points in S_v along the way). This information is trivially available if v is a leaf. So suppose that we have already computed this information for level $\ell+1$, and we want to compute it for level ℓ.

We claim that it suffices to show that the $M_{L_vR_v}$ matrix can be computed in $O(|S_v|^2)$ work and $O(\log n)$ time for each node v at level ℓ. This claim would imply an $O((\log n)^2)$ time, $O(n^2)$ work bound for Phase 1, as follows. That the time bound follows from the claim is obvious (we would spend a logarithmic amount of time per level, and there is a logarithmic number of levels). The work bound would follow from the fact that there are 2^ℓ nodes v at level ℓ, each having $|S_v| = n/2^\ell$, and hence the total work at level ℓ would be

$$O(2^\ell (n/2^\ell)^2) = O(n^2/2^\ell).$$

Summing over all levels ℓ gives $O(n^2)$ total work. We next prove the claim by showing that the $M_{L_vR_v}$ matrix can indeed be computed in $O(|S_v|^2)$ work and logarithmic time.

Let u (resp., w) be the left (resp., right) child of v in T. Let Y denote $R_u \cup L_w$, that is, Y consists of the horizontal projections of the points of S_v on the vertical line V_j that separates the region I_u from the region I_w. Since $M_{L_uR_u}$ is already available at u, we can easily obtain from it M_{L_vY}. Similarly, we obtain M_{YR_v} from $M_{L_wR_w}$. Now, simply observe that the conditions for Lemma 3 are satisfied, with L_v playing the role of X and R_v the role of Z. That is, we can obtain $M_{L_vR_v}$ from M_{L_vY} and M_{YR_v} in $O(|S_v|^2)$ work and logarithmic time. This completes the proof.

Remark: The astute reader may have observed that the above procedure can be modified so as to compute the L_v-to-S_v and S_v-to-R_v chain lengths information. This would involve only a logarithmic factor of additional work, and would exploit the kind of monotonicity depicted in Figure 1(a) by using the lower-dimensional parallel matrix searching algorithm of [5]. However, this would still leave us far from having solved our problem: We would still need something like Phase 2 below, since we cannot afford to multiply "non-square" Monge matrices — as of now, it is not known how to optimally (max, +)-multiply two non-square Monge matrices (for example, the best parallel algorithm for multiplying a $1 \times k$ Monge matrix with a $k \times k$ one in logarithmic time takes $O(k \log k)$ work [5]). Observe how Phase 2 below will satisfy the size requirements of Lemma 3, as expressed in the requirement that $|X| = c_1|Y| \le c_2|Z|$ for some positive constants c_1 and c_2.

3.2.2 Phase 2

Whereas Phase 1 involved a bottom-up computation in T, Phase 2 will involve a top-down computation, starting at the root and proceeding one level at a time until we reach the leaves. The purpose of the computation at a typical level ℓ is more

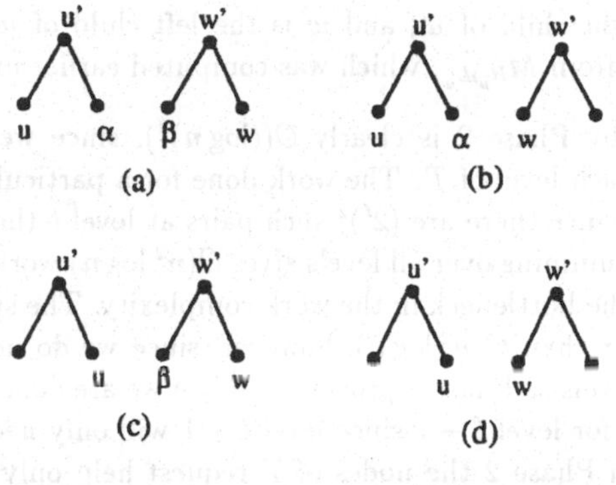

Figure 4: Illustrating the four cases of Phase 2.

ambitious than in Phase 1: We now seek, for every pair of nodes u, w at level ℓ such that u is to the left of w, the computation of the $M_{R_uL_w}$ chain lengths matrix (u is to the left of w iff it is in the subtree of the left child of the lowest common ancestor of u and w). The key idea is to get help from the parents of u and w, which we call u' and (respectively) w'. If $u' = w'$ then the desired information is trivially available, so suppose that $u' \neq w'$. We distinguish four cases, which are illustrated in Figure 4.

Case 1: u is the left child of u', and w is the right child of w'. Let α be the right child of u', β be the left child of w' (see Figure 4(a)). Since Phase 1 computed $M_{L_\alpha R_\alpha}$, we can obtain from it $M_{R_uR_\alpha}$, then $M_{R_uR_{u'}}$. Similarly, we obtain $M_{L_{w'}L_w}$ from $M_{L_\beta R_\beta}$ which was computed in Phase 1. Now, $M_{R_{u'}L_{w'}}$ is already available because Phase 2 is already done with processing the pair u', w' (recall that Phase 2 processes the levels from the root down). We use Lemma 3 to obtain the matrix $M_{R_uL_{w'}}$ from $M_{R_uR_{u'}}$ and $M_{R_{u'}L_{w'}}$, with R_u playing the role of X, $R_{u'}$ playing the role of Y, and $L_{w'}$ playing the role of Z. Finally, we use Lemma 3 again, this time to obtain the desired matrix $M_{R_uL_w}$ from $M_{R_uL_{w'}}$ and $M_{L_{w'}L_w}$.

Case 2: u is the left child of u', and w is the left child of w'. Let α be the right child of u' (see Figure 4(b)). From the $M_{L_\alpha R_\alpha}$ matrix which was computed in Phase 1, we obtain the $M_{R_uR_{u'}}$ matrix. Observe that $M_{R_{u'}L_{w'}}$ was already obtained earlier in Phase 2: get from it the $M_{R_{u'}L_w}$ matrix. We use Lemma 3 to obtain $M_{R_uL_w}$ from the matrices $M_{R_uR_{u'}}$ and $M_{R_{u'}L_w}$.

Case 3: u is the right child of u', and w is the right child of w'. Let β be the left child of w' (see Figure 4(c)). From the $M_{R_{u'}L_{w'}}$ matrix which was computed earlier in Phase 2, obtain the $M_{R_uL_\beta}$ matrix. From the $M_{L_\beta R_\beta}$ matrix which was computed in Phase 1, we obtain the $M_{L_\beta L_w}$ matrix. We use Lemma 3 to obtain $M_{R_uL_w}$ from $M_{R_uL_\beta}$ and $M_{L_\beta L_w}$.

Case 4: u is the right child of u', and w is the left child of w' (see Figure 4(d)). Obtain $M_{R_u L_w}$ from $M_{R_{u'} L_{w'}}$ which was computed earlier in Phase 2.

The time taken by Phase 2 is clearly $O((\log n)^2)$, since we take a logarithmic amount of time at leach level of T. The work done for a particular pair u, w at level ℓ is $O((n/2^\ell)^2)$, and since there are $(2^\ell)^2$ such pairs at level ℓ the total work done at that level is $O(n^2)$. Summing over all levels gives $O(n^2 \log n)$ work for Phase 2. Hence it is Phase 2 that is the bottleneck in the work complexity. The space taken by Phase 2 is still $O(n^2)$ rather than $O(n^2 \log n)$, however, since we do not need to store the matrices for all the levels as Phase 2 proceeds: When we are done with level ℓ, we can discard the matrices for level $\ell - 1$ since level $\ell + 1$ will only need information from level ℓ (recall that in Phase 2 the nodes of T request help only from their parents, not from their grandparents or from higher up in the tree T).

3.3 Computing the Actual Chains

In this subsection, we discuss how to obtain the matrix P which contains the n trees of longest chains, and how to pre-process the longest chain trees, so that each tree can support a longest chain query between any point in S and the point of S at the root of that tree.

First we sketch how the algorithm in the previous subsection can be modified so as to compute the P matrix as well. For each $M_{R_u L_w}$ matrix computed by that algorithm, we compute a companion $P_{R_u L_w}$ matrix whose significance is that, for $p \in R_u$ and $q \in L_w$, $P_{R_u L_w}(p, q)$ is the first point of S that lies on a longest p-to-q chain (it is undefined if no such point of S exists). Note that only points of S can be "parents". It is quite easy to modify the computation of an M_{XZ} so that it also produces P_{XZ}: If M_{XZ} is obtained by using Lemma 3, then P_{XZ} can be obtained from P_{XY} or P_{YZ} as a "byproduct" of this computation. For example, if q dominates p and if $M_{XZ}(p, q) = M_{XY}(p, t) + M_{YZ}(t, q)$, then we distinguish two cases for obtaining $P_{XZ}(p, q)$: If $P_{XY}(p, t)$ is undefined then $P_{XZ}(p, q) = P_{YZ}(t, q)$ (which could also be undefined), otherwise $P_{XZ}(p, q) = P_{XY}(p, t)$. When the modified algorithm finishes computing $P_{R_u L_w}$ for all leaves u, w (at the end of Phase 2), it is easy to obtain the matrix P: If $S_u = \{p_i\}$, $R_u = \{p_i'\}$, $S_w = \{p_j\}$, $L_w = \{p_j'\}$, then we set $P(p_i, p_j)$ equal to $P_{R_u L_w}(p_i', p_j')$ if the latter is defined; otherwise, we set $P(p_i, p_j)$ equal to p_j if p_j dominates p_i, and set $P(p_i, p_j)$ to be undefined if p_j does not dominate p_i. From now on, we assume that the matrix P is available. Note that this matrix is a description of n longest chain trees, each rooted at a point of S.

We pre-process each longest chain tree so that the following type of queries can be quickly answered: Given a node p in the tree and a positive integer i, find the i-th node on the path from p to the root of the tree. Such queries are called *level-ancestor queries* by Berkman and Vishkin [6], who gave efficient parallel algorithms for pre-processing rooted trees so that the level-ancestor queries can be answered quickly. The work of Berkman and Vishkin [6, 7] shows (implicitly) that a level-ancestor query can be handled sequentially in constant time, after a logarithmic time and linear work pre-processing in the CREW PRAM model. The pre-processing of the longest chain trees is done by simply applying the result of Berkman and Vishkin to each of the n trees, in totally $O(\log n)$ time and $O(n^2)$ work.

For the sake of processor assignment in reporting chains, we also need to compute the number of points of S on the actual longest chain which is to be reported. Suppose a longest chain between points p and q in S is to be reported. The number of points of S on such a p-to-q chain can be obtained from the depth of p in the longest chain tree rooted at q; it is known that the depths can be computed within the required complexity bounds by using the Euler Tour technique [20].

To report an actual longest chain between points p and q in S, we do the following (without loss of generality, we assume that q dominates p). First, we go to the longest chain tree rooted at (say) q, and find the number of nodes on the path in that tree from node p to the root q. Let that number be k. The p-to-q path in the tree corresponds to a longest chain from p to q, which we must report. We do so by performing, in parallel, $k-1$ level-ancestor queries, using node p and integers $1, 2, \ldots, k-1$. Each query is handled by one processor in $O(1)$ time. These queries find each point on the p-to-q chain. Finally, we report the k points of that chain in parallel, by assigning to k processors the task of reporting those k points (one point per processor).

4 Further Remarks

The following open problems remain:

- Give an $O(n^2)$ time sequential algorithm for the weighted case.
- Give an $O(n^2)$ time sequential algorithm for the three dimensional version of the problem (unweighted).
- For the three dimensional version of the problem, give an NC parallel algorithm that uses a quadratic (to within a logarithmic factor) number of processors.

Finally, using the methods we developed here in combination with other ideas, we can improve the processor complexity of the layers of maxima problem: We can achieve the same $O((\log n)^2)$ time complexity as in [1] with $O(n^2/(\log n)^3)$ processors, instead of the $O(n^2/\log n)$ processors used in [1].

References

[1] A. Aggarwal and J. Park. "Notes on Searching in Multidimensional Monotone Arrays (Preliminary Version)," *Proc. 29th Annual IEEE Symposium on Foundations of Computer Science*, 1988, pp. 497–512.

[2] A. Apostolico, M. J. Atallah, and S. E. Hambrusch. "New Clique and Independent Set Algorithms for Circle Graphs," *Discrete Appl. Math.*, Vol. 36, 1992, pp. 1–24.

[3] A. Apostolico, M. J. Atallah, L. L. Larmore, and H. S. McFaddin. "Efficient Parallel Algorithms for String Editing and Related Problems," *SIAM J. Comput.* 19 (5), 1990, pp. 968–988.

[4] M. J. Atallah and D. Z. Chen. "Parallel Rectilinear Shortest Paths with Rectangular Obstacles," *Computational Geometry: Theory and Applications*, 1, 1991, pp. 79–113.

[5] M. J. Atallah and S. R. Kosaraju. "An Efficient Parallel Algorithm for the Row Minima of a Totally Monotone Matrix," *Proc. 2nd ACM-SIAM Symp. on Discrete Algorithms*, San Francisco, California, 1991, pp. 394–403. (Accepted for publication in *J. of Algorithms.*)

[6] O. Berkman and U. Vishkin. "Finding Level-Ancestors in Trees," Tech. Rept. UMIACS-TR-91-9, University of Maryland, 1991.

[7] O. Berkman and U. Vishkin. Personal communication.

[8] R. P. Brent. "The Parallel Evaluation of General Arithmetic Expressions," *J. of the ACM*, Vol. 21, No. 2, 1974, pp. 201–206.

[9] B. M. Chazelle. "Optimal Algorithms for Computing Depths and Layers," *Proc. of the 20th Allerton Conference on Communications, Control and Computing*, 1983, pp. 427–436.

[10] R.B.K. Dewar, S.M. Merritt, and M. Sharir. "Some Modified Algorithms for Dijkstra's Longest Upsequence Problem," *Acta Informatica*, Vol. 18, No. 1, 1982, pp. 1–15.

[11] E.W. Dijkstra. "Some Beautiful Arguments Using Mathematical Induction," *Acta Informatica*, Vol. 13, No. 1, 1980, pp. 1–8.

[12] S. Even, A. Pnueli, and A. Lempel. "Permutation Graphs and Transitive Graphs," *Journal of the ACM*, Vol. 19, No. 3, 1972, pp. 400–410.

[13] M.L. Fredman. "On Computing the Length of Longest Increasing Subsequences," *Discrete Mathematics*, 1975, pp. 29–35.

[14] F. Gavril. "Algorithms for a Maximum Clique and a Maximum Independent Set of a Circle Graph," *Networks*, 1973, pp. 261–273.

[15] F. Gavril. "Algorithms on Circular-Arc Graphs," *Networks*, 1974, pp. 357–369.

[16] U.I. Gupta, D.T. Lee, and Y.-T. Leung. "Efficient Algorithms for Interval Graphs and Circular Arc Graphs," *Networks*, 1982, pp. 459–467.

[17] W.-L. Hsu. "Maximum Weight Clique Algorithms for Circular-Arc Graphs and Circle Graphs," *SIAM J. on Computing*, 1985, pp. 224–231.

[18] A. Pnueli, A. Lempel, and S. Even. "Transitive Orientation of Graphs and Identification of Permutation Graphs," *Canadian Journal of Math.* 23, 1, 1971, pp. 160–175.

[19] D. Rotem and U. Urrutia. "Finding Maximum Cliques in Circle Graphs," *Networks*, 1981, 1pp. 269–278.

[20] R. E. Tarjan and U. Vishkin. "An Efficient Parallel Biconnectivity Algorithm," *SIAM J. Comput.* 14 (4), 1985, pp. 862–874.

Towards a Better Understanding of Pure Packet Routing

Allan Borodin

Department of Computer Science, University of Toronto
and
Centre for Advanced Studies, IBM Canada

1 Introduction and Motivation

Routing is a fundamental field of study with origins in communications networks. The history and principal results concerning communications networks are carefully described in Pippenger [25]. Much of this work concerns "circuit switched" routing in which a path is selected and dedicated to a message (e.g. a telephone call) for the duration of the message's existence. With the advent of parallel and distributed computing, different forms of "packet" routing have emerged. In packet routing, a message is a packet or a sequence of packets being routed between two nodes of a network with (in its most general form) the only restriction being that at most one packet can traverse a network edge (or link) during a given "communication step".

The literature of routing and, in particular, packet routing is now so extensive that it is difficult for all but a few experts to appreciate the principal results and methodologies within a coherent framework. Fortunately, the understanding of packet routing has been greatly advanced by Leighton's excellent text [17] and the excellent survey articles by Valiant [33] and Leighton [18]. Still the field is not easy to comprehend and new results continue to come forth at an ever increasing rate. This should not be surprising given the importance and diversity of routing problems. Some of the many factors leading to this diversity are the following: the multiplicity of network models, the constantly changing (because of technology) models of a network node, different notions of performance (e.g. worst case or average case routing time, throughput, fault tolerance), different notions of cost (number of nodes or edges, area or volume in a layout).

In order to gain some focus, we shall restrict our attention to a concept of "pure packet routing" which we will define in this paper. In pure packet routing, a message is a single packet which must be routed from an origin node to a destination node in a fixed interconnection network. A network is simply a graph or multigraph with a set of "processor nodes" that can initiate, receive and transmit messages and possibly additional "switching nodes" that can only transmit packets. Thus in this "pure" form we do not include many important routing problems such as wormhole or cut through routing (where in both cases a message is a sequence of packets, called flits, which must occupy a path of nodes, see [7]), redundant or non redundant information dispersal strategies [27], and reconfigurable network routing schemes [21]. We do, however, include "hot potato" or deflection routing, which has very recently reemerged as an active research area.

Indeed the topic of hot potato routing [8], [9], [23], has forced us to more carefully consider the issue of "sorting based" routing and whether or not such routing falls within what we want to define as pure packet routing. Following the discussion in Bar-Noy, et al [2], let us consider the prototypical problem of 1-1 routing where each source node generates at most one packet and every destination node is the target of at most one packet. Since a sorting algorithm can be modified to yield a "sorting based" 1-1 routing algorithm (see the discussion in [17]), sorting is clearly a more general problem. What is a "non sorting based" routing algorithm and why bother with such a concept? Intuitively, sorting is usually thought of as a static computational problem while routing is often a dynamic problem and one that should only require simple switching decisions. Although our definition will not be able to rule out all sorting based methods, we can at least begin to articulate some desirable properties of a pure routing scheme. In order to do so we need to provide some details of our routing model and its application. For the purpose of analysis, we usually assume that the network (that is, each of its nodes) is operating in a synchronized mode relative to a global clock. However, in order to achieve simple and fast switching nodes it would be preferable if the node did not base its actions on the value of the clock. Not allowing the node access to the clock not only simplifies the node, but, moreover, makes it more plausible that the routing algorithm can work in the "dynamic mode" whereby packets are injected at different starting times.

Our model of a "finite state switching node" thus has the following properties:

Property 1: Upon initiating a packet, a processor node is allowed to possibly add some routing information to the packet in addition to the destination address. Thereafter nodes cannot alter the packets.

Property 2: On every step, the node must decide which packet to send on any given link as a (possibly randomized) function of the node address, the addressing and routing information of all the packets on its incoming queues, the sizes of its outgoing queues, the value of its finite state and the size of the network.

This description informally constitutes our concept of pure packet routing. Note, in particular, that we do not allow additional messages (beyond the packets to be routed). Krizanc [14] defines a similar concept of pure routing which in one sense is more restrictive than our definition. In particular, nodes cannot see the sizes of the outgoing queues. Our definition is closely related to the concept of pure routing as formulated in Maggs and Sitaraman [22]. Neither the Krizanc nor the Maggs and Sitaraman papers discuss the use of the clock or the possible altering of packets.

Researchers are typically interested in the following types of routing problems: worst case routing time of a static h-relation problem where all packets are initially injected at the same time and there are at most h packets initiated at any node and at most h packets destined for any node, average case routing time of a static random mapping where each injecting node chooses a destination at random, the throughput of the system or the routing time of a packet for the dynamic mode of routing under various assumptions on the injection rate. Of particular interest is the prototypical static problem of a 1-1 (partial) permutation (i.e. a 1-relation).

Although we will mainly discuss static problems, the restriction to pure routing algorithms is strongly motivated by good performance for the dynamic mode of packet injection. To make this motivation more precise, let us say that a routing algorithm dynamically routes (say) an h-relation in time T if when an adversary injects the packets constituting the h-relation at times $t_1, t_2, \cdots, t_r$, the ith packet will arrive at its destination by time $t_i + T$. Although our definition of pure routing does not preclude sorting based methods for static problems, it is our contention that on conventional networks sorting based methods are not readily modified so as to permit fast dynamic routing in our sense. On the other hand, we also contend that the pure routing methods discussed in this paper can be applied so as to yield dynamic routing with time bounds comparable to the stated static bounds.

In the next section, we review (for the most part following the chronology) some of the major results and techniques concerning pure packet routing. Notwithstanding the impressive but very incomplete list of substantial results outlined in section 2, we will see in section 3 that some of the most basic questions remain open. Stated briefly, it is very difficult to analyze simple routing strategies, there are almost no lower bound results and for some basic problems there is no known routing strategy with good dynamic performance (i.e. substantially faster than "the naive algorithm").

2 Some of the Things We Do Know

We will begin our history with the seminal paper of Valiant [31] and the improved analyses and extensions by Valiant and Brebner [35]. We start here at the cost of ignoring a number of more classical and basic ideas because these papers were the first to introduce the theory and algorithms research community to the topic of routing, and because they provide some foundational results of both theoretical and practical importance. The main result of Valiant and Brebner concerns static routing on the hypercube. An $N = 2^n$ node hypercube has nodes labelled by n bit vectors $(v_{n-1}v_{n-2}\cdots v_0)$ and edges of the form $\langle(v_{n-1}v_i\cdots v_0), (v_{n-1}\cdots \bar{v}_i \cdots v_0)\rangle$ for $0 \leq i \leq n-1$ where $\bar{v}_i = 1 - v_i$. Clearly the degree of each node of the 2^n node hypercube is n ($= \log N$) and the diameter is also n so that the best result possible for permutation routing is n steps. Motivated by standard technology constraints, almost all of the literature assumes "small" or even constant degree networks (with diameter at least $\log N$) although the advent of (say) optical switching may alter this assumption.

Consider what is perhaps the most naive algorithm for the hypercube, namely greedily routing the packet to its destination by correcting when necessary each dimension in some fixed order (say left to right). This Greedy algorithm seems to work very well "on average" but various natural and important permutations (e.g. the transpose permutation where the packet at origin $(v_{n-1} \cdots v_0)$ has destination $(v_{n/2-1} \cdots v_0 v_{n-1} \cdots v_{n/2})$) suffer the worst case behaviour of $2^{n/2} = \sqrt{N}$ steps. This is because (say) all packets with origins of the form $(v_{n-1} \cdots v_{n/2} 000 \cdots 0)$ will be routed through a "hotspot" node $(00 \cdots 0)$. Valiant introduced the idea of a two stage randomized algorithm, where in the first stage each packet is routed greedily to a random intermediate node and then in the second stage the packet is greedily routed from the intermediate node to its destination. The first stage is then a random mapping problem and the second stage is the reverse problem. A probabilistic analysis which applies to either stage shows that the expected number of packets which can intersect the path of a given packet is $\Theta(\log N)$. Since the routes of any two packets have the non repeating property (i.e. if their routes intersect and then diverge, they will not intersect again), it can be shown that with a reasonable scheduling rule the total delay in any stage is $\Theta(\log N)$ with high probability. Thus with high probability (depending only on the random choices for intermediate nodes) every 1-1 and indeed every $h = \log N$ relation can be routed in $\Theta(\log N)$ steps on the hypercube.

To simplify the analysis, Valiant and Brebner [35] assumed that stage one ended (for all packets) before beginning stage two. Thus, strictly speaking, the two stage algorithm is not a pure packet routing algorithm since the nodes might have to count to determine the completion of stage one. Valiant [34], however, observed that the hypercube analysis does easily extend to the case that each packet immediately begins stage two when its stage one is completed. It is also easy to see that a node can determine which stage a packet is in by inspecting the intermediate node (written in the packet by the originating node) and the link from which the packet entered the node. Thus the Valiant-Brebner randomized scheme can be viewed as an asymptotically optimal pure routing algorithm for the hypercube.

Valiant and Brebner also described and analyzed simple and efficient randomized algorithms for the 2-dimensional and 3-dimensional arrays. For example, in the 2-dimensional array, a packet proceeds along the origin row to a randomly chosen column, turns on this column to the destination row and then turns to the destination column. On a $N = \sqrt{N} \times \sqrt{N}$ node array, this algorithm routes every 1-1 permutation in expected time $\Theta(\sqrt{N})$ which is asymptotically optimal since this network has diameter $2\sqrt{N} - 2$. Similarly, there is a randomized pure routing algorithm with asymptotically optimal $\Theta(\sqrt[3]{N})$ expected time for the 3-dimensional array. All of the Valiant and Brebner randomized algorithms have the property that the expected maximum congestion at a node is $\Omega(\log N / \log\log N)$ and thus queues of at least that length must be provided.

The simplicity and performance of the randomized two stage hypercube scheme raises a number of immediate questions which provided the motivation for much of the routing research over the last decade. The most obvious question is whether or not there is a deterministic pure routing algorithm which achieves the same simplicity and comparable (or even better) performance on the hypercube and array networks. Another immediate question is what performance can be achieved using randomized or deterministic algorithms which have bounded (or no) queues. One also asks if there are networks of bounded degree and $\Theta(\log N)$ diameter for which optimal routing algorithms exist. Finally, can we exploit large degree so as to achieve (say) constant routing time? The remainder of this section describes the progress on such questions; the concluding section describes the lack of progress.

Greedy algorithms (i.e. those that route packets via an easily computed shortest path) are the canonical example of simple routing. Even though the packet path section is trivialized in Greedy routing, there is still great flexibility and importance in how such algorithms schedule packets that contend for the same link. Is there a fast deterministic Greedy routing strategy for some reasonable network model? Generalizing Greedy routing, Borodin and Hopcroft [4] define deterministic oblivious routing algorithms to be those algorithms in which the path taken by any packet is determined by the origin and destination of the packet independent of other packets being routed. For every degree d network, Borodin and Hopcroft [4] show that for any deterministic oblivious routing algorithm there is a permutation whose routing will result in $\Omega(\sqrt{N/d})$ node congestion (i.e. that many packets being routed through some node in the network) and hence the time to route this permutation will be at least $\Omega(\sqrt{N/d^3})$. (If a node can only send out a constant number of packets in any step the time lower bound is $\Omega(\sqrt{N/d})$.) Kaklamamis, Krizanc and Tsantilas [13] improve the time lower bound to $\Omega(\sqrt{N/d^2})$ by considering the congestion on some edge. Borodin and Hopcroft [4] then suggest a "simple" adaptive (= non oblivious) routing strategy, whereby on every step each node uses a maximum matching to route as many incoming packets as possible toward their destinations while sending the remaining packets away from their destinations. This turns out to be what we call hot potato or deflection routing, an idea whose origins (see Feige and Raghavan [8]) certainly precede the Borodin and Hopcroft suggestion. Prager [26] was able to show that a particular implementation of this maximum matching algorithm for the hypercube routes an interesting class of permutations in $\log N$ steps but, in general, there is no worst case analysis for the "simple algorithm".

Aleliunas [1] and Upfal [30] consider the Valiant [31] randomized two stage algorithm for bounded degree, $O(\log N)$ diameter networks, namely the d-way shuffle network and the butterfly, respectively. Let us just define the $N(\log N+1)$ node butterfly network (with $N = 2^n$) whose nodes are labelled $(v_{n-1}\cdots v_0, i)$ where $(v_{n-1}\cdots v_0)_2$ is the row address and i $(0 \leq i \leq \log N)$ is the column (or level) of the node. The butterfly edges are of the form $\langle(v_{n-1}\cdots v_0, i), (v_{n-1}\cdots v_0, i+1)\rangle$ and $\langle(v_{n-1}\cdots v_i\cdots v_0, i), (v_{n-1}\cdots \bar{v}_i\cdots v_0, i+1)\rangle$. The butterfly is one of the most important networks and is studied in terms of both end to end routing (where packets only originate at level 0 and are routed via increasing levels to level $\log N$) and "fully loaded" routing (where levels 0 and $\log N$ are identified and all nodes can originate and receive packets). For end to end routing there is a unique level-increasing path from every level 0 node to every level $\log N$ node and thus there is a unique (modulo the different scheduling options) Greedy algorithm. Following the type of analysis in Valiant and Brebner [35], it is possible to analyze the Greedy algorithm for a random mapping. Such an analysis would only yield an $O(\log^2 N)$ bound. Aleliunas [1] and Upfal [30] introduce a more sophisticated critical path analysis in order to show that the expected completion time (over random mappings) for the Greedy algorithm with "non predictive" scheduling is $O(\log N)$ steps (see Leighton [17]). Valiant's "two stage" approach can be used again to yield a randomized algorithm which performs every fully loaded routing problem in time $O(\log N)$ with high probability. The algorithm first sends a packet (i.e. via a level-increasing (mod N) path) to a random row in its origin column, then moves on this row to the destination column, and then finally greedily to the destination row. However, unlike the hypercube analysis, it is not clear if the desired timing can be proven when each packet immediately begins its next stage without waiting for the stage to complete for all packets. Thus, strictly speaking, the performance of the pure routing scheme is not fully analyzed.

Pippenger [24] and then Ranade [29] develop more sophisticated randomized algorithms in order to avoid the $\Omega(\log N/\log\log N)$ queue size of the previous algorithms. Leighton, Maggs and Rao [20] partition a routing algorithm into two components, path selection (i.e. which links to use) and scheduling (i.e. when to send a packet on its next link). For any N node "levelled network" (including butterflies, arrays, hypercubes) they show how a modification of Ranade's method can be used in the scheduling component so as to achieve a randomized, bounded queue algorithm with time bound $O(c + l + \log N)$ where c and l are respectively the edge congestion and maximum path length achieved by the path selection component. These bounded queue algorithms don't fit our model since they require additional messages. More recently, Maggs and Sitaraman [22] show how to "spread out the queue" along the Greedy paths being followed so that bounded queues can be achieved for end to end butterfly routing without the use of additional messages. Their fully loaded routing algorithm still requires additional messages (tokens) but is conceptually simpler than the previous algorithms.

Following Bassalygo and Pinsker's [3] use of "randomly wired" networks for offline routing, and Leighton's [15] AKS sorting based algorithm, Upfal [30] takes a creative approach to the problem of deterministic worst case routing by introducing the multibutterfly network model. As discussed in Leighton [18] the multibutterfly can be essentially viewed as a collection of "randomly wired" butterflies superimposed on each other. By using expander graphs between levels, this network can be constructively described. Upfal provides a deterministic adaptive algorithm that achieves worst case $O(\log N)$ complexity for both the end to end and fully loaded permutation problems. The analysis is rather complex and the hidden constant in the big O is quite large, but the multibutterfly and more primitive versions of it are of both great theoretical and practical importance.

Returning to the 2-dimensional array, Rajasekaran and Tsantilas [28] improve the Valiant and Brebner result [35] by presenting a nearly optimal $2\sqrt{N}+$

$O(\log N)$, constant buffer size randomized algorithm for the $N = \sqrt{N} \times \sqrt{N}$ mesh. Leighton [16] provides a very precise analysis for a random mapping using the Greedy algorithm on the mesh. Leighton shows that for most mappings, the Greedy algorithm routes each packet in time $D + O(\log N)$ using link queues of size at most 4 where D is the distance from the origin to the destination. Moreover, his analysis provides the only successful attempt at analyzing (without queueing-theoretic independence assumptions) routing on the array for stochastically generated dynamic routing. For the 3-dimensional $\sqrt[3]{N} \times \sqrt[3]{N} \times \sqrt[3]{N}$ mesh, Kaklamanis, Krizanc and Rao [11] describe a nearly optimal $3\sqrt[3]{N} + o(\sqrt[3]{N})$ constant buffer size randomized algorithm.

One of the most promising approaches towards simple and efficient routing algorithms on the basic hypercube and array networks is that of hot potato or deflection algorithms. In such algorithms, queues are not allowed so that each node treats a packet like a hot potato and immediately passes it along some link even if the chosen link represents a deflection away from the destination. Of course, as in the Borodin and Hopcroft [4] algorithm, the deflection strategy might be quite involved so that one really seeks a computationally simple deflection strategy. It should be noted that some deflection algorithms incur "livelock" in a subtle way and do not terminate. Hajek [9] was the first to prove the existence of a livelock free deterministic hot potato algorithm for the N node hypercube. The algorithm has worst case time $2k + \log N$ to deliver k packets. Any permutation is thus routed in $O(N)$ time. Feige and Raghavan [8] provide a number of interesting hot potato algorithms and analyses for both static and dynamic routing problems. In particular, they develop and analyze hot potato algorithms for the 2-dimensional and 3-dimensional mesh and toroidal networks. For example, for the 2-dimensional torus, Feige and Raghavan provide a $2\sqrt{N} + O(\log N)$ algorithm and a $O(\sqrt[3]{N} \log N)$ algorithm for the 3-dimensional torus for a random mapping. (These algorithms can be modified for meshes.) Feige and Raghavan [8] do not analyze the randomized algorithm that would result using Valiant's two stage idea. (In the case of hot potato routing, the algorithm surely cannot wait for the end of stage one since it would need queues to do so in addition to the possible use of a global clock.) Instead they adopt an idea from Leighton, et al. [20] and use random delays before starting the packets on their routes. These results have been extended in Kaklamanis, Krizanc and Rao [12]. For the N node hypercube, Feige and Raghavan provide an $O(\log N)$ average case analysis for a simple oblivious hot potato algorithm in which at every node a packet either makes progress or is returned along the same dimension it entered a node. In this algorithm, each node scans the incoming links in a fixed order and immediately assigns any packet encountered to an outgoing link. This one pass scheduling strategy provides a simple switching decision and is crucial for the analysis and absence of livelock for this algorithm. Bar-Noy, et al [2] provide a deterministic worst case time $O(N^{3/4})$ hot potato algorithm for the $\sqrt{N} \times \sqrt{N}$ mesh. A much more complicated hot potato algorithm achieves time $O(N^{1/2+\epsilon})$ for any $\epsilon > 0$. To our knowledge these are the first deterministic worst case pure routing algorithms for any of the previously discussed networks with bounded queues and proven performance asymptotically better than $O(N)$. Many deterministic sorting based algorithms, however, do perform well in the worst case. (For example, Batcher's classic $O(\log^2 N)$ algorithm can be implemented on many networks including the hypercube; Cypher and Plaxton [6] have the presently best deterministic sorter for the hypercube with performance $O(\log N \log\log N)$; Newman and Schuster [23] have a deterministic sorting based hot potato algorithm for the $\sqrt{N} \times \sqrt{N}$ mesh with time bound $O(\sqrt{N})$.) Some recent algorithms (see Han and Stanat [10] for a $O(\sqrt{N})$ algorithm on thc mesh) are surely not "sorting based" but do seem to require substantial use of a global clock.

Finally, we briefly discuss the potential use of large degree networks which may become feasible with (say) optical technology. For large degree networks, it is necessary to distinguish between single port and multi port models. Thus far, our model of pure routing on a d degree network has assumed a multi port model in which every node can receive d incoming packets and send d packets on each step. In particular, the Valiant and Brebner [35] hypercube analysis relies on this multi port model. To reflect more simple switching nodes, in a single port model processors can only send at most one packet per step. (A more restrictive definition would also only permit receiving at most one packet per step.) Obviously, for constant degree networks this restriction will only result in constant factor changes in the timing and queue sizes. Borodin, Raghavan, Schieber and Upfal [5] provide a number of results for large degree networks, considering the issues of oblivious vs adaptive, deterministic vs randomized, and single port vs multi port routing. For example, in the case of oblivious, randomized, single port routing on any degree $d \leq N/\log^3 N$ network there is a worst case permutation whose routing requires $\Omega(\log N/\log\log N)$ time. Thus large degree is mostly ineffective in this case since $O(\log N)$ can be achieved for constant degree networks. (In an oblivious randomized algorithm, the distribution of paths followed by a packet is independent of the other packets. Valiant's two stage algorithms satisfy this property.) For oblivious deterministic routing in both the single port and multi port models, simple array based networks are defined for all degrees which match the lower bounds provided by the node congestion and edge congestion results, by Borodin and Hopcroft [4] and, respectively Kaklamamis, Krizanc and Tsantilas [13]. In the case of adaptive, deterministic, multi port routing, there are pure routing algorithms for the "d-way multibutterfly" which achieve the asymptotically optimal diameter bound of $\log_d N = \log N/\log d$, which in the case of $d = N^\epsilon$ yields constant time algorithms.

3 Some of the Things We Don't Know

As stated in the introduction, the main embarrassment of the field is the absence of either lower bounds or "simple" and efficient deterministic algorithms for worst case permutation routing on the "classical" networks. (More precisely, we want simple algorithms which will also perform well in the case of dynamic routing.) Moreover in deterministic worst case routing, the allowable size of the queues seems to become a more critical issue (as compared to average case analysis for random mapping or worst case analysis in randomized algorithms). In particular, we do not know of a ("non sorting based") hypercube algorithm that performs better than the $O(\sqrt{N}/\log N)$ oblivious algorithm of Kaklamanis, et al [13] allowing unbounded ($= O(\sqrt{N})$) size queues. By their lower bound, this algorithm is optimal for oblivious routing. With bounded queues we do not know how to improve on a naive $O(N)$ algorithm (say using a Hamiltonian path in the network).

For the 2-dimensional array, the Bar-Noy, et al [2] hot potato algorithm provides the best known ("non sorting based") upper bound for deterministic worst case routing with bounded queues. With unbounded $O(\sqrt{N})$ queues, the most naive Greedy algorithm (i.e. proceed to the destination column and then turn) yields a $3\sqrt{N}$ step permutation bound since there is no delay along the row and a simple scheduling argument shows that the $\sqrt{N}$ packets heading for the same column cannot delay any packet by more than $\sqrt{N}$ steps. (The bound can be improved to $2\sqrt{N} - 2$ by using the farthest-to-go queueing discipline.) The 3-dimensional mesh seems to be a more challenging situation. Leighton [19] observes that the naive Greedy algorithm for the $\sqrt[3]{N} \times \sqrt[3]{N} \times \sqrt[3]{N}$ mesh results in a $\Omega(N^{2/3})$ delay for a natural permutation problem such as the bit reversal permutation. With bounded queues the best known upper bound may be the trivial N step solution again provided by a Hamiltonian path routing scheme.

Allowing the freedom to construct an appropriate network, is there a better choice than Upfal's multibutterfly? Can we construct an N node network and a deterministic worst case pure routing scheme which achieves $O(\log N)$ time with a "small" constant factor?

For randomized algorithms, the Valiant approach on (say) a hypercube forces every packet to have routing paths with expected distance $\log N$ even though the origin to destination distance D might be quite small. Is there a randomized (or deterministic) algorithm (for example) such that packets travel on paths of length $O(D)$ incurring delay $O(D)$? (In this regard, see Valiant [32] for a strong lower bound on randomized oblivious algorithms for networks with near optimal diameter.) Also, as mentioned before it would be beneficial to analyze the randomized Aleliunas/Upfal butterfly algorithm when the stages are not kept distinct.

For those who prefer lower bounds, routing provides a great source of problems. For example, can we prove worst case lower bounds for (deterministic or randomized) routing algorithms which only use minimal length paths? Can we prove lower bounds for hot potato algorithms? Can we at least prove lower bounds for minimal path length hot potato algorithms? In general, we have no techniques for proving lower bounds for adaptive algorithms. In addition to the previously stated lower bounds for oblivious algorithms, there are some results concerning restricted versions of deterministic source oblivious algorithms (i.e. the choice of the next link is determined by the destination independent of the origin). Krizanc [14] assumes that a node cannot see the size of the outgoing link queues, and shows for any N node bounded degree network that a bounded queue source oblivious algorithm requires $\Omega(N)$ worst case time. Maggs and Sitaraman assume single-port non predicative scheduling (i.e. as in FIFO scheduling, the choice of which packet the node attempts to forward is made without examining the destinations of the packets being queued) and show that for end to end routing on the butterfly that a bounded queue source oblivious algorithm requires $\Omega(N/q \log N)$ worst case time where q is a bound on the queue size. Can these results be extended to pure bounded queue oblivious algorithms? It is important to note that Leighton, et al. [20] show that for any network there is always an offline way to schedule packets achieving time $O(c + l)$ whenever there is an assignment of packet routes with edge congestion c and length l. Thus strong lower bounds in any network will undoubtedly require one to fully exploit the local and online nature of the model.

Compared to the case of static routing, there are relatively few results for dynamic routing. In particular, it is important to study the throughput (i.e. maximum injection rate so that the network does not become saturated) of various routing algorithms.

An unfulfilled goal of this paper is to find a definition of pure routing which captures the essence of "simplicity", permits provably efficient routing in a "practical sense", and does so in a way that allows static results to be extended to dynamic routing.

Acknowledgements

I am completely indebted to Prabhakar Raghavan for much of what I know about this subject and for his invaluable comments on this paper. I am also very indebted to Tom Leighton for many important suggestions. The paper has also greatly benefited from suggestions by Donald Chinn, Danny Krizanc, Quoc Tran Pham, Baruch Schieber, Hisao Tamaki, Thanasis Tsantilas, Martin Tompa and Gordon Turpin.

References

1. R. Aleliunas. Randomized parallel communication. In *Proc. ACM-SIGOPS Symposium on Principles of Distributed Systems*, 60–72, 1982.
2. A. Bar-Noy, P. Raghavan, B. Schieber, and H. Tamaki. Fast deflection routing for packets and worms. To appear in 1993 ACM PODC Conference.
3. L.A. Bassalygo and M.S. Pinsker. Complexity of an optimum non-blocking switching network without reconnections. *Problems of Information Transmission*, 9:64–66, 1974.
4. A. Borodin and J. Hopcroft. Routing, merging, and sorting on parallel models of computation. *Journal of Computer and System Sciences*, 30(1):130–145, 1985.
5. A. Borodin, P. Raghavan, B. Schieber, and E. Upfal. How much can hardware help routing? In *Proc. 25th Annual ACM Symposium on Theory of Computing*, 573–582, 1993.
6. R. Cypher and G. Plaxton. Deterministic sorting in nearly logarithmic time on the hypercube and related computers. In *Proc. 22nd Annual ACM Symposium on Theory of Computing*, 193–203, 1990.

7. W. Dally. Network and processor architecture for message-driven computers. In Robert Suaya and Graham Birtwhistle, editors, *VLSI and Parallel Computation*, chapter 3, 140–222. Morgan Kaufman Publishers, San Mateo, CA, 1990.
8. U. Feige and P. Raghavan. Exact analysis of hot-potato routing. In *Proc. 33rd Annual IEEE Symposium on Foundations of Computer Science*, 553–562, 1992.
9. B. Hajek. Bounds on evacuation time for deflection routing. *Distributed Computing*, 5:1–6, 1991.
10. T. Han and D. Stanat. "Move and smooth" routing algorithms on mesh-connected computers. In *Proc. 28th Allerton Conference*, 236–245, 1990.
11. C. Kaklamanis, D. Krizanc and S. Rao. Simple path selection for optimal routing on processor arrays. *SPAA 92*, 23-30.
12. C. Kaklamanis, D. Krizanc and S. Rao. Improved hot potato routing for processor arrays. *SPAA 93*, to appear.
13. C. Kaklamanis, D. Krizanc, and T. Tsantilas. Tight bounds for oblivious routing in the hypercube. *Math. Systems Theory*, 10:223–232, 1991.
14. D. Krizanc. Oblivious routing with limited buffer capacity. *Journal of Computer and System Sciences*, 43:317–327, 1991.
15. F.T. Leighton. Tight bounds on the complexity of parallel sorting. *IEEE Transactions on Computers*, C-34(4):344–354, 1985.
16. F.T. Leighton. Average case analysis of greedy routing algorithms on arrays. In *Proc. 2nd Annual ACM Symposium on Parallel Algorithms and Architectures*, 2–10, 1990.
17. F.T. Leighton. *Introduction to Parallel Algorithms and Architectures: Arrays, Trees, Hypercubes*. Morgan Kaufman Publishers, San Mateo, CA, 1992.
18. F.T. Leighton. Methods for message routing in parallel machines. In *Proc. 24th Annual ACM Symposium on Theory of Computing*, 77–96, 1992.
19. F.T. Leighton. Personal communication.
20. F.T. Leighton, B. Maggs, and S. Rao. Universal packet routing algorithms. In *Proc. 29th IEEE Symposium on Foundations of Computer Science*, 256–269, 1988.
21. H. Li and Q. Stout. Reconfigurable SIMD massively parallel computers. In *Proc. IEEE*, 79:429–443, 1991.
22. B. Maggs and R. Sitaraman. Simple algorithms for routing on butterfly networks with bounded queues. In *Proc. 24th Annual ACM Symposium on Theory of Computing*, 150–161, 1992.
23. I. Newman and A. Schuster. Hot-potato algorithms for permutation routing. Technical Report #9201, CS Department, Technion, Israel, 1992.
24. N. Pippenger. Parallel communication with limited buffers. In *Proc. 25th Annual IEEE Symposium on Foundations of Computer Science*, 127–136, 1984.
25. N. Pippenger. Communication networks. In J. Van Leeuwen, editor, *Handbook of Theoretical Computer Science, Vol. A: Algorithms, and Complexity*, 805–834. MIT Press, Cambridge, 1990.
26. R. Prager. An algorithm for routing in hypercube networks. M.Sc. thesis, University of Toronto, 1986.
27. M. Rabin. Efficient dispersal of information for security, load balancing, and fault tolerance. *Journal of the ACM*, 36(2):335–348, 1989.
28. S. Rajasekaran, and T. Tsantilas. Optimal routing algorithms for mesh-connected processor arrays. *Algorithmica*, 8:21–38, 1992.
29. A. Ranade. How to emulate shared memory. *Journal of Computer and System Sciences*, 42(3):307–326, 1991.

30. E. Upfal. Efficient schemes for parallel communication. *Journal of the ACM*, 31:507–517, 1984.
31. L. Valiant. A scheme for fast parallel communication. *SIAM Journal on Computing*, 11(2):350–361, 1982.
32. L. Valiant. Optimality of a two-phase strategy for routing in interconnection networks. *IEEE Transactions on Computers*, C-32(9):861-863, 1983.
33. L. Valiant. General purpose parallel architectures. In J. Van Leeuwen, editor, *Handbook of Theoretical Computer Science, Vol. A: Algorithms, and Complexity*, 943–972. MIT Press, Cambridge, 1990.
34. L. Valiant. Personal communication.
35. L. Valiant and G. Brebner. Universal schemes for parallel communication. In *Proc. 13th Annual ACM Symposium on Theory of Computing*, 263–277, 1981.

Tolerating Faults in Meshes and Other Networks

Richard Cole*

New York University

This talk describes methods for overcoming static faults in networks of processors. The scenario is that the faults are known apriori and their only effect is to stop the faulty unit (a processor or communication link) from operating. The problem is to simulate a computation on a fault-free network on a faulty network of the same type (e.g., a mesh or a butterfly) with small slowdown. The main focus of the talk is on simulations that result in only a constant factor slowdown. The methods of embedding and redundant computation are surveyed. Often, to tolerate more faults, the redundant computation approach is used recursively; unfortunately, each recursive level adds another multiplicative factor to the slowdown. This talk shows how this problem can be overcome, at least sometimes, by a new variant of redundant computation, called *multiscale simulation*. This method is applied to obtain substantially more efficient simulations on the mesh. In particular, for the case of random, independent faults, occuring with probability p, it is shown that there is a constant p, such that with high probability a faulty $N \times N$ mesh can simulate an equal sized fault-free mesh with only constant slowdown. (This is the first example of a constant slowdown simulation in the presence of constant probability faults for any family of networks). The multiscale simulation method was first reported in [1].

[1] R. Cole, B. Maggs, R. Satiraman. Multi-Scale Emulation: A technique for reconfiguring arrays with faults. *Twenty Fifth Annual ACM Symposium on the Theory of Computing*, 1993, 561–572.

* Richard Cole is supported in part by NSF grants CCR-8902221, CCR-8906949 and CCR9202900; this research was undertaken, in part, while he was visiting NEC Research Institute.

A Generalization of Binary Search

Richard M. Karp
University of California at Berkeley and
International Computer Science Institute, Berkeley, Ca. *

Abstract

Let f be a nondecreasing integer-valued function whose domain is the set of integers $[0..n]$. The (n, m) problem is the problem of determining f at all points of its domain, given that $f(0) = 0$ and $f(n) = m$. The paper [HM] determines the worst-case number of function evaluations needed to solve the (n, m) problem and gives one particular algorithm achieving the worst-case bound. We obtain the following further results concerning this problem:

- **A family of deterministic algorithms that minimizes the worst-case number of function evaluations needed to solve the (n, m)-problem;**
- **A deterministic algorithm that comes within one step of minimizing the worst-case number of parallel steps required to solve the (n, m)-problem, where a given number p of concurrent function evaluations may be performed in each parallel step. This result requires that $p \leq m$;**
- **A deterministic algorithm that minimizes the expected number of function evaluations when the function f is drawn from a probability distribution satisfying a natural symmetry property;**
- **A randomized algorithm that minimizes the worst-case expected number of function evaluations required to solve the $(n, 1)$-problem;**
- **Lower and upper bounds on the worst-case expected number of function evaluations required by a randomized algorithm to solve the (n, m)-problem for $m > 1$;**

All the algorithms presented in the paper are extremely simple.

The (n, m) problem is equivalent to the following natural search problem: given a table consisting of n entries in increasing order, and given keys $x_1 < x_2 < \ldots < x_m$, determine which of the given keys lie in the table. It is easily seen that the worst-case number of table entries that must be inspected in the search problem is equal to the worst-case number of function evaluations needed to solve the (n, m) problem.

*Research supported by NSF Grant No. CCR-9005448

1 Sequential and Parallel Algorithms that Minimize the Worst-Case Number of Function Evaluations

Let $t(n,m)$ be the worst-case number of function evaluations required to solve the (n,m)-problem. It is well known that $t(n,1) = \lceil \log_2 n \rceil$; several variants of binary search achieve this bound. We note that, when $m \geq n/2$, $t(n,m) = n-1$. Certainly, $t(n,m) \leq n-1$, since the (n,m)- problem can be solved by evaluating f at the points $1,2,\ldots,n-1$. To prove that $t(n,m) \geq n-1$, consider the nondecreasing function f such that $f(n) = m$ and, for $i = 0,1,\ldots,n-1$, $f(i) = \lceil i/2 \rceil$; no algorithm can determine f completely without evaluating it explicitly at all points in $[1..n-1]$.

The paper [HM] gives the following theorem, which determines $t(n,m)$ in general. Let k be the largest integer such that $2^k \leq n/m$.

Theorem 1 *For all $m \geq 2$, and $n \geq m$*

$$t(n,m) = km + \lceil \frac{n}{2^k} \rceil - 1$$

It is also shown in [HM] that there is an optimal algorithm for the (n,m) problem in which the first step is to evaluate $f(2^k)$. The following theorem generalizes this result.

Theorem 2 *Let i be any integer in the range $[0..n]$ such that either i or $n-i$ is a multiple of 2^k. Then there is an optimal algorithm for the (n,m)-problem in which the first step is to evaluate $f(i)$.*

Theorem 2 will follow from the discussion below, in which we give a simple optimal algorithm for the (n,m)-problem which proceeds in $k+1$ phases. We require some preliminary definitions. A set $[i..j]$ of consecutive integers is called an *interval*. The *length* of the interval $[i..j]$ is $j-i$. An interval of length 1 is called a *unit interval*. The interval $[i..j]$ is called a *dead interval* if $f(i) = f(j)$. An interval which is neither a unit interval nor a dead interval is called a *live interval*.

Function evaluations at the points $j_1 < j_2 < \ldots < j_t$ subdivide the interval $[0..n]$ into the subintervals $[0,j_1],[j_1,j_2],\ldots,[j_{t-1},j_t]$. These function evaluations suffice to determine f throughout its domain if and only if none of these subintervals is live. Thus, the task of any algorithm for the (n,m)-problem is to perform function evaluations that subdivide the interval $[0..n]$ into unit intervals and dead intervals. Note that, in any subdivision of $[0..n]$, at most m subintervals are live, since the function f has jumps at no more than m points.

Our algorithm proceeds in $k+1$ phases; within each phase, all the function evaluations can be carried out concurrently. In the first phase $\lceil \frac{n}{2^k} \rceil - 1$ function evaluations are performed so as to subdivide $[0..n]$ into $\lceil \frac{n}{2^k} \rceil$ subintervals of length at most 2^k. For $r = 2, 3, \ldots, k+1$, the rth phase starts with a subdivision in which each live interval is of length less than or equal to 2^{k+2-r}. Within each live subinterval $[i, j]$ of length greater than 2^{k+1-r} a function evaluation is made at the point $i + 2^{k+1-r}$, thus dividing the subinterval into two parts, each of which is of length less than or equal to 2^{k+1-r}. We refer to such a step as *halving the subinterval.* Once these function evaluations have been made, each live interval is of length at most 2^{k+1-r}, so we are ready for the next phase.

After phase $k+1$ a subdivision has been achieved in which each subinterval is either a unit interval or a dead interval; thus the computation is complete. The first phase requires $\lceil \frac{n}{2^k} \rceil - 1$ function evaluations and each subsequent phase requires at most m function evaluations, for a total of at most $km + \lceil \frac{n}{2^k} \rceil - 1$ function evaluations. Thus the algorithm is of optimal worst-case complexity.

Note that there is considerable freedom in choosing the locations of the function evaluations in the first phase and the order in which those function evaluations are performed. Thus, we get many different optimal sequential algorithms and, in particular, Theorem 2 follows.

We note in passing that our $k+1$-phase algorithm can easily be adapted to execute efficiently in parallel, provided that the degree of parallelism does not exceed m. Suppose that, for some $p \leq m$, the computation is to proceed in rounds, with at most p function evaluations per round. Since the $\lceil \frac{n}{2^k} \rceil - 1$ function evaluations to be performed in the first phase are known in advance, the first phase can be carried out in $\lceil \frac{\lceil \frac{n}{2^k} \rceil - 1}{p} \rceil$ rounds. After the first phase, each live subinterval is of size at most 2^k. The rest of the computation can be carried out in a series of rounds, in each of which the p live subintervals of greatest size (or all live subintervals if there are fewer than p) are halved. We leave it to the reader to verify that the number of rounds required for the rest of the computation is at most $\lceil \frac{mk}{p} \rceil$. Thus, the parallel algorithm terminates within $\lceil \frac{\lceil \frac{n}{2^k} \rceil - 1}{p} \rceil + \lceil \frac{mk}{p} \rceil$ rounds. Since $km + \lceil \frac{n}{2^k} \rceil - 1$ is a lower bound on the worst-case number of function evaluations required to solve the (n, m)-problem, any algorithm with degree of parallelism p requires at least $\lceil \frac{km + \lceil \frac{n}{2^k} \rceil - 1}{p} \rceil$ rounds in the worst case. It follows our parallel algorithm comes within one round of optimality.

2 An Algorithm that Minimizes the Expected Number of Function Evaluations

In this section we discuss the problem of minimizing the expected number of function evaluations required to solve the (n, m)-problem when the function f is drawn from a probability distribution. We shall exhibit an algorithm which is optimal for all probability distributions satisfying a natural symmetry property. For $i = 1, 2, \ldots, n$, let $d(i) = f(i) - f(i-1)$. Then $d(i)$ is a nonnegative integer, and $\sum_{i=1}^{n} d(i) = m$. Our symmetry property is that the joint distribution of the random variables $d(i)$ is invariant under all permutations of the index set $\{1,2,\ldots,n\}$. In particular, the following two probability distributions satisfy the symmetry property:

- The distribution in which all nondecreasing integer-valued functions satisfying $f(0) = 0$ and $f(n) = m$ are equally likely;

- The distribution in which all the probability is concentrated on the functions satisfying $f(0) = 0$, $f(n) = m$ and $f(i+1) - f(i) \leq 1$ for all i, and all such functions are equally likely.

Note that the symmetry property is inherited by subproblems: i.e., if f is drawn from a distribution satisfying the symmetry property and we observe $f(n_1)$, then the function f restricted to the domain $[0..n_1]$ or to the domain $[n_1..n]$ has a conditional distribution which satisfies the symmetry property.

Call an algorithm for the (n,m) problem *expected-time optimal* if, for every distribution satisfying the symmetry property, the algorithm minimizes the expected number of function evaluations.

Let $s(n)$ be the largest power of two less than or equal to $\frac{2n}{3}$.

Theorem 3 *There is an expected-time optimal algorithm for the (n,m)-problem. In this algorithm, the first function evaluation is made at the point $s(n)$.*

Note that, because the symmetry property is inherited by subproblems, the rule for choosing the first function evaluation in the overall problem also determines the rule for choosing subsequent function evaluations in the subproblems that arise recursively. Define the $[i..j]$ problem to be that of determining f throughout the domain $[i..j]$, given $f(i)$ and $f(j)$. If $f(i) = f(j)$ then no function evaluations are needed to solve the problem; if $f(i) \neq f(j)$ then the first function evaluation in the subproblem is made at $i + s(j-i)$, thus splitting the problem into the $[i..i+s(j-i)]$ and $[i+s(j-i)..j]$ problems. Let us denote this algorithm by $\mathcal{A}$. We shall prove that $\mathcal{A}$ is expected-time optimal.

Let us call an algorithm for the (n,m) problem *oblivious* if it can be represented by a rooted binary tree in which :

- Each node has a label of the form $[i..j]$, corresponding to the $[i..j]$ problem; such a node is said to be of *weight* $j-i$.

- The root has the label $[0..n]$;

- A node with label $[i..i+1]$ is a leaf;

- A node with label $[i..j]$, $j \geq i+2$, has children of the form $[i..k]$ and $[k..j]$, where $i < k < j$.

If node $[i..j]$ has children $[i..k]$ and $[k..j]$ then, whenever an instance of the $[i..j]$ problem is recursively generated, the algorithm solves the problem without function evaluations if $f(i) = f(j)$, and otherwise makes a function evaluation at k, thus generating instances of the $[i..k]$ and $[k..j]$ problems.

It is not clear *a priori* that, in seeking an expected-time optimal algorithm, one can restrict attention to oblivious algorithms; on the contrary, it seems that, in the problem of determining f throughout the domain $[i..j]$ the optimal choice of the first function evaluation might well depend on quantity $f(i+h) - f(i)$. However, we shall show that $\mathcal{A}$, which is oblivious, is expected-time optimal.

Consider an oblivious algorithm represented by the rooted tree T. Let $[i..j]$ be a node of T. The algorithm will make a first function evaluation in the problem $[i..j]$ if and only if $f(i) \neq f(j)$. Thus, the expected number of function evaluations required by the algorithm is the sum, over all nodes $[i..j]$ in the tree, of the probability that $f(i) \neq f(j)$.

Define $w(v)$, the weight of node v, as the number of leaves in the subtree rooted at v. By the symmetry property the probability that $f(i) \neq f(j)$ depends only on $j - i$, the weight of node $[i..j]$. Let $F(h)$ be the probability that $f(i) \neq f(j)$ when $j - i = h$. Then the oblivious algorithm represented by the tree T executes an expected number of function evaluations given by $\sum_v F(w(v))$, where v ranges over the internal nodes of T. The function F is obviously nondecreasing. We shall prove that it is concave. Consider a probability space of nondecreasing integer-valued functions over the domain $[1..n]$ satisfying the symmetry property. let $J(f)$, the *number of jumps* of f, be defined as $|\{i | d(i) \neq 0\}|$, where $d(i)$ is defined as $f(i) - f(i-1)$. Let $p(k)$ be the probability that $J(f) = k$, where f is drawn from the probability space. Noting that $f(i) = f(j)$ If and only if $d(i) = d(i+1) = \cdots = d(j-1)$, we see that the conditional probability that $f(i) = f(j)$, given that $J(f) = k$, is equal to the ratio of binomial coefficients $\frac{C_k^{n-j+1}}{C_k^n}$, and thus $F(h) = 1 - \sum_{k=1}^m p(k)\frac{C_k^{n-h}}{C_k^n}$, and $F(h) - F(h-1) = \sum_k p(k)\frac{C_k^{n-h+1} - C_k^{n-h})}{C_k^n} = \sum_k p(k)\frac{C_{k-1}^{n-h}}{C_k^n}$, from which it is evident that $F(h) - F(h-1)$ is a decreasing function of h. Thus F is concave.

It is proven in [GK] that, for every nondecreasing concave function F, the minimum value, over all rooted binary trees with n leaves, of $\sum_v F(w(v))$ is achieved by the tree in which, for all h, each internal node of weight h has children of weight $s(h)$ and $h - s(h)$. It follows that, for every probability space of functions satisfying the symmetry property, algorithm $\mathcal{A}$ requires the minimum expected number of function evaluations of any oblivious algorithm for determining f at all points of its domain.

We shall now prove that for all n and m, algorithm $\mathcal{A}$ is expected-time optimal among all algorithms, not just oblivious ones. Let $A(n)$ be the statement that, for all m and all probability distributions for the (n, m) problem satisfying the symmetry property, $\mathcal{A}$ is expected-time optimal. We shall prove by strong induction that $A(n)$ holds for all n. The base cases $n = 1, 2, 3$ are easily checked. Assume the result for all $n' < n$. Consider an algorithm for the (n, m) problem which makes its first function evaluation at n_1. Suppose the first function evaluation yields that $f(n_1) = m_1$. Then we are left with an instance of the (n_1, m_1)-problem and an instance of the $(n - n_1, m - m_1)$-problem. By induction hypothesis, regardless of the value of m_1, an expected-time optimal way to complete the computation is to apply the oblivious algorithm $\mathcal{A}$ to each of these subproblems. Thus, the oblivious algorithm which performs its first function evaluation at n_1 and then continues according to $\mathcal{A}$ minimizes the expected number of function evaluations, among all algorithms for the (n, m) problem which make their first function evaluation at n_1. It follows that, among all algorithms for the (n, m) problem, an oblivious algorithm minimizes the expected number of function evaluations. Since we know that $\mathcal{A}$ is expected-time optimal among oblivious algorithms, $\mathcal{A}$ must be expected-time optimal among all algorithms. This completes the proof of Theorem 3.

3 Randomized Algorithms

A *randomized algorithm* for the (n, m) problem is simply a probability distribution over deterministic algorithms for the problem. The complexity of a randomized algorithm is defined as the maximum, over all f, of the expected number of function evaluations required by the algorithm to determine f throughout its domain. $C(n, m)$, the *randomized complexity* of the (n, m)-problem, is defined as the minimum complexity of any randomized algorithm for the problem.

The following theorem determines the randomized complexity of the $(n, 1)$-problem.

Theorem 4 *Define k by the property that 2^k is the largest power of 2 less than or equal to n. For n even, $C(n, 1) = k + 2 - \frac{2^{k+1}}{n}$. For n odd, $C(n, 1) = \frac{(k+2)n-k-2^{k+1}}{n-1}$.*

Proof: The concept of an oblivious algorithm for the (n, m)-problem was introduced in Section 2. In the case $m = 1$, every deterministic algorithm which performs no redundant function evaluations (a redundant function evaluation is one whose outcome is determined by the results of previous function evaluations) is oblivious and can be represented by a rooted, oriented tree with n leaves, numbered from 1 to n in left-to-right order. Let the function f_i be defined by: $f_i(j)$ is equal to 1 if $j \geq i$ and 0 if $j < i$. Then the set of functions that can be presented to an algorithm for the $(n, 1)$-problem is $\{f_i : i = 1, 2, \ldots, n\}$. The number of function evaluations required by an oblivious algorithm on input f_i is equal to the distance from the root to the ith leaf of the tree that represents the algorithm.

The following statements are easily proved by induction on n:

Claim 1 In any rooted binary tree with n leaves, the average root-leaf distance is greater than or equal to $k + 2 - \frac{2^{k+1}}{n}$;

Claim 2 When n is odd the average distance from the root to an even-numbered leaf in any rooted oriented binary tree with n leaves is greater than or equal to $\frac{(k+2)n-k-2^{k+1}}{n-1}$.

The easy half of Yao's Lemma [Yao] tells us, that, for any probability distribution $\mathcal{D}$ over the set of functions $\{f_i : i = 1, 2, \ldots, n\}$, the randomized complexity of the $(n, 1)$-problem is greater than or equal to the minimum expected number of function evaluations required by a deterministic algorithm to solve the problem when the input function f is distributed according to $\mathcal{D}$. When n is even, take $\mathcal{D}$ to be the uniform distribution over the set $\{f_i : i = 1, 2, \ldots, n\}$. Then, by Claim 1 together with Yao's Lemma, $C(n, 1) \geq k+2-\frac{2^{k+1}}{n}$. When n is odd, take $\mathcal{D}$ to be the uniform distribution over the set $f_2, f_4, f_6, \ldots, f_{n-1}$ of even-indexed functions. Then, by Claim 2 together with Yao's Lemma, $C(n, 1) \geq \frac{(k+2)n-k-2^{k+1}}{n-1}$.

It remains to prove that the stated lower bounds on $C(n, 1)$ are also upper bounds.

When n is even, let the set $\{1, 2, \ldots, n\}$ be partitioned into pairs as follows: $(1, 2), (3, 4), \ldots, (n-1, n)$. It is easy to see that, for any choice of $n-2^k$ out of these $n/2$ pairs, there exists a rooted binary tree such that the leaves corresponding to the $2(n-2^k)$ elements of the $n-2^k$ selected pairs are at distance $k+1$ from the root, and the remaining $2^{k+1} - n$ leaves are at distance k from the root. If we draw a tree from the uniform distribution over this set of trees then, for all i, the expected distance from the root to leaf i is equal to $k+2-\frac{2^{k+1}}{n}$. Thus, if our randomized algorithm is specified as the uniform distribution over the oblivious algorithms corresponding to these trees, the expected number of function evaluations required for any input function f is $k+2-\frac{2^{k+1}}{n}$. This proves that $C(n, 1) \leq k+2-\frac{2^{k+1}}{n}$.

When n is odd, the proof is similar. We consider the following partition of the set $\{1, 2, \ldots, n\}$ into $\frac{n-1}{2}$ pairs and one singleton: $(1, 2), (3, 4), \ldots, (n-2, n-1), (n)$. It is easy to see that for any choice of $n-2^k$ out of the $\frac{n-1}{2}$ pairs in this partition, there exists a rooted binary tree such that the leaves corresponding to the $2(n-2^k)$ elements of the $n-2^k$ selected pairs are at distance $k+1$ from the root, and the remaining $2^{k+1} - n$ leaves are at distance k from the root. If we draw a tree from the uniform distribution over this set of trees then, for all $i \in \{1, 2, 3, \ldots, n-1\}$, the expected distance from the root to leaf i is equal to $\frac{(k+2)n-k-2^{k+1}}{n-1}$ and the distance to leaf n is uniformly k. Thus, if our randomized algorithm is specified as the uniform distribution over the oblivious algorithms corresponding to these trees, the expected number of function evaluations required for any input function f is at most $\frac{(k+2)n-k-2^{k+1}}{n-1}$. This proves that $C(n, 1) \leq \frac{(k+2)n-k-2^{k+1}}{n-1}$. □

We have not succeeded in determining the exact randomized complexity of the (n, m)-problem when $m > 1$. The worst-case complexity of the problem is clearly an upper bound, but the following lemma shows that, in general, the randomized complexity is less than the worst-case complexity.

We close by presenting two easy lower bounds.

Lemma 1

$$C(n,m) \geq \min(2m, n-1)$$

Proof: Consider the function f given by $f(i) = \min(\lceil i/2 \rceil, m)$. For all positive $i \leq \min(2m, n-1)$, $f(i-1) \neq f(i+1)$. No algorithm (deterministic or randomized) can determine f at such a point i unless it explicitly evaluates $f(i)$. Thus, every algorithm must evaluate f at the points $1, 2, \min(2m, n-1)$. □.

To prove our second lower bound, we again invoke the easy half of Yao's Lemma. Consider the distribution $\mathcal{D}$ in which all the probability is concentrated on the functions satisfying $f(0) = 0$, $f(n) = m$ and $f(i+1) - f(i) \leq 1$ for all i, and all such functions are equally likely. Then $C(n,m)$ is greater than the minimum expected number of function evaluations required to solve the (n,m)-problem when the input function is drawn from $\mathcal{D}$. By Theorem 2, this is the expected number of function evaluations required by Algorithm $\mathcal{A}$.

We shall study how this lower bound behaves asymptotically as n and m tend to infinity together with $m = \lfloor \alpha n \rfloor$, where α is a constant between 0 and 1/2. Then $F(h)$, the probability that $f(i) \neq f(i+h)$ tends to $1-(1-\alpha)^h$. Also, as n tends to infinity, almost all the nodes in the tree representing algorithm $\mathcal{A}$ have a weight which is a power of two and, for each i, the number of nodes of weight 2^i is asymptotic to $\frac{n}{2^i}$. Thus, the expected complexity of algorithm $\mathcal{A}$ on an input f drawn from $\mathcal{D}$ is asymptotic to $\sum_{i=1}^{\infty} \frac{1-(1-\alpha)^{2^i}}{2^i}$.

When $\alpha = 1/3$ the worst-case complexity is asymptotic to $\frac{5n}{6}$, the lower bound of Lemma 1 is $\frac{2n}{3}$ and the second lower bound is $.723n$. When $\alpha = 1/4$ the worst-case complexity is asymptotic to $\frac{3n}{4}$, the lower bound of Lemma 1 is $\frac{n}{2}$ and the second lower bound is $.627n$. Thus there is a substantial gap between the lower and upper bounds, and the randomized complexity of the (n,m)-problem is not yet well characterized.

Acknowledgement I thank Noga Alon, Amos Fiat and Donald Knuth for their helpful suggestions. I am particularly grateful to Amos Fiat for discussions that led to the proof of Theorem 4.

References

[GK] C.R. Glassey and R.M. Karp, "On the Optimality of Huffman Trees," *SIAM J. Applied Math*, Vol. 31, No. 2, pp. 368-378, September, 1976.

[HM] R. Hassin and N. Megiddo, "An Optimal Algorithm for Finding All the Jumps of a Monotone Step-Function," *J. Algorithms*, Vol. 6, No. 2, pp. 265-274, June, 1985.

[Yao] A.C.C. Yao, "Probabilistic Computation: Towards a Unified Measure of Complexity," Proc. 18th IEEE Symp. on Foundations of Computer Science, pp. 222-227, 1977.

Groups and Algebraic Complexity

Andrew C. Yao*

Department of Computer Science
Princeton University

Abstract

In recent years, concepts from group theory have played an important role in the derivation of lower bounds for computational complexity. In this talk we present two new results in algebraic complexity obtained with group-theoretical arguments. We show that any algebraic computation tree for the membership question of a compact set S in R^n must have height $\Omega(\log(\beta_i(S))) - cn$ for all i, where β_i are the Betti numbers. We also show that, to compute the sum of n independent radicals using logarithms, exponentiations and root-takings, at least n operations are required. (The second result was obtained jointly with Dima Grigoriev and Mike Singer.) This talk will be self-contained.

***This work was supported in part by NSF Grant CCR-9301430.**

Connected Component and Simple Polygon Intersection Searching*

(extended abstract)

Pankaj K. Agarwal[†] Marc van Kreveld[‡]

Abstract

Efficient data structures are given for the following two query problems: (i) Preprocess a set P of simple polygons with a total of n edges, so that all polygons of P intersected by a query segment can be reported efficiently, and (ii) Preprocess a set S of n segments, so that the connected components of the arrangement of S intersected by a query segment can be reported quickly. In both cases the data structure should return the labels of the intersected polygons or components, not their complete description. Efficient data structures are presented for the static case, the dynamic case, and an efficient on-line construction algorithm for the connected components is given.

1 Introduction

Consider the *colored segment intersection problem,* defined as follows: Given a set S of n segments in the plane and an m-coloring $\chi : S \rightarrow \{1, 2, \ldots, m\}$ of S, preprocess S into a data structure so that the colors of all segments in S intersected by a query segment can be reported efficiently. This problem is motivated by the fact that in several applications the input objects are not segments themselves, but are defined by a set of segments, and one wants to report the objects that intersect a query segment. (Here we assume that the objects are labeled and we just want to return the labels of the objects, not the segments defining the object). By coloring the segments of each object with the same color, we can reduce the problem to the colored segment intersection problem. The following examples illustrate the idea:

1. **Polygon intersection searching:** Let $P = \{P_1, \ldots, P_m\}$ be a set of simple polygons with a total of n edges. We wish to preprocess P, so that the polygons of P intersecting a query segment e can be reported quickly. (We do not want to report all the edges of polygons; we just want to report the indices of the polygons that intersect e.) By assigning the color i to the edges of P_i, one can reduce this problem to colored segment intersection searching.

***Part of this work was done while the second author was visiting the first author at Duke University on a grant of the Dutch Organization for Scientific Research (N.W.O.). The research of the second author was also supported by the ESPRIT Basic Research Action No. 3075 (project ALCOM) and an NSERC International Fellowship. The research of the first author was supported by National Science Foundation Grant CCR-91-06514.**

[†]Computer Science Department, Duke University, Durham, NC 27706, U.S.A.

[‡]School of Computer Science, McGill University, 3480 University St., Montréal, Québec, Canada, H3A 2A7.

2. **Connected component intersection searching:** Let $\mathcal{S}$ be a set of n segments in the plane. The *connected components* in the arrangement $\mathcal{A}(\mathcal{S})$ of $\mathcal{S}$ are the connected components of the planar graph induced by $\mathcal{S}$, which form a partition $\mathcal{S}^1, \ldots, \mathcal{S}^m$ of $\mathcal{S}$. We wish to preprocess $\mathcal{S}$, so that the connected components of $\mathcal{S}$ intersecting a query segment can be reported quickly. (For each $\mathcal{S}^i$, we do not want to report all the segments of the connected components; we just want to return the index i.) By coloring the segments of $\mathcal{S}^i$ with the color i, we can reduce the connected component searching problem to colored segment intersection searching.

The connected components and their labeling are an important concept in image processing, geographic information systems, etc.; see e.g. [5]. The *segment intersection searching problem*, where one wants to report all segments of $\mathcal{S}$ intersecting a query segment, is a special case of the above problem, because, by coloring each segment of $\mathcal{S}$ with a distinct color, we can reduce it to the colored segment intersection searching problem. Other colored intersection searching problems have been studied independently by Janardan and Lopez [8].

In the last few years a lot of work has been done on the segment intersection searching problem, see e.g. [2, 3, 6]. The best known solution is by Agarwal and Sharir [3], who showed that for any fixed $\epsilon > 0$, $\mathcal{S}$ can be preprocessed using $O(n^{1+\epsilon})$ space and time, so that all k segments of $\mathcal{S}$ intersecting a query segment can be reported in time $O(\sqrt{n}\log^{O(1)} n + k)$. The obvious extension to the colored segment intersection problem—just report the color of each intersected segment—is not efficient since the same color may be reported many times. The worst-case query time is $O((1+k)\sqrt{n}\log^{O(1)} n)$, where k is the number of colors reported, and this is much worse than the desired bound of $O(\sqrt{n}\log^{O(1)} n + k)$.

In this paper we present various algorithms for colored segment intersection searching, and on-line construction of the connected components. In Section 2, we present a dynamic solution for colored segment intersection searching for the case where the segments of $\mathcal{S}$ and the query segments are orthogonal. The query time is $O((k+1)\log^2 n)$, where k is the number of colors reported. The insertion or deletion of a segment takes $O(\log^2 n)$ amortized time. An algorithm for dynamic colored segment intersection searching does not immediately yield an algorithm for connected component intersection searching, because an update may affect several connected components of $\mathcal{S}$. We present an $O(n\log^2 n)$ time on-line algorithm to construct the connected components of n orthogonal segments. In the full paper we also show that static colored segment intersection searching can be solved using $O(n\log n)$ storage and $O(\log n + k)$ query time.

Section 3 solves the static version of the connected component searching problem for arbitrary segments. For any fixed $\epsilon > 0$, the data structure has size $O(n^{1+\epsilon})$ and a connected component intersection query can be answered in $O(\sqrt{n}\log^{O(1)} n + k)$ time, where k is the number of connected components intersected by the query segment. In the full paper we also present a variant of this data structure that can answer polygon intersection queries with the same performance.

In Section 4 we describe how the data structure for connected components can be modified so that new segments can be inserted efficiently. The semi-dynamic structure we obtain allows for insertions in $O(n^{1/2+\epsilon})$ amortized time. The amortized query time of the new structure is $O(n^{1/2+\epsilon}+k)$. The insertion of a segment is fairly expensive because before inserting a segment we need to determine the connected components that it intersects, which reduces to answering a query. If we allow the size of the data structure to be $O(n^{4/3+\epsilon})$, the update and query time can be improved to $O(n^{1/3+\epsilon})$ and $O(n^{1/3+\epsilon}+k)$. This leads to an $O(n^{4/3+\epsilon})$ time on-line incremental construction algorithm of the connected components.

An interesting feature of our dynamic algorithms is the *lazy* update of data structures. Part of the work is performed only later by the query algorithms. However, a query is answered correctly at all times. The idea is similar to the union-find structure with path compression [4]. A union performs its task correctly, but makes the structure less efficient. A find operation adjusts the structure so that later find operations can be performed more efficiently. This is exactly what happens in our solution for the maintenance of the connected components.

2 Orthogonal Colored Segment Intersection

In this section we study the dynamic orthogonal colored segment problem. We will store horizontal and vertical segments separately, and we only describe the data structure for the horizontal segments.

We apply the so-called *range restriction* technique, see e.g. [12]. It maintains a balanced binary tree on the y-coordinates of the segments in $\mathcal{S}$, of which each node v is associated with a subset $\mathcal{S}_v \subseteq \mathcal{S}$. At each node v, we preprocess the x-projection of $\mathcal{S}_v$ for *colored interval intersection searching*, i.e., we preprocess a collection of colored intervals into a data structure, so that the colors of all intervals containing a query point can be reported efficiently. The overall query and update time are a factor $O(\log n)$ higher than the query and update time of the secondary structure (see [12] for details). It thus suffices to describe an efficient data structure for the colored interval intersection searching.

Let Δ be a set of n colored intervals. We wish to preprocess Δ, so that an interval can be inserted to or deleted from the structure, and the colors of intervals in Δ containing a query point $x \in \mathbb{R}$ can be reported quickly. We maintain a dynamic interval tree $\mathcal{T}$ that supports insert and delete operations; see Mehlhorn [10, pp. 192–198]. Each node v of $\mathcal{T}$ is associated with a point x_v and a subset $\Delta_v \subseteq \Delta$ such that $x_v \in \delta$ for all $\delta \in \Delta_v$. Let C_v denote the set of colors of intervals in Δ_v. We store the following data structures at each node $v \in \mathcal{T}$.

- CLR_v: It stores C_v as a binary search tree (e.g. red-black tree). For a color i, let L_v^i (and R_v^i) denote the set of left (resp. right) endpoints of intervals in C_v. We store each of them as a balanced binary search tree at the node of CLR_v corresponding to the color i.
- TL_v: For each color in C_v, select the leftmost endpoint of the intervals of color i in Δ_v, and store the set of resulting points in a binary search tree.
- TR_v: For each color in C_v, select the rightmost endpoint of the intervals of color i in Δ_v, and store the set of resulting points in a binary search tree.

Answering a query: Let x be the query point. We want to report the colors of all intervals in $\mathcal{I}$ containing x. Let z be the rightmost leaf of $\mathcal{T}$ such that $x_z \leq x$. At each node v on the path from the root to z, we do the following: If $x > x_v$, then we visit TR_v from right-to-left and report the corresponding colors until we reach an endpoint r such that $r < x$. Otherwise we visit TL_v from left-to-right and report the corresponding colors until we reach an endpoint $l > x$. The total query time can be shown to be $O((1+k)\log n)$, where k is the number of colors reported.

Inserting an interval: Let $\delta = [l, r]$ be an interval that we want to add to Δ, and let i be the color of δ. We first insert δ in $\mathcal{T}$ using the standard insert procedure for interval trees; δ_v is stored at the highest node v of $\mathcal{T}$ such that $x_v \in \delta$. Then we update the secondary structures of v as follows. If $i \notin C_v$, we insert i in CLR_v, and the endpoints l in L_v^i and r in R_v^i. Finally, if l is the leftmost (r is the rightmost) endpoint among all the intervals in Δ_v of color i, we also update TL_v (resp. TR_v). The total amortized insert time can be shown to be $O(\log n)$.

An interval can be deleted from Δ in $O(\log n)$ amortized time using a similar procedure. Here we use CLR_v to find a new leftmost or rightmost endpoint if the present one is deleted.

For the original 2D problem the query and update time are a factor $O(\log n)$ higher, and hence, we obtain:

Theorem 2.1 *A set S of n colored orthogonal segments can be maintained in a data structure of size $O(n \log n)$, so that a segment can be deleted from or inserted in the structure in $O(\log^2 n)$ amortized time, and all k colors of segments intersecting an orthogonal query segment can be reported in time $O((k+1)\log^2 n)$.*

Corollary 2.2 *We can maintain a set P of rectilinear polygons with a total of n vertices in a structure of size $O(n \log n)$, so that a rectilinear m-gon can be inserted to or deleted from the data structure in time $O(m \log^2 n)$, and all k polygons intersecting a query orthogonal segment can be reported in time $O((k+1)\log^2 n)$.*

In the full paper, we show that the static case can be solved slightly more efficiently, using a different data structure. We state the result here, and refer to the full version for details (clearly, the result below implies the same result for orthogonal polygons and orthogonal connected components).

Theorem 2.3 *A set S of n (colored) orthogonal line segments in the plane can be preprocessed in time $O(n \log^2 n)$, into a data structure of size $O(n \log n)$, such that all k colors of segments intersecting a given orthogonal query segment can be reported in $O(\log n + k)$ time.*

Maintaining the connected components of orthogonal segments

The data structure as described above cannot be used for maintaining the connected components of $\mathcal{A}(S)$, because insertion of a segment can merge several connected components into one, therefore, if we label the segments in the ith connected component by color i, then the insertion of segment can change the colors of several segments. It will be very expensive to update the colors of all these segments.

We now briefly describe a variant of the above data structure which maintains the connected components of a set $\mathcal{S}$ of orthogonal segments under insertions and supports the following two queries:

REPORT-COMPONENT $(e, \mathcal{S})$: Report (the label of) all connected components in the arrangement of $\mathcal{S}$ that intersect e.

SAME-COMPONENT $(s_1, s_2, \mathcal{S})$: Decide whether the segments s_1 and s_2 are in the same connected component in $\mathcal{S}$, provided that they are in $\mathcal{S}$ themselves.

Let $\mathcal{S}_i \subseteq \mathcal{S}$ denote the set of segments of color i. We maintain $\mathcal{S}_1, \mathcal{S}_2, \ldots$ in a union-find data structure UF. Two subsets $\mathcal{S}_i, \mathcal{S}_j$ can be merged in $O(\log n)$ amortized time, and the color of a segment can be found in $O(1)$ time; see e.g. [4]. Using the structure UF we can determine in $O(1)$ time whether two query segments $s_1, s_2 \in \mathcal{S}$ lie in the same connected component of $\mathcal{A}(\mathcal{S})$.

Since we allow only insertions, we do not need the structure CLR_v. Moreover, we no longer require the colors of all endpoints in the secondary structures TL_v (and TR_v) to be different (we say that the color of an endpoint is i if it is an endpoint of a segment of color i), which enables us to perform lazy updates.

Next, we describe the procedure REPORT-COMPONENT for a query vertical segment e. In view of the above discussion on applying the range restriction technique, it suffices to describe how to report the colors of all intervals of Δ intersected by a query point x. To query the colored interval tree T, we use the same procedure as described above except that for each endpoint stored in TL_v, TR_v that we report, we perform a 'find' operation in UF with the corresponding segment and report its color. The query procedure also does part of the updating: if we encounter more than one endpoint of color i in TL_v (and TR_v), we discard all endpoints of color i, that we encountered, except the leftmost (resp. rightmost) endpoint. TL_v, TR_v may still have more than one endpoint of the same color, because we do not discard the endpoints that are not visited by the query algorithm. See the full version of the paper for details. The total amortized query time is $O(\log^2 n + k \log n)$.

To insert a new horizontal segment e, we first call the procedure REPORT-COMPONENT on the structure storing the vertical segments (recall that we construct a separate data structure for vertical segments), and determine the colors of all segments intersected by e. Let $i_1, \ldots, i_k$ be the reported colors. Next, we merge the subsets $\mathcal{S}^{i_1}, \ldots, \mathcal{S}^{i_k}$ into a single set $\mathcal{S}^{i_1}$, i.e., we perform k union operations on UF. Finally, we insert e to T as earlier. Notice TL_v, TR_v may now have more than one endpoint of the same color, because we did not update them. It can be shown that the total amortized update time is $O(\log^2 n)$. Hence, we obtain

Theorem 2.4 *We can store a collection $\mathcal{S}$ of n orthogonal segments into a data structure so that a new segment can be inserted in $O(\log^2 n)$ amortized time and the set of connected components intersecting a query segment can be reported in $O(\log^2 n + k \log n)$ amortized time. Given two segments in $\mathcal{S}$, we can determine in $O(1)$ time whether they are in the same connected component of $\mathcal{S}$.*

The above theorem immediately gives an on-line algorithm to compute the connected components of a set of n orthogonal segments. An optimal $O(n \log n)$ solution to the off-line problem was given by Imai and Asano [7].

Corollary 2.5 *There exists an on-line algorithm that computes the connected components of n orthogonal segments in $O(n \log^2 n)$ time.*

3 Connected Component Intersection Searching

In this section we consider the problem of preprocessing a set $\mathcal{S}$ of n (arbitrary) segments, so that the connected components in $\mathcal{A}(\mathcal{S})$ intersected by a query segment can be reported quickly. Notice that if the query object were a line, then the problem can be solved by replacing all segments of every component by the edges of their convex hull. Then the edges of all colored convex hulls can be stored for segment intersection queries. Any color is found at most twice, since any line intersects any convex polygon in at most two edges. When the query object is a segment, a solution is more difficult to obtain. However, the idea of replacing certain groups of segments by their convex hull edges will prove to be useful for segment queries too.

We first compute the connected components of $\mathcal{A}(\mathcal{S})$ in $O(n^{4/3} \log^{O(1)} n)$ time by computing the nonconvex faces of $\mathcal{A}(\mathcal{S})$ (i.e., the faces that contain endpoints of segments in $\mathcal{S}$); see e.g. [1]. The full paper contains more details. Let $\mathcal{S}^1, \ldots, \mathcal{S}^m$ be the partition of $\mathcal{S}$ formed by the connected components of $\mathcal{A}(\mathcal{S})$. For each $i \leq m$, we assign the color i to the segments of $\mathcal{S}^i$. We construct a segment tree $\mathcal{T}$ on $\mathcal{S}$. Each node v of $\mathcal{T}$ is associated with an x-interval δ_v and a vertical strip $I_v = \delta_v \times [-\infty, +\infty]$. A segment $e \in \mathcal{S}$ is associated with a node v if $\delta_v \subseteq e'$ and $\delta_{parent(v)} \not\subseteq e'$, where e' is the x-projection of e. For a node v, let $\mathcal{S}_v$ denote the set of segments associated with v, and let $n_v = |\mathcal{S}_v|$. Let $\mathcal{E}_v$ denote the set of segments associated with the proper descendants of v. The segments of $\mathcal{S}_v, \mathcal{E}_v$ will be referred to as *long* and *short* segments, respectively. We have: $\sum_{v \in \mathcal{T}} |\mathcal{S}_v| = O(n \log n)$, $\sum_{v \in \mathcal{T}} |\mathcal{E}_v| = O(n \log n)$. We clip the segments of $\mathcal{S}_v$ and $\mathcal{E}_v$ within I_v. At each node v, we store two secondary data structures:

(i) We preprocess $\mathcal{S}_v$ so that the colors of all segments of $\mathcal{S}_v$ intersected by a query *segment* can be reported, and

(ii) we preprocess $\mathcal{E}_v$ so that the colors of all segments intersected by a query *line* can be reported quickly.

Preprocessing long segments: Let $\mathcal{S}_v^1, \mathcal{S}_v^2, \ldots, \mathcal{S}_v^r$ be the connected components of $\mathcal{A}(\mathcal{S}_v)$. The segments of the same color (i.e., the segments of the same connected component in $\mathcal{A}(\mathcal{S})$) may split into several connected components of $\mathcal{S}_v$. For each $\mathcal{S}_v^i$, we choose an arbitrary segment $\gamma_v^i \in \mathcal{S}_v^i$ as a *representative* of $\mathcal{S}_v^i$. Let $\Gamma_v = \{\gamma_v^1, \ldots, \gamma_v^r\}$. The segments of Γ_v are pairwise disjoint, and they partition I_v into trapezoids. The structure associated with $\mathcal{S}_v$, TL_v, is a minimum height binary tree on the segments of Γ_v; the leaves of TL_v store the segments of Γ_v from left-to-right in the increasing order of their heights. For each node w, let Γ_{vw} denote the set of segments stored at the leaves of the subtree rooted at w. We store the set of colors of segments of Γ_{vw} in a linked list L_{vw}. At the ith leftmost leaf of the binary tree, we preprocess the lines containing the segments of $\mathcal{S}_v^i$ into a linear size data structure for segment intersection detection queries (i.e., whether a query segment intersects any of the lines) — preprocess the points dual to the lines containing the segments of S_v^i for determining whether a query double-wedge contains any of the points; see [9]. The total time spent in preprocessing S_v, and the size of TL_v, is $O(n_v \log n_v)$.

Preprocessing short segments: Let $\mathcal{F}_v = \{\mathcal{E}_v^1, \ldots, \mathcal{E}_v^t\}$ be the partition of $\mathcal{E}_v$ induced by the connected components of $\mathcal{A}(\mathcal{E}_v)$. We partition the connected components of $\mathcal{E}_v$ into three subsets:

- $\mathcal{F}_v^1$: A connected component $\mathcal{E}_v^i \in \mathcal{F}_v^1$ if there is a segment $e_i \in S$ that intersects a segment of $\mathcal{E}_v^i$ and the left boundary of I_v; either $e \in \mathcal{E}_v^i$, or $e_i \in S_w$ for some ancestor w of v. We will refer to e_i as the *leader* of $\mathcal{E}_v^i$. It can be shown that the leader of $\mathcal{E}_v^i$ can be computed efficiently. If two connected components have the same leader, we merge them into a single component.
- $\mathcal{F}_v^2$: A connected component $\mathcal{E}_v^i \in \mathcal{F}_v^2$ if $\mathcal{E}_v^i \notin \mathcal{F}_v^1$, and if S contains a segment e that intersects $\mathcal{E}_v^i$ and the right boundary of I_v. We refer to e_i as the *leader* of $\mathcal{E}_v^i$.
- $\mathcal{F}_v^3$: All the remaining connected components are in $\mathcal{F}_v^3$. The connected components of S_3 are also the connected components of the whole set S and they lie completely in the interior of I_v. Therefore, the color of all connected components in $\mathcal{F}_v^3$ is distinct.

We process each of the three subsets separately, so that the colors of the segments of each subset intersected by a query line can be reported efficiently.

- TS_v^1: Let e_i denote the leader of $\mathcal{E}_v^i$, and let p_i be the intersection point of e_i and the left boundary of I_v. Let $P_v^1 = \{p_i \mid \mathcal{E}_v^i \in \mathcal{F}_v^1\}$. Assume that the points in P_v^1 are sorted by their y-coordinates. TS_v^1 is a balanced binary tree whose leaves store P_v^i. For a node $w \in TS_v^1$, let $\mathcal{F}_{vw}^1 \subseteq \mathcal{F}_v^1$ be the set of connected components $\mathcal{E}_v^j$ such that p_j is stored at a leaf of the subtree rooted at w. For a color j, let Γ_{vw}^j denote the set of segments of color j, defined as $\Gamma_{vw}^j = \bigcup\{\mathcal{E}_v^i \mid \mathcal{E}_v^i \in \mathcal{F}_{vw}^1 \text{ and } \chi(\mathcal{E}_v^i) = j\}$. We compute the convex hull of Γ_{vw}^j, and let E_{vw}^j denote the set of edges in $CH(\Gamma_{vw}^j)$, and $E_{vw} = \bigcup_j E_{vw}^j$. We preprocess E_{vw} for line intersection queries using the algorithm described in [3]; all k segments of E_{vw} intersected by a query line can be reported in time $O(\nu \log^{O(1)} \nu + k)$, where $\nu = |E_{vw}|$.
- TS_v^2: Analogous to the previous structure, but for the components of $\mathcal{F}_v^2$.
- TS_v^3: For each connected component $\mathcal{E}_v^i \in \mathcal{F}_v^3$, we compute the convex hull of its segments. Let E_i denote the set of edges in the resulting convex hull. We preprocess all the edges for line intersection queries as above.

Answering a query: Let $e = \overline{pq}$ be the query segment, and let p be the left endpoint of e. We want to report the colors of segments in S intersecting e. We search with p in the segment tree T and find a leaf z such that $p \in I_z$. Let π_p denote the path from the root of T to z. Similarly, we compute the path π_q for q. At each node v

on π_p or π_q, we query the secondary structure constructed on S_v with $\hat{e} = e \cap I_v$ and report the colors of the segments in S_v intersected by $\hat{e}$. By searching with the endpoints $\hat{p}$ and $\hat{q}$ of $\hat{e}$ in TL_v (the tree constructed on long segments S_v), we compute $O(\log n)$ maximal subtrees that lie between the search paths to $\hat{p}$ and $\hat{q}$. At the root u of each such subtree, we report all colors in the associated lists L_{vu}. We also report the connected component S_v^j lying immediately above (and below) the upper (resp. lower) endpoint of $\hat{e}$; this can be tested by using the line-intersection detection structure constructed on S_v^j. It can be shown that the time spent in searching over all long segments is $O(\sqrt{n}\log^{O(1)} n + k_l \log^2 n)$, where k_l is the number of colors reported.

Next, let u be the node at which π_p and π_q bifurcate. For every descendant w of u that is also the right (left) child of a node on π_p (resp. π_q), but which is not on π_p (resp. π_q) itself, we report the colors of the segments in $\mathcal{E}_w$ intersected by the line ℓ containing e. Let λ be the intersection point of e and the left boundary of I_w. We search TS_w^1 with λ and determine the leaf ξ that stores the point lying immediately above λ. The subsets associated with the descendants of the nodes on the path from the root of TS_w^1 to ξ partition the connected components of $\mathcal{F}_w^1$ into $O(\log n)$ canonical subsets such that, for every connected component $\mathcal{E}_v^j$ within each subset, either p_j lies above λ, or it lies below λ for all of them. Let ζ be such a node. It is easily seen that ℓ intersects a segment of $\Gamma_{w\zeta}^j$ if and only if ℓ intersects an edge of $E_{w\zeta}^j$. Therefore we query the line-intersection detection structure for $E_{v\zeta}$ and report the colors of all segments (the segments of $E_{v\zeta}$) intersected by ℓ. Each color is reported at most twice at ζ. Next, we repeat a similar procedure with TS_w^2. Finally, we search TS_w^3 with ℓ and report all colors of the convex hull edges intersected by this line.

It can be shown that the overall time spent in querying the short segments is $O(\sqrt{n}\log^{O(1)} n + k_s \log^2 n)$, where k_s is the number of colors found. Thus the overall query time is $O(\sqrt{n}\log^{O(1)} n + k\log^2 n)$. By replacing all binary trees with trees of degree $O(n^\epsilon)$, for any $\epsilon > 0$, the query time can be improved to $O(\sqrt{n}\log^{O(1)} n + k)$ at a slight increase in the size of the data structure. Omitting all the details, we conclude

Theorem 3.1 *For any constant $\epsilon > 0$, a set of n line segments in the plane can be preprocessed in $O(n^{4/3}\log^{O(1)} n)$ time into a data structure of size $O(n^{1+\epsilon})$, so that all k connected components intersecting any query segment can be reported in $O(\sqrt{n}\log^{O(1)} n + k)$ time.*

In the full version we show that a modification of the above data structure can be used for simple polygon intersection searching. We only state the result here:

Theorem 3.2 *For any constant $\epsilon > 0$, a set $\mathcal{P}$ of simple polygons with a total of n edges can be preprocessed in time $O(n^{1+\epsilon})$ into data structure of size $O(n^{1+\epsilon})$, so that all k polygons intersecting a query segment can be reported in time $O(\sqrt{n}\log^{O(1)} n + k)$.*

4 Maintaining the Connected Components

Next we describe a semi-dynamic data structure for maintaining the connected components in the arrangement of a set S of arbitrary line segments in the plane under insertions. We will extend the data structure given in Section 3 using the ideas of the orthogonal case and some new ideas. The data structure supports three operations: inserting a new segment, and answering two types of queries REPORT-COMPONENT and SAME-COMPONENT as defined in Section 2.

Let $S_i \subseteq S$ denote the segments in the ith connected component of $\mathcal{A}(S)$. We assume that the color of the segments in the ith connected component is i (the connected components are indexed arbitrarily). As in Section 3, we will maintain $S_1, S_2, \ldots$ using a union-find data structure UF, so that the color of a segment can be found in $O(1)$ time and two sets can be merged in $O(\log n)$ amortized time. The basic data structure is the same as in Section 3. However, we have to modify the secondary structures stored at each node of $\mathcal{T}$ in order to facilitate the insert and query operations. The secondary structures (particularly the structures constructed on short segments) maintain weaker invariants, which enables us to postpone some of the changes that need to be made in the secondary structures as we insert a new segment; see below for details. We also maintain some additional structures. Due to lack of space, we only briefly describe the data structure and various operations. See the full version for details.

The overall data structure

We maintain a dynamic segment tree $\mathcal{T}$ on the segments in S. The overall structure of $\mathcal{T}$ is the same as in the previous section, i.e., each node $v \in \mathcal{T}$ stores a subset S_v of long segments and another set $\mathcal{E}_v$ of short segments. All the segments in $\mathcal{E}_v$ and S_v are clipped within the vertical strip I_v.

Storing long segments: As earlier, we construct a balanced binary tree TL_v on the representatives of each connected component of $\mathcal{A}(S_v)$. But TL_v now also supports insertion and deletion operations. Using the standard partial reconstruction method, a representative can be inserted to or deleted from TL_v in $O(\log^2 n)$ amortized time. Since the colors of representatives change dynamically, L_{vw} now stores (at least) one representative of each color in the set Γ_{vw} instead of the color itself (i.e., for each color present in Γ_{vw}, we pick an arbitrary segment of that color). The colors of these segments are obtained using UF. Moreover, we do not update the list L_{vw} as soon as the color of a segment in Γ_{vw} changes, so there may be more than one representative in L_{vw} of one color. We postpone the update of L_{vw} until w is visited by the REPORT-COMPONENT procedure. At each leaf of TL_v, storing the representative γ_i, we store S_v^i in a dynamic data structure for segment intersection detection queries; the query time is $O(m^{1/2+\epsilon})$ and amortized update time is $O(m^\epsilon)$, where $m = |S_v^i|$.

We preprocess the lines containing all segments of S_v into a dynamic data structure for segment intersection detection queries. We also assume that if a query segment intersects a line, the structure can return one of the lines intersected by the query segment. This data structure will be used to find the new leaders of short segments stored at descendants of v.

Storing short segments: Next, we describe the modifications required in the secondary structure storing the short segments. We maintain a partition $\mathcal{E}_v^1, \ldots, \mathcal{E}_v^t$ of $\mathcal{E}_v$ such that each $\mathcal{A}(\mathcal{E}_v^i)$ is a connected planar graph. However, unlike the static data structure, $\mathcal{A}(\mathcal{E}_v^i)$ is not necessarily a connected component of $\mathcal{A}(\mathcal{E}_v)$ (i.e., it may only be a subgraph of a connected component of $\mathcal{A}(\mathcal{E}_v)$). We will call each $\mathcal{E}_v^i$ a *group* of $\mathcal{E}_v$. The groups of $\mathcal{E}_v$ are partitioned into three subsets $\mathcal{F}_v^1, \mathcal{F}_v^2$ and $\mathcal{F}_v^3$, as earlier. Each group stored in $\mathcal{F}_v^1$ (and $\mathcal{F}_v^2$) satisfies the same condition as in the static structure, i.e., there is a segment $e_i \in \mathcal{S}$ that intersects the left (resp. right) boundary of I_v and a segment of $\mathcal{E}_v^i$; e_i is called the *leader* of $\mathcal{E}_v^i$. However, some of the groups satisfying this condition may be in $\mathcal{F}_v^3$. Intuitively, the reason is that when we insert a new segment e, it may intersect the left or right boundary of I_v and also a group in $\mathcal{F}_v^3$. It will be too expensive to detect all such groups of $\mathcal{F}_v^3$ and to move them to $\mathcal{F}_v^1$ or $\mathcal{F}_v^2$. Instead, we wait until v is visited by the REPORT-COMPONENT procedure and then all groups of $\mathcal{F}_v^3$ intersected by the query segment are checked whether they should be moved to $\mathcal{F}_v^1$ or $\mathcal{F}_v^2$. The groups of $\mathcal{F}_v^3$ maintain the following crucial property: If two groups $\mathcal{E}_v^i$ and $\mathcal{E}_v^j$ have the same color, then either they are connected to each other, or they are both connected to the boundary of I_v.

We now describe how to preprocess $\mathcal{F}_v^1$. Let p_j be the left endpoint of the leader of $\mathcal{E}_v^j \in \mathcal{F}_v^1$. We define the weight of p_j to be the number of segments in $\mathcal{E}_v^j$. Let $P_v^1 = \{p_j \mid \mathcal{E}_v^j \in \mathcal{F}_v^1\}$. We construct a weight balanced binary tree TS_v^1 on P_v^1. We define $\mathcal{F}_{vw}^1$ as earlier. Let $\Gamma_{vw} = \bigcup\{\mathcal{E}_v^j \mid \mathcal{E}_v^j \in \mathcal{F}_v^1\}$. We maintain a partition $\Gamma_{vw}^1, \ldots, \Gamma_{vw}^t$ of Γ_{vw} at w; all segments in each Γ_{vw}^i have the same color. However two different subsets may have the same color. The query procedure periodically merges some of the subsets Γ_{vw}^i of the same color. We store the (colored) segments of Γ_{vw} into a data structure based on the dynamic convex hull structure of Preparata [11] and the dynamic partition tree of Agarwal and Sharir [3], which supports the following operations:

(i) Report the edges of the convex hulls $CH(E_{vw}^i)$ intersecting a query line ℓ.

(ii) Insert a segment e of color i in $\mathcal{E}_{vw}$.

(iii) Merge two subsets E_{vw}^i and E_{vw}^j.

We can show that a query can be answered in time $O(m^{1/2+\epsilon} + k)$, and that the insert and merge operations require $O(m^\epsilon)$ amortized time, where k is the output size. This data structure allows to report the colors of segments in E_{vw} intersected by a query line.

$\mathcal{F}_v^2$ is processed in an analogous manner, and $\mathcal{F}_v^3$ is preprocessed in the same way as we preprocessed Γ_{vw}. We also maintain the segments of $\mathcal{E}_v$ for line intersection detection queries [3]. This completes the description of our data structure. It can be shown that the overall size of the structure is $O(n^{1+\epsilon})$.

Inserting a new segment

Let e be a segment that we want to add to $\mathcal{S}$. The insertion procedure consists of the following four steps:

1. Using the REPORT-COMPONENT procedure, determine the colors of all segments that e intersects. Let $i_1, \ldots, i_k$ be the colors of these segments. We

merge the sets $S^{i_1}, \ldots, S^{i_k}$ into a single set, say S^{i_1}, and add e to S^{i_1}. Set the color of all segments in S^{i_1} to i_1.

2. Insert e to the segment tree T; e is stored at $O(\log n)$ nodes as long and short segments.

3. Let v be a node where $\bar{e} = e \cap I_v$ is stored as short segment. We simply create a new group $\mathcal{E}_v^i$ consisting of only $\bar{e}$. We also add $\bar{e}$ to the line intersection detection structure. All the other updates (merging of groups intersecting $\bar{e}$) are postponed until v is visited by the REPORT-COMPONENT procedure.

4. If e is stored at a node v as a long segment, then update the secondary structure stored at v. Let $\bar{e} = I \cap e$. First, we add the line containing e to the segment intersection detection structure. Next, we delete from TL_v the representatives of all connected components of $\mathcal{A}(S_v)$ that intersect $\bar{e}$. (But we do not delete these representatives from the lists L_{vw}; this will be done when REPORT-COMPONENT visits w.) Merge the partition trees stored at the deleted leaves into a single structure, say Ψ. Add the line containing e to this structure. Finally, we add a new leaf ξ to TL_v, store Ψ at ξ, and add e to L_{vw} for every node w on the path from the root to ξ.

Answering a query

Let e be a query segment. We want to report the colors of all segments in S intersecting e. The query is answered as in the previous section. The only difference is that we use UF to find the current colors of segments. For example, the lists L_{vw} now store segments instead of colors, so we find the colors of these segments using UF. We do the same for short segments.

The query procedure also performs a number of updates which were not done while inserting a new segment. If the query procedure visits a node w and reports the colors of segments stored in L_{vw}, we delete all the representatives from L_{vw} that have been deleted from TL_v. Moreover, if we find two segments in L_{vw} of the same color, we arbitrarily delete one of them.

Suppose we report the colors of short segments stored at a node $v \in T$. If we find two subsets $\Gamma_{vw}^i, \Gamma_{vw}^j \subseteq \Gamma_{vw}$ of the same color at node w of TS_v^1 or TS_v^2, which intersect the line containing e, then we merge them. Next, if we find two groups $\mathcal{E}_v^i, \mathcal{E}_v^j$ of the same color in $\mathcal{F}_v^3$, then there must be a connection between them, since they are in the same overall connected component. So either they can be merged (when they are connected inside I_v), or they can be moved to $\mathcal{F}_v^1$ or $\mathcal{F}_v^2$ (if they are connected outside I_v). A leader of $\mathcal{E}_v^i$ is found using the segment intersection searching structure constructed on the lines supporting the long segments at ancestors of v. We omit the details from this version.

The analysis

The analysis of the insert and query procedures is rather involved, so we omit it from this version of the paper. The two main ingredients of the analysis are partial rebuilding of the trees (the primary segment tree and the secondary trees TL_v, TS_v^1 and TS_v^2), and the amortized analysis using the charging scheme. Roughly speaking, we assign $O(n^{1/2+\epsilon}) \cdot O(\log n)$ credits to each segment, which pays for various merge operations, for moving a group from $\mathcal{F}_v^3$ to $\mathcal{F}_v^1$, or the extra time spent by the query procedure. Omitting all the details we conclude

Theorem 4.1 *A set S of n segments can be stored in a data structure of size $O(n^{1+\epsilon})$ so that a new segment can be inserted in $O(n^{1/2+\epsilon})$ amortized time, and the set of all k connected components intersecting a query segment can be reported in $O(n^{1/2+\epsilon}+k)$ amortized time. Given two segments in S, we can determine in $O(1)$ time whether they are in the same connected component of S.*

If we allow more space, the update and query time can be improved. In particular, if we allow $O(n^{4/3+\epsilon})$ space, the amortized update and query time can be improved to $O(n^{1/3+\epsilon})$ and $O(n^{1/3+\epsilon}+k)$, respectively. An immediate corollary of this result is an $O(n^{4/3+\epsilon})$ time algorithm for computing the connected components of S on-line.

Corollary 4.2 *The connected components of n segments in the plane can be computed by an on-line algorithm in $O(n^{4/3+\epsilon})$ time.*

References

[1] Agarwal, P. K., Partitioning Arrangements of Lines: II. Applications, *Discr. & Comp. Geom.* 5 (1991), pp. 533–573.

[2] Agarwal, P. K., Ray Shooting and other Applications of Spanning Trees with Low Stabbing Number, *SIAM J. Computing* 21 (1992), 540–570.

[3] Agarwal, P. K., and M. Sharir, Applications of a New Space Partitioning Technique, *Discrete & Computational Geometry* 9 (1993), 11–38.

[4] Cormen, T. H., C. E. Leiserson, and R. L. Rivest, *Introduction to Algorithms*, MIT Press, Cambridge, 1990.

[5] Dillencourt M., H. Samet, and M. Tammiuen, A general approach to connected component labeling for arbitrary image representation, *JACM* 39 (1992), pp. 253–280.

[6] Dobkin, D. P., and H. Edelsbrunner, Space Searching for Intersecting Objects, *J. Algorithms* 8 (1987), pp. 348–361.

[7] Imai, H., and T. Asano, Finding the Connected Components and a Maximum Clique of an Intersection Graph of Rectangles in the Plane, *J. Algorithms* 4 (1984), pp. 310–323.

[8] Janardan, R., and M. Lopez, *Generalized Intersection Searching Problems, Int. J. Comp. Geom. & Appl.* 3 (1993), 39–70.

[9] Matoušek, J., Efficient Partition Trees, 8 (1992), pp. 315–334.

[10] Mehlhorn, K., *Data Structures and Algorithms 3: Multi-dimensional Searching and Computational Geometry*, Springer-Verlag, Heidelberg, 1984.

[11] Preparata, F.P., An optimal real time algorithm for planar convex hulls, *CACM* 22 (1979), pp. 402–405.

[12] Scholten, H. W., and M. H. Overmars, General Methods for Adding Range Restrictions to Decomposable Searching Problems, *J. Symb. Comp.* 7 (1989), pp. 1–10.

An Optimal Algorithm for Finding the Separation of Simple Polygons

Nancy M. Amato*

University of Illinois at Urbana-Champaign

Abstract. Given simple polygons P and Q, their *separation*, denoted by $\sigma(P,Q)$, is defined to be the minimum distance between their boundaries. We present an optimal $\Theta(N)$ algorithm for determining the separation of two simple polygons P and Q, where $|P|+|Q|=|N|$. The best previous algorithm for this problem is due to Kirkpatrick and has complexity $O(N\log N)$. In addition, a parallel version of our algorithm can be implemented in $O(\log N)$ time using $O(N)$ processors on a CREW PRAM. Our results are obtained by providing a unified treatment of the separation and the closest visible vertex problems for simple polygons.

1 Introduction

The problem of computing the minimum distance between two simple polygons P and Q has received much attention in the literature. The two most common variants of this problem are (i) finding the minimum distance between the boundaries of the two polygons (this problem is also known as finding the *separation* of the two polygons [DK90]), and (ii) finding the minimum distance between visible vertices on the boundaries of the two polygons, where two vertices $p \in P$ and $q \in Q$ are said to be visible if $\overline{pq}$ does not properly intersect P or Q.

Although the separation and the closest visible vertex problems seem to be closely related, it appears no benefit for either problem has been gained by this similarity in the past; in fact, the algorithms proposed for these two problems have differed greatly from one another. In this paper we show that there is no underlying geometric reason for these algorithmic differences by providing a unified treatment for both of these problems. In particular, we show that the general strategy used in [A92a] to optimally solve the closest visible vertex problem for simple polygons can indeed be used to optimally solve the separation problem as well. More specifically, although the geometric basis and the actual implementation of each step of the algorithm presented here differ significantly from those of the closest visible vertex algorithm of [A92a], the overall structure of the two algorithms is similar and we show that many of the ideas first appearing in algorithms for the closest visible vertex problem [CW83, AMSS89, A92a] prove valuable for the separation problem as well.

Computing the separation of two polygons P and Q, with $|P|+|Q|=N$, denoted by $\sigma(P,Q)$, has been studied extensively. The separation problem has been addressed

* Mailing address: University of Illinois at Urbana-Champaign, Coordinated Science Laboratory, 1308 W. Main St., Urbana, IL 61801. Email: amato@cs.uiuc.edu. This work was supported in part by the Joint Services Electronics Program (U.S. Army, U.S. Navy, U.S. Air Force) under contract N00014-90-J-1270 and NSF Grant CCR-89-22008.

sequentially in various cases: when both P and Q are convex their separation can be determined in $\Theta(\log N)$ time [CW83, E85, CD87, DK90], when only one polygon is convex their separation can be determined in $\Theta(N)$ time [CWS85], and when neither polygon is convex their separation can be found in $O(N \log N)$ time using a technique due to Kirkpatrick which traverses the contour (external skeleton) between the polygons in their generalized voronoi diagram [K79]. The parallel complexity of the separation problem has only been addressed in the special case in which both polygons are convex: [AG88] gives an algorithm for a CREW PRAM having $N^{1/k}$ processors that requires $O(k^{1+\epsilon})$ time, and in [DaK89] it is shown that time $O(\log N/(1+\log p))$ is sufficient on a CREW PRAM with p processors. Note that both of the above algorithms essentially achieve constant time if a linear number of processors is available. Since generalized voronoi diagrams can be constructed in $O(\log^2 N)$ time using $O(N)$ processors on a CREW PRAM [GOY89], the separation of two simple polygons, in the most general case, can be found within these same time and processor bounds by a naive parallelization of Kirkpatrick's sequential approach.

This paper studies, both sequentially and in parallel, the problem of determining $\sigma(P,Q)$ in the most general case, i.e., when P and Q are nonconvex simple polygons. A new sequential algorithm is presented that computes $\sigma(P,Q)$ optimally in $\Theta(N)$ time; this improves on the complexity of the technique proposed in [K79] by a factor of $O(\log N)$. (Optimality follows because [CWS85] establishes a linear lower bound even in the case when exactly one of the polygons is convex.) Our algorithm can be implemented in parallel; the complexity of our parallel version of the algorithm is $O(\log N)$ time using $O(N)$ processors on a CREW PRAM, which improves by a factor of $O(\log N)$ on the time complexity of the naive parallelization of Kirkpatrick's sequential algorithm. Therefore, the contribution of this paper is three-fold: (i) an optimal sequential algorithm for computing $\sigma(P,Q)$ for two nonconvex polygons P and Q; (ii) a unified treatment of the separation and closest visible vertex problems; (iii) a more efficient parallel algorithm for the separation problem.

We recall that Chazelle's linear time algorithm for triangulating a simple polygon can be adapted to determine whether two simple polygons P and Q intersect in time $O(N)$ [C90]; we will refer to this adapted algorithm as the (linear time) *intersection test*. Note that $\sigma(P,Q) = 0$ if and only if $P \cap Q \neq \emptyset$. Thus, if $\sigma(P,Q)$ can be computed in linear time when P and Q do not intersect, then this technique coupled with the intersection test will yield an optimal algorithm for finding the separation of any two simple polygons. In the remainder of the paper we explain how $\sigma(P,Q)$ can be determined in $\Theta(N)$ time if $P \cap Q \neq \emptyset$.

2 Nonintersecting Simple Polygons

As discussed above, a preprocessing step of the separation algorithm performs an intersection test: if $P \cap Q \neq \emptyset$, then $\sigma(P,Q) = 0$ and no further computation is required. Consequently, in this section we consider the case in which the simple polygons P and Q do not intersect, i.e., in this section it is assumed that $P \cap Q = \emptyset$.

As seems natural, the problem of determining the separation of two polygons P and Q, $\sigma(P,Q)$, is closely related to the problem of finding the closest visible vertex distance between P and Q, denoted by $CVV(P,Q)$. In fact, as we will see below, many of the techniques used to find $CVV(P,Q)$ can be adapted to our present problem of determining $\sigma(P,Q)$. Recall that $\sigma(P,Q)$ is either realized by a vertex pair or by a vertex and a edge, and note that in the former case $\sigma(P,Q) = CVV(P,Q)$ and a pair realizing $CVV(P,Q)$ also realizes $\sigma(P,Q)$. Thus, in order to extend the techniques for computing $CVV(P,Q)$ to the more general problem of finding $\sigma(P,Q)$, we must augment them to consider vertices and edges of the two polygons, rather than only considering vertices. Our approach is loosely modeled after the method of computing $CVV(P,Q)$ described in [A92a], which uses ideas that first appeared in the *CVV* algorithms of Wang and Chan [WC86] and Aggarwal *et al.* [AMSS89]; we note here that all of the above cited CVV algorithms are restricted to the case in which the two polygons are *nonintersecting*.

As in the *CVV* algorithms of [A92a], [AMSS89], and [WC86], we reduce the problem of computing $\sigma(P,Q)$ to solving a number of restricted versions of the problem. Specifically, the primitive operation is: compute $\sigma(P',Q')$ where P' and Q' are *linearly separable subchains* of P and Q, respectively. In this section we describe how to decompose the original problem into a set of these primitive operations, and in the next section we show how each of these operations can be solved in time linear in its size. Although the actual decomposition of the original problem into subproblems is necessarily different than that used in the above CVV algorithms, in all cases the decomposition is based on a sequence of line segments separating P and Q.

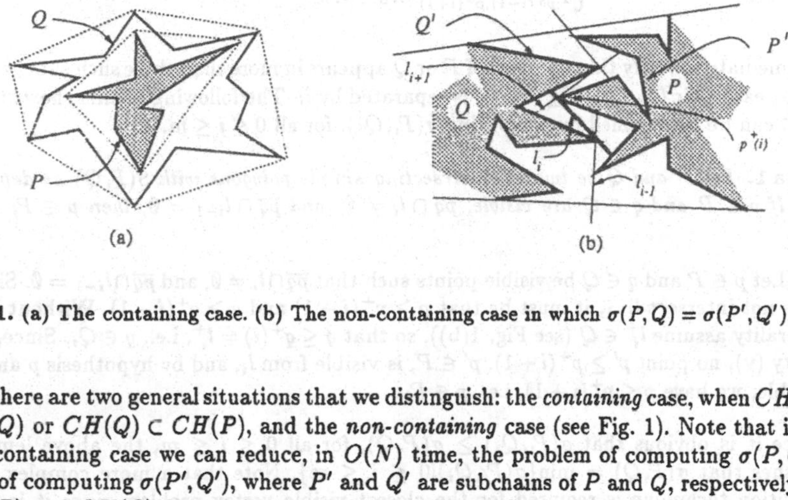

Fig. 1. (a) The containing case. (b) The non-containing case in which $\sigma(P,Q) = \sigma(P',Q')$.

There are two general situations that we distinguish: the *containing* case, when $CH(P) \subset CH(Q)$ or $CH(Q) \subset CH(P)$, and the *non-containing* case (see Fig. 1). Note that in the non-containing case we can reduce, in $O(N)$ time, the problem of computing $\sigma(P,Q)$ to that of computing $\sigma(P',Q')$, where P' and Q' are subchains of P and Q, respectively. Let $CH(P)$ and $CH(Q)$ denote the convex hulls of P and Q, respectively, and recall that the convex hull of a simple polygon with m vertices can be found in time $O(m)$ (see, e.g., [PS85]). Even though $CH(P)$ and $CH(Q)$ may intersect, it is a simple matter to verify that the convex polygon separation technique of Dobkin and Kirkpatrick [DK90] can be used to find the two lines tangent to $CH(P)$ and $CH(Q)$ in time $O(\log N)$. The facing portions of P and Q between these two tangents (P' and Q', respectively) contain all points of P and Q that could potentially realize $\sigma(P,Q)$. In the non-containing case, we assume that the subchains P' and Q' are indexed bottom-to-top, and that $|P'| = n_p$ and $|Q'| = n_q$; in the containing case, the vertices are also indexed consecutively, but there is no stipulation as to where the indexing begins.

In the following, the terms "highest" and "lowest" refer to positions of points (which follow the ordering of the indices of the vertices) on the polygons, and P_{p_i,p_j} denotes the subchain $(p_i, p_{i+1}, \ldots, p_j)$; Q_{q_i,q_j} is defined analogously. Following [A92a], a sequence of (intersecting) line segments $S(P,Q) = (l_0, l_1, \ldots, l_{m-1})$, where $l_i = (l_i^-, l_i^+)$, is called a *separator* of P and Q if, for $0 \leq i < m$, (i) $l_i \cap l_{i+1} \neq \emptyset$, (ii) $l_i \cap l_j = \emptyset$ for $j \notin \{i-1, i, i+1\}$, (iii) l_i does not intersect the interior of P or Q, (iv) l_i^- and l_i^+ lie on the boundaries of P and/or Q, and (v) each l_i is maximal in the following sense: if $l_i^+ \in P$ and $l_{i+1}^+ \in Q$, then no point of Q above l_{i+1}^+ is visible from l_i, and if $l_i^+ \in P$ and $l_{i+1}^+ \in P$, then l_{i+1} intersects the highest point (a vertex) of Q that is visible from l_i; analogous statements hold if $l_i^+ \in Q$ (see Fig. 1(b)). (In the containing case, all arithmetic is modulo m and we require $l_{m-1} \cap l_0 \neq \emptyset$.)

Note that property (v) ensures that the interior of each l_i, $0 \leq i < m$, intersects the boundary of P and/or Q; more specifically, this property guarantees that each l_i intersects the boundaries of both P and Q, either at an endpoint or an interior point. Let $p^+(i)$ and $q^+(i)$ denote the highest points of intersection of l_i with P and Q, respectively. We assign a subchain P_i of P to segment l_i, $0 \leq i < m$, as follows; a subchain Q_i of Q is assigned to l_i analogously.

$$P_i = \begin{cases} P_{p^+(i-1),p^+(i)} & \text{if } l_i^+ \in P \text{ or } i = m-1 \\ P_{p^+(i-1),p^+(i+1)} & \text{otherwise} \end{cases}$$

It is immediate to verify that no point of P or Q appears in more than three such subchains, and that each pair (P_i, Q_i), $0 \leq i \leq m$, is separated by l_i. The following lemma shows that $\sigma(P,Q)$ can be determined by computing $\sigma(P_i, Q_i)$, for all $0 \leq i \leq m$.

Lemma 1. *Let P and Q be two nonintersecting simple polygons with $S(P,Q)$ as defined above. If $p \in P$ and $q \in Q$ are visible, $\overline{pq} \cap l_i \neq \emptyset$, and $\overline{pq} \cap l_{i-1} = \emptyset$, then $p \in P_i$ and $q \in Q_i$.*

Proof. Let $p \in P$ and $q \in Q$ be visible points such that $\overline{pq} \cap l_i \neq \emptyset$, and $\overline{pq} \cap l_{i-1} = \emptyset$. Since $\overline{pq}$ does not intersect l_{i-1}, it must be that $p \geq p^+(i-1)$ and $q \geq q^+(i-1)$. Without loss of generality assume $l_i^+ \in Q$ (see Fig. 1(b)), so that $q \leq q^+(i) = l_i^+$, i.e., $q \in Q_i$. Since, by property (v), no point $p' \geq p^+(i+1)$, $p' \in P$, is visible from l_i, and by hypothesis p and q are visible, we have $p \leq p^+(i+1)$, i.e., $p \in P_i$. □

Since it is obvious that $\sigma(P_i, Q_i) \geq \sigma(P,Q)$, for all $0 \leq i \leq m$, the above lemma establishes that $\sigma(P,Q) = \min\{\sigma(P_i, Q_i) | 0 \leq i \leq m\}$. Note that a more complex decomposition technique is required for the closest visible vertex problem since it is not necessarily true that $CVV(P_i, Q_i) \geq CVV(P,Q)$. Assuming that $\sigma(P_i, Q_i)$ can be determined in $O(|P_i| + |Q_i|)$ time, as will be established in Section 3, we have the following theorem.

Theorem 1 *If P and Q are two nonintersecting simple polygons, where $|P| + |Q| = N$, then $\sigma(P,Q)$, and a pair of points realizing it, can be computed optimally in $\Theta(N)$ time.*

Proof. First, the technique of [A92a] is used to compute a separator $S(P,Q)$ satisfying properties (i-v) in $O(N)$ time; this technique constructs $S(P,Q)$, in both the containing and non-containing cases, by modifying a shortest path between two distinguished vertices in a simple polygon that lies between P and Q. Then, since each point of P and Q appears in at most three subchains, and $\sigma(P_i, Q_i)$ can be determined in $O(|P_i| + |Q_i|)$ time, for all $0 \leq i \leq m$, we see that $\sigma(P, Q) = \min\{\sigma(P_i, Q_i) | 0 \leq i \leq m\}$ can be found in $O(\sum_{i=0}^{m} |P_i| + |Q_i|) = O(N)$ time. This is optimal since it is shown in [CWS85] that $\Omega(N)$ time is required even if exactly one of P or Q is convex. □

3 Linearly Separable Polygonal Chains

We consider two disjoint polygonal *chains* P and Q that are separated by a line l; vertices and edges of P and/or Q may lie on the separating line, but no edge may cross the line. The algorithm for computing $\sigma(P, Q)$ presented in this section is loosely patterned after the techniques of [A92a] for determining $CVV(P, Q)$, where in both cases P and Q are linearly separable chains. For this reason, we briefly sketch the CVV algorithm of [A92a]. Without loss of generality we assume that l is vertical, Q lies to the left of l, $|P| = n_p$ and $|Q| = n_q$. Let $(p_1, p_2, \ldots, p_{n_p})$ and $(q_1, q_2, \ldots, q_{n_q})$ denote the sequences of vertices of P and Q, respectively, indexed bottom-to-top. We begin with some useful definitions.

Consider polygonal chain A and line l, where no edge of A crosses l. A point v is *A-visible* from a point w if the segment (v, w) does not properly intersect A. Similarly, a point v is *A-visible* from a line l if some point of l is A-visible from v. For each $a \in A$, $W(a)$

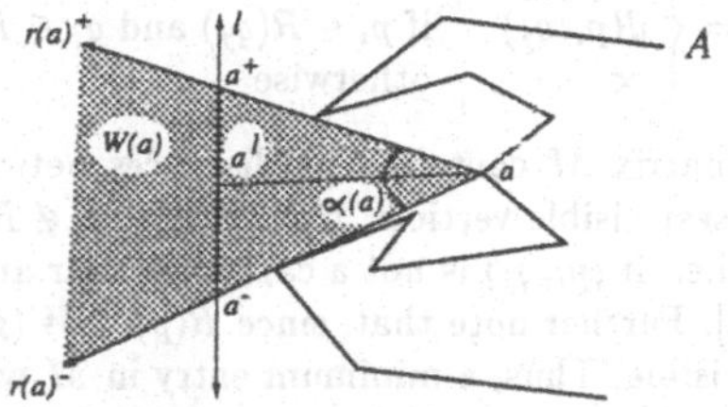

Fig. 2. The wedge $W(a)$ for vertex $a \in A$.

denotes the interior of the maximal wedge with apex a whose interior contains no vertex of A and all points of l A-visible from a. The upper and lower rays defining $W(a)$ are denoted by $r(a)^+$ and $r(a)^-$, respectively, and a^+ and a^- denote $r(a)^+ \cap l$ and $r(a)^- \cap l$, respectively. (If $r(a)^+ \cap l = \emptyset$ then $a^+ = +\infty$, and if $r(a)^- \cap l = \emptyset$ then $a^- = -\infty$.) The angle, in $W(a)$, between the rays $r(a)^-$ and $r(a)^+$ is denoted by $\alpha(a)$. Finally, let l_a be the line perpendicular to l that passes through a, and denote $l \cap l_a$ by a^l (see Fig. 2).

A CVV Algorithm. We first sketch the algorithm of [A92a] for computing $CVV(P, Q)$, when P and Q are linearly separable polygonal chains, and then discuss how this technique can be adapted so that it will determine the separation of P and Q. The following lemmas, which where established in [A92a] and [WC86], are used to form subchains P' and Q' of P and Q, respectively, so that $CVV(P', Q') = CVV(P, Q)$, and P' and Q' have a restricted form that can be exploited to calculate $CVV(P, Q)$ more easily.

Lemma 2 [A92a]. *Let P and Q be two polygonal chains that are separated by a line l, and let p be a vertex of P. If $p^l \notin W(p)$, then $CVV(p,Q) > CVV(P,Q)$, i.e., if p is not perpendicularly visible from l, then p cannot be a nearest vertex of P to Q. (Removal of these vertices creates a new chain monotone with respect to l.)*

Lemma 3 [WC86]. *Let P and Q be two polygonal chains that are separated by a line l and monotone with respect to it. If $\alpha(p) < 90°$, then $CVV(p,Q) > CVV(P,Q)$.*

After pruning P and Q according to Lemmas 2 and 3, a region $R() \subset W()$ is constructed for each remaining vertex of P and Q. (The structure for the $R()$ regions will be described in the separation algorithm; it suffices here to say that $R()$ is an unbounded region contained in $W()$.)

Lemma 4 [A92a]. *Let P and Q be two polygonal chains that are separated by a line l and monotone with respect to it, and $\alpha(p) \geq 90°$ and $\alpha(q) \geq 90°$, for all vertices $p \in P$ and $q \in Q$. If $p \notin R(q)$ or $q \notin R(p)$ then (p,q) cannot realize $CVV(P,Q)$, $p \in P$ and $q \in Q$.*

We next consider the $n_p \times n_q$ matrix M whose entries are defined as follows; if $n_p > n_q$ we instead consider the $n_q \times n_p$ matrix that is analogously defined. Let B be some constant that is greater than the maximal distance between P and Q. We use $q < R(p)$ or $q > R(p)$ to indicate that q lies *below* $R(p)$ or *above* $R(p)$, respectively.

$$M[i,j] = \begin{cases} B + n_q - j & \text{if } q_j < R(p_i) \text{ or } p_i > R(q_j) \\ d(p_i, q_j) & \text{if } p_i \in R(q_j) \text{ and } q_j \in R(p_i) \\ \infty & \text{otherwise} \end{cases}$$

By Lemma 4, the above matrix M contains the distances between all pairs of vertices that are candidates for closest visible vertices. Note that if $q_j \notin R(p_i)$ or $p_i \notin R(q_j)$, then $M[i,j] \geq B + n_q - j \geq B$, i.e., if (p_i, q_j) is not a candidate pair and (p_k, q_l) is a candidate pair, then $M[i,j] > M[k,l]$. Further note that, since $R(p) \subset W(p)$ and $R(q) \subset W(q)$, all candidate pairs (p,q) are visible. Thus, a minimum entry in M will yield $CVV(P,Q)$.

Following [AKMSW87], an $m \times n$ matrix is *monotone* if the (leftmost) minimum entry in its ith row lies below or to the right of the (leftmost) minimum entry in the $(i-1)$st row, for all $1 < i \leq m$. Furthermore, a matrix is *totally monotone* if each of its 2×2 submatrices (minors) is monotone. In [AKMSW87], it is shown that if every entry of an $m \times n$, $m \leq n$, totally matrix can be computed in constant time, then a minimum entry in every row can be found in $O(m)$ time. ([AMSS89] was the first to apply the totally monotone matrix techniques of [AKMSW87] to the CVV problem.)

Lemma 5 [A92a]. *If, for every $p \in P$ and $q \in Q$, the regions $R(p)$ and $R(q)$ are available, then the matrix M described above is totally monotone and each entry of M can be computed in constant time.*

Algorithm: $CVV(P,Q)$

1. Eliminate the vertices of P and Q that are not perpendicularly visible from l (Lemma 2).
2. Form $W(p)$ and $W(q)$ for all remaining vertices $p \in P$ and $q \in Q$. Eliminate those vertices that have $\alpha(p) < 90°$ or $\alpha(q) < 90°$ (Lemma 3).
3. Determine the $R()$ regions for all vertices remaining under consideration.

4. Find the smallest entry in the matrix M.

The complexity of the above algorithm is easily seen to be $O(N)$, where $N = |P| + |Q|$. Steps 1-3 can be accomplished by linear scans of the relevant subchains: Step 1 can be implemented by a single bottom-to-top scan of P or Q; Step 2 requires a bottom-to-top (to identify the lower rays of $W()$) and a top-to-bottom (to identify the upper rays of $W()$) scan of P or Q; Step 3 is accomplished (using the $W()$ regions) by a top-to-bottom and a bottom-to-top linear scan of both P and Q. Step 4 uses the $O(N)$ time algorithm of [AKMSW87] to find a minimum in each row, and then determines the minimum of the row minima.

The Separation Algorithm. We now return to the problem of determining the separation of P and Q. Recall that $\sigma(P,Q)$ is either realized by a vertex-vertex pair or a vertex-edge pair; i.e., $\sigma(P,Q) = \min\{CVV(P,Q), \sigma(P_e,Q_v), \sigma(P_v,Q_e)\}$, where P_v (Q_v) and P_e (Q_e) represent the vertices and (open) edges, respectively, of P (Q). Since $CVV(P,Q)$ can be found by the CVV algorithm described above, we now concentrate on the problem of determining $\sigma(P_e,Q_v)$; $\sigma(P_v,Q_e)$ can be found analogously. Our goal is to adapt the CVV algorithm of [A92a] sketched above to the present scenario; the modified algorithm that computes $\sigma(P_e,Q_v)$ will be referred to as the CVE (closest vertex edge) algorithm. We first note that Lemmas 2 and 3 hold for any point on the boundary of P. This fact is readily verified by realizing that there is nothing special about any interior point of an edge that distinguishes it from a vertex, indeed, such an "interior point" could in fact be a vertex. Thus, since $CVV(P,Q) \geq \sigma(P,Q)$, Lemmas 2 and 3, respectively, establish that we can eliminate all boundary points of P and vertices of Q that (i) are not perpendicularly visible from l (Lemma 2), and (ii) that have $\alpha() < 90^c$ (Lemma 3). The CVE algorithm has a pattern analogous to that of the CVV algorithm. Specifically, we form $W()$ and then $R()$ regions for each *vertex* of Q and each *edge* of P, and then we define a totally monotone matrix whose entires contain the distances between all candidates for closest vertex edge pair and use the algorithm of [AKMSW87] to find a minimum entry in this matrix.

Vertices of Q. Assuming that P and Q have been pruned according to Lemmas 2 and 3, we construct a new region $R(q) \subset W(q)$, for each remaining vertex $q \in Q$. (The $W()$ regions for each vertex of Q are defined as in the CVV algorithm.) Specifically, we associate with each vertex $q_i \in Q$, two (boundary) points of P, u_i^+ and u_i^-, as follows. Let $y(v)$ denote the y-coordinate of v. The point u_i^+ is the highest point of P that satisfies (i) $y(q_i) \leq y(u_i^+) \leq y(u_{i+1}^+)$, and (ii) $u_i^+ \in W(q_i)$ and $q_i \in W(u_i^+)$ (i.e., q_i and u_i^+ are visible); if no such point exists then $u_i^+ = q_i^l$. Similarly, the point u_i^- is the lowest point of P that satisfies (i) $y(q_i) \geq y(u_i^-) \geq y(u_{i-1}^-)$, and (ii) $u_i^- \in W(q_i)$ and $q_i \in W(u_i^-)$; if no such point exists then $u_i^- = q_i^l$. (The y-coordinate of $u_{n_q+1}^+$ (u_0^-) is assumed to be $+\infty$ ($-\infty$).) The region $R(q_i)$ is bounded above (below) by the segment (q_i, u_i^+) ((q_i, u_i^-)) and the horizontal ray originating at u_i^+ (u_i^-) (see Fig. 4(a)).

Edges of P. Our task now is to define, and construct, appropriate $W()$ and $R()$ regions for the *edges* of P. In the CVV algorithm, the $W()$ regions are used for two distinct purposes: (i) to eliminate those vertices v with $\alpha(v) < 90°$ and (ii) to help construct the $R()$ regions efficiently. (The fact that all vertices remaining under consideration have $\alpha() \geq 90°$ is used in the proof of Lemma 4, which is crucial in establishing the correctness of the CVV algorithm.) With these considerations in mind, we want to define and construct our

$W()$ regions for edges so that it is easy to eliminate all points p on the boundary of P that have $\alpha(p) < 90°$.

Consider an edge $e = (p_i, p_{i+1}) \in P$, and the wedges $W(p_i)$ and $W(p_{i+1})$ as defined in the *CVV* algorithm. Recall that $r(p)^+$ ($r(p)^-$) passes by the highest (lowest) vertex of P that is visible from p; denote these vertices by $h(p)$ and $l(p)$, respectively. In the special (and unusual) case in which $h(p_i) = h(p_{i+1})$ and $l(p_i) = l(p_{i+1})$, we can easily determine which points of e have $\alpha() \geq 90°$ as follows. Let C_e denote the circle with diameter $d(l(p_i), h(p_i))$ that is centered at the midpoint of $(l(p_i), h(p_i))$ (see Fig. 3(a)). It is easy to verify that all points of e external to C_e have $\alpha() < 90°$, and all points of e on the boundary or internal to C_e have $\alpha() \geq 90°$. After eliminating those points of e

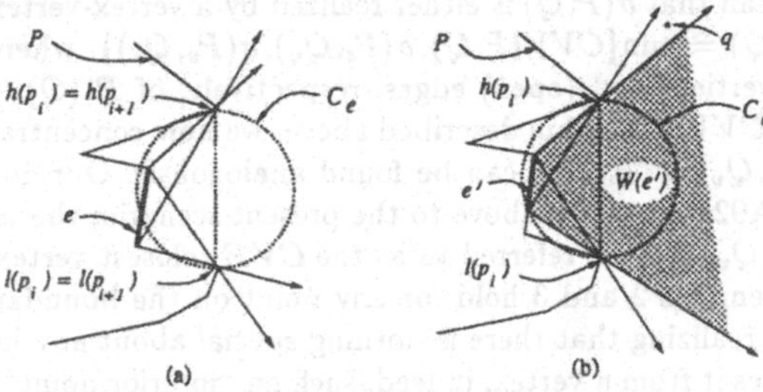

Fig. 3. (a) All points on the portion of e that is external to C_e have $\alpha() < 90°$. (b) The region $W(e')$ for that portion e' of e that has $\alpha() \geq 90°$.

with $\alpha() < 90°$, leaving at most one segment $e' = (p'_i, p'_{i+1})$, $W(e')$ is defined as follows: $W(e')$ is bounded above (below) by the ray originating at p'_{i+1} (p'_i) and passing by $h(p_i)$ ($l(p_i)$) (see Fig. 3(b)). Clearly, $W(e')$ contains all points of l that are visible from *every* point of e'. Let Q' denote the portion of Q that is external to $W(e')$. Although there may be points of Q' that are visible to some point of e', it is a simple matter to verify that $\sigma(e', Q') > \sigma(P, Q')$, i.e., any point of Q external to $W(e')$ that is visible to some point of e' is closer to some other point of P than it is to e'. Without loss of generality, consider a point $q \in Q'$ that lies above $W(e')$ but is visible from some point of e' (see Fig. 3(b)); in this case it is easy to see that $\sigma(P, q) \leq d(h(p_i), q) < d(e', q)$. This property of $W(e')$ will be referred to as *Prop-W*.

In general, however, for each $e = (p_i, p_{i+1}) \in P$, we do not have $h(p_i) = h(p_{i+1})$ and $l(p_i) = l(p_{i+1})$. Our solution to this problem is simply to partition each edge into a number of segments so that within each segment this special property holds. In the following we denote the partitioned set of edges of P_e by $P'_e = \{e_i | 1 \leq i \leq n_e\}$, where $e_i = (v_i^-, v_i^+)$ and $y(v_i^-) < y(v_i^+)$; later we will show that $n_e = |P'_e| = O(P)$.

Given the $W()$ regions for the partitioned set of edges, their $R()$ regions are defined similarly to those of the vertices $q \in Q_v$. We associate with each edge $e_i = (v_i^-, v_i^+) \in P'_e$ two *vertices*, u_i^- and u_i^+, of Q. The vertex u_i^+ is the highest vertex of Q that satisfies (i) $y(v_i^+) \leq y(u_i^+) \leq y(u_{i+1}^+)$, and (ii) $u_i^+ \in W(e_i)$ and $v_i^+ \in W(u_i^+)$. Similarly, the vertex u_i^- is the lowest vertex of Q that satisfies (i) $y(v_i^-) \geq y(u_i^-) \geq y(u_{i-1}^-)$, and (ii) $u_i^- \in W(e_i)$

and $v_i^- \in W(u_i^-)$. The region $R(e_i)$ is bounded above (below) by (v_i^+, u_i^+) $((v_i^-, u_i^-))$ and the horizontal ray originating at u_i^+ (u_i^-) (see Fig. 4(b)); if u_i^+ (u_i^-) does not exist, then $R(e_i)$ is bounded above (below) by the horizontal ray originating at v_i^+ (v_i^-), i.e., $u_i^+ = v_i^+$

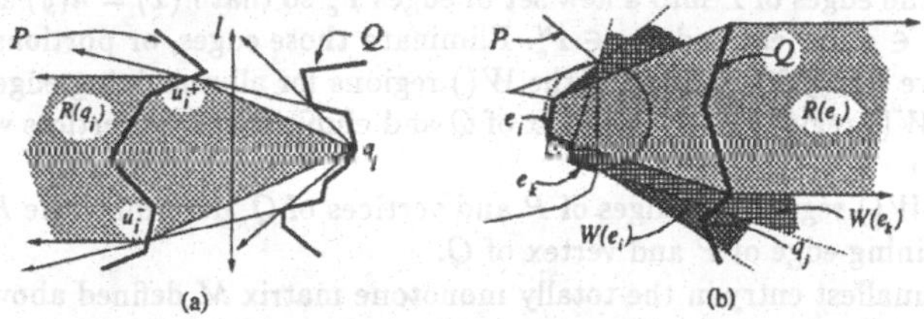

Fig. 4. (a) The region $R(q_i)$ for vertex $q_i \in Q_v$. (b)The region $R(e_i)$ for edge segment $e_i \in P_e'$.

$(u_i^- = v_i^-)$. The sufficiency of the $R()$ regions is established by the following lemma.

Lemma 6. *Let P and Q be two polygonal chains (or portions thereof) that are separated by a line l and monotone with respect to it, where $P_e' = \{e_i | 1 \leq i \leq n_e\}$ is the set of partitioned edges of P described above, and $\alpha(p) \geq 90°$ and $\alpha(q) \geq 90°$, for each point $p \in P_e'$ and each vertex $q \in Q_v$. If $p \notin R(q)$ or $q \notin R(e)$, for some $p \in e$, then (e, q) cannot realize $\sigma(P, Q)$, where $e \in P_e'$ and q is a vertex of Q.*

Proof. Assume $q_j \notin R(e_i)$ or $e_i \cap R(q_j) = \emptyset$. Clearly if some $p \in e_i = (v_i^-, v_i^+)$ and q_j are not visible, then (e_i, q_j) cannot realize $\sigma(P, Q)$; therefore we assume some $p \in e_i$ and q_j are visible. Without loss of generality, assume q_j lies below $R(e_i)$, the other cases are similar. If $q_j \notin W(e_i)$, then, by Prop-W, $d(e_i, q_j) > d(l(v_i^-), q_j) \geq \sigma(P, Q)$, where $l(v_i^-)$ is the lowest vertex of P visible from v_i^-. Thus, we now assume $q_j \in W(e_i)$ and $q_j \notin R(e_i)$ (see Fig. 4(b)); therefore, by definition of $R(e_i)$, there must be some edge $e_k \in P_e'$, $k < i$, such that q_j lies below $W(e_k)$. Again, by Prop-W, we have $d(e_k, q_j) > d(l(v_k^-), q_j) \geq \sigma(P, Q)$. Moreover, since $\alpha(x) \geq 90°$, for all $x \in e_k$, and both e_i and q_j lie outside $W(e_k)$, we have $d(e_i, q_j) > d(e_k, q_j) > \sigma(P, Q)$. □

The previous lemma establishes that we can use the $R()$ regions in the CVE algorithm in the same manner that they were used in the CVV algorithm. Specifically, we consider the $n_e \times n_q$ matrix M, defined as follows, where B is, once again, a constant greater than the maximal distance between P and Q. By Lemma 6, M contains the distances between all edge-vertex pairs that are candidates for $\sigma(P_e, Q_v)$, and thus a minimum entry in M will yield $\sigma(P_e, Q_v)$.

$$M[i,j] = \begin{cases} B + n_q - j & \text{if } q_j < R(e_i) \text{ or } e_i > R(q_j) \\ d(e_i, q_j) & \text{if } p \in R(q_j) \text{ and } q_j \in R(e_i), \text{ for some } p \in e_i \\ \infty & \text{otherwise} \end{cases}$$

Lemma 7. *If, for every $e \in P_e'$ and $q \in Q_v$, the $R()$ regions are available, then the matrix M described above is totally monotone and each entry can be computed in constant time.*

Proof. Omitted due to space limitations (see [A92b]).

Algorithm: $CVE(P_e, Q_v)$

1. Eliminate the portions of P and vertices of Q that are not perpendicularly visible from l (Lemma 2).
2. Partition the edges of P into a new set of edges P'_e so that $h(x) = h(y)$ and $l(x) = l(y)$ for all $x, y \in e$, for each edge $e \in P'_e$. Eliminate those edges, or portions thereof, that do not have $\alpha() \geq 90°$, and form the $W()$ regions for all remaining edges (Lemma 3). Form the $W()$ regions for each vertex of Q and eliminate those vertices with $\alpha() < 90°$ (Lemma 3).
3. Using the $W()$ regions for edges of P and vertices of Q, determine the $R()$ regions for each remaining edge of P and vertex of Q.
4. Find the smallest entry in the totally monotone matrix M defined above.

It is clear that all steps, with the exception of Step 2, can be accomplished with the same techniques and within the same time bounds as in the CVV algorithm, i.e., they are all accomplished in $O(N)$ time, where $N = |P| + |Q|$. We now explain how an edge $e = (p_i, p_{i+1}) \in P$ can be partitioned into a set of segments $S_e = \{s_j | 1 \leq j \leq k\}$, where $s_j = (u_j, u_{j+1})$, $u_1 = p_i$ and $u_{k+1} = p_{i+1}$, so that for all $s \in S_e$, and for all $x, y \in s$, $h(x) = h(y)$ and $l(x) = l(y)$. Without loss of generality we show how to partition e into a set of segments $S_e^+ = \{s_j | 1 \leq j \leq k' \leq k\}$ so that for all $s \in S_e^+$, and for all $x, y \in s$, $h(x) = h(y)$; a similar process is used to ensure that for all $s \in S_e$, and for all $x, y \in s$, $l(x) = l(y)$. Clearly, if $h(p_i) = h(p_{i+1})$ no subdivision is necessary. Next note that $h(p_i) \geq p_{i+1}$, and if $h(p_i) = p_{i+1}$ then $h(x) = p_{i+1}$ for all $x \in e$ and no subdivision is required (recall that we are dealing with open line segments since the vertices were tested separately). So we assume $h(p_i) > p_{i+1}$ and $h(p_i) \neq h(p_{i+1})$; it is easy to verify that $h(p_i) > h(p_{i+1})$, i.e., it is not possible that $p_{i+1} < h(p_i) < h(p_{i+1})$. Recall that the rays $r(p)^+$ (and thus the vertices $h(p)$) were determined by a top-to-bottom scan of P, and moreover, that this scanning process actually computed the successive convex hulls $C_i = CH(p_i \cup p_{i+1} \cup \ldots \cup p_{n_p})$, $1 \leq i \leq n_p$, i decreasing. Thus, after determining $h(p_{i+1})$ we have C_{i+1}, and note that $h(p_i) \in C_{i+1}$, and in particular $r(p_i)^+$ is the ray originating at p_i and tangent to C_{i+1}. Now it is clear that the scanning (up) of C_{i+1} to form C_i will actually determine the points on e which have different $h()$ values (see Fig. 5). Thus, once again, the same scanning process used in the CVV algorithm to form the $W()$ regions can be used in the CVE algorithm. We can also see that the number of new endpoints ("vertices") introduced in this manner is at most $n_p = |P|$, since a particular edge of C_i can only introduce one point as it disappears, and if it doesn't disappear it cannot introduce any new points. Thus, the total number of endpoints (and thus edges) we end up with after partitioning all edges is at most $3n_p$, i.e., the original n_p endpoints combined with an additional n_p from each of the scans of the partitioning process.

Theorem 2 *If P and Q are two linearly separable polygonal chains, where $N = |P| + |Q|$, then $\sigma(P, Q)$, and a pair of points realizing it, can be computed optimally in $\Theta(N)$ time.*

Proof. Since $CVV(P, Q)$ can be computed in $O(N)$ time, and the above discussion establishes that $\sigma(P_e, Q_v)$ and $\sigma(P_v, Q_e)$ can be determined in $O(N)$ time by the CVE algorithm, $\sigma(P, Q) = \min\{CVV(P, Q), \sigma(P_e, Q_v), \sigma(P_v, Q_e)\}$, can be found in $O(N)$ time as well. This is clearly optimal since it is shown in [CWS85] that $\Omega(N)$ time is required even if exactly one of P or Q is convex. □

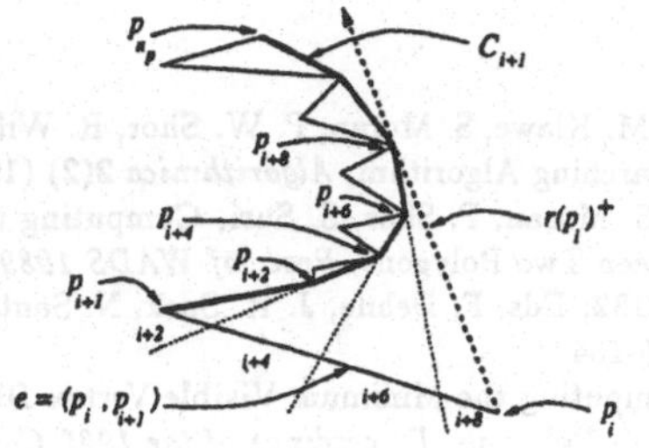

Fig. 5. The segments of $e = (p_i, p_{i+1})$ are labeled with the index of the highest vertex of P that is visible from them.

4 A Parallel Implementation

Recall that the separation of two simple polygons, in the most general case, can be found in $O(\log^2 N)$ time with $O(N)$ processors on a CREW PRAM by a naive parallelization of Kirkpatrick's sequential approach [K79]. As we have seen, the separation problem can be solved by the same primitive operations and techniques that were used to solve the CVV problem in [A92a]. Since [A92a] gives a parallel implementation of the CVV algorithm requiring $O(\log N)$ time and using $O(N)$ processors on a CREW PRAM, it is clear that $\sigma(P, Q)$ can be computed within these same time and processor bounds, even when P and Q are nonconvex simple polygons. Thus, a parallel implementation of the separation algorithm given here improves the time complexity by a factor of $O(\log N)$ over the naive adaptation of the parallel generalized voronoi diagram algorithm of [GOY89].

5 Conclusion and Open Problems

In this paper we have studied the separation problem in the scenario in which no preprocessing is allowed. However, separation problems, and the potentially simpler intersection detection problems (which simply decide whether or not two objects intersect), have also been studied in an environment which allows preprocessing. For example, when preprocessing is permitted, Dobkin and Kirkpatrick [DK90] give algorithms for determining the separation of two *convex* objects in two and three dimensions; the $O(\log N)$ algorithm for finding the separation of two convex polygons does not require any preprocessing as long the vertices of the polygons are stored in cyclic order, but the $O(\log^2 N)$ algorithm for computing the separation of two convex polyhedra requires a non-trivial amount of preprocessing. More recently, Mount [M92] describes an algorithm for determining whether or not two simple polygons intersect in $O(M \log^2 N)$ time, where M is the complexity (number of links) of a minimum link polygonal curve separating P and Q; this algorithm requires $O(N \log N)$ preprocessing. Thus, if $M < (N/\log^2 N)$, then Mount's intersection detection algorithm is more efficient than the test derived from Chazelle's linear time triangulation algorithm [C90]; note that $1 \leq M \leq N$. It would also be interesting to study the complexity of computing the separation of two simple polygons when preprocessing is allowed.

References

[AKMSW87] A. Aggarwal, M. Klawe, S. Moran, P. W. Shor, R. Wilber, Geometric Applications of a Matrix Searching Algorithm, *Algorithmica* **2**(2) (1986), pp. 195-208.

[AMSS89] A. Aggarwal, S. Moran, P. Shor, S. Suri, Computing the Minimum Visible Vertex Distance Between Two Polygons, *Proc. of WADS 1989*, and *Lecture Notes in Computer Science* **382**, Eds. F. Dehne, J. R. Sack, N. Santoro, Springer Verlag, Berlin, (1989), pp. 115-134.

[A92a] N. Amato, Computing the Minimum Visible Vertex Distance Between Two Nonintersecting Simple Polygons, *Proceedings of the 1992 Conference on Information Sciences and Systems* Vol. II, Princeton, NJ, (1992), pp. 800-805. (Also, Coordinated Science Laboratory Tech. Report, No. UILU-ENG-92-2206 (ACT 120), University of Illinois at Urbana-Champaign.)

[A92b] N. Amato, An Optimal Algorithm for Determining the Separation of Two Nonintersecting Simple Polygons, Coordinated Science Laboratory Tech. Report, No. UILU-ENG-92-2248 (ACT 125), University of Illinois at Urbana-Champaign, (1992).

[AG88] M. Atallah and M. Goodrich, Parallel Algorithms for Some Functions of Two Convex Polygons, *Algorithmica* **3** (1988), pp. 535-548.

[BE84] B. Bhattacharya and H. El Gindy, A New Linear Convex Hull Algorithm for Simple Polygon, *IEEE Inform. Theory* **c29** (1984), pp. 571-573.

[C90] B. Chazelle, Triangulating a Simple Polygon in Linear Time, *Proc. of 31st Annual Symposium on Foundations of Computer Science* (1990), pp. 220-230.

[CD87] B. Chazelle and D. Dobkin, Intersection of Convex Objects in Two and Three Dimensions, *Journal of the ACM* **34** (1987), pp. 1-27.

[CW83] F. Chin and C. Wang, Optimal Algorithms for the Intersection and the Minimum Distance Problems between Planar Polygons, *IEEE Trans. on Computers* **c32** (1983), pp. 1205-1207.

[CWS85] F. Chin, C. Wang, J. Sampson, An Unifying Approach for a Class of Computational Geometry Problem, *The Visual Computer - Internat. Journal of Computer Graphics* **1**(2) (1985), pp. 124-133.

[DaK89] N. Dadoun and D. Kirkpatrick, Optimal Parallel Algorithms for Convex Polygon Separation, Technical Report #89-21, Department of Computer Science, University of British Columbia, Vancouver BC, Canada (1989).

[DK90] D. Dobkin and D. Kirkpatrick, Determining the Separation of Preprocessed Polyhedra - A Unified Approach, *ICALP* (1990), pp. 400-413.

[E85] H. Edelsbrunner, On Computing the Extreme Distance Between Two Convex Polygons, *Journal of Algorithms* **6** (1985), pp. 213-224.

[GOY89] M. Goodrich, C. Ó'Dúnlaing, and C. Yap, Computing the Voronoi Diagram of a Set of Line Segments in Parallel, *Lecture Notes 382: WADS '89*, Springer-Verlag (1989).

[K79] D. Kirkpatrick, Efficient Computation of Continuous Skeletons, *Proc. of the 20th Annual Symposium on the Foundations of Computer Science* (1979), pp. 18-27.

[M92] D. Mount, Intersection Detection and Separators for Simple Polygons, *Proc. of the Eighth ACM Annual Symposium on Computational Geometry* (1992), pp. 303-311.

[PS85] F. Preparata and M. Shamos, *Computational Geometry*, Springer-Verlag, New York (1985).

[WC86] C. A. Wang and E. P. F. Chan, Finding the Minimum Visible Vertex Distance Between Two Nonintersecting Simple Polygons, *Proc. of the Second ACM Annual Symposium on Computational Geometry* (1986), pp. 34-42.

Balanced Search Trees Made Simple

Arne Andersson*

Department of Computer Science, Lund University, Box 118, S-221 00 Lund, Sweden

Abstract. As a contribution to the recent debate on simple implementations of dictionaries, we present new maintenance algorithms for balanced trees. In terms of code simplicity, our algorithms compare favourably with those for deterministic and probabilistic skip lists.

1 Introduction

It is well known that there is a huge gap between theory and practice in computer programming. While companies producing computers or cars are anxious to use the best technology available – at least they try to convince their customers that they do so – the philosophy is often different in software engineering; it is enough to produce programs that "work." Efficient algorithms for sorting and searching, which are taught in introductory courses, are often replaced by poor methods, such as bubble sorting and linked lists. If your program turns out to require too much time or space you advice your customer to buy a new, heavier and faster, computer.

This situation strongly motivates an extensive search for simple and short-coded solutions to common computational tasks, for instance worst-case efficient maintenance of dictionaries. Due to its fundamental character, this problem is one of the most well-studied in algorithm design. Until recently, all solutions have been based on balanced search trees, such as AVL-trees [1], symmetric binary B-trees [6] (also denoted red-black trees [8]), SBB(k)-trees [4], weight-balanced trees [11], half-balanced trees [12], and k-neighbour trees [9]. However, none of them has become the structure of choice among programmers as they are all cumbersome to implement. To cite Munro, Papadakis, and Sedgewick [10], the traditional source code for a balanced search tree "contains numerous cases involving single and double rotations to the left and the right".

In the recent years, the search for simpler dictionary algorithms has taken a new, and quite successful, direction by the introduction of some "non-tree" structures. The first one, *the skip list*, introduced by Pugh [14] is a simple and elegant randomized data structure. A worst-case efficient variant, *the deterministic skip list*, was recently introduced by Munro, Papadakis, and Sedgewick [10].

The important feature of the deterministic skip list is its *code simplicity*. As pointed out by Munro et al., the source code for insertion into a deterministic skip list is simpler (or at least shorter) than any previously presented code for a

* Arne.Andersson@dna.lth.se

balanced search tree. In addition, the authors claimed that the code for deletion would be simple, although it was not presented.

In this article we demonstrate that *the binary B-tree*, introduced by Bayer [5], may be maintained by very simple algorithms. The simplicity is achieved from three observations:

- The large number of cases occurring in traditional balancing methods may be replaced by two simple operations, called *Skew* and *Split*. Making a *Skew* followed by a *Split* at each node on the traversed path is enough to maintain balance during updates.
- The representation of balance information as one bit per node creates a significant amount of book-keeping. One short integer ($o(\log\log n)$ bits) in each node makes the algorithms simpler.
- The deletion of an internal node from a binary search tree has always been cumbersome, even without balancing. We show how to simplify the deletion algorithm by the use of two global pointers.

As a matter of fact, the coding of our algorithms is even shorter than the code for skip lists, both probabilistic and deterministic, are. In our opinion, it is also simpler and clearer. Hence, the binary search tree may compete very well with skip lists in terms of simplicity.

In Section 2 we present the new maintenance algorithms for binary B-trees, and in Section 3 we discuss their implementation. Section 4 contains a comparison between binary B-trees and deterministic skip lists. We also make a brief comparison with the probabilistic skip list. Finally, we summarize our results in Section 5.

2 Simple Algorithms

The binary B-tree, BB-tree, was introduced by Bayer in 1971 [5] as a binary representation of 2-3 trees [2]. Using the terminology in [4], we say that a node in a 2-3 tree is represented by a *pseudo-node* containing one or two binary nodes. Edges inside pseudo-nodes are *horizontal* and edges between pseudo-nodes are *vertical*. Only right-edges are allowed to be horizontal. In order to maintain balance during updates, we have to store balance information in the nodes. One bit, telling whether the incoming edge (or the outgoing right-edge) is horizontal or not, would be enough. However, in our implementation we chose to store an integer *level* in each node, corresponding to the vertical height of the node. Nodes at the bottom of the tree are on level 1.

Briefly, all representations of B-trees, including binary B-trees, red black trees, and deterministic skip lists, are maintained by two basic operations: joining and splitting of B-tree nodes. In a binary tree representation, this is performed by rotations and change of balance information. The reason why the algorithms become complicated is that a pseudo-node may take many different shapes, causing many special cases. For example, adding a new horizontal edge to a pseudo-node of shape ∘ or ↘ may result in five different shapes, namely

 , ,  , 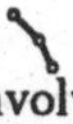 , or  . These possibilities give rise to a large number of different cases, involving more or less complicated restructuring operations.

Fortunately, there is a simple rule of thumb that can be applied to reduce the number of cases:

Make sure that only right-edges are horizontal
before you check the size of the pseudo-node.

Using this rule, the five possible shapes in the above will reduce to two, namely  and  , only the last one will require splitting.

In order to apply our rule of thumb in a simple manner, we define two basic restructuring operations (p is a binary node):

Skew (p): Eliminate horizontal left-edges below p. This is performed by following the right path from p, making a right rotation whenever a horizontal left-edge is found.

Split (p): If the pseudo-node rooted at p is too large, split it by increasing the level of every second node. This is performed by following the right path from p, making left rotations.

These operations are simple to implement as short and elegant procedures. They also allow a conceptually simple description of the maintenance algorithms:

Insertion

1. Add a new node at level 1.
2. Follow the path from the new node to the root. At each binary node p perform the following:
   (a) *Skew* (p)
   (b) *Split* (p)

Deletion

1. Remove a node from level 1 (the problem of removing internal nodes is discussed in Section 3).
2. Follow the path from the removed node to the root. At each binary node p perform the following:
   (a) If a pseudo-node is missing below p, i. e. if one of p's children is two levels below p, decrease the level of p by one. If p's right-child belonged to the same pseudo-node as p, we decrease the level of that node too.
   (b) *Skew* (p)
   (c) *Split* (p)

The algorithms are illustrated in Figures 1 and 2.

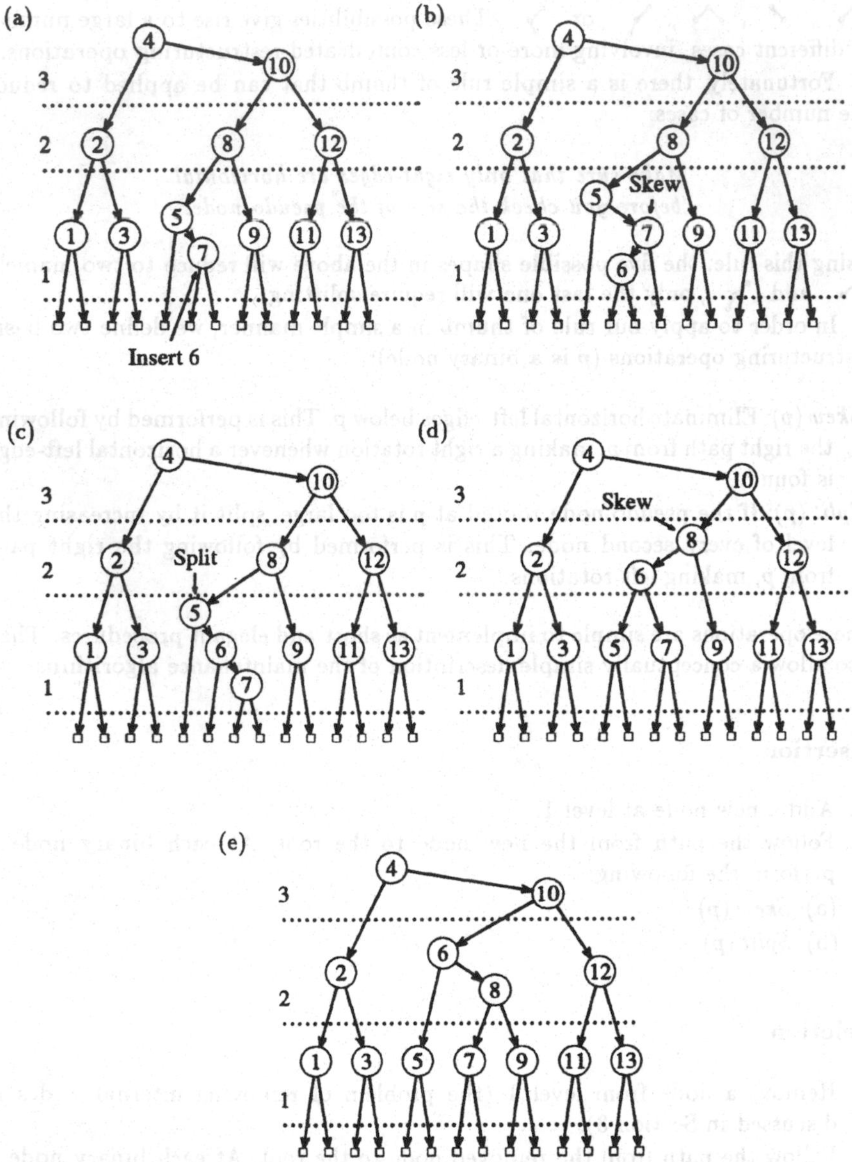

Fig. 1. ***Example of insertion into a BB-tree. The levels are separated by horizontal lines.***

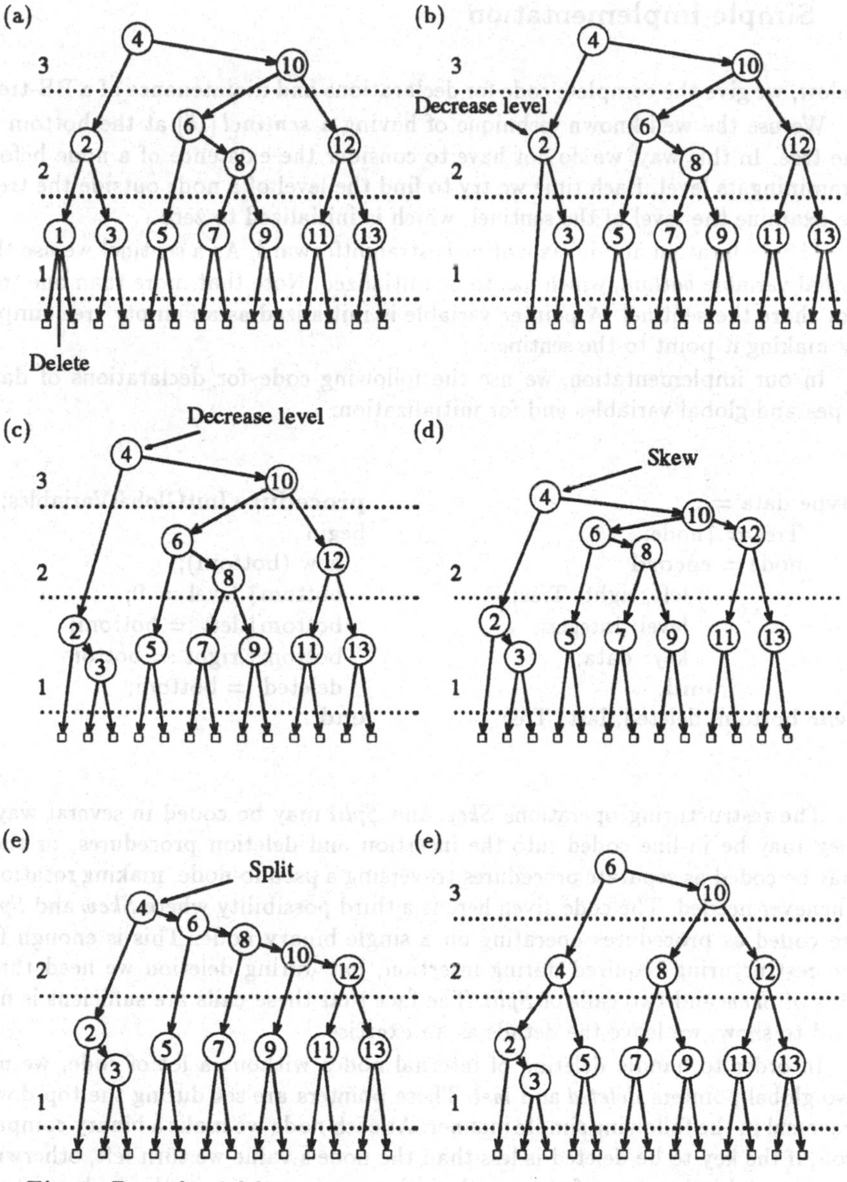

Fig. 2. ***Example of deletion.***

3 Simple implementation

Below, we give the complete code for declarations and maintenance of a BB-tree.

We use the well-known technique of having a *sentinel* [15] at the bottom of the tree. In this way, we do not have to consider the existence of a node before examining its level. Each time we try to find the level of a node outside the tree, we examine the level of the sentinel, which is initialized to zero.

The declaration and initialization is straightforward. As a sentinel we use the global variable *bottom*, which has to be initialized. Note that more than one tree can share the sentinel. A pointer variable is initialized as an empty tree simply by making it point to the sentinel.

In our implementation, we use the following code for declarations of data types and global variables and for initialization:

```
type data = ...;
    Tree = ↑node;
    node = record
              left, right: Tree;
              level: integer;
              key: data;
           end;
var bottom, deleted, last: Tree;
```

```
procedure InitGlobalVariables;
begin
  new (bottom);
  bottom↑.level := 0;
  bottom↑.left := bottom;
  bottom↑.right := bottom;
  deleted := bottom;
end;
```

The restructuring operations *Skew* and *Split* may be coded in several ways: they may be in-line coded into the insertion and deletion procedures, or they may be coded as separate procedures traversing a pseudo-node, making rotations whenever needed. The code given here is a third possibility where *Skew* and *Split* are coded as procedures operating on a single binary node. This is enough for the restructuring required during insertion, but during deletion we need three calls of *Skew* and two calls of *Split*. The fact that these calls are sufficient is not hard to show, we leave the details as an exercise.

In order to handle deletion of internal nodes without a lot of code, we use two global pointers *deleted* and *last*. These pointers are set during the top-down traversal in the following simple manner: At each node we make a binary comparison, if the key to be deleted is less than the node's value we turn left, otherwise we turn right (i.e. even if the searched element is present in the node we turn right). We let *last* point to each internal node on the path, and we let *deleted* point to each node where we turn right. When we reach the bottom of the tree, *deleted* will point to the node containing the element to be deleted (if it is present in the tree) and *last* will point to the node which is to be removed (which may be the same node). Then, we just move the element from *last* to *deleted* and remove *last*.

Altogether, insertion and deletion may be coded in the following way:

```
procedure Skew (var t: Tree);
var temp: Tree;
begin
 if t↑.left↑.level = t↑.level then
 begin { rotate right }
    temp := t;
    t := t↑.left;
    temp↑.left := t↑.right;
    t↑.right := temp;
 end;
end;

procedure Split (var t: Tree);
var temp: Tree;
begin
 if t↑.right↑.right↑.level = t↑.level then
 begin { rotate left }
    temp := t;
    t := t↑.right;
    temp↑.right := t↑.left;
    t↑.left := temp;
    t↑.level := t↑.level + 1;
 end;
end;

procedure Insert (var x: data;
    var t: Tree; var ok: boolean);
begin
 if t = bottom then begin
    new (t);
    t↑.key := x;
    t↑.left := bottom;
    t↑.right := bottom;
    t↑.level := 1;
    ok := true;
 end else begin
    if x < t↑.key then
      Insert (x, t↑.left, ok)
    else if x > t↑.key then
      Insert (x, t↑.right, ok)
    else ok := false;
    Skew (t);
    Split (t);
 end;
end;

procedure Delete (var x: data;
    var t: Tree; var ok: boolean);
begin
  ok := false;
  if t <> bottom then begin

    { 1: Search down the tree and }
    { set pointers last and deleted. }
    last := t;
    if x < t↑.key then
      Delete (x, t↑.left, ok)
    else begin
      deleted := t;
      Delete (x, t↑.right, ok);
    end;

    { 2: At the bottom of the tree we }
    { remove the element (if it is present). }
    if (t = last) and (deleted <> bottom)
    and (x = deleted↑.key) then
    begin
      deleted↑.key := t↑.key;
      deleted := bottom;
      t := t↑.right;
      dispose (last);
      ok := true;
    end

    { 3: On the way back, we rebalance. }
    else if (t↑.left↑.level < t↑.level-1)
    or (t↑.right↑.level < t↑.level-1) then
    begin
      t↑.level := t↑.level -1;
      if t↑.right↑.level > t↑.level then
        t↑.right↑.level := t↑.level;
      Skew (t);
      Skew (t↑.right);
      Skew (t↑.right↑.right);
      Split (t);
      Split (t↑.right);
    end;
  end;
end;
```

4 Discussion

Since the BB-tree has been known for a long time, an obvious question is whether our algorithms are actually simpler than known algorithms. A look in the original paper by Bayer [5] or in the textbook by Wirth [15] will show that this is the case.

First, by introducing a simple rule of thumb, we avoid all the various cases that occur in standard algorithms. Second, if we use just one bit in each node, we have to change the balance information in the two involved nodes after each rotation. This is not required in our case. Third, during deletion, the difference in level between a node and its child may become 0, 1, or 2. This is easily detected when the levels are stored explicitly in the nodes. If we use one bit in each node, we must use more bookkeeping, causing more (and quite tricky) code. Fourth, we have shown how to simplify the deletion of internal nodes. combining those observations, the implification becomes considerable.

We believe that our version of BB-trees is very suitable for classroom and textbook presentations. Since it fills less than one page, we can provide the student/reader with the entire code. Optimizing considerations, such as decreasing the number of bits used for balance information, are nice exercises for the interested student.

4.1 BB-trees versus deterministic skip lists

We start our comparison by examining the code for deterministic skip lists. Then, we discuss some other aspects.

Program code: Munro et al. presented a short and elegant source code for declaration, initialization, insertion, and search in a deterministic skip list. There also exist a preliminary version of the (quite complicated and long) deletion algorithm [13]. The code is given in C, but can easily be translated into Pascal. Doing so, we find that the total code for deterministic skip lists is about 40% longer than our code for BB-trees. The length was compared using the established method of counting lexicographical units (tokens) [7].

In addition, code for BB-trees appears to be simpler and cleaner. In our opinion, the occurrence of complicated pointer expressions, like $x\uparrow.d\uparrow.r\uparrow.r\uparrow.r\uparrow.key$ makes the code for deterministic skip lists harder to understand. The procedure for insertion into a DSL also contains a "hidden GOTO-statement" in the form of a **return** statement.

Recursion vs. non-recursion: A reader might argue that our comparison is "unfair" since we compare recursive code with non-recursive. To such an objection we answer

- Recursion is no disadvantage when aiming for simple and clear code.
- It does not seem likely that a recursive version of the DSL would be simpler than the BB-tree. At least, this remains to show.
- A stack of logarithmic size creates no problem in practice.

  The only possible drawback of recursion is that the execution time may become slightly longer. However, the subject treated here is to give simple and short algorithms, *not* to optimize code. If a minimal constant factor in execution time is very important, neither the binary B-tree nor the deterministic skip list is claimed to be the ideal choice.

Reserved key values: In the suggested implementation of deterministic skip lists two key values (*max* and *max + 1*) are used for special purposes. This has some drawbacks, which becomes evident when the DSL is to be used as a general-purpose abstract data type. If any of these keys are inserted or searched for, the program will fail. Furthermore, for some types of data, such as real numbers, it may be difficult to find suitable values for *max* and *max + 1*. Indeed, this "minor" problem may cause a great deal of confusion and irritation.

Note that a DSL does not necessarily require these reserved keys; they may be removed at the cost of more code.

Execution time: The purpose of this paper is not to minimize the execution time. However, for the sake of completeness we have run some experiments.

At the moment of testing, no code for deletion in a DSL was published, therefore we only compared the execution time for insertions and searches.

In our experiments, we used a Pascal version of the source code by Munro et al. [10]. For binary tree search we used the (non-recursive) algorithm discussed in [3].

The experiments were made in two environments: Sun-Pascal on a Sun SPARCstation 1 and Turbo-Pascal 6.0 on a Victor 2/86. We used two kinds of data: random real numbers and random strings. The strings where of type ***packed array*** [1..20] ***of char*** and consisted of randomly chosen capitol letters from the English alphabet. The dictionaries where of size 100 – 1000 on the PC and 100 – 10 000 on the Sun. (Trying to insert 2000 strings in a deterministic skip list, the PC run out of address space). In order to measure the cost of inserting/searching for *n* elements, the following experiment was made:

1. Start the clock.
2. Repeat ten times: Make *n* random insertions or search all *n* elements.
3. Stop the clock.
4. Repeat step 1 – 3 ten times, compute average time and standard deviation.

environment	input	n	insertion BB-tree	insertion DSL	search BB-tree	search DSL (1)	search DSL (2)
Turbo Pascal	real	100	1126	1587	752	1292	983
		200	1296	1825	856	1483	1100
		500	1548	2121	949	1741	1240
		1000	1724	2368	1039	1954	1368
	string	100	1230	1888	953	1539	1180
		200	1431	2154	1039	1784	1314
		500	1719	2520	1175	2067	1471
		1000	1922	2805	1259	2257	1594
Sun Pascal	real	100	109	58	13	15	15
		200	127	61	14	17	17
		500	152	66	16	20	20
		1000	174	70	17	21	22
		2000	198	72	19	24	25
		5000	221	85	23	30	30
		10000	241	89	26	32	33
	string	100	126	124	88	99	107
		200	150	130	92	103	112
		500	182	145	98	112	120
		1000	202	154	103	121	129
		2000	224	161	108	130	136
		5000	256	176	115	142	146
		10000	282	189	120	153	153

Table 1. *A comparison of execution time (in microseconds)*

The results of our experiment are given in Table 1. In all cases, the standard deviation was less than 6%.

From the table we conclude that searches are always faster in a BB-tree. For insertion we get varying results depending on environment. When using Turbo Pascal, it seems to be the case that the cost of recursion used by the BB-tree is compensated by the use of more comparisons and more calls to the memory allocation system used by the DSL. The only time when the DSL is faster is during insertions in the Sun Pascal environment.

We would like to point out that in many application element location is much more common that updates. In such cases, the BB-tree would probably consume less time than the DSL in any environment.

4.2 BB-trees versus the probabilistic skip list

So far, we have been concerned with worst-case efficient data structures. However, our binary tree implementation would also compete very well with the original, probabilistic, skip list. A demonstration program for this data structure has been announced by its inventor on the international network for anyone to fetch by anonymous ftp. Examining this source code, we find that the corresponding code for BB-trees is considerably shorter. In fact, a comparison of the number of tokens indicates that the total code of Pugh's program (including declaration, initialization, insertion, deletion, destroying a list, and search) is two to three times as long as the corresponding code for a BB-tree. The skip list code is also more "tricky". Of course, our comparison is not objective and therefore we leave it to the reader to judge.

It should be noted that also the probabilistic skip list uses a reserved key value with the same drawbacks as for the deterministic skip list.

5 Comments

The balanced binary search tree is not necessarily that complicated, although it took more than 30 years from its introduction to find out.

Finally, we hope that the search for simple maintenance algorithms for binary trees, skip lists, and other alternatives will help the mission of breaking the dominance of singly-linked lists and other poor data structures.

Acknowledgements

The comments of Thomas Papadakis and Kerstin Andersson has contributed to the presentation of this material.

References

1. G. M. Adelson-Velskii and E. M. Landis. An algorithm for the organization of information. *Dokladi Akademia Nauk SSSR*, 146(2):1259–1262, 1962.
2. A. V. Aho, J. E. Hopcroft, and J. D. Ullman. *Data Structures and Algorithms*. Addison-Wesley, Reading, Massachusetts, 1983. ISBN 0-201-00023-7.
3. A. Andersson. A note on searching in a binary search tree. *Software-Practice and Experience*, 21(10):1125–1128, 1991.
4. A. Andersson, Ch. Icking, R. Klein, and Th. Ottmann. Binary search trees of almost optimal height. *Acta Inormatica*, 28:165–178, 1990.
5. R. Bayer. Binary B-trees for virtual memory. In *Proc. ACM SIGIFIDET Workshop on Data Description, Access and control*, pages 219–235, 1971.

6. R. Bayer. Symmetric binary B-trees: Data structure and maintenance algorithms. *Acta Informatica*, 1(4):290–306, 1972.
7. S. D. Conte, H. E. Dunsmore, and V. Y. Shen. *Software Engineering Metrics and Models*. The Benjamin/Cummings Publishing Company Inc., 1986. ISBN 0-8053-2162-4.
8. L. J. Guibas and R. Sedgewick. A dichromatic framework for balanced trees. In *Proc. 19th Ann. IEEE Symp. on Foundations of Computer Science*, pages 8–21, 1978.
9. H. A. Maurer, Th. Ottmann, and H. W. Six. Implementing dictionaries using binary trees of very small height. *Information Processing Letters*, 5(1):11–14, 1976.
10. J. I. Munro, Th. Papadakis, and R. Sedgewick. Deterministic skip lists. In *Proc. Symp. of Discrete Algorithms*, pages 367–375, 1992.
11. J. Nievergelt and E. M. Reingold. Binary trees of bounded balance. *SIAM Journal on Computing*, 2(1):33–43, 1973.
12. H. J. Olivie. A new class of balanced search trees: Half-balanced binary search trees. *R. A. I. R. O. Informatique Theoretique*, 16:51–71, 1982.
13. Th. Papadakis. private communication.
14. W. Pugh. Skip lists: A probabilistic alternative to balanced trees. In *Proc. Workshop on Algorithms and Data Structures, WADS '89, Ottawa*, pages 437–449, 1989.
15. N. Wirth. *Algorithms and Data Structures*. Prentice-Hall, Englewood Cliffs, New Jersey, 1986. ISBN 0-13-022005-1.

Probing a Set of Hyperplanes by Lines and Related Problems

Yasukazu Aoki[1], Hiroshi Imai[1], Keiko Imai[2] and David Rappaport[3]

[1] Department of Information Science, University of Tokyo
Hongo, Bunkyo-ku, Tokyo 113, Japan

[2] Department of Information and System Engineering, Chuo University
Kasuga, Bunkyo-ku, Tokyo 112, Japan

[3] Department of Computing and Information Science, Queen's University
Kingston, Ontario, Canada K7L 3N6

Abstract. Suppose that for a set H of n unknown hyperplanes in the Euclidean d-dimensional space, a line probe is available which reports the set of intersection points of a query line with the hyperplanes. Under this model, this paper investigates the complexity to find a generic line for H and further to determine the hyperplanes in H. This problem arises in factoring the u-resultant to solve systems of polynomials (e.g., Renegar [13]). We prove that $d+1$ line probes are sufficient to determine H. Algorithmically, the time complexity to find a generic line and reconstruct H from $O(dn)$ probed points of intersection is important. It is shown that a generic line can be computed in $O(dn \log n)$ time after d line probes, and by an additional d line probes, all the hyperplanes in H are reconstructed in $O(dn \log n)$ time. This result can be extended to the d-dimensional complex space. Also, concerning the factorization of the u-resultant using the partial derivatives on a generic line, we touch upon reducing the time complexity to compute the partial derivatives of the u-resultant represented as the determinant of a matrix.

1 Introduction

Let H be a finite set of n hyperplanes $h_i = \{ x \in \mathbf{R}^d \mid a_i^T x = b_i \}$ $(i = 1, \ldots, n)$ in $\mathbf{R}^d$. The arrangement of H is denoted by $\mathcal{A}(H)$. A function $f_H: \mathbf{R}^n \to \mathbf{R}$ defined by

$$f_H(x) = \prod_{i=1}^{n} (a_i^T x - b_i)$$

is called the *defining polynomial* of the arrangement $\mathcal{A}(H)$ (Orlik and Terao [12]), and also plays an important role in the interior-point method for linear programming (Karmarkar [8], Sonnevend [15]).

In modern elimination theory for algebraic equations such as Lazard's method [10], some way of computing the u-resultant, which is the defining polynomial of n hyperplanes in $\mathbf{C}^d$, of a given system of algebraic equations is presented, and solutions are obtained by decomposing the u-resultant into linear factors, that is, determining the hyperplanes. This factorization is done by probing the hyperplanes by lines (Canny [3]) or computing the gradient on a generic line to the hyperplanes by partially using probes (Kobayashi, Fujise and Furukawa [9], Renegar [13], Murao [11]). Thus arises the problem of probing hyperplanes by lines, a problem which does not seem to have been well studied in computational geometry. In contrast with this, probing a convex polyhedron is a well-studied problem (Cole and Yap [4], Dobkin, Edelsbrunner and Yap [5]). For a survey of results in geometric probing see [14]. The theme of this paper is the previously unstudied problem of probing hyperplanes.

Let us define the problem in more detail. For simplicity, we consider the problem in real d-space $\mathbf{R}^d$ and all the hyperplanes in H are distinct. The cases of complex space and degenerate hyperplanes can be handled with slight modifications, which will be described in the full version of this paper. Suppose that a procedure to compute the defining polynomial $f_H(\boldsymbol{x})$ of the set H of n unknown hyperplanes in $\mathbf{R}^d$ is available in term of d variables $x_1, \ldots, x_d$ ($\boldsymbol{x} = (x_i)$), as in the case of the u-resultant. Using this procedure, for any point $\boldsymbol{p}$ in $\mathbf{R}^d$, we can determine whether $\boldsymbol{p}$ is on some hyperplane in H by checking if $f_H(\boldsymbol{p})$ is zero or not. This operation is called a *point probe*. For a line $\boldsymbol{p} + t\boldsymbol{q}$ passing through $\boldsymbol{p}$ with direction $\boldsymbol{q}$ ($\boldsymbol{q} \in \mathbf{R}^d$), $f_H(\boldsymbol{p} + t\boldsymbol{q})$ is a polynomial of degree at most n in terms of t. The points of intersection of the line with hyperplanes in H are computed by solving this univariate polynomial. This operation is called a *line probe*. It should be noted that a line probe obtains the points of intersection as a set, with no information regarding which point lies on which hyperplane. Concerning these fundamental operations, we assume that values of $f_H(\boldsymbol{p})$, $f_H(\boldsymbol{p} + t\boldsymbol{q})$ and the zeros, with their multiplicities, of a single-variable polynomial can be calculated exactly in a unit time (in section 5, we will discuss complexities related to compute $f_H(\boldsymbol{p}+t\boldsymbol{q})$). Also, fundamental operations $(+, -, \times, /)$ on real numbers are assumed to be executed exactly in a unit time. These assumptions allow us to focus on the number of necessary probes.

The following problems are considered when probing to reconstruct the set of hyperplanes H.

(1) Finding a generic line: Finding a *generic line*, a line intersecting all the hyperplanes in H at distinct points by line probes.

(2) Probing hyperplanes by lines: Constructing the hyperplanes in H by line probes.

(3) Computing the gradient of the defining polynomial: Computing the gradient $\nabla f_H(\boldsymbol{x})$ of $f_H(\boldsymbol{x})$ on a generic line efficiently.

Concerning (1), a randomly chosen line is generic with probability 1, but this is a probabilistic method [3, 9], and we want to devise a deterministic one. The algorithm by Renegar [13] is deterministic, but it uses $(d-1)n(n-1)/2+1$ line probes (a fixed direction) after finding a *generic point*, which is a point on no hyperplane in H, by $dn + 1$ point probes. This paper shows that, assuming a generic point is at hand, the number of line probes that are needed to determine a generic line is d in the worst case, and given at most dn probed points, a generic line can be computed in $O(dn \log n)$ time. If probed points are provided sorted on a probe line then the time complexity is reduced to $O(dn)$.

Concerning (2), Canny [3] shows that, given a generic line, performing d general line probes and $dn(n+1)/2$ special line probes which verify whether a probing line is included in some hyperplane in H, the hyperplanes can be reconstructed in $O(dn^2)$ time. This paper shows that, by executing, $2d$ probes from scratch the hyperplanes can be reconstructed in $O(dn \log n)$ time.

Thus in both cases of (1) and (2), by treating the problems from the viewpoint of computational geometry, the probing and time complexities are greatly improved. Our results can be readily extended to the d-dimensional complex space $\mathbf{C}^d$.

Concerning (3), in [9, 13], it is shown that, given a generic line, all the hyperplanes are recovered by computing the gradient of the defining polynomial at each intersection points of the generic line with the hyperplanes. Here, it is required to compute the gradient quickly. This paper presents an efficient way of computing the gradient when the u-resultant is represented as the determinant of a matrix ([9, 11]) by making use of the idea in [2, 7]. Our improvement reduces the time complexity by a factor of degree 2. This result is somehow a computational algebraic result, but would be interesting from the viewpoint of computational geometry since the defining polynomial of the arrangement may be utilized further in the field.

2 Preliminaries

We have already defined fundamental concepts such as point probe, line probe, generic point, and generic line for the set H of n distinct hyperplanes in $\mathbf{R}^d$. In connection with line probes, we need some more definitions. A line is *parallel* to a hyperplane if the normal vector of the hyperplane is orthogonal to the direction of the line. The direction of a line which is not parallel to any hyperplane in H is called a *generic direction.* Of course, the direction of a generic line is a generic direction. Finally, the *multiplicity* of a point in $\mathbf{R}^d$ with respect to H is defined to be the number of hyperplanes in H containing the point. As was assumed, by a line probe, intersection points of the line with hyperplanes in H are obtained with their multiplicities. If the multiplicities of intersection points of a probing line sum up to n, then the probing line is in a generic direction. When a probing line is contained in a hyperplane in H, we only obtain this fact and no other information. To avoid this problem, a generic point is useful, since any line passing the generic point is not contained in any hyperplane in H. Renegar [13] gives a simple solution to find a generic point by point probes using the moment curve $(1, t, t^2, t^3, \ldots, t^{d-1})$ $(t \in \mathbf{R})$ in $\mathbf{R}^d$.

Lemma 1 (Renegar [13]) *For $dn+1$ distinct points on the moment curve, at least one point is a generic point.*

Hence, with $dn+1$ point probes, a generic point q is obtained. To compare point probe and line probe and also to explain the approach in [13], we describe how to find a generic line by a similar approach. If the hyperplanes in H are homogeneous, that is, all the hyperplanes pass the origin, the direction q is not parallel to any hyperplane in H, and so is a generic direction. Even if the hyperplanes are not homogeneous, by means of line probes, a generic direction can be easily found, as stated in the following.

Lemma 2 *By considering $(d-1)n+1$ distinct directions from the origin to points on the moment curve, at least one direction is a generic direction.*

Then, we can find a generic line by line probes conceptually considering the projection of the whole arrangement to a hyperplane perpendicular to the generic direction.

Lemma 3 (Renegar [13]) *For a given generic direction q, considering $(d-1)n(n-1)/2+1$ distinct points $p_i(\neq q)$ on the moment curve, one of lines represented by p_i+tq ($t \in \mathbf{R}$) is a generic line.*

Thus, with $dn+1$ point probes (or $(d-1)n+1$ line probes) and $(d-1)n(n-1)/2+1$ line probes, a generic line can be found. Here, a crucial observation is that the numbers in Lemmas 1, 2 are tight in the worst case, which can be seen by considering a simple adversary, while the bound in Lemma 3 seems rather loose. We will show, in the next section, that much fewer line probes are sufficient to find a generic line.

Since a point probe is much less costly than a line probe, as may be seen in section 5, in the sequel, we assume that a generic point is pre-computed by point probes, and will not count for this complexity.

3 Tight Bounds for the Number of Necessary Probes

Making a given generic point as the origin, form the orthogonal $(x_1, x_2, \ldots, x_d)$-coordinate by choosing d independent directions in $\mathbf{R}^d$. Consider d line probes by these d axes. Denote by S_i the set of distinct intersection points of the x_i-axis with hyperplanes in H, where the multiplicities are ignored. Let $n_i = |S_i|$, where $n_i \leq n-$(the number of hyperplanes in H parallel to the x_i-axis). Denote the set of all probed points by S which is the disjoint union of S_i $(i = 1, \ldots, d)$. In total, we have $|S| = \sum_{i=1}^{d} n_i (\leq dn)$ probed points.

For a set S' of points and a hyperplane h in H, let $S' \cap h$ denote the set of points in S' which are contained in h. Since the probing lines are independent and are not contained in any hyperplane in H, we have the following lemma.

Lemma 4 *For any hyperplane h in H and $i = 1, \ldots, d$, $|S \cap h| \geq 1$ and $|S_i \cap h| \leq 1$.*

For $h \in H$, we represent the probed points by $I_h = \{ i \mid |S_i \cap h| = 1,\ i = 1, \ldots, d \}$. Then, for any $i \notin I_h$, the x_i-axis does not intersect with h, that is, the x_i-axis is parallel to h. Thus, h contains $|I_h|$ independent points in $S \cap h$ and is parallel to $d - |I_h|$ independent directions each of which is orthogonal to the affine hull of points in $S \cap h$. This implies that h is uniquely determined, in a reverse manner, if I_h and $|I_h|$ points, one from each S_i $(i \in I_h)$, are given.

Now, consider the set $\widetilde{H}$ of hyperplanes which are determined by I_h and $|I_h|$ points, one from each S_i ($i \in I_h$), in the above manner for all possible I_h and points in S_i ($i \in I_h$). Apparently, there are $\prod_{i=1}^{d}(n_i+1)-1$ possible ways of taking I_h and $|I_h|$ points, one from each S_i ($i \in I_h$), and, for each of these, a distinct hyperplane is uniquely determined. Note that $I_h \neq \emptyset$. The next lemma follows from the discussion thus far.

Lemma 5 *$H \subseteq \widetilde{H}$, and $|\widetilde{H}| = \prod_{i=1}^{d}(n_i+1)-1 < (n+1)^d$.*

We are now ready to prove the following theorem.

Theorem 1 *The number of line probes necessary to determine a generic line is at most d, and this bound is tight in the worst case.*

Proof. To show that d line probes are sufficient to find a generic line, we can use Lemma 5. First use line probes to obtain the superset $\widetilde{H}$. Then, we construct the arrangement $\mathcal{A}(\widetilde{H})$. (Although the size of $\mathcal{A}(\widetilde{H})$ is $O(n^{d^2})$ we will subsequently see how to make implicit use of $\mathcal{A}(\widetilde{H})$ without incurring a huge computational expense.) Given $\mathcal{A}(\widetilde{H})$, it is easy to compute a generic line for $\widetilde{H}$. Since $H \subseteq \widetilde{H}$, this line is also a generic line with respect to H.

We now prove this bound is tight in the worst case. For any set of at most $d-1$ line probes, there is a direction orthogonal to each probing line. Suppose there is a hyperplane in H which is orthogonal to this direction. Then, even if the hyperplane is moved orthogonally along the direction, the collection of probed points does not change. Hence, we cannot necessarily distinguish two different sets of hyperplanes by at most $d-1$ line probes. Furthermore, there may be a hyperplane not distinguishable by the first $d-1$ probes which can adversely effect any proposed generic line. □

Although the size of $\mathcal{A}(\widetilde{H})$ is $O(n^{d^2})$ we will subsequently see how to make use of the regular structure of $\widetilde{H}$ to obtain a generic line without incurring the huge computational expense of constructing the entire arrangement.

A further key result is proved leading to an algorithm to reconstruct H from probes. An efficient implementation of the algorithm is discussed in the next section.

Theorem 2 *The number of line probes necessary to determine H is at most $d+1$, and this bound is tight in the worst case.*

Proof. As in Theorem 1, execute d line probes, and further execute a line probe by using the computed generic line. Since this generic line is generic to both H and $\widetilde{H}$, each probed point on the line for H coincides with exactly one probed point for $\widetilde{H}$. Then, hyperplanes in $\widetilde{H}$ each of which uniquely contains a probed point on the generic line for H are exactly those in H.

The lower bound can be shown in a way similar to the proof of Theorem 1.
□

Although we have so far assumed that hyperplanes in H are distinct to one another, we can generalize the above theorems easily even to degenerate case. These bounds are a big improvement to those found in [3, 13].

4 Efficient Probing Algorithm for Hyperplanes

In this section, we give an efficient algorithm for finding a generic line and identifying the hyperplanes. A main idea is to make use of some regular structure of the superset $\widetilde{H}$ of H discussed in the preceding section, specifically its grid structure in the dual space. To find a generic line, this algorithm only needs d line probes. Using at most dn probed points of intersection, a generic line is computed in $O(dn \log n)$ time.

First, as in the preceding section, make a generic point as the origin, and choose an arbitrary orthogonal basis to form the $(x_1, \ldots, x_d)$-coordinate. Using each x_i-axis as a probing line, find all the points of intersection. Then, the superset $\widetilde{H}$ of H is formed from these probed points.

We now transform the original space (primal space) to the dual space so that the regular structure of the superset $\widetilde{H}$ may be more easily seen in the dual space. Since the hyperplanes in H do not pass the origin, each hyperplane may be represented as $\boldsymbol{a}^{\mathrm{T}}\boldsymbol{x} = 1$. We may then consider a dual transform which maps a point $\boldsymbol{a}$ in the primal space to a hyperplane $\boldsymbol{a}^{\mathrm{T}}\boldsymbol{x} = 1$ in the dual space. We will use the same coordinate system in the dual space. By this duality, a hyperplane in the primal space is mapped to a point in the dual space. Note that a hyperplane parallel to the x_i-axis in the primal space is mapped to a point in the hyperplane $x_i = 0$ in the dual space.

By this duality transform, each probed point on the x_i-axis is mapped to a hyperplane orthogonal to the same axis. By these hyperplanes corresponding to probed points and hyperplanes $x_i = 0$ $(i = 1, \ldots, n)$ in the dual space, $\prod_{i=1}^{d}(n_i + 1)$ grid points are formed in the dual space. Denote this set of at most $(n+1)^d$ grid points by $\widetilde{H}_D$. Hyperplanes in H are mapped to points in the dual space, whose set is denoted by H_D. Then, by the duality, the following lemma holds.

Lemma 6 *The dual of $\widetilde{H}$ is $\widetilde{H}_D$ minus the origin point, and $H_D \subseteq \widetilde{H}_D$.*

$\widetilde{H}_D$ has a nice regular structure as grid points. Although its total size is at most $(n+1)^d$, it is originally produced by d sets of at most $n+1$ parallel hyperplanes, and hence has a compact representation.

We will search for, L, a generic line to $\widetilde{H}$, passing through the origin in the primal space. Then this problem in the primal space is transformed to the problem, in the dual space, of finding a hyperplane orthogonal to the line L such that when moving the hyperplane orthogonally along the line no two grid points of $\widetilde{H}_D$ are hit simultaneously. Note that no points in $\widetilde{H}_D$ lie on L implying that there are no hyperplanes in H parallel to L.

Lemma 7 *A line L passing through the origin, with the property that every two points from $\widetilde{H}_D$ have distinct orthogonal projections onto L, is generic for $\widetilde{H}$.*

Let the set of n_i+1 distinct grid points in $\widetilde{H}_D$ incident to the dual x_i-axis be represented by their x_i coordinate as $\{x_i = \overline{x}_{i,j} \mid j = 1, \ldots, n_i+1\}$ with $\overline{x}_{i,0} < \overline{x}_{i1} < \cdots < \overline{x}_{i,n_i}$ (one of $\overline{x}_{i,j}$ $(j = 0, \ldots, n_i)$ is zero). For each $i = 1, \ldots, d$, define the width w_i of $\overline{x}_{i,j}$ $(j = 0, \ldots, n_i)$ to be $\overline{x}_{i,n_i} - \overline{x}_{i,0}$. Also, for each $i = 1, \ldots, d$, define the minimum gap g_i by $g_i = \min_{j=1,\ldots,n_i} (\overline{x}_{i,j} - \overline{x}_{i,j-1})$. We here assume that $n_i > 1$ for each i. Using these, we will try to find a line L, passing through the origin such that any hyperplane orthogonal to the line L contains at most one grid point in $\widetilde{H}_D$.

Theorem 3 *Let r_i be a real number with $r_i \geq 2$ $(i = 1, \ldots, d)$. Define the ith element q_i of $q \in \mathbf{R}^d$ for $i = 1, \ldots, d$ in this order by*

$$q_1 = \frac{4}{r_1}$$

$$q_i = \frac{g_{i-1}}{r_i w_i} q_{i-1} \quad (i = 2, \ldots, d)$$

Then, a line represented by tq $(t \in \mathbf{R})$ passing through the origin is a generic line in the primal space.

Proof. Consider two distinct grid points x' and x'' in $\widetilde{H}_D$. Let l be the minimum l such that the x_l-coordinate values of x' and x'' are different, that is, $x'_i = x''_i$ $(i = 1, \ldots, l-1)$ and $x'_l \neq x''_l$. Without loss of generality, we can assume $x'_l > x''_l$. Then, we have

$$\begin{aligned}
q^{\mathrm{T}} x' - q^{\mathrm{T}} x'' &= \sum_{i=l}^{d} q_i (x'_i - x''_i) \geq q_l (x'_l - x''_l) - \sum_{i=l+1}^{d} q_i |x'_i - x''_i| \\
&\geq q_l g_l - \sum_{i=l+1}^{d} q_i w_i = q_l g_l - \sum_{i=l+1}^{d} \left(w_i q_{l+1} \prod_{j=l+1}^{i-1} \frac{g_j}{r_{j+1} w_{j+1}} \right) \\
&\geq q_l g_l - q_{l+1} g_{l+1} \sum_{i=l+1}^{d} \left(\frac{1}{2^{i-l-1}} \prod_{j=l+2}^{i-1} \frac{g_j}{w_j} \right) \\
&\geq q_l g_l - q_{l+1} g_{l+1} \sum_{i=l+1}^{d} \frac{1}{2^{i-l-1}} \\
&> q_l g_l - 2 q_{l+1} g_{l+1} \geq q_l g_l \left(1 - \frac{g_{l+1}}{w_{l+1}} \right) \geq 0
\end{aligned}$$

Hence, two distinct grid points are not on the same hyperplane whose normal vector is q, and the theorem follows. □

Given $\{\overline{x}_{i,j}\}$, w_i and g_i $(i = 1, \ldots, d)$ are computed in $O(dn \log n)$ time by sorting $\overline{x}_{i,j}$ $(j = 0, \ldots, n_i)$ for each $i = 1, \ldots, d$. If probed points on a line in the primal space are given in a sorted order, these values can be computed in $O(dn)$ time. After computing these values, q is found in $O(d)$ time. Hence, we have the following theorem.

Theorem 4 *Given a generic point, from d line probes, a generic line can be found in $O(dn \log n)$ time, and, if intersection points on a probing line are given in a sorted order, the complexity is reduced to $O(dn)$.*

We now thus establish an efficient way of finding a generic line. We can generalize this result for the complex space $\mathbf{C}^d$. Concerning the problem of finding the hyperplanes in H by line probes, we have the following.

Theorem 5 *Given a generic point, from 2d line probes in total, all the hyperplanes in H can be determined in $O(dn \log n)$ time.*

Proof. In Theorem 3, q is computed for some r_i $(i = 1, \ldots, d)$ with $r_i \geq 2$. Let q_i be q in the theorem obtained for r_j $(j = 1, \ldots, d)$ defined by $r_j = 4$ for $j = i$ and $r_j = 2$ for $j \neq i$. Then, by the theorem, each line tq_i $(t \in \mathbf{R})$ is a generic line. Furthermore these generic lines hit the hyperplanes in H in the same order, because, for each q_i, by the proof of the theorem, we see that all grid points on $x_1 = \overline{x}_{1,j_1}$ are traversed before grid points on $x_1 = \overline{x}_{1,j_1+1}$ are done, and, among grid points on $x_1 = \overline{x}_{1,j_1}$, grid points on $x_2 = \overline{x}_{2,j_2}$ are traversed before grid points on $x_2 = \overline{x}_{2,j_2+1}$, and so on. Finally, since grid points on $d-1$ hyperplanes $x_i = \overline{x}_{i,j_i}$ $(i = 1, \ldots, d-1)$ have distinct x_d-coordinate values, grid points in $\widetilde{H}_D$ are passed over by the hyperplane with normal vector q_i in a unique order.

Hence, denoting the points of intersection of line tq_i and hyperplanes in H by $b_{i,j}q_i$ $(j = 1, \ldots, n)$ with $b_{i,1} < b_{i,2} < \cdots < b_{i,n}$ $(i = 1, \ldots, d)$, the jth hyperplane in H contains $b_{1,j}q_1, b_{2,j}q_2, \ldots, b_{d,j}q_d$.

Let A be a $d \times d$ matrix whose ith column vector is q_i. Defining α_i by $\alpha_1 = 1$ and $\alpha_i = \dfrac{g_{i-1}}{2w_i}\alpha_{i-1}$ $(i = 2, \ldots, d)$, A is expressed as follows.

$$A = \begin{pmatrix} \alpha_1 & 2\alpha_1 & 2\alpha_1 & \cdots & 2\alpha_1 \\ \alpha_2 & \alpha_2 & 2\alpha_2 & \cdots & 2\alpha_2 \\ \alpha_3 & \alpha_3 & \alpha_3 & \cdots & 2\alpha_3 \\ \cdot & \cdot & \cdot & \cdots & \cdot \\ \alpha_{d-1} & \alpha_{d-1} & \alpha_{d-1} & \cdots & 2\alpha_{d-1} \\ \alpha_d & \alpha_d & \alpha_d & \cdots & \alpha_d \end{pmatrix}$$

As is readily seen, A is nonsingular. Its inverse is given by

$$A^{-1} = \begin{pmatrix} -1/\alpha_1 & 0 & 0 & \cdots & 0 & 2/\alpha_d \\ 1/\alpha_1 & -1/\alpha_2 & 0 & \cdots & 0 & 0 \\ 0 & 1/\alpha_2 & -1/\alpha_3 & \cdots & 0 & 0 \\ \cdot & \cdot & \cdot & \cdots & \cdot & \cdot \\ 0 & 0 & 0 & \cdots & 1/\alpha_{d-1} & -1/\alpha_d \end{pmatrix}$$

Let B_j be a $d \times d$ diagonal matrix whose diagonal elements are $b_{1,j}, b_{2,j}, \ldots, b_{d,j}$. The ith column of AB_j is $b_{i,j}\boldsymbol{q}_i$, and, since A is nonsingular and $b_{i,j}$ ($i = 1, \ldots, d$) are not zeros, AB_j is nonsingular.

Let the jth hyperplane in H be $\boldsymbol{a}_j^{\mathrm{T}}\boldsymbol{x} = 1$ (recall that this hyperplane does not pass the origin). Then, $\boldsymbol{a}_j^{\mathrm{T}}AB_j = (1, 1, \ldots, 1)$. Since AB_j is nonsingular, $\boldsymbol{a}_j$ is determined uniquely from A and B_j, that is, the jth hyperplane in H can be uniquely determined by points $b_{1,j}\boldsymbol{q}_1, b_{2,j}\boldsymbol{q}_2, \ldots, b_{d,j}\boldsymbol{q}_d$ on the hyperplane.

All the hyperplanes in H can thus be obtained. About the time complexity, A^{-1} is simple enough as above, and, computing $\alpha_1, \ldots, \alpha_d$ once and computing $\boldsymbol{q}_1, \ldots, \boldsymbol{q}_d$ from them, we can compute $\boldsymbol{a}_j$ ($j = 1, \ldots, n$) in $O(dn \log n)$ time in total. □

5 Computing the Gradient of the *u*-Resultant

Due to the space limitations, we will only sketch the results on the u-resultant (e.g., see van der Waerden [16]). We illustrate our ideas through a small example taken from Lazard [10].

Consider a system of algebraic equations

$$x^2 + xy + 2x + y - 1 = 0$$
$$x^2 - y^2 + 3x + 2y - 1 = 0$$

which has a finite number (4 in this case) of solutions. The solution of this system is $(0, 1)$, $(1, -1)$, $(-3, 0)$ and a solution at infinity in the direction $(-1, 1)$. There are several methods to compute the u-resultant, and, by the algorithm in [10], the u-resultant is expressed as the determinant of a 4×4 matrix as follows:

$$f_H(u_1, u_2, u_3) = \begin{vmatrix} u_1 + u_2 - u_3 & 0 & 0 & 0 \\ -u_2 + u_3 & -u_1 + 2u_2 - u_3 & u_2 & 0 \\ u_3 & -2u_2 & u_1 - u_2 + u_3 & 0 \\ 0 & 0 & 0 & -u_2 + u_3 \end{vmatrix}$$

Since in this case the problem is small and well-behaved, we can directly expand the determinant and decompose it as follows:

$$f_H(u_1, u_2, u_3) = (u_1 + u_2 - u_3)(u_1 - 3u_2 + u_3)(u_1 + u_3)(u_2 - u_3)$$

A linear term $a_1u_1 + a_2u_2 + a_3u_3$ in the decomposition corresponds to a solution $(u_2/u_1, u_3/u_1)$ by the theory of u-resultant. Hence, four terms in the right-hand side in this case corresponds to $(1, -1)$, $(-3, 1)$, $(0, 1)$ and a solution at infinity in the direction $(-1, 1)$.

In this simple example, the determinant can be expanded and its decomposition can be found very easily. However, if the system becomes larger, expanding the determinant of a large matrix having d (in the above case, $d = 3$) indeterminates is very time-consuming, and, even in case the expansion is obtained, its direct decomposition is again very hard.

In general cases, $f_H(\boldsymbol{u})$ is given as the determinant of an $n \times n$ matrix $R(\boldsymbol{u})$ of d variables $u_1, \ldots, u_d$, and the ij-element of $R(\boldsymbol{u})$ is given by $r_{ij}(\boldsymbol{u}) = \boldsymbol{a}_{ij}^{\mathrm{T}} \boldsymbol{u}$ [9, 10]. As mentioned above, expanding the determinant of $R(\boldsymbol{u})$ takes much time. Roughly, it takes $O(n^{3+d})$ time by the interpolation method. However, to perform a line probe for line $\boldsymbol{p}+t\boldsymbol{q}$, we first compute the determinant $f_H(\boldsymbol{p}+t\boldsymbol{q})$ of $R(\boldsymbol{p}+t\boldsymbol{q})$ with a single variable t to obtain a single-variable polynomial, and then compute its zeros. Here, the determinant of $R(\boldsymbol{p}+t\boldsymbol{q})$ can be computed in $O(n^4)$ time, independent of d. Therefore, instead of expanding the original determinant with many variables once, which is almost impossible even for moderate-size problems, expanding the determinant of a single variable $d+1$ or $2d$ times by executing $d+1$ or $2d$ line probes, respectively, is definitely economical in total, and this is one of the most practical approaches to solve the problem. Probing by lines are thus useful.

Now, suppose all the solutions are distinct, or in other words in the corresponding arrangement there are no two identical hyperplanes (the general case will be discussed in the full version). Consider a generic line, represented by $\boldsymbol{p}+t\boldsymbol{q}$, and let $\boldsymbol{p}+t_i\boldsymbol{q}_1$ $(i = 1, \ldots, n)$ be points of intersection of the generic line with hyperplanes. Then, from the form of the defining polynomial,

$$\left(\left.\frac{\partial f_H(\boldsymbol{u})}{\partial u_1}\right|_{\boldsymbol{u}=\boldsymbol{p}+t_i\boldsymbol{q}}, \left.\frac{\partial f_H(\boldsymbol{u})}{\partial u_2}\right|_{\boldsymbol{u}=\boldsymbol{p}+t_i\boldsymbol{q}}, \ldots, \left.\frac{\partial f_H(\boldsymbol{u})}{\partial u_d}\right|_{\boldsymbol{u}=\boldsymbol{p}+t_i\boldsymbol{q}} \right)$$

is the direction of normal vector of the ith hyperplane [9, 13] (note that $\boldsymbol{p}+t_i\boldsymbol{q}$ is on only the ith hyperplane). From this, the ith solution of the original equation can be readily computed. Thus, computing the gradient of $f_H(\boldsymbol{u})$ with respect to $\boldsymbol{u}$ is a main step of this approach.

We here give an efficient way of doing this. A basic idea is from Baur and Strassen [2] and Iri [7].

Theorem 6 *Suppose that the determinant $|R(u(t))|$ of $R(u(t))$ with t as a variable is not vanishing. For $\boldsymbol{u} = \boldsymbol{u}(t) = \boldsymbol{p} + t\boldsymbol{q}$ with t as a variable,*

$$\left.\frac{\partial f_H(\boldsymbol{u})}{\partial u_l}\right|_{\boldsymbol{u}=\boldsymbol{u}(t)} = \mathrm{trace}\left[\frac{\partial R(\boldsymbol{u})}{\partial u_l}\left(|R(\boldsymbol{u}(t))|R(\boldsymbol{u}(t))^{-1}\right)\right]$$

where $\partial R(\boldsymbol{u})/\partial u_l$ is an $n \times n$ matrix whose ij-element is $\partial r_{ij}(\boldsymbol{u})/\partial u_l$, which is the lth element of $\boldsymbol{a}_{ij}$.

The proof is omitted in this version. Note that the trace in Theorem 6 can be computed in $O(n^3)$ time by summing up n^2 products of a real constant and a single-variable polynomial of degree at most $n-1$. The inverse $M(\boldsymbol{u}(t))^{-1}$ is computed just once, which takes $O(n^4)$ time by the interpolation method, and applying the above theorem for each i we have the following.

Theorem 7 *Given the defining polynomial $f_H(\boldsymbol{u})$ as the determinant of a matrix $R(\boldsymbol{u})$, the gradient of $f_H(\boldsymbol{u})$ with respect to $\boldsymbol{u}$ on a line $\boldsymbol{p}+t\boldsymbol{q}$ can be computed in $O(n^4)$ time.*

Acknowledgment

The work by the fourth author was done while he was visiting Department of Information Science, University of Tokyo during September, 1992, by the bilateral program between NSERC (Canada) and JSPS (Japan). The work by the second and third authors was partially supported by the Grant-in-Aid of the Ministry of Education, Science and Culture of Japan.

References

1. Aoki, Y., "*The Combinatorial Complexity of Reconstructing Arrangements,*" Doctoral Thesis, Department of Information Science, University of Tokyo, 1993.
2. Baur, W., and V. Strassen, "The Complexity of Partial Derivatives," *Theoretical Computer Science*, Vol.22 (1983), pp.317–330.
3. Canny, J., "A New Algebraic Method for Robot Motion Planning and Real Geometry," *Proceedings of the 28th IEEE Annual Symposium on Foundations of Computer Science*, 1987, pp.39-48.
4. Cole, R., and C. Yap, "Shape from Probing," *Journal of Algorithms*, Vol.8 (1987), pp.19–38.
5. Dobkin, D., H. Edelsbrunner, and C. K. Yap, "Probing Convex Polytopes," *Proceedings of the 18th Annual ACM Symposium on Theory of Computing*, 1986, pp.424–432.
6. Edelsbrunner, H., "*Algorithms in Combinatorial Geometry,*" Springer-Verlag, 1987.
7. Iri, M., "Simultaneous Computation of Functions, Partial Derivatives and Estimates of Rounding Errors — Complexity and Practicality," *Japan Journal of Applied Mathematics*, Vol.1, No.2 (1984), pp.223–252.
8. Karmarkar, N., "A New Polynomial-Time Algorithm for Linear Programming," *Combinatorica*, Vol.4 (1984), pp.373–395.
9. Kobayashi, H., T. Fujise and A. Furukawa, "Solving Systems of Algebraic Equations by a General Elimination Method," *Journal of Symbolic Computation*, Vol.5, No.3 (1988), pp.303–320.
10. Lazard, D., "Résolution des Systèmes d'Équations Algébriques," *Theoretical Computer Science*, Vol.15 (1981), pp.77–110.
11. Murao, H., *Development of Efficient Algorithms in Computer Algebra,*" Doctoral Thesis, Department of Information Science, University of Tokyo, 1992.
12. Orlik, P., and H. Terao, "*Arrangements of Hyperplanes,*" Springer-Verlag, 1991.
13. Renegar, J., "On the Worst-Case Arithmetic Complexity of Approximating Zeros of Systems of Polynomials," *SIAM Journal on Computing*, Vol.18, No.2 (1989), pp.350–370.
14. Skiena, S., "Interactive reconstruction via probing," *Proceedings of the IEEE*, Vol.80, No.9 (1992), pp.1364–1383.
15. Sonnevend, Gy., "An "Analytical Centre" for Polyhedrons and New Classes of Global Algorithms for Linear (Smooth, Convex) Programming," *Lecture Notes in Control and Information Sciences*, Vol.84, Springer-Verlag, 1986, pp.866–876.
16. van der Waerden, B. L., "*Moderne Algebra,*" Vol.II, Springer-Verlag, 2nd Edition, 1940.

A General Lower Bound on the I/O-Complexity of Comparison-based Algorithms*

Lars Arge, Mikael Knudsen and Kirsten Larsen

Aarhus University, Computer Science Department
Ny Munkegade, DK-8000 Aarhus C.**

Abstract. We show a general relationship between the number of comparisons and the number of I/O-operations needed to solve a given problem. This relationship enables one to show lower bounds on the number of I/O-operations needed to solve a problem whenever a lower bound on the number of comparisons is known. We use the result to show lower bounds on the I/O-complexity on a number of problems where known techniques only give trivial bounds. Among these are the problems of removing duplicates from a multiset, a problem of great importance in e.g. relational data-base systems, and the problem of determining the mode - the most frequently occurring element - of a multiset. We develop algorithms for these problems in order to show that the lower bounds are tight.

1 Introduction

In the study of complexity of algorithms, most attention has been given to bounding the number of primitive operations (for example comparisons) needed to solve a problem. However, when working with data materials so large that they will not fit into internal memory, the amount of time needed to transfer data between the internal memory and the external storage (the number of I/O-operations) can easily dominate the overall execution time.

In this paper we work in a model introduced by Aggarwal and Vitter [1] where an I/O-operation swaps B records between external storage and the internal memory, capable of holding M records. An algorithm for this model is called an I/O-algorithm. Aggarwal and Vitter [1] consider the I/O-complexity of a number of specific sorting-related problems, namely sorting, fast Fourier transformation, permutation networks, permuting and matrix transposition. They give asymptotically matching upper and lower bounds for these problems. The lower bounds are based on routing arguments, and in general no restrictions are made on the operations allowed in internal memory, except that records are considered to be atomic and cannot be divided into smaller parts. Only when the internal memory is extremely small, the comparison model is assumed.

* This work was partially supported by the ESPRIT II Basic Research Actions Program of the EC under contract No. 7141 (project ALCOM II).

** E-mail communication to large@daimi.aau.dk

In this paper we shall use the same model of computation, except that in general we will limit the permitted operations in the internal memory to comparisons. The main result of this paper is the following: Given an I/O-algorithm that solves an n-record problem P_n using $I/O(\overline{x})$ I/O's on the input $\overline{x}$, there exists an ordinary comparison algorithm that uses no more than $n \log B + I/O(\overline{x}) \cdot T_{\mathrm{merge}}(M - B, B)$ comparisons on input $\overline{x}$. $T_{\mathrm{merge}}(n, m)$ denotes the number of comparisons needed to merge two sorted lists, of size n and m respectively. While [1] shows lower bounds for a number of specific problems, our result enables one to show lower bounds on the number of I/O operations needed to solve a problem for any problem where a lower bound on the number of comparisons needed is known. Among these is sorting where we obtain the same lower bound as in [1]. We use the result to show lower bounds on a number of problems not formerly considered with respect to I/O-complexity. Among these are the problems of removing duplicates from a multiset, a problem of great importance in e.g. relational data-base systems, and the problem of finding the most frequently occurring element in a multiset. We develop algorithms for these problems in order to show that the lower bounds are tight.

The basic idea in the lower bound proofs in [1] is to count how many permutations can be generated with a given number of I/O-operations and to compare this to the number of permutations needed to solve a problem. This technique, however, is not generally applicable. For example only trivial lower bounds can be shown on the problems we consider in this paper using this technique. We use a different information theoretical approach, where we extend normal comparison trees with I/O-nodes. The proof of the main result then corresponds to giving an algorithm that transfers an I/O-tree that solves a given (I/O-) problem to a normal comparison tree with a height bounded by a function of the number of I/O-nodes in the original tree. An important property of the result is that it not only can be used to show worst-case I/O lower bounds, but that e.g an average lower bound on the number of comparisons needed also induces an average I/O lower bound.

In the next section, we formally define the model we will be working in. We also define the I/O-tree on which the main result in section 3 is based. In section 4, we discuss the generality of the I/O-tree model, and in section 5, we give optimal algorithms for two problems concerning multisets, namely determining the mode and removing duplicates. Finally some open problems are discussed in section 6.

2 Definition of the model and the I/O-tree

We shall consider n-record problems, where in any start configuration the n records - $x_1, x_2, \ldots, x_n$ - reside in secondary storage. The number of records that can fit into internal memory is denoted M and the number of records transferable between internal memory and secondary storage in a single block is denoted B ($1 \leq B \leq M < n$). The internal memory and the secondary storage device together are viewed as an extended memory with at least $M + n$ locations. The first M locations in the extended memory constitute the internal memory - we denote these $s[1], s[2], \ldots, s[M]$ - and the rest of the extended memory constitute secondary storage. The k'th *track* is defined as the B contiguous locations $s[M+(k-1)B+1], s[M+(k-1)B+2], \ldots, s[M+kB]$ in extended memory, $k = 1, 2, \ldots$. An I/O-operation is now an exchange of B records between the internal memory and a track in secondary storage.

An I/O-tree is a tree with two types of nodes: comparison nodes and I/O-nodes. Comparison nodes compare two records x_i and x_j in the *internal memory* using $<$ or $\leq$. i and j refer to the initial positions of the records and *not* to storage locations. A comparison node has two outgoing edges, corresponding to the two possible results of the comparison. An I/O-node performs an I/O-operation, that is, it swaps B (possibly empty) records in the internal memory with B (possibly empty) records from secondary storage. The B records from secondary storage must constitute a track (see figure 1).

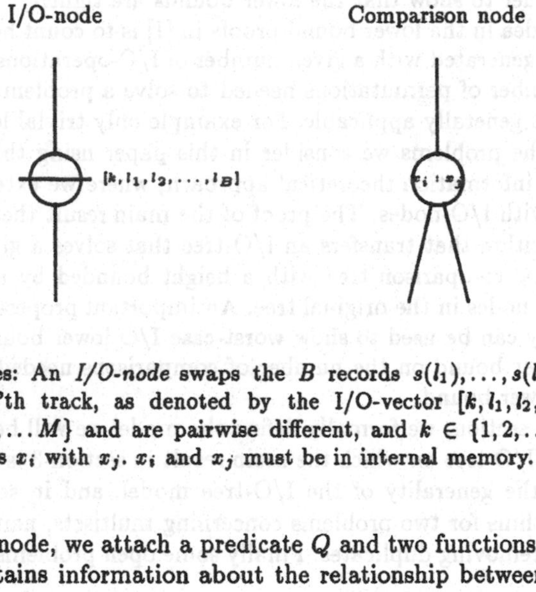

Fig. 1. Node-types: An *I/O-node* swaps the B records $s(l_1), \ldots, s(l_B)$ with the B records in the k'th track, as denoted by the I/O-vector $[k, l_1, l_2, \ldots, l_B]$, where $l_1, l_2, \ldots, l_B \in \{1, \ldots M\}$ and are pairwise different, and $k \in \{1, 2, \ldots\}$. A *comparison node* compares x_i with x_j. x_i and x_j must be in internal memory.

To each I/O-node, we attach a predicate Q and two functions π and π'. The predicate Q contains information about the relationship between the x_i's. We define the predicate recursively: First we attach a predicate P_k to each edge from a comparison node k. If the node made the comparison $x_i < x_j$ the predicate $x_i < x_j$ is attached to the left edge, and $x_i \geq x_j$ to the right edge. Similarly with $\leq$. We now consider a path S where we number the I/O-nodes $s_0, s_1, s_2, \ldots$ starting in the root and ending in the leaf.

Q_{s_i} is then defined as follows: $Q_{s_0} = True$

$$Q_{s_i} = Q_{s_{i-1}} \wedge P_1 \wedge P_2 \wedge \ldots \wedge P_l$$

where $P_1, P_2 \ldots P_l$ are the predicates along the path from s_{i-1} to s_i (see figure 2).

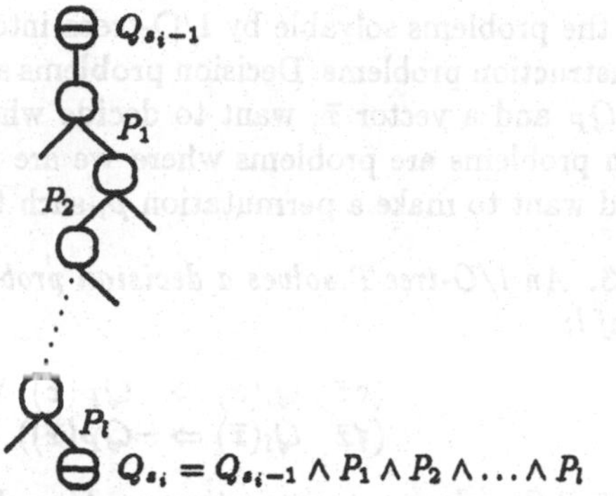

Fig. 2. The predicate Q_{s_i} is defined recursively from the predicate $Q_{s_{i-1}}$ and the predicates along the path between the two I/O-nodes.

The π's contain information about where in the extended storage the original n records - $x_1, x_2, \ldots, x_n$ - are placed. More formally, we have: $\pi : \{1, 2, \ldots, n\} \rightarrow \{1, 2, \ldots\}$, where $\pi(i) = j$ means that x_i is in the jth cell in the extended memory. Note that π is one-to-one. π' is the result of performing an I/O-operation in a configuration described by π, i.e., a track, consisting of B records, is swapped with B records from the internal memory (as denoted by the $(B+1)$-vector in figure 1). More formally, $\pi' = \pi$ except for the following:

$$
\begin{aligned}
&\pi'(\pi^{-1}(l_1)) &&= M + (k-1)B + 1 \\
&\pi'(\pi^{-1}(M + (k-1)B + 1)) &&= l_1 \\
&\vdots && \\
&\pi'(\pi^{-1}(l_B)) &&= M + kB \\
&\pi'(\pi^{-1}(M + kB)) &&= l_B
\end{aligned}
$$

π in an I/O-node is, of course, equal to π' in its closest ancestor, i.e., $\pi_{s_i} = \pi'_{s_{i-1}}$.

Definition 1. *An I/O-tree is a tree consisting of comparison and I/O-nodes. The root of the tree is an I/O-node where $\pi_{root}(i) = M + i$, i.e. corresponding to a configuration where there are n records residing first in secondary storage and the internal memory is empty. The leaves of the tree are I/O-nodes, again corresponding to a configuration where the n records reside first in secondary storage (possibly permuted with respect to the start configuration) and the internal memory is empty. This means that $\pi'_{leaf}(i) \in \{M+1, M+2 \ldots M+n\}$ for all i.*

Definition 2. *If T is an I/O-tree then $\text{path}_T(\overline{x})$ denotes the path $\overline{x}$ follows in T. $|\text{path}_T(\overline{x})|$ is the number of nodes on this path.*

We split the problems solvable by I/O-trees into two classes: decision problems and construction problems. Decision problems are problems where we, given a predicate Q_P and a vector $\overline{x}$, want to decide whether or not $\overline{x}$ satisfies Q_P. Construction problems are problems where we are given a predicate Q_P and a vector $\overline{x}$, and want to make a permutation ρ, such that $\rho(\overline{x})$ satisfies Q_P.

Definition 3. *An I/O-tree T solves a decision problem P, if the following holds for every leaf l:*

$$(\forall \overline{x} : Q_l(\overline{x}) \Rightarrow \quad Q_P(\overline{x})) \vee$$
$$(\forall \overline{x} : Q_l(\overline{x}) \Rightarrow \neg Q_P(\overline{x}))$$

An I/O-tree T solves a construction problem P, if the following holds for every leaf l:

$$\forall \overline{x} : Q_l(\overline{x}) \Rightarrow Q_P(x_{\pi_l'^{-1}(M+1)}, x_{\pi_l'^{-1}(M+2)}, \ldots, x_{\pi_l'^{-1}(M+n)})$$

It is important to note that an I/O-tree reduces to an ordinary comparison tree solving the same problem, if the I/O-nodes are removed. This is due to the fact that the comparison nodes refer to records (numbered with respect to the initial configuration) and not to storage locations.

3 The Main Result

Theorem 4. *Let P_n be an n-record problem, T be an I/O-tree solving P_n and let $I/O_T(\overline{x})$ denote the number of I/O-nodes in* $\text{path}_T(\overline{x})$*. There exists an ordinary comparison tree T_c solving P_n such that the following holds:*

$$|\text{path}_{T_c}(\overline{x})| \leq n \log B + I/O_T(\overline{x}) \cdot T_{\text{merge}}(M - B, B)$$

where $T_{\text{merge}}(n, m)$ denotes the number of comparisons needed to merge two sorted lists of length n and m, respectively.

Proof. We will prove the theorem by constructing the comparison tree T_c, but first we want to construct another I/O-tree T' that solves P_n from the I/O-tree T.

We consider a *comparison subtree* of T - an inner comparison tree of T with an I/O-node as the root and its immediately succeeding I/O-nodes as the leaves (see figure 3).

A characteristic of this tree is that, except from the I/O-nodes in the root and in the leaves, it only contains comparison nodes that compare records in the internal memory, i.e. comparisons of the form $x_i < x_j$ $(x_i \leq x_j)$ where $\pi(i), \pi(j) \in \{1, .., M\}$. In other words $Q_{i_1}, Q_{i_2}, \ldots, Q_{i_l}$ must be of the form $Q_i \wedge (x_{i_1} < x_{j_1}) \wedge (x_{i_2} \leq x_{j_2}) \wedge \ldots$ where $\pi(i_m), \pi(j_m) \in \{1, .., M\}$. Moreover, one and only one of the predicates $Q_{i_1}, Q_{i_2}, \ldots, Q_{i_l}$ is true for any $\overline{x}$ that satisfies Q_i.

We now build T' from T by inductively building comparison subtrees in T' from comparison subtrees in T starting with the "uppermost" comparison

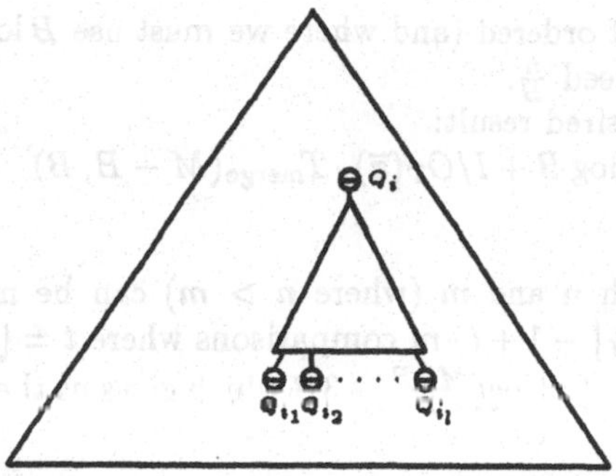

Fig. 3. Comparison subtree

subtree: The root of the new comparison subtree is the same as the root of the original comparison subtree. The internal comparison nodes are replaced with a tree that makes all the comparisons needed for a total ordering of records in internal memory. Finally, the leaves are I/O-nodes selected among the l I/O-nodes in the original subtree in the following way: If R is the predicate "generated" on the path from the root of T' to a leaf in the new subtree, the I/O-node with the predicate Q_{i_j} such that $R \Rightarrow Q_{i_j}$ is used. The choice of I/O-node is well-defined because the predicate R implies exactly one of the Q_{i_j}'s. If any of the leaves in the original comparison subtree are also roots of comparison subtrees, i.e., they are not the leaves of T, we repeat the process for each of these subtrees. Note that any of them may appear several times in T'. It should be clear that when T' is constructed in this way, it solves P_n. Furthermore, for all $\overline{x}$, $\text{path}_T(\overline{x})$ and $\text{path}_{T'}(\overline{x})$ contain the same I/O-nodes. This means that if the height of the comparison subtrees in T' is at most h, then the number of comparison nodes on $\text{path}_{T'}(\overline{x})$ is at most $h \cdot I/O_T(\overline{x})$. But then there exists an ordinary comparison tree T_c solving P_n, such that $|\text{path}_{T_c}(\overline{x})| \leq h \cdot I/O(\overline{x})$, namely the comparison tree obtained from T' by removing the I/O-nodes.

It is obvious that our upper bound on $|\text{path}_{T_c}(\overline{x})|$ improves the smaller an h we can get. This means that we want to build a comparison tree, that after an I/O-operation determines the total order of the M records in internal memory with as small a height as possible. After an I/O-operation we know the order of the $M - B$ records that were not affected by the I/O-operation - this is an implicit invariant in the construction of T'. The problem is, therefore, limited to placing the B "new" records within this ordering. If we, furthermore, assume that we know the order of the B records, then we are left with the problem of merging two ordered lists, this can be done using at most $T_{merge}(M - B, B)$ comparisons. We cannot in general assume that the B records are ordered, but because the I/O-operations always are performed on tracks and because we know the order of the records we write to a track, the number of times we can read

B records that are not ordered (and where we must use $B \log B$ comparisons to sort them) cannot exceed $\frac{n}{B}$.
Finally, we get the desired result:
$|\text{path}_{T_c}(\overline{x})| \leq \frac{n}{B} B \log B + I/O_T(\overline{x}) \cdot T_{\text{merge}}(M-B, B)$

□

Two lists of length n and m (where $n > m$) can be merged using *binary merging* [5] in $m + \lfloor \frac{n}{2^t} \rfloor - 1 + t \cdot m$ comparisons where $t = \lfloor \log \frac{n}{m} \rfloor$. This means that $T_{\text{merge}}(M-B, B) \leq B \log(\frac{M-B}{B}) + 3B$ which gives us the following corollary:

Corollary 5.

$$|\text{path}_{T_c}(\overline{x})| \leq n \log B + I/O_T(\overline{x}) \cdot \left(B \log(\frac{M-B}{B}) + 3B \right)$$

□

It should be clear that the corollary can be used to prove lower bounds on the number of I/O-operations needed to solve a given problem. An example is sorting, where an $n \log n - O(n)$ worst-case lower bound on the number of comparisons is known. In other words, we know that for any comparison tree (algorithm) T_c that sorts n records there is an $\overline{x}$ such that $|\text{path}_{T_c}(\overline{x})| \geq n \log n - O(n)$. From the corollary we get $n \log n - O(n) \leq n \log B + I/O_T(\overline{x}) \cdot (B \log(\frac{M-B}{B}) + 3B)$, hence the worst-case number of I/O-operations needed to sort n records is at least $\frac{n \log \frac{n}{B} - O(n)}{B \log(\frac{M-B}{B}) + 3B}$.
Note that no matter what kind of lower bound on the number of comparisons we are working with - worst-case, average or others - the theorem applies, because it relates the number of I/O's and comparisons for *each instance* of the problem.

4 Extending the Model

The class of algorithms for which our result is valid comprises algorithms that can be simulated by our I/O-trees. This means that the only operations permitted are binary comparisons and transfers between secondary storage and internal memory. It should be obvious that a tree, using ternary comparisons and swapping of records in internal memory, can be simulated by a tree with the same I/O-height, that only uses binary comparisons and no swapping (swapping only effects the $\pi's$). Therefore, a lower bound in our model will also be a lower bound in a model where swapping and ternary comparisons are permitted. Similarly, we can permit algorithms that use integer variables, if their values are implied by the sequence of comparisons made so far, and we can make branches according to the value of these variables. This is because such manipulations cannot save any comparisons.

The differences between our model and the model presented in [1] are, apart from ours being restricted to a comparison model, mainly three things. Firstly, Aggarwal and Vitter only assume that a transfer involves B contiguous records in secondary storage, whereas we assume that the B records constitute a track. Reading/writing across a track boundary, however, can be simulated by a constant number of "standard" I/O's. Hence, lower bounds proved in our model still apply asymptotically. Secondly, their I/O's differ from ours in the sense that they permit copying of records, i.e. writing to secondary storage without deleting them from internal memory. It can be seen that the construction in the proof of our theorem still works, if we instead of one I/O-node have both an I-node and an O-node that reads from, respectively writes to, a track. Therefore, our theorem still holds when record copying is permitted. Finally, they model parallelism with a parameter P that represents the number of blocks that can be transferred concurrently. It should be clear that we can get lower bounds in the same model by dividing lower bounds proved in our model by P.

It is worth noting that this parallel model is not especially realistic. A more realistic model was considered in [11] in which the secondary storage is partitioned into P distinct disk drives. In each I/O-operation, each of the P disks can simultaneously transfer one block. Thus, P blocks can be transferred per I/O, but only if no two blocks come from the same disk. Of course lower bounds in the Aggarwal and Vitter model also apply in the more realistic model. As using multiple disks is a very popular way of speeding up e.g. external sorting, extensive research has recently been done in this area [9] [10].

5 Optimal Algorithms

The common notion of $g(n, M, B) \in O(f(n, M, B))$ is that there exists a constant c, such that when n, M and B get sufficiently large, g is bounded by $c \cdot f$. Here, however, we want to bound g no matter what the values of B and M are, because B and M are parameters in our machine model. The intuition should be that first you choose a machine (i.e. B and M), and then you use it to solve a problem. For this reason, we define big-oh as follows:

Definition 6.

$$\overline{O}(f(n, M, B)) = \{g \mid \exists c \in R^+ : \forall M, B \in N : \exists n_0 \in N : \forall n \geq n_0 : \\ g(n, M, B) \leq c \cdot f(n, M, B)\}$$

Note that with this definition $\frac{n}{B} + M\sqrt{n} \in \overline{O}(\frac{n}{B})$ but on the other hand $n \notin \overline{O}(\frac{n}{B})$. This illustrates that the definition captures the idea that the leading term with respect to n is, in fact, the leading term with respect to all three variables, and that even though M and B are less important than n, they cannot be regarded as ordinary constants. Furthermore note that $\overline{\Omega}$ and $\overline{\Theta}$ can be defined similarly and that an extension of the definition to more variables describing the problem (other than n), is made by combining the above definition with the ordinary definition of big-oh in more variables.

Aggarwal and Vitter [1] show the following lower bound on the I/O-complexity of sorting:

$$\overline{\Omega}\left(\frac{n\log\frac{n}{B}}{B\log\frac{M}{B}}\right)$$

They also give two algorithms based on mergesort and bucketsort that are asymptotically optimal. As mentioned earlier our result provides the same lower bound.

An almost immediate consequence of the tight lower bound on sorting is a tight lower bound on set equality, set inclusion and set disjointness, i.e., the problems of deciding whether $A = B$, $A \subseteq B$ or $A \cap B = \emptyset$ given sets A and B. It can easily be shown (see e.g. [2]) that a lower bound on the number of comparisons for each of these problems is $n\log n - O(n)$. An optimal algorithm is, therefore, to sort the two sets independently, and then solving the problem by "merging" them.

In the following, we will look at two slightly more difficult problems for which our theorem gives asymptotically tight bounds.

5.1 Duplicate Removal

We wish to remove duplicates from a file in secondary storage - that is, make a set from a multiset. Before removing the duplicates, n records reside at the beginning of the secondary storage and the internal memory is empty. The goal is to have the constructed set residing first in the secondary storage and the duplicates immediately after.

A lower bound on the number of comparisons needed to remove duplicates is $n\log n - \sum_{i=1}^{k} n_i \log n_i - O(n)$, where n_i is the multiplicity of the ith record in the set. This can be seen by observing that after the duplicate removal, the total order of the original n records is known. Any two records in the constructed set must be known not to be equal, and because we compare records using only $<$ or $\leq$, we know the relationship between them. Any other record (i.e. one of the duplicates) equals one in the set. As the total order is known, the number of comparisons made must be at least the number needed to sort the initial multiset. A lower bound on this has been shown [7] to be $n\log n - \sum_{i=1}^{k} n_i \log n_i - O(n)$.

Combining a trivial lower bound of $\frac{n}{B}$ (we have to look at each record at least once) with an application of our theorem to the above comparison lower bound, we obtain:

$$I/O_{\text{Duplicate-Removal}} \in \overline{\Omega}\left(\max\left\{\frac{n\log\frac{n}{B} - \sum_{i=1}^{k} n_i \log n_i}{B\log\frac{M}{B}}, \frac{n}{B}\right\}\right)$$

We match this lower bound asymptotically with an algorithm that is a variant of merge sort, where we "get rid of" duplicates as soon as we meet them. We use a block (of B records) in internal memory to accumulate duplicates, transferring them to secondary storage as soon as the block runs full.

The algorithm works like the standard merge sort algorithm. We start by making $\lceil \frac{n}{M-B} \rceil$ runs; we fill up the internal memory $\lceil \frac{n}{M-B} \rceil$ times and sort the records, removing duplicates as described above. We then repeatedly merge c runs into one longer run until we only have one run, containing k records. $c = \lfloor \frac{M}{B} \rfloor - 2$ as we use one block for duplicates and one for the "outgoing" run. It is obvious that there are less than $\log_c(\lceil \frac{n}{M-B} \rceil) + 1$ phases in this merge sort, and that we in a single phase use no more than the number of records being merged times $\frac{2}{B}$ I/O-operations.

We now consider records of type x_i. In the first phase we read all the n_i records of this type. In phase j there are less than $\frac{\lceil n/(M-B) \rceil}{c^{j-1}}$ runs and we therefore have two cases:
$\frac{\lceil n/(M-B) \rceil}{c^{j-1}} \geq n_i$: There are more runs than records of the type x_i, this means that in the worst case we have not removed any duplicates, and therefore all the n_i records contribute to the I/O-complexity.
$\frac{\lceil n/(M-B) \rceil}{c^{j-1}} < n_i$: There are fewer runs than the original number of x_i's. There cannot be more than one record of the type x_i in each run and therefore the recordtype x_i contributes with no more than the number of runs to the I/O-complexity.
The solution to the equation $\frac{\lceil n/(M-B) \rceil}{c^{j-1}} = n_i$ with respect to j gives the number of phases where all n_i records might contribute to the I/O-complexity. The solution is $j = \log_c(\frac{\lceil n/(M-B) \rceil}{n_i}) + 1$, and the number of times the recordtype x_i contributes to the overall I/O-complexity is no more than:

$$n_i \left(\log_c(\frac{\lceil n/(M-B) \rceil}{n_i}) + 1 \right) + \sum_{j=\log_c(\frac{\lceil n/(M-B) \rceil}{n_i})+2}^{\log_c(\lceil n/(M-B) \rceil)+1} \frac{\lceil n/(M-B) \rceil}{c^j}$$

Adding together the contributions from each of the k records we get the overall I/O-complexity:

$$\frac{2}{B} \left[n + \sum_{i=1}^{k} \left(n_i \left(\log_c(\frac{\lceil n/(M-B) \rceil}{n_i}) + 1 \right) + \sum_{j=\log_c(\frac{\lceil n/(M-B) \rceil}{n_i})+2}^{\log_c(\lceil n/(M-B) \rceil)+1} \frac{\lceil n/(M-B) \rceil}{c^j} \right) \right]$$
$$\in \overline{O} \left(\max \left\{ \frac{n \log \frac{n}{B} - \sum_{i=1}^{k} n_i \log n_i}{B \log \frac{M}{B}}, \frac{n}{B} \right\} \right)$$

5.2 Determining the Mode

We wish to determine the mode of a multiset, i.e. the most frequently occurring record. In a start configuration, the n records reside at the beginning of the secondary storage. The goal is to have an instance of the most frequently occurring record residing first in secondary storage and all other records immediately after.

Munro and Raman [7] showed that $n \log \frac{n}{a} - O(n)$ is a lower bound on the number of ternary comparisons needed to determine the mode, where a denotes the frequency of the mode. This must also be a lower bound on the number of binary comparisons, thus, our theorem, again combined with a trivial lower bound of $\frac{n}{B}$, gives the following lower bound on the number of I/O-operations:

$$I/O_{\text{mode}} \in \overline{\Omega}\left(\max\left\{\frac{n\log\frac{n}{aB}}{B\log\frac{M}{B}}, \frac{n}{B}\right\}\right)$$

The algorithm that matches this bound is inspired by the distribution sort algorithm presented by Munro and Spira [8]. First, we divide the multiset into c disjoint segments of roughly equal size (a segment is a sub-multiset which contains all elements within a given range). We then look at each segment and determine which records (if any) have multiplicity greater than the segment size divided by B (we call this an B-majorant). If no segments contained an B-majorant, the process is repeated on each of the segments. If, on the other hand, there were any B-majorants, we check whether the one among these with the greatest multiplicity has multiplicity greater than the size of *the largest segment* divided by B. If it does, we have found the mode. If not, we continue the process on each of the segments as described above.

Both the division into segments and the determination of B-majorants can be done in a constant number of sequential runs through each segment. To determine B-majorants we use an algorithm due to Misra and Gries [6]. The division of a segment (of n records) into c disjoint new segments can be done so that either $\frac{3n}{\sqrt{\frac{M}{B}}}$ is a bound on the number of elements in a new segment or else all the records in the new segment are equal. This is done with a slightly modified version of an algorithm described by Aggarwal and Vitter [1]. Details of these algorithms will appear in a full paper.

As the number of I/O-operations at each level is proportional to n/B, the analysis reduces to bounding the number of levels. An upper bound on the size of the largest segment on level j must be $\frac{n}{(\frac{1}{3}\sqrt{M/B})^j}$. It follows that the algorithm can run no longer than to a level j where $\frac{n}{(\frac{1}{3}\sqrt{M/B})^j}/B \leq a$. Solving this inequality with respect to j gives us a bound on the number of levels of $\overline{O}\left(\frac{\log\frac{n}{aB}}{\log\frac{M}{B}}\right)$. This gives us the matching upper bound of $\overline{O}\left(\frac{n\log\frac{n}{aB}}{B\log\frac{M}{B}}\right)$.

6 Remarks and Open Problems

In the previous section we showed tight bounds on the problems of removing duplicates from a multiset, and determining the mode of a multiset. As mentioned in the introduction, previously known techniques [1] give only trivial lower bounds on these problems. On the other hand our theorem is also limited in the sense that there are problems for which it is useless. One example is the problem of permuting n records according to a given permutation π. An interesting and important problem "lying between" duplicate removal and permuting is multiset sorting. This problem is analyzed in [4], and lower bounds are given, both using our theorem and a (reduction-) variant of the technique from [1]. The obtained lower bounds are quite good, but we believe there is room for improvement.

Another interesting problem is to extend the model in which the lower bounds apply. Especially it would be interesting to extend our theorem to an I/O version of algebraic decision trees - thus allowing arithmetic. This would probably give interesting bounds on e.g. a number of computational geometry problems.

Acknowledgments

The authors thank Gudmund S. Frandsen, Peter Bro Miltersen and Erik Meineche Schmidt for valuable help and inspiration. Special thanks go to Peter Bro Miltersen and Erik Meineche Schmidt for carefully reading drafts of this paper and providing constructive criticism.

References

1. Aggarwal, A., Vitter, J.S.: The I/O Complexity of Sorting and Related Problems. Proceedings of 14th ICALP (1987), Lecture Notes in Computer Science 267, Springer Verlag, 467-478, and: The Input/Output Complexity of Sorting and Related Problems. Communications of the ACM, Vol 31 (9) (1988) 1116-1127.
2. Ben-Or, M.: Lower bounds for algebraic computation trees. Proceedings of 15th STOC (1983), 80-86.
3. Blum, M., Floyd, R.W, Pratt, V.R, Rivest, R.L, Tarjan, R.E.: Time bounds for selection. Journal of Computer and System Sciences, 7(4) (1972), 448-461.
4. Knudsen, M., Larsen, K.: I/O-complexity of comparison and permutation problems. M.Sc. Thesis, Aarhus University, November 1992.
5. Knuth, D.E.: The Art of Computer Programming, Vol 3: Sorting and Searching, Addison-Wesley (1973) (p. 205-206).
6. Misra, J., Gries, D.: Finding Repeated Elements. Science of Computer Programming 2 (1982), 143-152, North-Holland Publishing.
7. Munro, J.I., Raman, V.: Sorting Multisets and Vectors In-Place. Proceedings of 2nd WADS, Lecture Notes in Computer Science 519, Springer Verlag (1991), 473-479.
8. Munro, I., Spira, P.M.: Sorting and Searching in Multisets. SIAM Journal of Computing, 5 (1) (1976) 1-8.
9. Nodine, M.H., Vitter, J.S.: Optimal Deterministic Sorting on Parallel Disks. Brown University, CS-92-08, August 1992.
10. Vitter, J.S: Efficient Memory Access in Large-Scale Computation (invited paper). Proceedings of 8th STACS (1991), 26-41.
11. Vitter, J.S., Shriver, E.A.M.: Optimal Disk I/O with Parallel Block Transfer. Proceedings of 22nd STOC (1990), 159-169.

Point Probe Decision Trees for Geometric Concept Classes *

Esther M. Arkin[(1)] † Michael T. Goodrich[(2)] ‡ Joseph S. B. Mitchell[(1)] §
David Mount[(3)] ¶ Christine D. Piatko[(4)] Steven S. Skiena[(5)] ||

[(1)]Applied Math, SUNY, Stony Brook, NY 11794-3600

[(2)]Computer Science, Johns Hopkins, Baltimore, MD 21218

[(3)]Computer Science and UMIACS, University of Maryland, College Park, MD 20742

[(4)]NIST, Gaithersburg, Maryland 20899

[(5)]Computer Science, SUNY, Stony Brook, NY 11794-4400

Abstract

A fundamental problem in model-based computer vision is that of identifying to which of a given set of *concept classes* of geometric models an observed model belongs. Considering a "probe" to be an oracle that tells whether or not the observed model is present at a given point in an image, we study the problem of computing efficient strategies ("decision trees") for probing an image, with the goal to minimize the number of probes necessary (in the worst case) to determine in which class the observed model belongs. We prove a hardness result and give strategies that obtain decision trees whose height is within a log factor of optimal.

*These results grew out of discussions that began in a series of workshops on Geometric Probing in Computer Vision, sponsored by the Center for Night Vision and Electro-Optics, Fort Belvoir, Virginia, and monitored by the U.S. Army Research Office. The views, opinions, and/or findings contained in this report are those of the authors and should not be construed as an official Department of the Army position, policy, or decision, unless so designated by other documentation.

†Partially supported by NSF Grants ECSE-8857642 and CCR-9204585.

‡Partially supported by the NSF and DARPA under Grant CCR-8908092, by the NSF under Grants CCR-9003299, and IRI-9116843.

§Partially supported by grants from Hughes Research Laboratories, Boeing Computer Services, Air Force Office of Scientific Research contract AFOSR-91-0328, and by NSF Grants ECSE-8857642 and CCR-9204585.

¶Partially supported by NSF Grant CCR-89-08901 and by grant JSA 91-5 from the Bureau of the Census.

||Partially supported by NSF grant CCR-9109289 and New York Science and Technology Foundation grants RDG-90171 and RDG-90172.

1 Introduction

In computer vision, one is interested in devising algorithms that automatically interprets the contents of a digital image (a "scene"). In *model-based* computer vision, one is also given some information a priori about the objects to be searched for in the scene, specifically, a *library* S of k *models* (or "shapes"), partitioned into c *concept classes* (or "color classes"). The problem is to determine the concept class of each model that appears in the image.

For example, the library S may consist a set of alphabetic characters in many different fonts. The concept classes then may be defined to be sets of all characters that represent the same letter, regardless of font or style of print. We are more interested in determining whether the letter present is an "E" or an "F" than in what its precise font is. See Figure 1.

In this paper, we examine a fundamental instance of this model-based computer vision problem. Each model in the library S is given in a fixed position, orientation, and scale, and the given input image contains exactly one instance of one model. This situation arises, for example, after an image segmentation is applied to a scanned-in page of text, and also in recent approaches to model-based vision suggested by Arkin and Mitchell [1], Bienenstock et al. [3, 4] (for character recognition), Mirelli [10], and Papadimitriou [12]. Our problem is to determine the class of the object that is present in the scene by asking a sequence of "probe queries" of the following form: "Is there an object at location p in the scene?" We assume that there is an oracle that answers these probe queries, and we measure complexity in terms of the worst-case number of queries to the oracle before identifying the concept class of the model that is actually present in the image. In practice, such an oracle may be implemented as a local operator on a digitized image — e.g., as a measure of local texture or of gradient field.

A probing strategy is an interactive algorithm that can most naturally be thought of as a binary *decision tree*, in which each node, v, corresponds to a set of candidate models, $S(v)$, which in turn belong to a set of candidate concept classes. The root corresponds to the full set S, and each leaf corresponds to a set of models that all belong to the same concept class. Each internal node has an associated probe point that specifies the query that we ask the oracle at that particular stage of the identification. A path from the root to a leaf in the decision tree represents a possible outcome for a particular scene. An example is illustrated in Figure 2.

In this paper, we study the complexity of constructing minimum height decision trees for geometric concept classes. In other words, given a set S of geometric models, we want to construct (off line) a decision tree so that the worst-case number of probe queries needed to identify the concept class of the model present in the image is as small as possible. Any such decision tree has height at least $\lceil \lg c \rceil$. However, $\Omega(k)$ height decision trees are necessary for certain arrangements and colorings of models, even for two color classes.

Main Results. We formulate and study the geometric decision tree problem from the point of view of concept classes, proving several related results:

(1). Let $S = S_1 \cup S_2$ be a set of k non-degenerate aligned unit squares of two color classes. Then the problem of finding a minimum height decision tree to determine the color class is NP-complete. The same is true for the problem in which the models are of two color classes, are convex, and all share a point in common (but may be degenerate).

(2). Let $S = \cup_{i=1}^{c} S_i$ be a set of k simple polygons in the plane (having a total of n vertices), of classes $1, 2, \ldots, c$, whose arrangement is non-degenerate. Then, we can find, in polynomial time, a decision tree whose height is at most $2 \lg k$ times the height of an optimal decision tree that identifies the class to which the model in the scene belongs. This construction can be done in $O(n \log n + hc|A(S)|)$ time.

(3). Let $S = \cup_{i=1}^{c} S_i$ be a set of k simple polygons in the plane, of classes $1, 2, \ldots, c$, whose arrangement is possibly degenerate. Then, we can find, in polynomial time, a decision tree whose height is at most $4 \lg k$ times the height of an optimal decision tree that identifies the class to which the model in the scene belongs. The method uses a "double greedy" strategy for selecting pairs of probes to use in succession.

Relation with previous work. Most previous work on building decision trees [11] has focused on non-geometric instances of the problem. The *abstract* decision tree problem takes as input a finite *universal* set $X = \{1, \ldots, k\}$ and a family of subsets of X, $\mathcal{T} = \{T_1, T_2, \ldots, T_m\}$, representing the set of possible probes. Hyafil and Rivest [8] prove that it is NP-complete to construct a minimum height or a minimum external path length decision tree. Garey [6] presents dynamic programming algorithms for determining an optimal weighted decision tree. Our problem considered here can be viewed as the unweighted decision tree problem in which the set of possible "tests" is defined by the faces in the *arrangement* $A(S)$ determined by a set S of k geometric objects.

Arkin, et. al [2] recently studied the problem of building geometric decision trees, with the goal of identifying the specific model present in the scene (i.e., without considering the notion of color classes). They showed that, although optimal decision trees can be efficiently constructed for a *non-degenerate* set of k polygons that each contain the origin, the problem of constructing a minimum height decision tree is NP-complete if the models are possibly degenerate or if they do not share a common point. [2] also define a "greedy" heuristic that they prove to yield a decision tree of height at most $\lceil \lg k \rceil$ times that of an optimal decision tree. Further, there are examples in which the greedy heuristic attains the worst-case factor $\Theta(\lg k)$. The greedy decision trees of [2] can be arbitrarily bad, however, when the models are assigned color classes, and one is only interested in identifying the class to which a particular scene belongs.

Definitions and Notation. Let S be a set of k simple polygons (*"models"*) in the plane, having a total of n vertices. We assume that S is partitioned into c color classes, S_i, and we write $S = \langle S_1, S_2, \ldots, S_c \rangle$. Thus $S_i \cap S_j = \emptyset$ for all $i \neq j$, and $S = \cup_{i=1}^{c} S_i$.

A point in the plane designates a *probe* (or "test"). Each probe P can be identified with a subset of the set of objects that are "Hit" by the probe, $h_P(S) \subseteq$

S, and the remaining objects $m_P(S) = S \backslash h_P(S)$ are said to be "Missed" by the probe. If T is any subset of S then we extend the hit and missed subsets in the natural way so that $h_P(T) = h_P(S) \cap T$, and $m_P(T) = m_P(S) \cap T$. We say that a set of probes is *complete* for S, if for any two objects in different color classes there exists a probe that hits one of these objects and misses the other.

For a set $S = \langle S_1, S_2, \ldots, S_c \rangle$ of models and a complete set of probes, we desire a binary decision tree that discriminates between models of different classes without necessarily distinguishing models within the same class. Each node of the tree is associated with a subset of S and each nonleaf node is associated with a probe. A decision tree satisfies the following conditions:

- The subset associated with the root of the tree is S.
- Each leaf of the tree is associated with a (not necessarily proper) subset of a single color class.
- If T is the subset associated with some nonleaf node and P is the probe associated with this node, then the left child of this node is associated with $h_P(T)$, and the right child is associated with the missed set $m_P(T)$.

We let $A(S)$ denote the *arrangement* induced by S. All points within an edge or a cell of $A(S)$ intersect the same set of polygons of S, and therefore each point in an edge or a cell has the same discriminating power when it is used as a probe point. Thus, the set of possible probes can be identified with the faces (vertices, edges, and cells) of $A(S)$.

We say that $A(S)$ (or S) is *degenerate* if two distinct edges of polygons in S intersect in more than a single point (i.e., they intersect in a line segment).

2 The Non-Degenerate Case

One obvious lower bound on the height of a decision tree is $\lg c$, since there are at least c leaves. We now show another lower bound on the height of a decision tree, which applies whether the arrangement is degenerate or not. Let $s(\mathcal{C})$ be the minimum number of points needed to "stab" all models in the set $\mathcal{C}$, where a point *stabs* a model if it is in that model. Define $\bar{S}_i = \cup_{j \neq i} S_j$.

Lemma 2.1 *Let $S = \langle S_1, S_2, \ldots, S_c \rangle$ be a partitioning of a set of k simple polygons in the plane into c color classes. Assume that $A(S)$ is non-degenerate. Then, the height of any decision tree identifying the class to which the model in the scene belongs is at least* $\min_i\{s(\bar{S}_i)\}$.

Proof. We give an adversary argument. Consider the path in the probe tree from the root to a leaf, in which each probe is a "Miss". If the points along this path in which we probed do not stab all the models in two or more color classes, than we can not tell those classes apart at that leaf. □

With this lower bound, we can easily adapt a proof from Arkin *et al.* [2] that constructing an optimal decision tree is NP-complete, even for two color classes:

Theorem 2.2 *Let $S = \langle S_1, S_2\rangle$ be a set of k non-degenerate aligned unit squares of two color classes. Then the problem of constructing a minimum height decision tree for S is NP-complete.*

An easy case in which an optimal height decision tree can be constructed exactly is that in which there are two classes, one of which consists of a single model. (Refer to Figure 3.)

Lemma 2.3 *Let $S = \langle S_1, S_2\rangle$ be a set of k non-degenerate models, with $|S_1| = 1$. Then a minimum height decision tree for S is of height 1 or 2 and the tree can be constructed in $O(n \log n + |A(S)|)$ time.*

Proof. Clearly, it can be checked in time $O(|A(S)|)$ whether one probe point suffices. We show that at most *two* probes are needed to discriminate between the two classes: In the arrangement of all objects, consider a face f that is contained in the object of class S_1 and that has on its boundary some portion of the boundary of that object. (At least one such face exists, and can be found once the arrangement is built.) Probe in face f. If this is a "Miss", then we know the target object is of class S_2. If it is a "Hit", then probe in a face f' that is adjacent to f along some portion of the boundary of the S_1 object. Because of non-degeneracy, the set of models hit is identical to those hit by the previous probe, except for the S_1 object. If this second probe is a "Hit", then we conclude that an S_2 object is present; otherwise, we conclude S_1. This gives us an optimal decision tree. This tree can easily be constructed by a depth-first traversal of the arrangement, assuming it is given as a polygon arrangement (e.g., see Goodrich [7]), which can be constructed in $O(n \log n + |A(S)|)$ time by a method of Chazelle and Edelsbrunner [5]. □

Theorem 2.4 *Let $S = \langle S_1, S_2, \ldots, S_c\rangle$ be a partitioning of a set of k simple polygons in the plane into c color classes. Assume that $A(S)$ is non-degenerate. Then, we can find, in $O(n \log n + hc|A(S)|)$ time, a decision tree whose height is at most $2 \lg k$ times the height of an optimal decision tree identifying the class to which the model in the scene belongs.*

Proof. Our strategy is to approximate a minimum height decision tree that identifies which specific model is present (if any) among the models in $\bar{S}_i = \cup_{j \neq i} S_j$, where i is the class achieving the $\min_i\{s(\bar{S}_i)\}$. Probe along this tree, until reaching a leaf which corresponds to identifying a specific model. Since we identified the exact model we know which class it is in. (Clearly, this class is not S_i.) We have thus reduced our problem to that of deciding between one model, which is in a specific (and therefore known) color class other than S_i, and any of the models of color class S_i. But this problem can be done with only 2 more probes, by Lemma 2.3.

Let h be the height of the decision tree obtained by this method, and let h^* be the height of the minimum probe tree. We know by Lemma 2.1 that

$$h^* \geq \min_i\{s(\bar{S}_i)\}.$$

Also, by Theorem 5 in [2] we can build a decision tree that identifies which model is present among a set of k' non-degenerate models stabbed by $s \geq 2$ given points, and the tree is of height at most $s - 1 + \lceil \lg \lfloor k'/(s-1) \rfloor \rceil$. To this tree we add at most two more probes, as in Lemma 2.3. In our case we use $k' = |\bigcup_{j \neq i} S_j|$. ($k' < k$.) However, we can not find the exact stabbing number of $\bar{S}_i$, since that problem is NP-hard, so instead we approximate it to within a $\lg k'$ factor (e.g., by Lovasz [9]), allowing us to use $s \leq \lg k' \min_i\{s(\bar{S}_i)\}$. Thus,

$$h \leq \lg k' \min_i\{s(\bar{S}_i)\} + \lceil \lg \lfloor k'/(s-1) \rfloor \rceil + 1,$$

implying that

$$\frac{h}{h^*} \leq \lg k' + \frac{\lceil \lg \lfloor k'/(s-1) \rfloor \rceil + 1}{\min_i\{s(\bar{S}_i)\}}.$$

Now, if $\min_i\{s(\bar{S}_i\} \in \{1, 2\}$, then we can easily obtain a tree of height at most $\lg k' + 2$, since a $\lg k'$ height tree can identify exactly which model in $\bar{S}_i$ is present (if at all), and at most 2 additional probes complete the classification. If we assume that $\min_i\{s(\bar{S}_i\} \geq 3$, then we get

$$\frac{h}{h^*} \leq \lg k' + \frac{\lceil \lg \lfloor k'/(s-1) \rfloor \rceil + 1}{\min_i\{s(\bar{S}_i)\}} \leq 2 \lg k.$$

The bottleneck in the algorithm is implementing a greedy strategy for finding i. This can easily be done in $O(n \log n + hc|A(S)|)$ time by a series of depth-first traversals of the arrangement $A(S)$. □

In the next section we show that we can still get an $O(\lg k)$ approximation factor in the degenerate case, although the constant factor in this bound is not quite as good as in the non-degenerate case.

3 The Degenerate Case

In this section we give a set of strategies for dealing with the "degenerate" case when several models can share common edges. Such a "degenerate" situation can arise, for example, in character recognition, where different letters (such as "R", "E", and "D") can share common edges. (See Figure 4.)

Theorem 3.1 *Let $S = \langle S_1, S_2 \rangle$ be a set of k convex, possibly degenerate models of two color classes, with the property that all models (of both classes) have a point in common. Then the problem of constructing a minimum height decision tree for S is NP-complete.*

Proof. The problem is clearly in NP, since a complete set of candidate probe points can be concisely expressed by taking midpoints of segments joining vertices of polygons of S.

To prove the NP-hardness of determining the existence of a decision tree of height $\leq h$, we show a reduction from the set cover problem: Given a collection of subsets $C = \{C_1, C_2, \ldots C_m\}$ of $U = \{1, 2, \ldots, n\}$ and a number h, is there a subset $C' \subseteq C$ with $|C'| \leq h$, such that every element of U belongs to at least one member of C'?

Given an instance of the set cover problem, we create a regular $2m$–gon, M, with sides of unit length centered at the origin. Consider the sides of M to be indexed $j = 1, \ldots, 2m$. For each edge j, let $v_j \notin M$ be a point "just outside" edge j, and let Δ_j be the triangle determined by edge j and point v_j. Choose v_j such that the polygon $M \cup (\bigcup_j \Delta_j)$ is convex. The instance of our probe tree problem created will have n models of each color class, say red and blue, such that each element in U will have two corresponding models, one of each color. Each red (resp., blue) model corresponds to an element of $i \in U$, and is a convex polygon $R_i = M \cup \left[\bigcup_{\{j\ :\ i \in C_j\}} \Delta_j\right]$ (resp., $B_i = M \cup \left[\bigcup_{\{j\ :\ i \in C_j\}} \Delta_{m+j}\right]$).

We claim that a decision tree of height less than h exists if and only if there is a set cover of size h. Given h subsets that form a set cover, we use as probe points points in the corresponding (red) triangles Δ_j for j in the set cover. If one or more probes is a "Hit" then our model is of class S_1, (red), otherwise it is of class S_2, (blue).

Note that a probe point in M gives no information, since all models in S are present at such a point, thus we assume that any probetree will contain no such probes. The "only if" direction follows from the observation above and our lower bound of Lemma 2.1. □

3.1 The Framework

For a decision tree to exist, the only assumption we must make is that for any pair of models X and Y, there is some probe point p that distinguishes X from Y (either because $p \in X$ and $p \notin Y$, or vice versa). Because of a lack of nice geometric structure, we model the degenerate case in a more abstract fashion than we have used above.

Henceforth, we will refer to nodes of a decision tree by their associated sets. Consider some node in a decision tree associated with the set $T = \langle T_1, T_2, \ldots, T_c \rangle$. We will use lower-case, $\langle t_1, t_2, \ldots, t_c \rangle$ denote the respective cardinalities of the color classes. Define the ***weight*** of T to be its cardinality, $wgt(T) = |T|$. For each i, define the ***i-weight*** of T to be the sum of cardinalities except class i.

$$wgt_i(T) = \sum_{j \neq i} t_j = wgt(T) - t_i.$$

Observe that for any leaf in the decision tree, one of the weight functions is zero, namely the weight function of the only surviving color class.

One measure of the quality of a probe is how evenly it partitions the set of models among its children. Another way to think about this is, for each edge (T, T') in the decision tree, consider how many models in T have been eliminated from consideration in T'. We define the ***total elimination*** from T to T' to be

$$elim(T, T') = wgt(T) - wgt(T').$$

The *class most heavily eliminated* is the class i that maximizes $t_i - t'_i$. Finally, for any class i, define the *i-elimination* of the edge (T, T') to be the total elimination excluding class i,

$$elim_i(T, T') = wgt_i(T) - wgt_i(T') = elim(T, T') - (t_i - t'_i).$$

Of course, there must be a balance between the number of models eliminated from the left and right children of a node, since every model eliminated from one child will be present in the other child. In the classless case of [2] (or equivalently, where each object is in its own class) the greedy strategy is chosen to maximize the minimum elimination from each of the two children. In the presence of classes, this is not quite the right strategy, because it is quite acceptable to have one large color class.

3.2 The "Double-Greedy" Strategy

We now define the *double-greedy* heuristic for constructing decision trees. We select probes to be applied in ensembles of two consecutive probes, each of which is chosen by a greedy criterion (hence the term, double-greedy). Given a node T (with at least two color classes still active) the first probe of each ensemble is chosen exactly as in the standard greedy algorithm to maximize the minimum number of models eliminated along each of its two outgoing edges. That is, we select the probe P that maximizes

$$\min(elim(T, h_P(T)), elim(T, m_P(T))).$$

For the second probe, consider each of the two children, T' and T'' of T. For T', let i be the most heavily eliminated class by the previous probe. The second probe is chosen to maximize the minimum i-elimination for each of the two resulting grandchildren. That is, we select the probe P' that maximizes

$$\min(elim_i(T', h_{P'}(T')), elim_i(T', m_{P'}(T'))).$$

We do the same for T'' with whatever its most heavily eliminated class is.

Theorem 3.2 *Let $S = \langle S_1, S_2, \ldots, S_c \rangle$ be a set of k simple polygons in the plane, partitioned into c color classes, that are in a possibly degenerate arrangement. Let h^* denote the minimum height among all decision trees, and let h_g denote the height of the double-greedy decision tree. Then $h_g/h^* \leq 4 \lg k$.*

This theorem is proved by a variation of the argument appearing in [2] for standard decision trees. The argument is more complicated in this case because of the extra complexity of the heuristic, and the subtleties of how probes eliminate models between different classes. Consider the longest path in the double-greedy decision tree. The edges on this path can be partitioned into consecutive pairs since the probes have been chosen in ensembles. Let d denote the color class that is nonempty in the leaf of this path. At the root, the d-weight is at most k, and at the leaf it is zero. Observe that the d-weights decrease monotonically (not strictly) along the path. Classify an edge on this path as being ***light*** if the d-weight of the child is at most one half of the parent, and otherwise it is ***heavy***. Along any path there can be at most $\lg k$ light edges, and so for the remainder of the proof it suffices to consider only heavy edges.

For each node T at the head of an ensemble, let $m(T)$ denote the length of the longest subpath (of ensemble edge pairs) that reduces the d-weight by a factor of no more than one half. Let m denote the maximum value of this function over the entire path. Observe that the weight cannot be halved more than $\lg k$ times, and so $(m+2)\lg k$ is an upper bound on the height of the double-greedy decision tree. To complete the proof it suffices to show that for some constant C, $C \cdot m$ is a lower bound on the height of the optimum decision tree.

Consider the subpath that defines m. Consider any ensemble edge pair on this subpath. If d is the most heavily eliminated class from the first probe, then we ***mark*** the second edge of the pair, and otherwise we ***mark*** the first edge. The number of marked edges on the path is $m/2$.

Let W denote the d-weight at the start of the subpath. At the end of the path the d-weight is at least $W/2$, and so by the pigeonhole principal, along some marked edge (T, T') on the path the d-weight has decreased by at most $\dfrac{W-(W/2)}{m/2} = \dfrac{W}{m}$. Call the marked probe P that minimizes the decrease in d-weight the ***limiting probe***. We claim that after applying this probe, no later probe along an edge can decrease the d-weight by a factor greater than twice this amount, that is, $2W/m$. Before showing this, let us see why this suffices to complete the proof. At the end of the subpath the current d-weight is at least $W/2$. Clearly the height of the optimum tree to complete the discrimination of this subset of models is a lower bound on the height of the optimum tree for the entire problem. However, since no future probe can decrease the d-weight by more than $2W/m$, at least $\dfrac{W/2}{2W/m} = \dfrac{m}{4}$ probes are needed even in the optimum tree. Thus the optimum decision tree has height at least $m/4$, and this will complete the proof. (We illustrate the main ideas of this proof in Figure 5.) All that remains is to prove the following claim.

Lemma 3.3 *If the limiting probe P applied along edge (T, T') reduces the d-weight by E, then no later probe can reduce the d-weight by more than $2E$.*

Proof. We may assume that the probe P reduces the d-weight of T by less than half, since otherwise the lemma is trivially true. Thus, there are two cases.

If the limiting probe is applied along the second edge of an ensemble, then this probe was selected to reduce the d-weight by the largest amount possible, that is, to distinguish as many elements as possible from among all classes except d. Since later nodes involve subsets of T, any probe that distinguishes more than E models from these classes could have been applied at T to decrease the d-weight even more. The greedy choice of P implies that no such probe can exist.

If, on the other hand, the limiting probe was applied to the first edge of the ensemble, then by the definition of the double-greedy strategy we know that class d was not the most heavily eliminated class in this probe. Thus, the number of models eliminated from class d by this probe can be no more than half the number of models eliminated from all the other classes, and hence $elim(T, T') \leq 2E$. Now, suppose towards a contradiction that some later probe reduces the d-weight by more than $2E$. Such a probe must distinguish at least $2E$ total models from a subset of T, and hence could have been applied at T to eliminate more than $2E$ models, contradicting the choice of this probe. □

Acknowledgements. We thank Teresa Kipp and Vince Mirelli of the Center for Night Vision and Electro-Optics and all of the participants of the series of workshops on Geometric Probing in Computer Vision, sponsored by the U.S. Army Research Office.

References

[1] E. Arkin and J. Mitchell. Applications of combinatorics and computational geometry to pattern recognition. Technical report, Cornell University, 1990.

[2] E.M. Arkin, H. Meijer, J.S.B. Mitchell, D. Rappaport, and S.S. Skiena. Decision trees for geometric models. *Proc. Ninth ACM Symp. on Comput. Geometry*, 1993.

[3] E. Bienenstock, D. Geman, and S. Geman. A relational approach in object recognition. Technical report, Brown University, 1988.

[4] E. Bienenstock, D. Geman, S. Geman, and D.E. McClure. Phase II: Development of laser radar ATR algorithms. Contract No. DAAL02-89-C-0081, CECOM Center for Night Vision and Electro-Optics, 1990.

[5] B. Chazelle and H. Edelsbrunner. An optimal algorithm for intersecting line segments in the plane. *J. ACM*, 39:1–54, 1992.

[6] M. Garey. Optimal binary identification procedures. *SIAM J. Appl. Math.*, 23:173–186, 1972.

[7] M. T. Goodrich. A polygonal approach to hidden-line and hidden-surface elimination. *CVGIP: Graphical Models and Image Processing*, 54:1–12, 1992.

[8] L. Hyafil and R. Rivest. Constructing optimal binary decision trees is NP-complete. *Information Processing Letters*, 5:15–17, 1976.

[9] L. Lovász. On the ratio of optimal integral and fractional covers. *Discrete Mathematics*, 13:383–390, 1975.

[10] V. Mirelli. Computer vision is a highly structured optimization problem. Manuscript, Center for Night Vision and Electro-Optics, Fort Belvoir, VA, 1990.

[11] B. Moret. Decision trees and diagrams. *Computing Surveys*, pages 593–623, 1982.

[12] C. H. Papadimitriou. On certain problems in algorithmic vision. Technical report, Computer Science, UCSD, 1991.

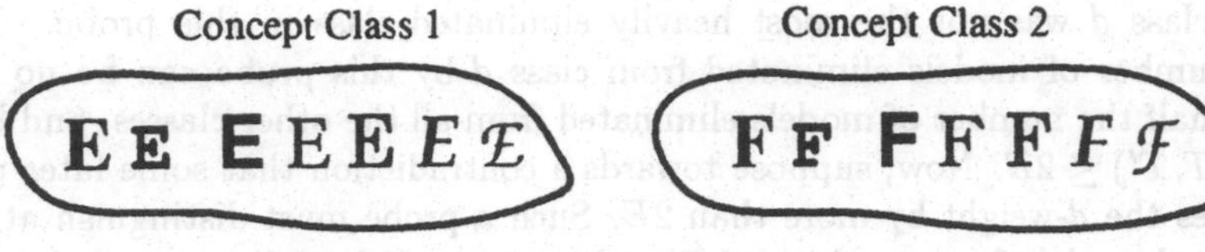

Figure 1. An example of two concept classes.

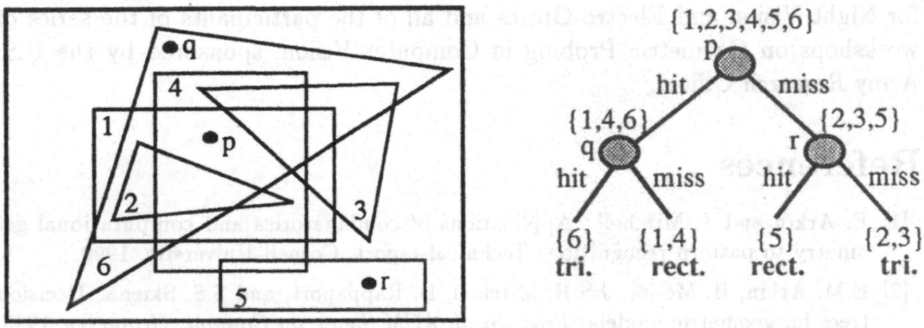

Figure 2. An instance of model-based computer vision and its decision tree for distinguishing a class of triangles from a class of rectangles.

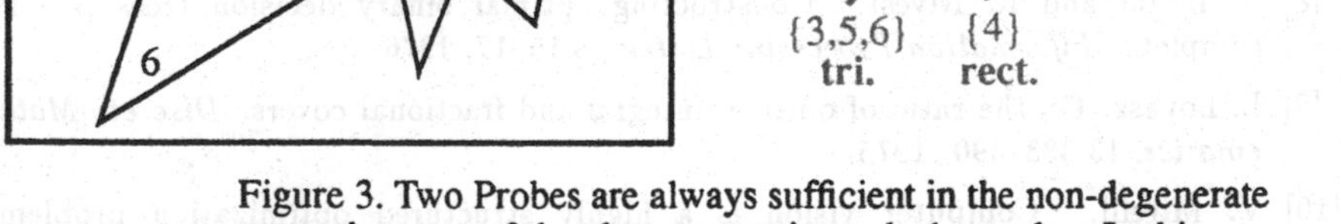

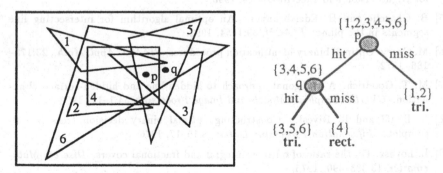

Figure 3. Two Probes are always sufficient in the non-degenerate case when one of the classes contains only 1 member. In this example we distinguish a class of triangles from a class containing a single rectangle.

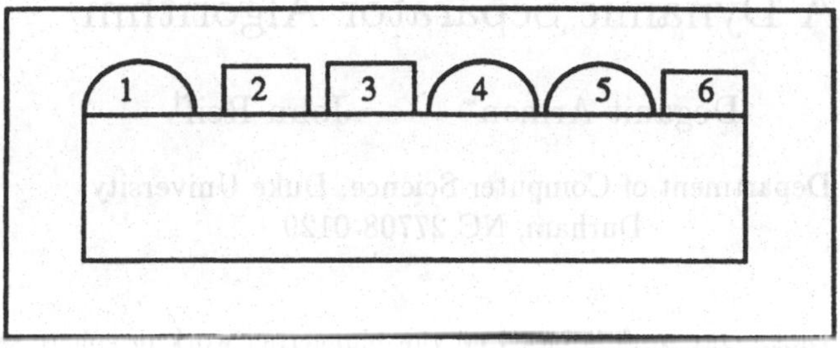

Figure 4. An example of a degenerate instance of model-based computer vision. In this case, each model is a rectangle with either a rectangular or hemispherical "tab" attached. Distinguishing these two classes requires a depth-3 tree even though all of the models can be "stabbed" by a single point.

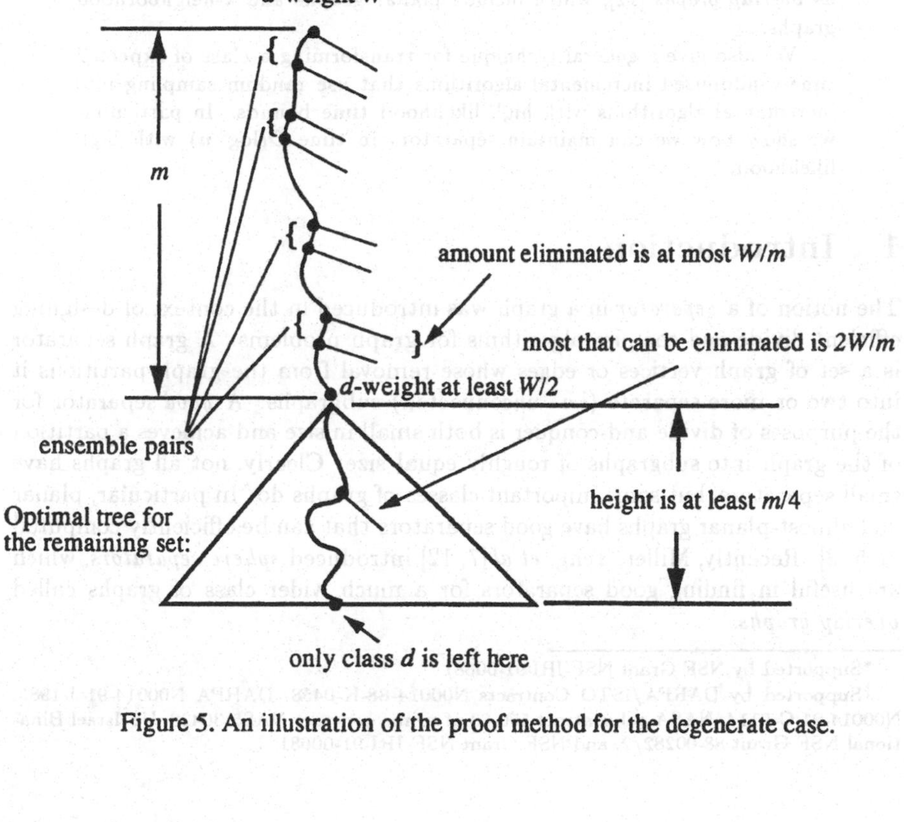

Figure 5. An illustration of the proof method for the degenerate case.

A Dynamic Separator Algorithm

Deganit Armon* John Reif†

Department of Computer Science, Duke University
Durham, NC 27708-0129

Abstract. Our work is based on the pioneering work in sphere separators of Miller, Teng, Vavasis *et al*, [8, 12], who gave efficient static algorithms for finding sphere separators of size $s(n) = O(n^{\frac{d-1}{d}})$ for a set of points in $\mathbf{R}^d$.

We present randomized, dynamic algorithms to maintain separators and answer queries about a dynamically changing point set. Our algorithms maintain a separator in expected time $O(\log n)$ and maintain a separator tree in expected time $O(\log^3 n)$. This is the first known polylog dynamic algorithm for finding separators of a large class of graphs known as *overlap graphs* [12], which include planar graphs and k-neighborhood graphs.

We also give a general technique for transforming a class of expected time randomized incremental algorithms that use random sampling into incremental algorithms with high likelihood time bounds. In particular, we show how we can maintain separators in time $O(\log^3 n)$ with high likelihood.

1 Introduction

The notion of a *separator* in a graph was introduced in the context of designing efficient divide-and-conquer algorithms for graph problems. A graph separator is a set of graph vertices or edges whose removal from the graph partitions it into two or more separate (i.e. unconnected) subgraphs. A *good* separator for the purposes of divide-and-conquer is both small in size and achieves a partition of the graph into subgraphs of roughly equal size. Clearly, not all graphs have small separators, but some important classes of graphs do. In particular, planar and almost-planar graphs have good separators that can be efficiently computed [6, 5, 3]. Recently, Miller, Teng, *et al* [7, 12] introduced *sphere separators*, which are useful in finding good separators for a much wider class of graphs called *overlap graphs*.

*Supported by NSF Grant NSF-IRI-91-00681

†Supported by DARPA/ISTO Contracts N00014-88-K-0458, DARPA N00014-91-J-1985, N00014-91-C-0114, NASA subcontract 550-63 of prime contract NAS5-30428, US-Israel Binational NSF Grant 88-00282/2, and NSF Grant NSF-IRI-91-00681.

Sphere separators are geometric entities, and overlap graphs are geometrically defined, derived from sets of points, and special neighborhoods of these points, in space. This makes them useful when solving problems in computational geometry, such as nearest neighbor queries ([4, 12]), in numerical analysis, such as nested dissection problems with large scale meshes in two and three dimensions [7], and path algebra problems [10, 11].

The basis for all these algorithms is a *separator based search structure*: a recursive structure which partitions the input graph or set of points using sphere separators [12]. This structure is a binary tree. Information about the separator is stored at the root, and the two partitioned subsets are recursively stored in the subtrees. The separators themselves are found using a randomized algorithm that uses sampling.

In this paper we describe a method for maintaining such a separator structure in the face of dynamically changing input.

1.1 Dynamizing Static Randomized Algorithms

Dynamic (or incremental) algorithms update their output of a solution to a problem when the input is dynamically modified. Usually it is not practical to recompute the solution "from scratch", so special data structures that can be updated at small cost are used. Dynamic algorithms are very useful in interactive applications, including network optimization, VLSI, and computer graphics. Many dynamic data structures have been devised to deal with problems in computational geometry [9, 2].

We describe a useful technique for enabling randomized algorithms that use sampling, to cope with dynamically changing inputs. This technique may be applied to a wide range of randomized algorithms. Throughout the rest of this paper we will refer to algorithms with fixed input as *static algorithms*.

Consider a simple static algorithm A which, given an input set of size n, uses only information from a small subset Σ of size σ from the input in its calculations. Many efficient algorithms use such random sampling. Let the time taken by A be $T(n)$. Now consider a a point p being presented to the algorithm to be inserted into the input set. Since the constructed data structure depends only on a small sample of the input set, the probability that p would be included Σ is small: $\frac{\sigma}{n+1}$ to be exact. With probability $1 - \frac{\sigma}{n+1}$ the same output would be produced regardless of whether p is in the input set. Thus we invoke A and rebuild the structure only with probabilty $\frac{\sigma}{n+1}$, and otherwise insert p into the existing structure. Since the cost of invoking A is at most $T(n+1)$, the cost of processing the new input point p is at most $\frac{\sigma}{n+1}T(n+1)$.

The dynamic maintenance of the data structure proceeds inductively. At each step, the following induction hypothesis holds:

- After a sequence of any number of updates to its input, the dynamic data structure output by the algorithm will have the same probability distribution as a data structure output by the static algorithm, had the same input been presented to it.

With each new insertion, with probability $\frac{\sigma}{n+1}$ the entire data structure may be completely rebuilt by the static algorithm, using unbiased independent random sampling. After each step, the induction still holds. Section 3 gives more details of this approach.

In reality, static algorithms are more costly. The static algorithms for finding sphere separators [12] use recursive randomized sampling. When constructing a separator structure [4, 12], there is a cost involved in building the tree even if the new input has no effect on the separator at the root. Nonetheless, this general idea is the basis for designing the dynamic algorithms in this paper, and analysis in all cases is similar.

1.2 Description of Our Model

Using a randomized algorithm, we maintain a separator based search structure for a set of points P in $\mathbf{R}^d$ which is dynamically changing. Our model assumes an adversary, who knows our randomized, dynamic algorithm but not the random choices which it makes. The adversary presents a sequence of points which are then fed to the algorithm, one at a time. Note that this is not a fully interactive model, as the adversary does not generate new points once the algorithm is running, and cannot take advantage of knowledge of choices the algorithm has already made. This feature of the model will be important only when making deletions to the point set. Associated with each point p is one of the following requests:

1. INSERT p into the search structure.
2. DELETE p from the search structure.
3. Answer a QUERY about p.

Queries presented to the algprithm request information about the location of a point with respect to the data structure, i.e. whether the point is inside or outside the separator, or where it is located in the separator tree. A query point may or may not be in P. If the application only allows queries to be answered about points in P (as in [4, 13]), the point can be temporarily inserted, then deleted from the structure.

1.3 The Replicant Paradigm: Terminology

When dealing with randomized algorithms we can only speak of expected bounds which can be attained with some probability. One method of attaining the expected bounds with high probability is to repeat a process a number of times. In Section 4 we present a general technique for transforming expected time bounds to high likelihood time bounds through the use of multiple, independent processes. This section introduces the terminology relevant to this technique.

The algorithm will maintain many independent processes, which carry out the same task. Each independent process and its dynamically maintained data structure is called a *replicant*[1]. Since the processes use random sampling, the data structures will be different, despite being derived from the same input. At any given time, a replicant may be in one of two states:

1. The replicant is *activated*. In this state, the information in the replicant data structure is current and can be used to process queries.

2. The replicant is in *retirement*. In this case, the replicant is being rebuilt, and the information it stores may not be current. A replicant in retirement may not be used to answer queries.

Retirement for replicants in our algorithm is a temporary state. Each replicant repeatedly alternates between periods of activation and retirement. A period of activation begins at an *incept date* and ends at a *retirement date*. An activation period's length is the replicant's *longevity*.

1.4 Organization of Our Paper

Section 2 gives preliminary definitions and related results. In section 3 we describe the process of converting a randomized algorithm based on sampling to a dynamic randomized algorithm, and give the details of the process of inserting and deleting a point from a separator structure. Finally, in section 4 we discuss the use of replicants to transform expected time bounds to high likelihood time bounds.

2 Definitions and Related Results

2.1 Neighborhood Systems

Let $P = \{p_1 \ldots p_n\}$ be a finite set of generally positioned points in $\mathbf{R}^d$. A *d-dimensional neighborhood system* in $\mathbf{R}^d$ is a set of balls $B = \{B_1 \ldots B_n\}$, with ball B_i centered at p_i. B is a *k-neighborhood system* if each B_i contains at most k points from P. B is a *k-ply neighborhood system* if each p_i is contained in at most k balls.

Given any set of points P in $\mathbf{R}^d$, we can easily construct a k-neighborhood system by centering at each point the largest ball that will contain $k+1$ points from P. A k-ply neighborhood system is more difficult, but the following lemma from [1] relates the two:

Lemma 2.1 *Each k-neighborhood system in* $\mathbf{R}^d$ *is* $\tau_d k$*-ply, where* τ_d *is the* kissing number *in d dimensions, i.e. the maximum number of nonoverlapping unit balls in* $\mathbf{R}^d$ *that can be arranged so that they all touch a central unit ball.*

[1] In the description that follows we used terminology and quotes [in square brackets] borrowed from the movie "Blade Runner".

Note that τ_d is a number independent of n which depends only on the dimension. While its exact value is not known for all d, the known bounds are $2^{.2075d(1+o(1))} \leq \tau_d \leq 2^{.401d(1+o(1))}$ [12].

The ***k*-nearest neighborhood digraph** of the points $P = \{p_1, \dots, p_n\}$ in $\mathbf{R}^d$ is a graph $G_k = (P, E)$ with vertex set P and directed edge set $E = \{(p_i, p_j) | p_j$ is one of the k nearest neighbors of $p_i\}$. We will call the 1-nearest neighborhood graph the *nearest neighborhood graph*.

Note that since we require the points to be in general position, the k nearest neighbors of a point are unique. The outdegree of every vertex in the graph is k, and by Lemma 2.1, the indegree is bounded by $k\tau_d$.

2.2 Sphere Separators of Points in $\mathbf{R}^d$

Let $s(n)$ be a positive, monotone function of n, such that $s(n) < n$. Given a finite set of points $P = \{p_1, \dots, p_n\}$ in $\mathbf{R}^d$, and a constant $0 < \delta \leq 1$, a $(d-1)$-sphere S in $\mathbf{R}^d$ is an *(s, δ)-separator* of P if it partitions P into two subsets, P_I and P_E (the points on the interior and the exterior of S respectively) such that

1. $|P_I| \leq \delta n$, $|P_E| \leq \delta n$, and

2. the *induced separator* of the point set is of size at most $s(n)$.

The induced separator set is defined by a neighborhood system $\mathcal{B}$ of P, and is the set of points whose associated balls intersect the sphere separator.

An *(s, δ)-separator tree* of P is a binary tree, at the root of which is stored information about an (s, δ)-separator of P. Points in the interior of the sphere are stored in a recursive structure in the left subtree, and points on its exterior are stored in a recursive structure in the right subtree. Recursion stops when a subtree contains less than a prespecified number of points. At each internal node of the tree, the sphere separator S' is required to be a good, that is, if the set of points separated by the parent of the node is n', then S' is required to be an $(s(n'), \delta)$-separating sphere. Note that since we only store the separators, the space $f(n)$ required to store the separator tree satisfies, for sufficiently large n, the recurrence $f(n) = f(n_0) + f(n_1) + s(n)$ where $n_0 + n_1 = n$. Thus if $s(n) \leq O(n^\beta)$, where $0 \leq \beta \leq 1$, the space is $f(n) = O(n)$ [4]. In this paper, wherever s, δ are determined by context, we will simply call the (s, δ)-separating sphere a *good* separating sphere.

2.3 Graph Separators

An *(α, k)-overlap graph* is a graph induced by a neighborhood system $\mathcal{B}$ of a set of points P. The vertex set of the graph is P, and an edge exists between two vertices p_i and p_j if $\alpha B(p_i) \cap B(p_j) \neq \emptyset$ and $\alpha B(p_j) \cap B(p_i) \neq \emptyset$, where, if B be a ball centered at p of radius r then αB is a ball centered at p with radius αr. Note that k-neighborhood graphs are a special case of overlap graphs.

The separator of a neighborhood system $\mathcal{B}$ naturally defines a separator set for an overlap graph G induced by $\mathcal{B}$. A vertex of G is included in the separator set if any of its incident edges crosses the sphere separator. It is shown in [12] that a good sphere separator will induce a good separator for the overlap graph. The separator tree for an overlap graph is defined analogously to the separator tree of the point set.

2.4 Static Algorithms for Finding a Sphere Separator

Our work is based on the pioneering work in sphere separators done by Miller, Teng, et al [4, 8, 7, 12]. They showed that for any (generally positioned) set of n points $P = \{p_1, \ldots, p_n\}$ in $\mathbf{R}^d$, there is a sphere S which (s^*, δ^*)-separates P, with separator size $s^* = O(n^{\frac{d-1}{d}})$ and a splitting ratio $\delta^* = \frac{d+1}{d+2}$ [8].
Furthermore, Teng [12] showed a randomized, linear time algorithm based on sampling for finding such a separator. His result states that:

Lemma 2.2 [12] *Let $P = \{p_1, \ldots, p_n\}$ be any set of n generally positioned points in $\mathbf{R}^d$, and let ϵ be a constant $0 < \epsilon < \frac{1}{d+2}$. Let Σ be a random sample of points in P of size $\sigma(n) = O(\frac{d}{\epsilon^2}(\log \frac{n}{\epsilon} + \log \eta))$. Then there is a randomized, linear time algorithm SPHERE-SEPARATOR(Σ), with success probability of $1 - \frac{1}{\eta}$, which yields a (s, δ)-separator of P with separator size $s \leq n^\beta, \beta = \frac{d-1}{d} + 2\epsilon$, and splitting ratio $\delta = \frac{d+1}{d+2} + \epsilon$.*

In particular, if we choose $\eta = O(\log n)$, then with a sample of size $\sigma(n) = O(\log n)$ the algorithm will have a probability of success $1 - \frac{1}{\log n}$. Note that the "goodness" of the sphere separator obtained by SPHERE-SEPARATOR(Σ) is only slightly non-optimal as compared with the existence results cited above.

To avoid time bounds which are exponential in d, their algorithm employs a method of recursive sampling. Sample sizes are carefully selected to be as stated in Lemma 2.2, and smaller samples are used to verify the "goodness" of the initial sample. For our purposes, it suffices to know that a random sample of the input points of size $\sigma(n)$ provides us a good sphere separator with probability $1 - \frac{1}{\sigma(n)}$. In what follows, we will take $\sigma(n)$ to be $O(\log n)$.

3 Dynamic Construction of Sphere Separator Trees

For a dynamically changing finite set of points P, we give algorithms for dynamic maintenance of a sphere separator of P, and for dynamic maintenance of a separator-based search structure. The algorithm accepts a sequence of requests from an adversary. Each request is a pair <point, action>, where an action may be to INSERT or DELETE the input point from P, or answer a QUERY about the input point, e.g. determine whether it is inside or outside the sphere separator of P, or search for it in the separator structure.

We show the following result:

Theorem 3.1 *Let $0 < \epsilon < \frac{1}{d+2}, \delta = \frac{d+1}{d+2} + \epsilon, \beta = \frac{d-1}{d} + 2\epsilon$, and $s \leq n^\beta$. Dynamic maintenance of an (s,δ)-sphere separator of n points in $\mathbf{R}^d$ with INSERT and DELETE operations, as well as queries about the separator, can be done in incremental expected time $O(\log n)$ per request. Dynamic maintenance of a sphere-separator-based search tree with INSERT, DELETE and QUERY requests can be done in $O(\log^3 n)$ incremental expected time per request.*

Recall that the separator of a set of n points is determined by examining a subset Σ of the points of size $\sigma(n) = O(\log n)$. We use the static algorithm for finding a sphere separator as a subroutine in our dynamic algorithm. The algorithm keeps track of the points selected to be in the sample, Σ. If the entire separator tree is maintained, the algorithm keeps track of all samples used to find separators at all levels of the tree. Thus a single point may be tagged up to $O(\log n)$ times.

Our algorithm proceeds inductively. Given a separator search structure, the entire structure or a part thereof may be completely rebuilt using the static algorithm with each new insertion or deletion. Assume that after sequence of i updates the set of points in the separator structure is P_i. Key to the correctness of the algorithm is the induction hypothesis which states that the structure output by the dynamic algorithm after i updates has the same probability distribution as the structure output by a static algorithm given P_i as input.

3.1 Insertion

When a request for insertion arrives, the input to that stage of the algorithm is a set of $n-1$ points P' arranged in a separator structure, and one additional point p. Consider the situation where all n points are presented to the static algorithm. The only step of the static algorithm which is important for our purpose is selecting the set Σ which is used to determine the separating sphere. If $|\Sigma| = \sigma(n)$, the probability that the new point would have not been included in Σ is $1 - \frac{\sigma(n)}{n}$. Thus we perform a single Bernoulli trial, with probability of success $1 - \frac{\sigma(n)}{n}$.

Failure in this trial means that p needs to be included in the subset of points that will determine the separating sphere. If that is the case, we retire the current data structure (if we maintain the entire separator tree, we retire the entire tree at this point) and invoke the static randomized linear time algorithm of Lemma 2.2 on the entire set of n points. That is, we perform SPHERE-SEPARATOR(Σ) on the set $P' \cup \{p\}$.

Success in the Bernoulli trial means that p is to be inserted into the existing structure. That process is straightforward. Look at the separating sphere, and determine whether p lies inside or outside it. Without loss of generality, assume p is in the interior, so it is added to P_I. If the entire separator structure is maintained, then p is added recursively to the left subtree.

Using a sample of size $\sigma(n) = O(\log n)$ to calculate separating spheres, we get

Lemma 3.1 *Given a set P' of $n-1$ points and a sphere separator S which δ-splits them, the expected incremental time to insert a new point p into P' is $O(\log n)$.*

Proof: The probability of needing to recompute a sphere separator for $P' \cup \{p\}$ is the same as the probability of including p in the sample used in selecting a separator, which is $\frac{\sigma(n)}{n}$. With that probability, we invoke the static, linear time algorithm for computing a sphere separator for $P' \cup \{p\}$. Otherwise, we determine (in constant time) which side of S the point lies in, and add it to that set (again in constant time). The expected update time is thus

$$E[T(n)] \leq \frac{\sigma(n)}{n} O(n) + \left(1 - \frac{\sigma(n)}{n}\right) O(1)$$

$$\leq \frac{\sigma(n)}{n} O(n) + O(1) = O(\sigma(n)) = O(\log n)$$

□

Similarly, if the entire separator structure is maintained,

Lemma 3.2 *The expected time to perform an insertion of a new point p into an existing sphere separator tree of a set of $n-1$ points, is $O(\log^3 n)$.*

Proof: At the top level, the probability of recalculation of the sphere separator, which would cause us to rebuild the entire structure, is the same as the probability of including p in the sample used in selecting a separator, which is $\frac{\sigma(n)}{n}$. With that probability, we invoke the static algorithm which constructs a new separator tree $O(n \log n)$ time. Otherwise, the time required will be the expected time to rebuild a subtree. Since the separators found using the static algorithm are guaranteed to $\left(\frac{d+1}{d+2} + \epsilon\right)$-split the point set, the size of a subtree is at most δ, where $\delta = \frac{d+1}{d+2} + \epsilon$. Thus we get the following recurrence equation for the expected time to maintain the separator tree:

$$E[T(n)] \leq \frac{\sigma(n)}{n} O(n \log n) + \left(1 - \frac{\sigma(n)}{n}\right) [T(\delta n)]$$

$$\leq \frac{\sigma(n)}{n} O(n \log n) + T(\delta n) = O(\sigma(n) \log^2 n) = O(\log^3 n)$$

□

3.2 Deletion

Using very similar reasoning, the dynamic separator structure also supports point deletion. Each point in the structure which was used in the sample when finding the separator is tagged. When deleting a tagged point, the separator needs to be rebuilt. If we maintain the entire search structure, a point is tagged with the highest level number in which it was used to find a separator – its highest *involvement level.* Points not selected are not tagged. When a point is presented for deletion, it is located in the separator structure down to its involvement level, then the subtree from that level down is rebuilt. Note that it is at this point that the importance of the model comes into play. Had the adversary had access to the choices made by the algorithm, he would have been able to keep presenting it with tagged points for deletion. The proofs of the next two lemmas are similar to the proofs of lemmas 3.1 and 3.2.

Lemma 3.3 *Given a set P of n points and a sphere separator S which δ-splits them, the expected time to delete a point p from P is $O(\log n)$.*

Lemma 3.4 *The expected time to perform a deletion of a point p from an existing sphere separator tree of a set of n points, is $O(\log^3 n)$.*

4 High Likelihood Time Bounds

In the previous section, we showed what the *expected* time bounds were for our dynamic algorithms. In this section, we'll show how to transform these expected time bounds to high likelihood time bounds, using the replicant paradigm. We define *high likelihood bounds* to be $1-\frac{1}{n^\alpha}$ for some $\alpha \geq 0$ where n is the input size. There are known techniques for transforming static (non-incremental) algorithms from expected time bounds to high likelihood bounds. The main idea is to repeat the execution of the algorithm a sufficient number of times to guarantee high likelihood time bounds.

We use multiple, independent processes, called replicants. Each replicant maintains its own dynamic data structure as described in Section 3, which is either active, or in retirement (being rebuilt). At all times when the replicant is active, its data structure satisfies the induction hypothesis specified in Section 3. As a result, a QUERY point presented to the algorithm will be processed by an activated replicant, and that replicant will process the query in $O(1)$ time (if we just query about the separator) or $O(\log n)$ time (if we actually search the entire separator tree).

In order for this idea to work, we need to know that, with high likelihood, there will be enough replicants in an activated state at any stage of the algorithm. The probability of *all* replicants being in retirement at any given time will only be $\frac{1}{n^\alpha}$, a very low likelihood. As we shall see, we can guarantee that condition by carefully choosing the number of replicants.

Naturally, there is a tradeoff involved. Maintaining more data structures will require extra time and result in a slowdown of the algorithm. We can place the following bounds on the slowdown factor:

Let the time per update or request if no retirement occurs (i.e. during activation) be $A(n)$ with high likelihood. If retirement does occur, let the wait time until reactivation be $R(n)$ with high likelihood. The expected time per update is then $T(n) = A(n)(1 - \frac{\sigma(n)}{n}) + R(n)\frac{\sigma(n)}{n}$.

Theorem 4.1 *Given an incremental, randomized algorithm with expected time $T(n)$, which uses a random sample of size $\sigma(n)$ of the input, and uses a static algorithm with time bound $R(n)$ as a subroutine when rebuilding its data structure, we can transform it to an incremental algorithm with high likelihood time bounds with a slowdown factor of $O(\frac{\sigma(n)R(n)\log n}{n})$.*

Proof:

Assume for simplicity that $\frac{n}{\sigma(n)R(n)} = o(1)$. This is true for the applications described in this paper.

As stated above, a replicant data structure needs to be rebuilt after a single update with probability $\frac{\sigma(n)}{n}$. A sequence of i updates are just i independent trials following a geometric distribution. Thus the expected *longevity* λ of a single replicant data structure is $\frac{n}{\sigma(n)}$.

We construct r independent replicants, each with its own data structure and with longevity $\lambda = \frac{n}{\sigma(n)}$. When a replicant is retired, the others don't wait for it to reconstruct its data structure. Rather, the reconstruction happens while activated replicants continue processing requests. As a result, we need to account for "catch-up" time before reactivation. Let the catch-up time for a retired replicant be $3R(n)$ (details below). The life of a replicant consists of periods of activation of expected length λ, followed by periods of retirement of length $3R(n)$.

Thus at any given time, the probability that a single replicant is activated is

$$\gamma = \frac{\lambda}{3R(n) + \lambda} = \frac{n}{3\sigma(n)R(n) + n}.$$

Since the r replicants are independent, the probability that at least one replicant is currently activated is $1 - (1-\gamma)^r$. The value of r for which this translates to a high likelihood event, i.e. an event with probability $1 - \frac{1}{n^\alpha}$, is

$$r = \frac{-\alpha \log n}{\log(1-\gamma)}$$

Since $\frac{n}{\sigma(n)R(n)} = o(1)$, as n increases $\gamma \to 0$, and $\log(1-\gamma) \to -\gamma$. Consequently,

$$r = \frac{-\alpha \log n}{\log(1-\gamma)} \to \frac{\alpha \log n}{\gamma} = \frac{\alpha \log n (3\sigma(n)R(n) + n)}{n} = O(\frac{\sigma(n)R(n)\log n}{n})$$

The only remaining problem is to make sure that catching up during retirement takes time $3R(n)$. The difficulty is that the while the reconstruction of the replicant data structure is performed, activated replicants continue processing input requests, so the retired replicant is in danger of falling farther and farther behind [*"accelerated decrepitude"*].

The solution to this problem of accelerated decrepitude is to proceed in stages through retirement toward activation. The first stage is rebuilding the structure with the new input point. Since restructuring will take $R(n)$ steps, we spread the rebuilding effort over the next $R(n)$ updates. As each new update request arrives, the replicant stores the request but otherwise ignores it. When this is done, the replicant is at most out of date by $R(n)$ updates.

In the second stage, the replicant processes this backlog, performing *two* updates at a time. For each new update performed by the activated replicants, the retired replicant will perform the next two updates from its backlog list. At the same time, new requests are still pouring in, and are stored for further reference. However, since the pace is quickened, when the original backlog is caught up with, the replicant is out of date by at most $R(n)/2$ requests. Continuing at this pace ["the light that burns twice as fast burns twice as bright"], after i catch-up stages the backlog will be of size $R(n)/2^{i-1}$. The retired replicant will be completely caught up after $2R(n)+1$ catch-up stages (Zeno's paradox does not apply here, because we are dealing with a discrete process). Along with the initial rebuilding phase, the total time from retirement to reactivation is $3R(n)+1$. This completes the proof. □

From this discussion, we get results relevant to the applications discussed in this paper. Note that since $R(n)$ is $O(n)$, and $\sigma(n) = O(\log n)$ in our application, the slowdown factor is only $O(\log n)$.

Corollary 4.1 *There exists a dynamic algorithm for maintaining the sphere separator of a set of dynamically changing points which works with high likelihood time bounds of* $T(n) = O(\log^3 n)$.

5 Applications and Further Work

Our results for dynamic separators can be applied to generate dynamic algorithms for a wide variety of combinatorial and numerical problems which have an underlying associated k-neighborhood graph. For example,

1. Computational geometry problems, such as dynamic k-nearest neighbor.

2. Solution of linear systems and inverses of dynamic matrices, with changes both in the numeric values of entries in the matrix, as well as modifications that change the sparsity structure of the matrix.

3. Monoid path problems on dynamic graphs, with vertex and edge insertion and deletion, as well as changes to edge labels.

References

[1] Cole R, Goodrich MT. Optimal parallel algorithms for polygon and point-set problems. Department of computer science, 88–14, Johns Hopkins University, 1988.

[2] Chiang Y, Tamassia R. Dynamic algorithms in computational geometry. Tech Report CS–91–24, Department of Computer Science, Brown University, 1991.

[3] Frederickson G. Planar graph decomposition and all pairs shortest paths. *JACM* 38:162–204, January 1991

[4] Frieze A, Miller GM, Teng S-H. Separator based Parallel divide and conquer in computational geometry. *Proceedings, Symposium on Parallel Algorithms and Architectures*, 1992.

[5] Gazit H, Miller GL A parallel algorithm for finding a separator in planar graphs. *Proceedings, Twenty-Eighth Annual Symposium on Foundations of Computer Science* 238–248, 1987.

[6] Lipton RJ, Tarjan RE. A separator theorem for planar graphs. *SIAM J of Appl Math*, 36:177–189, April 1979.

[7] Miller GL, Teng S-H, Thurston W, Vavasis SA. Automatic mesh partitioning. To appear, 1992.

[8] Miller GL, Teng S-H, Vavasis SA. A unified geometric approach to graph separators. *Proceedings, Thirty Second Annual Symposium on Foundations of Computer Science*, 538–547, 1991.

[9] Overmars M. The design of dynamic data structures. *Lecture Notes in Computer Science*, 156, 1983.

[10] Pan V, Reif JH. Extension of the parallel nested dissection algorithm to path algebra problems. *Proc Sixth Conference on Foundation of Software Technology and Theoretical Computer Science*, New Delhi, India, Lecture Notes in Computer Science, vol. 241, Springer -Verlag, 1986; also full version as Fast and Efficient Solution of Path Algebra Problems. *Journal of Computer and Systems Sciences* 38:494-510, June 1989.

[11] Pan V, Reif JH. Acceleration of minimum cost path calculations in graphs having small separator families. TR 1989.

[12] Teng S-H. Points, spheres and separators: a unified geometric approach to graph partitioning. PhD thesis, Carnegie-Mellon University, School of Computer Science, 1991. CMU- CS-91-184

[13] Vaidya PM. An $O(n \log n)$ algorithm for the all-nearest-neighbor problem. *Discrete and Computational Geometry* 4:101–105, 1989.

Online Load Balancing of Temporary Tasks

*Yossi Azar**
DEC SRC

Bala Kalyanasundaram†
Univ. of Pittsburgh

Serge Plotkin‡
Stanford Univ.

Kirk R. Pruhs§
Univ. of Pittsburgh

Orli Waarts¶
IBM Almaden

Abstract

We consider non-preemptive online load balancing problem under the assumption that tasks have *limited duration* in time. Each task has to be assigned immediately upon arrival to one of the machines, increasing the load on this machine for the duration of the task. The goal is to minimize the maximum load.

Azar, Broder and Karlin studied the *unknown duration* case where for each task there is a subset of machines capable of executing it; the increase in load due to assignment of the task to one of these machines depends only on the task and not on the machine. For this case, they showed an $O(n^{2/3})$-competitive algorithm and an $\Omega(\sqrt{n})$ lower bound, where n is the number of the machines. We close the gap by showing an $O(\sqrt{n})$-competitive algorithm.

We also consider the *related machines* case with unknown task duration. Here, a task can be executed by any machine and the increase in load depends on the speed of the machine and the weight of the task. For this case we show a 20-competitive algorithm and a lower bound of $3 - o(1)$.

Trying to overcome the $\Omega(\sqrt{n})$ lower bound for the case of unknown task duration, we study a variant of the load balancing problem for tasks with *known duration*. For this case we show an $O(\log nT)$-competitive algorithm, where T is the ratio of the maximum to minimum duration.

*DEC Systems Research Center.

†Dept. of Computer Science, University of Pittsburgh. Supported in part by NSF under grants CCR-9009318 and CCR-9202158.

‡Dept. of Computer Science, Stanford University. Research supported by U.S. Army Research Office Grant DAAL-03-91-G-0102, and by a grant from Mitsubishi Electric Laboratories.

§Dept. of Computer Science, University of Pittsburgh. Supported in part by NSF under grant CCR-9209283.

¶IBM Almaden Research Center.

1 Introduction

Online load-balancing can be viewed as a type of scheduling problem, where we have to assign each arriving task to one of the existing machines. Assigning a task to a machine raises the load on this machine by the amount that, in general, can depend both on the task and on the machine; the load is raised for the duration of the task. The goal is to minimize the maximum load. We consider the non-preemptive online load balancing problem, *i.e.* reassignment of tasks is not allowed.

The load-balancing problem naturally arises in many applications involving allocation of resources. In particular, many cases that are usually cited as applications for bin packing become load-balancing problems when one removes the assumption that the storage is permanent (*e.g.*, storing food in warehouses). As a concrete example (chosen for simplicity), consider the case where each "machine" represents a communication channel with bounded bandwidth. The problem is to assign each incoming request for bandwidth to one of the channels. Assigning a request to a certain communication channel increases the load on this channel, *i.e.* increases the percentage of the used bandwidth; the load is increased for the duration associated with the request. Note that the load balancing problem is different from the classical scheduling problem of minimizing the makespan of an online sequence of tasks with known running times (see [6, 11] for survey). Intuitively, in the load-balancing context, the notion of makespan corresponds to maximum load, and there is a new, orthogonal, notion of time. (See [1] for further discussion of the differences.)

Formally, each arriving task has an associated *load vector*, where the ith coordinate of this vector defines the increase in load due to this task if it will be assigned to the ith machine. Since the arriving tasks have to be assigned without knowledge of the future tasks, it is natural to evaluate the performance in terms of the *competitive ratio* [12], which in this case is the supremum, over all possible input sequences, of the maximum load achieved by the on-line algorithm to the load achieved by the optimal off-line algorithm. The competitive ratio may depend, in general, on the number of machines n, which is usually fixed, and should not depend on the number of tasks that may be arbitrary large.

Similar to the way it is done for scheduling problems, it is natural to categorize load balancing problems according to the properties of the load vectors. The simplest case is where the coordinates of each load vector are equal to some value that depends only on the task. It is easy to observe that the algorithm due to Graham [5], applied to load-balancing, leads to a $(2 - \frac{1}{n})$-competitive solution.

Azar, Broder, and Karlin [2] proposed to study a more complicated case, motivated by the problem of on-line assignment of network nodes to gateways. In this case, a task can represent a request of a network node to be assigned to a gateway; machines represent gateways. Since, in general, each node can be served by only a subset of gateways, this leads to a situation where each coordinate of a load vector is either ∞ or equal to a given value that depends only on the task. For this case,

Azar, Broder, and Karlin give an $\Omega(\sqrt{n})$ lower bound on the competitive ratio and described an $O(n^{2/3})$-competitive algorithm.

The work in [2] raises several questions. The obvious unresolved question is to find an $O(\sqrt{n})$-competitive algorithm for their case, which we will refer to as the *uniform-speed with assignment restrictions.* Other questions involve natural variations of this problem for which the $\Omega(\sqrt{n})$ lower bound for the competitive ratio does not apply. Two incomparable modifications are the *related machines* case, defined so that the ith coordinate of each load vector is equal to $w(j)/s_i$, where the "weight" $w(j)$ depends only on the task j and the "speed" s_i depends only on the machine i; and the *known duration* case, where the duration of each task is known upon its arrival.

This work answers the above three questions. In particular, the main results presented in this paper are as follows:

- A $2\lceil\sqrt{n}\rceil$-competitive algorithm, Robin-Hood, for the uniform-speed with assignment restrictions case studied by [2]. Their lower bound implies that this algorithm is optimal to within a constant multiplicative factor.
- A 20-competitive algorithm, Slow-Fit, for the related machines case with unknown task duration. We also show that any algorithm for this case can't be better than $(3 - o(1))$-competitive.
- An $O(\log nT)$-competitive algorithm for the known-duration case with unrestricted load vectors, where T is the ratio of the maximum to minimum duration.

The algorithm Slow-Fit is essentially identical to the algorithm of Aspnes, Azar, Fiat, Plotkin and Waarts for assigning permanent tasks [1]. However, their analysis is inapplicable for the case where tasks have limited duration. Our analysis is completely different and shows that Slow-Fit is 5-competitive if the maximum load is known. The $(3 - o(1))$ lower bound for the related machines case applies even if the maximal load of the optimal off-line algorithm is known in advance. This lower bound stands in contrast to the 2-competitive algorithm for the case where tasks are permanent and the optimal off-line maximal load is known in advance in [1] and the 2-competitive algorithm for the unknown-duration case where the machines are identical [5]. Both Slow-Fit and Robin Hood are simple, deterministic, and run in $O(n)$ per task assignment.

Our algorithm for the known duration case is an application of the virtual circuit routing algorithm of [1]. In the off-line setting, this problem can be stated as an integer LP and is known to be NP-hard; approximate solutions can be obtained, for example, by using techniques from [10, 8, 9]. These methods result in $O(\log T)$ approximations.

It is instructive to compare the results in this paper with the results for load-balancing of permanent tasks, *i.e.* tasks with infinite duration. The identical machines case (all coordinates of a load vector are equal) was considered by Graham [5], who showed an $(2-\frac{1}{n})$-competitive algorithm. The bound was improved in [4] by Bartal, Fiat, Karloff and Vohra to $2-\epsilon$ for a small constant ϵ. The uniform speed with assignment restriction case was considered by Azar, Naor, and Rom [3], who described an $O(\log n)$-competitive algorithm and a matching lower bound. The most general case, where there is no restriction on the load vectors (the unrelated machines case), was considered by Aspnes, Azar, Fiat, Plotkin and Waarts [1], who showed an $O(\log n)$-competitive algorithm. They also showed an 8-competitive algorithm for the related machines case. Notice that as the competitive factors show, the fact that tasks have duration can make the task of the online load balancing algorithm significantly more difficult, especially when this duration is unknown upon the arrival of the task.

2 Tasks with Unknown Duration

2.1 Uniform Speed with Assignment Restriction

In this section we describe an $O(\sqrt{n})$-competitive algorithm for the uniform speed with assignment restriction case, where the task durations are unknown. Formally, each task j has weight $w(j)$, and has to be assigned to one of the machines in the set $M(j)$; assigning this task to a machine rases the load on this machine by $w(j)$. First we present a version of the algorithm that assumes that n is a perfect square, that all tasks have unit weight, and that the online algorithm knows in advance the optimum load OPT achievable by the offline algorithm. The elimination of these assumptions is straightforward, and is deferred to the end of the section.

Let $\ell_g(t)$ denote the load on machine g at time t. A machine g is said to be ***rich*** at some point in time if at that time $\ell_g(t) \geq \sqrt{n}\ OPT$, and is said to be ***poor*** otherwise. A machine may alternate between being rich and poor over time. With each rich machine, we associate its ***windfall time***, which is the last instance it became rich. Let g be a machine that is rich at time t and has windfall time t_0. We define $Prev(g,t)$ as the set of machines that are rich at time t and have windfall time not later than t_0. Note that the machines in $Prev(g,t_0)$ are rich at t_0 and that $g \in Prev(g,t_0)$.

Algorithm Robin Hood (simplified version): If possible, assign the newly arrived task j to a poor machine. Otherwise, assign it to the rich machine with the most recent windfall time.[1]

Lemma 2.1 At most $\sqrt{n}$ machines can be rich at any point of time.

Proof: If at some point in time there were more than $\sqrt{n}$ rich machines, then at that

[1]Note that not only does Robin Hood always give to the poor in preference to the rich, but also invariably favors the nouveau riche over the noblesse.

point the aggregate load of the tasks in the system would be more than $n \cdot OPT$, contradicting the assumed value of OPT. ■

Theorem 2.1 The simplified version of Robin Hood, which assumes that the value of OPT is known in advance, is $2\sqrt{n}$-competitive.

Proof: The result clearly holds if g is poor at the considered instance of time. Otherwise, assume that at some time t, the load $\ell_g(t)$ exceeds $2\sqrt{n} \cdot OPT$. Let t_0 be g's windfall time. The fact that g was poor at time $t_0 - 1$ and $\ell_g(t) > 2\sqrt{n} \cdot OPT$, implies that in the interval $[t_0 - 1, t]$ the algorithm assigned to g a set S of tasks with aggregate load that exceeds $\sqrt{n} \cdot OPT$. Moreover, these tasks are all active at time t.

Consider the machines to which the optimal offline algorithm assigns tasks in S. Since g is rich throughout the interval $[t_0, t]$, each of these machines must be rich upon these tasks arrivals. Moreover, each of these machines must have been rich already at time t_0 since otherwise Robin Hood would have assigned the tasks to it. In other words, the offline algorithm assigns the tasks in S to machines in $Prev(g, t_0)$. By definition of OPT, the offline algorithm can assign at most OPT load to any machine at some point in time. In particular, each machine in $Prev(g, t_0)$ can be assigned at most OPT load due to the tasks in S. Since the aggregate load of the tasks in S exceeds $\sqrt{n}OPT$, we have $|Prev(g, t_0)| > \sqrt{n}$, which contradicts Lemma 2.1. ■

We now allow tasks with arbitrary nonnegative weights, allow the possibility that n may not be a perfect square, and remove the assumption that OPT is known in advance. We keep an estimate L for OPT satisfying $L \leq OPT$. A machine g is classified as rich at t if $\ell_g(t) \geq \sqrt{n}L$. The windfall time and $Prev(g, t)$ are defined as before.

Algorithm Robin Hood: Assign the first task to an arbitrary machine, and set L to the weight of the first task. When a new task j arrives, set:

$$L = \max\{L, w(j), \frac{1}{n}(w(j) + \sum_g \ell_g(t))\}$$

The last quantity is the aggregate weight of the tasks currently active in the system divided by the number of machines. Note that the recomputation of L may cause some rich machines to be reclassified as poor machines. If possible, assign j to some poor machine. Otherwise, j is assigned to the rich machine g with the most recent windfall time.

The following lemma is immediate from the fact that nL is the lower bound on the aggregate load of the currently active tasks.

Lemma 2.2 At most $\lceil\sqrt{n}\rceil$ machines can be rich at any point in time.

Lemma 2.3 At all times Robin Hood guarantees that $L \leq OPT$.

Proof: The proof is by induction on the number of the assigned tasks. After the first task $L \leq OPT$. For the inductive part, it suffices to consider only cases where L is increased. The claim follows from the fact that $w(j) \leq OPT$ and from the fact that $\frac{1}{n}(w(j) + \sum_g \ell_g(t)) \leq OPT$. ■

Theorem 2.2 The competitive factor of the algorithm Robin Hood is at most $2\lceil \sqrt{n} \rceil$.

Proof: We will show that the algorithm guarantees that at any point in time $\ell_g(t) \leq \lceil \sqrt{n} \rceil (L + OPT)$ for any machine g. The claim is immediate if g is poor at t. Otherwise, let S be the set of tasks that were assigned to g since g's windfall time t_0. As in the proof of Theorem 2.1, each task in S can only be assigned to machines in $Prev(g, t_0)$. Assume that $|Prev(g, t_0)| = k$, and note that $\frac{1}{k}\sum_{j \in S} w(j) \leq OPT$.

Let q be the task assigned to g between times $t_0 - 1$ and t_0 that caused g to become rich. We break the proof into cases. If $k \leq \lceil \sqrt{n} \rceil - 1$ then $\ell_g(t) \leq \sqrt{n}\, L + w(q) + \sum_{j \in S} w(j) \leq \lceil \sqrt{n} \rceil (L + OPT)$. The second inequality follows since $w(q) \leq OPT$. If $k = \lceil \sqrt{n} \rceil$, then $\ell_g(t_0) = L\sqrt{n}$, or we would get a contradiction to the fact that L is at least the aggregate weight of concurrent tasks in the system divided by n. Hence, $\ell_g(t) \leq \sqrt{n}\, L + \sum_{j \in S} w(j) \leq \lceil \sqrt{n} \rceil (L + OPT)$. ■

2.2 Related Machines with Unknown Task Duration

We now consider the related machines case where the task durations are not known upon their arrivals. We give an algorithm Slow-Fit that is 5-competitive for this problem if OPT is known and 20-competitive otherwise. Then we give a $(3 - o(1))$ lower bound on the competitive ratio for any solution to the problem.

2.2.1 Algorithm Slow-Fit

Slow-Fit is a minor variation of the algorithm Assign-R2 given in [1] for the corresponding problem with permanent tasks. It is interesting to note that the greedy algorithm is $\Theta(\log n)$-competitive for this problem even in the case that all tasks are permanent [1]. As in the previous section, we first describe a simplified version of Slow-Fit that assumes that the value of OPT is known in advance.

Let c be an integer constant (we will later choose $c = 5$ to minimize the resulting competitive factor). Recall that the load due to task j on a machine g_i is defined as $w(j)/s_i$, where $w(j)$ is the weight of the jth task, and s_i is the speed of machine g_i. We say that j is *assignable* to g_i if and only if $w(j)/s_i \leq OPT$ and that assigning j

to g_i will not cause the load ℓ_{g_i} to exceed $c \cdot OPT$. For convenience, we assume that $s_i \leq s_j$ for $i < j$.

Algorithm Slow-Fit: (simplified version) Assign the new task to the slowest *assignable* machine. Break ties by assigning to the machine with the smallest index.

Theorem 2.3 Provided $c \geq 5$, the simplified version of Slow-Fit guarantees that every task is assignable, which implies that the simplified Slow-Fit algorithm is c-competitive.

Proof: Suppose that at some point in time t_0, a task q arrives that is not *assignable*. We then reach a contradiction to the supposed value of OPT by showing that at some time in the past the aggregate weight of the tasks in the system was more than $\sum_{i=1}^{n} s_i OPT$. Starting from the current time, we will successively go backward in time and observe the load on certain sets of machines. Let G_i be a set of machines and t_i be the time under consideration. We define $J(G_i)$ to be the collection of tasks that are active at time t_i and that were assigned to machines in G_i by the online algorithm. Also, define $S(G_i) = \sum_{k \in G_i} s_k$.

At the start of ith iteration (during which we define G_{i+1} and t_{i+1}) of our backward journey in time the following four invariants hold:

1. The time t_i is defined and $t_i < t_j$ for $0 \leq j < i$.

2. The disjoint sets of machines G_j for $0 \leq j \leq i$ are defined.

3. At time t_i, each machine in G_i has a load that exceeds $(c-1)\, OPT$.

4. For $0 \leq j < i$, $S(G_{j+1}) \geq (c-3)S(G_j)$.

Initially, t_0 is the time at which the task q arrived and $G_0 = \{g_n\}$. Since the task q can not be assigned to the fastest machine g_n, and $OPT \geq w(q)/s_n$, we have $\ell_{g_n}(t_0) > (c-1)\, OPT$, which means that the cumulative weight of the tasks assigned to g_n that are active at t_0 exceeds $(c-1)\, s_n\, OPT = (c-1)\, S(G_0)\, OPT$.

Consider the assignment of tasks in $J(G_0)$ by the optimal offline algorithm. Let g_{m_0} be the slowest machine that is assigned a task in $J(G_0)$ by the offline algorithm. Construct the set $G_1 = \{g_{m_0}, g_{m_0+1}, \ldots, g_{n-1}\}$. The cumulative weight of tasks assigned to machine in G_0 by the offline algorithm is at most $s_n \cdot OPT$. Thus the cumulative weight of tasks assigned to machines in G_1 by the offline algorithm must be at least $(c-2)\, S(G_0)\, OPT$, or we would get a contradiction to $\ell_{g_n}(t_0) > (c-1)\, OPT$. Since the load on any machine does not exceed OPT in the offline assignment, we have $S(G_1)\, OPT \geq (c-2)\, S(G_0)\, OPT$. As a consequence we get

$$S(G_1) \geq (c-2)\, S(G_0) \geq (c-3)\, S(G_0),$$

which proves that all invariants except (3) hold at t_0.

Consider the task q_0 in $J(G_0)$ with the smallest weight. Say that q_0 was assigned at time t_1. According to our construction of the set G_1, we have $w(q_0) \leq s_k \cdot OPT$ for all $g_k \in G_1$. Since q_0 was assigned to g_n, all machines in G_1 were unassignable at t_1 with respect to this task. Hence, for each machine $g \in G_1$, $\ell_g(t_1) > (c-1)\ OPT$. This completes the proof of of the invariants for $i = 1$.

Now assume that t_i and G_i are defined, and that the above invariants hold. We will show how to construct t_{i+1} and G_{i+1} so that the invariants hold for $i+1$ as well. Using the 3rd invariant, we know that the load of each machine in G_i exceeds $(c-1)\ OPT$ at time t_i. Hence, the cumulative weight of tasks in $J(G_i)$ is at least $(c-1)S(G_i)OPT$. Consider how the optimal offline algorithm assigns tasks in $J(G_i)$. Let g_{m_i} be the slowest machine that gets a task in $J(G_i)$. We let

$$G_{i+1} = \{g_{m_i}, g_{m_i+1}, \ldots, g_{m_{i-1}-1}\}.$$

(Notice that G_{i+1} is empty if $m_i \geq m_{i-1}$.) Since all the tasks in $J(G_i)$ are active at t_i, the cumulative weight of the tasks from $J(G_i)$ assigned to machines in $\bigcup_{0 \leq k \leq i} G_k$ by the offline algorithm is bounded by $\sum_{0 \leq j \leq i} S(G_j)\ OPT$, which is at most $2\ S(G_i)\ OPT$ by the 4th invariant provided $c \geq 5$. The rest of the tasks in $J(G_i)$ have to be assigned to machines in G_{i+1} by the offline algorithm; the cumulative weight of these tasks is at least $(c-3)\ S(G_i)\ OPT$. Since the load on any machine never exceed OPT in the offline assignment, it must be the case that $S(G_{i+1})\ OPT \geq (c-3)\ S(G_i)\ OPT$. As a consequence we get

$$S(G_{i+1}) \geq (c-3)\ S(G_i).$$

Consider the task q_i with the smallest weight in $J(G_i)$. Say q_i arrived at time t_{i+1}. Since q_i was not assigned to any machine in G_{i+1}, all these machines were unassignable at t_{i+1}. The task q_i has the smallest weight of the tasks in $J(G_i)$, and hence $w(q_i)/s_k \leq OPT$ for any machine $g_k \in G_{i+1}$. Thus, we have:

$$\forall g_k \in G_{i+1} : \ \ell_{g_k}(t_{i+1}) > (c-1)\ OPT.$$

This proves that the invariants now hold for $i+1$.

Consider the last iteration, say k, where the set G_k includes g_1. We know that the cumulative weight of tasks in $J(G_k)$ is at least $(c-1)S(G_k)OPT$ at time t_k. Recall that $\sum_{0 \leq i \leq k} S(G_i) \leq 2S(G_k)$. The contradiction follows from the fact that at any point in time, the maximum task weight that can be handled by the system is at most

$$\sum_{0 \leq i \leq k} S(G_i)\ OPT \leq 2S(G_k)OPT < (c-1)S(G_k)OPT.$$

■

We now show how to modify SLOW-FIT so that it will not require advance knowledge of OPT.

Algorithm SLOW-FIT: After the arrival of the first task q set $L = w(q)/s_n$. Note that L is our estimate of OPT. We then repeat the following: (1) Apply the the preliminary version of SLOW-FIT, which uses L as the value of OPT, until it fails. (2) We set $L = 2L$, and return to step 1, ignoring any tasks encountered to this point. So when the preliminary version of SLOW-FIT starts up again it will behave as if the load on every machine is zero.

Theorem 2.4 The competitive factor of Slow-Fit is at most 4c, provided $c \geq 5$.

Proof: Notice that at any point in time $OPT \geq L/2$. During the last iteration of the preliminary version of SLOW-FIT the maximum load on any machine is at most cL, and hence at most $2cOPT$. Since L was doubled on each iteration, the cumulative load on any machine for all of the prior iterations is at most cL, or at most $2cOPT$. Therefore, the load on any machine is at most $4cOPT$. ∎

2.2.2 Lower Bound

In the previous section we show that there is a 5 competitive algorithm for the related machines case if OPT is known. Here we show that any on-line algorithm has competitive ratio of at least $(3 - o(1))$ for the related machine case even if the value of OPT is known.

Theorem 2.5 The competitive factor c of any on-line algorithm for the related machines case satisfies $c \geq 3 - o(1)$.

The proof is omitted due to lack of space. Roughly speaking, the lower bound is based on the idea of creating a sequence phases, where the weight of the tasks is exponentially increasing from phase to phase. A phase consists of identical tasks, and hence the online algorithm can not determine which ones should be assigned to slower machines. After the online algorithm assigns the tasks in a single phase, the adversary keeps only the tasks that were assigned to faster machines than necessary, and terminates all the rest of the tasks.

3 Tasks with Known Duration

In this section we consider the case where the duration of a task is known upon its arrival. With this additional assumption we can overcome the $\Omega(\sqrt{n})$ lower bound of [2] for the uniform speed with assignment restriction case, and design an algorithm whose competitive ratio is $O(\log nT)$, where T is the ratio of maximum to minimum task duration. In fact our algorithm works for the more general case of *unrelated machines*. Formally, each task j has an associated load vector $\vec{p}(j)$, and an associated duration value $d(j)$. All these values are specified upon the task's arrival.

Our algorithm is based on the on-line route allocation algorithm of [1]. In the on-line route allocation problem, we are given a graph $G = (V, E)$ with $|V| = N$ and $|E| = m$. Requests for route allocation arrive as tuples $(s_i, t_i, \{p_{i,e}\})$, where $s_i, t_i \in V$ and for all e, $p_{i,e} \in \mathcal{R}^+$. Request i is satisfied by assigning it to a permanent route P_i from s_i to t_i. Let $\mathcal{P} = \{P_1, P_2, \ldots, P_k\}$ be the routes assigned to requests 1 through k by the on-line algorithm, and let $\mathcal{P}^* = \{P_1^*, P_2^*, \ldots, P_k^*\}$ be the routes assigned by the off-line algorithm. Given a set of routes $\mathcal{P}$, define the load after the first j requests are satisfied by

$$\ell_e(j) = \sum_{i \leq j : e \in P_i} p_{i,e} \tag{1}$$

and let $\lambda(j) = \max_{e \in E} \ell_e(j)$. Similarly, define $\ell_e^*(j)$ and $\lambda^*(j)$ to be the corresponding quantities for the routes produced by the off-line algorithm. For simplicity we will abbreviate $\lambda(k)$ as λ and $\lambda^*(k)$ as λ^*. The goal of the on-line algorithm is to produce a set of routes $\mathcal{P}$ that minimizes λ/λ^*.

The on-line route allocation algorithm of [1] is $O(\log N)$-competitive, where N is the number of vertices in the given graph. We will reduce our problem of on-line load balancing of tasks with known duration to the on-line route allocation problem, where $N = nT$. Thus, we can apply the algorithm of [1] to achieve $O(\log N) = O(\log nt)$ competitive ratio.

We will assume that T is known in advance. It is easy to modify the algorithm to remove this assumption. For convenience of exposition, assume that the minimum task duration is 1, and tasks' arrival and departure times are integers. (In other word, the system is discrete, and 1 is the time unit.) The assumption that the system is discrete can be easily eliminated without paying more than a constant factor in the competitive ratio. Our description of the algorithm proceeds in two steps. First we describe an $O(\log nT')$-competitive algorithm, where T' is the duration of the whole sequence of tasks; that is, the time from the arrival of the first task to the departure of the last task. Later we will show an $O(\log nT)$-competitive algorithm, where T is the maximum duration of a task. The second algorithm that uses the first algorithm as a subroutine.

We will translate each task to a request for allocating a route. We construct a directed graph G consisting of T' layers, each of which consists of $n + 2$ vertices, numbered $1, \ldots, n + 2$. We refer to vertices $1 \ldots, n$ in each layer as the *common* vertices of the layer; to vertex $n + 1$ as the *source* of the layer; and to vertex $n + 2$ as the *sink* of the layer. There is an arc from each common vertex to the corresponding vertex in the next layer. That is, for each $i, 1 \leq i \leq n$ and $k, 1 \leq k \leq T' - 1$, there is an arc from vertex i in layer k to vertex i in layer $k+1$. In addition, there is an arc from each source to each common vertex in its layer, and an arc from each common vertex to the sink in its layer.

Denote by $v(i,k)$ vertex i in layer k. The arc from $v(i,k)$ to $v(i,k+1)$ will represent machine i during the time interval $[k,k+1)$, and the load on the arc will correspond to the load on machine i during this interval. Thus, minimizing the maximum load on an arc in the on-line route allocation problem corresponds to minimizing the maximum machine load over all machines and times in the load balancing problem.

Recall that in the load balancing problem, each task j has an associated load vector $\vec{p}(j)$ and a duration value $d(j)$, both are specified upon its arrival. Denote by $a(j)$ the arrival time of task j. We will convert each new task j in the on-line load balancing problem to a request for allocating a route in G from the source of layer $a(j)$ to the sink of layer $a(j)+d(j)$. The loads $p_{j,e}$ are defined as follows: for arcs $v_{i,k}$ to $v_{i,k+1}$ for $a(j) \le k \le a(j)+d(j)-1$, we set $p_{j,e} = p_i(j)$; we set $p_{j,e}$ is 0 for the arcs out of the source of layer $a(j)$ and for the arcs into the sink of layer $a(j)+d(j)$, and is ∞ for all other arcs. Clearly the only possible way to route the jth request is through the arcs $v_{i,k}$ to $v_{i,k+1}$ for $a(j) \le k \le a(j)+d(j)-1$ for some i. This route raises the load on the participating arcs by precisely $p_i(j)$ which corresponds to assigning the task to machine i. Thus, the routing algorithm for permanent routes provides a load balancing algorithm for tasks with known durations so that the load on the arc from $v(i,k)$ to $v(i,k+1)$ corresponds to the load on machine i at interval $[k,k+1)$. Thus, we can use the $O(\log N)$-competitive on-line route allocation algorithm of [1] to get an on-line load balancing algorithm which is $O(\log nT')$-competitive (recall that the number of vertices in G is $O(nT')$).

Next we construct an $O(\log nT)$-competitive algorithm, where T is the maximum duration of a task. Partition the tasks into groups according to their arrival times. Group k contains all tasks that arrive in the time interval $[(k-1)T, kT)$. Clearly each task in group k must depart by time $(k+1)T$, and the overall duration of tasks in a certain group is at most $2T$. Now, to assign tasks to machines, for each group use a separate copy of the algorithm described in the previous step. That is, assign tasks belonging to a certain group independently of the assignments of tasks in other groups. Clearly, for each group, the ratio between maximal load of the on-line and off-line algorithms is at most $O(\log nT)$. Moreover, at each instant of time there are tasks from at most two groups. Thus, the maximal on-line load at any given time is at most twice the maximal on-line load of a single group. The off-line load is at least the largest load of the off-line algorithms over all groups, and hence the resulting algorithm is $O(\log nT)$-competitive.

Theorem 3.1 The above on-line load balancing algorithm for unrelated machines with known tasks duration is $O(\log nT)$-competitive.

References

[1] J. Aspenes, Y. Azar, A. Fiat, S. Plotkin, and O. Waarts. On-line machine scheduling with applications to load balancing and virtual circuit routing. In *Proc. 23rd Annual ACM Symposium on Theory of Computing*, May 1993.

[2] Y. Azar, A. Broder, and A. Karlin. On-line load balancing. In *Proc. 33rd IEEE Annual Symposium on Foundations of Computer Science*, pages 218–225, 1992.

[3] Y. Azar, J. Naor, and R. Rom. The competitiveness of on-line assignment. In *Proc. 3rd ACM-SIAM Symposium on Discrete Algorithms*, pages 203–210, 1992.

[4] Y. Bartal, A. Fiat, H. Karloff, and R. Vohra. New algorithms for an ancient scheduling problem. In *Proc. 24th Annual ACM Symposium on Theory of Computing*, 1992.

[5] R.L. Graham. Bounds for certain multiprocessing anomalies. *Bell System Technical Journal*, 45:1563–1581, 1966.

[6] R.L. Graham, E.L. Lawler, J.K Lenstra, and A.H.G. Rinnooy Kan. Optimization and approximation in deterministic sequencing and scheduling: a survey. *Annals of Discrete Mathematics*, 5:287–326, 1979.

[7] T. Leighton, F. Makedon, S. Plotkin, C. Stein, É. Tardos, and S. Tragoudas. Fast approximation algorithms for multicommodity flow problem. In *Proc. 23th ACM Symposium on the Theory of Computing*, pages 101–111, May 1991.

[8] J.K. Lenstra, D.B. Shmoys, and É. Tardos. Approximation algorithms for scheduling unrelated parallel machines. *Math. Prog.*, 46:259–271, 1990.

[9] S. Plotkin, D. Shmoys, and É. Tardos. Fast approximation algorithms for fractional packing and covering problems. In *Proc. 32nd IEEE Annual Symposium on Foundations of Computer Science*, pages 495–504, October 1991.

[10] P. Raghavan and C.D. Thompson. Provably good routing in graphs: Regular arrays. In *Proc. of 17th ACM Symp. on Theory of Computing*, May 1985.

[11] D. Shmoys, J. Wein, and D.P. Williamson. Scheduling parallel machines on-line. In *Proc. 32nd IEEE Annual Symposium on Foundations of Computer Science*, pages 131–140, 1991.

[12] D.D. Sleator and R.E. Tarjan. Amortized efficiency of list update and paging rules. *Comm. ACM*, 28(2):202–208, 1985.

Connected Domination and Steiner Set on Asteroidal Triple-Free Graphs

Hari Balakrishnan **Anand Rajaraman**
C. Pandu Rangan*
Department of Computer Science and Engineering
Indian Institute of Technology, Madras 600 036, India.

Abstract

An *asteroidal triple* is a set of three independent vertices such that between any two of them there exists a path that avoids the neighbourhood of the third. Graphs that do not contain an asteroidal triple are called *asteroidal triple-free (AT-free) graphs*. AT-free graphs strictly contain the well-known class of cocomparability graphs, and are not necessarily perfect. We present efficient polynomial-time algorithms for the minimum cardinality connected dominating set problem and the Steiner set problem on AT-free graphs. These results, in addition to solving these problems on this large class of graphs, also strengthen the conjecture of White, et. al. [9] that these two problems are algorithmically closely related.

Keywords: Design of algorithms, asteroidal-triple free graphs.

1 Introduction

An independent set of vertices x,y,z is an *asteroidal triple* if between every two of them there exists a path that avoids the neighbourhood of the third. An *asteroidal triple-free (AT-free) graph* is a graph that contains no asteroidal triples. AT-free graphs were first considered by Lekkerkerker and Boland [8] in the early 1960s.The following facts about the AT-free graphs are well-known (for the definitions of the italicized terms, see [7]).

- **G is an *interval graph* if and only if it is *chordal* and AT-free [8].**
- **AT-free graphs are not necessarily *perfect*. Consider C_5, for example, which is AT-free but not perfect.**
- **Perfect AT-free graphs strictly contain the *cocomparability graphs*.**

*E-mail : rangan@iitm.ernet.in

- AT-free graphs have a characterization in terms of forbidden subgraphs [3].

A *dominating set* S of a graph $G = (V, E)$ is defined as a set of vertices $S \subseteq V$ such that every vertex in V is either in S or is adjacent to some vertex in S. S is a *connected dominating set* of G if and only if S dominates G and the subgraph induced by S is connected.

For a subset R of V, we define a set S to be a *Steiner set* if

i. S is a subset of V,

ii. the subgraph induced by $(R \cup S)$ in G is connected and

iii. S is a set with least cardinality satisfying (i) and (ii).

We call the vertices of S *Steiner vertices* with respect to R. A spanning tree on the Steiner set is called a *Steiner tree*. We will henceforth denote an instance of the Steiner set problem by (G, R).

It is known that the minimum connected dominating set problem and minimum Steiner set problem are NP-complete for general graphs [6, 2]. Polynomial times have been reported in the literature for some special classes of graphs such as interval graphs, permutation graphs, strongly chordal graphs and distance hereditary graphs. [4, 1, 9, 5]. We mention here two special cases of the Steiner tree problem - if $|R| = 2$, then this reduces to the well-known *shortest path* problem; if $R = V$, then this reduces to the *minimum spanning tree* problem (in the weighted case). Polynomial algorithms are known for both these special cases on general graphs.

In this paper, we present algorithms to solve the Minimum Cardinality Connected Dominating Set (MCDS) and Steiner set problems on AT-free graphs. Throughout this paper, $P = u \cdots v$ denotes a (chordless) path P between vertices u and v. Depending on the context, P could also stand for the set of vertices on the path. $N_G(v)$ and $N_G[v]$ denote the open and closed neighbourhoods of the vertex v respectively. We often denote the graph induced by a subset S of V by S itself, instead of by $G(S)$. $dom(S)$ denotes the set of vertices dominated by an arbitrary $S \subseteq V$. Note that a set D is a dominating set of $G = (V, E)$ if and only if $dom(D) = V$.

2 Dominating Pairs in AT-free Graphs

A pair of vertices (u, v) is said to be a *dominating pair* of a graph G if every $u \cdots v$ path in G dominates all vertices of G. Corneil et. al. [3] have stated a property of AT-free graphs in terms of dominating pairs. We state their result below as a theorem :

Theorem 2.1 ([3]) *Every connected AT-free graph contains a dominating pair of vertices.*

We now present a simple algorithm to find all the dominating pairs in an arbitrary graph. This is needed as a preprocessing step in our later algorithms. It is clear that a pair of vertices (u, v) is a dominating pair of G if and only if for any vertex $x \in V$, u and v lie in different connected components of $G - N[x]$. (Otherwise, there would exist a $u \cdots v$ path that did not pass through the closed neighbourhood of x). The algorithm proceeds by removing $N[x]$ from G for every vertex x in V, and finding the connectedness status of every other pair of vertices in the resulting reduced graph. We declare all pairs (u, v) which lie in different components in every such reduced graph to be the dominating pairs of G.

Algorithm *Dominating Pairs*

Input : An AT-free graph $G = (V, E)$. $|V| = n$, $|E| = m$.
Output : A list of all the dominating pairs of G.
A, A' : $array[1..n, 1..n]$ of boolean entries.

begin
1. initialize $A[i, j]$ to *true* for $i = 1, 2, \ldots, n$ and $j = 1, 2, \ldots, n$;
2. for all vertices $v \in V$ do
3. construct $G' = G - N[v]$;
4. do a DFS of G' to find out its connected components;
5. for $i = 1, 2, \ldots, n$ and $j = 1, 2, \ldots, n$ do
   if vertices i and j are in the same connected component of G' then
      $A'[i, j] \leftarrow$ *false*
   else
      $A'[i, j] \leftarrow$ *true*;
6. $A[i, j] \leftarrow A[i, j] \wedge A'[i, j]$;
7. for $i = 1, 2, \ldots, n$ and $j = 1, 2, \ldots, n$
   if $A[i, j] =$ *true* then (i, j) is a dominating pair

end.

Theorem 2.2 *Algorithm* Dominating Pairs *above correctly finds all the dominating pairs of an AT-free graph G in $O(n^3)$ time.*

Proof. Correctness follows from the discussion above. We now prove the complexity bound. Step 1 takes time linear in n. Steps 3 and 4 can be also executed in linear time for every iteration. Each execution of step 5 takes $O(n^2)$ time. Hence each execution of the loop in step 2 takes $O(n^2)$ time; so step 2 requires $O(n^3)$ time for execution. Step 7 takes time linear in n. Hence the overall complexity is dominated by step 2 and is $O(n^3)$. □

3 Connected Domination on AT-free Graphs

In this section, we consider the problem of computing a minimum connected dominating set ($MCDS$) of an AT-free graph. This problem has been extensively studied on various classes of graphs [4, 2]. Arvind and Pandu Rangan [1] give an $O(m + n \log n)$ algorithm for this problem on the class of permutation graphs. No results are currently known for this problem on the class of cocomparability graphs. We present an efficient algorithm on AT-free graphs, which are a large superset of cocomparability graphs. This algorithm can be used on any AT-free graph, regardless of whether it is perfect or not.

Let (u, v) be a dominating pair of G and let $X = N(u)$ and $Y = N(v)$. For each $x \in X$, let $A_x \subseteq V$ be the set of all vertices in $V - \{u\}$ such that if $a \in A_x$, then $\{a, x\}$ dominates all vertices in X. Similarly, for each $y \in Y$ let $B_y \subseteq V$ be the set of all vertices in $V - \{v\}$ such that if $b \in B_y$, then $\{b, y\}$ dominates all vertices in Y. Define Γ as follows :

$$\begin{aligned}\Gamma = \{P \mid P = u \cdots v, &\quad \text{or} \\ P = u \cdots by, &\quad \text{for } y \in Y, b \in B_y \text{ or} \\ P = xa \cdots v, &\quad \text{for } x \in X, a \in A_x \text{ or} \\ P = xa \cdots by, &\quad \text{for } x \in X, y \in Y, a \in A_x, b \in B_y \;\}\end{aligned}$$

Theorem 3.1 *Every path $P \in \Gamma$ such that $|P|$ is minimum in Γ is an $MCDS$ of G.*

Proof. Let $P \in \Gamma$. We show first that P is a connected dominating set. Clearly P is connected. To show that $dom(P) = V$, we have to consider four cases :

Case 1: $P = u \cdots v$. In this case clearly $dom(P) = V$, since (u, v) is a dominating pair.

Case 2: $P = u \cdots by$, for $y \in Y$, $b \in B_y$. Note that $Q = Pv$ is a dominating path since (u, v) is a dominating pair. But $\{b, y\}$ dominates every vertex dominated by v, and so P is also a dominating path.

Case 3: $P = xa \cdots v$, for $x \in X$, $a \in A_x$. This case is symmetric to Case 2.

Case 4: $P = xa \cdots by$. In this case, note that $Q = uPy$ is a dominating set; but $\{a, x, b, y\}$ dominates every vertex dominated by $\{x, y\}$ and so P is a dominating path.

Next, we show that given any MCDS S, we can convert it into an MCDS S' such that $S' \in \Gamma$. Once again, we consider four cases:

Case 1: $u, v \in S$. In this case clearly $S' \in \Gamma$.

Case 2: $u \notin S, v \in S$. In this case since S dominates u, some vertex $x \in X$ is in S. Since S is connected, it has a $x \cdots v$ path P.
Subcase 2(a): If $S \neq P$, then consider $S' = P \cup \{u\}$. Clearly, S' contains a (u, v) path and is therefore a dominating set of G with size smaller than or equal to that of S, satisfying $S' \in \Gamma$.
Subcase 2(b): In this case, $S = P$. If the vertex a adjacent to x in P belongs to A_x, clearly $S \in \Gamma$. If not, then there is some $x' \in X$ such that $\{x, a\}$ does not dominate x'. So x' must be dominated by some other vertex a' on the $x \cdots v$ path P. Replace $\{x, a\}$ in S ($= P$) by $\{u, x'\}$ to obtain S' from S. Clearly, S is connected because u is adjacent to x', x' is adjacent to a' ($\in S$), and S itself was originally connected. Now, S' is a $u \cdots v$ path, and so $S' \in \Gamma$. Also note that $|S'| = |S|$; this shows that the required transformation is possible in this case as well.

Case 3: $u \in S, v \notin S$. This case is symmetric to Case 2.

Case 4: $u \notin S, v \notin S$. In this case consider x, x', a, a' as in case 2(b) and replace $\{x, a\}$ by $\{u, x'\}$ to get S'' which satisfies the conditions of Case 3. Using the same technique as in Case 3, this can be transformed into an *MCDS* $S' \in \Gamma$.

So in all cases we can transform S to $S' \in \Gamma$ as desired. This proves the lemma. □

We present below an algorithm to find an MCDS of an AT-free graph G which relies on the above theorem.

Algorithm *Connected Domination*
Input: An AT-free graph $G = (V, E)$. $|V| = n, |E| = m$.
Output: An MCDS for G.

begin
1. Find a dominating pair (u, v) of G.
2. $X \leftarrow N(u), Y \leftarrow N(v)$.
3. for all $x \in X$ do
   construct A_x.
   Let $A \leftarrow \bigcup_{x \in X} A_x$.
4. for all $y \in Y$ do
   construct B_y.
   Let $B \leftarrow \bigcup_{y \in Y} B_y$.
5. do an All Pairs Shortest Paths on G and store the results in array D.
   i.e. $D[i, j]$ = length of a shortest (i, j) path.
6. Let

$l_1 \leftarrow D[u, v]$
$l_2 \leftarrow \min_{a \in A} D[a, v] + 1$
$l_3 \leftarrow \min_{b \in B} D[u, b] + 1$
$l_4 \leftarrow \min_{a \in A, b \in B} D[a, b] + 2$

7. $l \leftarrow min(l_1, l_2, l_3, l_4)$.
8. l gives the size of the MCDS. We can also easily construct the MCDS with the above information.

end.

Theorem 3.2 *Algorithm* Connected domination *correctly computes the MCDS of an AT-free graph G in $O(n^3)$ time using $O(n^2)$ space.*

Proof. The correctness of the algorithm follows from the previous theorem. The only thing to note is that we add 1 or 2 to l_2, l_3 and l_4 in step 6 to take care of the fact that the actual dominating path must contain a vertex from X or Y or both, as stated in the previous theorem.

We now prove the complexity bound. Step 1 can be done in $O(n^3)$ time using Algorithm *Dominating Pairs*. Step 2 takes linear time. Each iteration of steps 3 and 4 takes $O(n^2)$ time, and so steps 3 and 4 require $O(n^3)$ time. Step 5 takes $O(n^3)$ time using the standard All Pairs Shortest Paths algorithm. Step 6 takes $O(n^2)$ time because there are totally $O(n^2)$ pairs of distances to be compared. Step 7 takes constant time. Step 8 can be executed in $O(m + n)$ time using a BFS to find the actual shortest path given the endpoints of the path. Hence the overall time complexity of the algorithm is $O(n^3)$.

Computing the dominating pair requires $O(n^2)$ space (Theorem 2.2). X and Y take $O(n)$ space as do each A_x and B_y. Observe that we can avoid storing each A_x (B_y) explicitly by constructing A (B) directly, thus requiring only $O(n)$ space. The All Pairs algorithm requires $O(n^2)$ space, and the subsequent BFS requires $O(m + n)$ space. Hence the overall space requirement is $O(n^2)$. □

The above algorithm computes the minimum cardinality connected dominating set of an AT-free graph in $O(n^3)$ time. We have exploited the property that there exists an $MCDS$ that induces a path (or a cycle) in the algorithm. On permutation graphs, *every* minimum *weighted* connected dominating set induces a path or a cycle. This property has been exploited in [1] to solve the weighted version of the problem on permutation graphs in $O(m + n \log n)$ time. However on AT-free graphs it is not necessary that there be a minimum *weighted* connected dominating set that induces a path or a cycle. (See Figure 1.) Hence the same approach may not be extendable to the weighted case on AT-free graphs.

4 Steiner Set on AT-free Graphs

In this section we consider the problem of computing the Steiner set of an AT-free graph G, with respect to a set $R \subseteq V$ of vertices. The method used

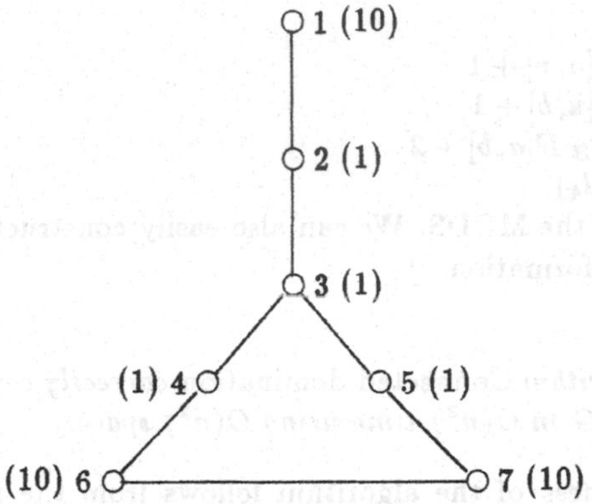

Figure 1: The weighted $MCDS = \{2,3,4,5\}$ is not a path or a cycle (numbers in brackets are the weights of the vertices).

to solve this problem is similar to that used to solve the $MCDS$ problem in the previous section. It has already been observed that these two problems are algorithmically closely related [9] in the sense that they are either both polynomial or both NP-complete on all classes of graphs investigated so far. However, their precise relationship to each other is unknown. Arvind and Pandu Rangan [1] provide an $O(m+n\log n)$ algorithm for this problem on the class of permutation graphs. We present an $O(n^3)$ algorithm for the Steiner set problem on AT-free graphs.

We denote by R the set of vertices with respect to which the Steiner set is to be computed. Denote by $R_1, R_2, \ldots, R_k$ the k connected components of R. Let P be a dominating path in G between any dominating pair of vertices. Number the vertices of this path 1,2, ..., p, where $|P| = p$. Let the number assigned to any vertex v on the path be $number(v)$. Define
$min_R(P) = \{v | v \in P,\ v \in R$ or $(v,r) \in E$ for some $r \in R$, and $number(v)$ is minimum$\}$, and
$max_R(P) = \{v | v \in P,\ v \in R$ or $(v,r) \in E$ for some $r \in R$, and $number(v)$ is maximum$\}$.

We define an R-dominating pair of an AT-free graph with respect to any $R \subseteq V$ to be a pair of vertices (u,v) such that every path from u to v dominates all vertices of R. This is a generalization of the concept of a dominating pair defined previously, which is just a V-dominating pair of G.

Construct an auxiliary graph G_R from G in the following manner :
1. Replace all the vertices of each connected component R_i by a single vertex r_i. with the adjacency of r_i equal to the union of the adjacencies of all the vertices in R_i.

2. Choose any dominating path P corresponding to some dominating pair. Compute $u = min_R(P)$ and $v = max_R(P)$.
3. Delete all vertices of P that are not numbered in the range $[number(u), number(v)]$. Let P_R be the corresponding path obtained.
4. Delete all vertices of $G - R$ that are not adjacent to any vertex of P_R.
5. Delete all vertices of $G - R$ that are adjacent to either u or v.

The graph thus obtained is $G_R = (V_R, E_R)$, which has certain interesting properties.

Lemma 4.1 *G_R is AT-free.*

Proof. Straightforward. For if not, there exists an asteroidal triple (x, y, z) in G_R. Since G_R is constructed from G by only deleting certain vertices and edges, and adding no new edges, (x, y, z) is an asteroidal triple in G as well – a contradiction. □

Lemma 4.2 *$u = min_R(P)$ and $v = max_R(P)$ form an R-dominating pair in G_R.*

Proof. It is clear that the subpath P_R of $P = x \cdots y$ between u and v is a connected dominating set of G_R (this follows directly from the construction of G_R), and therefore dominates all vertices of R. If (u, v) is not an R-dominating pair of G_R, then there is some path P' between u and v, and some vertex $r \in R$ such that r is adjacent to no vertex on P'. Now consider the path $Q = (x \cdots u) \cup P' \cup (v \cdots y)$. Since (x, y) is a dominating pair of G, r must be adjacent to Q somewhere – moreover, this adjacency *must* be in the P'-segment of Q, since u and v are the smallest and largest numbered vertices on P to which any vertex of R is adjacent. This completes the proof. □

4.1 The Algorithm

We now present the actual algorithm for the Steiner set.

Algorithm *Steiner-Set*

Input: An AT-free graph $G = (V, E)$ and a set $R \subseteq V$ of vertices.

Output: A minimum cardinality Steiner set of G with respect to R.

begin

1. Construct the graph G_R as explained previously.
2. Let $u = min_R(P)$ and $v = max_R(P)$.
3. Let $X = N(u) \cap R$ and $Y = N(v) \cap R$.
4. As in the *MCDS* algorithm, construct sets A_x and B_y.
   Use these to construct the sets A and B.
5. Construct a weighted graph $G' = (V', E')$ from G_R in the following manner:
   $wt(w) = 0$ if $w \in R$, and 1 otherwise.
6. As in the *MCDS* algorithm, compute the shortest path among the four

different kinds of paths using the All Pairs Shortest Paths algorithm.

7. Let P_m be the minimum weighted path.

8. Output $S = P_m - R$

end.

4.2 Proof of correctness

Theorem 4.1 *Algorithm Steiner-Set presented above correctly computes the minimum cardinality Steiner set of an AT-free graph G with respect to a set $R \subseteq V$ of vertices.*

Proof. **Let S be a Steiner set. If S is not already in a form that can be the output of the above algorithm, we show how it can be transformed to such a form, S'. Note that from the algorithm, (u, v) is an R-dominating pair of G_R, and that no two vertices of R are adjacent (since we have replaced each connected component of R by a single vertex). It is evident that this replacement does not affect the Steiner set – the same set is the output here as well. Since the weight of only those vertices in R is 0, and the others 1, it follows that a shortest weighted path will be one that minimises the number of vertices not in R, as we require. From the method of construction of the output set, S' in the above algorithm, it is apparent that $R \cup S'$ is *connected*. Define $T' = R \cup S'$. To show minimality, we distinguish the following cases :**

Case 1. **u and v are both in S.**
In this case, the minimum weight path between u and v in $R \cup S$ sets a lower limit on the size of the Steiner set. The above algorithm attains this limit, and so is correct.

Case 2. $u \notin S$, $v \in S$.
We show that S can be transformed to a set S' of same cardinality that the above algorithm can produce. We distinguish two sub-cases:
Sub-case 2(a). **Consider $T = R \cup S$. If in T, $r \in X$ is adjacent to some $a \in A_r$, we are done, because the above algorithm definitely considers a minimum weight A_r to v path.**
Sub-case 2(b). **$r \in X$ is adjacent to no $a \in A_r$.**
Since T is connected, r is adjacent to some $a' \notin A$. In addition, $a' \notin R$, since each r is maximally connected, and so $wt(a') = 1$. The set X consists only of vertices of R (by construction); if r is the only vertex in X, we are done since this sub-case doesn't apply at all. Therefore, there is a vertex $r' \in X$, not adjacent to r, which is connected to S in some manner, say, to vertex $s \in S$. Replace $a' \in S$ by u to obtain S' from S. Now, $T' = R \cup S'$ is connected and also dominates all the vertices of R, since (u, v) is an R-dominating pair of G_R. In addition, we have replaced one vertex of unit weight (a') by another (u);

therefore $|S'| = |S|$. In our algorithm we find a minimum weight $u \cdots v$ path, which sets a lower limit on the size of the Steiner set – a limit which we actually attain. Hence the required transformation is possible in this case too.

Case 3. $u \in S$, $v \notin S$.
This is symmetric to *Case 2* considered above, with u and v interchanged.

Case 4. Both u and $v \notin T'$.
In this case, as in cases 2 and 3, replace a by u and b by v to get a Steiner set of the same cardinality as S'. Formally, we can replace one of them and argue as in Case 2 (or 3).

Thus, in all cases, we can transform any Steiner set into one capable of being output by the above algorithm. This completes the proof. □

4.3 Complexity Analysis

We now analyse the complexity of our *Steiner-Set* algorithm.
Step 1 takes $O(n^3)$ time since we require to determine a dominating pair. Steps 2 and 3 can be done in linear time. As in the $MCDS$ algorithm, Step 4 requires $O(n^3)$ time. Step 5 is trivially linear, while Step 6 can be done in $O(n^3)$ time using the standard algorithm. Once again, the vertices on the path can be obtained in $O(m+n)$ time using a BFS, as we did in the $MCDS$. Thus, we see that the overall time complexity of the algorithm is $O(n^3)$. The algorithm can be implemented using $O(n^2)$ space, since we only require to store all graphs in the adjacency matrix form. The precise details are similar to those explained in the $MCDS$ algorithm.

Theorem 4.2 *The minimum cardinality Steiner set problem can be solved on AT-free graphs in $O(n^3)$ time, using $O(n^2)$ space.*

Proof. Follows from the discussion above. □

5 Conclusion

In this paper, we have presented $O(n^3)$ algorithms on AT-free graphs for the problems of connected domination and Steiner set. To the best of our knowledge, no algorithms have so far appeared in the literature for solving these problems on any non-trivial class of non-perfect graphs. As already noted, AT-free graphs are not necessarily perfect. These algorithms also generalize the algorithms presented in [1] for connected domination and Steiner sets on permutation graphs. In addition, this is the largest known class of graphs that strictly contains both interval and permutation graphs on which these problems have been solved. However, we have not been able to generalize the algorithms for the weighted case, which remain open. We also note that the algorithms presented here are actually applicable to a class of graphs *larger* than the class of AT-free graphs – more specifically, to the class of graphs that contain a dominating pair of vertices. This class is a proper superset of the class of AT-free graphs (e.g., C_6 belongs to this class, but is not AT-free). Our results also strengthen the conjecture in [9] that the connected domination and Steiner set problems are algorithmically closely related. However, the precise relationship between these two problems is unknown in general. We also feel that AT-free graphs have nice structural properties which can be exploited to design algorithms for several problems which are NP-complete on general graphs.

References

[1] K. Arvind and C. Pandu Rangan. Connected domination and Steiner set on weighted permutation graphs. *Info. Proc. Letts.*, 41:215–220, 1992.

[2] C.J. Colburn and L.K. Stewart. Permutation graphs: Connected domination and Steiner trees. *Discrete Math.*, 86:145–164, 1990.

[3] D.G. Corneil, S. Olariu, and L. Stewart. Asteroidal triple-free graphs. Technical Report 262/92, University of Toronto, June 1992.

[4] D.G. Corneil and L.K. Stewart. Dominating sets in perfect graphs. *Discrete Math.*, 86:179–189, 1990.

[5] A. D'Atri and M. Mascarini. Distance hereditary graphs, Steiner trees and connected domination. *SIAM J. Computing*, 17:521–538, 1988.

[6] M.R. Garey and D.S. Johnson. *Computers and Intractability: A Guide to the Theory of NP-completeness.* Freeman, San Fransisco, CA, 1979.

[7] M.C. Golumbic. *Algorithmic Graph Theory and Perfect Graphs.* Academic Press, 1980.

[8] C.G. Lekkerkerker and J.C. Boland. Representation of a finite graph by a set of intervals on a real line. *Fundamenta Mathematicae*, 51:45–64, 1962.

[9] K. White, M. Farber, and W.R. Pulleybank. Steiner trees, connected domination and strongly chordal graphs. *Networks*, 15:109–124, 1985.

The Complexity of Finding Certain Trees in Tournaments

R. Balasubramanian, Venkatesh Raman and G. Srinivasaraghavan

The Institute of Mathematical Sciences
C. I. T. Campus, Madras, India 600113.

Abstract. A tournament T_n is an orientation of the complete graph on n vertices. We continue the algorithmic study initiated by Hell and Rosenfeld[5] of recognizing various directed trees in tournaments. Hell and Rosenfeld considered orientations of paths, and showed the existence of oriented paths on n vertices finding which in T_n requires $\Theta(n \lg^{\alpha} n)$ "edge probes" where $\alpha \leq 1$ is any fixed non-negative constant. Here, we investigate the complexity of finding a vertex of prescribed outdegree (or indegree). In particular, we show, by proving upper and lower bounds, that the complexity of finding a vertex of outdegree $k(\leq (n-1)/2)$ in T_n is $\Theta(nk)$. We also establish an $\Omega(n^2)$ lower bound for finding a vertex of maximum outdegree in T_n. These bounds are in sharp contrast to the $O(n)$ bounds for selection in the case of transitive tournaments.

1 Introduction and Definitions

A tournament is an orientation of the complete graph. We denote by T_n, a tournament on n vertices $1, 2, \ldots n$. Between a pair of vertices v and w if $v \rightarrow w$, we say v beats w and we say w beats v otherwise. If this relation is transitive in the tournament, then the tournament is called *transitive*. Clearly, the vertices of a transitive tournament are linearly ordered by the relation "beats to" and there is a unique (up to isomorphism) transitive tournament on n vertices. The number of vertices beaten by a vertex v is the outdegree of v in T_n, and is also called the *score* of v. A *regular tournament* of degree k is a tournament in which the score of every vertex is k. Such a tournament has $(2k+1)$ vertices.

Hell and Rosenfeld[5] initiated the algorithmic study of finding certain oriented trees in tournaments. They considered orientations of a path on n vertices which are known as *generalized hamiltonian paths.* They showed that a directed (hamiltonian) path on n vertices in T_n can be found in $\Theta(n \log n)$ "probes". Their measure of cost is a probe involving a pair of vertices (v, w) to determine the direction in which the edge from v to w is oriented (or equivalently to determine whether v beats w). In the case of transitive tournaments, this measure corresponds to a comparison between two elements of the linear order, and finding the directed hamiltonian path corresponds to getting the linear order (i.e. sorting). Hell and Rosenfeld also showed that antidirected hamiltonian paths (paths in which the orientations of successive edges are different) can be found in T_n in $\Theta(n)$ probes. They gave a spectrum of oriented paths that require $\Theta(n \log^{\alpha} n)$ probes where α is any fixed nonnegative constant less than 1. They (see also [4]) ask whether there are oriented paths in T_n finding which requires $\Omega(n^2)$ probes, or even $\omega(n \log n)$ probes.

Here, we first observe that in a transitive tournament, any generalized hamiltonian path can be found in $\Theta(n \log n)$ probes. In fact, we establish the exact complexity (which depends on the orientations of the edges of the path) of finding such paths, the maximum value of which turns out to be $\Theta(n \log n)$. So, the question of Hell and Rosenfeld can be thought of as asking "whether transitivity helps in finding generalized hamiltonian paths in tournaments". (It didn't, for the specific oriented paths they considered.)

We, then consider a problem analogous to (and generalization of) finding the k-th smallest element in a linear order. That is, the problem of finding a vertex (if exists) with score k (for any fixed k). This corresponds to finding a spanning oriented tree in T_n in which one vertex has exactly k outgoing edges and $n-k-1$ incoming edges. We will call such an oriented tree a ***k*-star**. We investigate the problem of finding a k-star in T_n. We show, by proving lower and upper bounds, that finding a k-star in T_n requires $\Theta(n\ min(k, n-1-k))$ probes in the worst case. This is in sharp contrast to the transitive case, where finding such a vertex is equivalent to finding the k-th smallest element in a linear order and can be done in $O(n)$ time for any k. So, our result implies that transitivity makes a significant difference for finding certain trees in tournaments. Finally, we also show that finding the maximum score in a general tournament requires $\Omega(n^2)$ probes in the worst case.

The next section deals with the generalized hamiltonian paths in transitive tournaments. In the subsequent section, upper and lower bounds are proved for finding the k-star in T_n. We also give algorithms and lower bounds for finding vertices with score *k or less*. Section4 proves the $\Omega(n^2)$ lower bound for finding the maximum score in a general tournament. The final section presents some concluding remarks.

2 Finding Generalized Hamiltonian Paths in Transitive Tournaments

We follow the notations of Alspach and Rosenfeld[1]. A generalized hamiltonian path is a permutation $a_1, a_2, ...a_n$ of the vertices $1, 2, ...n$. The type of the path is characterized by the sequence $\sigma_{n-1} = e_1, e_2, ...e_{n-1}$ where $e_i = +1$ if $a_i \rightarrow a_{i+1}$ and $e_i = -1$ otherwise. We also say that σ_{n-1} is realized in T_n. In [6], it was conjectured that for $n \geq 8$, all possible σ_{n-1}'s are realizable in T_n. Forcade[2] proved this when n is a power of 2. For general n, certain types of paths are proved to be realizable ([3], [6], [7], [8]).

It is easy to see that all possible σ_{n-1}'s are realizable in a transitive tournament on n vertices. Here we concentrate on algorithms to actually find generalized hamiltonian paths in transitive tournaments. It is also easy to see (by induction on n) that once we get the linear order using $O(n \log n)$ probes by appealing to any $O(n \log n)$ sorting algorithm, there is enough information to realize any type of generalized hamiltonian path. However we observe that we can find any type of path without actually getting the complete linear order. The complexity of the algorithm depends on the type of σ_{n-1} (for example, if $e_i = (-1)^i$, such a path can be found using $O(n)$ probes).

Notice that every path can be uniquely determined by its blocks $B_1, B_2, ...B_l$, where a block is a maximal subsequence of consecutive elements having the same sign. We prove the following theorem.

Theorem 1. *Let $\sigma = e_1, e_2, ...e_{n-1} = B_1, B_2, ...B_l$ be a generalized hamiltonian path and let b_i be the length of B_i for $1 \leq i \leq l$. Then σ can be found in a transitive tournament on n vertices using $\Theta(\sum_{i=1}^{l} b_i \log b_i)$ probes.*

Proof. Let T be the transitive tournament on n vertices. We will prove a slightly stronger result: Given any vertex v in T, we will find a σ path in T in which $v \epsilon B_1$ using $\Theta(\sum_{i=1}^{l} b_i \log b_i)$ probes. The proof is by induction on l. The base case ($l = 1$) follows by sorting. For the induction step, without loss of generality, let the sign of the e_i's in B_1 be $+1$. Take v and any other $b_1 - 1$ other vertices of T and find among them, the directed hamiltonian path $B = a_1, a_2, ...a_{b_1}$ using $O(b_1 \log b_1)$ probes. Find a path characterized by $B_2, B_3, ...B_l$ in $T - \{a_1, a_2, ...a_{b_1 - 1}\}$ in which the vertex a_{b_1} is in the first block B_2, using $O(\sum_{i=2}^{l} b_i \log b_i)$ probes by the induction hypothesis. Let c_1 be the starting vertex of $B_2, B_3, ...B_l$. If $c_1 = a_{b_1}$, then concatenating B with $B_2 B_3 ...B_l$ gives the desired path. Otherwise $a_{b_1} \rightarrow c_1$. But, since $a_{b_1 - 1} \rightarrow a_{b_1}$, due to transitivity, $a_{b_1 - 1} \rightarrow c_1$. Hence concatenating the path $a_1, a_2, ...a_{b_1 - 1} c_1$ with $B_2 B_3 ...B_l$ gives the desired path. It can also be seen that the above bound is optimal since to find each block B_i in T requires $\Omega(b_i \log b_i)$ probes. □

Since $\sum_{i=1}^{l} b_i = n$, it follows that the bound on the number of probes in the above theorem is $\Theta(n \log n)$. Observe that the complexity of finding generalized paths considered by Hell and Rosenfeld is the same in transitive and general tournaments. And the question of whether there exists types of path requiring $\omega(n \log n)$ probes can be phrased as whether transitivity helps in finding generalized paths. In the next section we show that transitivity indeed helps in finding certain other trees in tournaments.

3 Finding a vertex with score k

In this section, k is any fixed integer between 1 and $n - 1$ and T_n is any tournament on n vertices.

In a transitive tournament, for every k between 0 and $n - 1$, there is a unique vertex with score k. However, in a general tournament, there could be several (or even 0) vertices with a prescribed score. We begin by proving the following structure result on T_n.

Lemma 2. *In any T_n, there are at most $2k + 1$ vertices with score less than or equal to k.*

Proof. Let m be the number of vertices in T_n with score less than or equal to k. Consider the subtournament on these m vertices. In this subtournament, let S be the sum of the scores (with respect to this subtournament) of the m vertices. Then it is clear that

$$mk \geq S = \frac{m(m-1)}{2}$$

from which it follows that $m \leq 2k + 1$. □

Note that by using exactly the same argument, one can conclude that there are at most $2k+1$ vertices with indegree at most k in T_n. Note also that the bound on m is independent of n.

In fact, it can be seen that seen than given any $2k+1$ vertices in a tournament, there exists a vertex with score (or indegree) at most k. Also for every k, and $n \geq 2k+3$ there exists a T_n that does not have a vertex with score less than or equal to k. One can simply consider a regular tournament of degree $k+1$ on $2k+3$ vertices and add the required number of extra vertices. All edges to the $2k+3$ vertices are outgoing from the extra vertices.

3.1 Upper Bounds – Algorithms

As mentioned earlier, finding a vertex with score k can be thought of as finding a k-star in T_n. We show that finding such a vertex in T_n requires $\Theta(n\ min(k, n-1-k))$ probes. For $k \leq (n-1)/2$, the same bound holds for finding vertices with score less than or equal to k. However, for $k > (n-1)/2$, a vertex with score less than or equal to k can be found faster using $\Theta((n-k-1)(n-k-2))$ probes.

Theorem 3. *In any T_n, a vertex with score (or indegree) k or its absence can be found using $O(n\ min(k, n-1-k))$ probes.*

Proof. It suffices to show that for $k \leq (n-1)/2$, a vertex with score k can be found using $O(nk)$ probes. Exactly the same strategy will work for finding a vertex with indegreee k (due to the note immediately after Lemma 2). Hence for $k > (n-1)/2$, to find a vertex with score k, one can look for a vertex with indegree $(n-1-k)$ and the bound follows.

In what follows, we assume that $k \leq (n-1)/2$ and T_n is any arbitrary tournament on n vertices. We exploit the fact that the number of vertices with score less than or equal to k is at most $2k+1$ in any subtournament, and is independent of n. So, we probe several subtournaments and eliminate vertices which are vertices with score larger than k in that subtournament. Thus we would have eliminated all but at most $2k+1$ of them at the end, and then we check the score of these vertices by probing them with every other vertex.

We describe the algorithm recursively. Initially we pick any $2k+1$ vertices and by probing every pair of them using $(2k+1)k$ probes, we determine and mark all vertices with score at most k. These are the only candidates among these vertices for the final output. In an intermediate stage of the algorithm, we would have considered some m vertices of which we would have marked the only candidates (at most $2k+1$) for vertices with score at most k. Initially $m = 2k+1$. If $m < n$, then pick a new vertex and probe it with each marked vertex, updating the scores of the marked vertices and the new vertex (its initial score was 0). Remove the mark from any vertex which gets a score of more than k. Mark the new vertex if its score is at most k. Update m to $m+1$. Clearly from Lemma 2 applied to the subtournament consisting of the marked vertices and the new vertex, the number of marked vertices remain at most $2k+1$, and the algorithm continues until m becomes n. During this process, we would have eliminated only those vertices with score more than k. Now, probe the marked vertices with every other vertex, updating the scores of the marked vertices.

Remove any marked vertex with the score more than k. All the remaining marked vertices are the vertices with score at most k. From these vertices, we can output a vertex of score k or declare that there is none.

Initially the number of probes performed is $(2k+1)k$. At each intermediate stage, at most $(2k+1)$ probes are performed, and there are $(n-2k-1)$ stages. Finally the algorithm performs at most $(2k+1)(n-2k-2)$ probes to confirm the scores of the marked vertices. So, the total number of probes performed by the algorithm is $(2k+1)(2n-3k-3)$ (which is $O(nk)$) and the theorem follows. □

Notice that we could find *all* vertices of scores at most k even with their scores in $O(n\ min(k, n-k))$ probes. It also follows from the above theorem that, finding all vertices of indegree k or more (with individual indegrees) can be found in $O(n\ min(k,(n-k)))$ probes.

However, for $k \geq (n-1)/2$, if we are interested in only *some* vertex with score k *or less*, (or equivalently some vertex with indegree i or more, for some $i \leq (n-1)/2$,) then that can be done using at most $2(n-1-k)(n-2-k)+1$ probes, as the following theorem shows.

Theorem 4. ***Let $k \geq (n-1)/2$. Then a vertex with score k or less can be found using at most $2(n-1-k)(n-2-k)+1$ probes.***

Proof. **We will find a vertex of indegree $i = (n-1-k)$ or more in $2i(i-1)+1$ probes.**

The observation following Lemma 2 assures us that any subtournament of size $2i+1$ guarantees a vertex with indegree i or more. Thus our strategy is simply to take any arbitrary $(2i-1)$ vertices and probe every pair of them using $(2i-1)(i-1)$ probes, keeping track of the indegrees. If a vertex has reached indegree i or more already, we are through. Otherwise, each of these vertices must have indegree exactly $(i-1)$. Now take a new vertex and start probing with these vertices of indegree $i-1$. If in i such probes, none of them gets indegree i, then the new vertex gets indegree i. The total number of probes used is at most $(2i-1)(i-1)+i = (2i)(i-1)+1$. □

For $k \geq (n-1)/2$, it can be verified that $2(n-1-k)(n-2-k)+1$ is less than $(2n-2k+1)(3k-n-3)$; the second term is the number of probes made by the previous algorithm. It is not surprising that for bigger k, we can find a vertex with score at most k faster, since there are several (at least 1) such vertices in any T_n.

3.2 Lower Bounds – Adversary Arguments

In this subsection, we show that the algorithms developed in the previous subsection are asymptotically optimal. We use adversary arguments to prove our lower bound results.

Theorem 5. ***For $k < (n-1)/2$, finding a vertex with score at most k in T_n requires $\Omega(nk)$ probes in the worst case.***

Proof. Since $k \leq (n-1)/2 - 1$, we have $n \geq (2k+3)$. Consider any algorithm to find a vertex of score at most k in a tournament on n vertices. Given a k, the adversary constructs a T_n with no vertex of score at most k as follows. It choses some $(2k+3)$ vertices and constructs a regular tournament of degree $(k+1)$ among them. We will call these vertices regular. The adversary directs the edges between the remaining $(n-2k-3)$ (non-regular) vertices and the regular vertices, towards the regular vertices. The edges between the non-regular vertices are arbitrary.

The adversary strategy is to force any algorithm into involving every vertex in at least $k+1$ probes. For that, it initially answers questions to the algorithm consistent with the tournament it has just constructed. Suppose the algorithm halts by involving a vertex, say x, in at most k probes and answers that there is no vertex with score at most k (if it answers otherwise, the algorithm is obviously wrong). Then there are at least $n-1-k(>k)$ vertices such that the orientations of the edges between these vertices and x are unknown to the algorithm. The adversary makes the orientation of all of these edges incoming to x, thereby declaring x as one with score at most k proving the algorithm wrong. Clearly this action of the adversary is consistent with its previous answers. Thus the adversary forces any algorithm to perform at least $n(k+1)/2$ probes in the worst case. □

Thus we see that there are oriented trees, determination of which in T_n requires more than $\omega(n \log n)$ probes (even $\Omega(n^2)$) probes. One might conclude that this is due to the fact that there are tournaments without vertices of score k for any small k. However, in the case of generalized hamiltonian paths, it is conjectured[6] that every tournament (on $n \geq 8$ vertices) realizes all types of generalized hamiltonian paths. We show that even if we know that a tournament has a vertex with score at most k, finding such a vertex requires $\Omega(nk)$ probes.

Theorem 6. *For $k < (n-1)/2$, finding a vertex with score at most k requires $\Omega(nk)$ probes, even if it is known that such a vertex exists in the tournament.*

Proof. Here the adversary constructs a regular tournament of degree $3k/4$ on $3k/2+1$ vertices. It directs the rest of the vertices towards the regular vertices. The orientations between the non-regular vertices are arbitrary. Only the regular vertices have score at most k. The adversary initially answers the probes according to this tournament. Its strategy is to force the algorithm to involve at least $k/2$ of the regular vertices in at least $n/8$ probes. It keeps track of the number of probes in which each regular vertex is involved. For the first $k/2$ vertices that are involved in $n/8$ probes, it will start reversing the direction between these vertices and some $k/4+1$ non-regular vertices which are yet to be probed with these vertices. This will make the score of these vertices more than k (their initial scores were $3k/4$). This action of adversary is consistent with its previous answers. Also, by this action, the score of a non-regular vertex may get decreased by at most $k/2$; Hence, none of the non-regular vertices can get a score less than or equal to k, since their scores were at least $3k/2+1$ initially. Furthermore, since $n-(3k/2+1)-(n/8-1) \geq k/4+1$, the algorithm has not probed for the orientations between the regular vertex (involved in $n/8$ probes) and at least $(k/4+1)$ non-regular vertices, whose answers the adversary can reverse.

Suppose the algorithm declares a vertex as having score less than or equal to k without involving at least $(k/2-1)$ other regular vertices in at least $n/8$ probes. If the declared vertex is non-regular then the algorithm is obviously wrong. If it is regular, but is involved in at least $n/8$ probes, then the adversary would have increased its score to more than k, and hence the algorithm is wrong. If it is involved in less than $n/8$ probes, then the adversary, by using the above strategy, can make the score of the declared vertex more than k, thereby proving the algorithm wrong. Thus the adversary can force any algorithm into making $\Omega(nk)$ probes before finding a vertex of score at most k. □

Clearly the above proof can be modified to prove the same $\Omega(nk)$ bound to find a vertex of score exactly k. Simply make the score of the $(k/2)$-th vertex involved in $n/8$ probes to be exactly k. This also proves that $\Omega(n(n-k))$ lower bound to find a vertex of score exactly k if $k \geq (n-1)/2$. Thus we have that the algorithm discussed in Theorem 3 is asymptotically optimal.

However, for $k \geq (n-1)/2$, we saw in Theorem 4 that we can find a vertex of score *at most* k, using at most $2(n-k-1)(n-k-2)+1$ probes. The following theorem proves that even that bound is asymptotically optimal.

Theorem 7. *For $k \geq (n-1)/2$, finding a vertex with score at most k requires at least $(n-1-k)(n-k)/2$ probes in the worst case.*

Proof. We will show that finding a vertex of indegree $i = (n-1-k)$ or more requires at least $i(i+1)/2$ probes in the worst case.

The adversary keeps track of the indegrees of each vertex and will increase the maximum indegree only when it is forced to (i.e. when the orientation between two vertices with the maximum indegree is probed). Its strategy simply is as follows: whenever the algorithm probes a pair (x, y) with indegree of x less than indegree of y, then the adversary answers that y beats x, thereby increasing the indegree of x. If the indegrees of x and y are equal, then the adversary answers arbitrarily thereby increasing the indegree of one of them.

Suppose the algorithm has found a vertex v_i of indegree i (or more). Initially the indegree of v_i was 0. So the algorithm must have probed v_i with some other vertex v_{i-1} of indegree $i-1$ before v_i got indegree i. In general, for the vertex v_i to get indegree $j+1$ from j, v_i must have been probed with a distinct vertex v_j of indegree j or more (for all j from 0 to $i-1$.).

Therefore, the number of probes made by the algorithm so far is

$$\sum_{v \epsilon T_n} indegree(v) \geq \sum_{j=0}^{i} indegree(v_j) \geq \sum_{j=0}^{i} j = i(i+1)/2$$

and the theorem follows. □

4 Finding the maximum score

Clearly, a vertex with the maximum score can be found by probing all possible edges in the tournament using $n(n-1)/2$ probes. We show that this bound is, indeed, optimal upto lower order terms.

Theorem 8. *Finding a vertex with the maximum score in a general tournament T_n requires at least $n(n-1)/2-(n-1)/2-1$ probes.*

Proof. Let $k = \lfloor (n-1)/2 \rfloor$. The adversary constructs T_n such that it has a regular subtournament on $2k+1$ vertices (called the regular vertices) and any edge not in the subtournament being directed away from the regular vertices. Its answers to the probes of any algorithm are consistent with this tournament.

We claim that unless the algorithm probes at least $n(n-1)/2-k-1$ edges, the adversary can flip the direction of some unprobed edges to prove the algorithm wrong. Suppose the algorithm has probed less than $n(n-1)/2-k-1$ edges. Consider the unprobed edges. There are at least $(k+2)$ of them.

Suppose two of them, say $a \leftarrow b$ and $c \leftarrow d$ are vertex disjoint. If the algorithm declares any vertex other than a as the answer, then the adversary can reverse the edge $a \leftarrow b$ while retaining the original directions of every other unprobed edge, thereby making a as the only vertex with the maximum score $(k+1)$. Similarly, if the algorithm declares any vertex other than c as the answer, it can make c as the only vertex with the maximum score. Thus, since a and c are distinct, the adversary can always prove the algorithm wrong.

Hence for the algorithm to be correct, all the unprobed edges must share a vertex, say x. Suppose x is not a regular vertex. If the algorithm answers x as the answer, then it is obviously wrong since the score of x in T_n is 0. If, however, the algorithm answers a vertex other than x, then the adversary reverses the directions of all the unprobed edges. This makes x the only vertex with the maximum score of at least $k+2$. Suppose x is a regular vertex. Since there are at least $k+2$ unprobed edges, and both the indegree and the outdegree of x are at most $k+1$ in T_n, there must exist edges $a \rightarrow x$ and $x \rightarrow b$. Now if the algorithm declares any vertex other than b as the one with the maximum score then the adversary just reverses the direction of the edge $x \rightarrow b$ while retaining the original directions of all the other unprobed edges, making b the only vertex of maximum score $k+1$. Instead if the algorithm declares b as the answer then reversing the direction of the edge $a \rightarrow x$ without disturbing any other, would make x the vertex of maximum score $k+1$. The adversary can thus always prove the algorithm wrong.

The lower bound now follows. □

5 Concluding Remarks

We investigated the problems of finding a vertex with a fixed/maximum score in a general tournament. The problem is interesting since, in a transitive tournament, this problem corresponds to the well studied problem of selection from a linear order. We proved upper and lower bounds to show that finding a vertex with score k requires $\Theta(n\ min(k, n-k))$ edge probes. We also showed that finding a vertex with the maximum score reqiures $\Omega(n^2)$ probes. This is in sharp contrast to the transitive case where selections can be performed in $O(n)$ time.

Hell and Rosenfeld[5] ask whether there is a generalized hamiltonian path whose recognition requires $\omega(nlogn)$ edge probes. Our result gives an example of a class of oriented trees finding which requires $\omega(n \log n)$ probes in a general tournament.

Perhaps, there is a connection between these classes of oriented trees, which can be exploited to answer their question using the lower bound result of this paper. In this context, we note that the existence of certain generalized hamiltonian paths are known [3, 1, 2], and some of those proofs have been converted into algorithms[5]. Alspach and Rosenfeld [1] proved the existence of generalized hamiltonian paths $B_1, B_2, \ldots B_k$ in which the length of $B_i \geq i$. Converting the proof into an algorithm, requires finding a vertex with score $\leq (n-1)/2$ in T_n, which requires $\Omega(n^2)$ probes in the worst case, as we have shown in this paper.

References

1. **B. Alspach and M. Rosenfeld, Realization of certain generalized paths in tournaments, *Discrete Math.*, 34 (1981), 199-202.**
2. **R. Forcade, Parity of paths and circuits in tournaments, *Discrete Math.*, 6 (1973) 115-118.**
3. **B. Grünbaum, Antidirected Hamiltonian paths in tournaments, *J. combinatorial Theory* (B) 11 (1971) 249-257.**
4. **P. Hell, Problem 4.1, Problem 4.2 in Problem Session 4, *Graphs and Orders*, (ed) I. Rival, D. Reidel Publishing Company. (1985) 543.**
5. **P. Hell and M. Rosenfeld, The Complexity of Finding Generalized Paths in Tournaments, *Journal of Algorithms*, 4 (1983) 303-309.**
6. **M. Rosenfeld, Antidirected Hamiltonian paths in tournaments, *Journal of Combinatorial Theory* (B), 12 (1972), 93-99.**
7. **M. Rosenfeld, Antidirected Hamiltonian paths in tournaments, *Journal of Combinatorial Theory* (B), 16 (1974), 234-256.**
8. **M. Rosenfeld, Generalized Hamiltonian paths in tournaments, *Annals of Discrete Mathematics*, 20 (1984), 263-269.**

Spirality of Orthogonal Representations and Optimal Drawings of Series-Parallel Graphs and 3-Planar Graphs* (Extended Abstract)

Giuseppe Di Battista† Giuseppe Liotta‡ Francesco Vargiu§

Abstract. An orthogonal drawing of a graph is a planar drawing such that all the edges are polygonal chains of horizontal and vertical segments. Finding the planar embedding of a planar graph such that its orthogonal drawing has the minimum number of bends is a fundamental open problem in graph drawing. This paper provides the first partial solution to the problem. It gives a new combinatorial characterization of orthogonal drawings based on the concept of spirality and provides a polynomial-time algorithm for series-parallel graphs and biconnected 3-planar graphs.

1 Introduction

A graph drawing algorithm receives as input a graph and produces as output a drawing that "nicely" represents such a graph; several references on the graph drawing area can be found in [5]. Most of the existing graph drawing algorithms can be roughly splitted into the following two main steps: (1) A planar embedding of the given graph is found by a *planarization algorithm*, possibly by inserting dummy vertices for crossings. The planar embedding is usually described through the cyclic ordering of the edges incident at each vertex. Planarization algorithms are implemented by using variations of the classical planarity testing algorithms. (2) Once a planar embedding has been found, a *representation algorithm* is applied to produce the final drawing. Such algorithm is selected depending on the requirements of the application and on the graphic standard. It can be targeted to minimize the global area of the drawing, to have as few bends as possible along

*Work partially supported by Progetto Finalizzato Sistemi Informatici e Calcolo Parallelo of the CNR and by Esprit BRA of the EC Under Contract 7141 Alcom II

†Dipartimento di Informatica e Sistemistica, Università di Roma "La Sapienza", via Salaria 113, I-00198 Roma, Italia.

‡Dipartimento di Informatica e Sistemistica, Università di Roma "La Sapienza". Part of this work has been done when this author was visiting the Department of Computer Science of McGill University

§Dipartimento di Informatica e Sistemistica, Università di Roma "La Sapienza" and Database Informatica Spa.

the edges, to emphasize symmetries, etc.

The representation algorithm produces a drawing within the planar embedding computed by the planarization algorithm. However, the choice of the embedding can deeply affect the results obtained by the representation algorithm. In Fig. 1 we show two different planar embeddings of the same series-parallel graph. Besides each planar embedding we show the ***orthogonal drawing*** (edges are mapped to polygonal chains of horizontal and vertical segments) with the minimum number of bends that can be constructed preserving that embedding. In the first drawing we have 13 bends, in the second we have 7 bends.

Thus, it raises naturally the problem of chosing, in the planarization algorithm, the "best" embedding from the point of view of the representation algorithm. Although the problem is quite natural there are only a few contributions on this topic, because of the exponential number of embeddings a planar graph (in general) has. Recently, it has been tackled the problem of contructing straight-line upward drawings of series-parallel digraphs (an upward drawing is such that all the edges follow monotonically the vertical axis). In [1] it is shown that fixed-embedding drawing strategies can lead to drawings with exponential area, and variable-embedding algorithms are needed to achieve optimal area.

In this paper we deal with the classical problem of constructing orthogonal drawings with the minimum number of bends along the edges. Tamassia [11] has proposed a very elegant representation algorithm that solves the problem in polynomial time for graphs with a fixed embedding. That also disproves a conjecture of Storer [10] that suspected the problem to be NP-hard. Linear time heuristics for the same problem have been proposed by Tamassia and Tollis in [13, 14]. Such heuristics guarantee at most $2n + 4$ bends for a biconnected graph with n vertices. Recently, Kant [7] has proposed efficient heuristics with better bounds for triconnected 4-planar graphs and general 3-planar graphs (a graph is k-planar if it is planar and each vertex has degree at most k). Observe that orthogonal drawings make sense only for 4-planar graphs. Tamassia, Tollis and Vitter [15] have given lower bounds for the problem and the first parallel algorithm. A brief survey on orthogonal drawings is in [12]. However, all the above papers work within a fixed embedding. The problem of finding the planar embedding that leads to the minimum number of bends is not known to be NP-hard or not and has been explicitely mentioned as open by several authors.

We provide a first partial solution to the problem. First, we introduce and study the new concept of ***spirality***, that is a measure of how an orthogonal drawing is "rolled up". Second, we give a combinatorial characterization relating the number of bends of an orthogonal drawing and its spirality. Third, we show that the problem of finding the planar embedding that leads to the orthogonal drawing with the minimum number of bends can be solved in polynomial time for series-parallel graphs. Our technique exploits both the properties of the spirality and a variation of the $SPQR$ trees [3, 4]: a data structure that implicitely represents all the planar embeddings of a planar graph. Observe that series-parallel graphs arise in a variety of problems such as scheduling, electrical networks, data-flow analysis, database logic programs, and circuit layout. Also, they play a very special role in planarity problems [3, 4]. Finally, we give a polynomial time

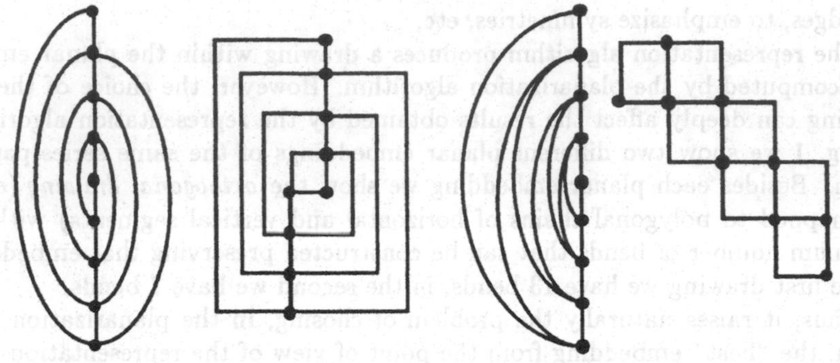

Figure 1: Two embeddings of a series-parallel graph that give raise to two optimal orthogonal drawings with a different number of bends.

algorithms for biconnected 3-planar graphs. Because of space limitations, in this extended abstract we omit proofs. They can be found in the full paper.

2 Preliminaries

We assume familiarity with connectivity and planarity [6, 9].

Two planar drawings Γ' and Γ'' of a planar graph are *equivalent* when, for each vertex v: (1) Γ' and Γ'' have the same circular clockwise ordering of the edges incident on v and of the faces around v; (2) Γ' and Γ'' have the same external face. An *embedding* of a planar graph G corresponds to an equivalence class of planar drawings of G. A planar graph G with a given embedding Ψ is an *embedded graph.* A circular list of edges and vertices of G that corresponds to a face of any drawing of Ψ is a *face* of G. The face of G that corresponds to the external face of any drawing of Ψ is the *external face* of G. The adjacency list of each vertex v of an embedded graph is a clockwise ordered sequence of the alternate faces and edges around v.

A planar drawing of a planar graph G such that all edges of G are mapped to polygonal chains of horizontal and vertical segments is an *orthogonal drawing* of G. A planar graph has an orthogonal drawing iff it is 4-planar [10, 11].

Two equivalent orthogonal drawings of an embedded graph are *shape equivalent* when: (1) for each vertex v, consecutive edges in the adjacency list of v form the same angle in the two drawings; (2) for each edge (u, v), following from u to v the polygonal chain representing (u, v), we have the same (possibly empty) sequence of left and right turns in the two drawings.

A *labelled embedded graph* H is an embedded graph where: (1) Each edge e is associated to a (possibly empty) sequence of L- and R-symbols. When walking along $e = (u, v)$, such a sequence is read in two different ways depending on the direction. When walking from u to v it is read as it is; when walking from v to u it is read by reversing the sequence and by exchanging L- and R-symbols. (2) Each face f in the adjacency list of vertex v has (possibly) a label in $\{R, L, LL\}$. An *orthogonal graph* H is a labelled embedded graph that describes a class of shape equivalent orthogonal drawings; it is defined as follows. Each edge (u,v) of H is associated to a (possibly empty) sequence of L- and R-symbols, each specifying a left or right turn that is found following (u,v) from u to v. For each vertex v of H, the label associated to each face f in the adjacency list of v specifies what happens when we go through v, walking clockwise around f (i.e. having f at one's right) in any drawing of the shape equivalent class. Namely: (1) f is labelled L if we turn left; (2) f is labelled R if we turn right; (3) f is labelled LL when v has degree one and we have to come back ("two turns left are enough to come back"); (4) f is not labelled when we go straight.

We denote by $deg(v)$ the degree of vertex v. The following property is proved by means of elementary geometric considerations (see also [11]).

Property 1 *A labelled embedded graph H is an orthogonal graph iff the following two conditions hold. (1) For each vertex v of H, let $|R_v|$ ($|L_v|$) be the number of R-symbols (L-symbols) of the labels associated to the faces that appear in the adjacency list of v. We have $|R_v| - |L_v| = 2deg(v) - 4$. (2) For each face f of H suppose to walk clockwise around f. Let $|R_e|$ ($|L_e|$) be the total number of R-symbols (L-symbols) that are encountered by traversing the edges of f. Let $|R_f|$ ($|L_f|$) be the number of R-symbols of the labels associated to f in the adjacency lists of vertices belonging to f. We have $|R_e| + |R_f| - |L_e| - |L_f| = \pm 4$; where we take the plus sign when f is an internal face and the minus sign when f is the external face.*

Property 2 *Let C be a simple cycle of an orthogonal graph. The number of right turns minus the number of left turns encountered by walking clockwise around C is four.*

An *orthogonal representation* of a graph G is an orthogonal graph with the same vertices and edges of G. An *optimal orthogonal representation* of G is an orthogonal representation of G with the minimum number of bends. An *optimal drawing* of G is an orthogonal drawing of the shape equivalence class described by an optimal orthogonal representation of G.

Let G be an acyclic digraph with one source and one sink. A *split pair* of G is either a separation pair of G or a pair of adjacent vertices. A *split component* of a split pair $\{u,w\}$ is either an edge (u,w) or a maximal subgraph G_{uw} of G such that $\{u,w\}$ is not a split pair of G_{uw}. Vertices u and w are the *poles* of the split component. In the following we often denote G_{uw} a split component whose poles are u and w. An acyclic digraph is an *st-digraph* if it has one source s, one sink t and contains edge (s,t). An *st-orientation* of an undirected graph G is an orientation of the edges of G such that the resulting directed graph is an st-digraph. Each 2-connected graph can be oriented into an st-digraph [6]. We define the planarity and connectivity properties of st-digraphs as the planarity and connectivity properties of their underlying undirected graphs. In the following we consider only orthogonal st-digraphs and orthogonal representations of st-digraphs such that the edge (s,t) is on the external face.

Let G be an embedded st-digraph and $G_{uw} \subset G$ be a split component of G. The edges of G that are incident on a pole v ($v = u, w$) of G_{uw} and that belong (do not belong) to G_{uw} are called *internal edges* (*external edges*) of v with respect to G_{uw}; the number of such edges is called *internal degree* (*external degree*) of v with respect to G_{uw}.

Let f' and f'' be the two faces of G "surrounding" G_{uw} (sharing both u and w). Let P_{uw} be the directed path of f' from u to w entirely composed by edges and vertices of G_{uw}. Suppose that, when going clockwise around f' (i.e. having f' at one's right), P_{uw} is traversed from u to w; face f' is called ***right face*** of G_{uw}. Face f'' is called ***left face*** of G_{uw}. Poles s and t divide the external face f_e of G into two simple paths. The ***right path*** is the one traversed from s to t when going clockwise around f_e; the ***left path*** is the other one.

A ***series-parallel graph*** is recursively defined as follows. A simple cycle with three edges is a series-parallel graph. The graph obtained by splitting an edge of a series-parallel graph into two edges is a series-parallel graph. The graph obtained by inserting an edge between a split pair $\{u, w\}$ of a series-parallel graph such that u and w are not adjacent is a series-parallel graph. A series-parallel graph is also called ***SP-graph***.

3 Souls and spirality of orthogonal graphs

Let H be an orthogonal *st*-digraph and $H_{uw} \subset H$ be a split component of H. A pole v ($v = u, w$) of H_{uw} is called ***bridge pole*** when its internal degree with respect to H_{uw} is one; v is called ***nonbridge pole*** when its internal degree with respect to H_{uw} is greater than one.

In order to define the spirality of H_{uw} we add to H some dummy vertices called ***alias vertices*** of the poles u, w. Several cases are possible. (1) Let v be a nonbridge pole of H_{uw} with external degree one; the ***alias vertex*** v' of v is defined as follows. Let $e_{out} \equiv (x, v)$ be the external edge of v with respect to H_{uw}; e_{out} is substituted by two new edges (x, v'), (v', v) and a new vertex v'. The labelling of (x, v') is the same as the one of e_{out}, (v', v) has no labels, and the two faces around v' have no labels in the adjacency list of v'. Intuitively, we have cut the first segment of the drawing of e_{out} into two pieces. (2) Let v be a nonbridge pole of H_{uw} with external degree two, let e'_{out} and e''_{out} be the external edges of v; the ***alias vertices*** of v are a pair $v'v''$ of dummy vertices obtained by cutting e'_{out} and e''_{out}, respectively. They are defined as above for v'. (3) Let v be a bridge pole of H_{uw}; the ***alias vertex*** v' of v coincides with v.

Definition 1 *Let H be an orthogonal st-digraph and let $H_{uw} \subset H$ be a split component of H. Let u' (w') be an alias vertex of u (w). Let P_{uw} be any undirected (in the following, paths will be, unless otherwise specified, undirected) simple path in H_{uw} from u to w. A* soul $S_{u'w'}$ *of H_{uw} is the simple path obtained by concatenating edge (u', u), path P_{uw}, and edge (w, w').*

Let v_1 and v_2 be vertices of H and $P_{v_1v_2}$ be a simple path between v_1 and v_2. We associate to $P_{v_1v_2}$ an integer $n(P_{v_1v_2})$ defined as the number of right turns minus the number of left turns found when going along $P_{v_1v_2}$ from v_1 to v_2.

Lemma 1 *Let H be an orthogonal st-digraph and let $H_{uw} \subset H$ be a split component of H. Let u' be an alias vertex of u and let w' be an alias vertex of w. Let $S'_{u'w'}$ and $S''_{u'w'}$ be two souls of H_{uw}; we have $n(S'_{u'w'}) = n(S''_{u'w'})$.*

Definition 2 *Let H be an orthogonal st-digraph and $H_{uw} \subset H$ be a split component of H. The spirality $\sigma_{H_{uw}}$ of H_{uw} is defined as follows; three cases are possible, depending on the number of alias vertices of u and w. (1) Both u and w have just one alias vertex, say u' and w', respectively. Let $S_{u'w'}$ be a soul of H_{uw}; $\sigma_{H_{uw}} = n(S_{u'w'})$. (2) Pole u has just one alias vertex, say u'; w has two alias vertices, say w' and w''. Let $S_{u'w'}$ and $S_{u'w''}$ be two souls of H_{uw}; $\sigma_{H_{uw}} = (n(S_{u'w'})+n(S_{u'w''}))/2$. (3) Pole u has two alias vertices, say u' and u''; w has two alias vertices, say w' and w''. Suppose u' and w' are on the same face of H. Let $S_{u'w'}$ and $S_{u''w''}$ be two souls of H_{uw}; $\sigma_{H_{uw}} = (n(S_{u'w'})+n(S_{u''w''}))/2$.*

Property 3 *If the poles of H_{uw} satisfy condition of Case 2 of Definition 2, then $2\sigma_{H_{uw}}$ is an odd integer number; if the poles of H_{uw} satisfy either condition of Case 1 or condition of Case 3 of Definition 2, then $\sigma_{H_{uw}}$ is an integer number.*

Property 4 *Let H and H' be two orthogonal st-digraphs that are orthogonal representations of the same st-digraph G; let G_{uw} be a split component of G. Suppose that: (1) the orthogonal representation H_{uw} of G_{uw} is the same in H and in H'; (2) the label associated to the right (left) face of H_{uw} in the adjacency list of each pole of H_{uw} that is not a bridge pole is the same in H and in H'. Then $\sigma_{H_{uw}}$ has the same value in H and in H'.*

Let G and G' be two embedded *st*-digraphs describing two different embeddings of the same *st*-digraph. If they are different only in the embedding of one split component they are ***split different***. Let $G_{uw} \subset G$ and $G'_{uw} \subset G'$ be two different embedded subgraphs of the same split component. Let H and H' be two orthogonal *st*-digraphs that are orthogonal representations of G and G', respectively, and let $H_{uw} \subset H$ and $H'_{uw} \subset H'$ be the orthogonal *st*-digraphs that are orthogonal representations of G_{uw} and G'_{uw}. Suppose that the labels that (possibly) appear between the external edges of u and w stay the same in H and in H'. Orthogonal *st*-digraphs H and H' are said to be ***split different***.

Let H and H' be two split different orthogonal *st*-digraphs that are orthogonal representations of the same *st*-digraph G. Suppose they are split different in a split component with poles u and w, such that $deg(u) \geq 3$ and $deg(w) \geq 3$. Let H_{uw} and H'_{uw} be the orthogonal representations of such split component in H and in H', respectively. Suppose $\sigma_{H_{uw}} = \sigma_{H'_{uw}}$. We define operation ***substitution*** in H of subgraph H_{uw} with H'_{uw} as follows: (1) It gives as result a labelled embedded *st*-digraph H''; (2) H'' has the same vertices and edges of G; (3) Each vertex of H'' that is not in the above split component has the same adjacency list it has in H; (4) If u (w) is a bridge pole, then it has in H'' the adjacency list it has in H; (5) If u (w) is a nonbridge pole, then it has in H'' the adjacency list it has in H'; (6) Any other vertex of H'' has the same adjacency list it has in H'.

Theorem 1 *Let H and H' be two orthogonal st-digraphs that are split different in a split component with poles u and w, such that $deg(u) \geq 3$ and $deg(w) \geq 3$. Let H_{uw} and H'_{uw} be such a split component in H and in H', respectively. Suppose $\sigma_{H_{uw}} = \sigma_{H'_{uw}}$. Let H'' be a labelled embedded st-digraph obtained by substituting in H the subgraph H_{uw} with H'_{uw}. H'' is an orthogonal st-digraph.*

Let H' and H'' be two orthogonal *st*-digraphs, such that H' and H'' are split different in the same split component. Let H'_{uw} and H''_{uw} be such split components in H' and in H'', respectively. The orthogonal graph H'_{uw} is *optimal within the spirality* $\sigma_{H'_{uw}}$ if it doesn't exist any H''_{uw} such that $\sigma_{H'_{uw}} = \sigma_{H''_{uw}}$ and such that H'_{uw} has more bends than H''_{uw}.

Theorem 2 *Let G be an st-digraph and let H be an optimal orthogonal representation of G; each split component of H is optimal within its spirality.*

Let G be an *st*-digraph and let $G_{uw} \subset G$ be a split component of G. G_{uw} is *odd* if one of its poles is a bridge pole or it has external degree one, while the other pole is a nonbridge pole with external degree two (the total number of alias vertices of u and w with respect to G_{uw} is odd); else G_{μ} is *even* (the total number of alias vertices of u and w with respect to G_{uw} is even).

Let H be an orthogonal representation of an *st*-digraph G, and let $H_{uw} \subset H$ be an orthogonal representation of a split component G_{uw} of G. Suppose G_{uw} is odd; H_{uw} satisfies Case 2 of Definition 2 and, according to Property 3, its spirality is a fractional number. Otherwise, suppose G_{uw} even; in this case H_{uw} satisfies either Case 1 or Case 3 of Definition 2 and, according to Property 3, its spirality is an integer number.

Property 5 *Let k be any integer number. If G_{uw} is odd, it always exists an orthogonal representation H of G such that the orthogonal representation $H_{uw} \subset H$ of G_{uw} has spirality $(2k+1)/2$. If G_{uw} is even, it always exists an orthogonal representation H of G such that the orthogonal representation $H_{uw} \subset H$ of G_{uw} has spirality k.*

Theorem 3 *Let G be an st-digraph with n vertices and let H be an optimal orthogonal representation of G; let $H_{uw} \subset H$ be the orthogonal representation of a split component of G; $|\sigma_{H_{\mu}}| \leq 3n - 2$.*

Let G be an *st*-digraph with n vertices and let $G_{uw} \subset G$ be a split component of G. Let H be an optimal orthogonal representation of G, and let $H_{uw} \subset H$ be the orthogonal representation of G_{uw}.

Corollary 1 *If G_{uw} is odd, the spirality of H_{uw} is a number $(2k+1)/2$ such that $-3n+2 \leq k \leq 3n-3$. If G_{uw} is even, the spirality of H_{uw} is an integer k such that $-3n+2 \leq k \leq 3n-2$.*

4 The SPQ^*R trees

In this section we propose a variation of the SPQR trees (presented in [3, 4]).

Let G be an acyclic embedded digraph with exactly one source s and exactly one sink t; a *SPQ^*R tree* T of G is a rooted ordered tree describing a recursive decomposition of G with respect to its split pairs, and it will be used to synthetically represent all the embeddings of G with s and t on the external face. Nodes of T are of four types: S, P, Q^* and R. Each node μ has an associated acyclic digraph called *skeleton* of μ and denoted by *skeleton*(μ). Non-root nodes of T are called *internal nodes*. Tree T is recursively defined as follows:

Chain case: If G consists of a simple path from s to t then T is a single Q^*-node μ whose skeleton is G itself.

Series case: If G is 1-connected, let $c_1, \ldots, c_{k-1}$ ($k \geq 2$) be the cutvertices of G such that no cutvertex has degree less than three; $c_1, \ldots, c_{k-1}$ partition G into its blocks $G_1, \ldots, G_k$ in this order from s to t. The root of T is an S-node μ. Graph *skeleton*(μ) is the chain $e_1, \ldots e_k$, where edge e_i goes from c_{i-1} to c_i, $c_0 = s$ and $c_k = t$.

Parallel case: If s and t are a split pair for G with split components $G_1, \ldots, G_k$ ($k \geq 2$), encountered in this order when going around s in clockwise order, the root of T is a P-node μ. Graph *skeleton*(μ) consists of k parallel edges from s to t, denoted $e_1, \ldots, e_k$.

Rigid case: If none of the above cases applies, let $\{s_1, t_1\}, \ldots, \{s_k, t_k\}$ be the maximal split pairs of G ($k \leq 1$), and for $i = 1..k$, let G_i be the union of all the split components of $\{s_i, t_i\}$. The root of T is an R-node μ. Graph *skeleton*(μ) is obtained from G by replacing each subgraph G_i with edge e_i from s_i to t_i.

We call *pertinent graph* of μ the graph whose decomposition tree is the subtree rooted at μ.

Let G be an *st*-digraph, it is always possible to define an SPQ^*R tree T of G such that: (1) The root of T is a P-node with two children; one of them is the edge (s,t); (2) Each internal P-node has children R, S, or Q^*-nodes; (3) Each S-node has two children; (4) Each Q^*-node has children edges of G.

We call T *canonical decomposition tree* of G. Since in the following we refer only to canonical decomposition trees, we briefly call them *decomposition trees*. We will denote H_μ the orthogonal representation of the pertinent graph of μ.

Let G be an *st*-digraph with n vertices, let T be a decomposition tree of G, and let μ be an internal node of T. Let $G_\mu \subset G$ be the pertinent graph of μ with poles u and w.

An orthogonal representation H_μ of G_μ is *feasible* if there exists an orthogonal representation of G having H_μ as a subgraph.

Let H_μ be a feasible orthogonal representation of G_μ; let $\mathcal{H}$ be the infinite class of orthogonal representations of G such that: (1) each element of $\mathcal{H}$ has H_μ as a subgraph; (2) the labels associated to the right (left) face of H_μ in the adjacency lists of poles u and w are the same for each element of $\mathcal{H}$. $\mathcal{H}$ is called *fitting class* of H_μ.

Let $\{\lambda_u, \rho_u, \lambda_w, \rho_w\}$ be a set of labels in $\{R, L\}$. The set $\{\lambda_u, \rho_u, \lambda_w, \rho_w\}$ is *sound* if there exists a fitting class $\mathcal{H}$ of H_μ such that the labels associated to the left (right) face of H_μ in the adjacency lists of poles u and w are λ_u (ρ_u) and λ_w (ρ_w) for each element of $\mathcal{H}$. Note that a feasible orthogonal representation H_μ and a set of labels that are sound for H_μ identify a fitting class of H_μ. Also, by Property 4, H_μ and a set of labels that are sound for H_μ are enough to evaluate the spirality of H_μ.

Let $< H_\mu, \lambda_u, \rho_u, \lambda_w, \rho_w, \sigma_{H_\mu}, c_{H_\mu} >$ be a 7-tuple such that: (1) H_μ is a feasible orthogonal representation of G_μ; (2) $\{\lambda_u, \rho_u, \lambda_w, \rho_w\}$ is a set of labels that is sound for H_μ; (3) σ_{H_μ} is the spirality of H_μ, computed according to the labels listed above; (4) c_{H_μ} is the number of bends of H_μ; it is called *cost*.

A 7-tuple $< H_\mu, \lambda_u, \rho_u, \lambda_w, \rho_w, \sigma_{H\mu}, c_{H_\mu} >$ where H_μ is optimal within the spirality $\sigma_{H\mu}$ is an *optimal 7-tuple* of μ for spirality $\sigma_{H\mu}$. An *optimal set* of

μ is a set of optimal 7-tuples of μ such that: (1) For each optimal 7-tuple, $-3n+2 \leq \sigma_{H_\mu} \leq 3n-2$; (2) No two optimal 7-tuples have the same σ_{H_μ}; (3) There is an optimal 7-tuple for each value of spirality that any orthogonal representation of G_μ can have in any optimal orthogonal representation of G.

Note that, by Property 3, the cardinality of an optimal set is at most $2(3n-2)+1$ if G_μ is even, at most $2(3n-2)$ if G_μ is odd.

Property 6 ***There exists an optimal orthogonal representation H of G such that the split components of H are in the 7-tuples of optimal sets of nodes of T.***

The *cost function* of G_μ associates each value of spirality σ of the optimal set to the cost of the orthogonal representation of the 7-tuple that has spirality σ. Note that the cost function is an invariant of a split component.

5 Series-parallel graphs

Let H be an orthogonal st-digraph, let T be a decomposition tree of H, and let μ be an internal S-node of T, with children μ_1 and μ_2; let $H_\mu \subset H$ be the pertinent orthogonal st-digraph of μ and let $H_{\mu_i} \subset H_\mu$ be the pertinent orthogonal st-digraph of μ_i $(i=1,2)$.

Lemma 2

$$\sigma_{H_\mu} = \sigma_{H_{\mu_1}} + \sigma_{H_{\mu_2}}.$$

Let μ be an internal P-node of T with three children μ_1, μ_2, μ_3. Let e_i be the edge of H_{μ_i} incident on pole u, and let e_{out} be the external edge of u with respect to H_μ. In the following lemma we suppose that edges e_{out}, e_3, e_2, e_1 appear in this order in the adjacency list of u.

Lemma 3

$$\sigma_{H_\mu} = \sigma_{H_{\mu_1}} + 2 = \sigma_{H_{\mu_2}} = \sigma_{H_{\mu_3}} - 2.$$

Let μ be an internal P-node of T with two children μ_1 and μ_2. Let f_1 and f_2 be the right face and the left face of H_μ, respectively. Suppose that the vertices and edges of H_μ belonging to f_1 are also vertices and edges of H_{μ_1}. We denote with $|RL^i_v|$ the number of R-symbols minus the number of L-symbols of the label associated to face f_i $(i=1,2)$ in the adjacency list of pole v $(v=u,w)$. We denote also with $extdeg_\mu(v)$ the external degree of v in the pertinent graph of μ and with $intdeg_\mu(v)$ the internal degree of v in the pertinent graph of μ.

Lemma 4

$$\sigma_{H_\mu} = \sigma_{H_{\mu_i}} + k_u(|RL^i_u|) + k_w(|RL^i_w|)$$

where

$$k_v(v=u,w) = \begin{cases} 1 & \textit{if } extdeg_\mu(v)=1 \textit{ and } intdeg_{\mu_i}(v)=1 \\ 1/2 & \textit{if } (extdeg_\mu(v)=2 \textit{ or } extdeg_{\mu_i}(v)=2) \textit{ and } i=1 \\ -1/2 & \textit{if } (extdeg_\mu(v)=2 \textit{ or } extdeg_{\mu_i}(v)=2) \textit{ and } i=2 \end{cases}$$

Let μ be the root of T; let μ_1 and μ_2 be the children of μ such that the pertinent graph of μ_2 is edge (s,t). Let $H_1 \subset H$ be the pertinent orthogonal st-digraph of μ_1 and let $H_{st} \subset H$ be the pertinent orthogonal st-digraph of μ_2.

Let f be the internal face of H containing (s,t) and let $|RL_v|$ be the number of R-symbols minus the number of L-symbols that are associated to f in the adjacency list of pole v $(v = s,t)$.

Lemma 5 *The spirality σ_{H_1} of H_{μ_1} and the spirality $\sigma_{H_{st}}$ of H_{st} are related by $\sigma_{H_1} - \sigma_{H_{st}} + k_s|RL_s| + k_t|RL_t| = 4$ if, going clockwise around f, edge (s,t) is traversed from t to s; $-\sigma_{H_1} + \sigma_{H_{st}} + k_s|RL_s| + k_t|RL_t| = 4$ if, going clockwise around f, edge (s,t) is traversed from s to t;*

$$\text{where } k_s\ (k_t) = \begin{cases} 1 & \text{if } s\ (t) \text{ is a bridge pole in } H_1 \\ 0 & \text{otherwise} \end{cases}$$

Based on the above characterization we can state the following lemma.

Lemma 6 *There exists an algorithm that computes an optimal set of a P-node or of an S-node or of a Q^*-node μ of the decomposition tree of a series-parallel graph in $O(n)$ time.*

In the following we call *procedure P-ORSet, procedure S-ORSet, procedure Q^*-ORSet* the algorithms of the previous lemma.

6 Extension to 3-planar graphs

The above results on how to compute an optimal set for each node of the decomposition tree of a series-paralle graph can be extended to 3-planar graphs. For such graphs the problem to solve is how to compute an optimal set of a R-node. This is done by using the following characterization:

Lemma 7 *The cost function of a split component of a 3-planar graph is nondecreasing, piecewise linear, and convex.*

We sligthly modify the algorthm by Tamassia [11] in order to compute an orthogonal representation optimal within a given value of spirality for a rigid component of an st-digraph. Namely, we transform the problem into a min-cost flow problem on a network such that the costs of the arcs are nondecreasing convex piecewise linear functions. This allows the use of a polynomial time algorithm to solve the problem [8].

Lemma 8 *There exists an algorithm that computes an optimal set of an R-node μ of the decomposition tree of a 3-planar graph in $O(n^3 \log n)$ time.*

In the following we call *procedure R-ORSet* the algorithm of the previous lemma.

7 Optimal orthogonal representations of SP-graphs and 3-planar graphs

Let G be an SP-graph or a 3-planar graph with n vertices, and let T be a decomposition tree of G. The following procedure receives as input the root of T and computes the optimal sets of all the nodes of T, by performing a preorder visit of T and applying the procedures described above.

procedure Fill-tree(μ)

if μ **is not a** Q^***-node, then for each children** μ_i **of** μ**, apply Procedure Fill-tree(**μ_i**)**
if μ **is not the root of** T
case (μ) **of**
Q^*-node: apply Procedure Q^*-ORSet(μ)
P-node: apply Procedure P-ORSet(μ)
S-node: apply Procedure S-ORSet(μ)
R-node: apply Procedure R-ORSet(μ)
end procedure.

Lemma 9 *Procedure Fill-tree computes an optimal set for all the internal nodes of T in* $O((|P| + |Q| + |S|n + |R|n^2 \log n)n)$ *time, where* $|P|$ *is the number of internal P-nodes of T,* $|S|$ *is the number of S-nodes of T,* $|R|$ *is the number of R-nodes of T, and* $|Q|$ *is the number of* Q^**-nodes of T.*

For each edge (s,t) of G we compute an st-orientation of G and an optimal orthogonal representation of G such that (s,t) is on the external face. An optimal orthogonal representation of G is the orthogonal representation with the minimum number of bends among all the computed ones.

We give first a procedure that receives as input a decomposition tree T of an st-digraph G and produces as output an optimal orthogonal representation of G. In the procedure we exploit the result of Lemma 5, considering all the admissible values of parameters $|RL_s|$, $|RL_t|$ in any orthogonal representation of G.

procedure H-from-T
Let μ be the root of T, let μ_1 and μ_2 be the children of μ; suppose the pertinent graph of μ_2 is edge (s,t)
Step 1: Apply Procedure Fill-tree(μ)
Step 2: By considering all possible assignments to $|RL_s|$, $|RL_t|$, find a 7-tuple t_1 in the optimal set of μ_1 and a 7-tuple t_2 in the optimal set of μ_2 such that: (1) the spirality in t_1 and the spirality in t_2 satisfy Lemma 5; (2) the sum of the costs is the lowest among all possible choices of the two 7-tuples
Step 3: Let $t_1 =< H_{\mu_1}, \lambda_{s_1}, \rho_{s_1}, \lambda_{t_1}, \rho_{t_1}, \sigma_{H_{\mu_1}}, c_{H_{\mu_1}} >$ and let $t_2 =< H_{\mu_2}, \lambda_{s_2}, \rho_{s_2}, \lambda_{t_2}, \rho_{t_2}, \sigma_{H_{\mu_2}}, c_{H_{\mu_2}} >$. Define H as follows. Associate to each vertex of H but s and t the same adjacency list it has in H_{μ_1} Associate to edge (s,t) the same sequence of L- and R-symbols it has in H_{μ_2}. Let e_1 and e_2, be the (possibly coincident) edges of H_{μ_1} incident on s such that: (1) they are on the external face f_{e_1} of H_{μ_1}; (2) when going clockwise around f_{e_1}, e_2, s, and e_1 appear consecutively. Associate to s in H the same adjacency list it has in H_{μ_1} and insert edge (s,t) between e_1 and e_2. Analogously for vertex t. Let f_e be the external face of H and let f be the internal face of H that contains (s,t). Associate labels to f_e and to f in the adjacency list of pole v $(v = s,t)$ according to the value of $|RL_v|$
end procedure.

Lemma 10 *Let G be an st-digraph with n vertices and let T be a decomposition tree of G. Procedure H-from-T gives as output an optimal orthogonal representation of G in $O((|P|+|Q|+|S|n+|R|n^2\log n)n)$ time, where $|P|$ is the number of internal P-nodes of T, $|S|$ is the number of S-nodes of T, $|R|$ is the number of R-nodes of T, and $|Q|$ is the number of Q^*-nodes of T.*

Theorem 4 *There exists an algorithm computes an optimal drawing of an SP-graph with n vertices and m edges in $O(n^3m)$ time; it computes an optimal drawing of a 3-planar graph with n vertices and m edges in $O(n^4m\log n)$ time.*

References

[1] P. Bertolazzi, R.F. Cohen, G. Di Battista, R. Tamassia, and I.G. Tollis, "How to Draw a Series-Parallel Digraph," Proc. 3rd Scandinavian Workshop on Algorithm Theory, 1992.

[2] G. Di Battista and R. Tamassia "Algorithms for Plane Representations of Acyclic Digraphs," Theoretical Computer Science, vol. 61, pp. 175-198, 1988.

[3] G. Di Battista and R. Tamassia "Incremental Planarity Testing," Proc. 30th IEEE Symp. on Foundations of Computer Sciene, pp. 436-441, 1989.

[4] G. Di Battista and R. Tamassia "On Line Planarity Testing," Technical Report CS-89-31, Dept. of Computer Science, Brown Univ. 1989.

[5] P. Eades and R. Tamassia, "Algorithms for Automatic Graph Drawing: An Annotated Bibliography," Technical Report CS-89-09, Dept. of Computer Science, Brown Univ. 1989.

[6] S. Even "Graph Algoritms," Computer Science Press, Potomac, MD, 1979.

[7] G. Kant "A New Method for Planar Graph Drawings on a Grid," Proc. IEEE Symp. on Foundations of Computer Science, 1992.

[8] E. L. Lawler "Combinatorial Optimization: Networks and Matroids," Holt, Rinehart and Winston, New York, Chapt. 4, 1976.

[9] T. Nishizeki and N. Chiba, "Planar Graphs: Theory and Algorithms," Annals of Discrete Mathematics 32, North-Holland, 1988.

[10] J. A. Storer "On Minimal Node-Cost Planar Embeddings," Networks, vol. 14, pp. 181-212, 1984.

[11] R. Tamassia "On Embedding a Graph in the Grid with the Minimum Number of Bends," SIAM J. Computing, vol. 16, no. 3, pp. 421-444, 1987.

[12] R. Tamassia, "Planar Orthogonal Drawings of Graphs," Proc. IEEE Int. Symp. on Circuits and Systems, 1990.

[13] R. Tamassia and I.G. Tollis "Efficient Embedding of Planar Graphs in Linear Time," Proc. IEEE Int. Symp. on Circuits and Systems, Philadelphia, pp. 495-498, 1987.

[14] R. Tamassia and I.G. Tollis "Planar Grid Embedding in Linear Time," IEEE Trans. on Circuits and Systems, vol. CAS-36, no. 9, pp. 1230-1234, 1989.

[15] R. Tamassia, I. G. Tollis, and J. S. Vitter "Lower Bounds and Parallel Algorithms for Planar Orthogonal Grid Drawings," Proc. IEEE Symp. on Parallel and Distributed Processing, 1991.

Separating the Power of EREW and CREW PRAMs with Small Communication Width*

Paul Beame[1] Faith E. Fich[2] Rakesh K. Sinha[1]

[1] Computer Science and Engineering, University of Washington, Seattle, WA 98195, USA

[2] Computer Science, University of Toronto, Toronto, Ontario, Canada M5S 1A4

Abstract. We prove that evaluating a Boolean decision tree of height h requires $\Omega(\frac{h}{\log^* h})$ time on any EREW PRAM with communication width one. Since this function can be easily solved in time $O(\sqrt{h})$ on a CREW PRAM with communication width one, this gives a separation between the two models. Our result can be extended to show a separation whenever the EREW PRAM has communication width $m \in o(\sqrt{h})$.

1 Introduction

Parallel random access machines (PRAMs) have been the model of choice for describing parallel algorithms and analyzing the parallel complexity of problems. Depending on whether or not we allow more than one processor to concurrently read from or write to a memory cell, we obtain different models of PRAMs and complexity classes associated with them. The three most popular models are the CRCW (concurrent read and write), CREW (concurrent read, but exclusive write), and EREW (exclusive read and write) PRAMs.

A basic issue in parallel complexity theory is to understand the relative power of different variants of PRAMs. Cook, Dwork, and Reischuk [CDR86], by an elegant argument, showed a separation between the powers of CRCW and CREW PRAMs. They proved that the OR of n bits, which can be easily computed in constant time on a CRCW PRAM, requires $\Omega(\log n)$ time on any CREW PRAM. This result was improved by Kutylowski [Kut91] and Dietzfelbinger et. al [DKR90] who determined the exact complexity of OR.

*Beame and Sinha's research supported by NSF/DARPA under grant CCR-8907960 and NSF under grant CCR-8858799. Fich's research supported by the Natural Science and Engineering Research Council of Canada and the Information Technology Research Centre of Ontario.

Snir [Sni85] proved that the problem of searching a sorted list is more difficult on the EREW PRAM than on the CREW PRAM. Gafni, Naor, and Ragde [GNR89] extended this result to a problem defined on a full domain. Because they use Ramsey theory, both the lower bounds rely crucially on the problems having an extremely large domain relative to the number of inputs (at least doubly exponential in the number of inputs). Essentially, they show that there is a large subset of the domain for which the state of the computation at each point depends only on the relative ordering of the input values.

Unfortunately, a lower bound proof that relies crucially on a problem's extremely large domain is not very satisfying. It may say more about the difficulty of handling very large numbers rather than the inherent difficulty of solving the problem on reasonable size domains. An open question that remains is whether or not there is a function $f : \{0,1\}^n \longrightarrow \{0,1\}$ that can be computed more quickly by a CREW PRAM than by an EREW PRAM.

Fich and Wigderson [FW90] have made some progress by resolving this question in a special case when there is a restriction imposed on where in shared memory processors can write. The EROW PRAM is an EREW PRAM in which each processor is said to "own" one shared memory cell and that is the only shared memory cell to which it is allowed to write. In a sense, the choice of processor that may write into a given cell at a given point in time is oblivious of the input. Processors are still allowed to read from any shared memory cell. The CROW PRAM is the CREW PRAM restricted in the same manner. Fich and Wigderson proved that the EROW PRAM requires $\Omega(\sqrt{\log n})$ time to compute a Boolean function that requires only $O(\log\log n)$ time on the CROW PRAM. The CROW PRAM never requires more than a constant factor more time than the CREW PRAM to compute any function defined on a complete domain (although it may drastically increase the number of processors) [Nis91]. However, the restriction to the owner write model with a single memory cell per processor seems much more drastic for exclusive read machines. A fast simulation of EREW PRAMs by EROW PRAMs seems unlikely. In fact, we do not see any obvious way to extend the lower bound proof of Fich and Wigderson to even allow more than one writable memory cell per processor.

This leaves open the following question: is there a separation between CREW and EREW PRAMs for any function $f : \{0,1\}^n \rightarrow \{0,1\}$ in the presence of non-owner writes? We cannot answer the question in general but we can when the amount of shared memory through which processors can communicate is small. The *communication width* of a PRAM [VW85] is defined as the number of shared memory cells that are available for both reading and writing. (A separate read-only memory is used to store the input.) We denote by EREW(m), CREW(m), and CRCW(m) the respective PRAM models with communication width m. We show that a special case of the problem considered by Fich and Wigderson can be solved in time $O(\sqrt{\log n})$ by a CREW(1) PRAM, but requires $\Omega(\frac{\log n}{\log^* n})$ time on any EREW(1) PRAM. It is easy to see that the sequential time complexity of this problem is $\log_2 n$, which is almost matched by our lower bound for the EREW(1) PRAM. For higher communication width we can prove a lower bound of $\Omega(\frac{\log n}{m+\log^* n})$ on any EREW(m) PRAM.

We would like to extend our separation between CREW and EREW PRAMs to larger communication width. Our hope is that some of the techniques developed for small communication widths will turn out to be useful even for the general case. For example, the technique in the lower bound result for the OR function on CREW(1) PRAMs [VW85, Bea86] is very similar to the technique that Kutylowski [Kut91] eventually used in his optimal bound for the OR on general CREW PRAMs.

Our lower bound proof consists of three parts. First we show that any EREW(1) PRAM running for a short time can only have a small number of processors doing useful work. We then determine the time complexity of our problem on CREW(1) and CRCW(1) PRAMs with limited numbers of processors. This also implies an $\Omega(\log^{2/3} n)$ time bound for the EREW(1) PRAM. Finally, we show that there are subproblems (obtained via restrictions) on which the number of processors doing useful work is drastically reduced. Applying this result recursively, we obtain a nearly optimal EREW(1) PRAM lower bound.

2 A Bound on the Number of Processors Doing Useful Work

For a function defined on n variables, we assume that a PRAM starts with its input stored in n read-only memory cells. We assume that it has m other shared read/write memory cells we call common cells. The output will be the contents of a designated common cell at the end of the computation. Computations proceed in steps. At every step, each processor may read a memory cell, perform some local computation, and then write into a common cell. We do not place any restrictions on the number of processors, word size, or the computational power of individual processors.

For any EREW or CREW PRAM computing a function f, we say that a processor p *writes by time* t if, on some input to f, p writes into some memory cell during the first t steps. Since the shared memory is the only means of communication, we can assume that for any PRAM running for t steps only the processors that write by time t are involved in the computation.

Using results in [Bea86], it can be shown that, for any CREW(1) PRAM, the number of processors that write by time t is in $2^{O(t^2)}$. For EREW(1) PRAMs, there is a considerably smaller upper bound.

Lemma 1 Consider any EREW(1) PRAM computing a function with domain $\{0,1\}^n$. For all $t \geq 0$, at most $2^{t+1}+2^t-3 \leq 2^{t+2}$ processors write by time t.

Proof: For any processor P, time t, and input x, let $P^t(x)$ be the set of input variables that P reads during the first t steps on input x. Since a processor reads at most one cell per step, $|P^t(x)| \leq t$.

For $1 \leq j \leq t$, let $R(j)$ be the set of processors that do not read the common cell on any input for the first $j-1$ steps, but do read it on some input at step j.

Consider any processor $P \in R(j)$ and suppose that, on input $x \in \{0,1\}^n$, P reads the common cell at step j. At step j, processor P must decide whether to read the common cell based on the values of the variables in $P^{j-1}(x)$. The fraction of all inputs that agree with x on these variables is at least $2^{-(j-1)}$, since $|P^{j-1}(x)| \leq j-1$. On all these inputs, P reads the common cell in step j. At most one processor can read the common cell in step j on any particular input, so there are at most 2^{j-1} processors in $R(j)$.

Similarly, if $W(j)$ is the set of processors that do not read the common cell on any input during the first j steps, but do write there on some input at step j, then $|W(j)| \leq 2^j$. Thus the number of processors that write by time t is bounded above by

$$\sum_{j=1}^{t}|R(j)|+\sum_{j=1}^{t}|W(j)| \leq \sum_{j=1}^{t}2^{j-1}+\sum_{j=1}^{t}2^j \leq 2^t-1+2^{t+1}-2. \quad \square$$

It is also possible to construct an EREW(1) PRAM computing a function with domain $\{0,1\}^{2^t}$ such that there are at least 2^{t-2} processors that write by time t. Thus the bound in Lemma 1 is optimal, to within a small constant factor.

Lemma 2 There is an EREW(1) PRAM computing a function of $\{0,1\}^{2^t}$ for which there are at least 2^{t-2} processors that write during step t.

Proof: View the problem of selecting a processor to write during step t as a competition between processors that is arbitrated by the input vector. A processor is a potential winner if there is some setting of the input bits that could cause it to write during step t. Let $b_1 = 1$, $b_j = \sum_{i=1}^{j-1}[b_i + 1] < 2^j$, $k_1 = 1$, and $k_j = \sum_{i=1}^{j-1} k_i = 2^{j-2}$, for $j \geq 2$. We claim that for any j, there is an EREW(1) PRAM algorithm for selecting a winning processor from among k_j potential winning processors that runs for j steps, does not access the common cell, and is arbitrated by only b_j bits of input. Notice that this is enough to prove the lemma as we can modify the algorithm to make the winning processor write at step j.

The claim is proved by induction on j. The case $j = 1$ is trivial. For larger values of j, consider an input of length b_j which is partitioned into disjoint groups $X_1, X_2, \ldots, X_{j-1}$ of length $b_1, b_2, \ldots, b_{j-1}$, respectively, as well as one extra group of $j-1$ bits: $y_1, y_2, \ldots, y_{j-1}$. Let $G_1, \ldots, G_{j-1}$ be disjoint sets containing $k_1, \ldots, k_{j-1}$ processors, respectively, for a total of k_j. By our induction hypothesis, for each $i < j$, we can select a winner from among G_i during the first i steps based on the input in X_i. The winner from G_i reads $y_1, y_2, \ldots, y_{j-i}$ in steps $i+1, i+2, \ldots, j$ respectively. This processor is a winner in step j if and only if $y_1 = y_2 = \cdots = y_{j-i-1} = 0$ and $y_{j-i} = 1$. It is easy to verify that read conflicts never occur. □

In contrast, for a CREW(1) PRAM, the number of processors that write by time t can be in $2^{\Omega(t^2)}$. See, for example, the algorithm given in the proof of Theorem 3.

3 A CREW(1) Upper Bound for Evaluating Decision Trees

We now define a Boolean function and show that it can be computed in $O(\sqrt{\log n})$ time on a CREW(1) PRAM. In the next two sections, we will prove that this function requires significantly more time to be computed on an EREW(1) PRAM. Specifically, we interpret the $n = 2^h - 1$ input variables as the labels of the internal nodes in a complete Boolean decision tree D_h of height h, taken in some fixed (e.g. breadth-first) order. The leaves of D_h that are left children are labelled 0; those that are right children are labelled 1. Given an input, proceed down from the root, going left when a node labelled by a variable with value 0 is encountered and going right when a node labelled by a variable with value 1 is encountered. The value of the function $F_h : \{0,1\}^{2^h-1} \rightarrow \{0,1\}$ is the label of the leaf node that is reached.

There is a trivial sequential algorithm that computes F_h in h steps. It is unknown whether one can do better than this on an EREW(1) PRAM. However, the following lemma shows that it can be computed substantially faster on a CREW(1) PRAM.

Theorem 3 If $\binom{t}{2} \geq h$, then there is a CREW(1) PRAM that computes F_h in t steps.

Proof: For each of the 2^h root-leaf paths in the decision tree D_h, we assign a group of $t-1$ processors. Exactly one of these root-leaf paths is the correct path. In the following algorithm, the common cell will contain the values of the labels of the first $\binom{j+1}{2}$ nodes on the correct root-leaf path at the end of step $j+1$.

The jth processor in each group is active for the first $j+1$ steps. During the first j steps, it reads the j variables labelling nodes $\binom{j}{2}+1$ through $\binom{j+1}{2}$ on its root-leaf path. At step $j+1$, it reads the common cell, which contains the values of the labels of the first $\binom{j}{2}$ nodes on the correct root-leaf path. (When $j=1$, the common cell contains no information.) At this point, the jth processor in each group knows whether or not its root-leaf path agrees with the correct root-leaf path at the first $\binom{j+1}{2}$ nodes. Among the processors whose paths agree, a prespecified one (e.g. the jth processor in the leftmost of these groups) appends the bits that it has read to the previous contents of the common cell. Thus, at the end of step $j+1$, the common cell contains the values of the labels of the first $\binom{j+1}{2}$ nodes along the correct root-leaf path.

To compute F_h, we modify the algorithm slightly so that at the last step, instead of appending bits to the common cell, a processor writes down the value of the leaf determined by the h internal nodes on its path, i.e. the leaf node in D_h that is reached. □

4 Lower Bounds for Processor-limited PRAMs

In this section, we show that to compute F_h quickly we need to have many processors doing useful work even on a CRCW(1) PRAM. Using our bounds from section 2 on the number of processors doing useful work, this will give an $\Omega(h^{2/3})$ lower bound for the EREW(1) PRAM. In the next section, we will improve this bound to a nearly optimal $\Omega(\frac{h}{\log^* h})$ by combining the techniques of this section with a new restriction technique.

For any restriction r that sets some input variables to 0 or 1, let $r(F_h)$ be the function F_h with restriction r applied to it. Define the *depth* of r, $d(r)$, to be the minimum depth of any node v in D_h, the underlying decision tree of F_h, such that the path from the root to v is consistent with r and the subtree rooted at v does not contain any variables set by r. Note that $F_{h-d(r)}$ is a subfunction of $r(F_h)$.

Define the *history* of the common cell on any input to be the sequence of values that it takes on that input. Our lower bounds proceed by fixing the history of the common cell. We use the following result of Vishkin and Wigderson [VW85].

Lemma 4 [VW85] For any CRCW(1) PRAM running for t steps, there is a restriction r which sets at most $\binom{t+1}{2}$ variables such that the history of the common cell for the first t steps is the same for all inputs consistent with r.

Lemma 5 For any integer $h \geq \binom{t+1}{2}$ and any CRCW(1) PRAM running for t steps, there is a restriction r' of depth $\binom{t+1}{2}$ such that the history of the common cell for the first t steps is the same for all inputs to F_h consistent with r'.

Proof: By Lemma 4, there is a restriction r which sets at most $\binom{t+1}{2}$ variables such that the history of the common cell for the first t steps is the same for all inputs consistent with r.

We define r' by tracing a path of length $\binom{t+1}{2}$ from the root of F_h as follows. If r sets the current variable then let r' set it to be consistent with r and take the appropriate branch. Otherwise, take the branch which leads to the subtree with fewer variables from among those that r sets. In each step, we are taking care of at least one variable that r sets. So, after $\binom{t+1}{2}$ steps, we will have a subtree with none of its variables set, whose path from the root of the original tree is consistent with r. We set all variables outside this subtree in some manner consistent with r. Clearly, the resulting restriction has depth at most $\binom{t+1}{2}$. From Lemma 4, it follows that the common cell has the same history for the first t steps for all inputs consistent with this restriction. □

In particular, this implies that the CREW(1) PRAM algorithm to compute F_h given in the proof of Theorem 3 is within one step of optimal.

Theorem 6 If a CRCW(1) PRAM computes F_h in t steps, then $\binom{t+1}{2} \geq h$.

Proof: Consider any CRCW(1) PRAM that computes F_h in t steps. Suppose, for contradiction, that $\binom{t+1}{2} \leq h-1$. Then by Lemma 5, there is a restriction r' of depth $\binom{t+1}{2}$ such that the history of the common cell for the entire computation is the same for all inputs to F_h consistent with

r'. Since $d(r') \leq h-1$, F_1 is a subfunction of $r'(F_h)$. Since F_1 is a not a constant function neither is $r'(F_h)$ and thus F_h has not been computed in t steps which is a contradiction. □

The following theorem shows that, with a limited number of processors, a larger lower bound may be obtained.

Theorem 7 Any CRCW(1) PRAM with p processors that computes F_h requires more than $\frac{2h}{3(1+\sqrt{\log p})}$ steps.

Proof: By induction on h. Suppose some CRCW(1) PRAM with p processors computes F_h in T steps. Then, by Theorem 6, $\binom{T+1}{2} \geq h$.

Let $t = \sqrt{\log p}$. If $\binom{t+1}{2} \geq h$ (which always holds in the base case when $h = 0$), then $T(t+1)/2 \geq h$, so $T \geq 2h/(t+1) > 2h/3(1+\sqrt{\log p})$. Therefore, assume that $\binom{t+1}{2} < h$.

By Lemma 5, there is a restriction r' of depth $\binom{t+1}{2}$ that fixes the common cell for the first t steps. Consider the computations of the CRCW(1) PRAM on all inputs consistent with r'. Since each input variable has at most two different values, and the value of the common cell is fixed at each time step, it follows that each processor is in one of at most 2^i states at the end of step $i < t$. Now the state of a processor determines which memory cell it will read next. Thus at most $p\sum_{i=0}^{t-1} 2^i < 2^t p$ different input variables are read during the first t steps by all processors on all these inputs.

Let v be any node of depth $d(r')$ such that the path from the root to v is consistent with r' and the subtree rooted at v does not contain any variables set by r'. Consider the $2^t p$ nodes at distance $t + \log p$ from v. There is at least one node w such that the subtree rooted at w contains input variables that no processor can possibly read in the first t steps. Let r'' be a restriction that extends r' by setting the variables labelling all ancestors of w so as to cause the path from the root of D_h to w to be followed. Only the variables labelling nodes in the subtree rooted at w are left unset. All remaining variables are set arbitrarily. The restriction r'' has depth $\binom{t+1}{2} + t + \log p = 3\binom{t+1}{2}$. Thus the functions $F_{h-d(r'')}$ and $r''(F_h)$ are identical, up to the renaming of variables. Since no processors have read any input variables of this subfunction at time t, it follows from the induction hypothesis that at least $\left[2h - 6\binom{t+1}{2}\right]/3(t+1)$ additional steps are required to compute this subfunction. Therefore $T \geq t + \left[2h - 6\binom{t+1}{2}\right]/3(t+1) = \frac{2h}{3(t+1)}$. □

We note that this lower bound is asymptotically optimal even for a CREW(1) PRAM.

Corollary 8 For all integers $1 \leq p \leq 2^h$, the complexity of F_h on a CRCW(1) or CREW(1) PRAM with p processors is $\Theta(h/\sqrt{\log p})$.

Proof: The lower bound follows from Theorem 7. For the upper bound, notice that the algorithm in Theorem 3 shows that, with p processors, a CREW(1) can evaluate a decision tree of height $\log p$ in time $O(\sqrt{\log p})$. To compute F_h with p processors on a CREW(1), we simply apply this algorithm sequentially $O(h/\log p)$ times to obtain a running time of $O(h/\sqrt{\log p})$. □

Using the bounds of section 2 we have:

Corollary 9 Any EREW(1) PRAM computing F_h must run for at least $\frac{1}{4}h^{\frac{2}{3}}$ steps.

Proof: Suppose the EREW(1) PRAM runs for T steps. Then, by Lemma 1, we can assume that it has at most $p = 2^{T+2}$ processors. Now, applying Theorem 7, it follows that $T \geq \frac{2h}{3(1+\sqrt{T+2})}$. Solving, we obtain $T \geq \frac{1}{4}h^{2/3}$, as required. □

5 A Near Optimal EREW(1) Lower Bound

We strengthen the arguments of the previous section to prove a nearly optimal $\Omega(\frac{h}{\log^* h})$ lower bound on the time for an EREW(1) PRAM to compute F_h. The key to the improvement of Lemma 5 for the EREW(1) PRAM is to take advantage of the fact that we can select a large subset of inputs on which very few processors ever do useful work. We then recursively apply the argument of the previous section to obtain a better lower bound.

Lemma 10 For any integers $T, h \geq 2T+2$, and any EREW(1) PRAM computing F_h, there is a restriction r of depth $2T + 2$ such that at most $2T$ processors write by time T on inputs consistent with r.

Proof: From Lemma 1, we can assume that the EREW(1) PRAM has at most 2^{T+2} processors. For each processor P, let $s(P)$ denote the set of input variables that P reads during the first T steps of computation on any input to F_h, but before it reads the common cell. Define $S = \bigcup_P s(P)$. As in the proof of Theorem 7, since each input variable has at most two different values, $|s(P)| < 2^T$ and thus $|S| < 2^{2T+2}$.

Consider the 2^{2T+2} nodes at depth $2T + 2$ from the root of D_h. For at least one such node v, none of the nodes in the subtree rooted at v are labelled by variables in S. Set the variables labelling ancestors of v so that the path from the root to v is followed in D_h. All variables labelling nodes in the subtree rooted at v are left unset. Set all other variables outside this subtree arbitrarily. Let r be the resulting restriction.

From now on, consider only those inputs consistent with r. The subtree rooted at v does not contain any input variables from S, so no processor reads any unset variable until after it has read the common cell. Since the PRAM is exclusive read, for each step $j \leq T$, there is at most one processor that does not read the common cell on any input for the first $j-1$ steps but does read the common cell on some input at step j. Hence, at most T processors read the common cell in the first T steps. Similarly, at most T processors from among those that have not read the common cell may write in the first T steps. So, altogether, there are at most $2T$ processors that can write into the common cell on inputs consistent with r. □

Let $A_2 = 3$ and, for $T \geq 3$, let $l_T = \lceil \log T \rceil$ and

$$\begin{aligned} A_T &= (2T+2) + \left\lceil \frac{T}{l_T} \right\rceil (A_{l_T} + 2l_T + 1), \\ &\leq (2T+2) + (1 + \frac{T}{\log T})(A_{\log T+1} + 2\log T + 3). \end{aligned}$$

It can be verified that $A_T \in O(T \log^* T)$.

Lemma 11 For any integers $T, h \geq A_T$, and any EREW(1) PRAM computing F_h, there is a restriction r of depth A_T such that the history of the common cell for the first T steps is the same for all inputs to F_h consistent with r.

Proof: The proof is by induction on the value of T.

For $T = 2$, $A_T = 3 = \binom{T+1}{2}$ and the claim follows from Lemma 5.

For larger values of T, first use Lemma 10 to obtain a restriction r_0 of depth $2T+2$ such that at most $2T$ processors write by time T on inputs consistent with r_0. Break the T steps of the computation into $\lceil T/l_T \rceil$ subintervals each of size at most l_T. It is sufficient to prove that, for $i \leq l_T$, there is an extension r_i of r_0 of total depth $2T+2+i(A_{l_T}+2l_T+1)$ so that none of these $2T$ processors read any variables left unset by r_i during the first i subintervals on all inputs to F_h consistent with r_i. This is proved by induction on i.

The case $i = 0$ is trivial, so suppose $i \geq 1$. Assume that a restriction r_{i-1} with the desired properties exists. Without loss of generality, we may also assume that r_{i-1} sets all variables labelling nodes outside of some tree of height $h - d(r_{i-1})$. Then $r_{i-1}(F_h)$ is the same as $F_{h-d(r_{i-1})}$ up to the renaming of variables. Since $l_T < T$, it follows from the induction hypothesis that there is a restriction of depth A_{l_T} such that the history of the common cell for the first l_T steps is the same for all inputs to $F_{h-d(r_{i-1})}$ consistent with this restriction. None of the $2T$ processors have read any variables left unset by r_{i-1} during the first $i-1$ subintervals on inputs to F_h consistent with r_{i-1}. Therefore there is also a restriction r'_i that extends r_{i-1} with depth $d(r_{i-1})+A_l$, such that the history of the common cell for the first i subintervals is the same for all inputs to F_h consistent with r'_i.

As in the proof of Theorem 7, each of the $2T$ processors can read at most $2^{l_T}-1$ input variables during the ith subinterval. Thus there is an extension r_i of r'_i with depth $d(r'_i) + \lceil\log(2T)\rceil + l_T = 2T + 2 + i(A_{l_T} + 2l_T + 1)$ so that none of the processors have read any inputs of F_h left unset by r_i during the first i subintervals. □

We can use this lemma to derive the near optimal lower bound for the EREW(1) PRAM.

Theorem 12 Any EREW(1) PRAM computing F_h must run for $\Omega(h/\log^* h)$ steps.

Proof: Suppose there is an EREW(1) PRAM computing F_h that runs for T steps, where $A_T \le h-1$. By Lemma 11, there is a restriction r of depth A_T such that the answer in the common cellis the same for all inputs to $r(F_h)$. However, there are two inputs in $r(F_h)$ that reach the same leaf, but differ in the value of that leaf. These inputs should have different answers. Thus $A_T \ge h$. Since $A_T \in O(T\log^* T)$, it follows that $T \in \Omega(\frac{h}{\log^* h})$. □

Theorem 12 and Theorem 3 immediately give our main separation theorem.

Theorem 13 There is a function on n Boolean variables that can be solved in $O(\sqrt{\log n})$ time on a CREW(1) PRAM but requires $\Omega(\frac{\log n}{\log^* n})$ time on every EREW(1) PRAM.

Since an EREW(1) PRAM can simulate an EREW(m) PRAM with a slowdown of $O(m)$, Theorem 12 immediately gives an $\Omega(\frac{h}{m\cdot\log^* h})$ lower bound for EREW(m) PRAMs. An extension of the techniques in this paper can be used to obtain a stronger lower bound.

Theorem 14 Any EREW(m) PRAM computing F_h must run for $\Omega(\frac{h}{m+\log^* h})$ steps.

This gives a separation between CREW(1) and EREW(m) PRAMs when m is small.

Theorem 15 For all $m \in o(\sqrt{\log n})$, there is a function on n Boolean variables that can be solved asymptotically faster on a CREW(1) PRAM than on any EREW(m) PRAM.

References

[Bea86] P. Beame. *Lower Bounds in Parallel Machine Computation.* PhD thesis, Department of Computer Science, University of Toronto, 1986. Also appears as Technical Report TR 198/87.

[CDR86] Steven A. Cook, Cynthia Dwork, and Rudiger Reischuk. Upper and lower time bounds for parallel random access machines without simultaneous writes. *SIAM Journal on Computing*, 15(1):87–97, February 1986.

[DKR90] M. Dietzfelbinger, M. Kutylowski, and R. Reischuk. Exact time bounds for computing Boolean functions on PRAMs without simultaneous writes. In *Proceedings of the 1990 ACM Symposium on Parallel Algorithms and Architectures*, pages 125–135, Crete, Greece, June 1990.

[FW90] Faith E. Fich and Avi Wigderson. Towards understanding exclusive read. *SIAM Journal on Computing*, 19(4):717–727, 1990.

[GNR89] E Gafni, J Naor, and P Ragde. On separating the EREW and CREW PRAM models. *Theoretical Computer Science*, 68(3):343–346, 1989.

[Kut91] M. Kutyloswki. The complexity of Boolean functions on CREW PRAMs. *SIAM Journal on Computing*, 20(5):824–833, 1991.

[Nis91] Noam Nisan. CREW PRAMs and decision trees. *SIAM Journal on Computing*, 20(6):999–1007, December 1991.

[Sni85] M. Snir. On parallel searching. *SIAM Journal on Computing*, 14:688–708, 1985.

[VW85] U. Vishkin and A. Wigderson. Trade-offs between depth and width in parallel computation. *SIAM Journal on Computing*, 14(2):303–314, May 1985.

Triply-Logarithmic Upper and Lower Bounds for Minimum, Range Minima, and Related Problems with Integer Inputs

O. Berkman[1] Y. Matias[2] P. Ragde[3]

[1] King's College London[†]
[2] AT&T Bell Laboratories[‡]
[3] University of Waterloo[§]

Abstract. We consider the problem of computing the minimum of n values, and several well-known generalizations (prefix minima, range minima, and all-nearest-smaller-values (ANSV) problems) for input elements drawn from the integer domain $[1..s]$ where $s \geq n$. Recent work [4] has shown that parallel algorithms that are sensitive to the size of the input domain can improve on more general parallel algorithms. The cited paper demonstrates an $O(\log\log\log s)$-step algorithm on an n-processor PRIORITY CRCW PRAM for finding the prefix-minima of n numbers in the range $[1..s]$. The best known upper bounds for the range minima and ANSV problems were previously $O(\log\log n)$ (using algorithms for general input). This was also the best known upper bound for computing prefix minima or even just the minimum on the COMMON CRCW PRAM; this model has a $\Theta(\log n/\log\log n)$ time separation from the stronger PRIORITY model when using the same number of processors. In this paper we give simple and efficient algorithms for all of the above problems. These algorithms all take $O(\log\log\log s)$ time using an optimal number of processors and $O(ns^{\epsilon})$ space on the COMMON CRCW PRAM. We also prove a lower bound demonstrating that no algorithm is asymptotically faster as a function of s, by showing that for $s = 2^{2^{\Omega(\log n \log\log n)}}$ the upper bounds are tight.

1 Introduction

Let $A = (a_1, \ldots, a_n)$ be an array of input elements. Denote by $MIN(i,j)$ the minimum over $a_i, \ldots, a_j$. We consider the following problems:

- The *minimum* problem: find $MIN(1,n)$.
- The *prefix minima* problem: find $MIN(1,i)$ for all i, $1 \leq i \leq n$.

[†] Dept. of Computer Science, King's College London, The Strand, London WC2R 2LS, England.

[‡] AT&T Bell Laboratories 600 Mountain Ave., Murray Hill, NJ 07974-0636, USA.

[§] Dept. of Computer Science, University of Waterloo, Waterloo, Ontario, Canada N2L 3G1.

– The *range minima* problem: build a data structure that will permit a constant-time answer to any query $MIN(i,j)$ for any $1 \leq i < j \leq n$.
– The *all nearest smaller values (ANSV)* problem: find for all i, $1 \leq i \leq n$, the maximum j, $j < i$, such that $a_j < a_i$ (the "left match" of a_i) and the minimum k, $k > i$, such that $a_k < a_i$ (the "right match" of a_i).

In this paper we consider the above problems when the elements of A are drawn from the integer domain $[1..s]$ where $s \geq n$. We show:

Theorem 1. *Each of the above problems can be solved on the* COMMON *CRCW PRAM in* $O(\log\log\log s)$ *time using* $n/\log\log\log s$ *processors and* $O(ns^{\epsilon})$ *space.*

Theorem 2. *Any n-processor* PRIORITY *CRCW PRAM algorithm for computing the minimum (and thus any algorithm for the other three problems) takes* $\Omega(\log\log\log s)$ *time for any* s, $s \geq 2^{2^{\Omega(\log n \log\log n)}}$.

1.1 The model of computation

The model of parallel computation used in this paper is the concurrent-read concurrent-write (CRCW) parallel random access machine (PRAM). The CRCW PRAM model employs synchronous processors, all having access to a shared memory with allowed concurrent access. There are several sub-models of CRCW PRAM regarding the conflict resolution rule in case of a concurrent writing. In the COMMON model, several processors may attempt to write simultaneously at the same location only if they write the *same* value; COMMON thus forbids write conflicts. Following [13, 20], Boppana [7] gave a lower bound of $\Omega(\log n/\log\log n)$ for computing the *Element Distinctness* problem on an n-processor COMMON. This problem can be solved in constant time on models that allow write conflicts. Such models include: (i) TOLERANT, where if two or more processors attempt to write to the same cell in a given step then the contents of that cell does not change; (ii) the stronger ARBITRARY, in which a concurrent writing results in an arbitrary winner among the writing processors; and (iii) the yet stronger PRIORITY in which a write conflict is resolved by having the processor with highest priority succeed. Algorithms running on PRIORITY or ARBITRARY might not be transferable to COMMON without a significant slow-down or loss of efficiency.

A parallel algorithm is said to be *optimal* if its time-processor product is (asymptotically) equal to the lower bound on the time complexity of any sequential algorithm for the problem. A primary goal in parallel computation is to design optimal algorithms that also run as fast as possible.

1.2 Related Work

We review below previous and related results for the four problems considered in this paper.

Sequential algorithms Gabow, Bentley, and Tarjan [14] gave a linear-time preprocessing algorithm for range minima that results in constant-time query retrieval. The ANSV problem has a simple linear-time algorithm using a stack.

Bounds on general inputs Using n processors in the deterministic comparison model, the minimum-finding problem, and hence all four problems have an $\Omega(\log\log n)$ time lower bound [26]. This lower bound is matched for each of the four problems by optimal COMMON CRCW PRAM $O(\log\log n)$ time algorithms: [24] for minimum, [22] for prefix minima, and [5] for range minima and ANSV.

Input from restricted domains [13] gives an optimal constant-time algorithm on COMMON for finding the minimum for integers in the domain $[1..n^k]$ for a constant k. [4] gives a prefix-minima algorithm on PRIORITY (and thus also a minimum finding algorithm) that runs in $O(\log\log\log s)$ time and $O(n)$ operations, using $O(ns^\epsilon)$ space for input restricted to integers in the domain $[1..s]$ where $s \geq n$. For the case $s = n$, [4] gives a PRIORITY algorithm that takes $O(\log^* n)$ time and $O(n)$ operations, using $O(n)$ space.

Lower bounds The following lower bounds were proved using Ramsey-theoretic arguments. [25] gave an $\Omega(\sqrt{\log n})$ lower bound on searching in a sorted table of size n with an EREW PRAM. An $\Omega(\sqrt{\log n})$ lower bound on sorting n items with an n-processor PRIORITY CRCW PRAM is given in [19]. This paper also gives an $\Omega(\log\log n)$ lower bound for finding the minimum among n numbers on PRIORITY assuming that the numbers are drawn from a domain of size at least doubly exponential in n. An $\Omega(\sqrt{\log n})$ lower bound on deciding element distinctness of n items with an n-processor COMMON CRCW PRAM is given in [20]. This was improved in [7] to the optimal result $\Omega(\log n/\log\log n)$. [23] gave an optimal $\Omega(\log\log n)$ lower bound on merging two sequences of length n with an $n\log^{O(1)} n$-processor PRIORITY CRCW PRAM. Finally, Grolmusz and Ragde [16] gave an $\Omega(\log\log\log n)$ lower bound on simulating a n-processor ARBITRARY PRAM on an n-processor COLLISION PRAM. This was improved in [8] to $\Omega(\log\log n)$.

1.3 Our results

Upper bounds We present the first optimal triply-logarithmic time algorithms for the range-minima and ANSV problems. The algorithms assume the COMMON model of computation. For the minimum and prefix-minima problems, the improvement is on the model of computation; PRIORITY was used in previous triply-logarithmic algorithms.

Our algorithms are simple and no large constants are hidden in the "big-Oh" notations. It should also be noted that our algorithms improve upon previous $O(\log\log n)$ time algorithms even for very large domain s; in particular, $\log\log\log s = O(\log\log\log n)$ for $s = n^{(\log n)^{(\log\log n)^{O(1)}}}$.

Applications It is hardly necessary to mention applications of minimum and prefix minima. We mention two recent papers that need range minima search: [1] on scaled string matching, and [21] on parallel triconnectivity. Other applications of the algorithm are given in [14]. Our new ANSV algorithm implies an optimal $O(\log\log\log s)$ time using $O(ns^\epsilon)$ space algorithm for triangulating a monotone polygon whose coordinates are taken from the domain $[1..s]$, $s \geq n$.

Previous optimal parallel algorithms for triangulating a monotone polygon are those of [5] and [15]. Their running times are $O(\log\log n)$ and $O(\log n)$ using the COMMON CRCW PRAM and the CREW PRAM respectively. Both assume that coordinates have unrestricted domain.

Lower bounds Few techniques exist to show general lower bounds for parallel computation. One of the most useful ones has been the application of powerful methods from Ramsey theory. Intuitively, a Ramsey-like theorem states that in some large and possibly complex universe, there exists a subuniverse with some simpler or more regular structure. To prove a lower bound on the complexity of a problem, it is often possible to take an arbitrary program which may exhibit complex behavior when considered over all inputs, and apply Ramsey theory to show that there exists a subdomain of inputs on which the program behaves in very simple ways. In effect, the program is reduced to operating in a structured fashion, or with a restricted set of operations. Ad-hoc techniques can then be used to prove a lower bound on the running time of the program on this subdomain. In this fashion, each of the above mentioned lower bounds were proved.

One of the drawbacks of these uses of Ramsey theory is the fact that, in order to show that the subdomain exists, the domain size must be a very rapidly growing function of n. The possibility thus exists that, if inputs are taken from the domain $[1..s]$, where s may be polynomial or even singly or doubly exponential in n, then algorithms may exist which beat these lower bounds. As an analogy, consider the case of sequential sorting; Radix sort will, for suitably restricted domains, give an $O(n)$ algorithm.

The challenge, then, is to either reduce the domain size required in the lower bounds, or to produce algorithms with better running times on moderate sized domains. [2] improves both the asymptotic result and the domain size for the sorting bound mentioned above by proving an $\Omega(\log n/\log\log n)$ lower bound on computing parity with a PRIORITY CRCW PRAM. This implies the same lower bound for sorting with domain size 2. [9] has obtained the same lower bound as [7] for element distinctness but with a domain size that is doubly exponential in n.

The domain-sensitive lower bound implied by Theorem 2 above cannot be improved without further restriction on s. This represents a modest beginning to the search for lower-bound techniques that work on problems defined over small domains.

The rest of this paper is organized as follows. In Section 2 we present constant-time non-optimal algorithms for each of the four problems. In Section 3 we present the optimal algorithms. The lower bound is given in Section 4. Concluding remarks and open problems are given in Section 5.

2 Constant-Time Non-Optimal Algorithms

We begin with an algorithm for finding the minimum. This algorithm is then used as a subroutine for an algorithm that solves both the prefix-minima and ANSV problems. Finally, the prefix minima algorithm is used to get an algorithm for the range minima problem.

2.1 Minimum

The following lemma and algorithm demonstrate a basic step which appears (in different forms) in some of the algorithms below.

Lemma 3. *Let $A = (a_1, \ldots, a_n)$ be an array of elements drawn from the domain $[1..s]$. The algorithm below finds the minimum in A in $O(1)$ time using $n \log s$ processors and $O(s)$ space.*

Step 1 (Data Structure). Build a complete binary tree T_s whose leaves are the numbers $[1..s]$. We assume that the space allocated for the tree is initialized to zero. It will be shown at the end of the algorithm how to get rid of this assumption.

Step 2 (Processor Allocation). Allocate $\log s$ processors to each element a_i, $1 \leq i \leq n$: a processor for each ancestor of the leaf in T_s whose value is a_i.

Step 3 (Marking). Each processor assigned to the ancestor v of a leaf a_i writes '1' in a variable attached to v, for $i = 1, \ldots, n$.

Step 4 (Information Gathering). The $\log s$ processors of each a_i are assigned to the ancestors of the leaf a_i as in Step 3. A processor that is assigned to ancestor v of a_i which is a right sibling reads the variable of the left sibling of v. Element a_i is the minimum in A if and only if none of its processors has read a '1'. The minimum can therefore be found in constant time by simple OR computations.

To handle the case that the input is not initialized to zero we add a step between Step 2 and Step 3 that initializes to zero only those locations which are being read in Step 4. The complexity of such a step is the same as that of Step 4. Lemma 3 follows.

Lemma 4. *There is an algorithm for finding the minimum that runs in $O(1)$ (more precisely, $O(\frac{1}{\epsilon})$) time using $n \log s$ processors and $O(ns^\epsilon)$ space for any constant ϵ, $0 < \epsilon < 1$.*

2.2 Prefix Minima and All Nearest Smaller Values

Lemma 5. *Let $A = (a_1, \ldots, a_n)$ be an array of elements drawn from the domain $[1..s]$. The algorithm below solves both the prefix-minima and the ANSV problems with respect to A in $O(1)$ time using $n \log^2 s$ processors and $O(ns^\epsilon)$ space for any constant ϵ, $0 < \epsilon < 1$.*

Step 1 (Data Structure). Build a complete binary tree T_A whose leaves are the elements of A.

Step 2 (Processor Allocation). Allocate $\log^2 s$ processors to each leaf a_i of T_A: $\log s$ processors for each ancestor of leaf a_i in T_A (note that a_i has $\log n \leq \log s$ ancestors).

Step 3 (Minima Computation). Find the minimum over the leaves of each internal node v of T_A, using the algorithm of Lemma 4. Let r denote the number of leaves of v. Then this step with respect to v uses $r \log s$ processors, which is indeed the number of processors allocated to v; it takes $O(1)$ time and uses $O(rs^\epsilon)$ space. In case the minimum is not unique, we compute the rightmost minimum. The total space used is $O(ns^\epsilon)$.

Step 4 (Prefix- and Suffix-Minima Computation). For each internal node v compute prefix minima and suffix minima over an array $L(v)$ that contains the leaves of v:
Consider a leaf l of (the subtree rooted at) v. Let $LS_v(l)$ be the set of left siblings of the nodes which are on the path from l to v. The leaves in arrays $L(u)$ of nodes u in $LS_v(l)$, together with l itself, represent exactly all the leaves in the prefix of l in $L(v)$. Therefore, the minimum over the prefix of l in $L(v)$ is the minimum over $\{\min(L(u)) : u \in LS_v(l) \cup l\}$. Since $|LS_v(l)| \leq \log n$ we can find this minimum in constant time with $(\log n)^2$ processors (out of the $\log^2 s$ processors allocated to l) using the algorithm of [24]. Note that the prefix minima computed with respect to the root is actually prefix minima with respect to A. This concludes the computation of prefix minima.
The next steps complete the computation of ANSV.

Step 5 (Find the Nodes whose Subtrees Contain the Matches). Each leaf a_i finds its lowest ancestor that has the left match of a_i among its leaves. Finding the lowest ancestor that has the right match of a_i is similar. For this consider the (at most) $\log n$ nodes which are left siblings of the ancestors of a_i. Among these $\log n$ nodes, we find the lowest node whose minimum is smaller than a_i. This problem can be restated as the problem of finding the leftmost '1' in an array of 0's and 1's and can therefore be done in $O(1)$ time with $\log n$ processors using the algorithm of [13]. The parent of this node is the lowest ancestor of a_i that has a_i's left match among its leaves.

Step 6 (Merge Left Child's Suffix Minima with Right Child's Prefix Minima). For each node v do the following. Let u and w be the left and right children of v, respectively, and let S_u and P_w denote their respective suffix minima and prefix minima (computed in Step 4 above). We merge S_u (which is a non-decreasing array) with the reverse of array P_w (P_w itself is non-increasing) into an array $A(u, w)$. Denoting by r the number of leaves of v, this can be done in $O(1)$ time with $r \log s$ processors (which is the number of processors allocated to v) and $O(rs^\epsilon)$ space, using the integer merging algorithm of [6]. The overall space used is $O(ns^\epsilon)$.

Step 7 (Find Left and Right Matches for all Elements). Let v be the lowest ancestor of a_i which contains a_i's left match; let u and w be its left and right children, respectively, and let r_1 be the number of leaves of u and w. Let j be the index of a_i in P_w (which is also its index in $L(w)$), and let k be the index of a_i in $A(u, w)$. Then out of the first $k-1$ elements of $A(u, w)$ (these $k-1$ elements constitute the elements of u and w that are smaller than a_i), $r_1 - j$ are elements of w, and thus $(k-1)-(r_1-j) = k - r_1 + j - 1$ are elements of u. It follows that the $(k - r_1 + j - 1)$'st element of u is the left match of a_i. Finding the right match of a_i is similar.

Lemma 5 follows.

2.3 Range Minima

Lemma 6. *Let $A = (a_1, \ldots, a_n)$ be an array of elements drawn from the domain $[1..s]$. The preprocessing algorithm below solves the range minima problem with respect to A in $O(1)$ time using $n \log^3 s$ processors and $O(ns^\epsilon)$ space for any*

given ϵ, $0 < \epsilon < 1$. Following this preprocessing, each range minimum query can be answered in constant time by one processor.

Preprocessing

Step 1 (Data Structure). **Build a complete binary tree T_A whose leaves are the elements of A.**

Step 2 (Processor Allocation). **Allocate $\log^3 s$ processors to each leaf a_i of T_A: $\log^2 s$ processors for each ancestor of a_i in T_A (note that a_i has $\log n \leq \log s$ ancestors).**

Step 3 (Prefix- and Suffix-Minima Computation). **For each internal node v compute prefix minima and suffix minima over an array $L(v)$ that contains the leaves of v. This is done using the algorithm for prefix minima (and a similar algorithm for suffix minima) of Lemma 5. The time is $O(1)$ using $r \log^2 s$ processors (which is the number of processors allocated to v) and $O(rs^\epsilon)$ space, where r is the size of v. The overall processor and space complexities are $O(n \log^3 s)$ and $O(ns^\epsilon)$ respectively.**

Query Retrieval **To answer a query $MIN(i,j)$ we find the lowest common ancestor v of a_i and a_j. $MIN(i,j)$ is then the minimum between the following two minima: (1) the minimum over the suffix of a_i with respect to the left child of v; and (2) the minimum over the prefix of a_j with respect to the right child of v. These two minima are computed in the preprocessing algorithm above. We note that since T_A is a complete binary tree, the computation of the lowest common ancestor of a_i and a_j can be done in $O(1)$ time using a single processor (see [17]).**

Lemma 6 follows.

3 Optimal Algorithms

We present optimal algorithms for the ANSV problem and the range-minima problem. Since range minima is a generalization of prefix minima, this also implies optimal algorithms for the problems of finding the minimum and prefix minima.

3.1 All Nearest Smaller Values

We divide the input into $n/\log^2 s$ subarrays of size $\log^2 s$ each and apply the optimal doubly logarithmic ANSV algorithm of [5] to each subarray. This takes $O(\log\log\log s)$ time using $n/\log\log\log s$ processors. We now solve the ANSV problem with respect to an array of $n/\log^2 s$ minima, one minimum from each subarray. This is done in $O(1)$ time using n processors and $O(ns^\epsilon)$ space using Lemma 5 and can be implemented in $O(\log\log\log s)$ time using $n/\log\log\log s$ processors and $O(ns^\epsilon)$ space. Finally, using this data we reduce, in $O(1)$ time, the problem of finding nearest smaller values for all elements into (at most) $2n/\log^2 s$ merging problems each of size $2\log^2 s$. The details of this reduction

are given in [5] and are thus omitted from this manuscript. We solve each such merging problem using the optimal doubly-logarithmic algorithm for merging of [18]. This takes $O(\log\log\log s)$ time using $\log^2 s/\log\log\log s$ processors for each merging problem and $O(\log\log\log s)$ time using $n/\log\log\log s$ processors overall.

Triangulating a monotone polygon Using ideas from [5], the above optimal triply-logarithmic ANSV algorithm, and the optimal triply-logarithmic merging algorithm of [6] we get the following:

Theorem 7. ***A monotone polygon whose coordinates are taken from the domain $[1..s]$, $s \geq n$, can be triangulated in $O(\log\log\log s)$ time using an optimal number of processors and $O(ns^\epsilon)$ space.***

3.2 Range Minima

We divide the input into $n/\log^3 s$ subarrays of size $\log^3 s$ each and preprocess each subarray so that range-minima queries within the subarray can be answered in $O(1)$ time. This can be done using the optimal doubly logarithmic range-minima algorithm of [5] and takes $O(\log\log\log s)$ time using $n/\log\log\log s$ processors and linear space for all subarrays. Next we apply the algorithm of Lemma 6 to an array of $n/\log^3 s$ minima, a minimum from each subarray. This takes $O(1)$ time using n processors and $O(ns^\epsilon)$ space and enables answering a range minimum query with respect to this array in $O(1)$ time. It is easy to see that using this data each range minimum query can be answered in constant time.

4 The Lower Bound

The lower bound given here follows the general outlines of other PRAM lower bounds [11, 12, 16, 20]. The input to a PRAM will be an n-tuple of positive integers $(x_1, x_2, \ldots, x_n)$, where x_i is drawn from the domain $[1..s]$ and is initially stored in the local memory of processor P_i. (Since memory is unbounded, this is equivalent to the situation where the input variables are stored in shared memory, one to a cell.) The output of the PRAM will be in the local memory of processor P_1 at time T. One step of a PRAM consists of a parallel write followed by a parallel read.

It is useful to slightly modify the PRIORITY PRAM. We disallow overwriting of memory – that is, a cell may be written into only once. To compensate, we allow each processor to simultaneously read $t-1$ cells at step t, providing that those cells, if they were written into at all, were written into at steps $1, 2, \ldots, t-1$ respectively. One can prove easily (see [11]) that for infinite memory, this does not decrease the power of the PRAM. This is a technical convenience that makes the proof slightly easier.

Theorem 8. *Any* PRIORITY *CRCW PRAM requires $\Omega(\log\log\log s)$ steps to find the maximum of n numbers in the domain $[1..s]$, when $s = 2^{2^{\Omega(\log n \log\log n)}}$.*

Proof. Given a PRIORITY CRCW PRAM algorithm that claims to solve the maximum problem, we proceed to construct a set of "allowable" inputs for each step. This set is chosen to restrict the behavior of the machine so that its state of knowledge can be easily described. As long as the set of allowable inputs for step t is sufficiently rich, we can show (based on our characterization of the state of knowledge of the machine) that there exists an allowable input on which the machine cannot answer correctly after t steps. In order to fully describe the set of allowable inputs after step t, we will require some additional sets, which are described below.

- A set $\mathcal{U}_t$ of *free* variables. These are variables to which no fixed value has been assigned. We denote the total number of variables in $\mathcal{U}_t$ as v_t. Intuitively, after t steps the algorithm has succeeded in determining only that the maximum is one of the free variables. In other words, the free variables are the candidates that the algorithm has to work with (whether or not the algorithm is explicitly structured in this fashion).
- A set S_t of positive integers. In any allowable input, the values given to the free variables will have distinct values chosen from S_t.
- A set $\mathcal{M}_t$ of *fixed* variables. Any variable that is not free will be fixed. A fixed variable has the same value in any allowable input. It is set to some value that is smaller than any value in S_t. Intuitively, either the algorithm has determined that the variables in $\mathcal{M}_t$ are not the maximum, or we as adversary have given that information away.

Any input for which all the variables in $\mathcal{M}_t$ have their assigned fixed values and all the variables in $\mathcal{U}_t$ have values in S_t is an allowable input for step t. We can now state two inductive hypotheses which will be shown to hold by construction.

Hypothesis 1: The state of each processor and each memory cell at each step up to and including step t, considered over the domain of allowable inputs for step t, is a function of at most one free variable. For a given processor or memory cell, this variable, if it exists, is the same over all allowable inputs. We say that the processor or memory cell ***knows*** that variable.

Because of Hypothesis 1, the choice of which cell processor P_i reads at a given step t (again, considered over the domain of allowable inputs for step t) is also a function of the one free variable that P_i knows. This is called the ***read access function*** of P_i. A read access function should be considered as a function of some variable z that can take on values from S_t; a processor uses the read access function by substituting as an argument the value of the free variable it knows. Similarly, the write access function of P_i (the choice of where the processor writes) is a function of that one free variable.

Hypothesis 2: For every step $t' \leq t$ and over all allowable inputs, a processor either does not write at step t' or always writes. Any read or write access function at step t', considered as a function over S_t, is either constant or 1–1; any two such functions used before or at step t are either identical, or have disjoint ranges.

Given these hypotheses, if at any time there are at least two free variables in $\mathcal{U}_t$ and at least $v_t + 1$ values in S_t, then the algorithm cannot answer after step t. This is because processor 1 cannot distinguish two cases: the case when the variable it knows is set to the second highest value in S_t and all other free variables have lower values and the case when one other free variable is set to the highest value in S_t. We must attempt to carry out the construction so as to keep the set of free variables and the domain size as large as possible. When we can no longer maintain two free variables, the construction will stop, yielding a lower bound on T; we can then extract an initial value for s which allows the construction to continue for that many steps.

The proof proceeds by induction on t. For the base case, we set $S_0 = \{1, 2, \ldots, s\}$, $\mathcal{U}_0 = \{x_1, \ldots, x_n\}$, $\mathcal{M}_o = \phi$, and $v_0 = n$; the hypotheses are trivially satisfied. For the inductive step, suppose the situation as described above holds through step t. We describe how to maintain the inductive hypotheses by defining $\mathcal{U}_{t+1}$, $\mathcal{M}_{t+1}$, and S_{t+1}. Initially, let $S_{t+1} = S_t$; we will change S_{t+1} by removing values, based on what the PRAM algorithm does at step $t+1$.

We will find it useful to borrow a technique from [16]. Lemmas 9 and 10 were used there to restrict the manner in which processors may communicate with each other by restricting the domain S_{t+1}. The importance of the lemmas lies in the relatively small reduction in domain size. Similar lemmas with greater reduction were given in [11].

Lemma 9. *If $f_1, f_2, \ldots, f_k$ are functions with common domain S, where $|S| = k!q^{k+1}$, then there exists a subdomain S' of size q such that when $f_1, \ldots, f_k$ are restricted to S', each function is either constant or 1-1.*

Proof. A theorem of Erdős and Rado ([10]) states that in any family of at least $\ell!k^{\ell+1}$ (not necessarily different) sets of size at most ℓ, there is a *sunflower* formed by k sets; that is, a collection of k sets whose pairwise intersection is equal to its intersection. With each element $e \in S$, associate the set of ordered pairs $A_e = \{(r, f) | f \in \{f_1, \ldots, f_k\}, f(e) = r\}$. There are $k!q^{k+1}$ such sets, and so there exists a sunflower of size q among them.

Let the elements corresponding to the sets in the sunflower be $e_1, e_2, \ldots e_q$. If we set $S' = \{e_1, e_2, \ldots e_q\}$, the desired property is obtained. Consider an ordered pair (r, f_i) in the sunflower. If this pair is in the center of the sunflower (that is, in all the sets A_e, $e \in S'$), it follows that $f_i(e) = r$ for all $e \in S'$, and f_i is constant over S'. If (r, f_i) is in a petal (that is, it is in the set A_{e_j} and in no other set), then $f_i(e_j) = r$ but for no other e_k does $f_i(e_k) = r$. Since there was nothing special about our choice of r, we conclude that f_i is 1–1 over S'. ∎

Let us define the value of a write function to be 0 if the processor does not wish to write, and apply Lemma 9 to the set of all read and write access functions used at step $t+1$. This restricts S_{t+1}. Remember that each processor uses t read access functions and one write access function at step $t+1$; this is a total of $k = n(t+1)$ functions. We overestimate the domain reduction necessitated by Lemma 9 by assuming an initial domain size of $(kq)^{k+1}$ reduced to q. Once we have applied Lemma 9 to a given f, if it is 1–1, then there is at most one value in S_{t+1} on which it does not write. We can remove that value from S_{t+1}, thereby ensuring that processors using f always write. At this point, then, the size of S_{t+1} is $\dfrac{s_t^{1/(n(t+1)+1)}}{n(t+1)} - n(t+1)$.

Lemma 10. *If f, g are two 1-1 functions with common domain S, $|S| = 4q$, then there exists a subdomain S' of size q such that f and g, restricted to S', are either identical or have disjoint ranges.*

Proof. If f, g have the same value for q elements in S, then let S' be those elements. As a result, f and g are identical when restricted to S'. Otherwise, remove all such elements from S. Form a graph whose nodes are the elements of S; there is an edge between a and b if $f(a) = g(b)$. This graph consists of disjoint cycles and thus is 3-colourable; choose any independent set of size q and let S' be this set. It follows that f and g have disjoint ranges when restricted to S'. ■

We apply Lemma 10 to all pairs consisting of one read or write access function used before step $t+1$ and one function used at step $t+1$. There are $n(t+1)(t+2)/2$ functions in the first category and $n(t+1)$ functions in the second category; each application reduces the size of S_{t+1} by a factor of 4. This ensures that Hypothesis 2 holds after step $t+1$.

It remains to ensure that Hypothesis 1 holds after step $t+1$. There are two ways in which it can be violated: if a cell that knows one free variable is written into by a processor knowing another free variable, the state of that cell after step $t+1$ may be a function of two free variables. Also, if a processor knowing one free variable reads a cell knowing another free variable, the state of that processor may be a function of two free variables.

Let us construct a graph whose nodes are the free variables; there is an edge between x_i and x_j if a processor knowing x_j learns something about x_i (in the sense described above). Each processor can contribute at most $t+1$ edges to this graph, since it reads at most t cells and writes into at most one cell at step $t+1$. Turán's theorem [3] states that in any graph with v vertices and e edges, there exists an independent set of size $\frac{v^2}{v+2e}$. Hence in our graph there is an independent set of size $\frac{v_t^2}{v_t+2n(t+1)} \geq \frac{v_t^2}{3n(t+1)}$.

If there are j variables not in this independent set, then we choose the j smallest values of S_{t+1}, fix the variables to those values in an arbitrary fashion, and remove those values from S_{t+1}, thus ensuring Hypothesis 1. The two inductive hypotheses are now satisfied. The resulting recurrence equations (slightly simplified) are:

$$v_{t+1} \geq \frac{v_t^2}{3n(t+1)}$$

$$s_{t+1} \geq \frac{s_t^{1/n(t+1)+1}}{n(t+1)2^{(t+1)^2(t+2)n^2}} - n(t+1)$$

It is now not difficult to obtain the following inequalities by estimation, and to prove them using induction on t (for n sufficiently large).

$$v_t \geq \frac{n}{2^{2^{3t}}}$$

$$s_t \geq \frac{s^{n^{-3t}}}{2^{n^{2t}}}$$

Since the process can continue as long as there are at least two free variables, the bound on v_t ensures $T \geq \frac{1}{3}\log\log n$. If the domain size after step T is to be at least n, then s need only be as large as $2^{n^{4\log\log n}} = 2^{2^{\Omega(\log n \log\log n)}}$. A simple calculation shows that $T \geq \frac{1}{3}\log\log\log s$ for n sufficiently large. ∎

5 Conclusions

We have shown that the minima, prefix-minima, range-minima, and ANSV problems, with input elements taken from the integer domain $[1..s]$, $s \geq n$, can all be solved in $O(\log\log\log s)$ time using $n/\log\log\log s$ processors (optimal speedup) on the COMMON CRCW PRAM. As an application, we obtain an algorithm with the same bounds for the problem of triangulating a monotone polygon whose coordinates are taken from the integer domain $[1..s]$. The algorithms presented are simple and no large constants are hidden in the asymptotic complexity bounds.

We also gave a matching lower bound of $\Omega(\log\log\log s)$ for $s = 2^{2^{\Omega(\log n \log\log n)}}$. Thus, our algorithms cannot be improved when expressed solely in terms of the domain size. If the range of the domain is further restricted, however, improvements are possible. [12] gave a technique which could be applied to find the minimum of n integers from the range $[1..n^k]$ in $O(k)$ time on a COMMON CRCW PRAM; [4] gave an $O(\log^* n)$ time algorithm on PRIORITY for computing the prefix-minima when $s = O(n)$. This shows that $t = \Theta(\log\log\log s)$ does not give the correct tradeoff between domain size and computation time for all values of s. More work is needed to discover upper and lower bounds for parallel minimum computation that are tight for all s.

[4] gives an algorithm for merging sorted lists of length n from the domain $[1..2n]$ in time $\alpha(n)$, where $\alpha(n)$ is the very slowly growing functional inverse of Ackermann's function. The technique presented here does not seem to be powerful enough to deal with the problem of merging, since fixing values very quickly constrains the adversary. The technique in [9] allows processors to learn more than one variable, but is only good for moderately large (doubly exponential in n) domains, and its applicability to other problems remains unclear.

References

1. A. Amir, G. M. Landau, and U. Vishkin. Efficient pattern matching with scaling. In *Proc. of the First Annual ACM-SIAM Symposium on Discrete Algorithms*, pages 344–357, 1990.
2. P. Beame and J. Håstad. Optimal bounds for decision problems on the CRCW PRAM. In *Proc. of the 19th Ann. ACM Symp. on Theory of Computing*, pages 83–93, 1987.
3. C. Berge. *Graphs and Hypergraphs*. North-Holland, 1973.
4. O. Berkman, J. JáJá, S. Krishnamurthy, R. Thurimella, and U. Vishkin. Some triply-logarithmic parallel algorithms. In *Proc. of the 31st IEEE Annual Symp. on Foundation of Computer Science*, pages 871–881, 1990. To appear in *SIAM J. of Comput.* as 'Top-bottom routing around a rectangle is as easy as computing prefix minima'.
5. O. Berkman, B. Schieber, and U. Vishkin. Optimal doubly logarithmic parallel algorithms based on finding all nearest smaller values. *Journal of Algorithms*, 14:344–370, 1993.

6. O. Berkman and U. Vishkin. On parallel integer merging. Technical Report UMIACS-TR-90-15.1 (revised version), Institute for Advanced Computer Studies, University of Maryland, College Park, Maryland 20742, USA, 1990. To appear in *Information and Computation*.
7. R. B. Boppana. Optimal separations between concurrent-write parallel machines. In *Proc. of the 21st Ann. ACM Symp. on Theory of Computing*, pages 320–326, 1989.
8. S. Chaudhuri. *Lower Bounds for Parallel Computation*. PhD thesis, Rutgers University, 1991.
9. J. Edmonds. Lower bounds with smaller domain size on concurrent write parallel machines. In *Proc. 6th Annual IEEE Conference on Structure in Complexity Theory*, 1991.
10. P. Erdős and R. Rado. Intersection theorems for systems of sets. *J. London Math. Soc.*, 35:85–90, 1960.
11. F. E. Fich, F. Meyer auf der Heide, and A. Wigderson. Lower bounds for parallel random-access machines with unbounded shared memory. In *Advances in Computing Research*. JAI Press, 1986.
12. F. E. Fich, P. L. Ragde, and A. Wigderson. Relations between concurrent-write models of parallel computation (preliminary version). In *Proceedings 3rd Annual ACM Symposium on Principles of Distributed Computing*, pages 179–189, 1984.
13. F. E. Fich, P. L. Ragde, and A. Wigderson. Simulations among concurrent-write PRAMs. *Algorithmica*, 3:43–51, 1988.
14. H. N. Gabow, J. L. Bentley, and R. E. Tarjan. Scaling and related techniques for geometry problems. In *Proc. of the 16th Ann. ACM Symp. on Theory of Computing*, pages 135–143, 1984.
15. M. T. Goodrich. Triangulating a polygon in parallel. *Journal of Algorithms*, 10:327–351, 1989.
16. V. Grolmusz and P. L. Ragde. Incomparability in parallel computation. In *Proc. of the 28th IEEE Annual Symp. on Foundation of Computer Science*, pages 89–98, 1987.
17. D. Harel and R. E. Tarjan. Fast algorithms for finding nearest common ancestors. *SIAM J. Comput.*, 13(2):338–355, 1984.
18. C. P. Kruskal. Searching, merging, and sorting in parallel computation. *IEEE Trans. on Comp*, C-32:942–946, 1983.
19. F. Meyer auf der Heide and A. Wigderson. The complexity of parallel sorting. In *Proc. of the 26th IEEE Annual Symp. on Foundation of Computer Science*, pages 532–540, 1985.
20. P. L. Ragde, W. L. Steiger, E. Szemerédi, and A. Wigderson. The parallel complexity of element distinctness is $\Omega(\sqrt{\log n})$. *SIAM Journal on Disceret Mathematics*, 1(3):399–410, Aug. 1988.
21. V. Ramachandran and U. Vishkin. Efficient parallel triconnectivity in logarithmic parallel time. In *Proc. of the 3rd Aegean Workshop on Parallel Computing, Springer LNCS 319*, pages 33–42, 1988.
22. B. Schieber. *Design and analysis of some parallel algorithms*. PhD thesis, Dept. of Computer Science, Tel Aviv Univ., 1987.
23. B. Schieber and U. Vishkin. On finding lowest common ancestors: simplification and parallelization. *SIAM J. Comput.*, 17(6):1253–1262, 1988.
24. Y. Shiloach and U. Vishkin. Finding the maximum, merging, and sorting in a parallel computation model. *Journal of Algorithms*, 2:88–102, 1981.
25. M. Snir. On parallel searching. *SIAM J. Comput.*, 14:688–707, 1985.
26. L. G. Valiant. Parallelism in comparison problems. *SIAM J. Comput.*, 4:348–355, 1975.

Parallel Construction of Quadtrees and Quality Triangulations

Marshall Bern* David Eppstein† Shang-Hua Teng‡

Abstract

We describe efficient PRAM algorithms for constructing unbalanced quadtrees, balanced quadtrees, and quadtree-based finite element meshes. Our algorithms take time $O(\log n)$ for point set input and $O(\log n \log k)$ time for planar straight-line graphs, using $O(n+k/\log n)$ processors, where n measures input size and k output size.

1 Introduction

A crucial preprocessing step for the finite element method is mesh generation, and the most general and versatile type of two-dimensional mesh is an unstructured triangular mesh. Such a mesh is simply a triangulation of the input domain (e.g., a polygon), along with some extra vertices, called *Steiner points*. Not all triangulations, however, serve equally well; numerical and discretization error depend on the *quality* of the triangulation, meaning the shapes and sizes of triangles. A typical quality guarantee gives a lower bound on the minimum angle in the triangulation.

Baker et al. [2] first proved the existence of quality triangulations for arbitrary polygonal domains; their grid-based algorithm produces a triangulation with all angles between 14° and 90°. Chew [7] also bounded all angles away from 0° using incremental constrained Delaunay triangulation. Both of these algorithms, however, produce triangulations in which all triangles are approximately the size of the smallest input feature; hence, many more triangles than necessary may be generated, slowing down both the mesh generation and finite element procedures. Bern et al. [4] used adaptive spatial subdivision, namely quadtrees, to achieve guaranteed quality with a small number of triangles (within a constant factor of the optimal number). Modifications yield other desirable properties, such as small total length [10] and no obtuse triangles [13]. Mitchell and Vavasis

*Xerox Palo Alto Research Center, 3333 Coyote Hill Rd., Palo Alto, CA 94304.

†Department of Information and Computer Science, University of California, Irvine, CA 92717. Supported by NSF grant CCR-9258355. Work performed in part while visiting Xerox Palo Alto Research Center.

‡Department of Mathematics, Massachusetts Inst. of Technology, Cambridge, MA 02139. Work performed in part while at Xerox Palo Alto Research Center.

generalized this approach to three dimensions [14]. Subsequently, Ruppert [15] built on Chew's more elegant algorithm to achieve the same theoretical guarantee on the number of triangles. For more mesh generation theory, see our survey [5].

In this paper, we give parallel algorithms for mesh generation. The finite element method is often performed on parallel computers, but parallel mesh generation is less common. (We are unaware of any theoretical papers on the subject and only a few practical papers, for example [18].) In some applications, a single mesh is generated and used many times; in this case the time for mesh construction is not critical and a relatively slow, sequential algorithm would suffice. In other applications, especially when the physics or geometry of the problem changes with time, a mesh is used once and then discarded or modified. Then a parallel mesh generator would offer considerable speed-up over a sequential generator.

We parallelize the quadtree-based methods of Bern et al. [4]. The grid-based method of Baker et al. [2] and a grid-based modification of Chew's algorithm [7] both parallelize easily, but as mentioned above these may produce too many triangles. It is currently unknown whether Ruppert's method [15] has an efficient parallel version.

A *quadtree* [16] is a recursive partition of a region of the plane into axis-aligned squares. One square, the *root*, covers the entire region. A square can be divided into four *child* squares, by splitting it with horizontal and vertical line segments through its center. The collection of squares then forms a tree, with smaller squares at lower levels of the tree.

It may seem that quadtrees are easy to construct in parallel, a layer a time. In practice this idea may work well, but it does not provide an asymptotically efficient algorithm because the quadtree may have depth proportional to its total size. We use the following strategy instead. We first find a "framework", a tree of quadtree squares such that every internal node has at least two nonempty children. This tree guides the computation of the complete quadtree. We then *balance* the quadtree so that no square is adjacent to a square more than twice its side length. For polygonal inputs, we further refine and rebalance the quadtree so that edges are well separated. Finally, we perform local "warping" as in [4] or [13], to construct a guaranteed-quality triangulation. Figure 1 shows a mesh computed by a variant of our sequential algorithm.

1.1 Input assumptions

We assume that the input is a point set or planar straight-line graph, with n vertices. The coordinates of the points or vertices are fixed-point binary fractions, strictly between 0 and 1, that can be stored in a single machine word. We assume the ability to perform simple arithmetic and Boolean operations on such words in constant time per operation, including bit shift operations. Finally, we assume the ability to detect the highest order nonzero bit in the binary representation of such a number in constant time.

These assumptions are similar to a model in which the input coordinates are machine-word integers, as in the work of Fredman and Willard [12, 17]. Our

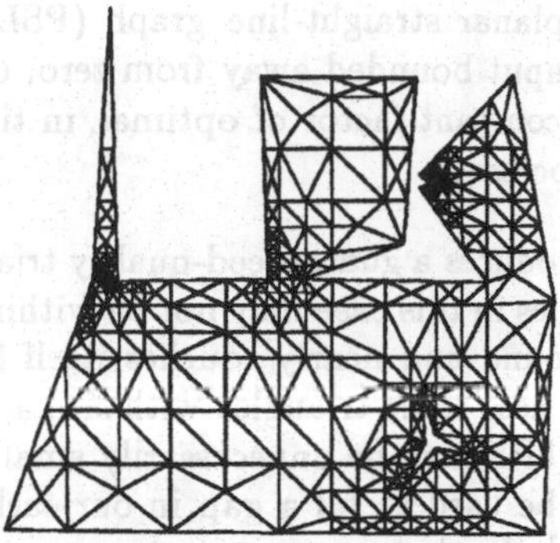

Figure 1. A mesh derived from a quadtree (courtesy Scott Mitchell).

description in terms of fixed point fractions is somewhat more convenient for our application, but equivalent in expressive power. If we were to use a real number model instead—as is more usual in computational geometry—we would incur a slight expense in time. The key operation of finding the highest bit can be simulated by binary search in time $O(\log\log R)$ per operation, where R is the ratio between the largest and smallest numbers tested by the algorithm. For our purposes $R = O(2^k)$ where k is the output size, so the penalty for weakening the model is a factor of $O(\log k)$ time.

Our parallel algorithms use the EREW PRAM model of exclusive access to shared memory [11]. Since our processor bound depends on k, the size of the output triangulation, which is not known in advance, we need a mechanism for allocating additional processors within the course of the computation. We assume that a single allocation step, in which each of the N processors already allocated asks to be replaced by some number of additional processors, can be performed in $O(\log N)$ time.

1.2 New results

We describe EREW PRAM algorithms to perform the following tasks.

- We construct a balanced or unbalanced quadtree for an arbitrary point set in time $O(\log n)$ with $O(n + k/\log n)$ processors.
- We triangulate a point set with total edge length $O(1)$ times the minimum and all angles bounded between 36° and 80°, using a total number of points within a constant factor of optimal, in time $O(\log n)$ using $O(n+k/\log n)$ processors.
- We triangulate a point set with total edge length $O(1)$ times the minimum and all angles less than 90°, using $O(n)$ Steiner points, in time $O(\log n)$ using $O(n)$ processors.

- We triangulate a planar straight-line graph (PSLG) with all angles except those of the input bounded away from zero, using a total number of triangles within a constant factor of optimal, in time $O(\log k \log n)$ using $O(n + k/\log n)$ processors.

Our last algorithm produces a guaranteed-quality triangulation of a polygon, but the number of triangles in this case may not be within a constant of optimal. If the polygon wraps around and nearly touches itself from the outside (as in Figure 1), our algorithm uses some triangles with size approximately the size of this outside "feature", which may be unnecessarily small.

Our results can also be used to fill a gap in our earlier paper [4], noted by Mitchell and Vavasis [14]. In that paper we gave sequential algorithms for triangulation of point sets, polygons, and planar straight-line graphs, all based on quadtrees. We claimed running times of $O(n \log n + k)$ in all cases, but gave a proof only the case of point sets. There turned out to be some complications in achieving this bound for polygons and PSLGs. (A straightforward implementation achieves $O(k \log n)$.) However, the ideas given here also improve sequential quadtree-based triangulation, yielding the first guaranteed-quality triangulation algorithms with the optimal running time of $O(n \log n + k)$.

2 Unbalanced quadtrees

We first describe how to generate a quadtree for a set of points in the plane. In the resulting quadtree, there are no restrictions on the sizes of adjacent squares, but no leaf square may contain more than one point. The root square of the quadtree will be the semi-open unit square $[0, 1)^2$. Thus the corners of quadtree squares have coordinates representable in our fixed-point format.

The sides of all squares in the quadtree have lengths of the form 2^{-i}, and for any square of side length 2^{-i} the coordinates of all four corners are integral multiples of 2^{-i}. We define a square with bottom left corner (x, y) and size 2^{-i} to contain a point (x', y') if $x \leq x' < x + 2^{-i}$ and $y \leq y' < y + 2^{-i}$. Given two input points (x, y) and (x', y'), we define their ***derived square*** to be the smallest such square containing both points. The size of this square can be found by comparing the high order bits of $x \oplus x'$ and $y \oplus y'$, and the coordinates of its corners can be found by masking off lower order bits.

Given a point (x, y), where x and y are k-bit fixed point fractions, we define the ***shuffle*** $Sh(x, y)$ to be the $2k$-bit fixed point fraction formed by alternately taking the bits of x and y from most significant to least significant, the x bit before the y bit. We represent $Sh(x, y)$ implicitly by the pair (x, y). We can compare two numbers $Sh(x, y)$ and $Sh(x', y')$ in this implicit representation in constant time, using arithmetic and high bit operations separately on x and y.

The first step of our algorithm will be to sort all the input points by the values of their shuffled coordinates. This can be done in $O(\log n)$ time with $O(n)$ EREW processors [8]. From now on we assume that the points occur in this sorted order.

Lemma 1. *The set of points in any square of a quadtree rooted at* $[0,1)^2$ *form a contiguous interval in the sorted order.*

Proof: The points in a square of size 2^{-i} have the same i most significant shuffled-coordinate bits, and any pair of points with those same bits shares a square of that size. If a point (x,y) is outside the given square, one of its first i bits must differ. If that bit is zero, (x,y) will appear before all points in the square. If the bit is one, (x,y) will appear after all points in the square. Hence it is impossible for (x,y) to appear before some points and after some others in the sorted order. ∎

Lemma 2. *If more than one child of quadtree square s contains an input point, then s is the derived square for two adjacent points in the sorted order.*

Proof: By Lemma 1, the points in s form an interval, which can be divided into two to four smaller intervals corresponding to the children of s. Then s is the derived square for any pair of points in two different smaller intervals. ∎

Consider the desired unbalanced quadtree as an abstract rooted tree. If we remove all leaves of this tree that do not contain input points, and contract all remaining paths of nodes having one child each, we obtain a tree T_F in which all internal nodes have degree two or more. We call T_F the *framework* and use it to construct the quadtree.

By Lemma 2, the nodes of T_F exactly correspond to the derived squares for adjacent points in the sorted order, and the structure of T_F corresponds to the nesting of intervals induced by the derived squares, as in Lemma 1. So for each adjacent pair of points, we compute their derived square and note its side length. (Some squares may be derived in as many as three ways, but we can eliminate this problem later.) The nesting of intervals is computed by finding, for each derived square, the first larger one to its right and to its left in the sorted order. This is an all-nearest-larger-values computation, taking $O(\log n)$ time with $O(n/\log n)$ work [3].

Once the framework T_F is computed, we construct the quadtree T_Q. Each edge in T_F corresponds to a path of perhaps many squares in T_Q, with the number of squares determined by the relative sizes of T_F squares. So we perform a processor allocation step in which each framework edge e requests $O(p_e/\log n)$ processors, where p_e is e's number of squares. Now all remaining squares (leaves and children of path squares), can be constructed in $O(\log n)$ time. The total number of processors is $O(n+k/\log n)$ where k is the complexity of the resulting (unbalanced) quadtree.

Theorem 1. *Given n input points, we can compute a quadtree with k squares, in which each point is alone in its square, in time* $O(\log n)$ *using* $O(n+k/\log n)$ *EREW processors.* ∎

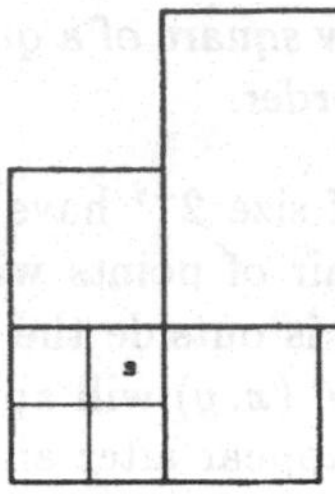

Figure 2. The squares forced by square s.

3 Balancing the quadtree

The quadtree-based mesh generation algorithms of [4, 10, 13] impose a *balance condition*: no leaf square is adjacent to another leaf square smaller than half its size. (Some variants of these algorithms impose stronger conditions; the techniques given here generalize.) These algorithms also need *cross pointers* between squares of the same size sharing a common side.

We proceed in two stages, starting from the unbalanced quadtree T_Q of Section 2. In the first stage, we produce a tree of squares T'_Q in which some non-leaf squares may have fewer than four child squares. Tree T'_Q, however, will satisfy the balance condition above; it will also include cross pointers.

The balance condition is ensured if, for each square s, three other squares (not necessarily leaves) exist: the two squares adjacent to both s and its parent, twice as large as s, and the square sharing a corner with both s and its parent, four times as large as s. See Figure 2. (If any of these squares protrudes from $[0,1)^2$, then an exception is granted.) The parents of these forced squares exist either by the same rule applied to the parent of s, or because they are themselves ancestors of s. The three forced squares do not force other new squares in an unbounded chain of requirements; the nine squares forced by these three either already exist or are forced by the parent of s. (We do not require the siblings of the forced squares, as this could result in such an unbounded chain. This is why we allow a square to have fewer than four children.)

How do we add forced squares in parallel all over T_Q? For each side length in T_Q, we compute a list of squares (say top left corners) containing: all current squares, all squares of that size that we must create, and all squares that should be cross-linked if they exist or are created. Then we simply sort these lists using shuffled or unshuffled coordinates. Each list contains $O(1)$ copies of any top left corner; these determine all actions necessary to produce T'_Q (creating squares and adding cross pointers). Each quadtree level has at most $O(n)$ squares, so this algorithm takes only $O(\log n)$ time, but requires $O(k)$ processors and hence $O(k \log n)$ total work.

In the second stage, we turn T'_Q back into a quadtree by simply splitting all squares, leaves as well as internal nodes without all four children. This preserves the balance condition and restores the required number of children per parent.

Due to this extra split, the resulting balanced quadtree T_B may have up to four times as many squares as the quadtree produced by the sequential balancing algorithm. Cross pointers for T_B can be found recursively by examining squares' siblings as well as the children of squares cross-linked to their parents.

The two stages together give us the balanced quadtree T_B, but we may no longer know which squares contain which input points. To obtain this important information, we group the squares of T_B into blocks of $O(n)$ squares each and then process each block separately. Within a block, we compute a framework (that is, a compressed quadtree as above) for the input points along with the bottom left corners of the new squares. We find the squares of this framework that contain one input point, and from them determine the nearest input point above and to the right of each T_B square corner. For each T_B square, we can then test its closest above-right point to see whether it lies within the square. These steps also take $O(\log n)$ time and $O(k \log n)$ work.

The following theorem shows how to combine the "sorting algorithm" just described with a top-down algorithm in order to achieve optimal efficiency.

Theorem 2. *An unbalanced quadtree can be balanced and cross-linked in time $O(\log n)$ using $O(n + k/\log n)$ processors.*

Proof: Consider the following top-down method. Assume a virtual processor per square of the unbalanced quadtree. This processor waits until all the balancing and cross-linking is done for its parent's level. Then it creates nearby squares to enforce the balance condition, coordinates with neighboring squares so that no new square is created multiple times, and adds all necessary cross-links, in $O(1)$ time. It also tests whether its square contains the (at most one) input point contained in its parent. This results in an algorithm which has time t bounded by the number of levels in the quadtree, and total work $O(k)$. By Brent's lemma [6], only $O(k/t)$ actual processors are required to simulate all virtual processors in time $O(t)$.

To combine the two algorithms, we apply the sorting algorithm only at certain levels. We choose a set of levels, equally spaced and $\ell = \lceil \log_2 n \rceil$ apart. Once the spacing is fixed there are ℓ ways of making this choice, each giving rise to a set of levels disjoint from other such sets, so for some such set there are a total of only $O(k/\log n)$ squares. The appropriate choice of set can be determined from the framework tree. Once we have added the required squares in this set of levels, we apply the parallelization of the top-down algorithm in $O(\log n)$ time and $O(k)$ total work. The scheduling required for Brent's lemma can be performed using information on the number of squares per level, again computed from the $O(n)$-size framework tree. ∎

4 Mesh generation for point sets

The balanced quadtree computed in the last section can be modified by a set of "warping" steps to give a triangulation of the input point set, with no angles smaller than about 20°, as in [4]. These warping steps are local, involving only $O(1)$ squares each, and hence can obviously be performed in constant parallel time with optimal work. In this section, we go on to solve two slightly harder problems: (1) approximate minimum-weight no-small-angle triangulation, and (2) approximate minimum-weight nonobtuse triangulation.

Eppstein [10] showed how to sequentially compute triangulations of point sets with these guarantees: all angles between 36° and 80°, total edge length within a constant factor of the minimum, and total number of triangles within a constant of the minimum for any angle-bounded triangulation. The algorithm again uses local warping, trivial to parallelize, but the quadtree must also satisfy some stronger conditions than the ones given directly by Theorems 1 and 2:

- The coordinate axes of the quadtree must be rotated so that the diameter, of length d, connecting the farthest pair of points, is horizontal.
- A row of equal-size smaller squares that contain the input is cut from the root square. These squares have side length proportional to $\max\{d', d/n\}$, where d' is the maximum distance of any point from the diameter.
- The points must be *well-separated* from each other, meaning that for some specified constant c, if a point is in a square with side ℓ then its nearest neighbor must be at least distance $c\ell$ away. (In [4], $c = 2\sqrt{2}$.)

We relax these requirements somewhat to simplify our parallel algorithm. Rather than computing the exact diameter, it suffices to find some line segment with length within a factor of, say, .95 of the diameter, and rotate the points so that line segment forms an angle of $O(1/n)$ with the horizontal axis. Such a line segment can be found by projecting the point set onto $O(1)$ different axes, taking the extrema of each projected point set, and choosing the pair forming the longest segment. The rotation can be performed in our integer model by treating our inputs as complex numbers with integer coordinates, and multiplying by another such number chosen appropriately. The result will be a set of points with the correct orientation but scaled by a factor of $O(n)$, which corresponds to using $O(\log n)$ additional bits to represent each coordinate value.

We scale and translate the rotated point set to fit in the rectangle $[1/4, 3/4) \times [2^{-i}, 2^{-i+1})$, where i is the smallest integer with $2^{-i} \geq 1/n$ for which this is possible. The row of squares will then have side lengths equal to 2^{-i+1}. These steps may all be performed in $O(\log n)$ time with $O(n/\log n)$ processors.

It remains to ensure the separation of points. We first compute *near neighbors* (approximate nearest neighbors), using the balanced quadtree constructed by Theorem 2. We simply examine the $O(1)$ squares around each square containing an input point. If all those squares are empty, we need not find a near neighbor for the input point; otherwise we take the near neighbor to be any point in a neighboring square. We can now add notations to the framework tree of Section 2 specifying the desired size of the quadtree square containing a point with a near neighbor; the size is, say, one-eighth of the distance to the near neighbor. A suitable quadtree can then be computed as in Theorems 1 and 2.

Theorem 3. *Given a set of n points in the plane, we can compute a triangulation with k triangles, total length $O(1)$ times the minimum possible, in which all angles measure between 36° and 80°, in time $O(\log n)$ using $O(n + k/\log n)$ EREW processors. The output size k is $O(m + n)$, where $m = m(\epsilon)$ is the minimum number of Steiner points required to triangulate the input point set with no angle smaller than fixed constant ϵ.* ∎

We now consider the second problem: nonobtuse triangulation. Sequential quadtree based methods can triangulate a point set with all acute angles and only $O(n)$ Steiner points [4]. Moreover, the total edge length can be made to approximate that of the minimum-weight triangulation [10]. The following theorem extends this result to the parallel case. To save space, we omit the algorithm and proof; these will appear in the journal version of this paper.

Theorem 4. *A set of n input points can be triangulated with $O(n)$ triangles, all angles less than 90°, and total length $O(1)$ times the minimum possible, in time $O(\log n)$ using $O(n)$ EREW processors.*

5 Mesh generation for planar straight-line graphs

For most finite-element mesh generation applications, the input is not a point set but rather a polygonal region. We discuss here the most general input, a planar straight-line graph (PSLG). Our triangulation algorithm handles simple polygons as a special case, but the output complexity may be larger than necessary due to input features that are near to each other in Euclidean distance but far in geodesic distance.

Several new complications arise with PSLG input. First, we must modify the input to eliminate any pre-existing acute (below 90°) corners [4], and this should be done without introducing new points or edges too close to existing ones. Second, we must subdivide the edges of the PSLG where they cross the sides of quadtree squares. Third, we require that vertices not only be well-separated from other vertices, but also from other edges, and that edges be well-separated from each other; this means that each (piece of an original) edge is contained in a sufficiently small square that no vertex or other edge passes nearby (less than a specified constant—such as 3—times the square's side length).

To accomplish these goals, we take the following approach. We compute vertex-to-vertex and vertex-to-edge approximate nearest neighbors using an initial balanced quadtree T_B. This information enables us to cut off the acute corners and build a second quadtree T'_B in which vertices are well-separated from edges and other vertices. We subdivide the edges into pieces in this quadtree, and then split the squares of the quadtree to create a third quadtree T''_B in which each square contains only $O(1)$ pieces of edges. We can then—finally—compute approximate nearest neighbor information for pieces of edges, which we use to construct the final quadtree T^*_B, in which everything is well-separated.

We now flesh out the steps mentioned above. Approximate vertex-vertex nearest neighbors proceeds as in Section 4. The following lemma (easy geometric proof omitted) shows how to find a nearby non-incident edge for each vertex of the planar straight-line graph.

Lemma 3. *If the nearest edge e to a vertex v has distance d from v, then either some vertex v' is within distance $\sqrt{2}d$ from v, or e is visible to v through an angle of at least $90°$.*

So for each vertex of the PSLG, we need only determine which edge is visible along each of the four axis directions, and choose the nearest of these four edges. This horizontal and vertical ray-shooting problem is exactly that solved by the ***trapezoidal decomposition***, which can be constructed in time $O(\log n)$ using $O(n)$ CREW processors [1]; this algorithm can be simulated on an EREW machine with logarithmic slowdown [11].

Now we cut off acute angles around each vertex at a distance proportional to the vertex's nearest neighbor as in [4, 15], so that cut-off triangles do not contain other parts of the input, two such triangles on the same edge match up, and the new cutting edges are not unnecessarily short. We can ignore the cut-off triangles for the remainder of the algorithm, simply reattaching them (with appropriately subdivided cutting edges) after the final warping step. For the remainder of the algorithm, we can assume as in [4, 15] that all angles measure at least $90°$, and consequently each vertex has bounded degree.

We can now compute a quadtree T'_B in which each vertex lies in a square of size proportional to the nearer of its nearest neighbors (vertex and edge). Vertices will be well-separated from other vertices and edges, but two PSLG edges may yet cross the same square. We subdivide the PSLG edges into pieces contained in each quadtree square so that we will be able to discover these problems and correct them. In a sequential algorithm, we could subdivide simply by walking from one end of each edge to the other, keeping track of the crossed squares, but in a parallel algorithm we must do this differently.

We apply the accelerated centroid technique [9] to T'_B. This produces a binary tree T_C, the ***accelerated centroid tree***, which has logarithmic height and linear size; each subtree of T_C corresponds to a subtree of T'_B. To find the subdivisions of a given PSLG edge, we test it against T_C's root node r. Node r corresponds to a square in the quadtree T'_B, and we simply test whether the edge misses the square, is contained in the square, or crosses the square's boundary. If one of the first two cases occurs, we continue recursively to the left or right child of r. If the third case occurs, we split the segment in pieces and continue in parallel for each piece to the appropriate child of r. We process all edges in parallel, one level of T_C at a time. Each level can be handled in $O(\log k)$ time by $O(k/\log k)$ EREW processors. Between levels we reallocate processors and split the lists of edges being tested, resulting in time $O(\log k)$ and linear work.

This edge-subdivision step of our algorithm takes $O(\log^2 k)$ time and a total of $O(k \log k)$ work. But in each of these bounds, a factor of $O(\log k)$ can be reduced to $O(\log n)$ by the following trick: partition the quadtree T'_B into $O(k/n)$ subtrees, each of $O(n)$ squares, and perform the edge partition algorithm described above in parallel for each edge in each subtree. The number of levels in each centroid decomposition tree is thus reduced from $O(\log k)$ to $O(\log n)$, but $O(\log k)$ time per level is still required for processor reallocation.

We now look at the arrangement of edges in a single square of quadtree T'_B. We assert that each cell in this arrangement has only $O(1)$ complexity. If the cell is not convex, then there is a vertex of the PSLG on its boundary, and the bound on angles then implies the assertion. If the cell is convex, PSLG edges on its boundary must either meet at a nearby PSLG vertex or be nearly parallel. The fact that vertices are well-separated in T'_B then implies the assertion.

We consider in parallel each pair of edge pieces that bound a common cell. By the assertion above, there are only $O(k)$ such pairs over all of T'_B. For each pair, we determine the further quadtree subdivision that would be necessary to separate the edges in that square. We combine and balance all such subdivisions with sorting (as in Section 3) to produce a new balanced quadtree T''_B in time $O(\log n)$ and work $O(k \log n)$. We again use a centroid tree to determine the sub-pieces of edge pieces; each quadtree square now intersects only $O(1)$ sub-pieces.

Finally, for each square we search $O(1)$ nearby squares to determine how much further splitting is necessary. We again subdivide and rebalance to construct a last quadtree T^*_B, in which all PSLG edges are finally well-separated. Local warping then offers the following result. Bern et al. [4] achieve an angle bound of 18° sequentially; the same specific bound can be achieved in parallel, only with a constant factor more triangles.

Theorem 5. *Given a planar straight-line graph with n vertices, we can compute a triangulation with k triangles, in which all new angles are bounded away from 0°, in time $O(\log k \log n)$ using $O(n + k/\log n)$ EREW processors. The output size k is $O(n + m)$, where $m = m(\epsilon)$ is the minimum number of Steiner points required to triangulate the given PSLG with no new angle smaller than fixed constant ϵ.* ∎

6 Conclusions

We have given a theoretical study of parallel two-dimensional mesh generation. We believe this area deserves further research, both theoretical and practical.

The triangulations we construct are within a constant factor of the optimal complexity, but it might be of some practical interest to improve the constant factors. In particular, our balancing method wastes a factor of four; is it possible to compute the minimum balanced quadtree for a set of n points in time $O(\log n)$? Another interesting problem that we are leaving open is the parallel computation of approximate geodesic nearest neighbors. Efficient algorithms for this problem (both vertex-vertex and vertex-edge) would extend our PSLG methods to simple polygons with a stronger bound on the total number of triangles.

References

[1] M.J. Atallah, R. Cole, and M.T. Goodrich. Cascading divide-and-conquer: a technique for designing parallel algorithms. *SIAM J. Comput.* 18 (1989) 499–532.

[2] B. Baker, E. Grosse, and C. Rafferty. Nonobtuse triangulation of polygons. *Discrete Comput. Geom.* 3 (1988) 147–168.

[3] O. Berkman, D. Breslauer, Z. Galil, B. Schieber, and U. Vishkin. Highly parallelizable problems. *21st Symp. Theory of Computing* (1989) 309–319.

[4] M. Bern, D. Eppstein, and J.R. Gilbert. Provably good mesh generation. *31st Symp. Found. Comput. Sci.* (1990) 231–241. To appear in *J. Comp. Sys. Sci.*

[5] M. Bern and D. Eppstein. Mesh generation and optimal triangulation. In *Euclidean Geometry and the Computer*, World Scientific, 1992.

[6] R.P. Brent. The parallel evaluation of general arithmetic expressions. *J. ACM* 21 (1974) 201–206.

[7] L.P. Chew. Guaranteed-quality triangular meshes. TR-89-983, Cornell, 1989.

[8] R. Cole. Parallel merge sort. *SIAM J. Comput.* 17 (1988) 770–785.

[9] R. Cole and U. Vishkin. Optimal parallel algorithms for expression tree evaluation and list ranking. *3rd Aegean Workshop on Computing*, Springer LNCS 319 (1988).

[10] D. Eppstein. Approximating the minimum weight triangulation. *3rd Symp. Discrete Algorithms* (1992) 48–57. To appear in *Discrete Comput. Geom.*

[11] D. Eppstein and Z. Galil. Parallel algorithmic techniques for combinatorial computation. *Ann. Rev. Comput. Sci.* 3 (1988) 233–283.

[12] M.L. Fredman and D.E. Willard. Blasting through the information-theoretic barrier with fusion trees. *22nd Symp. Theory of Computing* (1990) 1–7.

[13] E.A. Melissaratos and D.L. Souvaine. Coping with inconsistencies: A new approach to produce quality triangulations of polygonal domains with holes. *8th Symp. Comput. Geom.* (1992) 202–211.

[14] S.A. Mitchell and S.A. Vavasis. Quality mesh generation in three dimensions. *8th Symp. Comput. Geom.* (1992) 212–221.

[15] J. Ruppert. A new and simple algorithm for quality 2-dimensional mesh generation. *4th Symp. Discrete Algorithms* (1993) 83–92.

[16] H. Samet. The quadtree and related hierarchical data structures. *Computing Surveys* 16 (1984) 188-260.

[17] D.E. Willard. Applications of the fusion tree method to computational geometry and searching. *3rd Symp. Discrete Algorithms* (1992) 286–295.

[18] R.D. Williams. Adaptive parallel meshes with complex geometry. Tech. Report CRPC-91-2, Center for Research on Parallel Computation, Cal. Tech.

PATTERN MATCHING FOR PERMUTATIONS*

Prosenjit Bose† **Jonathan F. Buss**‡ **Anna Lubiw**‡

jit@muff.cs.mcgill.ca jfbuss@uwaterloo.ca alubiw@uwaterloo.ca

Abstract

Given a permutation T of 1 to n, and a permutation P of 1 to k, for $k \leq n$, we wish to find a k-element subsequence of T whose elements are ordered according to the permutation P. For example, if P is $(1, 2, \ldots, k)$, then we wish to find an increasing subsequence of length k in T; this special case can be done in time $O(n \log \log n)$ [CW]. We prove that the general problem is NP-complete. We give a polynomial time algorithm for the decision problem, and the corresponding counting problem, in the case that P is *separable*—i.e. contains neither the subpattern $(3, 1, 4, 2)$ nor its reverse, the subpattern $(2, 4, 1, 3)$.

1. Introduction

The *pattern matching problem for permutations* is the following: We are given a permutation $T = (t_1, t_2, \ldots, t_n)$ of 1 to n, which we call the *text*, and a permutation $P = (p_1, p_2, \ldots, p_k)$ of 1 to k, $k \leq n$, which we call the *pattern*. We wish to know whether there is a length k subsequence of T, say $T' = (t_{i_1}, t_{i_2}, \ldots, t_{i_k})$, with $i_1 < i_2 < \ldots < i_k$, such that the elements of T' are ordered according to the permutation P—i.e. $t_{i_r} < t_{i_s}$ iff $p_r < p_s$. If T does contain such a subsequence, we will say that T *contains* P, or that P *matches into* T.

Herb Wilf spoke at the 1992 SIAM Discrete Math meeting about counting pattern matchings of permutations. He asked whether there is an algorithm to decide the pattern matching problem for permutations that runs faster than exponential time.

This note contains one positive and one negative result. We show that the general decision problem is NP-complete, and the counting problem is #P-complete. We give a polynomial time algorithm for the decision and counting problems in case the pattern P contains neither the pattern

* work supported in part by NSERC

† Dept. Computer Science, McGIll University, Montreal, Canada, H3A 2A7

‡ Dept. Computer Science, University of Waterloo, Waterloo, Canada, N2L 3G1

(3, 1, 4, 2) nor the pattern (2, 4, 1, 3). Such a pattern P is called *separable*, for reasons given in section 3.

There has been considerable interest in counting the number of permutations T that do *not* contain a smaller permutation P. Knuth [K, section 2.2.1] showed that the number of permutations of $1, \ldots, n$ without the pattern (3, 1, 2) is equal to the nth Catalan number. In [L, p. 27] Lovász considered permutations without the pattern (2, 1, 3). Rotem [R] considered permutations without the either of the patterns (2, 3, 1) or (3, 1, 2). Simion and Schmidt [SS] go further, counting the number of permutations avoiding all the patterns in any subset of the permutations on 1, 2, 3.

2. The NP-Completeness Result

Theorem 2.1. The pattern matching problem for permutations is NP-complete.

Proof. We will reduce 3-SATISFIABILITY to this problem. Suppose we have an instance of 3-SATISFIABILITY with variables $v_1, \ldots, v_l$ and clauses $C_1, \ldots, C_m$, where each clause C_i is the disjunction of three literals, and each literal is a variable or the negation of a variable.

We will first construct a pattern and a text with repeated elements, and later show how to eliminate the need for the repetitions. When the pattern has some element repeated, then copies of the repeated element must match to the copies of some repeated element in the text. In other words, using the notation of the introduction, we add the condition that $t_{i_r} = t_{i_s}$ iff $p_r = p_s$. For example, the pattern (1, 3, 1, 2) matches in the text (1, 3, 2, 6, 3, 4, 5) but does not match in the text (1, 4, 2, 3).

The pattern P will have a substring (i.e., contiguous subsequence) P_V for the variables, and a substring P_{C_i} for each of the clauses. Similarly, the text T will have a substring T_V for the variables, and a substring T_{C_i} for each clause. We will join these substrings of P and of T one after another, with "guards" in-between to ensure that elements of one of the substrings of the pattern can only match to elements of the corresponding substring of the text. The construction of these guards will be described later.

Construct $P_V = (1, 2, \ldots, l+1)$, and $T_V = (2, 1, 4, 3, 6, 5, \ldots, 2l, 2l-1, 2l+1)$. Note that a match of P_V into T_V will force i in P_V to match with either $2i$ or $2i-1$ in T_V, for $1 \leq i \leq l$. The former will correspond with setting the variable v_i TRUE, and the latter with setting it FALSE. Also, $l+1$ in P_V must match with $2l+1$ in T_V; these values will separate the variables from other values in the other substrings.

Consider the clause C_i containing variables $v_{i_1}, v_{i_2}, v_{i_3}$ in either positive or negated form, with $i_1 < i_2 < i_3$. Construct $P_{C_i} = (l+i+1, i_1, i_2, i_3, l+i+1)$. For $j = 1,2,3$, let t_j be $2i_j$ if variable v_{i_j} appears in C_i positively, and $2i_j - 1$ otherwise; and let f_j be the other of the two values. Let $b_i = 2l + 7i$. There are seven truth assignments to the variables that make the clause true; these assignments correspond to the seven sequences $s_{i,1} = (t_1, t_2, t_3)$, $s_{i,2} = (t_1, t_2, f_3)$, $s_{i,3} = (t_1, f_2, t_3)$, ..., $s_{i,7} = (f_1, f_2, t_3)$. The text string T_{C_i} comprises these seven sequences, each bracketed by a pair of equal values; i.e., T_{C_i} is the sequence

$$(b_i+1, s_{i,1}, b_i+1, b_i+2, s_{i,2}, b_i+2, b_i+3, s_{i,3}, b_i+3, \ldots, b_i+7, s_{i,7}, b_i+7).$$

For example, consider the clauses $C_1 = (v_1 \vee v_2 \vee \overline{v_4})$ and $C_2 = (\overline{v_2} \vee v_3 \vee \overline{v_5})$ over variables $\{v_1 \ldots, v_5\}$. The corresponding patterns and text strings are

$$\begin{aligned}
P_V &= (1,2,3,4,5,6),\\
T_V &= (2,1,4,3,6,5,8,7,10,9,21,11,13),\\
P_{C_1} &= (7,1,2,4,7),\\
T_{C_1} &= (18,2,4,7,18,19,2,4,8,19,20,2,3,7,20,21,2,3,8,21,\\
&\qquad 22,1,4,7,22,23,1,4,8,23,24,1,3,7,24),\\
P_{C_2} &= (8,2,3,5,8),\\
T_{C_2} &= (25,3,6,9,25,26,3,6,10,26,27,3,5,9,27,28,3,5,10,28,\\
&\qquad 29,4,6,9,29,30,4,6,10,30,31,4,5,9,31).
\end{aligned}$$

In any matching of P_{C_i} into T_{C_i} that respects a matching of P_V into T_V, the occurrences of $l+i+1$ must match with one of the bracket pairs, and thus (i_1, i_2, i_3) must be one of the sequences $s_{i,k}$.

Claim 2.2. The set of clauses $\{C_i\}$ is satisfiable if and only if there is a simultaneous matching of P_V into T_V and each P_{C_i} into T_{C_i}.

Proof. A matching of P_V into T_V corresponds to a truth-value assignment to the variables. By the above construction, this matching can be extended to a matching of P_{C_i} into T_{C_i} if and only if the corresponding assignment satisfies C_i. ∎

Two issues remain: how to put guards between the segments of P and of T to ensure that they remain separate; and how to eliminate the need for repeated elements in P and T.

We consider the guards first. Let P be the sequence $(l+m+1, P_V, l+m+1, P_{C_1}, l+m+1, P_{C_2}, l+m+1, \ldots, l+m+1, P_{C_m})$, followed by $3m+1$

copies of $l+m+1$. Let T be the sequence $(2l+7m+8, T_V, 2l+7m+8, T_{C_1}, 2l+7m+8, T_{C_2}, 2l+7m+8, \ldots, T_{C_m})$, followed by $3m+1$ copies of $2l+7m+8$. Observe that any matching of P into T must match the element $l+m+1$ of P to the element $2l+7m+8$ of T, because there are $4m+2$ copies of $l+m+1$ in P, and no value except $2l+7m+8$ appears more than $4m+1$ times in T. Since there are the same number of occurrences of $l+m+1$ in P as there are of $2l+7m+8$ in T, they do perform as guards, ensuring that P_V matches into T_V and each P_{C_i} matches into T_{C_i}.

Finally, let us turn to the question of eliminating repeated elements. It is easier to describe the construction if we allow the pattern and the text to each be a sequence of distinct numbers, not necessarily integers. Then to obtain a permutation, simply replace each element by its ordinal.

Instead of using a singleton in P_V for each variable, which must map to one of two possible values in T_V, we will use a pair in P_V which must map to one of two possible pairs in T_V. Occurrences of the variable in clauses will then be modelled as values intermediate to the corresponding pair, thus avoiding the need to duplicate values. Construct $P_V = (1, 2, \ldots, 2l-1, 2l, 2l+1)$, and $T_V = (3, 4, 1, 2, 7, 8, 5, 6, \ldots, 4l-1, 4l, 4l-3, 4l-2, 4l+1)$. Observe that if P_V matches into T_V, then the pair $2i-1, 2i$ must match with either $4i-1, 4i$ or $4i-3, 4i-2$, which we will interpret as setting variable v_i TRUE or FALSE respectively.

Consider a clause C_i containing variables $v_{i_1}, v_{i_2}, v_{i_3}$ in either positive or negated form, with $i_1 < i_2 < i_3$. Construct $P_{C_i} = (2l+2i+1, u_{i,1}, u_{i,2}, u_{i,3}, 2l+2i)$, where $u_{i,j}$ is a number strictly between $2i_j - 1$ and $2i_j$, larger than all values previously chosen in that interval. For $j = 1, 2, 3$, let $\hat{t}_j$ be the open interval $(4i_j - 1, 4i)$ if variable v_{i_j} appears in C_i positively, and the interval $(4i_j - 3, 4i_j - 2)$ otherwise; and let $\hat{f}_j$ be the other of the two intervals. Let $t_{j,1}$, $t_{j,2}$, $t_{j,3}$ and $t_{j,4}$ be values in the interval $\hat{t}_j$, and let $f_{j,1}$, $f_{j,2}$ and $f_{j,3}$ be values in the interval $\hat{f}_j$; all these values are to be distinct from each other and larger than all values previously chosen in their respective intervals. Let $b_i = 4l + 14i$.

For the seven satisfying assignments of C_i we choose the seven sequences $s_{i,1} = (t_{1,1}, t_{2,1}, t_{3,1})$, $s_{i,2} = (t_{1,2}, t_{2,2}, f_{3,1})$, $s_{i,3} = (t_{1,3}, f_{2,1}, t_{3,2})$, ..., $s_{i,7} = (f_{1,3}, f_{2,3}, t_{3,4})$. The text segment T_{C_i} is the sequence

$$(b_i+2, s_{i,1}, b_i+1, b_i+4, s_{i,2}, b_i+3, b_i+6, s_{i,3}, b_i+5, \ldots, b_i+14, s_{i,7}, b_i+13).$$

In any matching of P_{C_i} into T_{C_i} that respects a matching of P_V into T_V, the occurrences of $2l+2i+1$ and $2l+2i$ must match with one of the bracket pairs, and thus (i_1, i_2, i_3) must be one of the sequences $s_{i,k}$.

Claim 2.3. The set of clauses $\{C_i\}$ is satisfiable iff there is a matching of P_V into T_V and each P_{C_i} into T_{C_i}.

It remains to implement the guards without the use of duplicate elements. Note that the parts of P so far $(P_V, P_{C_1}, \ldots, P_{C_m})$ form a permutation of $1, \ldots, 2l+5m+1$, and the corresponding parts of T form a permutation of $1, \ldots, 4l+21m+1$. Augment P to a permutation of $1, \ldots, 4l+21m+2$ by inserting the decreasing subsequence of $2l+16m+1$ values $4l+21m+2, \ldots, 2l+5m+2$, using the first $m+1$ values as guards—the first before P_V, the second before P_{C_1}, etc.—and putting the remaining values at the end of P. Similarly, augment T to a permutation of $1, \ldots, 6l+37m+2$ by inserting the decreasing subsequence of $2l+16m+1$ values $6l+37m+2, \ldots, 4l+21m+2$, using the first $m+1$ values as guards, and putting the remaining $2l+12m$ values at the end of T. The last element of P is thus $2l+5m+2$, and the last value of T is $4l+21m+2$, and each of P and T has $2l+15m$ larger values.

If the last element of P maps to the last element of T, then, since there are an equal number of larger elements in P as in T, these large elements act as guards. If the last element of P maps to a different one of the large trailing elements of T, then there are insufficiently many larger elements in T to accommodate the larger elements of P. The final possibility is that the last element of P maps to one of the smaller $4l+21m+1m$ values of T. But then the trailing $2l+15m$ values of T cannot be used in the match, leaving only $4l+22m+1$ values—not enough to match P. ∎

Corollary 2.4. Counting the number of matchings of a pattern into a text is #P-complete.

Proof. In the reduction above, the matchings of P into T are in one-to-one correspondence with the satisfying assignments of the original formula. ∎

3. A Good Algorithm for Separable Patterns

A pattern P is said to be *separable* if it contains neither the subpattern $(3,1,4,2)$ nor the subpattern $(2,4,1,3)$. We begin with a lemma that clarifies the sense of this name "separable", and a discussion of the relationship between separable permutations and a class of graphs called P_4-free graphs.

3.1. Separable Permutations

A *separating tree* for a permutation $(p_1, \ldots, p_k)$ of $1, \ldots, k$ is an ordered binary tree T with leaves $(p_1, \ldots, p_k)$ in that order, such that for each node V, if the leaves of the subtree rooted at V are $p_i, p_{i+1}, \ldots, p_{i+j}$, then the set of numbers $\{p_i, p_{i+1}, \ldots, p_{i+j}\}$ is a *subrange* of the range $1, \ldots, k$—i.e. is of the form $\{l, l+1, \ldots, l+m\}$, for some l, m with $1 \leq l \leq k, 0 \leq m \leq k-l$. This subrange is called *the range* of the node V. If V is a node of the tree with left child V_L and right child V_R then the above condition implies that either the range of V_L just precedes the range of V_R, or vice versa. In the first case V is called a *positive* node, and in the second case a *negative* node.

Lemma 3.1. A pattern $P = (p_1, \ldots, p_k)$ is separable iff it has a separating tree.

The separating tree need not be unique. For example, the permutation $(4,5,3,1,2,6)$ has the two separating trees $((((4,5),3),(1,2)),6)$ and $(((4,5),(3,(1,2))),6)$.

A graph-theoretic formulation of this result is well-known: From any permutation T of $1, \ldots, n$, a graph can be defined with vertices $1, \ldots, n$, and with an edge (i,j) for $i < j$ iff i appears after j in the permutation. The graphs formed this way are *permutation graphs*. It is easy to see that the permutation containing the pattern $(3,1,4,2)$ or $(2,4,1,3)$ is equivalent to the graph containing an induced path on 4 vertices—a P_4. In general, graphs that do not contain an induced path on 4 vertices are called *P_4-free* graphs. Lemma 3.1 follows from the result that a graph is a P_4-free graph iff it is a *cograph*—a graph constructible from single vertices by the operations of disjoint union and complementation [CLS]. The use of a tree to represent this construction is explicit in [CLS]. This characterization of P_4-free graphs also implies that any P_4-free graph is a permutation graph. A linear time algorithm to recognize P_4-free graphs was given in [CPS]. This provides a linear time algorithm to recognize separable permutations.

Since we begin with a permutation rather than a graph, the results we need are significantly simpler, and thus it seems worthwhile—especially for those not already familiar with cographs—to give a direct proof of Lemma 3.1 and a direct algorithm to recognize separable permutations.

One final note on the relationship to graphs: The problem we are solving here could be expressed in terms of graphs. However, we are *not* solving the subgraph isomorphism problem for a cograph in a permutation graph, since the permutation pattern matching problem depends on more than just the unlabelled graphs than can be derived from the permutations.

Proof of Lemma 3.1. If the permutation does contain the subpattern $(3,1,4,2)$ or $(2,4,1,3)$ then there is no separating tree, since no two consecutive elements of these patterns form a range.

It remains to prove that any permutation either has a separating tree or has a subpattern $(3,1,4,2)$ or $(2,4,1,3)$. Suppose the permutation has neither of the subpatterns. By induction, it suffices to prove that there is a consecutive pair p_i, p_{i+1} forming a range.

Look at p_1 and p_2. Suppose without loss of generality that $p_1 < p_2$. If they form a range, fine; otherwise there is some $i > 2$ for which $p_1+1 = p_i$. Let j be the smallest index such that the substring $R = (p_2, p_3, \ldots, p_j)$ contains all the values in the range $p_i, p_i + 1, \ldots, p_2$. It is possible that $i = j$. We have $p_1 < p_i \leq p_j < p_2$. If the set of values in R forms a range then by induction R contains a consecutive pair forming a range, since the patterns $(3,1,4,2)$ and $(2,4,1,3)$ are excluded.

Thus we may assume that R does not form a range. There are two possibilities: (1) R contains a small value—i.e. there is an l in $2, 3, \ldots, j$ with $p_l < p_1$; or (2) R contains a large value, but not all values before it—i.e. there is an l in $2, 3, \ldots, j$ and an $m > j$ with $p_2 < p_m < p_l$. In case (1) the subsequence p_1, p_2, p_l, p_j has the pattern $2,4,1,3$; and in case (2) the subsequence p_2, p_l, p_j, p_m has the pattern $2,4,1,3$. ∎

Separable permutations can be viewed as permutations sortable in a particular way. This is somewhat analogous to the characterization by Knuth [K, section 2.2.1] that permutations without the pattern $2,3,1$ are exactly the permutations sortable using one stack; and to the further characterizations by Tarjan [T], by Even and Itai [EI], and by Pratt [P] (see also [G, chapter 7]).

Separable permutations are exactly the permutations that can be sorted in the following way. Imagine a straight segment of railway track, with the elements of the permutation lined up on the track, uncoupled, so that any element can be moved back and forth though without changing the relative ordering of elments. Imagine a segment in the middle of the track that can be rotated 180°, thus reversing the order of the elements on the middle segment. Using this device, we may move any consecutive subsequence of elements onto the middle section of track and reverse their order. There is one restriction: a sequence of elements to be rotated on the middle section must first be coupled together and must remain coupled forever afterwards. Separable permutations are exactly the permutations that can be sorted using this method. An efficient algorithm for sorting in this way is described below. It provides a test for whether a permutation is separable. (Large railway switching yards usually have these devices, implemented as a huge rotatable disc, though they generally have many segments of track converging radially on the disc, and of course do not permit an unlimited number of cars on the disc at one time.)

Lemma 3.1 provides a linear time algorithm to test if a permutation P is separable: Use a stack S whose elments are subranges $l, l+1, \ldots, l+m$ of the range $1, \ldots, k$. In general, to add a range r to S, see if the top element of S forms a larger range together with r—i.e. the union of the two sets of numbers is a range. If so, pop the top element of S, form the combined range, and recursively add it to S; and if not, then push the range r onto S. The algorithm proceeds by scanning through P once, making each element of S into a singleton range and adding that range to S, as just described. If, at the end of P, the stack S contains a single range then the permutation is separable, and otherwise it is not. P's separating tree T can easily be recovered.

3.2. Pattern Matching with Separable Permutations

We will now describe the algorithm to match a separable permutation pattern $P = (p_1, \ldots, p_k)$ into a text permutation $T = (t_1, \ldots, t_n)$. In fact, the number of such matches will be computed. The technique is dynamic programming, and the problem is subdivided based on P's separating tree, T. For each node V of T, try to match the substring corresponding to the subtree rooted at V—say, $(p_l, \ldots, p_m)$—into the text T. More specifically, solve one subproblem for each node V as above, and for each substring $(t_i, \ldots, t_j)$ of T, and each subrange $a, a+1, \ldots, b$ of the range $1, \ldots, n$. For each such subproblem, count the number of matches, $M(V, i, j, a, b)$, of the pattern $(p_l, \ldots, p_m)$ into the text $(t_i, \ldots, t_j)$, *using* t_i, and using text values in the range $a, \ldots, b$ *including* a. Note that there are $O(kn^4)$ such subproblems, since the tree T has $O(k)$ nodes, and there are $O(n)$ choices for each of i, j, a, b. After solving the subproblems we can recover the number of matchings by summing $M(\text{Root}, i, n, a, n)$ over all values of i and a.

We will show how to compute $M(V, i, j, a, b)$ based on the values of M for the children of node V. (Note that solutions to the subproblems for the leaves of the tree T are immediately obtainable.) Let V_L and V_R be the left and right children of V, respectively. If node V is a positive node, in the sense defined above—i.e. the range of values for V_L precedes the range of values for V_R, then

$$M(V, i, j, a, b) = \sum \{M(V_L, i, h-1, a, c-1) \cdot M(V_R, h, j, c, b) : \\ i < h \leq j, a < c \leq b\}$$

On the other hand, if V is a negative node then

$$M(V, i, j, a, b) = \sum \{M(V_L, i, h-1, c, b) \cdot M(V_R, h, j, a, c-1) : \\ i < h \leq j, a < c \leq b\}$$

Thus it takes $O(n^2)$ arithmetic steps to compute $M(V,i,j,a,b)$ from previously computed values. Altogether, the running time of the dynamic programming algorithm is $O(kn^6)$.

Rather than just computing the number of matchings, the algorithm can easily be augmented to compute a matching (if there is one) or to compute a list of all matchings—in the latter case the running time will have an additional factor of the number of matchings. For the case of recovering one matching, it is enough to record for each $M(V,i,j,a,b)$ that is non-zero, values of h and c in the above formulas that cause it to be non-zero.

4. Open Problems

How many separable permutations on $1,\ldots,k$ are there?

References

[CLS] D.G. Corneil, H. Lerchs, L. Stewart Burlingham, Complement-reducible graphs, Discrete Applied Math. 3, 163–174, 1981.

[CPS] D.G. Corneil, Y. Perl, L.K. Stewart, A linear recognition algorithm for cographs, SIAM J. Computing 14, 926–934, 1985.

[CW] M.-S. Chang and F.-H. Wang, Efficient algorithms for the maximum weight clique and maximum weight independent set problems on permutation graphs, Information Processing Letters 43, 293–295, 1992.

[EI] S. Even and A. Itai, Queues, stacks and graphs, in *Theory of Machines and Computations*, ed. Z. Kohavi and A. Paz, Academic Press, New York, 71–86, 1971.

[G] M.C. Golumbic, *Algorithmic Graph Theory and Perfect Graphs*, Academic Press, New York, 1980.

[K] D.E. Knuth, *The Art of Computer Programming, Vol. 1: Fundamental Algorithms*, 2nd edition, Addison-Wesley, 1973.

[L] L. Lovász, *Combinatorial Problems and Exercises*, North-Holland, 1979.

[P] V.R. Pratt, Computing permutations with double-ended queues, parallal stacks, and parallel queries, Fifth ACM Symposium on Theory of Computing, 268–277, 1973.

[R] D. Rotem, Stack-sortable permutations, Discrete Math, 33, 185–196, 1981.

[SS] R. Simion and F.W. Schmidt, Restricted permutations, European J. Combinatorics 6, 383–405, 1985.

[T] R. Tarjan, Sorting using networks of queues and stacks, J. Assoc. Computing Machinery 19, 341–346, 1972.

Filling Polyhedral Molds*

Prosenjit Bose Marc van Kreveld Godfried Toussaint

Abstract

In the manufacturing industry, finding an orientation for a mold that eliminates surface defects and insures a complete fill after termination of the injection process is an important problem. We study the problem of determining a favorable position of a mold (modeled as a polyhedron), such that when it is filled, no air bubbles and ensuing surface defects arise. Given a polyhedron in a fixed orientation, we present a linear time algorithm that determines whether the mold can be filled from that orientation without forming air bubbles. We also present an algorithm that determines the most favorable orientation for a polyhedral mold in $O(n^2)$ time. A reduction from a well-known problem indicates that improving the $O(n^2)$ bound is unlikely for general polyhedral molds. But we give an improved algorithm for molds that satisfy a local regularity condition that runs in time $O(nk \log^2 n \log\log(n/k))$, where k is the number of local maxima. Finally, we relate fillability to certain known classes of polyhedra.

1 Introduction

A well-known technique used in the manufacturing of goods is injection molding. A mold, as defined in [3], refers to the whole assembly of parts that make up a cavity into which liquid is poured to give the shape of the desired component when the liquid hardens. Given a mold (modeled as a polyhedron), establishing whether there exists an orientation that allows the filling of the mold using only one pin gate (the pin gate is the point from which the liquid is injected into the mold) as well as determining an orientation that allows the most complete fill are two major problems in the field of injection molding. These problems seem difficult and to date only heuristics have been proposed as solutions to these two problems [3, 10, 17]. In fact, until now, determining the favorable position for filling a mold was considered a cut-and-try process [17].

An initial study of the geometric and computational aspects of mold filling in two dimensions has been carried out by Bose and Toussaint [2], who show that for any mold modeled by a simple polygon P with n vertices, one can decide in $O(n)$ time whether a given orientation allows for a complete fill (the point from which a polygon is filled is always the highest point with respect to the direction of gravity). They also presented a linear time algorithm that determines whether

*Research supported in part by NSERC PGSB scholarship, an NSERC international fellowship, and NSERC-OGP0009293 and FCAR-93ER0291. School of Computer Science, McGill University, 3480 University St., Montréal, Québec, Canada H3A 2A7.

or not a polygonal mold is 1-fillable and if the mold is 1-fillable, the algorithm provides all the orientations that allow the 1-fillability. Furthermore, in $O(n \log n)$ time they find the orientation that minimizes the number of venting holes needed to avoid air bubbles (a venting hole is a point from which air, but no liquid, is allowed to escape). This problem is equivalent to finding the orientation that minimizes the number of local maxima of P. Finally, they related fillability to certain known classes of polygons.

Figure 1: Injection molding of a star-shaped object using one filling hole and two additional venting holes.

In this paper, we study the three-dimensional aspects of the mold filling problem. We show that given a mold, represented by a polyhedron with n vertices, and given a direction of gravity, we can determine in $O(n)$ time whether or not the mold can be filled without forming air pockets (Section 3). Then, we show that in $O(n^2)$ time, we can find all the orientations of a polyhedron that allow 1-filling. This algorithm also finds the orientation that minimizes the number of venting holes needed to ensure a complete fill when the polyhedron is not 1-fillable (Section 4). The above problem is equivalent to finding the orientation that minimizes the number of local maxima in the positive z-direction assuming that gravity points in the negative z-direction. The pin gate is placed at the global z-maximum of the polyhedron and the venting holes are placed at the local z-maxima. See Figure 1. These results were discovered independently in [8].

A *pseudo-lower bound* on the complexity of this problem can be given by a reduction from the problem $\exists: A+B=C$ to mold filling. The problem $\exists: A+B=C$ is defined as follows: Given three sets A, B and C of n real numbers each, decide if there exist $a \in A$, $b \in B$ and $c \in C$ such that $a+b=c$. The best known algorithm for $\exists: A+B=C$ uses $O(n^2)$ time. Several other geometric problems also reduce to this problem, such as: 'Given a set of n points in the plane, are there three collinear points?' and 'Given a set of n rectangles in the plane, do they cover a given rectangle R completely?' [9]. In the full version of this paper [1], we reduce the rectangle covering problem to the filling problem, and for the reduction from $\exists: A+B=C$ we refer to [9]. Our reduction takes $O(n \log n)$ time. Thus it seems difficult to improve upon the quadratic bound for the filling problem, even to determine whether there is an orientation of a polyhedron with only one local maximum.

We also consider the relation between mold filling and certain classes of polyhedra, such as weakly monotonic or star-shaped polyhedra (Section 5).

An interesting question that arises is whether one can improve the $O(n^2)$ time bound for restricted classes of polyhedra. Indeed, we show that if the polyhedron is in some sense not too irregular, then we can determine in $O(nk \log^2 n \log\log(n/k))$ time the orientation that minimizes the number of venting holes needed to ensure a complete fill. Here k is the number of local maxima in the optimal orientation (Section 6). The main idea behind this improvement is that the restriction imposed on the polyhedron, under a suitable transformation, leads to a set of *fat convex polygons* in the plane, and we use the fact that a set of fat convex polygons has small union size.

Both the general $O(n^2)$ time algorithm and the improved algorithm for restricted polyhedra are fairly simple and should perform well in practice. The latter algorithm provides the correct result for any polyhedron, but a guarantee on the asymptotic running time can only be given for polyhedra with the restriction.

2 Notation and Preliminaries

The boundary of a polyhedron partitions the space into two disjoint domains, the *interior* (bounded) and the *exterior* (unbounded). We will denote the open interior of the polyhedron P by $int(P)$, the boundary by $bd(P)$. and the open exterior by $ext(P)$. The boundary and interior are considered part of the polyhedron; that is, $P = int(P) \cup bd(P)$. We only consider *simple* polyhedra and polygons in this paper. Furthermore, facets and edges are assumed to be closed, i.e., they include their boundary. If we intersect a polyhedron with an arbitrary plane, the result is a collection (possibly empty) of simple polygons with holes (or line segments or points) lying on the plane. A polygon in this collection will be referred to as a *sectional* polygon. Notice that a sectional polygon divides the polyhedron into two simple polyhedra. Thus in this sense a sectional polygon is the three dimensional equivalent to a chord or diagonal in a polygon.

It will be convenient to have the set of all directions in space be represented by two planes. Although this is not standard, it will help simplify the exposition. Let the plane $z = -1$, denoted by $DP^{(-)}$, represent all directions with a negative z-component. Let the plane $z = 1$, denoted by $DP^{(+)}$, represent all directions with a positive z-component. We do not consider the horizontal directions. This assumption simplifies our discussion but is not an inherent limitation of our methods. A point q in $DP^{(-)}$ or $DP^{(+)}$ represents the direction $\vec{oq}$, where o represents the origin of E^3. Given a direction d, represented by $\vec{oq}$, we define $opp(d)$ to be the opposite direction. Thus, $opp(d)$ is pointing in the direction of the vector $\vec{qo}$.

As in the 2-dimensional problem, the point on the boundary of a polyhedral mold through which the liquid is injected into the polyhedron is called the *pin gate*.

A *venting hole* is a point from which only air and no liquid is allowed to escape. We assume that neither the liquid being poured into the mold, nor the air in the mold are compressible. Finally, we assume that air cannot bubble out through the liquid.

When liquid is injected into a polyhedron, the level of the liquid rises in the direction opposite that of gravity. When the level of the liquid reaches the pin gate, we say the polyhedron is *maximally injected.* A region containing air in a maximally injected polyhedron is called an *air pocket.* A polyhedron is said to be *1-fillable* if there exists a pin gate and direction of gravity such that when the polyhedron is injected through the pin gate, there are no air pockets when the polyhedron is maximally injected. We make the following observations.

Observation 1 *A polyhedron P in 3-space is said to be* 1-fillable in direction $-z$ *provided that for every point inside P there is a $+z$-monotone path from it to the z-maximum of P. Thus, a polyhedron is* 1-fillable *if there is an orientation of P in which it is 1-fillable.*

We extend the notion of fillability in the following two ways. A simple polyhedron P in 3-space is said to be *k-fillable* from a fixed orientation with gravity in the negative z-direction provided that there are k points on the boundary of P such that for every point inside P there is a $+z$-monotone path to at least one of the k points. Thus a polyhedron is *k-fillable* provided that there exists an orientation of P in which it is k-fillable.

A simple polyhedron P in 3-space is *k-fillable* with re-orientation provided that the polyhedron can be re-oriented and filled from a new pin-gate after partial filling from an initial direction and pin-gate. The k refers to the number of times that the polyhedron needs to be re-oriented before it is completely filled. We assume that after the completion of a partial filling, the liquid that is injected into the polyhedron hardens. Notice that both definitions are equivalent when $k = 1$. Unless stated otherwise, we will always refer to *k-fillable* as filling from a fixed orientation.

3 The Decision Problem

In this section we will present an $O(n)$ time algorithm to decide whether a polyhedron P is 1-fillable given an orientation of the polyhedron.

Let P be a simple polyhedron of which all facets are triangulated, and let v be an arbitrary vertex of P. We define P_v to be the union of the facets incident to v. Let $f_1, \ldots, f_m$ be the sequence of facets of P_v such that f_i and f_{i+1} are incident to an edge denoted e_i, and f_m and f_1 are incident to an edge e_m. If v is incident to m facets, and if P has a triangulated boundary, then P_v has $2m$ edges and $m+1$ vertices. Let S_v be a sphere centered at v, such that S_v only intersects the m edges incident to v, and no other facets, edges or vertices of P.

Definition 1 *A vertex v is a* convex vertex of P *provided that there exists a plane h_v, with $v \in h_v$, such that $S_v \cap h_v$ does not intersect the interior of P.*

Let h_v^- and h_v^+ denote the closed half-spaces below and above the plane h_v, containing the vertex v. Let h_v° be the closed half-space bounded by the plane h_v with normal d, containing the vertex v and where $\circ \in \{-,+\}$ is the opposite of the sign of the z-component in d. Recall that we assume, for simplicity, that d is not a horizontal direction.

Definition 2 *A vertex v is a* local maximum *of P in direction d provided that $P_v \subset h_v^\circ$.*

The following theorem is used to establish the linear time decision algorithm (a proof is given in the full paper [1]).

Theorem 1 *A polyhedron P is 1-fillable if and only if the orientation of P has precisely one local maximum in direction $+z$.*

From this theorem, we see that given a polyhedron P and a direction of gravity g, to test 1-fillability of P with respect to g, we need only determine the number of local maxima with respect to gravity. We can determine if a vertex is a local maximum in time linear in the degree of the vertex. This immediately gives us a linear time algorithm to determine whether or not a polyhedron is 1-fillable from a fixed orientation.

Theorem 2 *Given a polyhedron P, we can determine in $O(n)$ time whether or not the polyhedron is 1-fillable with respect to a given direction of gravity.*

4 Determining all Directions of Fillability

In this section we will give an $O(n^2)$ time algorithm to find the orientation of a given polyhedron P that minimizes the number of venting holes needed in order to ensure a complete fill from that orientation. This orientation is equivalent to the orientation that minimizes the number of local maxima. The algorithm has two stages. In the first stage, the fillability problem is transformed to a planar problem for a set of convex (possibly unbounded) polygons that cover the plane. In the second stage, the following problem is solved: Given a set of n convex polygons in the plane, find the point that is covered by a minimum number of them.

Let P be a polyhedron with n vertices, and assume that P is given by its incidence graph (see [7]). First, we triangulate every facet of P (see [4, 16]). We choose an initial orientation of P such that no edge of P is vertical. Let v be any vertex of P. We extract the description of P_v from the description of P in time proportional to its size. Let $f_1, \ldots, f_m$ be the sequence of facets incident to v, such that f_i and f_{i+1} are incident to an edge e_i of P_v (and f_m and f_1 are incident to an edge e_m). Let $w_1, \ldots, w_m$ be the sequence of endpoints corresponding to $e_1, \ldots, e_m$, see Figure 2.

Suppose that v is a convex vertex. We define the *cone C_v of v* to be the unbounded polyhedron consisting of v as its only vertex, m half-lines $E_1, \ldots, E_m$ starting at v, which contain the edges $e_1, \ldots, e_m$, respectively, and m unbounded facets incident to E_i and E_{i+1} $(1 \leq i \leq m-1)$, or E_m and E_1. Since C_v need not be a convex polyhedron, but its only vertex is convex, we say that C_v is a *semi-convex cone.* Let CC_v be the convex hull of C_v, which clearly is a *convex cone.* The half-lines that are the edges of CC_v are a subset of the edges of C_v; we denote them by $E_{i_1}, \ldots, E_{i_j}$, where $1 \leq i_1 < \cdots < i_j \leq m$. Finally, we define the *inverted cone* IC_v of the convex cone CC_v as follows. Let $h_{i_1}, \ldots, h_{i_j}$ be the set of planes that pass through v and are perpendicular to $E_{i_1}, \ldots, E_{i_j}$. Let $H_{i_1}, \ldots, H_{i_j}$ be the closed half-spaces bounded by $h_{i_1}, \ldots, h_{i_j}$ such that they contain $E_{i_1}, \ldots, E_{i_j}$, respectively. Then IC_v is the convex region that is bounded by $H_{i_1} \cap \cdots \cap H_{i_j}$. Notice that if CC_v is a sharp cone then IC_v is a blunt cone, and vice versa.

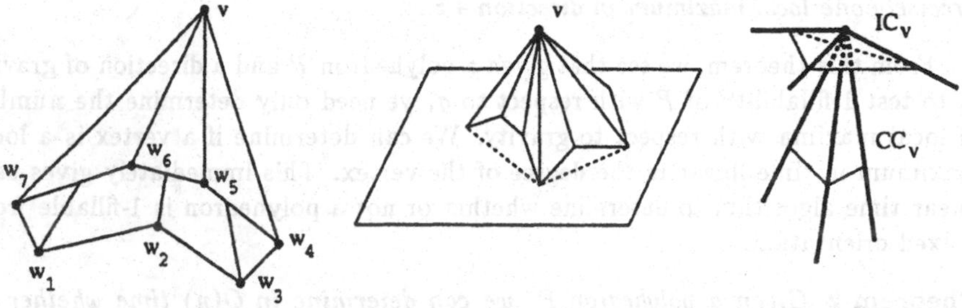

Figure 2: Left: P_v. Middle: the convex hull CC_v of C_v. Right: the convex cone CC_v and the inverted cone IC_v.

Each convex vertex of the polyhedron P defines a convex region in $DP^{(-)}$ and/or $DP^{(+)}$, which corresponds to the directions with respect to which it is a local maximum. Hence, P gives rise to $O(n)$ convex regions in these planes. It follows that a direction for which P has the smallest number of local maxima corresponds to some point in the plane that is covered by the smallest number of convex regions. The following lemma relates the inverted convex cones to the direction planes, $DP^{(-)}$ and $DP^{(+)}$.

Lemma 1 *For every convex vertex v of a polyhedron P such that v coincides with the origin o and direction $d = \vec{oq}$ where q is a point on one of the direction planes, it holds that v is a local maximum in (non-horizontal) direction opp(d) if and only if $q \in IC_v \cap DP^{(-)}$ or $q \in IC_v \cap DP^{(+)}$.*

We first determine if v is a convex vertex. This is the case if and only if v is an extremal point in the set $\{v, w_1, \ldots, w_m\}$. This is equivalent to the problem of determining if v can be separated from $\{w_1, \ldots, w_m\}$ by a plane, which in turn is equivalent to linear programming [6]. Therefore we can determine if v is convex by linear programming in linear time (see [7, 14]). If v is not a convex vertex, then v is not a local maximum for any direction, and we stop considering v. Otherwise, let h_v be a plane that contains v and has $w_1, \ldots, w_m$ to one side of it. Such a plane is returned by the linear programming test. Let h'_v be a plane parallel to h_v which intersects the edges $e_1, \ldots, e_m$. The intersection of h'_v with P_v is a simple polygon $\bar{P}_v$ with m vertices (corresponding to $e_1, \ldots, e_m$) and m edges (corresponding to $f_1, \ldots, f_m$). We compute the convex hull of $\bar{P}_v$ in linear time [13, 15]. Let us denote the convex hull by $CH(\bar{P}_v)$. Let $\bar{e}_{i_1}, \ldots, \bar{e}_{i_j}$ be the sequence of vertices of $CH(\bar{P}_v)$, where $1 \leq i_1 < \cdots < i_j \leq m$. These vertices correspond to the edges $e_{i_1}, \ldots, e_{i_j}$ of P_v. We have in fact computed the edges adjacent to v on the convex hull of P_v. This information gives us the description of the convex cone CC_v of v in linear time. Furthermore, the inverted cone IC_v can also be computed in linear time.

Translate IC_v such that v coincides with the origin o. Let $Q_v^{(-)}$ be the convex polygon $IC_v \cap DP^{(-)}$ and let $Q_v^{(+)}$ be $IC_v \cap DP^{(+)}$. Either $Q_v^{(-)}$ is a bounded convex polygon and $Q_v^{(+)}$ is empty, or vice versa, or both $Q_v^{(-)}$ and $Q_v^{(+)}$ are unbounded convex polygons. The convex polygons have the following meaning: v is a local maximum in a non-horizontal direction *opp*(d) if and only if the half-line starting at the origin o in direction d intersects the interior of one of the polygons $Q_v^{(-)}$ or $Q_v^{(+)}$. We compute the convex polygons $Q_v^{(-)}$ and $Q_v^{(+)}$ for all vertices v of P, giving sets $\mathcal{Q}^{(-)}$ and $\mathcal{Q}^{(+)}$ of at most n convex polygons in each of the planes $DP^{(-)}$ and $DP^{(+)}$, respectively. The total complexity of the polygons in $\mathcal{Q}^{(-)}$ and $\mathcal{Q}^{(+)}$ is $O(n)$. The question: 'Is P 1-fillable?' or 'Is there an orientation of P such that it has only 1 maximum?' translates to the question: 'Is there a point in $DP^{(-)}$ or $DP^{(+)}$ that is covered by only one convex polygon?' Similarly, the question of k-fillability translates to deciding whether there exists a point that is covered by only k convex polygons. We therefore have established the following result:

Lemma 2 *In $O(n)$ time, one can transform the problem of k-fillability to the problem of finding a point in the plane covered by only k convex polygons.*

The next step in the algorithm involves solving the following problem: 'Given a set $\mathcal{Q}$ of n convex, but not necessarily bounded, polygons in the plane, with total complexity $O(n)$, find a point that is covered by the minimum number of polygons in $\mathcal{Q}$.' Our algorithm constructs the subdivision induced by $\mathcal{Q}$, and associates to each cell the number of polygons that contain it.

The subdivision induced by $\mathcal{Q}$ without the numbering can be constructed deterministically in $O(n \log n + A)$ time by the algorithm of Chazelle and Edelsbrunner[5], where A is the total number of intersection points of all polygons in $\mathcal{Q}$. Clearly $A = O(n^2)$, and we obtain a planar subdivision $\mathcal{S}$ with $O(n^2)$ vertices, edges and cells. It is straightforward to traverse the cells of the subdivision, and determine the one covered by the minimum number of polygons.

Returning to the k-fillability problem, the above algorithm finds the direction d such that the polyhedron has the minimum number of local maxima, if we apply it to both the set $\mathcal{Q}^{(-)}$ of convex polygons in the plane $DP^{(-)}$ and $\mathcal{Q}^{(+)}$ in the plane $DP^{(+)}$. We thus conclude:

Theorem 3 *Given a simple bounded polyhedron P with n vertices in 3-space, one can find in $O(n^2)$ time an orientation for P such that P is fillable with the minimum number of venting holes.*

Remark: The solution presented above makes no use of the fact that P is topologically equivalent to a sphere. The algorithm works equally well for a polyhedron of higher genus. In practice, this is relevant because a plastic cup with an ear is modeled by a polyhedron that is a torus.

5 Fillability of Certain Classes of Polyhedra

In this section, we investigate the relationship between the notion of fillability and certain known classes of restricted polyhedra. These results are relevant to the manufacturing industry because in practice many objects are not modeled by polyhedra of arbitrary shape complexity.

A polygon P is monotonic in direction l if for every line L orthogonal to l that intersects P, the intersection $L \cap P$ is a line segment (or point) [16, 19]. We generalize this notion to 3-dimensions to obtain a family of monotone polyhedra. We define the class as follows.

Definition 3 *A polyhedron P is* weakly monotonic *in direction l if there exists a direction l such that the intersection, of each plane orthogonal to l that intersects P, is a* simple polygon *(or a line segment or point). The direction l is referred to as the direction of monotonicity.*

Theorem 4 *A weakly monotonic polyhedron P is 1-fillable if it is oriented such that gravity points in the direction of monotonicity.*

Two points inside a polyhedron are said to be *visible* if the line segment between them does not intersect the exterior of the polyhedron. A point p is *weakly visible* from a facet f if there is a point x on f such that p is visible from x.

A polyhedron P is *open-facet visible* if there is a facet f in P such that all points p are visible from some point x on f that is not on the boundary of the facet.

Let P be an open-facet visible polyhedron. Without loss of generality, let f_1 be the open facet from which the polyhedron is weakly visible. Let d denote the direction of the interior normal to the facet.

Theorem 5 *An open-facet visible polyhedron P is 1-fillable if it is oriented such that d points in the direction of gravity.*

Corollary 1 *Every polyhedron that is weakly visible from the interior of a sectional polygon is 2-fillable with re-orientation.*

A *star-shaped polyhedron* is a polyhedron that contains at least one point x from which all points of the polyhedron are visible (see Figure 1). The set of points from which all points are visible is known as the *kernel* of the star-shaped polyhedron. A point in the kernel of a star-shaped polyhedron can be computed in $O(n)$ time using linear programming [7, 14]. This implies that in $O(n)$ time, a sectional polygon can be found from which the star-shaped polyhedron is weakly visible. However, a star-shaped polyhedron may not be 1-fillable (a ball with spikes in all directions). In fact, if a star-shaped polyhedron is filled from one fixed orientation, it may need $\Omega(n)$ venting holes.

Theorem 6 *A star-shaped polyhedron is not necessarily 1-fillable but can always be 2-filled with re-orientation in $O(n)$ time.*

6 An Improved Algorithm for Restricted Polyhedra

We next consider the issue of improving our $O(n^2)$ time algorithm for polyhedra that satisfy certain regularity conditions. We place a local condition on each vertex to ensure that the covering problem we obtain can be solved more efficiently. Our new algorithm runs in $O(nk \log^{O(1)} n)$ time, where k is the minimum number of local maxima of P. If this number is small compared to n, the new algorithm improves considerably upon the previous algorithm. The local restriction is such that the convex polygons that are obtained for the covering problem are *fat*, that is, the ratio of the diameter to the width of each polygon is bounded from above by a constant.

Definition 4 *A convex vertex v of a polyhedron P in 3-space is* non-flat *provided that there exists a positive constant β and plane h through v such that $int(C_v) \subset h^-$ or $int(C_v) \subset h^+$, and every edge e incident to v makes an angle at least β with h. A bounded polyhedron P in 3-space is* non-flat *provided that each convex vertex v of P is non-flat.*

We will show that any non-flat polyhedron P yields—using an adapted transformation to the covering problem—a set of fat convex polygons. In [1], a different restriction on convex vertices is given that also gives fat polygons. We omit the description here.

The second problem is that of computing regions that are covered not too often without having to compute the full subdivision of the convex polygons (which may have quadratic complexity even for fat objects). We will show that one can compute the regions that are at most k-covered in $O(nk \log^2 n \log\log(n/k))$ time, if the convex polygons are fat.

Let P be a non-flat polyhedron with n vertices. As before, we triangulate all facets of P, and for each vertex $v \in P$, we test whether v is convex. If so, we compute the convex cone CC_v and the inverted convex cone IC_v. However, instead of letting all non-horizontal directions be represented by the planes $z = -1$ and $z = 1$, we let all directions be represented by an axis-parallel cube. Let DC be the axis-parallel cube centered at the origin o and with edge length 2. The inverted cone may intersect all six facets of DC, or its interior may contain a whole facet. For any facet F of DC, consider the set $\mathcal{Q}$ of clipped convex polygons $IC_v \cap F$, see Figure 3. The proof of the following lemma can be found in the full paper [1].

Lemma 3 *Any clipped convex polygon $Q \in \mathcal{Q}$ can be extended to a fat convex polygon that coincides with Q inside the facet F.*

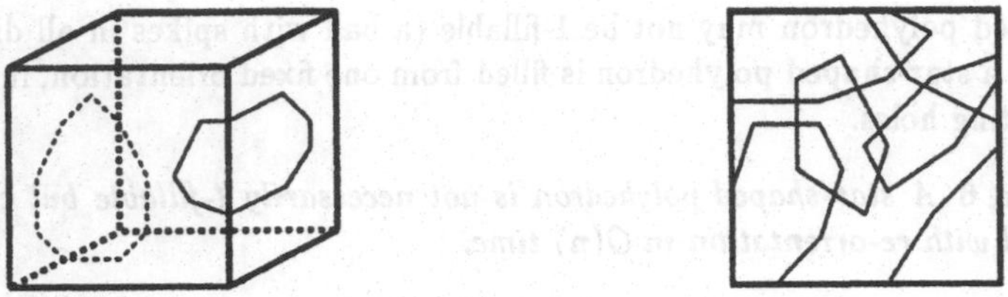

Figure 3: Left: the intersection of the cube with two convex cones. Right: one facet of the cube intersected by several convex cones.

A set of n fat convex polygons with total complexity $O(n)$ has the property that the boundary of the union has close to linear complexity, whereas it may be quadratic for non-fat convex polygons. A more general result, which we require for our algorithm, is:

Theorem 7 ([11], see also [12]) *Given a set $\mathcal{Q}$ of n fat convex polygons with total complexity $O(n)$, the total complexity of all cells in the subdivision induced by $\mathcal{Q}$ which are covered by at most k polygons of $\mathcal{Q}$ is $O(nk \log\log(n/k))$.*

If we clip each polygon in the set $\mathcal{Q}$ with a square F, then the theorem still holds. For some value of k, we will test whether every point in F is at least k-covered. If not, then we know that the polyhedron that gave rise to $\mathcal{Q}$ has less than k local maxima for some orientation. If for each of the six facets of DC every point is at least k-covered, then we know that the polyhedron has at least k local maxima for every orientation.

For a facet F, a set $\mathcal{Q}$ of polygons with total complexity $O(n)$, and an integer $k \geq 1$, we test whether there exists a point inside F that is covered by at most k polygons of $\mathcal{Q}$ (see also Sharir [18]). This is done by computing the subdivision defined by $\mathcal{Q}$ using divide-and-conquer. However, we only compute edges and vertices of regions that are covered by at most k polygons of $\mathcal{Q}$. With every region, an integer is associated that represents the number of polygons that cover the region. The regions for which the integer is greater than k are associated with the special integer ∞.

We partition $\mathcal{Q}$ into two subsets $\mathcal{Q}_1$ and $\mathcal{Q}_2$ such that the total complexity of the polygons in each of the subsets is approximately the same. Recursively, we compute the subdivisions S_1 and S_2 of the regions that are covered by at most k polygons of $\mathcal{Q}_1$ and $\mathcal{Q}_2$, respectively. Then we merge these two subdivisions S_1 and S_2 by a plane sweep. Keep all regions covered by at most k polygons, and for all regions covered at least $k+1$ polygons, associate the integer ∞. Then remove all boundaries of regions with ∞.

If the polygons in $\mathcal{Q}_1$ and $\mathcal{Q}_2$ have complexity $O(n)$, then the subdivisions $\mathcal{S}_1$ and $\mathcal{S}_2$ have complexity $O(nk \log\log(n/k))$ by Theorem 7. The number of intersection points of edges of $\mathcal{S}_1$ and edges of $\mathcal{S}_2$ is $O(nk \log\log(n/k))$, since such an intersection point is obtained from two edges that bound a region that is at most k times covered. Hence, any intersection point lies on the closure of a cell that is at most $2k$ times covered. By the complexity result of Theorem 7, the plane sweep takes $O(nk \log n \log\log(n/k))$ time. We conclude that the divide-and-conquer algorithm requires $O(nk \log^2 n \log\log(n/k))$ time in total (see also [18]).

We show how to use the above algorithm for k-fillability of a polyhedron P. Let $F_1, \ldots, F_6$ be the six facets of the cube DC, and let $\mathcal{Q}_1, \ldots, \mathcal{Q}_6$ be the six sets of polygons on these facets. Let $j = 1$ and determine for each F_i and $\mathcal{Q}_i$ whether there exists a point on F_i that is covered by only j polygons of $\mathcal{Q}_i$. If the answer is no for all F_i, then we double j and try again. If the answer is yes for some facet F_i, then traverse the subdivisions of $F_1, \ldots, F_6$ that we computed, and find the region with smallest associated integer $k \leq j$. Any point in this region corresponds to an orientation of the polyhedron P for which it has k local maxima. Any other orientation gives as least as many maxima. The time taken by the algorithm is $O(\sum_{b=0}^{\lceil \log_2 k \rceil} n \cdot 2^b \log^2 n \log\log(n/2^b)) = O(nk \log^2 n \log\log(n/k))$ time.

Theorem 8 *Let P be a non-flat polyhedron with n vertices. An orientation for P can be found such that P can be filled using k venting holes, which is the minimum possible for P, and the algorithm takes $O(nk \log^2 n \log\log(n/k))$ time.*

7 Conclusions and Open Problems

This paper presented the first algorithm to compute an orientation of a polyhedron that minimizes the number of venting holes needed to guarantee a complete fill without forming air pockets. For a polyhedron with n vertices, our algorithm runs in $O(n^2)$ time and should be easy to implement.

We gave a second algorithm that improves upon the first one in most practical situations and, under some local conditions of the polyhedron, we showed that the second algorithm performs asymptotically better than the first one. If the minimum number of venting holes required to fill the polyhedron is k, our algorithm runs in $O(nk \log^2 n \log\log(n/k))$ time. In practice, k will usually be small compared to n.

There are several directions for further research. For the second algorithm, we imposed restrictions on the polyhedron to be able to prove the stated running time. It may, however, be possible to weaken the restrictions that we imposed and still prove the same time bound. Also, a different approach may yield other algorithms that improve upon the $O(n^2)$ time bound for other types of restricted polyhedra. Finally, our algorithms require $O(n^2)$ and $O(nk \log\log(n/k))$ storage, respectively. It may be possible to improve upon this.

References

[1] Bose, P., M. van Kreveld, and G. Toussaint, *Filling polyhedral molds.* Tech. Rep. No. SOCS 93.1, School of Computer Science, McGill University, 1993.

[2] Bose, P., and G. Toussaint, *Geometric and computational aspects of injection molding.* Tech. Rep. No. SOCS 92.16, School of Computer Science, McGill University, 1992.

[3] Bown, J., *Injection Moulding of Plastic Components.* McGraw-Hill, England, 1979.

[4] Chazelle, B., Triangulating a simple polygon in linear time. *Proc. 31st IEEE Symp. Found. Comp. Science* (1990), pp. 220–230.

[5] Chazelle, B., and H. Edelsbrunner, An optimal algorithm for intersecting line segments in the plane. *J. ACM* 39 (1992), pp. 1–54.

[6] Dobkin, D., and S. Reiss, The Complexity of Linear Programming. *Theoretical Computer Science* 11 (1980), pp. 1–18.

[7] Edelsbrunner, H., *Algorithms in Combinatorial Geometry.* Springer-Verlag, Berlin, 1987.

[8] Fekete, S., and J. Mitchell, *Geometric aspects of injection molding.* Manuscript, 1993.

[9] Gajentaan, A., and M.H. Overmars, *$O(n^2)$ difficult problems in computational geometry.* Tech. Report, Dept. of Computer Science, Utrecht University, 1993, to appear.

[10] Hui, K.C., and S.T. Tan, Mould design with sweep operations - a heuristic search approach. *C.A.D.*, 24-2: 81-90, 1992.

[11] van Kreveld, M., *Fat partitioning, fat covering and the union size of polygons.* Tech. Rep. No. SOCS 93.2, School of Computer Science, McGill University, 1993. Extended abstract in these proceedings.

[12] Matoušek, J., N. Miller, J. Pach, M. Sharir, S. Sifrony, and E. Welzl, Fat triangles determine linearly many holes. *Proc. 32nd IEEE Symp. Found. Comp. Science* (1991), pp. 49–58.

[13] McCallum, D. and D. Avis, A linear time algorithm for finding the convex hull of a simple polygon. *Inform. Process. Lett.* 8, (1979), pp. 201–205.

[14] Megiddo, N., Linear-time algorithms for linear programming in R^3 and related problems. *SIAM J. Comp.* 12 (1983), pp. 759–776.

[15] Melkman, A., On-line construction of the convex hull of a simple polyline. *Inform. Process. Lett.* 25, (1987), pp. 11–12.

[16] Preparata, F.P., and M.I. Shamos, *Computational geometry – an introduction.* Springer-Verlag, New York, 1985.

[17] Pribble, W.I., Molds for reaction injection, structural foam and expandable styrene molding. In: *Plastics mold engineering handbook*, J.H. DuBois and W.I. Pribble (Eds.), Van Nostrand Reinhold Company Inc., New York, 1987.

[18] Sharir, M., On k-sets in arrangements of curves and surfaces. *Discr. & Comp. Geom.* 6 (1991), pp. 593–613.

[19] Toussaint, G., Movable Separability of Sets. In: *Computational Geometry*, North-Holland, New York, (1990), pp. 335–375.

Deferred-Query — An Efficient Approach for Problems on Interval and Circular-Arc Graphs

Extended Abstract

Maw-Shang Chang, Sheng-Lung Peng and Jenn-Liang Liaw

Department of Computer Science and Information Engineering
National Chung Cheng University
Ming-Shiun, Chiayi, Taiwan 621

Abstract. An efficient approach, called deferred-query, is proposed in this paper to design $O(n)$ algorithms for the domatic partition, optimal path cover, Hamiltonian path, Hamiltonian circuit and matching problems on a set of sorted intervals. Using above results, the optimal path cover, hamiltonian path and hamiltonian circuit problems can also be solved in $O(n)$ time on a set of sorted arcs.

1. Introduction

A graph $G = (V,E)$ is an interval graph if its vertices can be put in a one-to-one correspondence with a family I of intervals on the real line such that two vertices are adjacent in G if and only if their corresponding intervals have nonempty intersection. I is called an interval model for G. If no interval in I is properly contained in some other interval in I, then G is called a proper interval graph. For an interval family I, $G(I)$ denotes the graph constructed from I. Booth and Lueker [BL] gave a linear algorithm for deciding whether a given graph is an interval graph and constructing, in the affirmative case, the required interval family. Interval graphs have been used in many practical applications. In particular, they have applications in genetics, archaeology, biology, psychology, sociology, engineering, computer scheduling, traffic control, storage information retrieval and others (See [G] and [R]).

A graph G is a circular-arc graph if there is a one to one correspondence between the vertices of G and the arcs of a circular arc family F such that two vertices of G are adjacent if and only if their corresponding arcs overlap. The circular arc family is called an arc model of the graph. For an arc family F, $G(F)$ denotes the graph constructed from F. If there exists a point on the circle such that no arc of F passes through, then $G(F)$ is an interval graph. Circular–arc graphs are rich in combinatorial structures. Various characterization and optimization problems on circular–arc and interval graphs have been studied (some of them listed in [G]).

A set of vertices D is a dominating set of a graph $G = (V,E)$ if every vertex in $V - D$ is adjacent to a vertex in D. The domatic number of G, denoted $d(G)$, is the maximum number of pairwise disjoint dominating sets in G. The domatic number was defined and studied in [CH]. In particular, they showed that $d(G) \geq \delta + 1$ for any graph G, where δ is the minimum degree of vertices in G. G is domatically full if $d(G) = \delta + 1$. Farber [F] showed that strongly chordal graphs are domatically full. Interval graphs are domatically full since they are special strongly chordal graphs. The domatic partition problem is to partition V into $d(G)$ disjoint dominating sets. The domatic partition problem of a graph is NP–hard on general graphs [GJ]. Bertossi [B1] solved this problem in $O(n^{2.5})$ time, where n is the number of vertices in G, for interval graphs and $O(nlogn)$ for proper interval graphs. Recently, $O(m+n)$ time algorithms were given by [LHC] and [RR] independently, where m is the number of edges in G.

A path cover for a graph G is a set of vertex disjoint paths that cover all the vertices of G. An *optimal path cover* (OPC) of G is a path cover of minimum cardinality. Like many other graph problems, this problem is NP–hard for arbitrary graphs [GJ]. Arikati and Rangan [AR] showed that the optimal path cover problem can be solved in $O(m+n)$ time on interval graphs. Linear algorithm on circular–arc graphs was proposed by Bonuccelli and Bovet [BB2]. Their algorithm was pointed out to be incorrect in [HSC] and [MMS] independently.

A path (circuit) in G is simple if no vertex occurs more than once. A *Hamiltonian path* (HP) in a graph G is a simple path including all vertices. A *Hamiltonian circuit* (HC) in a graph G is a simple circuit including all vertices. The problems of finding an HP and an HC are NP–complete for some classes of graphs, such as bipartite graphs [K2], edge graphs [B2], planar cubic 3–connected graphs [GJT], and some other classes. Keil [K1] showed that an HC could be constructed for a general interval model (with no restriction that it be proper) in time and space $O(m+n)$. Manacher, Mankus and Smith [MMS] showed an optimal $O(nlogn)$ time algorithm for finding an HP and an HC in a set of intervals. Whether HCP can be solved in polynomial time on circular–arc graphs was listed as an open problem in [J]. Recently, Hsu, Shih and Chern solved the problem in $O(n^2logn)$ time [HSC].

A set of pairwise disjoint edges is called a matching. A matching of maximum cardinality is called a maximum matching. There is an algorithm which solves this problem on interval graphs in $O(m+n)$ time [MJ].

Some researchers on interval graphs are interested in the complexity of problems on interval graphs given a set of sorted intervals [MMS]. We refer it as a sorted interval model. In Section 2, we present a new approach, called deferred–query, to design efficient algorithms for the domatic partition, optimal path cover, Hamiltonian path, Hamiltonian circuit and maximum matching problems on sorted interval models. Using this approach all above problems can be solved in $O(n)$ time. Note that all the previous results for the above problems remain $O(m+n)$ or $O(nloglogn)$ time even on sorted interval models.

A circular–arc graph can be recognized in $O(n^3)$ time [T]. An arc model F with endpoints sorted of a circular–arc graph can be built as a byproduct of this recognition algorithm. Hence, researchers who study circular–arc graphs sometimes assume that a set of endpoints sorted arcs is given [GP, HT and MN]. In Section 3, we show how to reduce the HC, HP and OPC problems on circular–arc graphs to a constant number of the same problems on interval graphs. Hence, the HP, HC and OPC problems on circular–arc graphs can be solved in $O(n)$ time on a set of sorted arcs.

2. Algorithms for Interval Graphs

In this Section, we show how to design efficient algorithms for the optimal path cover, Hamiltonian path and Hamiltonian circuit problems by deferred–query approach. For other problems, please refer to [CPL]. Let $I = \{v_1, v_2, ..., v_n\}$ be a family of n intervals corresponds to graph $G = (V,E)$. Without loss of generality, we assume endpoints are restricted to distinct integers between 1 to $2n$. We number the intervals of I from 1 to n in the increasing order of their right endpoints. That is, for any two intervals v_i and v_j, $b_i < b_j$ if $i < j$. We say that $v_i < v_j$ if $i < j$.

2.1. The Optimal Path Cover (OPC) Problem

We first present a greedy algorithm for the OPC problem by using deferred–query approach. This algorithm output the same result as the algorithm given by Arikati and Rangan [MMS,AR] does. This new algorithm visits intervals one by one, creates, extends or merges paths depending on the position of the left point of the currently visited interval. The algorithm uses a data

structure L to keep the right endpoints of the path ends of all paths. Initially, there is only one path $P_1 = v_1$ and $L = \{b_1\}$. Assume that v_i is the interval visited currently and let $T = \{b \mid a_i < b, b \in L\}$. Then we consider two cases:
Case 1: $T = \phi$. In this case, the algorithm creates a new path $P_z = v_i$ and insert b_i to L, i.e. $L = L \cup \{b_i\}$.
Case 2: $T \neq \phi$. The algorithm find b_k = MIN T. Suppose b_k is not the only element of T. Let b_h be successor of b_k in T (the smallest number that is larger than b_k in T). Let P_x and P_y be the two paths ending at v_k and v_h respectively. Since v_i overlaps with path end of P_x and path start of P_y, we can form a new path $P_x = P_x * v_i * P_y$. Remove b_k from L. On the other hand, if $T = \{b_k\}$, let $P_y = P_x * v_i$ and $L = (L - \{b_k\}) \cup \{b_i\}$.

It is not hard to see that the algorithm described above produces the same result as the algorithm given by Arikati and Rangan does [AR]. The main difference between them is that we defer the decision to determine to which path an interval should go until the right endpoint of the interval is visited in the new algorithm. This is the spirit of deferred–query approach. Another important issue is how to manipulate L. We use a more efficient data structure for L such that the insertion, deletion and finding the successor can be done in constant amortized time. This data structure is possible because every endpoint related to the operations on the data structure has been visited.

We now show how the algorithm manipulate L. The algorithm first scans endpoints to find out sequences of consecutive left endpoints. A sequence of consecutive left endpoints is a sequence of left endpoints without any right endpoints among them. For a sequence of consecutive left endpoints, we make a set and this set is associated with the right endpoint that is closest to the largest left endpoint in the set. What we mean by saying that a set is associated with a right endpoint is that both of the operations of finding the set from the right endpoint and finding the right endpoint from the set can be done in constant time. For example, the right endpoint b_i will be associated with the set of consecutive left endpoints $\{a_j \mid b_{i-1} < a_j < b_i\}$. If $b_i = b_{i-1} + 1$, then the consecutive left endpoint set associated with b_i is the empty set. Elements in L are stored in a list in the increasing order. Since there are at most n right endpoints in L, we can easily find the position in L of a right endpoint. By using pointers, the successor and predecessor of a right endpoint in L can be found in a constant time. In the following algorithm, the definition of L is slightly modified to contain the right endpoint of all path ends and the right endpoints of the intervals that have not been visited by the algorithm yet. The detailed algorithm is as follows:

Algorithm GP(I)
Input. A family of interval $I = \{v_1, v_2, ..., v_n\}$.
Output. An optimal path cover of G(I).
Method.

1. Scan the endpoints and create a set for each sequence of consecutive left points. Each right endpoint b_i, $1 \leq i \leq n$, is associated with a consecutive left endpoint set. That is, every $b_i, 1 \leq i \leq n$, is associated with the consecutive left endpoint set $\{a_j \mid b_{i-1} < a_j < b_i\}$.
   $L = \{b_1, b_2, ..., b_n\}$; $z = 0$;
2. For $i = 1$ to n do
   - Find the set that contains a_i and let this set be associated with b_k.
   - If $b_k = b_i$ (a_i is in the set that is associated with b_i, the right endpoint of v_i), then
     - $z = z + 1$; $P_z = v_i$; $P = P \cup \{P_z\}$;
   - else
     - Let b_h be the successor of b_k in L.
     - If $b_h = b_i$, then
       - Let P_x be the path containing v_k;
       - $P_x = P_x * v_i$;
       - Unite the set associated with b_k to the set associated with b_i and destroy the set associated with b_k;
       - Delete b_k from L;
     - else /* merge two paths. */
       - Let P_x be the path ending at v_k;
       - Let P_y be the path ending at v_h;
       - $P_y = P_x * v_i * P_y$; $P = P - \{P_x\}$;
       - Unite the set associated with b_k to the set associated with b_h and destroy the set associated with b_k;
       - Delete b_k from L.
       - If $i \neq n$ then
         - Unite the set associated with b_i to the set associated with b_{i+1} and destroy the set associated with b_i;

It is clear that these disjoint sets union and find operations can be implemented by Gabow and Tarjan's static disjoint set union algorithm where the union tree is a path [GT]. According to Gabow and Tarjan [GT], it can be implemented in $O(p+q)$ time and $O(q)$ space where p is the number of union and find operations and q is the total number of elements in the sets. Besides, the operations on L can be implemented in $O(1)$ time by using the double linking list on an array.

Theorem 1 Given a sorted interval model I of an interval graph G, Algorithm GP(I) computes an optimal path cover of G in $O(n)$ time and $O(n)$ space.

2.2. The Hamiltonian Path and Hamiltonian Circuit Problem

If Algorithm GP(I) produces only a path, then it is an HP. Otherwise, there doesn't exist any HP and HC in the interval graph G. It is straightforward to verify that the output of Algorithm GP(I) is equivalent to that of Algorithm FIND–PATH [MMS]. Therefore, if Algorithm GP(I) produces exactly one path, we can apply Algorithm FIND–CIRCUIT [MMS] to determine whether G has an HC. The time complexity of Algorithm FIND–CIRCUIT is $O(n)$. Therefore, given a sorted interval model of an interval graph G, we can solve the HP and HC problem in $O(n)$ time by using Algorithm GP(I) and Algorithm FIND–CIRCUIT of Manacher, Mankus and Smith [MMS]. This answer the question, posted by Manacher, Mankus and Smith [MMS], can the HC problem be solved in $O(n)$ time on sorted interval models of interval graphs?

3. Algorithms for Circular–Arc Graphs

In this section, we give an $O(n)$ algorithm for the optimal path cover and Hamiltonian path problem on circular–arc graphs. For the Hamiltonian circuit problem, please refer to [C]. Denote an arc x that begins at endpoint p and ends at endpoint q in the clockwise direction by (p, q). We call p the *head* (or counterclockwise endpoint, denoted by $h(x)$) and q the *tail* (or clockwise endpoint, denoted by $t(x)$) of the arc (p,q). A point on the circle is said to be contained in arc (p,q) if it falls within the interior of (p, q). An arc x of F is said to be *contained in* another arc y if every point of x is in arc y. Without loss of generality, assume (1) all arc endpoints are distinct, (2) no arc covers the entire circle and (3) an arc does not include its two endpoint, i.e. it is an open arc of the circle.

Denote the set of paths output by Algorithm GP(I) by *GP*(I). The paths in *GP*(I) are called *greedy paths.* Define the *path–end* (resp. *path–start*) of a path Q, denoted by *path–end*(Q) (resp. *path–start*(Q)), to be the last (resp. first) interval of Q. Note that the path–start of a greedy path is the smallest interval in the path. Define $L(Q)$ of a greedy path Q to be the collection of intervals in Q which are larger than *path–end*(Q). For a greedy path Q output by Algorithm GP, define the subset of intervals in Q, called a *greedy connecting set* and denoted by $C(Q)$, by the following algorithm.

Algorithm C(Q)
Input: Q, a greedy path output by Algorithm GP(I).
Output: $C(Q)$, *greedy connecting set* of path Q and $R(Q)$, the set of path components of $G(Q - C(Q))$.
Method.
1. Let C denote the current set of connectors and R denote the current collection of paths. Initially, $R = \{Q\}$, $C = \phi$.
2. Repeat the following process until $L(S) = \phi$ for every path S in R:
   Let $L^* = \cup_{S \in R} L(S)$, q^* be the largest interval in L^* and S^* be the path containing q^*. Delete q^* from path S^*. Let S_1 and S_2 be the two paths created by deleting q^* from S^*. Let $R = (R - \{S^*\}) \cup \{S_1, S_2\}$ and $C = C \cup \{q^*\}$.
   Note that both S_1 and S_2 are not empty. Let *path–end*(S_1) be smaller than *path–end*(S_2). If s_1 and s_2 are any two intervals of $S_1 - L(S_1)$ and S_2 respectively, then s_1 does not overlap with s_2.

Lemma 1 [HSC] The following statements are all true:
(1) $L(S)$ is empty for every path S in $R(Q)$.
(2) $|R(Q)| = |C(Q)| + 1$.
(3) For any two paths S_i and S_j of $R(Q)$, any two intervals s_i and s_j of S_i and S_j respectively, s_i does not overlap with s_j.
(4) $L(Q) \subseteq C(Q)$.

Algorithm C(Q) can be implemented to run in $O(n)$ time and space on sorted interval models. Let $C(GP(I)) = \cup_{Q_i \in GP(I)} C(Q_i)$ and $R(GP(I)) = \cup_{Q_i \in GP(I)} R(Q_i)$. Then, we have the following corollary.

Corollary 1.1 $|R(GP(I))| = |C(GP(I))| + |GP(I)|$.

For a graph $G = (V,E)$, let $\tau(G)$ be the minimum cardinality of a node disjoint path cover of G. If we find a subset D of V such that $G(V–D)$, the induced subgraph of $V - D$, has $|D| + k$ $(k > 0)$ connected components, then, following pigeonhole principle, $\tau(G) \geq k$. Following Corollary 1.1, we can find a subset $C(GP(I))$ of I such that $|R(GP(I))| = |C(GP(I))| + |GP(I)|$. Each path of $R(GP(I))$ is a connected component of $G(I - C(GP(I)))$. This proves the correctness of Algorithm GP(I).

Let $B(z) = \{x \mid x \in I \text{ and } x \text{ contains } z\}$. For $v \in B(z)$, define

$B(z,v) = \{x \mid x \in B(z) \text{ and } x > v\}$ and $U(z,v) = \{x \mid x \in B(z) \text{ and } x \leq v\}$. Then, we have the following lemma.

Lemma 2 If $GP(I) = \{Q\}$, $v \in B(\textit{path-start}(Q))$, $GP(I-\{v\}) = \{Q_1, Q_2\}$ and $\textit{path-end}(Q_1) < \textit{path-end}(Q_2)$, then $B(\textit{path-start}(Q),v) \subseteq C(Q_1)$.

Lemma 3 If $GP(I) = \{Q\}$, $v \in B(\textit{path-start}(Q))$, $GP(I-\{v\}) = \{Q'\}$ and $\textit{path-end}(Q') < \textit{path-end}(Q)$, then $B(\textit{path-start}(Q),v) \subseteq C(Q')$.

Lemma 4 Let $v \in I$, $GP(I-\{v\}) = \{Q\}$ and $v \in B(\textit{path-start}(Q))$. If $r \in I$ and $r < v$, then $|GP(I-\{r\})| = 1$.

Let I_h be a set of intervals that $|GP(I_h)| = 1$. D is called a *connecting set* of I_h if $I_h - D$ has $|D| + 1$ connected components. Apparently, each connected component has a hamiltonian path. For an nonempty connecting set D of I_h, define $C^*(I_h,D)$, a subset of I_h, by the following algorithm.

Algorithm C*(I_h,D)
Input. I_h, a set of intervals where $GP(I_h) = \{Q\}$ and D, a connecting set of I_h.
Output. $T(I_h,D)$, a flag. $T(I_h,D) = \text{true}$ if $C(Q) \cap D \neq \phi$. $C^*(I_h,D)$, a subset of I_h that $G(I_h - C^*(I_h,D))$ has at least $|C^*(I_h,D)|$ path components. $R^*(I_h,D)$, the set of path components of $G(I_h - C^*(I_h,D))$.
Method.
1. Initially, $R^* = \{Q\}$, $C^* = \phi$ and $T = \text{false}$.
2. Repeat the following process until either $T = \text{true}$ or $L(S)$ is empty for every path S in R^*:
   Let $L^* = \bigcup_{S \in R^*} L(S)$, s^* be the largest interval in L^* and S^* be the path in R^* containing s^*. Delete s^* from path S^*. Let W_1 and W_2 be the two paths formed by deleting s^* from S^* and $\textit{path-end}(W_1)$ is smaller than $\textit{path-end}(W_2)$. Let $C^* = C^* \cup \{s^*\}$ and $R^* = (R^* - \{S^*\}) \cup \{W_1, W_2\}$. If $s^* \in D$, then let $T = \text{true}$.
   Note that both W_1 and W_2 are not empty. For any interval s, $s \in W_2$, s is not adjacent to all intervals in $W_1 - L(W_1)$.
3. If $T = \text{true}$, then repeat the following process until $L(S)$ is empty for every path S, $S \in (R^* - \{S_1\})$ where S_1 is the path in R^* with the smallest path–end:
   Let $L^* = \bigcup_{S \in R^*,\, S \neq S_1} L(S)$, s^* be the largest interval in L^* and S^* be the path in R^* containing s^*. Delete s^* from path S^*. Let Y_1 and Y_2 be the two paths formed by deleting s^* from S^* and $\textit{path-end}(Y_1)$ is smaller than $\textit{path-end}(Y_2)$. Let $C^* = C^* \cup \{s^*\}$ and $R^* = (R^* - \{S^*\}) \cup \{Y_1, Y_2\}$.

Note that both Y_1 and Y_2 are not empty. For any interval s, $s \in Y_2$, s is not adjacent to all intervals in $Y_1 - L(Y_1)$.

4. Let $W = S_1 \cap D$, $C^* = C^* \cup W$. Deleting W from S_1 forms $|W| + 1$ path components. Let R^* include all of them and exclude S_1. Output T, C^* and R^*.

Lemma 5 Let Q_1 be the path in $Q - C(Q)$ with the smallest path end where $GP(I_h) = \{Q\}$. All of the following statements are true. (1) If $L(Q) = \phi$, then $T(I_h,D) =$ false and $C^*(I_h,D) = D$. (2) If $T(I_h,D) =$ true, then $|R^*(I_h,D)| = |C^*(I_h,D)| + 1$. (3) If $T(I_h,D) =$ false, then $C(Q) \cap D = \phi$ and $C^*(I_h,D) = D \cup C(Q)$. (4) If $T(I_h,D) =$ false and $path\text{–}end(Q_1) \notin D$, then $|R^*(I_h,D)| = |C^*(I_h,D)| + 1$. (5) If $T(I_h,D) =$ false and $path\text{–}end(Q_1) \in D$, then $|R^*(I_h,D)| = |C^*(I_h,D)|$. (6) For any two paths S_i and S_j of $R^*(I_h,D)$, any two intervals s_i and s_j of S_i and S_j respectively, s_i does not overlap with s_j. (7) $L(Q) \subseteq C^*(I_h,D)$.

3.1. Solving the OPC Problem on Circular–Arc Graphs

We first describe a procedure that converts F to a set of intervals. For arc x, $x \in F$, $B(x)$ denotes the set of arcs which contain arc x. For arc u, the *clockwise* (resp. *counterclockwise*) *conversion with respect to arc u and* $\mathcal{B}$ where $\mathcal{B}$ is a subset of $B(u)$ will convert F to a set of $|F|$ intervals, denoted by $I_c(\mathcal{B},u)$ (resp. $I_{cc}(\mathcal{B},u)$), as follows:

1. Starting from point $h(u)$ (resp. $t(u)$), label F by clockwise (resp. counterclockwise) labeling. That is, $h(u)$ (resp. $t(u)$) is labeled by 1.
2. For each arc $x \notin \mathcal{B}$, x is converted to interval $I_c(\mathcal{B},u,x) = (h(x), t(x))$ if $t(x) \geq h(x)$ (resp. $I_{cc}(\mathcal{B},u,x) = (t(x), h(x))$ if $h(x) \geq t(x)$) and to interval $I_c(\mathcal{B},u,x) = (h(x), 2n+t(x))$ (resp. $I_{cc}(\mathcal{B},u,x) = (t(x), h(x)+2n)$) otherwise.
3. For each arc x in $\mathcal{B}$, x is converted to interval $I_c(\mathcal{B},u,x) = (h(x)-2n, t(x))$ (resp. $I_{cc}(\mathcal{B},u,x) = (t(x)-2n, h(x))$).
4. Re–label the endpoints of the intervals such that every endpoint is labeled by a number between 1 to $2n$. Note that the label of the left endpoint of $I_c(\mathcal{B},u,u)$ (resp. $I_{cc}(\mathcal{B},u,u)$) will be $|\mathcal{B}| + 1$ after the re–labeling. Number intervals from 1 to n in the increasing order of the right endpoints.
5. Let $I_c(\mathcal{B},u) = \{I_c(\mathcal{B},u,x) \mid x \in F\}$ (resp. $I_{cc}(\mathcal{B},u) = \{I_{cc}(\mathcal{B},u,x) \mid x \in F\}$). Let i be an interval of $I_c(\mathcal{B},u)$ (resp. $I_{cc}(\mathcal{B},u)$). Use $A_c(\mathcal{B},u,i)$ (resp. $A_{cc}(\mathcal{B},u,i)$) to denote the arc x where $I_c(\mathcal{B},u,x) = i$ (resp. $I_{cc}(\mathcal{B},u,x) = i$).

Lemma 6 Both $G(I_c(\mathcal{B},u))$ and $G(I_{cc}(\mathcal{B},u))$ are subgraphs of $G(F)$.

We have shown how to convert F to a set of intervals. Careful implementation of this converting process takes $O(n)$ time and space. The following is an $O(n)$ algorithm for the optimal path cover problem on circular–arc graphs.

Algorithm OPC
Input: F, a set of arcs with endpoints sorted.
Output: NP, a minimum set of disjoint paths which covers F.

1. Find an arc u that contains no other arc of F. Starting from the head of u, number the endpoints of F from 1 to $2n$ in the clockwise order. Number arcs from 1 to n in the increasing order of tails. That is, $i < j$ if $t(i) < t(j)$. Let $B_u = B(u)$. Convert F to $I_c(B_u, u)$. Call Algorithm GP($I_c(B_u, u)$). If $|GP(I_c(B_u, u))| = 1$, then let $NP = GP(I_c(B_u, u))$ and stop.
2. Let P be the first path (the path with the smallest path–end) output by Algorithm GP($I_c(B_u, u)$) and $v = A_c(B_u, u, \textit{path–end}(P))$. If $u = v$, then no arc of F contains $t(u)$. Thus, $G(F)$ is an interval graph and the problem can be solved in $O(n)$ time and space [CHA92]. If $t(v) = 2n$ or $t(v) < t(u)$, then $|GP(I_c(B_u, u))| = 1$ and the algorithm should already terminate at Step 1. Hence, in the following steps, we assume that $t(u) < t(v) < 2n$. Let $B_v = \{x \in F \mid I_c(B_u, u, x)$ contains $I_c(B_u, u, v)\}$. Convert F to $I_{cc}(B_v, v)$. Call Algorithm GP($I_{cc}(B_v, v)$). If $|GP(I_{cc}(B_v, v))| = 1$, then let $NP = GP(I_{cc}(B_v, v))$ and stop.
3. Let Q be the first path (the path with the smallest path–end) output by Algorithm GP($I_{cc}(B_v, v)$). Call both Algorithm C(P) and Algorithm C(Q). Call Algorithm C*($Q, C(P)$)). If $T(Q, C(P))) =$ true or $\textit{path–end}(Q_1) \notin C(P)$, then let $NP = GP(I_{cc}(B_v, v))$ and stop.
4. Let $P_1, P_2, \ldots, P_k$ be the k paths of $P - C(P)$ sorted in the increasing order of their path ends in $I_c(B_u, u)$ where $k = |C(P)| + 1$, arc $y = A_c(B_u, u, \textit{path–end}(P_1))$ and $B_y = \{x \in F \mid I_c(B_u, u, x)$ contains $I_c(B_u, u, y)\}$. Convert F to $I_{cc}(B_y, y)$. Call Algorithm GP($I_{cc}(B_y, y)$). If $|GP(I_{cc}(B_y, y))| < |GP(I_{cc}(B_v, v))|$, then let $NP = GP(I_{cc}(B_y, y))$. Otherwise, let $NP = GP(I_{cc}(B_v, v))$.

Theorem 2 Given a set F of n sorted arcs, the optimal path cover and Hamiltonian path problems on $G(F)$ can be solved in $O(n)$ time and space.

References.

[AR] S. Rao Arikati and C. Pandu Rangan, Linear algorithm for optimal path cover problem on interval graphs, *Inform. Process. Lett.* 35 (1990), 149–153.

[B1] A. A. Bertossi, On the domatic number of interval graphs, *Inform. Process. Lett.* **28** (1988) 275–280.

[B2] A. A. Bertossi, The edge Hamiltonian path problem is NP–complete, *Inform. Process. Lett.* **13** (1981) 157–159.

[BB1] A. A. Bertossi and M. A. Bonuccelli, Hamiltonian Circuits in Interval Graph Generalizations, *Information Process. Lett.* **23** (1986) 195–200.

[BB2] M. A. Bonuccelli and D. P. Bovet, Minimum Node Disjoint Path Covering for Circular–Arc Graphs, *Information Process. Lett.* **8** (1979) 159–161.

[BL] K. S. Booth and G. S. Leuker, Testing for consecutive ones property, interval graphs and graph planarity using PQ–tree algorithms, *J. Comput. system Sci.* **13** (1976), 335–379.

[C] M. S Chang, $O(n)$ Algorithms for the Hamiltonian Cycle and Node Disjoint Path Cover Problems on Circular–Arc Graphs, Preprint.

[CH] E. J. Cockayne and S. T. Hedetniemi, Towards a theory of domination in graphs, *Networks* **7** (1977), 247–261.

[CPL] M. S. Chang, S. L. Peng and J. L. Liaw, Deferred–Query, An Efficient Approach for Some Problems on Interval Graphs, Preprint.

[F] M. Farber, Domination, independent domination, and duality in strongly chordal graphs, *Disc. Appl. Math.* **7** (1984), 115–130.

[GT] H. N. Gabow and R. E. Tarjan, A linear–time algorithm for a special case of disjoint set union, *J. Comput. System Sci.* **30** (1985), 209–221.

[GJ] M. R. Garey and D. S. Johnson, *Computers and Intractability: A Guide to the Theory of NP–Completeness*, W.H. Freeman, San Francisco, CA, 1979.

[GJT] M. R. Garey, D. S. Johnson and R. E. Tarjan, The planar Hamiltonian circuit problem is NP–complete, *SIAM J. Comput.* **5** (1976), 704–714.

[G] M. C. Golumbic, *Algorithmic Graph Theory and Perfect Graphs*, Academic Press, New York, 1980.

[GP] M. C. Golumbic and P. L. Hammer, Stability in Circular–Arc Graphs, *J. of Algorithms* **9** (1988) 314–320.

[HSC] W. L. Hsu, W. K. Shih and T. C. Chern, An $O(n^2 logn)$ Time Algorithm for the Hamiltonian Cycle Problem, *SIAM J. on Compt.*

[HT] W. L. Hsu and K. H. Tsai, Linear Time Algorithms on Circular–Arc Graphs, *Information Process. Lett.* **40** (1991) 123–129.

[J] D. S. Johnson, The NP–Complete Column: an Ongoing Guide, *J. of Algorithms* **6** (1985) 434–451.
[K1] J. M. Keil, Finding Hamiltonian Circuits in Interval Graphs, *Inform. Process. Lett.* **20** (1985), 201–206.
[K2] M. S. Krishnamoorthy, An NP–hard problem in a bipartite graphs, *SIGACT News* **7** (1), (1976) 26.
[LHC] T. L. Lu, P. H. Ho and G.J. Chang, The domatic number problem in interval graphs, *SIAM J. Disc. Math.* **3**.(1990), 531–536.
[MM] G. K. Manacher and T. A. Mankus, Determining the domatic number and a domatic partition of an interval graph in time $O(n)$ given its sorted model, Technique report, Submitted to *SIAM J. Disc. Math.* 1991.
[MMS] G. K. Manacher, T. A. Mankus and C. J. Smith, An Optimum $O(nlogn)$ Algorithm for Finding a Canonical Hamiltonian Path and a Canonical Hamiltonian circuit in a set of intervals, *Inform. Process. Lett.* **35** (1990), 205–211.
[MN] S. Masuda and K. Nakajima, An Optimal Algorithm for Finding a Maximum Independent Set of a Circular–Arc Graph, *SIAM J. Comput.* **17** (1988) 41–52.
[MJ] A. Moitra and R. C. Johnson, A parallel algorithm for maximum matching on interval graphs, 1989 *International Conference on Parallel Processing*, III 114–120.
[PC] S. L. Peng and M. S. Chang, A new approach for domatic number problem on interval graphs, *Proceedings of National Computer Symposium* 1991 R.O.C., 236–241.
[R] F. S. Roberts, *Graph theory and its Applications to problems of society*, SIAM, Philadelphia, PA, 1978.
[RR] A. Srinivasa Rao and C. Pandu Rangan, Linear algorithm for domatic number problem on interval graphs, *Inform. Process. Lett.* **33** (1989), 29–33.
[T] A. Tucker, An Efficient Test for Circular–Arc Graphs, *SIAM J. Comput.* **9** (1980) 1–24.
[V] P. Van Emde Boas, Preserving order in a forest in less than logarithmic time and linear space, *Inform. Process. Lett.* **6** (1977), 80–82.

On the Complexity of Graph Embeddings
(Extended Abstract)

Jianer Chen *   Saroja P. Kanchi **   Arkady Kanevsky ***

Texas A&M University, College Station TX 77843, USA

Abstract. It is known that embedding a graph G into a surface of minimum genus $\gamma_{\min}(G)$ is *NP*-hard, whereas embedding a graph G into a surface of maximum genus $\gamma_M(G)$ can be done in polynomial time. However, the complexity of embedding a graph G into a surface of genus between $\gamma_{\min}(G)$ and $\gamma_M(G)$ is still unknown. In this paper, it is proved that for any function $f(n) = O(n^\epsilon)$, $0 \leq \epsilon < 1$, the problem of embedding a graph G of n vertices into a surface of genus at most $\gamma_{\min}(G) + f(n)$ remains NP-hard, while there is a linear time algorithm that approximates the minimum genus embedding either within a constant ratio or within a difference $O(n)$. A polynomial time algorithm is also presented for embedding a graph G into a surface of genus $\gamma_M(G) - 1$.

1 Introduction

Minimum genus $\gamma_{\min}(G)$ of a graph G is defined to be the smallest integer k, such that G has a 2-cell embedding into an orientable surface of genus k. Maximum genus $\gamma_M(G)$ of a graph G is defined to be the largest integer k such that G has a 2-cell embedding into an orientable surface of genus k.

Embedding a graph into topological surfaces is a fundamental, yet very difficult, problem. The computational complexity of constructing the embeddings of a graph into surfaces of different genus is not well-understood. Not much progress had been made until very recently. Algorithms have been developed for embedding a graph into the minimum genus surface as well as into the maximum genus surface. It was demonstrated by Furst, Gross, and McGeoch [5] that a maximum genus embedding of a graph can be constructed in polynomial time. The algorithm presented in [5] is based on a characterization of the maximum genus of a graph given by Xuong [14]. On the other hand, research shows that constructing minimum genus embeddings of a graph is more difficult. For the class C_g of graphs whose minimum genus is bounded by a constant g, Filotti, Miller, and Reif [4] derived an $O(n^{O(g)})$ time algorithm for determining the minimum genus of a graph, which was improved recently by Djidjev and Reif [2] who developed an algorithm of time $O(2^{O(g)!}n^{O(1)})$. The celebrated work of Robertson and Seymour [12] gives an $f(g)n^2$ time algorithm for determining the

* Supported by the National Sciecne Foundation under Grant CCR-9110824.

** Supported by Engineering Excellence Award from Texas A&M University.

*** Supported in part by the NATO Scientific Affairs Division under collaborative research grant 911016.

minimum genus of a graph in the class C_g, where $f(g)$ is a very fast growing function. A result for general graphs was obtained recently by Thomassen [13], who showed that the following problem is *NP*-complete: given a graph G and an integer k, is $\gamma_{\min}(G) \leq k$? Note that this question was one of the remaining basic open problems, listed by Garey and Johnson [6].

At the end of their paper [5], Furst, Gross, and McGeoch posed several open problems, one of them is asking the complexity of embedding a graph into a surface of genus k, where $\gamma_{\min}(G) < k < \gamma_M(G)$.

In the present paper, we will provide a partial answer to the above question. Our main results are: 1) for any function $f(n) = O(n^\epsilon)$, where $0 \leq \epsilon < 1$ is a fixed constant, constructing an embedding of a graph G of n vertices into a surface of genus at most $\gamma_{\min}(G) + f(n)$ is still *NP*-hard; 2) a polynomial time algorithm for embedding a graph G into a surface of genus $\gamma_M(G) - 1$; and 3) a linear time algorithm that, given a graph G of n vertices, constructs an embedding $\Pi(G)$ of G such that either the genus of $\Pi(G)$ is less than $\gamma_{\min}(G) + O(n)$ or the ratio between the genus of $\Pi(G)$ and the minimum genus $\gamma_{\min}(G)$ is bounded by a constant.

We point out that our first two results are not simple consequences of the process of locally altering an embedding of a graph and decreasing the embedding genus. In fact, the process of decreasing embedding genus by locally altering the embedding is not always possible. As demonstrated by Gross and Rieper [9], there are non-minimum genus embeddings of some graphs, on which no local alteration will decrease the embedding genus. It is in fact unknown whether there exist a region R between the minimum genus and the maximum genus and a polynomial time algorithm $\mathcal{A}_R$ such that given an embedding of a graph G into a surface of genus k, where k is within the region R, the algorithm $\mathcal{A}_R$ constructs a genus $k-1$ embedding of the graph G.

Our first result of the intractability of embedding a graph G into a surface of genus at most $\gamma_{\min}(G) + f(n)$, where $f(n) = O(n^\epsilon)$, $0 \leq \epsilon < 1$, is based on a simple graph operation, the bar amalgamation. The contribution of the second result is to demonstrate that given a graph G, there exists a special maximum genus embedding of G, which can be constructed in polynomial time, such that it is always possible to decrease the embedding genus by locally altering the embedding. Our third result is based on a simple combinatorial analysis, and it is in contrast to the first result in some sense: the first result demonstrates that it is hard to approximate the minimum genus of a graph within a difference $O(n^\epsilon)$, $0 \leq \epsilon < 1$, while the third result claims that we can always approximate the minimum genus of a graph easily either within a constant ratio or within a difference $O(n)$.

The paper is organized as follows. In Section 2, we introduce the necessary background on theory of graph embeddings. The intractability of embedding a graph G into a surface of genus at most $\gamma_{\min}(G) + f(n)$, where $f(n) = O(n^\epsilon)$, $0 \leq \epsilon < 1$, is demonstrated in Section 3. A polynomial time algorithm for constructing an embedding of genus $\gamma_M(G) - 1$ for a graph G is presented in Section 4. The approximability of graph minimum genus is discussed in Section 5.

2 Preliminaries

It is assumed that the reader is somewhat familiar with topological graph theory. For further description, see Gross and Tucker [8].

A *graph* may have multiple adjacencies or self-adjacencies. A graph is a *simple graph* if it has neither multiple-adjacency nor self-adjacency. An *embedding* must have the "cellularity property" that the interior of every face is simply connected. The closed orientable surface of genus k is denoted S_k.

A *rotation* at a vertex v is a cyclic permutation of the edge-ends incident on v. Thus, a d-valent vertex admits $(d-1)!$ rotations. A list of rotations, one for each vertex of the graph, is called a *rotation system*.

An embedding of a graph G in an orientable surface induces a rotation system, as follows: the rotation at vertex v is the cyclic permutation corresponding to the order in which the edge-ends are traversed in an orientation-preserving tour around v. Conversely, by the Heffter-Edmonds principle [10, 3], every rotation system induces a unique embedding of G into an orientable surface. This bijectivity enables us to study graph embeddings based on graph rotation systems, a more combinatorial structure.[4] In particular, if $\Pi(G)$ is a rotation system of a graph G, we will denote by $\gamma(\Pi(G))$ the genus of the corresponding graph embedding of G.

A rotation system of a graph can be represented by a doubly-connected-edge-list (DCEL) in which given a vertex v and an edge e incident on v, in constant time we can find the edge following e in the rotation system at v [11]. Moreover, there is a linear time algorithm to trace the boundary walks of all faces in a rotation system [1, 11]. Thus, given a rotation system $\Pi(G)$ of a graph $G=(V,E)$, the genus of $\Pi(G)$ can be calculated in time $O(|E|)$ using the Euler polyhedral equation

$$|V|-|E|+|F|=2-2\gamma(\Pi(G))$$

where $|F|$ is the number of faces in the embedding $\Pi(G)$.

We now describe the effect of inserting a new edge into an embedded graph. Let $\Pi(G)$ be an embedding of a connected graph G. Suppose that we insert a new edge $e=[u,v]$ into the embedding $\Pi(G)$, where u and v are vertices of G. There are two possible cases.

If the edge-ends u and v of e are inserted between two corners of the same face F, then the new edge e splits the face F into two faces. More precisely, if the boundary walk around the face F in $\Pi(G)$ is of the form $u\alpha v\beta u$, where α and β are subwalks, then the new edge e splits the boundary walk of F into two walks: $u\alpha veu$ and $v\beta uev$, resulting in two new faces. Since both the number of faces and the number of edges are increased by 1, with the number of vertices unchanged, by the Euler polyhedral equation, the embedding genus remains the same.

[4] We will interchangeably use the phrases "an embedding of a graph" and "a rotation system of a graph".

If the edge-ends u and v of e are inserted between corners of two different faces F_1 and F_2, then both these faces are merged by e into one larger face. In particular, suppose that the edge e runs from the corner of u in face boundary walk $u\alpha u$ of F_1 to the corner of v in face boundary walk $v\beta v$ of F_2, then the merged face has boundary walk $uev\beta veu\alpha u$. In this case, since the number of faces is decreased by 1 and the number of edges is increased by 1, with the number of vertices unchanged, by the Euler polyhedral equation, the embedding genus is increased by 1.

Finally, we point out that inserting the edge e into the embedding $\Pi(G)$ never decreases the embedding genus.

3 Intractability of $f(n)$-close-min embeddings

Let G be a graph of n vertices and let $\Pi(G)$ be an embedding of G into the surface S_k of genus k. The embedding $\Pi(G)$ is an *$f(n)$-close-min* embedding of G if $k \leq \gamma_{\min}(G) + f(n)$.

The main result of this section is to show that constructing an $f(n)$-close-min embedding of a graph is *NP*-hard, provided that $f(n) = O(n^{\epsilon})$, where $0 \leq \epsilon < 1$ is a fixed constant.

Let G and G' be two graphs, and v and u be vertices of G and G', respectively. The ***bar-amalgamation*** of G and G' at vertices v and u, denoted $G_v * G'_u$, is the result of running a new edge $[v, u]$ (called the "bar") from the vertex v of G to the vertex u of G'. When the vertices v and u are irrelevant, we simply write $G * G'$ instead of $G_v * G'_u$.

Similarly, suppose that $\Pi_1(G)$ and $\Pi_2(G')$ are rotation systems of G and G', respectively, then a ***bar-amalgamation*** of $\Pi_1(G)$ and $\Pi_2(G')$ at vertices v and u is a rotation system of the graph $G_v * G'_u$ obtained by inserting the vertex u somewhere into the rotation system $\Pi_1(G)$ at v and inserting the vertex v somewhere into the rotation system $\Pi_2(G')$ at u and then taking the union of the two adjusted rotation systems. Note that every rotation system of $G_v * G'_u$ can be obtained in this way from two rotation systems of G and G', respectively.

Let G be a graph and let H be a subgraph of G. Let $\Pi(G)$ be a rotation system of G. A rotation system $\Pi'(H)$ of H can be obtained from $\Pi(G)$ by deleting all edges that are not in H. For example, suppose that v is a vertex of H and that the rotation system of $\Pi(G)$ at v is $u_1 u_2 \cdots u_k$, where $[u_{i_1}, v]$, $[u_{i_2}, v]$, $\cdots$, $[u_{i_t}, v]$ are edges in H, $1 \leq i_1 < i_2 < \cdots < i_t \leq k$, and all other edges $[u_p, v]$ are not in H, $p \in \{1, \cdots, k\} - \{i_1, i_2, \cdots, i_t\}$, then the corresponding rotation system of $\Pi'(H)$ at v will be $u_{i_1} u_{i_2} \cdots u_{i_t}$. The rotation system $\Pi'(H)$ of H will be called an ***induced rotation system*** of H from the rotation system $\Pi(G)$ of G. Note that if $\Pi(G_1 * G_2)$ is a bar-amalgamation of two rotation systems $\Pi_1(G_1)$ and $\Pi_2(G_2)$, where G_1 and G_2 are two arbitrary graphs, then $\Pi_1(G_1)$ and $\Pi_2(G_2)$ are the induced rotation systems of G_1 and G_2, respectively, from the rotation system $\Pi(G_1 * G_2)$.

Theorem 1. *Let $\Pi_1(G_1)$ and $\Pi_2(G_2)$ be rotation systems for graphs G_1 and G_2, respectively, and let $\Pi(G_1 * G_2)$ be a bar-amalgamation of $\Pi_1(G_1)$ and $\Pi_2(G_2)$. Then*

$$\gamma(\Pi(G_1 * G_2)) = \gamma(\Pi_1(G_1)) + \gamma(\Pi_2(G_2))$$

We can extend the definition of bar-amalgamation on two graphs to a set of more than two graphs. A *bar-amalgamation* of r graphs G_1, G_2, $\cdots$, G_r, written $G_1 * G_2 * \cdots * G_r$, is recursively defined as follows.

1. If $r = 1$, the bar-amalgamation of the graph G_1 is itself,
2. If $r > 1$, a bar-amalgamation of the graphs G_1, G_2, $\cdots$, G_r is a bar-amalgamation of G' and G_r, where G' is a bar-amalgamation of the $r-1$ graphs G_1, $\cdots$, G_{r-1}.

The definition of bar-amalgamation for rotation systems can be similarly extended to a set of more than two graph rotation systems.

The following theorem can be proved by simple induction based on Theorem 1.

Theorem 2. *Let $\Pi_i(G_i)$ be a rotation system of a graph G_i, $1 \le i \le r$, and let $\Pi(G_1 * \cdots * G_r)$ be a bar-amalgamation of $\Pi_1(G_1)$, $\cdots$, $\Pi_r(G_r)$. Then*

$$\gamma(\Pi(G_1 * \cdots * G_r)) = \sum_{i=1}^{r} \gamma(\Pi_i(G_i))$$

Corollary 3. *Let G_1, $\cdots$, G_r be graphs and let $\Pi(G_1 * \cdots * G_r)$ be a rotation system of a bar-amalgamation $G_1 * \cdots * G_r$ of G_1, $\cdots$, G_r. Then*

$$\gamma(\Pi(G_1 * \cdots * G_r)) = \sum_{i=1}^{r} \gamma(\Pi_i(G_i))$$

where $\Pi_i(G_i)$ is the induced rotation system of G_i from $\Pi(G_1 * \cdots * G_r)$, $1 \le i \le r$.

Corollary 4. *Let G_1, $\cdots$, G_r be graphs and let G' be a bar-amalgamation of G_1, $\cdots$, G_r. Then*

$$\gamma_{\min}(G') = \sum_{i=1}^{r} \gamma_{\min}(G_i)$$

Now we are ready for the main theorem of this section. Recall that an $f(n)$-close-min embedding of a graph G of n vertices is an embedding of G of genus at most $\gamma_{\min}(G) + f(n)$.

Theorem 5. *Let $f(n) = O(n^\epsilon)$ be a function, where $0 \le \epsilon < 1$ is a fixed constant. Then the following problem is NP-hard.*

$f(n)$-Close-Min Embedding

Input: *A graph G of n vertices*
Output: *An $f(n)$-close-min embedding of G*

Proof. By Thomassen's result [13], it is easy to see that the following problem is *NP*-hard:

Minimum-Genus Embedding

Input: A graph G of n vertices
Output: A minimum genus embedding of G

We present a polynomial time reduction from Minimum-Genus Embedding problem to $f(n)$-Close-Min Embedding problem.

Let k be an integer such that $\epsilon < \frac{k}{k+1}$. Then for sufficiently large n, we have $f(n) < n^{\frac{k}{k+1}}$. Thus

$$f(n^{k+1}) \leq n^k - 1$$

Given a graph G of n vertices and m edges, without loss of generality, we may assume that $f(n^{k+1}) \leq n^k - 1$. Let n^kG be a graph that is an arbitrary bar amalgamation of n^k copies of G. Then the number of vertices of n^kG is $N = n^{k+1}$ and the number of edges of n^kG is $M = n^k m + n^k - 1 \leq n^{k+1}m$. The graph n^kG can be obviously constructed from G in time $O(M) = O(n^{k+1}m)$. Moreover, by Corollary 4

$$\gamma_{\min}(n^kG) = n^k \cdot \gamma_{\min}(G)$$

A solution of $f(n)$-Close-Min Embedding problem on the graph n^kG is an $f(n)$-close-min embedding $\Pi(n^kG)$ of n^kG, which has genus at most $\gamma_{\min}(n^kG) + f(N)$. Therefore,

$$\begin{aligned}\gamma(\Pi(n^kG)) &\leq \gamma_{\min}(n^kG) + f(N) \\ &= n^k\gamma_{\min}(G) + f(n^{k+1}) \\ &\leq n^k\gamma_{\min}(G) + n^k - 1 \qquad (1)\end{aligned}$$

On the other hand, if we let $\Pi_1(G), \cdots, \Pi_{n^k}(G)$ be the n^k induced rotation systems of G from $\Pi(n^kG)$, then by Corollary 3

$$\gamma(\Pi(n^kG)) = \sum_{i=1}^{n^k} \gamma(\Pi_i(G)) \qquad (2)$$

Combining Equations (1) and (2) and noticing that the genus of $\Pi_i(G)$ is at least as large as $\gamma_{\min}(G)$ for all $1 \leq i \leq n^k$, we conclude that at least one induced rotation system $\Pi_i(G)$ of G achieves the minimum genus $\gamma_{\min}(G)$. This rotation system of G can be found by calculating the genus for each induced rotation system $\Pi_i(G)$ from $\Pi(n^kG)$ and selecting the one with the smallest genus. This can be accomplished in time $O(M) = O(n^{k+1}m)$.

This completes the proof that the Minimum-Genus Embedding problem can be reduced to the $f(n)$-Close-Min Embedding problem in polynomial time. Since the Minimum-Genus Embedding problem is *NP*-hard, the $f(n)$-Close-Min Embedding problem is also *NP*-hard. □

4 Graph embeddings of genus $\gamma_M(G) - 1$

In this section, we present a polynomial time algorithm for constructing a genus $\gamma_M(G) - 1$ embedding of a graph G. The reduction of genus is achieved by partially altering a special maximum genus embedding.

We say that two edges in a graph are *adjacent* if they have an endpoint in common.

Lemma 6. *Let H be a spanning subgraph of a graph G and let $\Pi(H)$ be an embedding of H. Then for any two adjacent edges e_1 and e_2 in $G - H$, there is a way to insert e_1 and e_2 into $\Pi(H)$ and obtain an embedding of $H \cup \{e_1, e_2\}$ of genus larger than $\gamma(\Pi(H))$.*

Let H be a subgraph of a graph G. An *adjacency matching* in H is a partition of edges of H into groups of one or two edges, called *1-groups* and *2-groups*, respectively, such that two edges in the same 2-group are adjacent. If edges e_1 and e_2 belong to the same 2-group, we say that e_1 and e_2 are *matched*. If an edge e belongs to a 1-group, we say that e is *unmatched*. A *maximum adjacency matching* in H is an adjacency matching that maximizes the number of 2-groups.

Let T be a spanning tree of a graph G. The edge complement $G - T$ will be called a *co-tree*. The number of edges in any co-tree is known as the *cycle rank* of G, denoted $\beta(G)$.

The *deficiency* $\xi(G,T)$ of a spanning tree T is the number of unmatched edges in a maximum adjacency matching in the co-tree $G - T$. The *deficiency* $\xi(G)$ of a graph G is the minimum $\xi(G,T)$ over all spanning trees T of G. A spanning tree T is a *Xuong tree* if $\xi(G,T) = \xi(G)$. Xuong [14] obtained a characterization of maximum genus of a graph in terms of deficiency of the graph.

Theorem 7 (Xuong). *Let G be a connected graph. The maximum genus of G is given by the formula*

$$\gamma_M(G) = \frac{1}{2}(\beta(G) - \xi(G))$$

¿From now on we fix a Xuong tree T of a graph G, and fix a maximum adjacency matching

$$\mathcal{M} = \left[\{s_1\}, \cdots, \{s_{\xi(G)}\}, \{e_1, e_1'\}, \cdots, \{e_{\gamma_M(G)}, e'_{\gamma_M(G)}\}\right]$$

in the co-tree $G - T$, where each $\{s_i\}$, $1 \le i \le \xi(G)$, is a 1-group, and each $\{e_j, e_j'\}$, $1 \le j \le \gamma_M(G)$, is a 2-group.

We briefly describe Furst-Gross-McGeoch's algorithm [5] for constructing a maximum genus embedding of the graph G based on the maximum adjacency matching $\mathcal{M}$.

Arbitrarily embed the Xuong tree T into the 2-sphere (i.e., the plane) of genus 0. Then insert each pair $\{e_i, e_i'\}$ of matched edges into the embedding. Finally, insert all unmatched edges s_i into the embedding. By Lemma 6, we can

always insert a pair $\{e_i, e_i'\}$ of matched edges and increase the embedding genus by at least 1. Furthermore, inserting a pair $\{e_i, e_i'\}$ cannot increase genus by more than 1 since otherwise by Lemma 6 we would be able to insert the $\gamma_M(G)$ pairs $\{e_1, e_1'\}, \cdots, \{e_{\gamma_M(G)}, e'_{\gamma_M(G)}\}$ and obtain an embedding of genus larger than $\gamma_M(G)$. Similarly, any way of inserting the unmatched edges will not increase the embedding genus. Therefore, from the planar embedding of the Xuong tree T, we insert each pair $\{e_i, e_i'\}$ of matched edges and increase the embedding genus by exactly 1, then arbitrarily insert the unmatched edges. The resulting embedding must be a maximum genus embedding of the graph G.

The above process suggests that if we insert a pair $\{e_i, e_i'\}$ of matched edges, for some i, in a way that the embedding genus is not increased, then we would obtain a $\gamma_M(G)-1$ genus embedding of G. This is not true, however. First of all, if the pair $\{e_i, e_i'\}$ is inserted without increasing the embedding genus, then it is possible that a later insertion of a pair $\{e_j, e_j'\}$, $j > i$, increases the embedding genus by 2, or an insertion of some unmatched edge increases the embedding genus, thus, again resulting in a maximum genus embedding of the graph G. Moreover, we must make sure that there exists a pair $\{e_i, e_i'\}$ that can always be inserted without increasing the embedding genus.

In the following, we will demonstrate that there is a way to insert the edges in the co-tree $G-T$ such that inserting each pair $\{e_i, e_i'\}$ of matched edges increases the embedding genus by exactly 1 except for the last pair $\{e_{\gamma_M(G)}, e'_{\gamma_M(G)}\}$, which is inserted without increasing the embedding genus. Moreover, all unmatched edges s_i are inserted without increasing the embedding genus. Such an insertion method obviously results in a $\gamma_M(G) - 1$ genus embedding of G.

Let T_1 be any spanning tree of the graph G and let e be an edge in the co-tree $G - T_1$. There is a unique simple cycle in the graph $T_1 \cup \{e\}$, called the *fundamental cycle* of e with respect to the tree T_1. If we then delete an edge e' in $T_1 \cup \{e\}$ from the fundamental cycle of e, we obtain a new spanning tree T_1' of G. We say that the spanning tree T_1' is obtained from T_1 by *swapping* the edges e and e'.

Lemma 8. *Let C_i and C_j be the fundamental cycles of unmatched edges s_i and s_j with respect to the Xuong tree T, respectively, $i \neq j$. Then the cycles C_i and C_j are vertex disjoint.*

We define a sequence of spanning subgraphs of the graph G.

$$G_k = T \cup \{s_1, \cdots, s_{\xi(G)}\} \cup \bigcup_{i=1}^{k} \{e_i, e_i'\} \qquad k = 0, 1, \cdots, \gamma_M(G) \tag{3}$$

Note that G_0 is the Xuong tree T plus all unmatched edges, and $G_{\gamma_M(G)}$ is the graph G itself. Moreover, all graphs G_k have the same vertex set as the graph G.

Lemma 9. *The maximum genus of the graph G_k is k, for $0 \leq k \leq \gamma_M(G)$.*

Let $\Pi(G)$ be an embedding of a graph G and let F be a face in $\Pi(G)$. We say that F is a *spanning face* of $\Pi(G)$ if all vertices of G appear on the boundary walk of the face F.

Theorem 10. *For $0 \leq k \leq \gamma_M(G)$, the graph G_k has a maximum genus embedding with a spanning face. Moreover, for $k \geq 1$, a maximum genus embedding of G_k with a spanning face can be constructed in linear time from a maximum genus embedding of G_{k-1} with a spanning face.*

Now we are ready to present the algorithm. The graphs G_k, $0 \leq k \leq \gamma_M(G)$, in the algorithm are as defined in Equation (3).

ALGORITHM Max-1 Embedding

INPUT: A graph G
OUTPUT: An embedding of G of genus $\gamma_M(G) - 1$

1. Construct a Xuong tree T of G and find a maximum adjacency matching $\mathcal{M}$ in the co-tree $G-T$
$$\mathcal{M} = \left[\{s_1\}, \cdots, \{s_{\xi(G)}\}, \{e_1, e_1'\}, \cdots, \{e_{\gamma_M(G)}, e'_{\gamma_M(G)}\}\right]$$
2. Construct a maximum genus embedding $\Pi_0(G_0)$ of the graph G_0 with a spanning face.
3. For $1 \leq k \leq \gamma_M(G) - 1$, construct a maximum genus embedding $\Pi_k(G_k)$ of the graph G_k with a spanning face from the maximum genus embedding $\Pi_{k-1}(G_{k-1})$ of the graph G_{k-1} with a spanning face.
4. Insert the last pair $\{e_{\gamma_M(G)}, e'_{\gamma_M(G)}\}$ of matched edges into the rotation system $\Pi_{\gamma_M(G)-1}(G_{\gamma_M(G)-1})$ *without* increasing the embedding genus.

The only step that needs explanation is Step 4. Let F_0 be a spanning face in the maximum genus embedding $\Pi_{\gamma_M(G)-1}(G_{\gamma_M(G)-1})$ of the graph $G_{\gamma_M(G)-1}$. Let $e_{\gamma_M(G)} = [v, u]$ and $e'_{\gamma_M(G)} = [v, w]$. Since all vertices of G are on the boundary walk of the face F_0, we can insert the edge $e_{\gamma_M(G)} = [v, u]$ into the embedding $\Pi_{\gamma_M(G)-1}(G_{\gamma_M(G)-1})$ so that $e_{\gamma_M(G)}$ splits the face F_0 into two faces F' and F'' without increasing the embedding genus. Now the vertex w must be either on the boundary walk of face F' or on the boundary walk of face F'' because w was also on the boundary walk of face F_0. Moreover, *both* faces F' and F'' have the vertex u on their boundary walks. Therefore, we are always able to insert the edge $e'_{\gamma_M(G)} = [u, w]$ so that $e'_{\gamma_M(G)}$ splits either face F' or face F'', without increasing the embedding genus.

Since the embedding $\Pi_{\gamma_M(G)-1}(G_{\gamma_M(G)-1})$ is a maximum genus embedding of the graph $G_{\gamma_M(G)-1}$, which has maximum genus $\gamma_M(G) - 1$, we conclude that the embedding of the graph G constructed by the Max-1 Embedding algorithm has genus $\gamma_M(G) - 1$.

Step 1 of the algorithm can be done in time $O(nmd\log^6 n)$, by Furst-Gross-McGeoch's algorithm [5], where n, m, and d are the number of vertices, the number of edges, and the largest vertex-degree of the graph G, respectively. Step 2 and Step 4 can be done in time $O(m)$. By Theorem 10, each execution of the loop in Step 3 can be done in time $O(m)$. Thus, the total time of Step 3 is bounded by $O(m^2)$. Therefore, the time complexity of the above algorithm is dominated by Step 1, which is $O(nmd\log^6 n)$.

5 On approximating minimum genus

Let $\Pi(G)$ be an embedding of a graph G. The *approximation ratio* of $\Pi(G)$ to the minimum genus $\gamma_{\min}(G)$ is defined to be the value $\gamma(\Pi(G))/\gamma_{\min}(G)$.

Let $\mathcal{C}$ be a class of graphs. We say that an algorithm $\mathcal{A}$ *approximates the minimum genus within a constant ratio* for the class $\mathcal{C}$ if there is a constant c such that for any graph G in $\mathcal{C}$, the algorithm $\mathcal{A}$ constructs an embedding $\Pi(G)$ whose approximation ratio to the minimum genus is bounded by c.

Theorem 5 indicates that there is no polynomial-time algorithm that can construct an n^ϵ-close-min embedding for an arbitrary graph unless $P = NP$. That is, no polynomial-time algorithm can approximate the minimum genus of an arbitrary graph within a difference n^ϵ unless $P = NP$. In this section, using a simple combinatorial derivation, we show a result in contrast to Theorem 5: there is a linear time algorithm that, given a graph G of n vertices, constructs an embedding $\Pi(G)$ of G such that either $\Pi(G)$ is an $O(n)$-close-min embedding of G or the approximation ratio of $\Pi(G)$ to the minimum genus is bounded by a constant.

Lemma 11. *Let G be a simple graph of n vertices. Then*

$$2n + 6\gamma_{\min}(G) > \beta(G)$$

Lemma 11 gives a trivial linear time algorithm to approximate the minimum genus to a constant ratio for a large class of graphs.

Lemma 12. *Let $\delta > 0$ be a fixed constant. Then there is a linear time algorithm $\mathcal{A}_\delta$ that approximates the minimum genus within a constant ratio for all simple graphs $G = (V, E)$ satisfying the condition $|E| \geq (3 + \delta)|V|$.*

Proof. Let $G = (V, E)$ be a simple graph, and let $m = |E|$ and $n = |V|$. If the graph G satisfies the condition $m \geq (3 + \delta)n$, then by Lemma 11, we have

$$2n + 6\gamma_{\min}(G) > \beta(G) = m - n + 1$$

Therefore, $\gamma_{\min}(G) > (\delta n)/6$.

Now pick an arbitrary embedding $\Pi(G)$ of the graph G. Since $\Pi(G)$ has at least one face, by the Euler polyhedral equation, we have $\gamma(\Pi(G)) \leq \beta(G)/2$. By Lemma 11, we have

$$\gamma(\Pi(G)) < n + 3\gamma_{\min}(G)$$

This gives immediately

$$\frac{\gamma(\Pi(G))}{\gamma_{\min}(G)} < \frac{n}{\gamma_{\min}(G)} + 3 < \frac{6}{\delta} + 3 \tag{4}$$

because $\gamma_{\min}(G) > (\delta n)/6$.

Since the embedding $\Pi(G)$ is arbitrary, it can be constructed in linear time by simply picking an arbitrary rotation system. Moreover, the genus of the rotation system $\Pi(G)$ can be computed in linear time by tracing all faces then applying the Euler polyhedral equation. □

Theorem 13. *There is a linear time algorithm $\mathcal{A}$ such that given a simple graph G of n vertices, the algorithm $\mathcal{A}$ constructs an embedding $\Pi(G)$ of G that is either a $(4n)$-close-min embedding or an embedding whose approximation ratio to the minimum genus is bounded by* 4.

Proof. Let $G = (V, E)$ be an arbitrary simple graph where $|V| = n$ and $|E| = m$. Let $\Pi(G)$ be an arbitrary embedding of G. The embedding $\Pi(G)$ can be constructed by picking an arbitrary rotation system of G. We show that the embedding $\Pi(G)$ satisfies the conditions in the theorem.

If $m \geq 9n = (3+6)n$, then by Lemma 12 and its proof, specifically, Equation (4), the approximation ratio of $\Pi(G)$ to the minimum genus is bounded by $6/6 + 3 = 4$.

On the other hand, if $m < 9n$, then the cycle rank $\beta(G)$ of G is at most $8n$. Since the genus of an embedding of the graph G cannot be larger than $\beta(G)/2$, we conclude that the genus of the embedding $\Pi(G)$ is bounded by $4n$. Thus, $\Pi(G)$ is a $(4n)$-close-min embedding. □

To eliminate the restriction of simpleness in Theorem 13, we note that a graph can be "simplicialized" by removing self-loops and multiple edges without changing the number of vertices. This observation gives us the following theorem.

Theorem 14. *There is a linear time algorithm $\mathcal{A}$ such that given a graph G of n vertices, the algorithm $\mathcal{A}$ constructs an embedding $\Pi(G)$ of G that is either a $(4n)$-close-min embedding or an embedding whose approximation ratio to the minimum genus is bounded by* 4.

Proof. Given a graph G of n vertices, we first delete all self-loops, and delete all edges but one in each multiple adjacency. Let the resulting graph be G_1. The graph G_1 is a simple graph of n vertices and can be easily constructed from the graph G in linear time. Moreover, it is easy to see that the graphs G and G_1 have the same minimum genus.

By Theorem 13, we can construct in linear time an embedding $\Pi_1(G_1)$ of the graph G_1 such that either $\gamma(\Pi_1(G_1)) \leq \gamma_{\min}(G_1) + 4n$ or $\gamma(\Pi_1(G_1))/\gamma_{\min}(G_1) \leq 4$. Let $\Pi(G)$ be an embedding of the graph G that is obtained from the embedding $\Pi_1(G_1)$ by adding back those deleted self-loops and multiple edges *without increasing the genus*. It is easy to see that the embedding $\Pi(G)$ can be constructed from the embedding $\Pi_1(G_1)$ in linear time. Since $\gamma(\Pi(G)) = \gamma(\Pi_1(G_1))$

and $\gamma_{\min}(G) = \gamma_{\min}(G_1)$, we conclude that either $\Pi(G)$ is a $(4n)$-close-min embedding of G or the approximation ratio of $\Pi(G)$ to the minimum genus is bounded by 4. □

References

1. J. Chen, J. L. Gross and R.G. Rieper, Overlap matrices and imbedding distributions, *Discrete Mathematics*, to appear.
2. H. Djidjev and J. Reif, An efficient algorithm for the genus problem with explicit construction of forbidden subgraphs, *Proc. 23rd Annual ACM Symposium on Theory of Computing*, (1991), pp. 337-347.
3. J. Edmonds A combinatorial representation for polyhedral surfaces. *Not. Am. Math. Soc.* **7** (1960), pp. 646.
4. I. S. Filotti, G. L. Miller, and J. H. Reif, On determining the genus of a graph in $O(v^{O(g)})$ steps, *Proc. 11th Annual ACM Symposium on Theory of Computing*, (1979), pp. 27-37.
5. M. L. Furst, J. L. Gross, and L. A. McGeoch, Finding a maximum-genus graph imbedding, *J. of ACM* **35-3**, (1988), pp. 523-534.
6. M. R. Garey and D. S. Johnson, *Computers and Intractability, A Guide to the Theory of NP-Completeness*, W. H. Freeman, 1979.
7. J. L. Gross and M. L. Furst, Hierarchy of imbedding-distribution invariants of a graph, *J. of Graph Theory* **11-2**, (1987), pp. 205-220.
8. J. L. Gross and T. W. Tucker, *Topological Graph Theory*, Wiley-Interscience, 1987.
9. J. L. Gross and R. G. Rieper, Local extrema in genus-stratified graphs, *J. Graph Theory* **15-2**, (1991) pp. 159-171.
10. L. Heffter, Uber das Problem der Hachbargeibeite, *Math. Annalen* **38**, (1891), pp. 447-508.
11. F. P. Preparata and M. I. Shamos, *Computational Geometry: An Introduction*, Springer-Verlag, 1985.
12. N. Robertson and P. D. Seymour, Graph minors – a survey, *Survey in Combinatorics*, I. Anderson ed., Cambridge University Press, (1985), pp. 153-171.
13. C. Thomassen, The graph genus problem is *NP*-complete, *J. of Algorithms* **10**, (1989), pp. 568-576.
14. N. H. Xuong, How to determine the maximum genus of a graph, *J. Combinatorial Theory B* **26**, (1979), pp. 217-225.

Algorithms for Polytope Covering and Approximation

Kenneth L. Clarkson
AT&T Bell Laboratories
Murray Hill, New Jersey 07974
e-mail: clarkson@research.att.com

Abstract

This paper gives an algorithm for *polytope covering*: let L and U be sets of points in R^d, comprising n points altogether. A *cover* for L from U is a set $C \subset U$ with L a subset of the convex hull of C. Suppose c is the size of a smallest such cover, if it exists. The randomized algorithm given here finds a cover of size no more than $c(5d \ln c)$, for c large enough. The algorithm requires $O(c^2 n^{1+\delta})$ expected time.[1] More exactly, the time bound is

$$O(cn^{1+\delta} + c(nc)^{1/(1+\gamma/(1+\delta))}),$$

where $\gamma \equiv 1/\lfloor d/2 \rfloor$. The previous best bounds were $cO(\log n)$ cover size in $O(n^d)$ time.[MS92b] A variant algorithm is applied to the problem of approximating the boundary of a polytope with the boundary of a simpler polytope. For an appropriate measure, an approximation with error ϵ requires $c = O(d/\epsilon)^{d-1}$ vertices, and the algorithm gives an approximation with $c(5d^3 \ln(1/\epsilon))$ vertices. The algorithms apply ideas previously used for small-dimensional linear programming.

1 Introduction

Applications in graphics, design, and robotics lead to the problem of approximating a surface by a simpler one. The results here apply to the special case of approximating a surface that is the boundary of a convex body.

If P is a convex polytope (polyhedron) in R^d, then for $\epsilon > 0$, the points within ℓ_1 distance ϵ of the boundary of P lie between the boundaries of two nested convex polyhedra A and B with $A \subset P \subset B$. Thus the problem of finding a polytope P' with few facets whose boundary is close to that of P is reducible to that of separating two nested convex polyhedra by a polytope with few facets. In this way, an *approximation* problem leads naturally to a

[1] In this paper, δ will denote any fixed value greater than zero.

separation problem. Unfortunately, finding a minimal-facet separating polytope for arbitrary nested polyhedra is NP-hard[DJ90], and we hope only to find a polynomial-time algorithm that gives a separating polytope whose number of facets is within a small factor of the smallest possible number. Such an algorithm is given in §3.

It is not hard to show that if P' is any polytope separating A and B, then there is a coarsening A' of A that also separates A and B, such that A' has at most d times as many facets as P'. What is a "coarsening"? Recall that A can be described as the intersection of a family of halfspaces $\mathcal{H}(A)$: $A = \cap_{H \in \mathcal{H}(A)} H$. A coarsening of A is a polytope of the form $\cap_{H \in \mathcal{H}'} H$, where $\mathcal{H}' \subset \mathcal{H}(A)$. Thus in trying to find a simple separating polytope, we give away only a factor of d by restricting ourselves to coarsenings. This simple but crucial observation was made by Mitchell and Suri[MS92b].

Using projective duality, the problem of finding a coarsening with few facets is linear-time equivalent to finding a *covering* with few points: let L be a set of ℓ points in convex position, and let U be a set of u points, both in R^d. A *cover* for L from U is a set $C \subset U$ with L a subset of the convex hull of C. Seeking C with as few points as possible is equivalent to seeking a coarsening of A contained in B with as few facets as possible.

This paper gives an algorithm for covering such that the number of points in the returned cover is $O(d \log c)$ times the optimal number c. This implies an algorithm for separation that finds a polytope with $O(d^2 \log c)$ times as many facets as optimal. In the version of the approximation problem discussed above, the optimal cover size $c = O(d/\epsilon)^{d-1}$, independent of the combinatorial complexity of the polytope being approximated; hence the algorithm given below for covering can be applied to find an approximation within a factor of $5d^3 \ln(1/\epsilon)$ of optimal in number of vertices, as discussed in §3.

This dependence on $\ln c$ (and so $\ln(1/\epsilon)$) contrasts with previous results on this problem for $d > 2$, where the corresponding factor is $\log n$.[MS92b] Moreover, the algorithm given here is faster, sometimes much faster, than the previous $O(n^d)$; the new algorithm requires, in its simplest form,

$$O(c^2 \ell \log c + cu) \log(u/c)$$

expected time: linear in $n \equiv \ell + u$ for fixed c, and always $n^3 \log^{O(1)} n$. (A more complicated form of the algorithm has expected running time $cn^{1+\delta}n$ for $d = 3$.)

The algorithm is Las Vegas, and makes random choices; the expectation is with respect to the algorithm's behavior and independent of the input points.

The algorithm applies ideas used previously for linear programming[Cla88].

2 The covering algorithm

We will assume that the points of L and U are in general position: no $d+1$ are on the same hyperplane. For points p and q, say that p and q *see* each other, or are *visible* to each other, if the (relatively open) line segment $\overline{pq}$ does not meet P. Say that q and a facet F of a polytope P see each other if every $p \in F$ sees q. (Note that q sees F if and only if it sees at least one point of the relative interior of F.) A point q sees F if and only if q is in the halfspace bounded by the hyperplane through F and not containing P.

Let conv S denote the convex hull of the point set S.

Fact 2.1 *Let F be a facet of a polytope P. A point q sees F if and only if F is not a facet of* conv $P \cup \{q\}$. *Moreover, q is contained in P if and only if q sees no facets of P.*

The algorithm needs answers to *visibility queries*: given a set S of n points in general position, and point p, the answer to such a query is a facet of conv S visible to p, or the answer that $p \in \operatorname{conv} S$ and no such facet exists. Visibility queries can be answered in $O(n)$ time using linear programming (e.g.,[Cla88]); with $m^{1+\delta}$ preprocessing, for any fixed $\delta > 0$, queries can be answered in $O(n(\log^{2d+1} n)/m^{\gamma})$ time, where $\gamma \equiv 1/\lfloor d/2 \rfloor$.[MS92a]

Let $C \subset U$ be some optimum cover of L, so $|C| = c$. The statement of the algorithm assumes that the optimal cover size c is known; this is no loss of generality, as discussed at the end of this section.

The development of the algorithm begins with Fact 2.1 above: for $R \subset U$, if L is not contained in conv R, then there is a facet F of conv R that is visible to some point $p \in L$. Moreover, F must be visible to some point $q \in C$, since otherwise F is a facet of conv $R \cup C$, which implies $p \notin \operatorname{conv} R \cup C \supset \operatorname{conv} C$. Let U_F be the set of points of U that see F. Then there is a point of C in the set U_F.

Lemma 2.2 below says that when R is a *random* subset of U, then U_F contains few points. Hence some information about C has been obtained: a small known set contains one of its members.

This may motivate the following algorithm outline, that closely follows an algorithm for linear programming[Cla88]: let each $p \in U$ have a weight w_p, with $w_p := 1$ for all $p \in U$ initially. Let $w(V)$ denote $\sum_{p \in V} w_p$ for $V \subset U$. Repeat the following: choose random $R \subset U$, by choosing each $p \in U$ independently to be in R with probability $1 - (1 - w_p/w(U))^r \leq r w_p / w(U)$, where $r = c4d \ln c$. For each point $p \in L$ in turn, make visibility queries with respect to R; if there are no facets of conv R visible to a point $p \in L$, output R as a cover and quit. If there is such a facet F, and $w(U_F) \leq w(U)/2c$, then double the weights of the points of U_F, so $w_p \mathrel{*}= 2$ for $p \in U_F$. This completes the loop.

We turn to the analysis of this algorithm. Say that an iteration of the loop is *successful* if the weights were changed: there was a facet F visible to a point in L and with $|U_F| \leq w(U)/2c$. Call a facet of conv R an *L-facet* if it is seen by a point in L.

For a time bound, we need to know the chance of a successful iteration.

Lemma 2.2 *Given that an L-facet is found, the probability that an iteration will be successful is at least* $1/2$.

Proof. We show that with high probability, every facet of conv R is seen by few points of U; this follows easily from ancient results[HW87, Cla87], but is included for completeness. Suppose F is a potential facet of conv R: an oriented simplex with d vertices in U, with at least j points $U_F \subset U$ on its positive side. The convex hull of R will have F as a facet if and only if its d vertices are in R, and points U_F that see F are not in R. The probability of this event for given F is

$$\prod_{p\in \mathrm{vert}\, F} (1-(1-\frac{w_p}{w(U)})^r) \prod_{p\in U_F} (1-\frac{w_p}{w(U)})^r \le \prod_{p\in \mathrm{vert}\, F} \frac{rw_p}{w(U)} e^{-rw(U_F)/w(U)}$$
$$\le \frac{r^d}{w(U)^d} e^{-rj/w(U)} \prod_{p\in \mathrm{vert}\, F} w_p.$$

Let $\binom{U}{d}$ denote the set of subsets of U of size d. Since each $V \in \binom{U}{d}$ gives two F, it remains to bound

$$\sum_{V\in\binom{U}{d}} \prod_{p\in V} w_p,$$

subject to the conditions $\sum_{p\in U} w_p = w(U)$ and $|U| = n$. It's not hard to show that this expression is maximized when all $w_p = w(U)/n$, and so the probability that any facet of conv R is seen by more than j points of U is

$$\frac{r^d}{w(U)^d} e^{-rj/w(U)} \binom{n}{d} (w(U)/n)^d \le \frac{r^d}{n^d} \frac{(en)^d}{d^d} e^{-rj/w(U)}$$
$$= \left(\frac{er}{d}\right)^d e^{-rj/w(U)},$$

which is less than $1/2$ for $j = n/2c$, $r \ge c(4d \ln c)$, and c large enough. □

Lemma 2.3 *The expected number of iterations is at most* $1 + 8c\lg(u/c)$.

Proof. The proof follows arguments of Littlestone[Lit87] and Welzl[Wel88]. By Lemma 2.2, the number of "successful" iterations, in which the weights of U_F are changed, is on average at least half the total number of iterations. Consider the total weight $w(U)$. At each successful iteration, $w(U)$ increases by a factor of $1 + 1/2c < e^{1/2c} < 2^{3/4c}$. After I successful iterations, $w(U) \le u2^{3I/4c}$. On the other hand, as noted above, U_F contained a member of C, and so that member is doubled in weight. Hence after I successful iterations, $w(C) \ge \sum_{p\in C} 2^{z_p}$, where $\sum_{p\in C} z_p = I$, and so by the convexity of the exponential function, $w(C) \ge c2^{I/c}$. Since $w(C) \le w(U)$, the algorithm does at most $4c\lg(u/c)$ successful iterations, or $8c\lg(u/c)$ iterations on average. □

Theorem 2.4 *Let $\gamma \equiv 1/\lfloor d/2 \rfloor$, and let δ be any fixed value greater than zero. For known c, a cover of size no more than $c(4d \ln c)$ can be found in*

$$O(u^{1+\delta} + c\ell^{1+\delta} + c(\ell c)^{1/(1+\gamma/(1+\delta))} + (uc)^{1/(1+\gamma/(1+\delta))})$$

expected time, using sophisticated data structures. A simpler algorithm requires $O(\ell c \log c + u)c \log(u/c)$ expected time.

Proof. First consider the time needed for finding sets U_F and reweighting their points, over the whole algorithm. This requires answering $O(c \log(u/c))$ range queries on average, on a set of u points; with a simple algorithm, this requires $O(uc \log(u/c))$ time. Using sophisticated data structures, however, $O(u^{1+\delta} + (uc)^{1/(1+\gamma/(1+\delta))})$ expected time can be achieved, by trading off preprocessing time for query time.[Mat92]

It remains to bound the time required for answering visibility queries during each iteration; this requires answering no more than ℓ visibility queries on a set of points with expected size $r = O(c \log c)$. Using linear-time linear programming, this step requires $O(\ell c \log c)$ expected time per iteration. By again trading off preprocessing for query time, the queries can be answered in $O(\ell^{1+\delta} + (\ell c)^{1/(1+\gamma/(1+\delta))})$ time.[MS92a] This is multiplied by $O(c \log(u/c))$ for the bound.

(This has ignored any correlation between $|R|$ and $|U_F|$; it is quite likely that $|R| < 5r$, so if the algorithm is changed to make an iteration successful only if this holds, the change in $E|U_F|$ will be slight.) □

If the size c of an optimal cover is not known, the algorithm can postulate $c = (5/4)^i$ for $i = \lg d, \lg d + 1, \ldots$, and stop execution for a given value $(5/4)^i$ if the number of successful iterations exceeds the proven bound for covers of that size. In this way the cover returned is no more than 5/4 as big as that for known c, and the work is dominated asymptotically by the work for the returned cover.

3 Approximation of polytopes

This section considers a polytope approximation problem: using the ℓ_1 distance measure, and given a set S of n points, find a polytope Q such that every point of $P \equiv \operatorname{conv} S$ is within ϵ distance of some point of Q, and every point of Q is within ϵ of some point of P. That is, suppose B is the set of points no farther than ϵ from the origin. We seek Q such that $Q \subseteq P + B$ and $P \subseteq Q + B$, with Q having as few vertices as possible. Here $P + B \equiv \{p + b \mid p \in P, b \in B\}$.

(An alternative approximation problem is simply to scale P by some $1 + \epsilon$, and solve the resulting separation problem. For very "flat" polytopes, this problem is plainly quite different from the one above, and arguably less useful.)

First, with small loss we need consider only a finite set of possible vertices for an approximating polytope. Let E be the set of $2d$ extreme points of B: these have coordinates all zero, except one that is either ϵ or $-\epsilon$.

Lemma 3.1 *If Q is an approximating polytope, there is another approximating polytope Q' that has vertices in $S+E$, and with at most d times as many vertices as Q.*

Proof. Since $Q \subset P+B = \operatorname{conv} S+E$, by a slight extension of Caratheodory's theorem, one can pick an arbitrary $v \in Q$ such that each vertex of Q is a convex combination of d points of $S+E$ together with v. For each vertex of Q, pick such points in $S+E$ and include them in a set C. So $Q' \equiv \operatorname{conv} C$ has no more than d times as many vertices as Q, and Q' is an approximating polytope since $P \subset Q+B \subset Q'+B$ and $Q' \equiv \operatorname{conv} C \subset \operatorname{conv} S+E = P+B$. □

We now have a covering problem: choose small $C \subseteq S+E$ such that $P \subset B+\operatorname{conv} C$. (Again, note that $\operatorname{conv} C \subset P+B$.)

To solve this problem, change the algorithm of the last section slightly: while the set U here is $S+E$, and the set L is S, we seek an "L-facet" that is a facet not of $\operatorname{conv} R$, but of $\operatorname{conv} R+E$, for $R \subset S+E$. The set U_F is computed not as the points of U that see F, but rather as the points $p \in U$ such that there is some $e \in E$ such that $p+e$ sees F.

With these changes, and with $r = c5d^3 \ln(1/\epsilon)$, the algorithm and its analysis are analogous to that for the covering problem; the only change in the analysis is the bound on potential L-facets: rather than $2\binom{r}{d}$, it is $2\binom{2dr}{d}$, since facets have vertices in $R+E$, not just R.

The estimate $c = O(d/\epsilon)^{d-1}$ assumes that P is contained in an ℓ_∞ ball of radius one. Consider the regular grid of points with coordinates that are integral multiples of $2\epsilon/d$; every point of the boundary of P is within ℓ_∞ distance ϵ/d of such a point, and so within ℓ_1 distance ϵ. Thus C can be taken to be the set of such grid points within ℓ_∞ distance ϵ/d. Letting $A(P)$ denote the surface area of P, the number of such grid points is $A(P)(d/2\epsilon)^{d-1}(1+O(\epsilon/d))$, and since P is contained in a cube of side length 2, $A(P) \leq d2^d$, yielding a bound of $d(d/\epsilon)^{d-1}$ for c.

4 Concluding remarks

Of course, the most interesting open question is whether the $\log c$ factor in the performance ratio can be reduced, as well as the d factors. The bound $4d \ln c$ can be easily sharpened to $2d \ln(Kc \ln c)$, for a small constant K.

In three dimensions, the bounds reduce to $O(cn^{1+\delta})$; this can readily be sharpened to $cn \log^{O(1)} n$.

Related ideas yield an output-sensitive algorithm for extreme points, requiring $O(an)$ time to find the a extreme points of a set of n points. This result will be reported elsewhere.

Acknowledgements. I'm grateful to Pankaj Agarwal, Michael Goodrich, and Subhash Suri for helpful comments. Of course, they aren't to blame.

References

[Cla87] K. L. Clarkson. New applications of random sampling in computational geometry. *Discrete and Computational Geometry*, 2:195–222, 1987.

[Cla88] K. L. Clarkson. A Las Vegas algorithm for linear programming when the dimension is small. In *Proc. 29th IEEE Symp. on Foundations of Computer Science*, pages 452–456, 1988. Revised version: Las Vegas algorithms for linear and integer programming when the dimension is small (preprint).

[DJ90] G. Das and D. Joseph. The complexity of minimum nested polyhedra. In *Canadian Conference on Computational Geometry*, 1990.

[HW87] D. Haussler and E. Welzl. Epsilon-nets and simplex range queries. *Discrete and Computational Geometry*, 2:127–151, 1987.

[Lit87] N. Littlestone. Learning quickly when irrelevant attributes abound: A new linear-threshold algorithm. In *Proc. 28th IEEE Symp. on Foundations of Computer Science*, pages 68–77, 1987.

[Mat92] J. Matoušek. Reporting points in halfspaces. *Computational Geometry: Theory and Applications*, pages 169–186, 1992.

[MS92a] J. Matoušek and O. Schwartzkopf. Linear optimization queries. In *Proc. Eighth ACM Symp. on Comp. Geometry*, pages 16–25, 1992.

[MS92b] J. Mitchell and S. Suri. Separation and approximation of polyhedral objects. In *Proc. 3rd ACM Symp. on Discrete Algorithms*, pages 296–306, 1992.

[Wel88] E. Welzl. Partition trees for triangle counting and other range searching problems. In *Proc. Fourth ACM Symp. on Comp. Geometry*, pages 23–33, 1988.

Global Strategies for Augmenting the Efficiency of TSP Heuristics*

Bruno Codenotti[1], Giovanni Manzini[1], Luciano Margara[2] and Giovanni Resta[1]

[1] IEI-CNR, Via S.Maria, 46 56100-Pisa (Italy)
[2] Dipartimento di Informatica, Università degli studi di Pisa (Italy)

Abstract. In this paper we introduce two different techniques, *clustering* and *perturbation*, which use global information on TSP instances to speed-up and improve the quality of the tours found by heuristic methods. This global information is related to two correspondent features of any instance, namely *distribution* and *sensitivity*. The performance of our techniques has been tested and compared with known methods. To this end, we performed a number of experiments both on test instances, for which the optimal tour length is known, and on uniformly distributed instances, for which the comparison is done with the Held-Karp lower bound. The experimental results show that our techniques are competitive with the most efficient known methods. It turns out that the viewpoint used in this paper is very satisfactory and deserves further examination.

1 Introduction

Given N cities $i = 1, \ldots, N$, separated by distances d_{ij} the (euclidean) Traveling Salesman Problem — TSP from now — consists of finding the shortest closed path visiting each city exactly once. The solution of very large instances of the TSP has challenged several authors over the last few years [1, 2, 7, 10]. The results have been quite satisfactory. In fact, by using e.g. the Lin-Kernighan method — LK from now — it is possible to face TSP instances with thousands of cities and obtain, within a reasonable time, tours which are very close to the optimal one. On the other hand we are still far from solving the general problem of evaluating the performance of local search heuristics and capturing the mathematical properties of the correspondent local optima. For some preliminary results, see [8].

The problem of gathering some global information on an instance of the TSP seems to be central; in fact all the methods which avoid this are characterized by either a significant loss of precision or a running time penalty.

In this paper, we try to develop an adaptive framework based on two global parameters, the *problem distribution* and the *problem sensitivity* whose eval-

* This work has been partially supported by the CNR through the "Progetto Finalizzato Sistemi Informatici e Calcolo Parallelo. Sottoprogetto 2". G. Manzini and G. Resta have been partially supported by an INDAM postdoc fellowship.

uation suggests a number of new heuristics and/or improvements of existing methods.

Problem Distribution. The problem distribution approach consists of introducing a function which measures the density of cities in any given zone. This suggests to partition the plane into clusters determined by the relative maxima of the above *density function*.

The techniques for solving the TSP can be subdivided into constructive - or direct - and iterative methods. The former class essentially identifies methods which construct a solution in polynomial time, while the latter class corresponds to methods which iteratively construct a solution from another one, e.g., by local search, and do not in general run in guaranteed polynomial time. A popular approach consists of using the outcome of a direct method as the starting point for an iterative one. Here we adopt a completely different strategy: we use an iterative method to perform one step of the process of tour construction which is guided by the clustering procedure. We thus combine the power of iterative methods with the low-cost of direct ones. This is possible because, roughly speaking, we use an iterative method on a problem of substantially smaller size and with special properties.

From the algorithmic viewpoint, the clustering notion leads to the design of both iterative and direct methods. For the iterative methods, one can extend the solution by adding, e.g. to the convex hull, one cluster at a time rather than one city at a time. As a by-product we also have that the LK heuristic becomes much faster once used inside our technique. Furthermore, we use the density function with different parameter setting to obtain "different level descriptions" of the problem depending on the value of specific parameters of the density function (see Sect. 2). This gives rise to a direct method that starts constructing a solution over a set of maxima with fixed cardinality. Then the subsolution over the maxima obtained with a certain value of the parameter is extended to another subsolution over a larger set of maxima, obtained modifying the parameter of the density function. At each step we obtain solutions over sets of cities given by the maxima of the density function with different parameter setting. When c is small enough, the solution found over the maxima is also a solution for the TSP.

Problem Sensitivity. There have been several attempts to formalize some intuitions on the structure of the local optima found by local search procedures (see e.g. [8]). We use the notion of problem sensitivity as the theoretical background for devising a perturbation strategy to be used in the iteration of local search procedures.

Sensitivity has been widely recognized as one important measure of hardness for computational problems. It consists of measuring how the output of the problem changes upon slight changes in the input. For the TSP this property could be interpreted as the relationship between the optimal tour of a given problem instance and the optimal tour of a slightly different one, e.g. with a small modification in the matrix of distances. It could also be used to analyze the changes of the local optima. This general viewpoint leads to the idea of

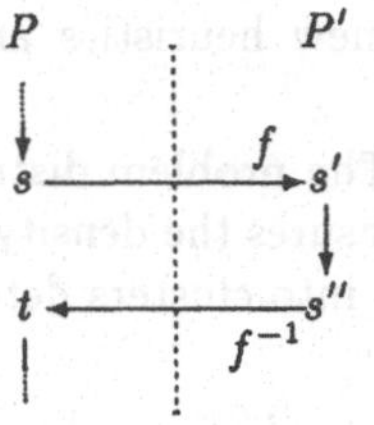

Fig. 1. We start with a local optimum s for the original problem P. We apply a transformation f to s obtaining a solution s' for a perturbated problem P'. In general, s' it is not a local optimum for P'; starting from s' we obtain a local optimum s''. Inverting the trasformation f we convert s'' into a solution t for P, which, again, is not necessarily a local optimum, so that local search can be re-applied.

moving from a given instance to another one as shown in Fig. 1.

More precisely, given a local optimum s for a problem P, we consider a "perturbated" problem P' and a one-to-one correspondence f between the cities in P and P'. We then construct a solution s' for P' as $s' = f(s)$. s' needs not to be a local optimum for P' so one can apply local search — with s' as starting point — to get a local optimum s'' for P'. This can then be mapped onto a solution $t = f^{-1}(s'')$ for P which again is not — in general — a local optimum for P. Thus local search can start again.

The key feature of this approach is the fact that we perturb problem instances rather than solutions so that we can map local optima onto good solutions which can be improved by local search. More in general, this new viewpoint can be a first step towards the development of novel intuitions about how to speed-up local search and how to *escape* from local optima.

The rest of this paper is organized as follows. In Sect. 2, we present the clustering technique; in Sect. 3 the perturbation method; in Sect. 4 the experimental results and in Sect. 5 some concluding remarks.

2 Clustering Techniques

In this section we present a clustering technique for the euclidean TSP which suggests an iterative method (CLR1), a constructive method (CLR2). We introduce the notion of *mass density*, which allows us to develop iterative and direct methods, by analyzing each problem instance in terms of the distribution of the cities.

Euclidean TSP allows us to define a density function $\sigma_c(x, y)$ which gives some information about the concentration of cities in a particular region of the plane. Let $P = \{p_1, \ldots, p_n\}$ be a set of points distributed on the plane and (x_i, y_i) be the coordinates of p_i, $i = 1, .., n$. Then the density function $\sigma_c(x, y)$ is defined as $\sigma_c(x, y) = \sum_{i=1}^{n} f_i(x, y)$, where $f_i(x, y) = e^{-t}$, $t = (\frac{x-x_i}{c})^2 + (\frac{y-y_i}{c})^2$.

We can modify the form of the above function by working on the value of c. It can be easily shown that, as c grows, then each city contributes more to the value of $\sigma_c(x,y)$, for x and y belonging to a larger and larger region of the plane. Given a euclidean graph G, we denote by $\#M_c$ the number of maxima of the function $\sigma_c(x,y)$ and by $(x_{i,c}, y_{i,c})$ the coordinates of the ith maximum.

Let $P = \{p_1, \ldots, p_n\}$ be a set of points distributed on the plane. Let $\sigma_c(x,y)$ and $\sigma_d(x,y)$ be two density functions defined on P. If $d > c > 0$, we have $\#M_c \geq \#M_d$. The definition of the density function suggests a generalization of the notion of convex hull of a set of points. Consider the set of points $M_c = \{(x_{i,c}, y_{i,c}) : 1 \leq i \leq \#M_c\}$, for which the density function attains its maxima, then the generalized convex hull is the convex hull formed on M_c

We now describe an iterative clustering technique which exploits the notion of density function to localize some regions of the plane containing *clusters* of cities.

Algorithm CLR1.

1. **Compute a subsolution s by using a convex hull (or a generalized convex hull) procedure.**
2. **Repeat the following 5 steps until a complete solution S is obtained.**
   a. **Find all the maxima M_c of the density function $\sigma_c(x,y)$ restricted to the cities not yet inserted in s.**
   b. **Choose the maximum $M \in M_c$ closest to the subsolution we are building.**
   c. **Insert in the subsolution all the cities which are placed within a certain distance d_M from M.**
   d. **Use local search to reduce the cost of the subsolution.**
   e. **If all the maxima have been considered then goto step a, else goto step b.**

CLR1 depends on two parameters, namely the value of c and the values of d_M. CLR1 produces solutions which can be substantially different based on the corresponding value of c.

A crucial step of algorithm CLR1 consists of deciding when a city belongs to a certain cluster. Let M be a maximum of the density function, and let S_d be the subset of cities which lie within distance d from M. We map M onto a real number r_M such that $r_M = \min\{d : \sum_{q \in S_d} \exp[-(\frac{x_M - x_q}{c})^2 - (\frac{y_M - y_q}{c})^2] \geq p \cdot \sigma_c(M)\}$, where $0 < p < 1$. We call p the *cluster rebuilding coefficient.*

For c small enough, the number of maxima of the density function approaches the number of cities, and each maximum $(x_{i,c}, y_{i,c})$ converges to a different city. We show now a constructive algorithm which exploits these properties.

Algorithm CLR2.

1. **Set $c = c_0$. Find all the maxima M_c of the density function.**
2. **Find a solution s_c over M_c by using a constructive method (e.g., convex hull + cheapest insertion method).**
3. **Let f be a function such that $f(c) < c$. Compute $d = f(c)$.**

4. **Find all the maxima M_d of the density function.**
5. **Extend the solution s_c over M_c to a solution s_d over M_d by using the polynomial time algorithm described at the end of this section. Let $s_c = s_d$ and $c = d$.**
6. **If a stop criterium occurs, then extend s_c to the set of the cities in order to obtain the solution.**
7. **Goto step 3.**

3 The Method of Perturbation

The most accurate local search heuristic for the TSP (LK heuristic) finds solutions whose lengths are, on the average, 2.1% off the optimum. The only known technique which allows one to find tours which are significantly shorter than those found by single applications of local search is *Iterated Local Search* (ILS). ILS works as follows:

ILS.
1. **Find an initial solution s by using local search.**
2. **Do the following for a given number M of iterations.**
   2.1. **Perturb s obtaining a new solution t.**
   2.2. **Run local search on t obtaining u.**
   2.3. **If $length(u) < length(s)$, set $s = u$.**
4. **Return s.**

The effectiveness of ILS mainly depends on two factors: the perturbation strategy used in Step 2.1 and the local search procedure used in Step 2.2. In practice, ILS finds solutions much shorter than those found by *repeated local search*, which consists of running the local search procedure for a certain number of times, starting from independently chosen initial tours.

ILS has been introduced by Martin, Otto, and Felten [11]. They initially tested this technique on the euclidean TSP by using 3-Opt local search procedure in Step 2.2. To perturb the solution s - Step 2.1 - they remove 4 arcs from s and replace them in order to obtain a non-sequential move. More precisely, they first replace two arcs belonging to s by the two arcs which disconnect the tour. Then they repeat the same operation in order to reconnect the tour (see [11]). This kind of perturbation move is called *Double Bridge*.

They also introduce (see step 2.3 above) the possibility of accepting solutions with increasing cost with a certain probability, using a technique which is similar to simulated annealing. Johnson [7] has investigated the improvement of ILS when 3-Opt local search is replaced by the more powerful LK heuristic in Step 2.2. From now on, DB-ILS will denote the ILS procedure using the Double Bridge move.

In this section we present a new approach for iterating local search. Local search heuristics take a solution s as input and yield a local optimum w which is shorter than s. Experimental results show that the quality of w strongly depends on the quality of s. As an example, local search applied to an initial solution obtained by a direct method, e.g., multiple fragment heuristic [2], finds solutions which are better than those obtained by applying the same local search procedure to random initial tours. ILS takes advantage of this experimental evidence by applying local search to solutions which are obtained by perturbing a previously found local optimum.

Intuitively, one has to look for a perturbation strategy which neither produces a too long tour t nor keeps substantially the same structure as s. In fact, in the former case, local search applied to t is not likely to find a solution shorter than s, and, in the latter case, local search stops soon and does not make any significant improvement.

Informally, a good perturbation strategy should produce a solution t such that: (*i*) the length of t is close to the length of s, and (*ii*) t is quite different from any local optimum (in particular from s). If t satisfies properties (*i*) and (*ii*), then it is conceivable that local search, applied to t, yields a local optimum u better than s.

Here we present a new perturbation strategy which produces perturbated solutions satisfying properties (*i*) and (*ii*). Our basic idea consists of working on *perturbated problems* instead of perturbated solutions. More precisely, we perform Step 2.1 as follows.

Step 2.1.

a. **Find a new set P' of cities by applying a small random perturbation to P.**
b. **Starting from the solution s for P, find the correspondent solution s' for P'**
c. **Find a new solution s'' by applying a local search procedure to P' starting from s'.**
d. **Starting from the solution s'' for P', find the correspondent solution t for P.**

Let π be a permutation of $1, \ldots, n$, $P = \{p_1 \ldots, p_n\}$ be a set of cities and $s = p_{\pi(1)}, \cdots p_{\pi(n)}$ be a solution for P.

In Step *a* we find a perturbated problem P'. This can be done, for example, moving $p_1 \ldots, p_n$ by ϵ or removing from P a certain number k of cities. In Step *b* we find a solution s' for P' starting from s. As an example, if we perturb P by moving the cities by ϵ, s' is equal to s.

Since P' is different from P, s' is not a local optimum for P'. In Step *c* we apply a local search procedure to s' obtaining a local optimum s'' for P'. In Step *d*, starting from s'', we find the correspondent solution t for P. Now, t is not a local optimum for P. Experimental results show that the solution t is quite different from s and has a cost close to the cost of s. This perturbation strategy allows the local search procedure to continue its work even though a local optimum for P has already been found.

Local search runs temporarily on a perturbated problem P'. When no further improvement can be done on P', local search turns again to the original problem P. The intuition behind this approach is visualized in Fig. 1.

We propose two different ways to introduce perturbations, which we call *ε-move* and *k-remove*.

ε-move Given a euclidean TSP $P = \{p_1, \ldots, p_n\}$, this strategy produces a new problem $P' = \{p'_1, \ldots, p'_n\}$ such that $dist(p_i, p'_i) = \epsilon_i$, $\forall i$, $1 \leq i \leq n$, where $dist(a, b)$ stands for the euclidean distance between the points a and b. We choose the value of ϵ_i equal to 1/4 of the sum of the distances of p_i from its neighbors in the actual tour. This allows us to perturb the problem according to the quality of the solution found so far. Each city moves towards a randomly chosen direction. One can readily verify that any solution s for P is also a feasible solution for P', and viceversa.

k-remove This strategy produces a new graph P' simply by removing from P a certain number $k << n$ of cities. Any solution s for P can be translated into a solution s' for P' by disregarding the removed cities. The inverse operation is slightly more complicated. In fact we have to take into account the requirements expressed by properties (i) and (ii). We adopt the following strategy. Each city $p_{\pi(i)}$ is inserted between two other cities $p_{\pi(j)}, p_{\pi(j+1)}$ belonging to s'', chosen at random among the first m neighbors of $p_{\pi(i)}$.

Once a perturbated solution s' has been found (Step 2.1), ILS runs local search on s' in order to obtain a local optimum s'' for the perturbated problem, and then a new local optimum u for the original problem. At this point, ILS accepts or rejects u depending on its length. If u is shorter than s, then ILS sets $s = u$. We perform this step by allowing a low probability acceptance for tours u longer than s. Let $\delta = (length(u) - length(s))/length(u)$. We accept solution u with probability $p = c_1 \cdot e^{-c_2\delta}$, where c_1 and c_2 are suitable constants. Experimental results show that this mechanism enables local search to find shorter tours than those found by using the criterium described in Step 2.3.

4 Implementation Issues and Experimental Results

Both the clustering technique and the perturbation method have been inplemented and tested. The code for ILS with ε-move and k-remove is very efficient and can be executed over very large instances. We are currently working on making the code for the clustering technique more efficient. The implementation currently available allows us to work on up to 3,000 city instances for CLR1 and up to 300 cities for CLR2.

This section is organized as follows. We first give some ideas on the techniques used to speed-up the code and some details on the implementation of ILS, and then we discuss the experimental results.

It is well known that the most successful local search procedures for the TSP are the *2-Opt*, *3-Opt*, and *LK* algorithms (see [2][9] for a complete description of these methods). Unfortunately, it does not exist any theoretical estimate of the quality of the solutions found by these procedures. However, they have been extensively tested for many years and their performance *in practice* is well known. For example, for random Euclidean instances with 1,000 cities, it is known that on the average 2-Opt finds solutions that are within 6.4% off the Held-Karp lower bound[3], and for the 3-Opt and LK procedures this percentage reduces to 3.5% and 2.1%, respectively [7].

The naive implementations of 2-Opt, 3-Opt and LK algorithms take $O(N^2)$, $O(N^3)$, and $O(N^5)$ time, respectively, where N is the number of cities. However, by using appropriate data structures and by taking advantage of certain geömetric properties of the tours, it is possible to substantially reduce the running time of these algorithms [2]. Further significant reductions of the time complexity are possible by implementing "approximate" algorithms that find slightly longer tours but are much faster than the original versions. We underline that all these "tricks" are *necessary* in order to apply, within reasonable time limits, local search procedures to TSP instances with 1,000 cities and more. This is particularly true for ILS algorithms where local search procedures are applied many times.

So far, we have implemented efficiently 2-Opt and 3-Opt. In both cases we have obtained running times that are roughly proportional to the number of cities, and we are currently working on the efficient implementation of LK algorithm.

In our first experiments we have implemented the ILS algorithms based on the ϵ-move strategy, and have performed local search at steps 2.1.c and 2.2 using 3-Opt. One of the basic operation of all local search procedures consists of finding all cities within a certain radius of a given city. In order to perform efficiently this near neighbor search, our algorithms execute a preprocessing stage in which the following data structures are created:

1. a bidimensional array `near[][]` such that `near[i][·]` is the list of the 50 cities closer to city `i`, sorted by increasing distance;
2. a bidimensional array `dist[][]` such that `dist[i][j]` is the distance between cities `i` and `near[i][j]`.

Unfortunately, at step 2.1.c of the ILS algorithm based on the ϵ-move strategy this data structure cannot be used. In fact, local search at step 2.1.c is performed on a *perturbated* problem P' that changes at each iteration. Therefore, the distances between cities are different from the distances contained in `dist[][]` (however, note that the distances change only slightly).

For each perturbated problem P' it would be necessary to compute a new pair of arrays `aux_near[][]` and `aux_dist[][]`. In order to speed up the algorithm we have used the data contained in `near[][]` also for the perturbated problems P'. Clearly, for P' the cities contained in `near[i][]` are simply a set of cities close to city `i`, but in general they are *not* the *more close* cities to city `i`. This implies that in general the solution s'' found at step 2.1.c it is not a local optimum. However, since s'' must be transformed into a solution t for the problem P, the length of s'' is not critical and also a non-optimal solution can be accepted.

[3] Held and Karp [6] have proposed an iterative technique based on minimum spanning trees which produces sharp lower bounds on the optimal tour length.

The distances between cities `i` and `near[i][j]` (for the problem P') are stored in `aux_dist[i][j]`. For these values we utilize a "lazy evaluation" scheme: distances are computed only when required by the algorithm. This guarantees that each distance is computed only once, and that useless distances are not computed.

The resulting algorithm has been tested on random euclidean 1,000 city instances. Setting $M = 1,000$ the ILS algorithm finds solutions whose average percentage excess over the Held-Karp lower bound is 1.67%. A single iteration of the algorithm — that is, a single execution of step 2 — takes on the average 1.49 seconds on a VAX-8750.

We have tried to reduce the running time of this algorithm by using a faster local search procedure at step 2.1.*c*. As we have already pointed out, the length of the solution s'' found at step 2.1.*c* is not critical. For this reason we have tested a modified ILS algorithm in which the local search at step 2.1.*c* is performed by using the 2-Opt procedure. The modified algorithm is much faster: on random euclidean 1,000 city instances a single iteration of step 2 takes on the average 0.52 seconds. Surprisingly enough, the new algorithm finds shorter solutions: the average percentage excess over the Held-Karp lower bound is 1.60%. All experimental results reported in the following refer to this modified ILS algorithm.

The implementation of the k-remove strategy is more straightforward. The data contained in the arrays `near[][]` and `dist[][]` can be utilized also for the perturbated problem P'; we only need to handle properly the cities that have been removed. We have tested the algorithm using both 2-Opt and 3-Opt procedures for the local search at step 2.1.*c*. It turns out that using 3-Opt the algorithm is 17% slower and the solutions are only slightly shorter.

The experimental results reported in the following section have been obtained using 2-Opt at step 2.1.*c*, and at step 2.1.*a* the perturbated problem P' is obtained by removing k cities from P, where $k = \lceil \text{number of cities}/200 \rceil$. The parameter 200 has been chosen on the basis of several tests performed with different values of k.

Table 1 and 2 contain a comparison between the most successful direct (iterative) methods and CLR2 (CLR1). Table 3 gives the results obtained for DB-ILS, ϵ-move, and k-remove on random euclidean instances with up to 10,000 cities (the quality of the tours is compared with the Held-Karp lower bound).

Table 4 provides the results — for DB-ILS, ϵ-move, k-remove— on instances for which the optimal tour length is known. Figs. 2 and 3 give the performance of these algorithms as a function of the number of iterations and the actual running time, respectively, for a 10,000 city instance. Note that the comparisons between DB-ILS, ϵ-move, and k-remove have been performed by using 3-Opt local search. A comparison by using LK will be performed as soon as efficient LK code will be available.

The analysis of the experimental results provides the following indications:

(*i*) The clustering technique leads to direct and iterative methods that outperform — from the viewpoint of the tours quality — all the known correspondent techniques for small instances.

Table 1. Average percentage excess over the Held-Karp lower bound on optimal tour length for direct methods.

Cities	CLR2	Christof.	CH + Cheap. ins.	Near.Neigh.
100	5.1	8.8	13.2	29.5
200	6.1	8.8	14.0	29.6
300	7.1	8.9	14.2	27.4
500	7.5	9.2	15.2	27.2
1,000	9.0	9.9	15.6	27.4
2,000	9.2	9.9	16.7	26.6
3,000	9.2	10.2	17.8	26.2

Table 2. Average percentage excess over the Held-Karp lower bound on optimal tour length for iterative methods. Note that DB-ILS and k-remove have been used to iterate local search 1,000 times.

Cities	*Iterative methods*				*ILS methods*	
	CLR1	LK	3-Opt	2-Opt	DB-ILS	k-remove
100	0.7	1.9	3.0	6.6	0.61	0.65
200	1.0	2.0	3.2	6.7	0.99	1.04
300	1.1	2.2	3.5	6.9	0.73	1.25

(*ii*) Our perturbation techniques give tours whose quality is
- similar to the quality of the tours found by DB-ILS for small instances ($< 1,000$ cities),
- far better than the quality of the tours found by DB-ILS for large instances ($\geq 1,000$ cities).

(*iii*) For what concerns the running time, for instances greater or equal to 5,000
- ϵ-move is slightly slower than DB-ILS,
- k-remove is faster — by a factor of 2 — than DB-ILS.

Table 3. Average percentage excess over the Held-Karp lower bound for iterated local search methods after 500 and 1,000 iterations, respectively. Column ***3-Opt*** gives the average percentage excess over the Held-Karp lower bound for the initial solution found by using 3-Opt local search. For each method, t denote the average running time in seconds on a VAX 8750 for a single iteration.

Cities	3-Opt	DB-ILS			k-remove			ϵ-move		
		500	1000	t	500	1000	t	500	1000	t
100	3.55	0.61	0.03	0.61	0.65	0.02	0.65	0.84	0.84	0.02
500	3.70	1.28	0.24	1.15	1.79	0.14	1.55	1.51	1.36	0.19
1,000	3.77	1.90	0.62	1.71	1.85	0.43	1.49	1.80	1.60	0.52
5,000	3.99	2.92	7.06	2.78	2.71	4.79	2.50	2.10	2.00	7.21
10,000	3.88	3.39	21.48	3.09	2.67	12.78	2.63	2.22	2.06	23.91

Table 4. Average percentage excess over the optimal tour length for known problems after 500 and 1,000 iterations, respectively.

Problems	DB-ILS		*k*-remove		ϵ-move	
	500	1000	500	1000	500	1000
LIN 318	0.43	0.43	0.43	0.43	1.03	1.03
ATT 532	0.55	0.55	0.50	0.50	2.50	2.42
PR 1173	1.36	1.29	2.05	1.90	1.81	1.63
PR 2392	1.43	1.43	1.65	1.39	1.31	1.12

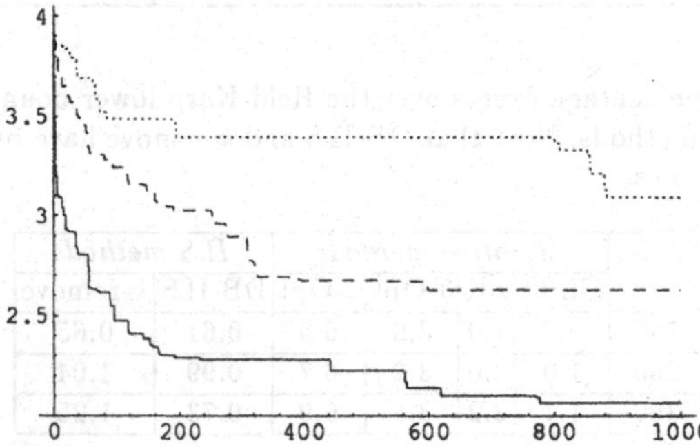

Fig. 2. Average percentage excess over the Held-Karp lower bound versus number of iterations for a 10,000 cities instance. Dotted, dashed and solid lines represent the behaviour of DB-ILS, *k*-remove, and ϵ-move, respectively.

5 Conclusions and Further Work

In this paper we have proposed some alternatives to existing techniques for iterating local search procedures for the TSP. The behaviour of our approach has been experimentally evaluated both on random instances and on test problems. We have shown in the previous section that our results can be favourably compared to those provided by known methods.

Further work to be done includes the application of the perturbation technique to LK method, the development of more efficient code for the clustering methods, and the implementation of the different methods on parallel machines. Once we will have an efficient implementation of LK method, we will be able to compare ϵ-move and *k*-remove with DB-ILS by performing local search with LK heuristics. In addition, we are currently exploring the features of different perturbation techniques. We hope that the idea of working on perturbated instances could lead to new clues on the structure of local optima.

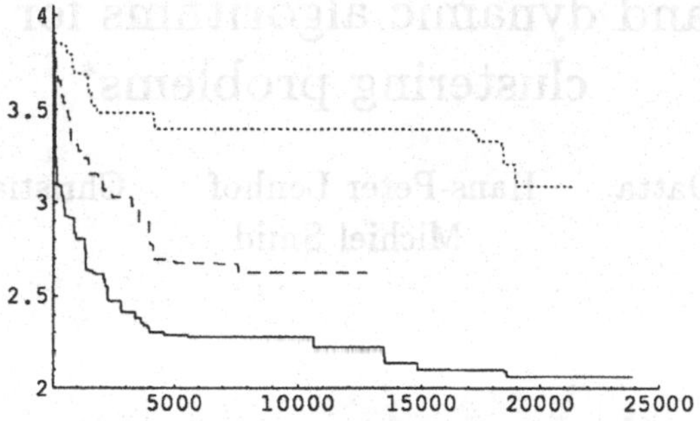

Fig. 3. Average percentage excess over the Held-Karp lower bound versus running time, in seconds, for a 10,000 cities instance. Dotted, dashed and solid lines represent the behaviour of DB-ILS, k-remove, and ϵ-move, respectively. Note that, since the experiments were performed by executing up to 1,000 iterations, the graph corresponding to k-remove stops after less than 15,000 seconds because this time bound is enough to accomplish this number of iterations.

References

1. J. L. Bentley. Experiments on traveling salesman heuristics. *Proc. 1st Symp. on Discrete Algorithms*, 91–99, 1990.
2. J. L. Bentley. Experiments on geometric traveling salesman heuristics. *AT&T Bell Laboratories, Technical Report No. 151*, August 1990.
3. B. Codenotti and L. Margara. Efficient clustering technique for the traveling salesman problem. *I.C.S.I. Technical Report 92-036, International Computer Science Institute, Berkeley, CA 94704*, June 1992.
4. N. Christofides. Worst-case analysis of a new heuristic for the traveling salesman problem. *Management sciences research report No. 388, Carnegie-Mellon University*, February 1976.
5. B. L. Golden, L. D. Doyle, W. Stewart JR. Approximate traveling salesman algorithm. *Oper. Res.*, (28):694–711, 1980.
6. M. Held and R. Karp. The traveling salesman problem and minimum spanning trees. *Oper. Res.* 18:1138–1162, 1970.
7. D. S. Johnson. Local optimization and the traveling salesman problem. *Proc. 17th Colloq. on Automata, Languages, and Programming, Lecture Notes in Computer Science 443.* , 446–461, 1990.
8. D. S. Johnson, C. H. Papadimitriou, and M. Yannakakis. How easy is local search ? *J. Comput. System Sci.* 37(1):79-100, 1988.
9. S. Lin and W. Kernighan. An effective heuristic algorithm for traveling salesman problem. *Oper. Res.*, (21):493–515, 1973.
10. E. Lawler, J. Lenstra, A. Rinnoy Kan, and D. Shmoys. *The traveling salesman problem.* John Wiley and Sons, 1985.
11. O. Martin, S. W. Otto, and W. Felten. Large-step markov chains for the TSP incorporating local search heuristics. *Oper. Res. Lett.*, (11):219–224, 1992.
12. D. Rosenkrantz, R. Stearns and P. Lewis II. An analysis of several heuristics for the TSP. *SIAM J. Comput.*, (6), 1977.

Static and dynamic algorithms for k-point clustering problems*

Amitava Datta Hans-Peter Lenhof Christian Schwarz
Michiel Smid

1 Introduction

We consider clustering problems of the following type. Given a set S of n points in d-dimensional space and an integer k between one and n, find a subset of S of size k that minimizes some closeness measure. As an example, we may want to minimize the perimeter of the convex hull of the k points. This measure was considered by Dobkin et al. [4]. Other measures were considered by Aggarwal et al. [1]. To be more precise, they gave algorithms for finding k points such that their diameter, or their enclosing square, or the perimeter of their enclosing rectangle is as small as possible. Smid [9] also considered the case of minimizing the enclosing square.

Eppstein and Erickson [5] give a general framework for solving such k-point clustering problems. They start by computing for each point its $\Theta(k)$ nearest neighbors, where the constant depends on the problem. Then they use this information to reduce the original problem to $O(n/k)$ subproblems for only $O(k)$ points each. Every single subproblem is solved by some other algorithm $\mathcal{A}$ for the k-point clustering problem in question. (In this reduction, the parameter k remains the same, but the size of the point set is reduced.) If $T(n,k)$ resp. $S(n,k)$ denote the time resp. space complexity of algorithm $\mathcal{A}$ running on a set of size n, then the entire running time of their algorithm is bounded by $O(n \log n + nk + (n/k)T(O(k),k))$ if $d = 2$ and $O(nk \log n + (n/k)T(O(k),k))$ if $d > 2$. Moreover, their algorithm uses space $O(n \log n + nk + S(O(k),k))$ if $d = 2$ and $O(nk + S(O(k),k))$ if $d > 2$.

In this paper, we improve the results of [5] by generalizing techniques that were designed for closest pair problems. Using the search technique of [6], we also reduce the problem to $O(n/k)$ subproblems for $O(k)$ points each. Our reduction, however, is more direct and it circumvents the necessity to compute $\Theta(k)$ neighbors for each point. For any dimension $d \geq 2$, the resulting algorithm has a running time of $O(n \log n + (n/k)T(O(k),k))$ and it uses space $O(n + S(O(k),k))$. Hence, our algorithm uses strictly less space than the one in [5], and our time bound does not exceed that of [5].

Eppstein and Erickson also consider the problem of maintaining the optimal k-point subset if points are inserted. In the planar case, their result is a data structure of size $O(n \log n + S(O(k),k))$ with an insertion time of $O(\log^2 n + k \log n + T(O(k),k))$.

*This work was supported by the ESPRIT Basic Research Actions Program, under contract No. 7141 (project ALCOM II). Authors' address: Max-Planck-Institut für Informatik, W-6600 Saarbrücken, Germany.

They mention that for higher dimensions their method gives results that are only slightly better than brute force.

We give a data structure that, for any dimension $d \geq 2$, maintains the optimal k-point subset in $O(\log n + T(O(k), k))$ time per insertion, using only $O(n + S(O(k), k))$ space.

Eppstein and Erickson mention that no fully dynamic solutions, i.e., solutions that maintain the optimal solution under insertions and deletions of points, are known. We show that the technique of [8] can be generalized to give such a fully dynamic data structure. It uses $O(n \log^d(n/k) + S(O(k), k))$ space and it has an amortized update time of $O(\log n \log^{d-1}(n/k) + \log^d(n/k) \log\log n + T(O(k), k) \log^d(n/k))$.

This paper is organized as follows. In Section 2, we define the class of problems that we can solve and we give the general algorithm for solving them. In order to apply this general algorithm, we need a variant of a grid. If we use a standard grid, then we need the non-algebraic floor function to identify the grid-cell that contains a given point. In Section 3, we introduce a degraded grid that has basically the same properties as a standard grid, but for which we do not need the floor function. In this way, we get algorithms that fall inside the algebraic decision tree model. The notion of degraded grid we use is simpler than the one in [6]. In Section 4, we use the search method of [6] to construct a degraded grid, such that each grid box contains $O(k)$ points and at least one box contains at least k points. This grid is needed to reduce the k-clustering problem for the n points of our input set S to $O(n/k)$ subproblems for $O(k)$ points each. In Section 5, we give several applications of our general algorithm. In Section 6, we give the data structure that maintains the optimal k-point subset under insertions. Section 7 gives a data structure that supports both insertions and deletions. See also Tables 1 and 2.

2 A general approach

Let S be a set of n points in d-dimensional space and let k be an integer such that $1 \leq k \leq n$. A d-dimensional axes-parallel rectangle of the form $[a_1 : b_1) \times [a_2 : b_2) \times \ldots \times [a_d : b_d)$, where a_i and b_i, $1 \leq i \leq d$, are real numbers is called a *box*. If $b_i = a_i + \delta$ for all i, then the box is called a *δ-box*. The closure of a box, i.e., the product of d closed intervals $[a_i : b_i]$ is called a *closed box*. Throughout this paper, we will use the following notations:

- μ denotes a function that maps a set V of points in $\mathbb{R}^d$ to a real number $\mu(V)$, the *measure* of V.
- $P(S, k)$ denotes the problem of finding a subset of S of size k whose measure is minimal among all k-point subsets.
- $\mu_{opt}(S)$ denotes this minimal measure.
- S_{opt} denotes a k-point subset of S such that $\mu(S_{opt}) = \mu_{opt}(S)$.
- $\mathcal{A}$ denotes an algorithm that solves problem $P(S, k)$.
- $T(n, k)$ resp. $S(n, k)$ denote the time resp. space complexity of algorithm $\mathcal{A}$.

Assumption 1 *There exists a closed $\mu_{opt}(S)$-box that contains S_{opt}.*

Assumption 2 *There exists an integer constant c such that for any $\delta < \mu_{opt}(S)/c$, any closed δ-box contains less than k points of S.*

Lemma 1 *Let δ be a real number. Assume there exists a closed δ-box that contains at least k points of S. Then, $\mu_{opt}(S) \leq c\delta$ and there exists a closed $(c\delta)$-box that contains the optimal solution S_{opt}.*

We show how to reduce problem $P(n, k)$ to $O(n/k)$ subproblems $P(S', k)$ for subsets S' of size $O(k)$. Each of these subproblems is then solved using algorithm $\mathcal{A}$. First, we need the following

Definition 1 *Let δ be a positive real number, let $\alpha \leq \beta$ be positive integers, and let R be a collection of δ-boxes such that*

1. *each box in R contains at least one point of S,*
2. *each point of S is contained in exactly one box of R,*
3. *there is a box in R that contains at least α points of S,*
4. *each box in R contains at most β points of S.*

Then R is called an $(\alpha, \beta; \delta)$-covering of S.

Now we can give the algorithm.

Step 1. Compute a positive real number δ together with a $(k, 2^d k; \delta)$-covering R of S. In Section 4, we show that such a δ and such a covering R exist and that they can be found in $O(n \log n)$ time using $O(n)$ space. Moreover, it will be shown there how this collection can be stored in a data structure of size $O(n)$ such that point location queries can be solved in $O(\log n)$ time. This data structure can be built in $O(n \log n)$ time.

Step 2. Initialize $\mu_{opt} := \infty$ and $S_{opt} := \emptyset$.

Step 3. For each box $B \in R$, do the following:

3.1 Find all boxes in R that overlap the $(2c+1)\delta$-box that is centered at B. These boxes are found as follows: Let $(b_1, b_2, \ldots, b_d)$ be the "lower-left" corner of B. Then, in the data structure for R, locate the $(2c+1)^d$ points $(b_1 + \epsilon_1\delta, b_2 + \epsilon_2\delta, \ldots, b_d + \epsilon_d\delta)$, for $\epsilon_i \in \{-c, -c+1, \ldots, c-1, c\}$, $1 \leq i \leq d$.

3.2 Let S' be the set of points of S that are contained in the boxes that are found in Step 3.1. If $|S'| \geq k$, solve problem $P(S', k)$ using algorithm $\mathcal{A}$. Let S'_{opt} be the optimal k-point subset of S'. If $\mu(S'_{opt}) < \mu_{opt}$, then set $\mu_{opt} := \mu(S'_{opt})$ and $S_{opt} := S'_{opt}$.

Step 4. Output μ_{opt} and S_{opt}.

Theorem 1 *The algorithm correctly solves problem $P(S, k)$. Moreover, there is a constant c' such that the algorithm takes $O(n \log n + (n/k)T(c'k, k))$ time and uses $O(n + S(c'k, k))$ space.*

Proof: By Lemma 1, there is a closed $(c\delta)$-box that contains the optimal solution. It is clear that this box must be contained in the $(2c+1)\delta$-box that is centered at some box of R. The algorithm checks all these $(2c+1)\delta$-boxes. If there are less than k points in such a box, then it does not contain the optimal solution. ∎

3 Degraded grids

In the previous section we saw that we need a real number δ and a $(k, 2^d k; \delta)$-covering R for S. Moreover, we need a data structure for these boxes that support point location queries. Assume that the value of δ and a δ-box containing at least k points are known already. Then, of course, we can take a grid with mesh size δ containing this box, and take for R the set of non-empty grid cells. Then, however, we need the floor-function to find the cell that contains a given point. In this section, we introduce so-called degraded grids, that have basically the same properties as standard grids. We can build and search in a degraded grid, however, without using the floor-function.

To give an intuitive idea, in a standard δ-grid, we divide d-space into slabs of width δ. The grid is then defined by fixing an arbitrary point of $\mathbb{R}^d$ to be a lattice point of the grid. So, if e.g. $(0, \ldots, 0)$ is a lattice point, then for $1 \leq i \leq d$, a slab along the i-th axis consists of the set of all points in d-space that have their i-th coordinates between $j\delta$ and $(j+1)\delta$ for some integer j. In a degraded δ-grid, we also have slabs. The difference is that slabs do not necessarily start and end at multiples of δ. Moreover, slabs have width at least δ, and slabs that contain points of S have width exactly δ. That is, while a δ-grid may be defined independently of the point set by fixing an arbitrary point of $\mathbb{R}^d$ to be a lattice point, the degraded δ-grid is defined in terms of the point set stored in it. We will now make this precise.

Definition 2 *Let S be a set of n real numbers and let δ be a positive real number. Let $a_1, a_2, \ldots, a_l$ be a sequence of real numbers such that (1) for all $1 \leq j < l$, $a_{j+1} \geq a_j + \delta$; (2) for all $p \in S$, $a_1 \leq p < a_l$; and (3) for all $1 \leq j < l$, if there is a point $p \in S$ such that $a_j \leq p < a_{j+1}$, then $a_{j+1} = a_j + \delta$.*

The collection of intervals $[a_j : a_{j+1})$, $1 \leq j < l$, is called a one-dimensional degraded δ-grid for S.

Definition 3 *Let S be a set of n points in d-space and let δ be a positive real number. For $1 \leq i \leq d$, let S_i be the set of i-th coordinates of the points in S. Let $[a_{ij} : a_{i,j+1})$, $1 \leq j < l_i$, be a one-dimensional degraded δ-grid for the set S_i. The collection of d-dimensional boxes $\prod_{i=1}^d [a_{ij_i} : a_{i,j_i+1})$, where $1 \leq j_i < l_i$, is called a d-dimensional* degraded δ-grid for S.

Lemma 2 *Let S be a set of n points in d-space and let δ be a positive real number. Assume the points of S are stored in an array S. Moreover, assume that for each $1 \leq i \leq d$, the elements of S_i are sorted, and each element of this set contains a pointer to the corresponding point in S.*

Then we can construct a d-dimensional degraded δ-grid for S in $O(n)$ time using $O(n)$ space. Moreover, we can preprocess this grid in $O(n)$ time, such that for any point p in S, we can report all ℓ points of S that are contained in the δ-box of p, in $O(\log n + \ell)$ time.

4 Constructing a degraded grid with $O(k)$ points per cell

In this section, we give the algorithm that computes the real number δ together with a corresponding $(k, 2^d k; \delta)$-covering R for S. We will use the following notations:

- Assume δ is a real number and R is a degraded δ-grid for S. Number the boxes of R (arbitrarily) $1, 2, \ldots, r = |R|$ and define n_i to be the number of points of S that are contained in the i-th box of R. Then we denote $M(R) = \max_{1 \le i \le r} n_i$.
- Let S' be a subset of S of size $2^d k$ with minimal L_∞-diameter among all $(2^d k)$-point subsets. Then, δ^* denotes the L_∞-diameter of S'.

Lemma 3 *Using these notations, the following holds:*

1. *For any $\delta \ge \delta^*$ and any degraded δ-grid R for S, we have $M(R) \ge k$.*
2. *For any $\delta \le \delta^*$ and any degraded δ-grid R for S, we have $M(R) \le 2^d k$.*

Proof: Let $\delta \ge \delta^*$ and let R be a degraded δ-grid for S. The set S' is contained in an axes-parallel square with sides of length δ^*. This square overlaps at most 2^d boxes of R. Since S' has size $2^d k$, there must be one box in R that contains at least k points of S. This shows that $M(R) \ge k$.

Let $\delta \le \delta^*$ and let R be a degraded δ-grid for S. Assume that $M(R) > 2^d k$. Then there is a box in R that contains more than $2^d k$ points of S. Since this box is the product of half-open intervals of length δ, there are $2^d k$ points in S with L_∞-diameter less than δ. Since $\delta \le \delta^*$, this contradicts the definition of δ^*. ■

The algorithm that is presented below searches for a real number δ together with a degraded δ-grid R for S such that $k \le M(R) \le 2^d k$. This grid is the $(k, 2^d k; \delta)$-covering we want. Lemma 3 implies that there is a δ for which such a covering exists, namely $\delta = \delta^*$. In fact, such a δ is contained in the set of all L_∞-distances between pairs of points in S. As in [6], we do a binary search in the larger set consisting of all possible differences $|p_i - q_i|$, where p and q are points of S and $1 \le i \le d$.

The algorithm maintains the following information:

- Arrays $A_1, \ldots, A_d$ of length n, where A_i contains the points of S sorted w.r.t. their i-th coordinates. For each $1 \le i \le d$, each point in A_i contains a pointer to its occurrence in A_1.
- For each $1 \le i \le d$ and $1 \le j < n$, we store with $A_i[j]$ an interval $[l_{ij} : h_{ij}]$, where l_{ij} and h_{ij} are integers, such that $j < l_{ij} \le h_{ij} + 1 \le n + 1$.

We define the set of *candidate differences* as follows. Let $p = (p_1, \ldots, p_d)$ and $q = (q_1, \ldots, q_d)$ be two distinct points in S, and let $1 \le i \le d$. Moreover, let j and j' be such that $A_i[j] = p$ and $A_i[j'] = q$. Assume w.l.o.g. that $j < j'$. Then $|q_i - p_i|$ is a candidate difference iff $l_{ij} \le j' \le h_{ij}$.

The algorithm makes a sequence of iterations. In each iteration, the number of candidate differences is decreased by a factor of at least one fourth. The algorithm maintains the following

Invariant: At each moment, the value of δ^* is contained in the set of candidate differences.

Initialization: Build the arrays $A_1, \ldots, A_d$. Then, for each $1 \le i \le d$ and $1 \le j < n$, store with $A_i[j]$ the interval $[l_{ij} : h_{ij}] = [j + 1 : n]$.

Iteration:

Step 1. For each $1 \le i \le d$ and $1 \le j < n$, such that $l_{ij} \le h_{ij}$, take the pair $A_i[\lfloor (l_{ij} + h_{ij})/2 \rfloor]$ and $A_i[j]$, and take the (positive) difference of their i-th coordinates.

Give this difference *weight* $h_{ij} - l_{ij} + 1$. This gives a sequence of at most $d(n-1)$ weighted differences.
Step 2. Compute a weighted median δ of these weighted differences.
Step 3. Construct a degraded δ-grid R for S, and compute $M(R)$. There are three possible cases.
3.1 If $k \leq M(R) \leq 2^d k$, then output δ and R, and stop.
3.2 If $M(R) < k$, then for each pair $A_i[\lfloor (l_{ij}+h_{ij})/2 \rfloor]$ and $A_i[j]$ selected in the first step such that the difference of their i-th coordinates is at most δ, set $l_{ij} := \lfloor (l_{ij}+h_{ij})/2 \rfloor + 1$. Go to Step 1.
3.3 If $M(R) > 2^d k$, then for each pair $A_i[\lfloor (l_{ij}+h_{ij})/2 \rfloor]$ and $A_i[j]$ selected in the first step such that the difference of their i-th coordinates is at least δ, set $h_{ij} := \lfloor (l_{ij}+h_{ij})/2 \rfloor - 1$. Go to Step 1.

Lemma 4 *The algorithm correctly maintains the invariant.*

Proof: After the initialization, the set of candidate differences equals the set of all $d\binom{n}{2}$ differences $|p_i - q_i|$. Therefore, the invariant holds initially. Consider one iteration. Assume that Case 3.2 applies, i.e., $M(R) < k$. Then, Lemma 3 implies that $\delta < \delta^*$. The algorithm only removes differences $|p_u - q_u|$ from the set of candidate differences that are at most equal to δ. Hence, at the end of the iteration, the invariant still holds. Case 3.3 can be treated similarly. ∎

Theorem 2 *In $O(n \log n)$ time and using $O(n)$ space, we can compute a real number δ and a degraded δ-grid R for S, such that $k \leq M(R) \leq 2^d k$.*

5 Applications

Minimum diameter k-point subset: In this problem, $\mu(V)$ is the L_2-diameter of the set V. Hence, we want to find k points that have a minimal diameter. In order to show that the algorithm of Section 2 can be applied, we only have to show that Assumptions 1 and 2 are satisfied. It is easy to see that both assumptions hold with $c = \lceil \sqrt{d} \rceil$. It follows that we can apply our general algorithm. Recall that we need an algorithm $\mathcal{A}$ that is called for subsets of size $\Theta(k)$. We take the algorithm of [5]. This algorithm solves the problem in time $T(n,k) = O(n^3 \log^2 n)$ using $S(n,k) = O(n)$ space in the planar case. For the d-dimensional case, the algorithm runs in time $T(n,k) = O(kn \log n + 2^{O(k)} n)$ and uses space $S(n,k) = O(kn)$.

Theorem 3 *Given a set S of n points in d-space and an integer $1 \leq k \leq n$, we can find a subset of size k with minimal diameter in $O(n \log n + nk^2 \log^2 k)$ time and $O(n)$ space, if $d = 2$, and in $O(n \log n + 2^{O(k)} n)$ time and $O(n + k^2)$ space, if $d > 2$.*

Minimum L_∞-diameter k-point subset: We want to find k points with minimal L_∞-diameter, i.e., $\mu(V)$ is the L_∞-diameter of the set V. This is the same as finding a smallest d-dimensional cube that contains at least k points of S. For μ the L_∞-diameter, Assumptions 1 and 2 hold with $c = 1$. We take the algorithm $\mathcal{A}$ from [5]. This algorithm solves the problem in $O(n^{d/2} \log^2 n)$ time using $O(n^{d/2})$ space.

Theorem 4 *Given a set S of n points in d-space and an integer $1 \leq k \leq n$, we can find a subset of size k with minimal L_∞-diameter in $O(n \log n + n \log^2 k)$ time using $O(n)$ space, if $d = 2$, and in $O(n \log n + nk^{d/2-1} \log^2 k)$ time using $O(n + k^{d/2})$ space, if $d > 2$.*

Minimum perimeter k-point subset: For this problem, the points are planar. We want to find k points whose convex hull has minimal perimeter. That is, $\mu(V)$ is the perimeter of the convex hull of V. Assumptions 1 and 2 hold with $c = 4$. We take the algorithm $\mathcal{A}$ from [4]. This algorithm has a running time of $O(n^3 k)$ and uses $O(nk)$ space.

Theorem 5 *Given a set S of n points in the plane and an integer $1 \leq k \leq n$, we can find a subset of size k with minimal perimeter in $O(n \log n + nk^3)$ time using $O(n+k^2)$ space.*

If we take for μ the L_∞-perimeter, then Assumptions 1 and 2 still hold with $c = 4$. Note that for this measure, we want to find k points such that the perimeter of their axes-parallel enclosing rectangle is minimal. We take for $\mathcal{A}$ the brute-force algorithm of [1]. This algorithm runs in time $O(n^3)$ and uses $O(n)$ space.

Theorem 6 *Given a set S of n points in the plane and an integer $1 \leq k \leq n$, we can find a subset of size k with minimal L_∞-perimeter in $O(n \log n + nk^2)$ time using $O(n)$ space.*

Minimum circumradius k-point subset: This is a problem in d-space again. We want to find a smallest ball that contains at least k points. Hence, we can take for $\mu(V)$ the diameter of the smallest ball that contains V. Assumptions 1 and 2 hold with $c = \lceil \sqrt{d} \rceil$. We take the algorithm $\mathcal{A}$ from [5]. This algorithm runs in time $O(n^d \log^2 n)$ and uses space $O(n^d \log n)$. In the planar case, the time and space bounds are both $O(n^2 \log n)$.

Theorem 7 *Given a set S of n points in d-space and an integer $1 \leq k \leq n$, we can find a subset of size k with minimal circumradius in $O(n \log n + nk \log k)$ time using $O(n + k^2 \log k)$ space, if $d = 2$, and in $O(n \log n + nk^{d-1} \log^2 k)$ time using $O(n + k^d \log k)$ space, if $d > 2$.*

6 Maintaining an optimal k-point subset under insertions

In this section, we consider the problem of maintaining the optimal solution S_{opt} if points are inserted into S.

Lemma 5 *Let B be a box that contains at least $(2c)^d k$ points of S. For $1 \leq i \leq d$, let m_i resp. M_i denote the minimal resp. maximal i-th coordinate of any point in $S \cap B$. Then there is an index i such that $M_i - m_i \geq 2\mu_{opt}(S)$.*

measure	dimension	insertion time	space
diameter	2	$\log n + k^3 \log^2 k$	n
diameter	$d > 2$	$\log n + 2^{O(k)}$	$n + k^2$
L_∞-diameter	2	$\log n + k \log^2 k$	n
L_∞-diameter	$d > 2$	$\log n + k^{d/2} \log^2 k$	$n + k^{d/2}$
perimeter	2	$\log n + k^4$	$n + k^2$
L_∞-perimeter	2	$\log n + k^3$	n
circumradius	2	$\log n + k^2 \log k$	$n + k^2 \log k$
circumradius	$d > 2$	$\log n + k^d \log^2 k$	$n + k^d \log k$

Table 1: Semi-dynamic solutions.

Proof: Assume the claim is false. Then there is a $\delta < 2\mu_{opt}(S)$ and a closed δ-box that contains all points of $S \cap B$. Partition this box into $(2c)^d$ closed subboxes with sides of length $\delta/(2c)$. Then one of these closed subboxes contains at least k points of S. This contradicts Assumption 2. ■

In [7], it is shown how a partition of d-space into at most n boxes can be maintained in $O(n)$ space, such that point location queries can be solved in $O(\log n)$ time, and such that a box can be split into two boxes in $O(\log n)$ amortized time. (The two new boxes must still be axes-parallel.) Using a technique described in [3], the time for a split operation can even be made worst-case. In the following insertion algorithm, we assume that we have a partition of d-space into boxes such that

- each point of S is contained in exactly one box of the partition,
- each box in the partition has sides of length at least $\mu_{opt}(S)$,
- each box in the partition contains at most $(2c)^d k$ points of S.

Moreover, we assume that this partition is stored using the method of [3]. With each box of this partition, we store a list of all points of S that are contained in it.

The insertion algorithm: We denote the current value of $\mu_{opt}(S)$ by μ_{opt}. Recall that S_{opt} denotes the optimal k-point subset of the current set S. Let $p = (p_1, p_2, \ldots, p_d)$ be the point to be inserted.

Step 1. Find all boxes of the partition that overlap the $(2\mu_{opt})$-box that is centered at p. These boxes are found by performing 3^d point location queries with the points

$$(p_1 + \epsilon_1\mu_{opt}, p_2 + \epsilon_2\mu_{opt}, \ldots, p_d + \epsilon_d\mu_{opt}), \quad \epsilon_1, \epsilon_2, \ldots, \epsilon_d \in \{-1, 0, 1\}.$$

Let S' be the set of points of S that are contained in these boxes.

Step 2. If $|S'| \geq k$, solve problem $P(S', k)$ using algorithm $\mathcal{A}$. Let S'_{opt} be the optimal k-point subset of S'. If $\mu(S'_{opt}) < \mu_{opt}$ then set $\mu_{opt} := \mu(S'_{opt})$ and $S_{opt} := S'_{opt}$.

Step 3. Output μ_{opt} and S_{opt}.

Step 4. Insert p into the box of the partition that contains it. If this box contains $(2c)^d k + 1$ points, then split this box into two boxes with sides of length at least μ_{opt} such that both new boxes contain at most $(2c)^d k$ points. (By Lemma 5, this is possible.)

Theorem 8 *The insertion algorithm correctly maintains the optimal solution of problem $P(S,k)$. Moreover, there is a constant c' such that the insertion time is bounded by $O(\log n + T(c'k,k))$ and the amount of space used is bounded by $O(n + S(c'k,k))$.*

Proof: Let $\mu_{opt} = \mu_{opt}(S)$ and $\mu'_{opt} = \mu_{opt}(S \cup \{p\})$. We know from Assumption 1 that the optimal solution for the set $S \cup \{p\}$ is contained in a closed μ'_{opt}-box. It is clear that if the optimal solution changes because of the insertion of p, then this optimal solution must be contained in the $(2\mu'_{opt})$-box which is centered at p. Since $\mu'_{opt} \leq \mu_{opt}$, it suffices to consider all points that are contained in the $(2\mu_{opt})$-box centered at p. The algorithm indeed considers all these points. This proves that the optimal solution is correctly maintained.

Next we show that the partition of d-space is correctly maintained. If a box in the partition is not split, it has sides of length at least $\mu_{opt} \geq \mu'_{opt}$. The two new boxes that arise because of a split operation have sides of length at least μ'_{opt}. It is clear that the other two requirements also hold.

It remains to prove the time and space bounds. The bound on the size of the data structure is clear. To insert a point, we perform 3^d point location queries, each taking $O(\log n)$ time. Then we solve a problem $P(S',k)$ for a subset S' of size at most $3^d(2c)^d k$. Finally, we may split a box. This takes $O(\log n + k)$ time. Hence, for an appropriate constant c', the entire insertion algorithm takes time $O(\log n + k + T(c'k,k)) = O(\log n + T(c'k,k))$. ∎

7 A fully dynamic data structure

The algorithm of the previous section only works for insertions: It is crucial that $\mu_{opt}(S)$ does not increase during insertions. In this section, we show that the method of [8] can be adapted such that the optimal k-point subset can be maintained under insertions and deletions.

Recall that for V a set of points, $\mu_{opt}(V)$ denotes the minimal measure of any k-point subset of V. If V has size less than k, then $\mu_{opt}(V) = \infty$. Let $1 \leq i \leq d$. The space $\mathbb{R}^i$ consists of 2^i quadrants. (For $i = 1$, the quadrants of $\mathbb{R}^i = \mathbb{R}$ are $(-\infty : 0]$ and $[0 : \infty)$.) Quadrants are assumed to be closed. We number them arbitrarily. For $1 \leq j \leq 2^i$, we denote the j-th quadrant by Q^i_j.

As in [8], we recursively define data structures of type i for $i = 0, 1, \ldots, d$. The data structure of type d maintains the optimal solution S_{opt} and its measure $\mu_{opt}(S)$. The data structure of type i stores a collection of 2^{d-i} sets. For $1 \leq j \leq 2^{d-i}$, the j-th set of this collection lies in $Q^{d-i}_j \times \mathbb{R}^i$.

We start with the data structure of type 0. For $1 \leq j \leq 2^d$, let V_j be a set of points that lies in the j-th quadrant of d-space, i.e., $V_j \subseteq Q^d_j$. (The 2^d quadrants can intersect in an arbitrary point of $\mathbb{R}^d$. W.l.o.g. we take this point to be the origin.) Let $n_j = |V_j|$ and $n = \sum_j n_j$. Some of the n_j's may be zero.

The data structure of type 0:

1. For $1 \leq j \leq 2^d$, the points of V_j are stored in the leaves of a balanced binary search tree T_j, sorted by their L_∞-distances to the origin. Points with equal L_∞-distance to the origin are stored in lexicographical order.
2. For $1 \leq j \leq 2^d$, let V'_j be the set of $\min((2c)^d k, |V_j|)$ smallest—i.e., leftmost—points in T_j. We store a variable η_0 having value $\eta_0 = \mu_{opt}(\cup_j V'_j)$.

The meaning of the variable η_0 will become clear later. We consider updates of the following type: If we insert or delete a point p that lies in the j-th quadrant of d-space, then we insert or delete p in the set V_j. If p lies on the boundary of several quadrants, then we insert or delete p in only one (arbitrary) set V_j.

Now let $0 < i \leq d$ and assume that the data structure of type $(i-1)$ has been defined already. We define the data structure of type i. For $1 \leq j \leq 2^{d-i}$, let V_j be a set of points that lie in $Q_j^{d-i} \times \mathbb{R}^i$. Let $n_j = |V_j|$ and $n = \sum_j n_j$. Some of the n_j's may be zero.

The data structure of type i: All points of the set $\bigcup_j V_j$ are stored in the leaves of one balanced binary search tree, sorted by their $(d-i+1)$-th coordinates. Points with equal $(d-i+1)$-th coordinate are stored in lexicographical order. In each node u of this tree, we store a hyperplane $x_{d-i+1} = \sigma_u$, where σ_u is the maximal $(d-i+1)$-th coordinate stored in its left subtree.

Each node u of this tree contains the following additional information.

1. If the subtree of u contains less than k points, then it contains a variable $\eta_i(u)$ with value ∞.
2. Assume the subtree of u contains at least k points. Let v resp. w be the left resp. right son of u.

   For $1 \leq j \leq 2^{d-i}$, let V_j^v resp. V_j^w be the subsets of V_j that are stored in the subtrees of v resp. w. Note that $V_j^v \subseteq Q_j^{d-i} \times (-\infty : \sigma_u] \times \mathbb{R}^{i-1}$, and $V_j^w \subseteq Q_j^{d-i} \times [\sigma_u : \infty) \times \mathbb{R}^{i-1}$. That is, w.r.t. an appropriate origin, we have $V_j^v \subseteq Q_j^{d-i} \times Q_1^1 \times \mathbb{R}^{i-1}$, and $V_j^w \subseteq Q_j^{d-i} \times Q_2^1 \times \mathbb{R}^{i-1}$.

   (a) We store in u a pointer to a data structure of type $(i-1)$ which stores the 2^{d-i+1} sets V_j^v and V_j^w, $1 \leq j \leq 2^{d-i}$. Let η_{i-1} be the variable that is stored with this data structure.

   (b) Let $\eta_i(v)$ resp. $\eta_i(w)$ be the variables that are stored with the nodes v resp. w. We store in u a variable $\eta_i(u)$ with value $\eta_i(u) = \min(\eta_i(v), \eta_i(w), \eta_{i-1})$.

Finally, the data structure of type i stores a variable η_i with value $\eta_i = \eta_i(r)$, where r is the root of the tree.

We consider updates of the following type: If we insert or delete a point p that lies in $Q_j^{d-i} \times \mathbb{R}^i$, then we insert or delete p in the set V_j. If p lies on the boundary of several regions, then we insert or delete p in only one (arbitrary) set V_j. Note that the data structure of type d stores one set of points. We show that this data structure for the set S stores the optimal solution to problem $P(S,k)$:

Lemma 6 *Consider the data structure of type d for the set S. The value η_d that is stored with this structure is equal to $\mu_{opt}(S)$.*

Proof: First note that all $\eta_i(\cdot)$-variables have value either ∞ or $\mu_{opt}(S')$ for some subset S' of S. Therefore, $\mu_{opt}(S) \leq \eta_d$. The data structure of type d contains type 0 structures as substructures. If we can show that the η_0-variable of one of these type 0 substructures has value $\mu_{opt}(S)$, then it follows that $\eta_d \leq \mu_{opt}(S)$ and, hence, $\eta_d = \mu_{opt}(S)$.

We show that such a type 0 substructure exists. Consider the optimal k-point subset S_{opt} of S. Note that $\mu(S_{opt}) = \mu_{opt}(S)$. We inductively define a sequence $u_1, u_2, \ldots, u_d$ of nodes having the following properties:

1. u_i is a node of a data structure of type $(d-i+1)$.
2. u_i is a node of the data structure of type $(d-i+1)$ that is pointed to by u_{i-1}.
3. The subtree of u_i contains all points of S_{opt}.
4. u_i is the highest node in its tree s.t. both its left and its right subtree contain points of S_{opt}.

Consider the data structure of type d. Then, u_1 is the highest node in its binary search tree such that both the left and the right subtree of u_1 contain points of S_{opt}. Let $1 < i \leq d$ and assume that $u_1, u_2, \ldots, u_{i-1}$ have been defined already. Then, u_i is the highest node in the binary tree of the data structure of type $d-i+1$ that is pointed to by u_{i-1}, such that both the left and the right subtree of u_i contain points of S_{opt}. Clearly, the nodes defined in this way have the four given properties.

Consider node u_d. This node belongs to the binary tree of a data structure of type 1, and it contains a pointer to a data structure D of type 0. We claim that the variable η_0 that is stored with D has value $\mu_{opt}(S)$. This will complete the proof of the lemma.

Let V_j, $1 \leq j \leq 2^d$, be the sets stored in D. Note that $S_{opt} \subseteq \bigcup_j V_j$. Moreover, the sets V_j, some of which may be empty, lie in different quadrants that are defined by the hyperplanes stored in the nodes $u_1, u_2, \ldots, u_d$. We assume w.l.o.g. that these hyperplanes intersect in the origin. Let V_j', $1 \leq j \leq 2^d$, be the subsets defining the value of η_0, i.e., $\eta_0 = \mu_{opt}(\bigcup_j V_j')$, cf. the definition of the data structure of type 0. We now briefly sketch the argument that shows that $S_{opt} \subseteq \bigcup_j V_j'$, which in turn implies $\eta_0 = \mu_{opt}(S)$. For more details, we refer to the full paper.

Assume there is a point p in $S_{opt} \setminus \bigcup_j V_j'$. Assume w.l.o.g. that all coordinates of p are nonnegative and that $p \in V_1$. Then $p \in V_1 \setminus V_1'$. Since $|V_1'| = \min((2c)^d k, |V_1|)$, this means $|V_1'| = (2c)^d k$. Let δ be the maximal L_∞-distance between any point of V_1' and the origin. Then, V_1' is contained in the closed box $[0:\delta]^d$. By a standard argument, this box contains a closed subbox with sides of lengths $\delta/(2c)$ which holds at least k points of S. Using Lemma 1, it follows that $\mu_{opt}(S) \leq c \cdot \delta/(2c) < \delta$. By the choice of the nodes $u_1, \ldots, u_d$, the partition of d-space into quadrants in the data structure D of type 0 is such that there is a point $q \in S_{opt}$ having L_∞-distance at least δ to p, a contradiction since by Assumption 1, S_{opt} is contained in a $\mu_{opt}(S)$-box. Hence, we have shown that $S_{opt} \subseteq \bigcup_j V_j'$. This completes the proof. ∎

The data structure of type i is similar to a range tree. To maintain it under insertions and deletions, we take the binary trees from the class of BB[α]-trees. First, we analyze its space complexity. Let $G(n,i)$ be the size of a data structure of type i storing n points. Then, $G(n,0) = O(n)$ and $G(n,i) = O(n + G(n,i-1)\log(n/k))$. This solves to $G(n,d) = O(n \log^d(n/k))$.

The update algorithm is virtually the same as in [8]. Rebalancing is done by means of rotations. As in [8], we can apply dynamic fractional cascading. We refer the reader to that paper for the details. If $U(n,i)$ denotes the amortized update time, then we can show that $U(n,1) = O(\log n + \log(n/k) \log\log n + T(O(k),k) \log(n/k))$, and $U(n,i) = O(\log n + U(n,i-1) \log(n/k) + \log^i(n/k) \log\log n + T(O(k),k)/k \log^i(n/k))$. It follows that $U(n,d) = O(\log n \log^{d-1}(n/k) + \log^d(n/k) \log\log n + T(O(k),k) \log^d(n/k))$. Note that during the update algorithm, we need an extra amount $S(O(k),k)$ of space, because we call the algorithm $\mathcal{A}$ for subsets of size $O(k)$.

measure	dimension	update time	space
diameter	2	$f(n,k,2)+k^3\log^2 k\log^2(n/k)$	$n\log^2(n/k)$
diameter	$d>2$	$f(n,k,d)+2^{O(k)}\log^d(n/k)$	$n\log^d(n/k)+k^2$
L_∞-diameter	2	$f(n,k,2)+k\log^2 k\log^2(n/k)$	$n\log^2(n/k)$
L_∞-diameter	$d>2$	$f(n,k,d)+k^{d/2}\log^2 k\log^d(n/k)$	$n\log^d(n/k)+k^{d/2}$
perimeter	2	$f(n,k,2)+k^4\log^2(n/k)$	$n\log^2(n/k)+k^2$
L_∞-perimeter	2	$f(n,k,2)+k^3\log^2(n/k)$	$n\log^2(n/k)$
circumradius	2	$f(n,k,2)+k^2\log k\log^2(n/k)$	$n\log^2(n/k)+k^2\log k$
circumradius	$d>2$	$f(n,k,d)+k^d\log^2 k\log^d(n/k)$	$n\log^d(n/k)+k^d\log k$

Table 2: Fully dynamic solutions. The update times are amortized. $f(n,k,d)$ denotes the function $\log n\log^{d-1}(n/k)+\log^d(n/k)\log\log n$.

Theorem 9 *There exists a data structure that maintains the optimal solution of problem $P(S,k)$ under insertions and deletions. For some constant c' this data structure uses $O(n\log^d(n/k)+S(c'k,k))$ space and it has an amortized update time of*

$$O(\log n\log^{d-1}(n/k)+\log^d(n/k)\log\log n+T(c'k,k)\log^d(n/k)).$$

References

[1] A. Aggarwal, H. Imai, N. Katoh and S. Suri. *Finding k points with minimum diameter and related problems.* J. Algorithms 12 (1991), 38-56.

[2] J.L. Bentley. *Decomposable searching problems.* IPL 8 (1979), 244-251.

[3] R.F. Cohen and R. Tamassia. *Combine and conquer.* Report CS-92-19, Brown University, Providence, 1992.

[4] D.P. Dobkin, R.L. Drysdale and L.J. Guibas. *Finding smallest polygons.* In: Advances in Computing Research, Computational Geometry, J.A.I. Press, London, 1983, 181-214.

[5] D. Eppstein and J. Erickson. *Iterated nearest neighbors and finding minimal polytopes.* Proc. 4th SODA, 1993, 64-73.

[6] H.P. Lenhof and M. Smid. *Enumerating the k closest pairs optimally.* Proc. 33rd Annual IEEE Symp. Foundations of Computer Science, 1992, 380-386.

[7] C. Schwarz, M. Smid and J. Snoeyink. *An optimal algorithm for the on-line closest pair problem.* Proc. 8th Symp. on Comput. Geom., 1992, 330-336.

[8] M. Smid. *Maintaining the minimal distance of a point set in polylogarithmic time.* Discrete Comput. Geom. 7 (1992), 415-431.

[9] M. Smid. *Finding k points with a smallest enclosing square.* Report MPI-I-92-152, Max-Planck-Institut für Informatik, Saarbrücken, 1992.

Scalable Algorithms for Bichromatic Line Segment Intersection Problems on Coarse Grained Multicomputers*

Olivier Devillers and Andreas Fabri

INRIA, B.P.93, 06902 Sophia-Antipolis cedex (France),
E-mail: Firstname.Name@sophia.inria.fr.

Abstract. We present output-sensitive scalable parallel algorithms for bichromatic line segment intersection problems for the Coarse Grained Multicomputer model. Under the assumption that $n \geq p^2$, where n is the number of line segments and p the number of processors, we obtain an intersection counting algorithm with a time complexity of $O(\frac{n \log n \log p}{p} + T_s(n \log p, p))$, where $T_s(m, p)$ is the time used to sort m items on a p processor machine. An additional $O(\frac{k}{p})$ time is spent on the reporting of the k intersections. As the sequential complexity is $O(n \log n)$ and $O(k)$ for counting and reporting, respectively, we obtain a speedup of $\frac{p}{\log p}$.

1 Introduction

The *bichromatic line segment intersection problems* can be stated as follows. Let R (resp. B) be two sets containing n red (resp. blue) non-intersecting line segments. The problem is to *count* or to *report* the intersections between red and blue segments. This problem is well understood in a sequential setting [4, 10] and its sequential time complexity is $\Theta(n \log n)$ for counting and $\Theta(n \log n + k)$ for reporting, where k is the number of intersections. In the parallel setting there is an optimal algorithm for the CREW-PRAM model solving the problem in time $O(\log n)$ with $O(n + \frac{k}{\log n})$ processors [7].

In this paper we consider the parallel complexity of this problem in the *coarse grained multicomputer* model, or $CGM(m, p)$ for short. In this model a set of p processors, each with $O(\frac{m}{p})$ local memory, are connected by some arbitrary interconnection network. The model is *coarse grained*, as the size $O(\frac{m}{p})$ of each local memory is defined to be considerably larger than $O(1)$. Throughout this paper we assume that $\frac{m}{p} \geq p$, unless otherwise stated. Interprocessor communication is done in *communication rounds.* An operation performed in a communication round is typically a sorting or a broadcasting step. Such operations are usually available as system calls on commercially available parallel machines, but nevertheless more expensive than local computations. Hence, the main objective in the design of algorithms for this model is to use a constant or small number of

* *This work was partially supported by the ESPRIT Basic Research Action Nr. 7141 (ALCOM II).*

communication rounds. Roughly, this is achieved by reorganizing the m data in memory in such a way that each processor can sequentially treat an independent subproblem of size $O(\frac{m}{p})$.

The coarse grained multicomputer model has recently been used to solve problems in computational geometry [6] and image processing [1]. The first paper uses a decomposition of the Euclidean space and the second paper a data compression technique. In this paper we present a further decomposition technique which can be applied to tree-like data structures, in our case segment trees. We further present a load balancing scheme to solve the 1-dimensional range query reporting problem. The combination of these techniques yields scalable parallel algorithms, which are independent of the communication network.

The main result of this paper can be stated as follows. Given two sets of n non-intersecting red and n non-intersecting blue line segments, we show how to solve the intersection counting problem on a $CGM(n \log p, p)$, $\frac{n}{p} \geq p$, in time $O(\frac{n \log n \log p}{p} + T_s(n \log p, p))$, where $T_s(m, p)$ is the time used to sort m items on a p processor machine. For reporting the k intersections we need a machine with $\max(k, n \log p)$ memory and it takes additional time $O(\frac{k}{p})$, that is our algorithm is output-sensitive. We further show how to obtain a trade off between the memory and the number of communication rounds.

Our algorithms are *scalable* in the following sense. Their running time is a function of two parameters, the problem size n and the number of processors p, but their speedup is only a function of p, not of n. This condition is crucial as the following example shows. Consider the simulation of a n-processor hypercube sorting n numbers with Batcher's bitonic sort algorithm on a p-processor hypercube with n memory. The runningtime is $O(\frac{n}{p} \log^2 n)$, the speedup is $\frac{p}{\log n}$ and, hence, the parallel sorting algorithm is only faster than the sequential algorithm for $n \leq 2^p$. This phenomenon can be observed whenever we simulate a non-optimal fine grained algorithm on a coarse grained machine.

This paper is organized as follows. In Section 2 we shortly discuss the underlying communication model. In Sections 3, 4 and 5, we present a binary search, a trapezoidal decomposition and a 1-dimensional range query reporting algorithm, which are interesting in their own right and which are used as building blocks for the bichromatic line segment intersection algorithms presented in Section 6.

2 Communication Model

The processors of a $CGM(m, p)$ communicate via an interconnection network in which each processor can exchange messages, of size $O(\log m)$ bits each, with any one of its immediate neighbors in constant time. Commonly used interconnection networks for CGM include 2D-mesh (e.g. Intel Paragon), hypercube (e.g. Intel iPSC/860) and the fat-tree (e.g. Thinking Machines CM-5). We refer the reader to [2, 3, 8, 9] for a more detailed discussion of architectures and algorithms.

We will now outline the four operations involving interprocessor communication which we use in this paper and give the time complexity of the operations for the hypercube. Note however that, as we restrict ourselves to these communication functions, the algorithms we present are architecture independent. That

is the reason why, in the sequel, complexities depend also on T_s and not only on the number of processors and the size of the data. Assume that the p processors of the $CGM(m,p)$ are numbered from 0 to $p-1$.
1) global sort: after completion, the data are distributed such that keys of data on processor p_i are greater than keys of data on p_j if $i > j$, and data are sorted on each processor. The time complexity, based on Batcher's bitonic sort [2], is $T_s(m,p) = O(\frac{m}{p}(\log m + \log^2 p))$, which is optimal for $m \geq p^{\log p}$. Note that for the hypercube better deterministic [5] and randomized [12] sorting algorithms exist, which, however, are not of practical use.
2) segmented broadcast: $q \leq p$ processors with numbers $j_1 < j_2 < \ldots < j_q$ are selected. Each such processor p_{j_i} broadcasts $O(\frac{m}{p})$ data from its local memory to the processors p_{j_i+1} to $p_{j_{i+1}-1}$. The time complexity is $T_{sb}(m,p) = \Theta(\frac{m}{p}\log p)$. If the data to broadcast are on processors not at the beginning of a segment, we first have to move them using monotonic routing operations. This can be done in the same time complexity as the segmented broadcast.
3) multinode broadcast: every processor (in parallel) sends one message to every other processor. The time complexity $T_b(p)$ for any network is $T_b(p) = \Theta(p)$.
4) total exchange: every processor (in parallel) sends a different message to every other processors. The time complexity is $T_x(p) = \Theta(p \log p)$.

3 Binary Search

The problem can be stated as follows: given a sequence S of n elements perform a binary search on S for each element of a set Q of n queries. This problem can easily be solved by sorting $S \cup Q$, if S and Q come from a totally ordered universe, but that is not the case in our application later on (point location in a planar subdivision). Essentially, it is the lack of a total order for the queries, which makes it necessary to perform n independent search processes.

The idea of the algorithm is to determine for the p subsequences S_i of size $\frac{n}{p}$ how many queries end up in each of them. Then, we make the appropriate number of copies of each subsequence, in order to balance the ratio of queries per subsequence. The details of the algorithm are as follows:

0. S_i, the ith subsequence of size $\frac{n}{p}$ of the sorted sequence S, and a set Q_i of $\frac{n}{p}$ queries are on processor p_i.
1. Perform a multinode broadcast, where each processor p_i sends the biggest element of S_i to all other processors. Let $\hat{S}$ denote this ordered sequence.
2. Each processors p_i performs the binary search on $\hat{S}$ for its $\frac{n}{p}$ queries Q_i. Let q_{ij} be the number of queries from Q_i which end up at entry $\hat{S}[j]$.
3. Perform a total exchange, where each processor p_i sends q_{ij} as message to processor p_j, $1 \leq j \leq p$. Each processor p_i then computes $q_i := \sum_{j=1}^{p} q_{ji}$, which is the overall number of queries which want to do binary search in S_i.
4. In order to loadbalance the number of queries per copy of a set S_i we need $\lceil \frac{q_i}{n} p \rceil$ copies of S_i. Perform a segmented broadcast to create them. Note that we end up with at most $2p-1$ copies, if $\frac{q_i}{n} = \frac{n}{p} + 1$, for $1 \leq i \leq p$. Hence, each processor has to treat at most two copies of sets S_i.

5. Perform a multinode broadcast, where each processor p_i sends q_i as message to all other processor. Given this information each processor generates the $O(\frac{n}{p})$ dummy queries, which are needed, for that the number of queries falling in S_i is an exact multiple of $\frac{n}{p}$. Globally sort the queries (now queries which ended up at $\hat{S}[j]$ in step 2 are on a processor holding a copy of S_j).
6. Locally continue the binary search.

The correctness of the algorithm is easy to see. The time complexity of the binary search can be analyzed as follows. The local binary search in Steps 2 and 6 takes time $O(\frac{n}{p}\log p)$ and $O(\frac{n}{p}\log n)$, respectively. The communication complexity is $2T_b(p)+T_x(p)+T_{sb}(n,p)+T_s(n,p)$, where the sorting step is the dominating term. We thus can state the following lemma.

Lemma 1. *Given a sorted sequence S of n elements and a set Q of n queries we can perform binary search on a CGM(n,p), $\frac{n}{p}\geq p$, in time $O(\frac{n\log n}{p}+T_s(n,p))$.*

4 Trapezoidal Decomposition

The *trapezoidal decomposition* of a set of n non-intersecting line segments in the plane is a planar subdivision and obtained as follows. Starting at each endpoint of each segment in S we draw two vertical rays, one upwards one downwards, each extending until it hits another segment from S. Various sequential algorithms compute this subdivision in optimal $\Theta(n\log n)$ time and $\Theta(n)$ space. In this section we present a less efficient sequential algorithm, its parallelization and its modification in order to obtain a scalable parallel algorithm with speedup $\frac{p}{\log p}$.

The idea of the non-optimal sequential algorithm is as follows. We build a segment tree [11] on the x-coordinates of the segment endpoints. Each node v represents an interval and stores the catalogue $S(v)$ of segments covering the interval of the node but not the interval of its parent. As the segments do not intersect we can order the segments in $S(v)$ by the relation *below*. As each segment is stored in at most $2\log n$ catalogues, the size of the segment tree is $O(n\log n)$. For each segment endpoint we traverse the tree and perform a binary search on the catalogue $S(v)$ for each node v on the path from the root to the appropriate leaf. The binary search gives us the closest segments above and below the query point in $S(v)$, that is we have to determine a kind of "minimum" of $\log n$ segments, to determine the closest segments in S. The time complexity is thus $O(n\log^2 n)$ and the space requirement is $O(n\log n)$.

We obtain an obvious parallelization of the query phase, by unfolding a loop of the sequential algorithm. Instead of letting a query s traverse the nodes of a path q of length $\log n$ in the tree, we create $\log n$ queries for each segment s. Each query consists of pairs (v,s) for each node v on the path q. Suppose that the nodes of T are uniquely numbered. For these $n\log n$ queries we perform a binary search on the concatenation of the catalogues $S(v)$ of all nodes $v\in T$ with increasing node numbers, which is a totally ordered sequence of length $O(n\log n)$. We finally determine the closest segment in S by choosing the closest among the $\log n$ segments, which were the closest in the catalogues on the path. To compute the catalogues we make $2\log n$ copies of each segment, each associated to a node v on the path for the left and right endpoint and check if it is in $S(v)$. We then sort them lexicographically by the number of the node v they are associated with and their order in $S(v)$.

Note that the sorting step and the binary search performed with $n \log n$ data give us a speedup of $\frac{p}{\log n}$, that is the parallel algorithm is not scalable. Using the fact, that our machine model is coarse grained, we now refine the algorithm in order to obtain a speedup of $\frac{p}{\log p}$. To do so, we partition the segment tree T as follows. Let T_0 be the subtree of T which consists of level 1 to $\log p$ and let T_i, $1 \leq i \leq p$, be the subtree of T rooted at the ith leaf of T_0. We make the following useful observations.

Lemma 2. *A segment with endpoints in T_i and T_j, $i < j$, appears only in catalogues $S(v)$ of nodes v in $T_i, T_{i+1}, T_{j-1}, T_j$ or in T_0. When v is a node of T_{i+1}, T_{j-1}, then v is the root. Each segment tree $T_1, \ldots, T_p$ has size $O(\frac{n}{p}\log n)$ and the size of T_0 is $O(n \log p)$. There are $\frac{n}{p}$ queries traversing tree T_i, $0 \leq i \leq p$.*

The consequence of this lemma is, that we do not have to loadbalance trees $T_1, \ldots, T_p$ on the p processors, as far as the levels $\log p + 1$ to $\log n$ are concerned. We map each tree T_i and the set of queries on processor p_i, $1 \leq i \leq p$. For levels 1 to $\log p$ we perform the above sketched parallel algorithm. We next give a more detailed description of the algorithm.

0. Initially, the segments are randomly distributed: $\frac{n}{p}$ on each processor.
1. Make a copy of each segment and choose the left endpoint as the reference point in one copy and the right endpoint in the other copy. Globally sort the segments by the x-coordinate of the reference point. Let S_i be the set of segments stored in the local memory of processor p_i, that is the segment with an endpoint in T_i.
2. Locally perform on each processor p_i the sequential algorithm to determine for the reference point of each segment in S_i the closest segment above and below among the segments in S_i.
3. Perform a multinode broadcast, where each processor p_i sends *root*(T_i) to all other processors. Each processor constructs the skeleton tree of T_0.
4. Locally traverse for each segment $s \in S_i$ the path to the root of T_0 starting at the root of T_i. Test for the right (left) child v of each node on this path, if s covers the interval associated to v, when the reference point of s is the left (right) segment endpoint. Create the pair (v, s) for each covering segment. Create the query pair (w,*ref*(s)) for each node w on the path q, where *ref()* returns the reference point of a segment.
5. Globally sort the at most $2n \log p$ pairs (v, s) lexicographically, with the relation "<" for the nodes and *below* for the segments.
6. Perform the binary search process as described in the previous section for the $n \log p$ queries, with the comparison operator ($<$,*below*).
7. Globally sort the queries by their x-coordinate, and determine for the $\log p$ copies which is the closest segment above and below.

The correctness of the algorithm follows immediately from Lemma 2 and its time complexity can be analyzed as follows. The local computation takes time $O(\frac{n}{p}\log n)$ for Step 2 and $O(\frac{n}{p}\log p)$ for Step 4. Constructing all $\log p$ levels of T_0 we need $O(n\log p)$. The communication complexity for Steps 1, 3, 5 and 7 is $T_s(n,p)+T_b(p)+T_s(n\log p,p)$. Step 6 is complete in $O(\frac{n\log p\log n}{p}+T_s(n\log p,p))$ from Lemma 1. We thus obtain the following result.

Theorem 3. *Given a set S of n non-intersecting line segments in the Euclidean plane, the trapezoidal decomposition can be computed on a CGM($n\log p, p$), with $\frac{n}{p}\geq p$, in time $O(\frac{n\log n\log p}{p}+T_s(n\log p,p))$.*

With a slight modification, namely by treating points as degenerated segments, we can solve the following point location problem, which will be used in Section 6. Given a set S of n non-intersecting line segments and a set Q of n query points in the Euclidean plane, the problem is to compute for each point $q\in Q$ the trapezoid of the trapezoidal decomposition of S containing q. We obtain the following corollary.

Corollary 4. *Given a set S of n non-intersecting line segments and a set Q of n points in the Euclidean plane, we can solve the point location problem for Q in the trapezoidal decomposition of S on a CGM($n\log p, p$), wih $\frac{n}{p}\geq p$, in time $O(\frac{n\log n\log p}{p}+T_s(n\log p,p))$.*

Instead of performing Steps 4-7 of the above algorithm on $n\log p$ data, we can also perform it r times on groups of $\lceil\frac{\log p}{r}\rceil$ levels of T_0, with $1\leq r\leq\log p$. We thus immediately obtain the following result about a possible trade off between memory and number of communication rounds. Note that $O(r\cdot T_s(n\frac{\log p}{r},p))=O(T_s(n\log p,p))$, if communication is as expensive as computation, but that in practice it means r times more setups of communication channels.

Corollary 5. *Given a set S of n non-intersecting line segments in the Euclidean plane and a set Q of n points, we can solve the point location problem for Q in the trapezoidal decomposition of S on a CGM($n\frac{\log p}{r}, p$), with $\frac{n}{p}\geq p$, in time $O(\frac{n\log n\log p}{p}+r\cdot T_s(n\frac{\log p}{r},p))$, for $1\leq r\leq\log p$.*

5 Range Query Reporting

Suppose we are given a set $X=\{x_1,\ldots,x_n\}$ of n points and a set $S=\{s_1,\ldots,s_n\}$ of n segments on the x-axis. The *range query reporting problem* is to report the pair (x,s), $x\in X$, $s\in S$, iff x is covered by s. We assume that the points are sorted by x-coordinate and evenly distributed over the memory of the p processors. We use $\bar{c}_i$ to denote the number of segments covering point x_i and c_i to denote the number of points covered by segment s_i. We further use k to denote the total number of point segment pairs to report. As in sequential these values can be computed by sorting all points and segment endpoints, followed by a parallel prefix, in time $O(\frac{n}{p}+T_s(n,p))$.

We now give three algorithms which are used depending on the number k of point segment pairs to report. k ranges between 0 and n^2. If $k \leq \frac{n}{p}$ or $k \geq np$, we can copy all involved segments to all processors. If $\frac{n}{p} \leq k \leq np$, we need a more elaborate loadbalancing scheme.

Case 1: $k \leq \frac{n}{p}$. As there are at most $\frac{n}{p}$ segments covering at least a single point, we can copy all these segments to all processors. Each processor reports for its $\frac{n}{p}$ points, performing binary search for the two endpoints of each segment. The communication complexity is $O(\frac{n}{p})$ and the reporting takes time $O(\frac{n}{p}\log n)$.

Case 2: $k \geq np$. As each processor has to report more than n point segment pairs, each processor can keep the full set of points and segments. The algorithm is as follows.

1. Make a copy of each segment and choose the left endpoint as the reference point in one copy and the right endpoint in the other copy. Globally sort the segments and the points of X by the x-coordinate of the reference point.
2. Copy this sorted sequence of size $3n$ to all processors.
3. Greedily group segments together in p groups G_i, such that $\sum_{s_j \in G_i} c_j = O(\frac{k}{p})$. This can be done, as $c_j \leq n \leq \frac{k}{p}$.
4. Each processor p_i reports point segment pairs for the segments in group G_i, that is the reporting is balanced. We do the reporting in linear time by scanning over the entire sorted sequence. Before reaching the beginning of G_i we add (delete) a segment when we encounter its left (right) endpoint and we skip points. After addition and before deletion we call a segment *active*. When we scan a point from G_i we report it with all active segments.

The communication complexity is $O(n)$, the grouping takes time $O(n)$, as it is a simple scan and the reporting takes time $O(n + \frac{k}{p})$ as an addition or deletion takes time $O(1)$.

Case 3: $\frac{n}{p} \leq k \leq np$. We need some more notation. As the points are ordered and distributed over the p processors we have p intervals. We say a segment *covers* an interval, if the endpoints of the segment lie to the left and right of the interval. It *partially covers* an interval, if at least one endpoint lies in the interval. The algorithm for the third case has two phases. In the first phase we report point segment pairs for segments which cover an interval. It can be outlined as follows. We first make an appropriate number of copies of the intervals (steps 1-3 of the below algorithm). We then make as many copies of a segment as it covers intervals (steps 4-7). In order to generate the copies in a loadbalanced way we have to rearrange them first. As there may be several copies of an interval, we have to distribute the set of segments destinated for an interval evenly. We finally do the reporting. We next describe each step of the algorithm.

1. Broadcast the p values of the intervals. Locally determine for each of the p intervals by how many of the $\frac{n}{p}$ local segments it is covered. Perform a total exchange of these values. Determine the number d_i of segments covering interval I_i.
2. Locally determine which of the $\frac{n}{p}$ segments are covering and compute the number e_i of intervals covered by segment s_i. Let $e := \sum_{i=1}^{n} e_i$. As a covering

segment covers $\frac{n}{p}$ points in each interval $e = \frac{kp}{n}$. As we are in the case $k \leq np$, we have $e \leq p^2$ and $e_i \leq p$.

3. Perform a segmented broadcast in order to make $\lceil \frac{d_i}{e} p \rceil$ copies of the $\frac{n}{p}$ points in the interval I_i.
4. We now generate the copies of segments in a balanced way. Group the segments in p groups $G_1, \ldots, G_p$, such that $g_i := \sum_{s_i \in G_j} e_i = O(p)$ for $j = 1, \ldots, p$, and move group G_i on processor p_i. This is possible since $e_i \leq p$.
5. Locally write up the $O(p)$ segment interval pairs.
6. Globally sort the e segment interval pairs by the interval. Doing so each group of $O(p)$ pairs is on a processor which holds a copy of the same interval.
7. Locally report all $\frac{k}{p}$ pairs of points and segments.

To analyze the time complexity we will first look at the local computation and then at the communication operations. Step 1 takes time $O(\frac{n}{p} \log p)$ using a segment tree with p leaves, Steps 2, 5 and 7 take each linear time, that is $O(\frac{n}{p})$, $O(p)$ and $\frac{k}{p}$, respectively. The communication complexity is $T_b(p) + T_x(p) + T_{sb}(n,p) + T_s(p^2, p)$.

In the second phase for the case $\frac{n}{p} \leq k \leq np$, we report the point segment pairs for the partially covering segments. The number of partially covering segments is $O(n)$, as it is bound by the number of segment endpoints. The idea of the algorithm is to loadbalance the intervals first by number of partially covering segments and to loadbalance them again by number of point segment pairs to report. The details of the algorithm are as follows.

1. Loadbalance the intervals by number of segments, that is make the appropriate number of copies of intervals using a segmented broadcast.
2. Globally sort the partially covering segments, which associates $O(\frac{n}{p})$ of them with an interval where they lie in.
3. Loadbalance these intervals by number of point segment pairs to report. Let $\bar{c}_i$ be the number of partially covering segments associated with the copy of an interval, which cover the point x_i. Group the points in each interval greedily, such that $\sum_{x_i \in G_j} \bar{c}_i = O(\frac{k}{p})$ for groups G_j, $1 \leq j \leq p$. This is possible as $\bar{c}_i \leq \frac{n}{p} \leq \frac{k}{p}$. Use a segmented broadcast to make the appropriate number of copies of the interval together with the $\frac{n}{p}$ associated segments,
4. Each processor p_j reports the point segment pairs for points in group G_j and the $\frac{n}{p}$ segments associated to the interval where G_j lies in. To do so the $O(\frac{n}{p})$ points and segments are sorted and scanned from left to right.

The communication complexity is $2T_{sb}(n,p) + T_s(n,p)$. The grouping can be done in time $O(\frac{n}{p})$ and the reporting in time $O(\frac{n}{p} \log n + \frac{k}{p})$. This completes the description of the second phase and the third case. We thus can conclude this section stating the following theorem.

Theorem 6. *Given a set of n points and n intervals on the x-axis we can solve the 1-dimensional range query reporting problem on a CGM*$(\max(n,k),p)$, *with $\frac{n}{p} \geq p$, in time $O(\frac{n \log n}{p} + \frac{k}{p} + T_s(n,p))$, where k is the size of the output.*

6 Bichromatic Segment Intersection Problems

Let R (resp. B) be a set of n non-intersecting red (resp. blue) line segments in the plane. The ***bichromatic segment intersection problems*** are to ***count*** or to ***report*** the intersections between red and blue segments. Before presenting the parallel algorithms we briefly sketch the underlying data structure, make two crucial observations and state the sequential compexity of the problem.

In the basic sequential algorithm a segment tree is built on the x-coordinates of the segment endpoints of red and blue segments. Each node of the segment tree represents a vertical slab and has catalogues for red long, red short, blue long and blue short segments. With respect to a given slab a segment is called ***short***, if it has an endpoint in the slab and ***long*** if it completely traverses the slab from left to right, and does not traverse the slab of the parent node. The following observations are crucial.

- An intersection between two segments arises in exactly one node of the segment tree, either as an intersection between a short and a long segment, or between two long segments.
- At each level of the segment tree, a segment is stored at most four times, namely in two short catalogues and two long catalogues.

The first observation allows to organize the counting or reporting of intersections by processing each slab independently. Inside a slab, the long red segments define a subdivision of the slab in strips (see Figure 1). In order to count or report the intersections of a blue segment in this slab, it is enough to locate its endpoints (after clipping to the slab). In the remainder of this section we only consider the intersections with blue segments and long red segments.

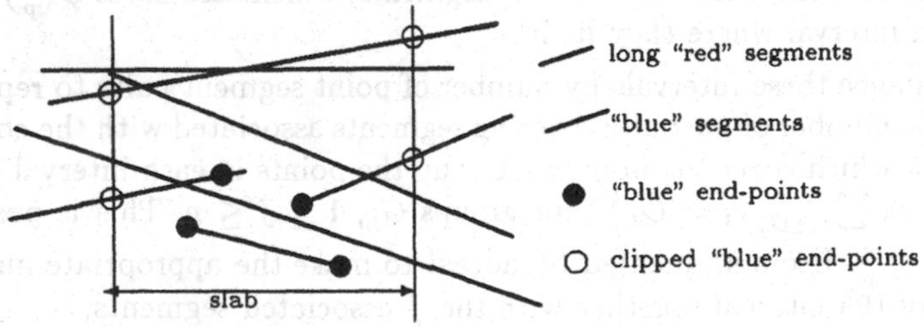

Fig. 1. A slab represented by a node of the segment tree.

The complexity of the basic algorithm, given in [4], is $O(n \log n)$ for space, $O(n \log^2 n)$ time for counting and $O(n \log^2 n + k)$ time for reporting the k intersections. In the same paper the space complexity is reduced to $O(n)$ by streaming and the time complexity to $O(n \log n)$ and $O(n \log n + k)$ respectively, using dynamic fractional cascading. Recently, Palazzi and Snoeyink presented a simple and very elegant method to get the same optimal bounds [10]. The basic algorithm is the same, but they presort the segments in order to reduce the time complexity and they store only one level of the segment tree at a time, to reduce the space complexity, using the second observation.

Lemma 7. *Let R (resp. B) be a set of n non-intersecting red (resp. blue) line segments in the plane. Then, the intersections between red and blue segments can be counted in $O(n \log n)$ time and reported in $O(n \log n + k)$ time using $O(n)$ space, where k is the number of intersections.*

6.1 The Counting Problem

We now describe the parallel counting algorithm for the coarse grained model. As for the trapezoidal decomposition we partition the segment tree T in trees T_0 and T_i, $1 \leq i \leq p$. We map the latter on processors p_i. Only tree T_0 is treated in parallel. The details of the algorithm are as follows.

1. Make a copy of each segment and choose the left endpoint as the reference point in one copy and the right endpoint in the other copy. Globally sort the red and blue segments by the x-coordinate of the reference point. Let S_i be the set of segments stored in the local memory of processor p_i.
2. Each processor p_i performs an optimal sequential algorithm for the set S_i. The algorithm constructs the segment trees T_i, $1 \leq i \leq p$, level by level.
3. Construct T_0 with long blue segments in the catalogues and perform the point location for the endpoints of long and short red segments using Corollary 4.
4. Construct T_0 with long red segments in the catalogues and perform the point location for the endpoints of short blue segments using Corollary 4.
5. Globally sort query pairs (s, v) lexicographically and subtract the ranks of neighbor pairs in this sorted sequence, which are the two endpoints of a segment.
6. Sum up the number of intersections.

The correctness results from the above discussion. The time complexity can be analyzed as follows. Segment trees T_i, $1 \leq i \leq p$, have size $O(\frac{n}{p} \log n)$. Running an optimal sequential algorithm on sets S_i, $1 \leq i \leq p$, takes time $O(\frac{n}{p} \log n)$ and space $O(\frac{n}{p})$, according to Lemma 7. The segment tree T_0 has size $O(n \log p)$ and there are the same number of queries. The point location can be done in time $O(\frac{n \log n \log p}{p} + T_s(n \log p, p))$ using Corollary 4 and sorting the queries lexicographically in Step 5 takes time $T_s(n \log p, p)$.

Again we can obtain a trade off between memory and number of communication rounds, by performing the point location algorithm in Steps 3 and 5 for groups of $\lceil \frac{\log p}{r} \rceil$ levels of T_0 in r rounds, with $1 \leq r \leq \log p$. We thus obtain

Theorem 8. *Let R (resp. B) be a set of n non-intersecting red (resp. blue) line segments in the plane. Then we can solve the bichromatic segment intersection counting problem on a CGM$(n\frac{\log p}{r}, p)$, $\frac{n}{p} \geq p$, in time $O(\frac{n \log n \log p}{p} + r \cdot T_s(n\frac{\log p}{r}, p))$, for $1 \leq r \leq \log p$.*

6.2 The Reporting Problem

In this section we will look at the bichromatic intersection reporting problem. The main difficulty is to loadbalance the writing up of the $O \leq k \leq n^2$ intersections. That is we want each processor to report not more than $O(\frac{\max(k,n)}{p})$ intersections. The algorithm has two phases. In the first phase we report the intersections for subtrees $T_1, \ldots, T_p$, in the second phase we report the intersections for subtree T_0.

The idea of the algorithm of the first phase is to make an appropriate number of copies of each subtree and to let each processor which has such a copy report a fraction of the intersections. The details of the algorithm are as follows.

1. Make a copy of each segment and choose the left endpoint as the reference point in one copy and the right endpoint in the other copy. Globally sort the red and blue segments by the x-coordinate of the reference point. Let S_i be the set of segments stored in the local memory of processor p_i.
2. Each processor p_i locally performs an optimal sequential intersection counting algorithm on set S_i, constructing the sub segment trees $T_1, \ldots, T_p$. Let k_i, be the number of intersections in subtree T_i, and let k be $\sum_{i=1}^{p} k_i$.
3. Make $\lceil \frac{k_i}{k} p \rceil$ copies of the sets S_i with $k_i \geq 1$, performing a segmented broadcast.
4. Each processor performs the intersection counting algorithm in order to determine which fraction of intersections to report. The details are as follows. For each level of the segment tree construct the catalogues interleaved with the query segment endpoints using an optimal sequential algorithm. For each catalogue segment s_j compute the number $\bar{c}_j$ of query segments intersecting s_j. Perform a scan on the concatenation of the catalogues of nodes of T_i, in order to determine the catalogue segments $\hat{s}_1, \hat{s}_2, \ldots$, which contribute to each $\frac{k}{p}$-th intersection in T_i. The processor holding the l-th copy of T_i reports all intersections where catalogue segments between $\hat{s}_{l-1}$ and $\hat{s}_l$ are involved.

Note that we only determine the catalogue segment which is involved in each $\frac{k}{p}$-th intersection, but not the $\frac{k}{p}$-th intersection itself. This does not disturb our intention to loadbalance, as each catalogue segment in a tree $T_1, \ldots, T_p$ has at most $\frac{n}{p}$ intersections. That is the maximal difference between the number of intersections to report on two arbitrary processors is $\frac{n}{p}$. As each processor stores only one level at a time step, we need at most $O(n)$ memory during the first phase.

The running time can be analyzed as follows. Steps 2 and 4 are sequential, the first taking time $O(\frac{n}{p} \log n)$ and $O(n)$ memory as the algorithm stores only one level of the segment tree at a time step. The counting part of Step 4 has the same complexity. The computation of $\bar{c}_j$ for catalogue segments s_j and the scan to determine when to start reporting take linear time, that is $O(\frac{n}{p})$, the reporting itself takes time $O(\frac{k}{p})$. The communication complexity is $T_s(n,p) + T_{sb}(n,p)$ for Steps 1 and 3 respectively.

In the second phase we report the intersections for subtree T_0. Once the point location of query points in the slabs is done, the reporting is easy. As the catalogue segments of the same color do not intersect, they can be reduced to *points* on an axis. The query segments can be reduced to intervals on this axis. We thus have to solve a 1-dimensional range query reporting problem for $n \log p$ points and intervals. Hence, we can do the bichromatic segment intersection reporting for subtree T_0 in time $O(\frac{n \log n \log p}{p} + \frac{k}{p} + T_s(n \log p, p))$ using Theorem 6. This completes the description of the algorithm and we can state

Theorem 9. *Let R (resp. B) be a set of n non-intersecting red (resp. blue) line segments in the plane. Then, we can solve the bichromatic segment intersection reporting problem on a* $CGM(\max(k, n \log p), p)$, $\frac{n}{p} \geq p$, *in time* $O(\frac{n \log n \log p}{p} + \frac{k}{p} + T_s(n \log p, p))$.

In this case the trade off between memory and number of communication rounds only makes sense, if $k \leq n \log p$, because the memory requirement is at least $O(k)$. The idea then is to perform the range query reporting algorithm on r groups of $\lceil \frac{\log p}{r} \rceil$ levels of T_0, with $1 \leq r \leq \log p$.

References

1. M.J. Atallah and F. Dehne and S.E. Hambrusch. *A coarse-grained, architecture-independent approach for connected component labeling. TR-90-008. Purdue University, 1993.*
2. K.E. Batcher. *Sorting networks and their applications. Proc. AFIPS Spring Joint Computer Conference,* pages 307–314, 1968.
3. D.P. Bertsekas and J. N. Tsitsiklis. *Parallel and Distributed Computation: Numerical Methods.* Prentice Hall, Englewood Cliffs, NJ, 1989.
4. B. Chazelle, H. Edelsbrunner, L. Guibas, and M. Sharir. Algorithms for bichromatic line segment problems and polyhedral terrains. Report UIUCDCS-R-90-1578, Dept. Comput. Sci., Univ. Illinois, Urbana, IL, 1989.
5. R. Cypher and C.G. Plaxton. *Deterministic sorting in nearly logarithmic time on the hypercube and related computers. ACM Symposium on Theory of Computing,* 193–203. ACM, 1990.
6. F. Dehne and A. Fabri and A. Rau-Chaplin. *Scalable parallel geometric algorithms for coarse grained multicomputers. ACM Symposium on Computational Geometry,* 1993.
7. M.T. Goodrich and S.B. Shauck and S. Guha. *Parallel Methods for Visibility and Shortest-Path Problems in Simple Polygons. Algorithmica,* 8, 461–486, 1992.
8. R.I. Greenberg and C. E. Leiserson. *Randomized Routing on Fat-trees. Advances in Computing Research,* 5:345–374, 1989.
9. F.T. Leighton. *Introduction to Parallel Algorithms and Architectures: Arrays, Trees, Hypercubes.* Morgan Kaufmann Publishers, San Mateo, CA, 1992.
10. L.P. Palazzi and J. Snoeyink. Counting and Reporting Red/Blue Segment Intersections. *Workshop on Algorithms and Data Structures 1993. Springer LNCS.*
11. F.P. Preparata and M.I. Shamos. *Computational Geometry: an Introduction.* Springer-Verlag, New York, NY, 1985.
12. J.H. Reif and L.G. Valiant. *A logarithmic time sort for linear size networks. J. ACM,* Vol.34, 1:60–76, 1987.

Persistence, Randomization and Parallelization: On Some Combinatorial Games and their Applications* (Abstract)

Paul F. Dietz[1] and Rajeev Raman[2]

[1] Department of Computer Science, University of Rochester, Rochester, NY 14627
[2] UMIACS, University of Maryland, College Park, MD 20742

Abstract. We explore a family of combinatorial "pebble" games associated with eliminating amortization from data structures in general, continuing a line of research initiated in [10, 21, 8]. These results are used to eliminate amortization from the general schemes of [13] for making linked data structures *partially* and *fully persistent* on a pointer machine, the *monotonic list labeling* problem [5, 28, 10] and the *fully persistent disjoint set union-find* problem [19].
An important motivation for eliminating amortization is that using amortized data structures in a parallel environment may severely degrade performance [12]. We demonstrate interesting aspects of this issue by introducing the problem of *parallel persistence.* The best solutions to this problem are obtained by replacing an amortized data structure by one of our new randomized ones and using fast load-balancing schemes.

1 Introduction

Data structures are often designed to be efficient in the *amortized* sense, i.e., the average time per operation to process a sequence of operations is small. However, the time taken by an efficient amortized data structure to process an individual operation in the sequence may be considerably more than the average for th sequence; in addition, an adversary can usually ensure that an expensive operation occurs at a "bad" time, making this variation unpredictable. Disadvantages of amortized algorithms include:

(a) Unsuitability in real-time and interactive environments [30, 2]; Wadler [30] gives the example of a tennis-playing robot stopping in mid-swing to garbage collect (classical garbage collection is efficient in the amortized sense).
(b) If an amortized data structure is shared in a synchronous parallel environment [12], all operations requested by one processor may be time-consuming, preventing that processor from completing and slowing down the entire task.
(c) Amortized data structures perform poorly when made *fully persistent* [26].

This brings us to the problem of *eliminating amortization* which is, loosely speaking, deriving from an amortized data structure a comparably efficient data structure with guaranteed time bounds for every operation. On the other hand, we view deriving a randomized data structure with good expected time bounds as being almost as good eliminating amortization, as randomized data structures do not suffer from the above problems. In situation (b) random variations in processor load are less pernicious in their effect on performance, and are also sometimes amenable to load balancing; while in situation (c) an expected time data structure will show good expected perfomance

* Paul Dietz acknowledges support from NSF grant CCR-8909667. Rajeev Raman did most of this research at the University of Rochester. Authors' e-mail addresses: dietz@cs.rochester.edu, raman@umiacs.umd.edu

when made persistent. Finally, a randomized algorithm that runs in close to expected time with high probability could be used in situation (a) if missing a deadline is not catastrophic.

Early results regarding eliminating amortization from data structures [23, 31, 12] were inspired by work on real-time Turing machines [14, 15, 20]. A different approach to eliminating amortization was intiated in [10, 21], who reduced the analysis of a worst-case data structure to analyzing a two-player combinatorial "pebble" game. The data structures obtained by these methods were somewhat complicated and had large constant factors. Dietz and Raman [8] obtained more results on pebble games of this type and applied them to obtain new worst-case data structures, some of which were quite simple.

In this paper we study more games from the pebble game family, obtaining tight bounds on the payoffs to the players (Section 2), and apply these results to obtain several new worst-case and expected-time data structures for the *monotonic list labeling* problem [5, 28, 10] (Section 3), for making linked data structures fully and partially persistent [13, 8] (Sections 4 and 6) and for the *fully persistent disjoint set-union problem* [19] (Section 5). Many of these new data structures are simple and appear to have constant factors comparable to the original amortized algorithms. We demonstrate the usefulness of randomization in eliminating amortization by deriving efficient randomized data structures from amortized ones — although earlier results [8] suggested that randomization might be useful in eliminating amortization, no examples of its use in any concrete data structuring problem were given. We study the interaction between amortization and parallelization in Section 7 and give parallel algorithms for making data structures persistent. On a completely different note, we indicate how these results can be used to save buffer space in a transaction processing system (Section 3.2).

Some of our algorithms make use of the random access features of the RAM model. Unless otherwise stated, our model allows an algorithm to manipulate words of $\Theta(\log m)$ bits if a sequence of m operations is to be processed. The following result was shown by Ajtai et al. [1] for the cell probe model of computation; however, as noted in [8], on a RAM their algorithm can be implemented using lookup tables that can be incrementally precomputed at a cost of $O(1)$ worst-case time and space per operation:

Lemma 1. *Let n be an integer and let $S \subseteq \{1, \ldots, n\}$ such that $|S|$ is $\log^{O(1)} n$. There is a data structure that supports insertions, deletions and predecessor queries (find the largest $i \in S$ such that $i \leq x$) on S in constant time. The data structure occupies space that is linear in the size of the subset.*

We say that an event occurs "with high probability" (w.h.p.) if it occurs with probability at least $1 - n^{-\alpha}$ for any pre-specified constant $\alpha > 0$. The following lemma can be proved using standard techniques (see, e.g.,[24]):

Lemma 2. *Let $X_1, \ldots, X_m$ be independent, identically distributed geometric random variables, $E[X_i] = 1/p$, $p \in (0, 1)$. Let $X = \sum X_i$. Then,*

$$\Pr(X > cm/p) < (ce^{1-c})^m.$$

2 "Pebble" Games

Pebble games are a family of two-player ames that arise in the context of eliminating amortization from data structures [10, 21, 8]. These games have the following general form: player I (increaser) and player D (decreaser) play each other on a set of n real- or integer-valued variables, all of which have value 0 initially. The players alternate moves for n rounds, player I increasing the value of some of the variables on his move, and player D decreasing the value of some variables on his. The payoff to player I, denoted by M, is the maximum value of any variable at the end of any round in the game.

Game 1: (Subtracting game)
Player I: Chooses $\vec{a}$, sets $\vec{x} \leftarrow \vec{x} + \vec{a}$.
Player D: Sets some x_j to $\max\{x_j - c, 0\}$, for some constant $c \geq 1$.

Game 2: (Zeroing game with costs)
Player D: Sets some variable x_j to 0. Let the value of x_j be y.
Player I: Picks $\vec{a}$ and sets $\vec{x} \leftarrow \vec{x} + (f(y) + 1)\vec{a}$.

Fig. 1. Games 1 and 2 are simple games, played on n real-valued variables $\vec{x} = (x_1, \ldots, x_n)$. $\vec{a}$ is a vector of non-negative real numbers such that $\sum_{i=1}^{n} a_i = 1$. For game 2, f is some real-valued function.

Game 3: (Graph Game)
Player I: Places a pebble on some vertex $v \in V$, and optionally makes v point to a different vertex without violating degree constraints.
Player D: Does the following at most (expected) $O(1)$ times: Zeroes some vertex $v \in V$, causing one pebble to be placed on each of the predecessors of v in G.

Fig. 2. Game 3 is the *graph* game, played on an n-vertex bounded in- and out- degree directed graph $G = (V, E)$, with degree bound b. Player I may only increment variables by a unit amount: this is viewed as player I placing pebbles on vertices.

These games were classified in [8] into *simple* and *graph* games. Figures 1 and 2 define the pebble games studied in this paper more precisely.

In the data structuring context, the variables represent the sizes of collections of objects, which the algorithm attempts to keep small (to make queries efficient, for instance) although updates may increase the sizes of the collections. The simple games were originally motivated by problems in bucketed data structures and the particular graph game we describe (Game 3) models Driscoll et al.'s *fat node* with *node copying* method for making linked data structures partially persistent [13]. Other variants could describe the related problems of full persistence [13] and dynamic fractional cascading [22].

An important special case of Game 2, where $f(y)$ is identically zero, was posed by [10], who proved tight bounds on M in the situation that player I may decide on his next move based on player D's last move ([21] studied a closely related game). In [8] in addition to the above *on-line* mode of interaction between the players, the *oblivious*[3] and *off-line* modes were introduced. In the oblivious mode, player I reveals moves one at a time but may not base his actions on player D's moves. In the off-line mode, player I reveals all his moves before the game starts. It was proved in [8] that by using randomization, player D could do much better on the average against an oblivious player I than in the on-line case.

Game 3 was posed in [8], where an on-line strategy for player D was described that achieved $M = O(1)$. Unfortunately, the strategy described was not efficiently implementable, and so did not lend itself to constructing efficient worst-case data structures. A different D-strategy from [8] could be implemented in $O(1)$ time per D-move but allowed M to be $\Theta(\log n)$

In this paper, we consider Game 1 in the on-line and oblivious settings, and prove tight bounds on M in both cases. For Game 2, we consider several different variants of the cost function f in the on-line setting. Applications of these results are given in Subsection 2.1 and Sections 3 and 4. We obtain new results on Game 3 in the oblivious setting using randomization. Some of these do not obtain bounds on M, but rather on

[3] The oblivious mode was called the *hidden on-line* mode in [8]

$\overline{M}$, a quantity we define as follows: before the game begins, player I fixes a vertex v and a round r but does not reveal these choices to player D. Then we define $\overline{M}$ to be the expected number of pebbles on vertex v at the end of round r.

2.1 Simple Games

We now describe the results we have obtained on Games 1 and 2.

Theorem 3. *For the on-line version of Game 1, if player D decrements a variable with the maximum value at each move, then $M \leq \ln n + 1$ even when $c = 1$, and this bound is tight to within a constant additive factor.*

Proof. Let $\vec{x}$ denote the vector of values $x_1, \ldots, x_n$. For any real z, define the function

$$S(z, \vec{x}) = \sum_{i=1}^{n} \max(x_i - z, 0) \tag{1}$$

Also let $m(\vec{x}) = \max_i\{x_i\}$. We let $\vec{x}^{(i)}$ denote the value of the variables after round i. Consider any strategy of player I against the above D-strategy that achieves $M = \mu > 1$, and let r be such that $m(\vec{x}^{(r)}) = \mu$. Let $p < r$ be the last time such that $m(\vec{x}^{(p)}) \leq 1$. We restrict our attention to rounds $p+1$ through r. We define times $t_0 < \ldots < t_s$ and numbers $m_0 \leq \ldots \leq m_s$ in the following way: We let $t_0 = p$, $t_s = r$, $m_0 = 1$, $m_s = \mu$, $\Delta_s = m_s - m_{s-1}$, and for $0 < j < s$ we let

$$\begin{aligned} t_j &= \max\{p < i < t_{j+1} \mid m(\vec{x}^{(i)}) \leq m_{j+1}\}, \\ m_j &= m(\vec{x}^{(t_j)}) \text{ and} \\ \Delta_j &= m_j - m_{j-1}. \end{aligned} \tag{2}$$

Clearly $0 \leq \Delta_j \leq 1$ for all j. Let $f_j = S(m_j - 1, \vec{x}^{(t_j)})$, for $0 \leq j \leq s$.

Lemma 4. *For $0 \leq j \leq s$, $f_{j+1} \leq (1 - \Delta_{j+1}) f_j$.*

Proof. Because no variable contributes more than 1 to f_j, and each variable that contributes to f_j contributes at most $1 - \Delta_{j+1}$ to $S(m_{j+1} - 1, \vec{x}^{(t_j)})$:

$$S(m_{j+1} - 1, \vec{x}^{(t_j)}) \leq (1 - \Delta_{j+1}) f_j.$$

Also, in round $t_j + 1$, player I can add at most 1 to $S(m_{j+1} - 1, \vec{x}^{(t_j)})$ on his move, while player D always subtracts 1 from this quantity since he must decrement a variable of value at least m_{j+1} on his move. Thus $S(m_{j+1} - 1, \vec{x}^{(t_j+1)}) \leq S(m_{j+1} - 1, \vec{x}^{(t_j)})$, and we have that:

$$S(m_{j+1} - 1, \vec{x}^{(t_j+1)}) \leq (1 - \Delta_{j+1}) f_j.$$

Whenever $t_{j+1} = t_j + 1$, the claim is true. Otherwise, we note that after each round i in the interval $[t_j + 1, t_{j+1}]$, $m(\vec{x}^{(i)}) > m_{j+1}$, and so player D must subtract 1 from $S(m_{j+1} - 1, \vec{x}^{(i)})$ while player I can add at most 1 to it. Thus $S(m_{j+1} - 1, \vec{x}^{(t_j+1)}) \geq S(m_{j+1} - 1, \vec{x}^{(t_{j+1})}) = f_{j+1}$ and the claim follows. □ **Lemma 4**

Clearly, $f_s \geq 1$, since at least one variable has value μ at time r. In addition, for all i, $S(0, \vec{x}^{(i)}) \leq n$, since this quantity does not increase if the maximum valued variable has value > 1 (if not, $S(0, \vec{x}^{(j)}) \leq n$ anyway), and so $f_0 \leq n$. From Lemma 4 we infer that:

$$\prod_{i=1}^{s} (1 - \Delta_i) \geq 1/n. \tag{3}$$

Since $\mu = 1 + \sum_{i=1}^{s} \Delta_i$, to find the maximum possible value of μ we have to maximize the quantity $\sum_{i=1}^{s} \Delta_i$ subject to equation 3. This maximum is achieved when all the Δ_i are equal to $(\mu-1)/s$. Substituting in equation 3 we obtain that $(1-(\mu-1)/s)^s \geq 1/n$. Since $(1-y)^{1/y} < e^{-1}$ for all $y > 0$, we see that μ must satisfy $e^{\mu-1} < n$, from which we find that $\mu < \ln n + 1$. □ **Theorem 3**

It is easy to show how to make M be arbitrarily close to H_n, the nth harmonic number, thus giving $H_n < M < \ln n + 1$. Dietz and Sleator [10] considered Game 2 for the case where $f(y)$ is identically zero and proved that M is essentially H_{n-1} in this game. Since the rules of their game can be viewed as strengthening player D's moves in Game 1 to permit him to set a variable to zero, rather than just decrement it, it is somewhat surprising that almost the same bound holds in each case.

We omit the proofs of the following theorems in this abstract:

Theorem 5. *In the oblivious version of Game 1, there is a strategy for player D that achieves $M \in \Theta(\log\log n)$ w.h.p., provided $c = 2 + \epsilon$ for some constant $\epsilon > 0$.*

Theorem 6. *When $f(y) = \alpha y$ in the on-line version of Game 2, for some constant $0 < \alpha < 1$, then if player D zeroes a variable with the maximum value at each move, then $M \in \Theta(n^\alpha)$.*

We consider other functions as well for Game 2. For example, when $f(y) = \log y$, we can show that $M \in \Theta(\log n \log\log n)$, and when $f(y) = \sqrt{y}$, we can show that $f(y) = \Theta(\log^2 n)$.

Application: Queue Selection

We consider the situation of a "transaction processing system" where a processor serves transactions arriving on n queues, each of which has some associated buffer space. Our model is as follows. Transaction records are large equal-sized packets each arriving as a stream of bytes along some queue. Every time unit, the processor removes a completed transaction record off one queue in order to serve it. In this time unit, fractions of packets appear at the n queues and are buffered there; the only assumption we make about how large a fraction appears at a queue in a time step is that the fractions should sum up to at most 1. (Without this assumption the buffer space needed at each queue cannot be bounded as a function of n alone.) We want to find the queue selection algorithm that minimizes the total amount of buffer space and guarantees that no record gets lost due to a buffer being full.

One approach would be to use a *round-robin* schedule. However, it is easy to see that this requires a buffer capable of holding $n-1$ records at each queue, as does *first-in-first-out*. However, by theorem 3, using the *longest queue first* strategy ensures that $O(\log n)$ buffer space per queue suffices. Furthermore, by using the strategy of theorem 5, we ensure that with very high likelihood, $O(\log\log n)$ buffer space per queue suffices. The *exhaustive longest queue first* algorithm picks the largest queue each time and empties it before changing queues. If the processor is $\alpha > 1$ times faster than the data rate, theorem 6 shows that $O(n^{1/\alpha})$ buffer space per queue suffices.

2.2 The Graph Game

We now give three randomized strategies for Game 3. Again, let d be the maximum in-degree of any node. The first strategy is as follows:

Algorithm 1 Whenever a vertex receives a pebble (either directly from an I-move or because player D zeroed one of its successors), it is zeroed with probability $1/2d$.

We can show that the number of pebbles at a vertex is bounded above by $2d$ plus a geometric random variable with expectation $2d$. Moreover, for different vertices these random variables are mutually independent. This in particular implies that $\overline{M}$ is $O(d)$. Also, the expected number of vertices zeroed is $O(1)$. A modification to the above algorithm makes M (instead of $\overline{M}$) be $O(1)$. The expected number of vertices zeroed is still $O(1)$.

Algorithm 2 In addition to randomly zeroing vertices as in Algorithm 1, we always zero a vertex that has at least $cd \log d$ pebbles, for some constant c to be chosen.

Theorem 7. *In Algorithm 2, the expected number of vertices zeroed is $O(1)$, at most $O(\log n)$ vertices are zeroed w.h.p., and M is $O(d \log d)$.*

Proof. (Sketch) If c is chosen to be large enough, then the probability that some vertex exceeds $cd \log d$ is at most $1/2d$. Since zeroing a vertex causes at most d other vertices to receive pebbles, the expected number of other vertices incremented is at most $1/2$. More careful analysis shows that w.h.p. $O(\log n)$ vertices are zeroed. Clearly vertices always have at most $cd \log d$ pebbles. □ **Theorem 7**

The next algorithm is complementary to Algorithm 2, in that $\overline{M}$ is $O(d)$, but player D zeroes exactly one vertex:

Algorithm 3 Zero the first vertex to be incremented. Thereafter, on each move zero one of: the vertex incremented by player I and the vertices incremented by player D's zeroing on the last move, making the choice uniformly at random.

We can show (proof omitted):

Theorem 8. *Algorithm 3 ensures that $\overline{M}$ is $O(d)$.*

3 Monotonic List Labeling

The monotonic list labeling problem [5, 28, 10] is the problem of maintaining a monotone labeling of the elements of a list L, $|L| = n$, with labels drawn from the set $\{1, \ldots, p(n)\}$, for some polynomial p, while performing insertions and/or deletions on L. Algorithms have been given which perform each operation in $O(\log n)$ amortized relabelings. This problem arises in the implementation of fully persistent arrays [6], and in the solution of the *list ordering* problem: perform queries of the form "does x precede y?" in a list while performing insertion and deletions.

One of us [7] has shown how to solve the problem in $O(\log n)$ worst-case relabelings per insertion. This algorithm can be used to simplify the real-time list ordering algorithm of [10]. The new algorithm for this problem is interesting in that it makes use of Game 1 to eliminate amortization; using the "zeroing" game from [10] is insufficient.

The relabeling algorithm works recursively. We show, for all $n = 2^k$ for some integer $k \geq 0$, how to label n items with $4n^4$ labels so that n insertions, each causing $O(k) = O(\log n)$ relabelings, can be performed.

The labeling algorithm will decompose the list L into sublists. An L that initially contains 2^k elements is called a ***level k list***. For $k > 1$, L is decomposed into two or more level $k-1$ lists. List elements will be moved between the sublists of L as insertions are performed into L. The labeling of a level k list can be thought of as the combination of the labelings of its level $k-1$ sublists. The labels in a level k list are integers in the range $[0, S(k) - 1]$, where $S(k) = 2^{4k+2}$. Let x be an element with label $l(x)$. Let x be a list element. The *label for the ith level*, $i > 0$ is defined to be

$$l(x, i) = l(x) \bmod S(i). \tag{4}$$

The level $i-1$ sublist position of x is defined to be

$$p(x,i) = \lfloor l(x,i)/S(i-1) \rfloor. \quad (5)$$

That is, $p(x,i)$ is the index of the level $i-1$ sublist containing element x in the list of sublists (empty or not) of the level i sublist containing x. Stated more concisely, the label space associated with a level k list L ($k > 1$) is decomposed into sixteen subintervals of equal size. Elements with labels in each subinterval comprise the level $k-1$ sublists of L. In this way, the labels can be thought of as inducing a 16-ary tree on the elements of L. Call this the *labeling tree* of L. A level k sublist will initially contain at most 2^k elements, and will be required to accept at most 2^k insertions. The algorithm will be such that by the time this many insertions are performed, all the elements of the sublist will have been relabeled so as to be removed from the sublist.

The initial label of the ith ($i \in [0, 2^k - 1]$) element of a level k list, $INIT(k,i)$, is defined to be

1. If $k = 0$ or 1 then $INIT(k,i) = i$.
2. Otherwise, let $n = 2^k$. L is decomposed into two nonempty level $k-1$ sublists L_0 and L_1, where L_0 contained the first $n/2$ elements of L and L_1 the last $n/2$ elements. Then,
   (a) For $i = 0, \ldots, n/2 - 1$ (the elements of L_0), $INIT(k,i) = INIT(k-1,i)$,
   (b) For $i = n/2, \ldots, n-1$ (the elements of L_1), $INIT(k,i) = INIT(k-1, i - n/2) + S(k-1)$

Lemma 9. *The labelings defined above satisfy* $INIT(k,0) < \ldots < INIT(k, 2^k - 1) < S(k)$.

Proof. Straightforward by induction on k. □

We next show how to perform $n = 2^k$ insertions into a list initially containing n elements so that each insertion causes $O(k)$ elements to be relabeled and so that all labels remain in the range $[0, S(k) - 1]$.

We now describe how to insert an element y after and element x in a list that initially contained $n = 2^k$ elements and has had $I < n$ insertions performed on it.

1. If $k = 0$ or 1 then insert y after x and relabel the entire list $0, \ldots, n + I$. This causes at most $2n \leq 4$ relabelings.
2. Otherwise, recursively insert y after x in the $(k-1)$–sublist. If this sublist has now had $n/4$ insertions performed into it, it becomes "active". Then, elements are incrementally moved between active sublists. Elements are moved according to these rules:
   (a) Elements are moved from a sublist four at a time, (or, if the sublist contains fewer than four elements, all are moved). Elements are moved from the right end of one sublist to the left end of a succeeding sublist.
   (b) "Moving" an element means placing it in a specified succeeding sublist. The i-th element placed into the pth level $k-1$ sublist ($0 \leq p < 15$) is given the label $INIT(j, 2^j - i) + pS(k-1)$.
   (c) Elements are moved between sublists as follows: sublist 0 moves to 1 and 4, 1 to 2 and 3, 4 to 5 and 6. Elements from sublists 8, 9 and 12 are moved similarly (with an offset of 8). Sublists 7 and 15 are unused.
   (d) When moving elements from sublist j_1 to sublists j_2 and j_3 ($j_2 < j_3$), the first $n/2$ elements moved are sent to sublist j_3; all remaining elements are moved to sublist j_2.

Theorem 10. *For any level i sublist, $1 \leq i < k$, no more than 2^i insertions are performed into the sublist before it is emptied.*

Proof. We show that if 2^{i+1} insertions are performed into a level $i+1$ list L, none of its sublists receive more than 2^i insertions.

Sublist 0 initially contain 2^i elements. It becomes active when 2^{i-1} insertions have been performed into it. At that point, it contains at most $3\,2^{i-1}$ elements. After that, each insertion into L causes the size of sublist 0 to decreaseby at least 3, so that at most 2^{i-1} additional insertions can be performed before it becomes empty. An identical argument applies to sublist 8.

For sublists 1, 4, 9 and 12, note that the algorithm finishes moving elements into these sublists before they become active. Each sublist has at most 2^i elements moved into it. After each becomes active, the same argument applies as for sublists 0 and 8.

For sublists 2, 3, 5, 6, 10, 11, 13 and 14, note that no elements are moved into these sublists until after at least 2^i insertions have been performed into L. Therefore, none of these sublists can experience more than 2^i insertions, and none need be emptied. □

Theorem 11. ***The insertion algorithm maintains a strictly monotonic integer labeling, and performs $O(\log n)$ relabelings per insertion.***

Proof. It is easy to see that the movement of list elements between sublists maintains monotonicity. The fact that sublists never overflow implies that there is always space at the bottom, in a level 1 sublist, to insert a new list element.

The number of relabelings per insertion is at most $16 \log_2 n$, since no more than four sublists of any list are active at the same time. □

The algorithm as described performs n insertions into a list of size n. We now show how to incrementally increase n, starting from a list of length $O(1)$.

Begin with a list of length at most $n = 2$. Repeatedly do the following. Let k be the order of the current list (initially 1).

1. After 2^{k-1} insertions have occured into the list, it becomes active.
2. On succeeding insertions, four elements are copied from the list into another list of order $k+1$, whose label set immediately follows the label set of the current list.
3. After the current list is emptied (in at most 2^{k-1} insertions), the next list becomes the current list.

Note that at most 2^{k-1} insertions can occur into the order $k+1$ list before it becomes the current list, so at most one list at this top level can be active at any time.

It is easily seen that the labels are $O(n^4)$. The size of the label space could be made smaller (for example, by eliminating the unused sublists 7 and 15, attaining $O(n^{3.81})$), but no attempt has been made in this paper to minimize the exponent.

The algorithm as described performs $O(\log n)$ relabelings per insertion. However, the algorithm appears to require $O(\log^2 n)$ time per operation, since $O(\log n)$ time is spent updating the tree of sublists whenever one list element is relabeled.

A solution to this problem is to group list elements into subgroups of size $O(\log^2 n)$. For this to work, we must have a labeling scheme for lists of length k^2 that can perform k^2 insertions, each in time $O(k)$. The list $x_0, \ldots, x_{k^2-1}$ is initially labeled $l(x_i) = (i \bmod k) + 2k\lfloor i/k \rfloor$ (that is, the elements are grouped into groups of k, with labels inside each group consecutive integers, and a gap of size k between each group). When an insertion occurs, two elements are removed from the group into which the insertion occured, and are placed at the front of the next group. Two elements are then removed from the end of that group, etc. At the end, elements are added to a group of size $\leq k+1$; when it reaches $\geq k$ elements a new group is started after it with a gap of size $k+1$. It is easy to see that there will be $O(k)$ groups of size $O(k)$, so insertions cause $O(k)$ relabelings. See figure 3. The rest of the data structure must also be modified. A list becomes active after only $n/8$ insertions (for $n \geq 8$). Elements are then moved six a time until the sublist is emptied. In this way, no sublist would ever has more than $3n/4$ elements. However, we in fact do not move the elements immediately, but rather keep a priority queue of sublists ordered by the number of "pending" elements that

Fig. 3. Relabeling in $O(k)$ time

should have been moved (this number increases by six each time an insertion occurs in the sublist or one of its siblings). On each insertion we find the sublist with the largest number of pending moves and move $c_1 \log n$ elements. By the analysis of Game 1, we know that no sublist ever has more than $O(\log^2 n)$ pending moves. If the constant c_1 is large enough and if the $O(\log^2 n)$ size sublists at the bottom are large enough, no upper level sublist ever has more than $n/8$ pending moves, and cannot overflow.

4 Full Persistence: General Schemes

A general scheme for persistence is one where an ephemeral data structure can be made persistent just by replacing, in the algorithms for manipulating the ephemeral data structure, an ephemeral read of a cell by an *access* step, which is a call to a subroutine that takes the name of the cell and the required version and returns the value of the cell in that version. Ephemeral writes are similarly replaced by *update* steps. Driscoll et al. [13] gave a general scheme for making any bounded in-degree linked data structure (hereafter called a "BID" data structure) fully persistent, with $O(1)$ amortized time for update steps and $O(1)$ time for access steps. Dietz [6] gave a general scheme for making arbitrary data structures fully persistent on a RAM in $O(\log\log m)$ amortized expected time per access or update step, where m is the number of versions of the data structure. Dietz and Raman [8] showed how to make any n-node BID data structure fully persistent with $O(\log\log m)$ worst-case time per access or update step. This algorithm was complex, as it used dynamic fractional cascading [22, 8] as a subroutine.

4.1 Deterministic Solutions

We first give a very simple implementation of a general scheme for making any BID data structure fully persistent in worst-case time. This represents a major simplification over the previous algorithm [8] and runs faster for non-constant d (the in-degree bound) as well. The improvement in the case of non-constant in-degree is useful in section 5. We use the *split node* data structure of [13] and "split" nodes according to the following simple rule:

Algorithm 4 After each update step proceed as follows: if no fat node has more than $3d + 3$ versions, then do nothing; otherwise move between $d + 4$ and $2d + 3$ version records from a fat node of maximum size among all fat nodes to a newly created fat node

Using theorem 3 it is easy to show that the size of the largest fat node is bounded by $O(d \log m)$, where m is the number of versions, and thus access steps take $O(\log\log m + \log d)$ time. It is possible to implement an entire update step in $O(\log\log m + \log d)$ time using simple data structures, and the algorithm can be simplified further when d is a constant (details in [9]). We thus obtain:

Theorem 12. *An n-node linked data structure with degree bound d can be made fully persistent with $O(\log\log m + \log d)$ worst-case time for both access and update steps and $O(1)$ worst-case space per update, where m is the number of versions of the data structure.*

We now speed up queries in the above algorithm at the expense of slowing down updates. The idea is to replace binary search among the versions in a fat node by queries to the data structure of Lemma 1. If the degree bound d is $\log^{O(1)} m$ then the number of versions in a fat node is also bounded by $\log^{O(1)} m$, by the above argument. If the versions in the *version list* [13] are monotonically labeled with distinct integers of $O(\log m)$ bits each, then binary searches among the versions in a fat node can be replaced by predecessor queries among the integers corresponding to these versions, which can be done in $O(1)$ time by Lemma 1. Creating a new version causes $O(1)$ insertions into the version list (as in [13]); maintaining the monotonicity of the labeling under these insertions can be done using the algorithm of of section 3. This algorithm relabels $O(\log m)$ versions per insertion, necessitating $O(\log m)$ updates to small priority queues. Each update can be performed in $O(1)$ worst-case time, giving a $O(\log m)$ slowdown for update steps. This gives the following (details in [9]):

Theorem 13. *An n-node linked data structure with degree bound* $\log^{O(1)} m$ *can be made fully persistent with* $O(1)$ *worst-case time for access steps,* $O(\log m)$ *worst-case time for update steps and* $O(1)$ *worst-case space per update on a RAM, where m is the number of versions of the data structure.*

4.2 A Randomized Solution

We now adapt Algorithm 1 to full persistence of BID data structures to give a randomized solution. Let d be the maximum sum of in and out-degrees of nodes in the ephemeral data structure. If a fat node receives a new version and has at $s \geq d$ versions, then with probability $1/2d$ it is split into two fat nodes, each with at most $\lceil s/2 \rceil$ versions. Sen [27] has independently discovered a similar method for the related problem of (static) fractional cascading. We can show that:

Lemma 14. *The size of each fat node x is bounded above by $O(d) + G_x$, where G_x is a geometric random variable with expectation $O(d)$. Moreover, the G_x's associated with different fat nodes whose version intervals are overlapping are mutually independent.*

Using this result and Lemma 2, we can show the following:

Theorem 15. *A BID data structures can be made fully persistent with $O(1)$ expected ($O(\log\log m)$ w.h.p.) time for each access step and $O(1)$ expected time for each update step. Moreover, a sequence of s accesses to different cells can be performed in $O(s)$ time w.h.p. if $s \in \Omega(\log m)$.*

5 Fully Persistent Union–Find

Italiano and Sarnak [19] considered the problem of making a fully persistent data structure for the disjoint set union-find problem. Buchsbaum [4] detected and corrected an error in their paper, leading to an algorithm which runs in $O(\log n)$ amortized time for query and update operations and $O(1)$ amortized space per operation. They also claimed, but have not yet given a complete proof, that the time bounds above could be made worst-case. The worst-case result requires a non-trivial modification of the *displaced storage of changes* idea of [13]. We now quickly sketch why we feel the necessary modifications will be difficult. The idea of displaced storage of changes is to not always store the change to an ephemeral node x in the corresponding persistent node $\bar{x}$, but rather in some possibly different node that lies on the access path to $\bar{x}$ in the new version. Driscoll et al. use this technique to get persistent search trees, where there is a *unique* access path from the root to an internal node in the search tree. Italiano and Sarnak use a forest of 2–3 trees as the ephemeral data structure in their paper, with the root containing the name of the set. A *find* ascends from the leaf representing the element to the root of the tree representing the set, meaning that there may be *many* access paths to a particular persistent node in a version. It appears prohibitively expensive to store the change to x along all access paths to $\bar{x}$ or to search for the displacement record along all paths to $\bar{x}$. Our results are as follows:

Theorem 16. *Provided m, the number of versions, is polynomial in n, there are data structures for the fully persistent disjoint set union-find problem with (worst-case) complexities: (1) Persistent finds take $O(\log n/\log k)$ time, and unions take $O(k \log n)$ time and $O(k)$ space, for any $2 \leq k \leq \log n$, on a RAM. (2) Persistent find and union take $O(\log n)$ time each and unions use $O(\log^{\epsilon} n)$ space, for any constant $\epsilon > 0$.*

We obtain these results by applying Theorems 12 or 13 to a variant of Blum's *k-UF* tree data structure [3]. In the new data structure the in-degree of any node is poly-logarithmic in n (Blum's data structure places no non-trivial upper bound on the in-degree). We augment Blum's algorithm for the *union* operation with the following "thinning" step; that is, the *union* operation is first performed as in [3], after which we do the following:

Algorithm 5 Take the node v with the largest in-degree in the entire forest, and if it has more than $2k+1$ children then "thin" it down as follows: take its rightmost $k+1$ children and make them children of a new node v'. Insert v' as a sibling of v. If v was the root of its tree then make a new root r and insert v and v' as the children of r.

The number of memory modifications of this algorithm is clearly $O(k)$ in the worst case. In addition, the depth of any tree is $O(\log_k n)$, and updates takes $O(k + \log_k n)$ time. Theorem 3 allows us to prove a bound of $O(k \log n)$ on the in-degree of any node. Applying Theorems 12 or 13 to this data structure now gives us the theorem.

6 Partial Persistence of BID Data Structures

Driscoll et al. gave a general scheme for making BID data structures partially persistent, so that accesses and updates took $O(1)$ time (amortized for updates). [8] achieved constant worst-case time for both on a RAM assuming that version names are small integers. Under the assumption that the operations requested of the data structure do not depend on the random choices of made by it, we obtain pointer machine algorithms with the following complexities for this problem, as a straightforward consequence of Algorithms 2 and 3:

Theorem 1 *$O(1)$ expected time ($O(\log n)$ time w.h.p.) for update steps and $O(1)$ worst-case time for accesses. (2) $O(1)$ worst-case time for updates and $O(1)$ expected ($O(\log \log n)$ w.h.p.) time for accesses.*

7 Parallel Persistence

We consider algorithms in which a p-processor parallel pointer machine [17] accesses an $n \geq p$ node BID data structure in synchrony. On each "tick" of the clock, each processor may modify one ephemeral record (including updating pointers). Queries are also assumed to use p processors, again accessing the data structure synchronously. We would like to make the data structure partially persistent. We assume that we only have to deal with batches consisting either of p updates or p accesses.

A direct application of Driscoll et al.'s node copying method unfortunately fails to achieve even amortized constant slowdown. The reason is that up to $\Theta(n)$ nodes may need to be copied after a particular update. If the copying induced by different processors occurs at different times, most of the processors may be idle. Since the copying takes place "sequentially" (i.e., copying one node may cause another to be copied), it appears difficult to utilize load-balancing schemes. In [8] a worst-case algorithm was given for the sequential partial persistence problem. It was shown there that copying the node with the largest number of versions ensures that each node always has $O(\log n)$ versions. We can prove that if we were to copy the p largest nodes each time, then no node would have more than $O(\log n)$ versions. Access steps would then take $O(\log\log n)$ time ($O(1)$ time if version numbers are small integers). The problem with this solution is that while in the case $p = 1$ it is quite easy to find the node with the most versions, parallelizing the sequential data structure used for this purpose is difficult and yields a running time of no better than $O(\log n)$.

On the other hand, replicating our randomized algorithms for sequential partial persistence yields good randomized parallel solutions. We assume, as before, that the operations requested of the data structure do not depend on the random choices made by it. Our first solution will use a modification of Algorithm 1 to replace the amortized one (d is again the bound on the in-degree):

Algorithm 6
(1) Whenever a node receives a new version, then with probability $1/2d$ it is placed onto a queue associated with the processor that modified that fat node (or that caused some successor of the fat node to be copied).
(2) On each iteration, the processor removes a fat node from the queue. If it has at least $2d$ versions, it is copied, and a version added to each predecessor fat node.

We can show that (a) the expected size of a fat node at the time it is copied is $O(d)$ and (b) no fat node has more than $O(d \log n)$ versions, w.h.p. If d is $\log^{O(1)} n$, access steps now take $O(\log\log n)$ time w.h.p. on a parallel pointer machine. On the other hand, if if version names are small integers, Lemma 1 can be used to improve the complexity of an access to $O(1)$ time w.h.p. on a PRAM. Thus we obtain:

Theorem 17. *Batches of p update steps can be processed in $O(1)$ worst-case time, and batches of access steps in $O(1)$ time w.h.p. on a CREW PRAM assuming that version names are small integers, or on a parallel pointer machine with $O(\log\log n)$ slowdown w.h.p.*

We now remove the restriction that the version names be small integers. We replicate Algorithm 3 on the p processors. Updates thus take $O(1)$ time per processor. For access steps, we proceed as follows: if an access step needs to search in a fat node of size s, we allocate $\lceil sp/S \rceil$ processors to this searching task, where S is sum of the sizes of all searching tasks in this round. Note that the expected value of S is $O(p)$ and so the expected number of processors allocated to a searching task is proportional to the size of the searching task, enabling it to be completed in $O(1)$ expected time. The allocation of processors to tasks in this fashion can be done in $O(\log^* p)$ time and $O(p)$ operations ([18, Theorem 4.6] or [16]). We thus obtain:

Theorem 18. *Batches of p updates can be performed in $O(1)$ time and $O(p)$ operations and batches of p accesses can be performed in $O(\log^* p)$ expected time and $O(p)$ expected operations on a CRCW PRAM.*

8 Open Problems

We have proved a tight $\Theta(\log\log n)$ bound for the oblivious version of the game with decrements (Game 1), when $c = 2 + \epsilon$, for some constant $\epsilon > 0$. We would like to prove something about the case when c is smaller. It would also be interesting to find a way of improving either the query or the update complexity to $O(1)$ in the algorithm for full persistence (perhaps at the cost of some degradation in the other). We would like to see a data structure for the fully persistent union-find problem with sub-logarithmic times for both queries and updates.

Acknowledgement: We thank Neal Young for a suggestion that simplified the proof of Theorem 18.

References

1. M. Ajtai, M. Fredman, and J. Komlós. Hash functions for priority queues. *I & C*, **63** (1984), pp. 217–225.
2. H. Baker. List processing in real time on a serial computer. *CACM*, **21** (1978), pp. 280–294.
3. N. Blum. On the single-operation worst-case time complexity of the disjoint set union problem. *SICOMP*, **15** (1986), pp. 1021–1024.
4. A. Buchsbaum. Personal communication.
5. P. F. Dietz. Maintaining order in a linked list. In *Proc. 14th STOC* (1982), pp. 122–127.
6. P. F. Dietz. Fully persistent arrays. In *Proc. WADS 89* (1989), LNCS 382, pp. 67–74.
7. P. F. Dietz. Monotonic list labeling with good worst case performance. manuscript, February 1991.
8. P. F. Dietz and R. Raman. Persistence, amortization and randomization. In *Proc. 2nd SODA* (1991), pp. 77–87. Revised version, Univ. of Rochester CS TR 353, 1991.
9. P. F. Dietz and R. Raman. On some combinatorial games and their applications. Univ. of Rochester CS TR 392, 1991.
10. P. F. Dietz and D. D. Sleator. Two algorithms for maintaining order in a list. In *Proc. 19th STOC* (1987), pp. 365–372.
11. M. Dietzfelbinger, A. Karlin, K. Mehlhorn, F. Meyer auf der Heide, H. Rohnhert and R. E. Tarjan. Dynamic perfect hashing: upper and lower bounds. In *Proc. 29th FOCS* (1988), pp. 524–531.
12. J. R. Driscoll, H. N. Gabow, R. Shrairman and R. E. Tarjan. Relaxed heaps: an alternative to Fibonacci heaps with applications to parallel computation. *CACM* **31** (1988), pp. 1343–1354.
13. J. R. Driscoll, N. Sarnak, D. D. Sleator and R. E. Tarjan. Making data structures persistent. *JCSS*, **38** (1989), pp. 86–124.
14. M. Fischer, A. Meyer and A. Rosenberg. Real-time simulation of multihead tape units. *JACM*, **19** (1972), pp. 590–607.
15. Z. Galil. String matching in real time. *JACM*, **28** (1981), 134–149.
16. J. Gil, Y. Matias and U. Vishkin. Towards a theory of nearly constant-time parallel algorithms. In *Proc. 32nd FOCS* (1991), pp. 698–710.
17. M. T. Goodrich and S. R. Kosaraju. Sorting on a parallel pointer machine with applications to set expression evaluation. In *Proc. 30th FOCS* (1989), pp. 190–196.
18. T. Hagerup. Fast parallel space allocation, estimation and integer sorting. TR MPI-I-91-106, MPI für Informatik, 1991. Preliminary version in *Proc. 23rd STOC*, 1991.
19. G. F. Italiano and N. Sarnak. Fully persistent data structures for disjoint set union problems. *Proc. WADS '91* (1991), LNCS 519, pp. 449–460.
20. B. Leong and J. I. Seiferas. New real-time simulations of multihead tape units. *JACM*, **28** (1981), pp. 166–180.
21. C. Levcopolous and M. H. Overmars. A balanced search tree with $O(1)$ worst-case update time. *Acta Inf.*, **26**, (1988), pp. 269–278.
22. K. Mehlhorn and S. Näher. Dynamic fractional cascading. *Algorithmica*, **5** (1990), pp. 215–241.
23. M. H. Overmars. *The design of dynamic data structures*, LNCS 156, Springer-Verlag, Berlin, 1983.
24. P. Raghavan. Lecture notes in randomized algorithms. TR RC 15340, IBM, 1989.
25. R. Raman. *Eliminating amortization: on data structures with guaranteed response time.* PhD thesis, U. of Rochester, 1991.
26. N. Sarnak. *Persistent data structures.* PhD thesis, New York Univ., 1986.
27. S. Sen. Fractional cascading revisited. In *Proc. SWAT '92* (1992), LNCS 621, pp. 212–220.
28. A. K. Tsakalidis. Maintaining order in a generalized linked list. *Acta Inf.*, **21** (1984), pp. 101–112.
29. P. van Emde Boas. Preserving order in a forest in less than logarithmic time and linear space. *IPL*, **6** (1977), pp. 80–82.
30. P. Wadler. Analysis of an algorithm for real-time garbage collection. *CACM*, **19** (1976), pp. 491–500.
31. D. E. Willard and G. S. Lueker. Adding range restriction capability to dynamic data structures. *JACM*, **32** (1985), pp. 597–617.

The K-D Heap: An Efficient Multi-dimensional Priority Queue

Yuzheng Ding† and Mark Allen Weiss‡

†Computer Science Department, University of California
Los Angeles, CA 90024, U.S.A.
‡School of Computer Science, Florida International University
Miami, FL 33199, U.S.A

Abstract. This paper presents the *k-d heap*, an efficient data structure that implements a multi-dimensional priority queue. The basic form of the k-d heap uses no extra space, takes linear time to construct, and supports instant access to the items carrying the minimum key of any dimension, as well as logarithmic time insertion, deletion, and modification of any item in the queue. Moreover, it can be extended to a multi-dimensional double-ended mergeable priority queue, capable of efficiently supporting all the operations linked to priority queues. The k-d heap is very easily implemented, and has direct applications.

1 Introduction

The *Priority queue* is one of the fundamental abstract data types widely used, and has been extensively studied. A classic priority queue consists of a collection of items, each of which has a priority drawn from a fully-ordered set. The basic operations on a priority queue are the insertion of new items and the retrieval and deletion of the item with highest priority. In some applications, it is also desired that the deletions and the priority changes of arbitrary items are allowed in a priority queue, and priority queues can be merged and/or split. There are also cases where the access to both the item of highest priority and that of lowest priority is necessary, i.e. the priority queue is required to be *double-ended.*

Implementations of priority queues are usually named *heaps,* in which priority is represented by the *keys* of the data items. If the smallest key represents the highest priority, the heap is called a *min-heap*; if the largest key represents the highest priority, the heap is a *max-heap.* Heaps are usually based on certain tree structures; if pointers are used to maintain such structures, it is an *explicit implementation*; if pointers are not used (in which case an indexed array is usually the alternative), it is an *implicit implementation.* Implicit implementations are generally preferable since they use no or less extra space, unless merging is supported, in which case it seems that implicit implementations suffer a linear number of data movements.

Since the invention of the first priority queue implementation, the *binary heap*, by Williams [19], many papers have been published on this issue, including [1, 3, 4, 5, 6, 7, 9, 10, 11, 12, 15, 16, 17, 18], to name just a few. In [7] a data structure is introduced that efficiently supports *all* the priority queue operations that have been discussed in literature, including merge and double-ended operations.

On the other hand, in practice, data objects are often associated with more than one priority category. In a database system, records can be indexed on different fields, and each index usually defines a priority relationship. Access based on any index should be supported as efficiently. It is not difficult to give more examples. For such applications, a ***multi-dimensional priority queue*** is desired.

In all these cases, if the classic priority queue data type is used, for each priority relation we have to build a queue; moreover, we have to maintain the links among the presences of the same object in different queues, and take appropriate operations on all the queues every time an operation based on one priority is required. This clearly consumes extra space and time, and is inconvenient to implement.

In this paper we present a data structure that implements a multi-dimensional priority queue efficiently. This structure, named the ***k-d heap***, can implement a priority queue on a set of k different priority relationships, and support integrated priority queue operations according to any priority relationship as efficiently as the same operations on standard one-dimensional priority queues. Since a single structure is maintained, there is no need for extra space to keep links, and extra operations to maintain different priority relationships. Using an implicit implementation, the k-d heap does not require *any* extra space. Moreover, the k-d heap can be implemented to support some or all the priority operations on any number of dimensions, and it contains previously studied heap structures, e.g. the double-ended heaps, as a special case.

A related work is the ***k-d tree*** [2], which extends binary search tree to multiple keys. The major drawback of the k-d tree is that both deletion and balancing are difficult. The k-d heap proposed in this paper follows similar order, but is automatically balanced, and simple to implement.

The rest of this paper is organized as follows. Section 2 presents the general structure of a k-d heap, and its properties. In section 3 we discuss the operations for the special case of a two-dimensional priority queue; this will be extended to the general case in section 4. Section 5 is devoted to further extensions to support non-standard operations. Section 6 discusses the improvement of the k-d heap structure for large values of k. Section 7 concludes the paper. Due to page limitations we are unable to include detailed complexity analysis; it can be found in [8].

2 The K-D Heap

We are given a set of *items*, each of which contains k *keys* $key_1, key_2, ..., key_k$, where key_i is drawn from fully-ordered set K_i. A (min) ***k-d heap*** H over this set

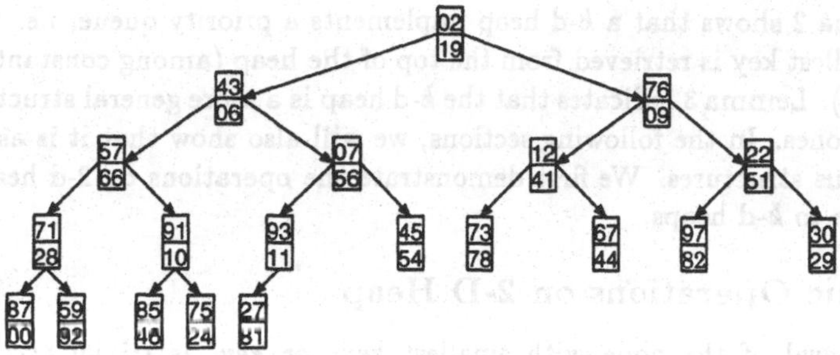

Fig.1 A 2-dimensional heap of 20 items

of items is a binary tree that satisfies the following conditions:

1. H is a complete binary tree (i.e. it is full except possibly at the rightmost positions of the bottom level);
2. H maintains *k-d heap order*:
   (a) the item at root has the smallest key_1 value in the tree;
   (b) for any node v other than the root, if the item at its parent w has the smallest key_i value in the subtree rooted at w, then the item at v has the smallest $key_{\mathrm{mod}(i,k)+1}$ value in the subtree rooted at v.

For simplicity we will not distinguish a *node* and the *item* at the node. We call k the *dimension* of the heap. An example of a 2-d heap is shown in Figure 1.

Define the *level* of a node to be the number of nodes on the path from the root to that node, and the *height* of the heap to be the largest level of its nodes. In practice, the height of a heap is usually much larger than the dimension. The k-d heap order can be re-expressed as the following:

Lemma 1 In a k-d heap, a node at level i has the smallest $key_{\mathrm{mod}(i-1,k)+1}$ in its subtree. ∎

We will refer to level i as a *level* for the key $key_{\mathrm{mod}(i-1,k)+1}$, and $key_{\mathrm{mod}(i-1,k)+1}$ the *key* of the level i. Also, if a node in a complete binary tree satisfies Lemma 1, we call it a *heap node*. A complete binary tree is a k-d heap if and only if every node is a heap node.

Clearly, a k-d heap can be implemented either implicitly or explicitly. Unless otherwise specified, we will assume an implicit implementation. It is easily seen that k-d heaps have the following properties.

Lemma 2 In a k-d heap, the node that has the smallest key_i value is among the highest i levels of the heap. ∎

Lemma 3 A 1-d heap is a *binary heap* [19]; a 2-d heap with $key_1 = -key_2$ for all nodes is a *min-max heap* [1]. ∎

Lemma 2 shows that a k-d heap implements a priority queue, i.e. the node with smallest key is retrieved from the top of the heap (among constant number of nodes[1]). Lemma 3 indicates that the k-d heap is a more general structure than previous ones. In the following sections, we will also show that it is as efficient as previous structures. We first demonstrate the operations on 2-d heaps, then generalize to k-d heaps.

3 Basic Operations on 2-D Heap

The retrieval of the node with smallest key_1 or key_2 is trivial according to Lemma 2. Insertion will add a new node to the end of the heap, however the new node might not be in accordance with the 2-d heap order at this position. Deletion of a node with the smallest key will leave a *hole* at one of the top two levels, and we can move the node from the end of the heap to fill this hole. The more general case is that an arbitrary node is deleted and the last node is used to replace it. The substitution might not be in accordance with the 2-d heap order at the new position. Therefore, the most essential operation is the *normalization* of an abnormal node in an otherwise 2-d heap ordered complete binary tree.

3.1 Normalization

Assume that in a 2-d heap, one node is altered. If every node was a heap node, then every node except the altered node and its ancestors is still a heap node. Moreover, the restoration of the heap order property at a node relates only to the descendants of the node. In other words, if a node is a heap node, it is still a heap node after any rearrangement among its descendants. Therefore, we restore the order from top to bottom so that restoration is permanent.

The first part of the normalization procedure is to restore the order among the altered node and its ancestors. The algorithm is outlined as follows.

algorithm normalize-up(*node*)
  for each *ancestor* of *node* starting from *root* **do**
    if the order relationship between *ancestor* and *node* is violated
    then swap *ancestor* and *node*.
end.

To see how this works, note that the swap of an ancestor and the altered node does not affect the order property on the ancestors at smaller levels, and after the swap, the heap order at this ancestor is restored since the altered node was the only problem. Therefore, after each iteration we convert one ancestor to a heap node (if it was not). Figure 2(a) shows a 2-d heap with an abnormal node and Figure 2(b) is the result of *normalize-up*() on it (a crossed out arrow indicates a comparison, but no swap).

[1] The cost of the search among these nodes can be easily eliminated by distributing it to the operations modifying the heap.

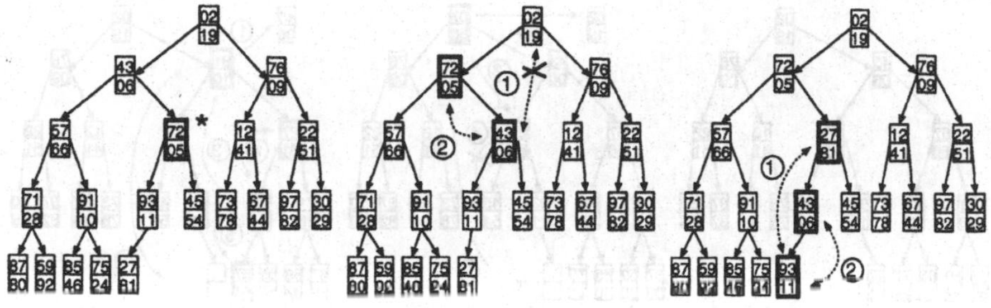

(a) Heap with abnormal node (b) After normalize-up() (c) After normalize-down()

Fig.2 Normalization of an abnormal node in 2-d heap

After such a procedure, all nodes except one are heap nodes, so we only need to restore the heap order in the subtree rooted at this abnormal node. The algorithm follows.

algorithm normalize-down(*node*)
  Let the key for the level of *item* be key_i.
  Find the node v with smallest key_i among *node*,
    its children and grandchildren.
  if v =*node* **then stop.**
  swap v and *node*.
  if v is a grandchild of *node* and
    has smaller $key_{\mathrm{mod}(i,2)+1}$ value than its parent
   then swap v and its parent.
  call normalize-down(v).
end.

This procedure is illustrated in Figure 2(c) where the abnormal node in Figure 2(b) is normalized towards its descendants. The procedure can be regarded as a procedure of pushing the abnormal node down the heap. Clearly, recursion is not necessary in implementation.

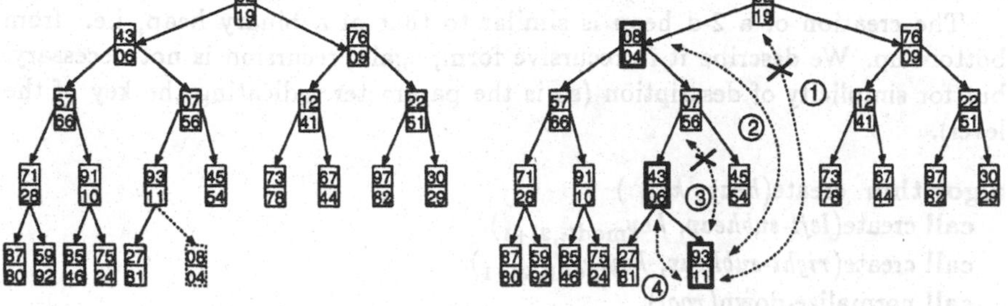

(a) Before insertion (b) After insertion

Fig.3 Insertion of an item into a 2-d heap

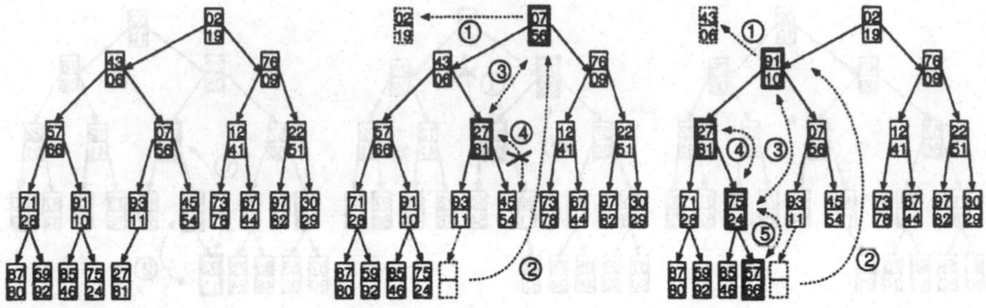

(a) The original 2-d heap (b) After deletion of min key1 (c) After deletion of min key2

Fig.4 Deletion of the item with minimum key in a 2-d heap

Now the normalization procedure is simply the following:

algorithm normalize(*node*)
  call normalize-up(*node*).
  call normalize-down(*node*).
end.

3.2 2-D Heap Operations

Based on the algorithms in the last subsection, the basic operations in 2-d heaps are easy.

To insert an item, we add it to the end of the heap, and normalize it. In this case, only *normalize-up*() is required. Figure 3 gives an example.

To delete the minimum for key_1, we replace the root with the last item in the heap, and normalize it. In this case, only *normalize-down*() is required. An example is shown in Figure 4(b), where the original heap is in Figure 4(a).

To delete the minimum for key_2, we first locate it among the first three items in the heap. Then we replace it with the last item in the heap and normalize it. Figure 4(c) shows how this is applied to the original heap.

To delete an arbitrary item (with known position), we simply replace it with the last item in the heap and normalize it. To modify an arbitrary item, we just do the modification and then normalize the item.

The creation of a 2-d heap is similar to that of a binary heap, i.e. from bottom up. We describe it in recursive form; again, recursion is not necessary, but for simplicity of description (so is the parameter indicating the key of the level).

algorithm create(*heap*, key_i)
  call create(*left-subheap*, $key_{\mathrm{mod}(i,2)+1}$).
  call create(*right-subheap*, $key_{\mathrm{mod}(i,2)+1}$).
  call normalize-down(*root*).
end.

Figure 5 shows an example.

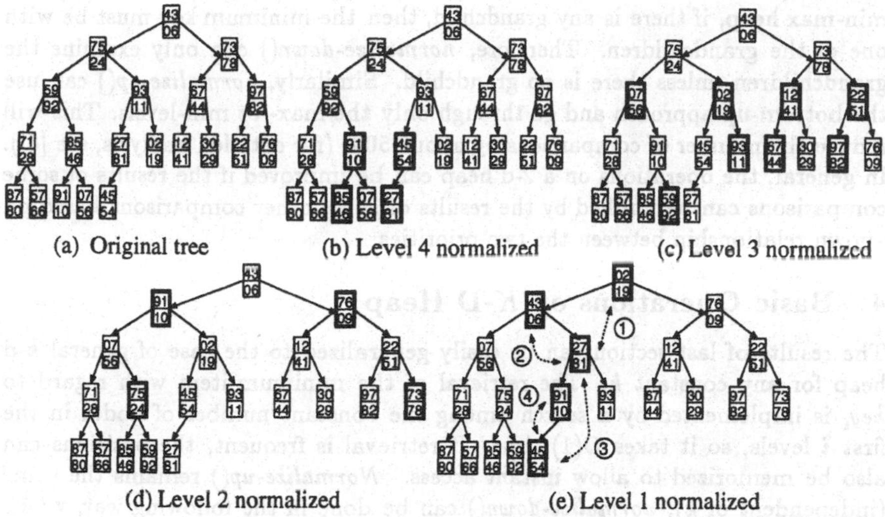

Fig.5 Creation of a 2-d heap

3.3 Complexity

It is easily seen that the cost of *normalize-up*() is linear in the level of the node it is applied to, and the cost of *normalize-down*() is linear in the height of the subtree rooted at the node to which the operation is applied. Therefore, except for creation, the complexity of any of the operations discussed in the last subsection is linear in the height of the heap. Creation is linear in the number of items; the proof is similar to that of binary heap, since *normalize-down*() is linear in the height of the heap. Thus we have

Theorem 1 A 2-d heap of n items can be created in $O(n)$ time. On such a heap, finding the item with minimum value of either key takes constant time; insertion, deletion of the item with minimum value of either key, deletion of an arbitrary item whose position is known, and modification of an arbitrary item take $O(\log n)$ time each in the worst case. ∎

The detailed analysis of the exact number of comparisons and data movements can be found in [8].

3.4 2-D Heap as Min-Max Heap

According to Lemma 3, a 2-d heap can implement a min-max heap if we set $key_2 = -key_1$ for all nodes, where key_1 is the original *min*-key, and key_2 is the new *max*-key. However, since the operations of 2-d heap are designed to handle the general case, where key_1 and key_2 can be totally unrelated, it is more efficient to use the min-max heap operations for this special case. For instance, when a node is normalized towards its descendant, in a general 2-d heap the minimum key can be with any of its children and grandchildren. However, in a min-max heap, if there is any grandchild, then the minimum key must be with one of the grandchildren. Therefore, *normalize-down*() can only examine the grandchildren unless there is no grandchild. Similarly, *normalize-up*() can use the bottom-up approach and go through only the max- or min-levels. This will reduce the number of comparisons by about 50% (for detailed analysis, see [8]). In general, the operations on a 2-d heap can be improved if the results of some comparisons can be implied by the results of some other comparisons based on known relationship between the two priorities.

4 Basic Operations on K-D Heap

The results of last section can be easily generalized to the case of general k-d heap for any constant k. The retrieval of the minimum item with regard to key_i is implemented by a search among the constant number of nodes in the first i levels, so it takes O(1) time. If retrieval is frequent, the positions can also be memorized to allow instant access. *Normalize-up*() remains the same (independent of k); *normalize-down*() can be done in the following way, which is a direct generalization of the 2-d case.

algorithm normalize-down(*node*)
  Let key_i be the key for the level of *node*.
  Find the node v with smallest key_i value
    in the descendants of *node* within k levels.
  if *node* has smaller key_i than v
    then stop
    else swap *node* and v.
  call normalize-up(v) for the subheap rooted at *node*.
  call normalize-down(v).
end.

Note that the call of *normalize-up*() is within a heap of k levels, so the cost is constant.

Creation is similar to the case of 2-d heap except the key for any level is determined with modulo k instead of 2. Therefore,

Theorem 2 A k-d heap of n items, where k is any given constant, can be created in $O(n)$ time. It takes constant time to retrieve the item with minimum value of any of the k keys, and it takes $O(\log n)$ time to do insertion, deletion of an item with minimum value of any of the k keys, and deletion or modification of an arbitrary item (whose position in the heap is known) on such a heap. ∎

Figure 6 shows a 3-d heap and the result of the deletion of the root.

It should be pointed out that the hidden constant for the deletion operations is exponential in k, which makes the operations impractical when k is very large. (On the other hand, insertion cost is independent of k, and creation cost is proportional to k^2 [8].) In Section 6 we will show that the exponential constant can be eliminated by a slight modification of the heap structure.

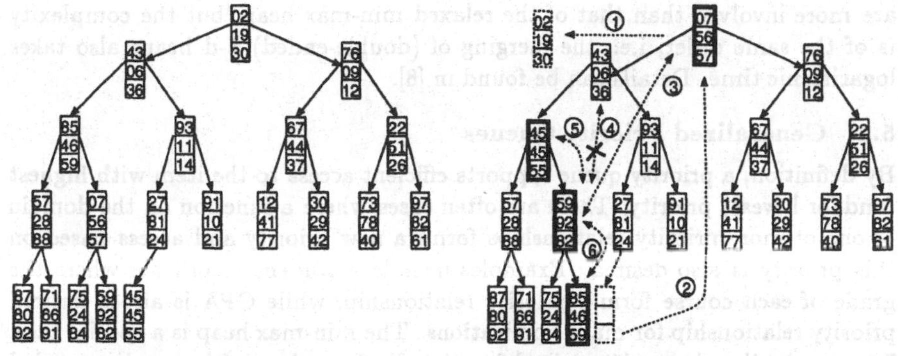

(a) Original 3-d heap (b) After deletion of min key1

Fig.6 Deletion of item with minimum key_1 in a 3-d heap

5 Extensions of K-D Heap

The basic k-d heaps can be extended to support other operations. We present some of them briefly; details are in [8].

5.1 Double-Ended K-D Heap

As we have demonstrated in Section 3.3.4, 2-d heaps can be used to implement min-max heaps. In general, if each item has k keys $key_1, key_2, ..., key_k$, we can (conceptually) include k more keys $key_1^-, key_2^-, ..., key_k^-$ where $key_i^- = -key_i$, where the negative sign represents the general functional mapping that reverses the order. Then, the $2k$-d heap becomes a double-ended k dimensional heap. Note that the new keys need not be stored; instead, the mapping can be encoded in the operations, so there is no need for extra space.

5.2 Mergeable K-D Heap

We assume explicit implementation of binary trees when merging is considered. Merging of the basic k-d heaps is difficult. As special cases, binary heaps (1-d heaps) take $O(\log^2 n)$ time to merge [16], and min-max heaps (a special type of 2-d heap) have an $\Omega(n)$ lower bound for merging [13].

On the other hand, in [7] it is shown that by introducing certain kind of relaxation to the min-max order, a structure similar to the binomial queue is possible, and thus merging becomes an $O(\log n)$ time operation. Specifically, the heap is decomposed into a set of units, each containing 2^i items for a unique i, in the form of a perfect binary tree plus a single item. For odd (even) i, the perfect binary tree is a *relaxed* min-max (max-min) heap, where the word *relaxed* means that some nodes are allowed not to obey the heap order. It was shown that two such units of the same size can be merged in constant time, therefore the entire heap is merged in logarithmic time. This technique can be generalized to the k-d heaps by a similar decomposition, and for a perfect binary tree of height i, the root has $key_{\text{mod}(i,k)+1}$ has its key for the relaxed order. The operations are more involved than that of the relaxed min-max heap, but the complexity is of the same order, i.e. the merging of (double-ended) k-d heaps also takes logarithmic time. Details can be found in [8].

5.3 Generalized Priority Queues

By definition, a priority queue supports efficient access to the item with highest (and/or lowest) priority. There are often cases where a function on the domain of one or more priority relationships forms a new priority and access based on this priority is also desired. Examples include a student record file where the grade of each course form a priority relationship, while GPA is also a desired priority relationship for many applications. The min-max heap is a special case. In general, all such pre-determined functionally formed priorities can be encoded into the operations, and explicit storage of the corresponding "keys" is not necessary. In this sense, k-d heaps can be used to implement such *generalized priority queues*. Double-ended operations is a special case of such generalization.

6 Improvement of K-D Heap for Large K Values

The k-d heap structure we have discussed so far maintains a very loose order relationship, in the sense that in any subheap, although the smallest key is at the root, the second smallest key can be anywhere among the first k levels of descendants. This results in the exponential constant for the deletion operations. When k is large, as is the case for double-ended implementations, this must be improved to make the k-d heap practical.

In this section we briefly describe an approach to solve this problem. The detailed analysis is given in [8].

Instead of building a complete binary tree for the heap, we allow each node at the levels i, $\mathrm{mod}(i, k) \neq 0$, to have at most one child, while the nodes at the levels i, $\mathrm{mod}(i, k) = 0$, still have up to two children. Figure 7(a) shows the modified structure for the 3-d heap shown in Figure 6(a). In order to allow efficient implicit implementation, every chain of k nodes can be packed into a single node, so that the underlying tree is still a complete binary tree, as shown in Figure 7(b).

Clearly, the *normalize-down* operation now only needs to examine at most $2k$ descendants to select the minimum. On the other hand, the height of the heap is about k times as large, so the constant factor for the deletion operations is reduced to $O(k^2)$. The cost of the insertion operation, however, is increased by a factor of k. It can be shown that the cost of heap creation is also slightly increased (but still $O(k^2n)$).

Although Figure 7 illustrates the modification for $k = 3$, it is only preferred when k is large. In practice, small values of k (for instance, 2 or 3) are more usual, and the original structure is more efficient.

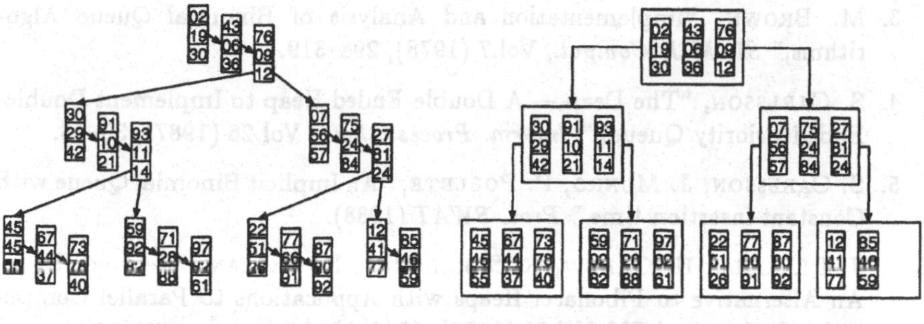

(a) The structure (b) The implementation

Fig.7 Improved k-d heap structure for large k

7 Conclusion

We have presented the k-d heap, a data structure that efficiently implements a multi-dimensional priority queue without using extra space. One form supports insertion, deletion of any minimum, and creation in $O(\log n)$, $O(2^k \log n)$, and $O(k^2 n)$, respectively, with particularly easy operations for $k = 2$, while another form gives times of $O(k \log n)$, $O(k^2 \log n)$, and $O(k^2 n)$. (Note that for typical values of k, the difference between 2^k and k^2 is not large). Moreover, the k-d heap can be extended to support double-ended and merging operations.

The implementation of the k-d heap is extremely simple. We have coded the operations in C, and they take about 120 lines totally. Therefore, it is very practical. A complete implementation with performance comparison is given in [8].

Several related problems are still open. It is possible to improve the bounds for insertion and minimum-deletion to $O(k \log n)$ by using only $O(n)$ additional space. We do not know if this can be achieved with only contant extra space.

Although the k-d heaps presented in Section 4 can be merged in logarithmic time via order relaxation, the same technique does not extend to the modified structure discussed in Section 6. Another interesting problem is to exploit the inter-priority relationships to improve the efficiency of k-d heap operations. In the case of double-ended priority queues, we have shown that such relationships yield significantly complexity reduction. The impact of other commonly presented relationships is also worth studying.

References

1. M. ATKINSON, J. SACK, N. SANTORO, T. STROTHOTTE, "Min-Max Heaps and Generalized Priority Queues," *Comm. ACM*, Vol.29 (1986), 996–1000.

2. J. BENTLEY, "Multidimensional Binary Search Trees Used for Associative Searching," *Comm. ACM*, Vol.18 (1975), 509–517.

3. M. Brown, "Implementation and Analysis of Binomial Queue Algorithms," *SIAM J. Comput.*, Vol.7 (1978), 298–319.

4. S. Carlsson, "The Deap — A Double-Ended Heap to Implement Double-Ended Priority Queues," *Inform. Process. Lett.*, Vol.26 (1987), 33–36.

5. S. Carlsson, J. Munro, P. Poblete, "An Implicit Binomial Queue with Constant insertion time," *Proc. SWAT* (1988).

6. J. Driscoll, H. Gabow, R. Shrairman, R. Tarjan, "Relaxed Heaps: An Alternative to Fibonacci Heaps with Applications to Parallel Computation," *Comm. ACM*, Vol.31 (1988), 1343–1354.

7. Y. Ding, M. Weiss, "The Relaxed Min-Max Heap: A Mergeable Double-Ended Priority Queue," *Acta Informatica*, Vol.30 (1993), to appear.

8. Y. Ding, M. Weiss, "Efficient Implementations of Multi-dimensional Priority Queues," School of Computer Science Technical Report, Florida International University, Feb. 1993.

9. R. Floyd, "Algorithm 245: Treesort," *Comm. ACM*, Vol.7 (1964), 701.

10. M. Fredman, R. Sedgewick, D. Sleator, R. Tarjan, "The Pairing Heap: A New Form of Self-Adjusting Heap," *Algorithmica*, Vol.1 (1986), 111–129.

11. M. Fredman, R. Tarjan, "Fibonacci Heaps and Their Uses in Improved Network Optimization Algorithms," *J. ACM*, Vol.34 (1987), 596–615.

12. G. Gambosi, E. Nardelli, M. Talamo, "A Pointer-free Data Structure for Merging Heaps and Min-Max Heaps," *Theoretical Computer Science*, Vol. 84 (1991), 107–126.

13. A. Hasham, J. Sack, "Bounds for Min-Max Heaps," *BIT*, Vol.27 (1987), 315-323.

14. D. Knuth, *The Art of Computer Programming*, Vol. 3, Addison-Wesley, Reading, MA, 1973.

15. S. Olariu, C. Overstreet, Z. Wen, "A Mergeable Double-ended Priority Queue," *The Computer Journal*, Vol.34 (1991), 423–427.

16. J. Sack, T. Strothotte, "An Algorithm for Merging Heaps," *Acta Informatica*, Vol.22 (1985), 171–186.

17. D. Sleator, R. Tarjan, "Self-Adjusting Heaps," *SIAM J. Comput.*, Vol.15 (1986), 52–69.

18. J. Vuillemin, " A Data Structure for Manipulating Priority Queues," *Comm. ACM*, Vol.21 (1978), 309–315.

19. J. Williams, "Algorithm 232: Heapsort," *Comm. ACM*, Vol.7 (1964), 347–348.

A Complete and Efficient Algorithm for the Intersection of a General and a Convex Polyhedron*

Katrin Dobrindt[1] and Kurt Mehlhorn[2] and Mariette Yvinec[13]

[1] INRIA, B.P.93, 06902 Sophia-Antipolis cedex, France.
e-mail : dobrindt@sophia.inria.fr

[2] Max-Planck-Institut für Informatik, Im Stadtwald 6600 Saarbrücken, Germany.
e-mail : mehlhorn@mpi-sb.mpg.de

[3] Laboratoire I3S, CNRS-URA 1376, 06902 Sophia-Antipolis, France.
e-mail : yvinec@sophia.inria.fr

Abstract. A polyhedron is any set that can be obtained from the open halfspaces by a finite number of set complement and set intersection operations. We give an efficient and complete algorithm for intersecting two three–dimensional polyhedra, one of which is convex. The algorithm is efficient in the sense that its running time is bounded by the size of the inputs plus the size of the output times a logarithmic factor. The algorithm is complete in the sense that it can handle all inputs and requires no general position assumption. We also describe a novel data structure that can represent all three–dimensional polyhedra (the set of polyhedra representable by all previous data structures is not closed under the basic boolean operations).

1 Introduction

A polyhedron is any subset of three–dimensional Euclidean space that can be obtained from the open halfspaces by a finite number of set complement and set intersection operations. Figure 1 shows some polyhedra. We give an algorithm to compute the intersection of a polyhedron P with a convex polyhedron C. The algorithm runs in time $O((|P| + |C| + |P \cap C|) \log(|P| + |C| + |P \cap C|))$ where $| \, |$ denotes the size of a polyhedron. The algorithm works for all inputs and not only for inputs in general position. The only previous algorithm with similar efficiency of Mehlhorn and Simon [MS85] applied only to regular[4] polyhedra in general position, i.e., a face of P and a face of C may intersect only if the sum of their affine hulls is the entire space. The intersection of two regular polyhedra in general position is again regular.

* The research of all three authors was partly supported by the ESPRIT Basic Research Actions Program, under contract No. 7141 (project ALCOM II). The research of the second author was also partially supported by the BMFT (Förderungskennzeichen ITS 9103). The paper is based on the first author's master's thesis [Dob90].

[4] All mathematical terms and notations are summarized in the appendix A.

The standard data structures for three–dimensional polyhedra, e.g. the quad–edge–structure of [EM85], the doubly–connected–edge–list of [PS85], and the half–edge–structure of [Män88], cannot represent all polyhedra. This implies that the class of representable (in any one of these data structures) polyhedra is not closed under the basic boolean operations intersection, union, and complement. The traditional remedy is to redefine the basic boolean operations (by adding a regularization step, cf. [Req80, Män88, Hof89]). We propose a new data structure (called the ***local–graphs–data–structure***) that can represent all three–dimensional polyhedra. Our data structure is based on the fundamental work of Nef [Nef78] (see also [BN88]) who studied the mathematical properties of polyhedra. The data structure stores a polyhedron as a collection of faces (vertices, edges, and facets); each face is described as the set of points comprising the face and its local graph. The local graph is a planar graph embedded into a sphere that captures the local properties of the polyhedron in the neighborhood of the face. The details are given in section 2.

The intersection algorithm is described in section 3. It first computes the intersections of all faces of P and all faces of C and then builds the local–graphs–data–structure for $P \cap C$. We employ different strategies for intersecting faces depending on the dimension of the faces involved.

In the extended abstract submitted to this conference we outlined an intersection algorithm based on the symbolic perturbation technique introduced by Edelsbrunner and Mücke [EM90] . We believe that the algorithm presented here is simpler. In the other algorithm, we first perturb the convex polyhedron C by moving its facets outwards by infinitesimal amounts. This brings the two polyhedra into general position. We then apply an extension of the intersection algorithm of Mehlhorn and Simon [MS85] to P and the perturbation $C(\varepsilon)$ of C. Finally, we let ε go to zero and obtain $P \cap C$ from $P \cap C(\varepsilon)$. The limit process is mathematically and algorithmically quite involved and so the overall algorithm is more complex than the algorithm presented here. However, if the exact output is not needed and $P \cap C(\varepsilon)$ suffices then the other algorithm is to be preferred.

2 The Local–Graphs–Data–Structure

We start with a brief review of Nef's theory of polyhedra [Nef78].

Definition 1. [Nef78] A ***polyhedron*** in $\mathbb{R}^3$ is a set $P \subseteq \mathbb{R}^3$ generated from a finite number of open halfspaces by set complement and set intersection operations.

Figure 1 shows some polyhedra. A face of a polyhedron is a maximal set of points which have the same local view of the polyhedron. The exact definition requires the concept of the local pyramid of a point.

Definition 2. A set $K \subseteq \mathbb{R}^3$ is called a *cone with apex 0*, if $K = \mathbb{R}^+ K$ and a *cone with apex x*, $x \in \mathbb{R}^3$, if $K = x + \mathbb{R}^+(K - x)$. Thus $(K - x)$ is a cone with apex 0. A cone is *polyhedral* if it is a polyhedron. A set $Q \subseteq \mathbb{R}^3$ is called a *pyramid with apex x*, if it is a a polyhedral cone with apex x.

Note that $\mathbb{R}^+$ does not include zero and thus a cone K with apex x may or may not include x. Furthermore, a cone can have more than one apex and the set of all apices of a cone is a flat.

Definition 3. [Nef78] Let $P \subseteq \mathbb{R}^3$ be a polyhedron and $x \in \mathbb{R}^3$. There is a neighborhood U_0 such that the cone $Q := x + \mathbb{R}^+((P \cap U) - x)$ is the same for all neighborhoods $U \subseteq U_0$. The cone Q is called the *local pyramid of the polyhedron P in the point x* and is denoted $Pyr_P(x)$.

Indeed, the cone Q is a polyhedron and thus a pyramid with apex x. It describes the local characteristics of the polyhedron in the neighborhood of the point x.

Definition 4. [Nef78] Let $P \subseteq \mathbb{R}^3$ be a polyhedron. A *face s* of P is a maximal (with respect to set inclusion) non-empty subset of $\mathbb{R}^3$ such that all of its points have the same local pyramid Q, i.e., $s = \{x \in \mathbb{R}^3 | Pyr_P(x) = Q\}$. Q is called the pyramid associated with the face and is denoted $Pyr_P(s)$ or simply $Pyr(s)$. The dimension of s is the dimension of the linear subspace of all apices of Q.

A face s_1 is *incident* to a face s_2 if $s_1 \subseteq clos\ s_2$. As usual we call a 0-dimensional face a *vertex*, a 1-dimensional face an *edge*, and a 2-dimensional face a *facet*. A face of dimension two or less is called *low-dimensional*. In Figure 2 an example in $\mathbb{R}^2$ is given: the polyhedron has one vertex v, two edges e_1 and e_2, and two 2-dimensional faces f and $int(cpl(f))$. We now list some basic properties of faces.

Fact 5. [Nef78]

a) *All faces of a polyhedron are polyhedra.*
b) *The linear subspace of all apices of the pyramid associated with a face is the affine hull of the face.*
c) *Faces are relatively open sets.*
d) $x \in \mathrm{Pyr}_P(x)$ *iff* $x \in P$.
e) *A polyhedron P has at most two 3-dimensional faces, namely* $\mathrm{int}(P) = \{x \in \mathbb{R}^3 | \mathrm{Pyr}_P(x) = \mathbb{R}^3\}$ *and* $\mathrm{ext}(P) = \{x \in \mathbb{R}^3 | \mathrm{Pyr}_P(x) = \emptyset\}$. *The boundary* $\mathrm{bd}\ P = \{x;\ \emptyset \neq P \cap U \neq U$ *for every neighborhood U of x*$\}$ *is equal to the union of the low-dimensional faces.*
f) *Let s be a face of the polyhedron P and let t be a face of s. Then t is the union of some faces of P (cf. Figure 2 for an illustration: the face f has one edge e that is the union of faces e_1, v, and e_2.)*
g) *A face of P is either a subset of P or disjoint from P.*

A polyhedron may have an arbitrary number of edges with the same affine hull but can have at most six facets with the same affine hull (since the local pyramid of a facet is either an open or a closed halfspace or a plane or the complement of a plane). A face is not necessarily connected or bounded and a polyhedron does not necessarily have faces of all dimensions. We are now ready to define the (abstract) representation of a polyhedron. We will later develop a concrete data structure for it.

Definition 6. For a polyhedron P let $rep(P)$ be the set $\{(s, Pyr_P(s));\ s$ is a low-dimensional face of $P\}$.

Every polyhedron different from the full space and the empty set has a low-dimensional face.

Lemma 7. *Let P and R be distinct polyhedra. Then $rep(P) \neq rep(R)$ or $\{P, R\} = \{\emptyset, \mathbb{R}^3\}$.*

Proof: Let P and R be distinct polyhedra. We may assume w.l.o.g. that $P \backslash R \neq \emptyset$. If $P = \mathbb{R}^3$ then there is nothing to show. Otherwise, let x and y be points with $x \in P \setminus R$ and $y \notin P$. Let z be the first point on the ray from x to y that belongs to the boundary of either P or R. We claim that $Pyr_P(z) \neq Pyr_R(z)$. If $x \neq z$ this follows from the observation that the open line segment with endpoints x and z is contained in P and is disjoint from R, and if $x = z$ this follows from $x \in P \setminus R$. Also, z belongs to a low-dimensional face of either P or R. Thus $rep(P) \neq rep(R)$. □

We propose to store a polyhedron as the collection of its low-dimensional faces together with their local pyramids. So we need data structures for local pyramids and faces.

Let $Pyr_P(x)$ be the local pyramid of point x and let $S(x)$ be a sphere with center x. The intersection of $S(x)$ with $Pyr_P(x)$ is a planar graph embedded into $S(x)$ that we denote $G_P(x)$. The nodes, arcs, and regions of this graph[5] correspond to the edges, facets, and three-dimensional faces of $Pyr_P(x)$ respectively. The graph $G_P(x)$ may consist of a single edge and no vertex. In this case we call the unique edge a *selfloop*. For each feature (=node, arc, or region) f of the graph we have a label *inP*(f) indicating whether the feature is contained in $Pyr_P(x)$. We also have such a label for the point x. We call $G_P(x)$ together with these labels the local graph of x and also use $G_P(x)$ to denote it. The following lemmata characterize local graphs and give a criterion to determine the dimension of the face containing its center. We use the phrase *to classify* x to mean to determine the dimension of the face containing x.

Lemma 8. *The following properties hold for every local graph.*

a) *Every arc of G is part of a great circle.*
b) *For every arc a of G there is a region r incident to a with* inP(a) $\neq$ inP(r).
c) *For every node v of G there is an arc or region f incident to v with* inP(v) $\neq$ inP(f).
d) *For every node v of G of degree 2 with two coplanar arcs a_1 and a_2 incident to it the three labels* inP(a_1), inP(v), *and* inP(a_2) *are not identical.*

Moreover, any graph G (embedded into a sphere) satisfying properties a to d above is the local graph $G_P(x)$ for some polyhedron P and some point x.

[5] We reserve the words vertex, edge, and facet for polyhedra.

Lemma 9. *Let G be the local graph of some point x with respect to some polyhedron P. The point x belongs to a 3-dimensional face of P iff G is empty and the unique region of G has the same label as x. It belongs to a facet of P if G consists of a single selfloop that, in addition, has the same label as x. It belongs to an edge of P if G has exactly two nodes and these nodes are antipodal and have the same label as x (there can also be an arbitrary number of arcs connecting the two nodes). In all other cases, x is a vertex of P.*

We now complete the description of the local-graphs-data-structure. A vertex is represented by its coordinates and its local graph. An edge is represented by the equation of its affine hull, an ordered sequence of open line segments comprising the points belonging to the edge, and the local graph of the edge. The endpoints of these line segments are the 0-dimensional faces of the edge. They correspond to vertices of P.

A facet is represented by the equation of its affine hull, the set of points belonging to the facet, its local graph, and its set of vertices and edges. The set of points of the facet is stored as a straight-line planar graph. For each region of this planar graph there is a label indicating whether the region belongs to the facet or not. The vertices (0-dimensional faces) of a facet are precisely the nodes of this planar graph. The edges (1-dimensional faces) of a facet partition the arcs of the planar graph. If two arcs belong to the same edge of the facet then the arcs are collinear (the converse is not true). We associate with each arc the edge containing it and with each edge (of the facet) the ordered sequence of arcs contained in the edge.

We also store cross links between the different occurrences of the same object. For example, with every edge e of a facet f we associate the set of edges and vertices of P comprising e. If $\overline{e}$ is an edge or vertex of P contributing to e then the item representing $\overline{e}$ in this set points to the arc in the local graph $G_P(\overline{e})$ corresponding to f. The arc points back to the edge of the facet. Similarly, every node x of a local graph $G_P(v)$ for a vertex v of P points to the edge segment corresponding to that node, ...

The *size* of a polyhedron P is defined as the size of its local-graphs-data-structure and is denoted $|P|$. It is proportional to the number of incidences between the faces of P.

For convex polyhedra we also use the *hierarchical representation* introduced by Dobkin and Kirkpatrick [DK82]. It requires linear space and can be constructed in linear time. It supports intersection queries with lines and planes in logarithmic time. An intersection query with a line returns the intersection (a line segment) and an intersection query with a plane returns an arbitrary point of the boundary of the intersection (if any). If the plane intersects the polyhedron in a vertex, edge, or facet then this fact is also reported.

3 The Intersection Algorithm

We show how to compute $P \cap C$, where P is an arbitrary polyhedron and C is a convex polyhedron, in time $O((|P|+|C|+|P \cap C|) \cdot \log(|P|+|C|+|P \cap C|))$. We refer to this bound as our *target time* and use N to denote $|P|+|C|+|P\cap C|$, i.e., the combined input and output size. A polyhedron is called *convex* if with any two points the entire line segment connecting the two points is contained in it. A convex polyhedron is not necessarily closed. Our algorithm is based on the following fact.

Fact 10. [Nef78] *Let P and C be polyhedra. Then every face of $P\cap C$ is the union of intersections of faces of P and C.*

If one of the polyhedra is convex then a slight ly stronger property holds (cf. Figure 3).

Lemma 11. *Let P be a polyhedron and let C be a convex polyhedron. Then every face of $P\cap C$ is the intersection of one face of C with the union of some faces of P.*

We use different strategies for intersecting faces depending on the dimension of the faces. In sections 3.2 to 3.4 we discuss how to intersect the vertices, edges, and facets of P with the faces (of all dimensions) of C. All three sections rely on a subroutine to intersect two local graphs that we introduce in section 3.1. In section 3.5 we show how to intersect the full–dimensional faces of P with the faces (of all dimensions) of C and finally in section 3.6 the information gained previously is put together to construct the local–graphs–data–structure for $P\cap C$.

3.1 Intersecting Two Local Pyramids

We show how to compute $G_{P\cap C}(x)$ from $G_P(x)$ and $G_C(x)$ for a point x and how to classify x in time $O((|G_P(x)|+|G_C(x)|)\cdot\log(|G_P(x)|+|G_C(x)|))$. First sweep a plane through the sphere centered at x and intersect the two graphs $G_P(x)$ and $G_C(x)$. Let $G(x)$ be the resulting graph. We also compute the label *inP* for each feature of $G(x)$ and for x. Since $G_C(x)$ is either trivial (if x is not a vertex of C) or corresponds to a convex cone, any arc of $G_P(x)$ can intersect at most two arcs of $G_C(x)$. Thus $|G(x)|=O(|G_P(x)|+|G_C(x)|)$ and hence $G(x)$ can be computed within the time bound stated above. An alternative strategy for computing $G(x)$ is to intersect any pair of features of the two local graphs. This takes time $O(|G_P(x)|\cdot|G_C(x)|)$ and is to be preferred if one of the two local graphs has constant size.

The graph $G(x)$ determines the local pyramid of x with respect to $P\cap C$ but it may contain spurious features, i.e., features violating one of the conditions (b) to (d) of Lemma 8. Simply remove all these features (ensure condition (b) first, then (c) and finally (d)) and obtain $G_{P\cap C}(x)$. Lemma 9 can then be used to classify the point x. Simplification and classification take time $O(|G(x)|)$.

3.2 Vertices

Let v be a vertex of P. Locate v with respect to C in time $O(\log|C|)$. This decides whether v belongs to C and determines the local graph $G_C(v)$. Next compute $G_{P\cap C}(v)$ as described in section 3.1 and classify x.

All of this takes time $O(|G_P(v)| \log N)$ except if v is also a vertex of C in which case it takes time $O((|G_P(v)| + |G_C(v)|) \cdot \log N)$. Summation over all vertices of P shows that this step stays within the target time.

3.3 Edges

Let e be an edge of P. First intersect $aff(e)$ with C in time $O(\log |C|)$ and obtain either an empty intersection or a line segment $\overline{uv}$ ($u = v$ is possible) contained in $aff(e)$. Intersect $\overline{uv}$ with e and obtain a number of subsegments of $\overline{uv}$. All endpoints of these subsegments except maybe u and v are vertices of P and hence were already treated in section 3.2. Next determine the local graphs $G_C(u), G_C(v)$, and $G_C(x)$ where x is an arbitrary point between u and v[6], and from this and the local graph of e compute $G_{P\cap C}(u), G_{P\cap C}(v)$, and $G_{P\cap C}(x)$.

The computation of the local graphs takes time $O(|G_P(e)|)$ except if either u or v is a vertex of either P or C. The vertices of P were already accounted for in section 3.2. For vertices u (or v) of C that are not also vertices of P the time required is $O((|G_P(e)| + |G_C(u)|) \cdot \log(|G_P(e)| + |G_C(u)|))$. Since any vertex of C can lie on at most one edge of P (recall that edges are relatively open sets) we conclude that the total cost of treating the edges of P is within our target.

3.4 Facets

Let s be a facet of P. It is given as a polygonal subset of $aff(s)$ and by its local graph $G_P(s)$. Our goal is to intersect s with all faces of C.

There is a simple but inefficient way to do so. $Aff(s)$ intersects $clos(C)$ in a convex polygonal region that we call R. We could explicitly compute R, then use plane sweep to intersect s and R, and finally compute the local graphs of all features of the intersection by the method of section 3.1. Unfortunately, this approach exceeds our target time[7]. We now describe an alternative approach that only looks at those parts of R that actually contribute to the output.

First intersect $aff(s)$ with C and determine an arbitrary point z of R. This has cost $O(\log |C|)$ per facet of P and hence total cost $O(|P| \cdot \log N)$. If R is empty then we are done.

If R is a vertex of C, say v, determine whether v belongs to s in time $O(|s|)$. If $v \in s$ then compute $G_{P\cap C}(v)$ and classify v in time $O(G_C(v))$: either $v \in ext(P \cap C)$ or v is a vertex of $P \cap C$. Since any vertex of C can lie in at most one facet of P the total cost of this case is $O(|C|)$.

Assume from now on that R is nonempty and not a vertex of C, i.e., either an edge or a facet of C or a planar cross section of C. If R is an edge of C we may also assume that R is not contained in the affine hull of any edge of P (otherwise, the required information was already computed in the previous

[6] $G_C(x)$ does not depend on the choice of x.

[7] Assume for example that C is a polygonal cylinder with n facets and that P consists of m disjoint rectangles contained in parallel planes orthogonal to the axis of C. Then the naive approach would take time $\Omega(m \cdot n)$ even if P and C are disjoint.

sections). The boundary of s consists of one or more polygons (not necessarily simple). The information gained in the previous sections permits to classify these polygons into three classes: those that intersect *bd* R, those that are contained in *bd* R, and those that either contain *bd* R or are disjoint from *bd* R and do not contain it. The last class can be split by locating an arbitrary point of R, e.g., the point z determined above, with respect to each polygon in the class. The classification process takes time $O(\log |C| + |s|)$ for the facet s and hence total time $O(|P| \log N)$.

We are now in a position to compute $\overline{s} = s \cap R$ in time $O((|s| + |\overline{s}|) \cdot \log(|s| + |\overline{s}|))$. Note that all intersections between *bd* s and R are known at this point and if *bd* R does not intersect *bd* s it is also known whether *bd* R contributes to *bd* $\overline{s}$ at all. Sort for each edge of R (that is intersected at least once) the intersections with *bd* s (this divides the edge into segments) and then explore *bd* $\overline{s}$ starting at the intersections between *bd* s and *bd* R. It is not too hard to see that only those parts of *bd* R need to be inspected that actually contribute to *bd* $\overline{s}$.

At this point we have computed the point set $\overline{s}$. We next compute the local graphs of all points in this set and on its boundary. Consider first a point $x \in$ *rel int*$(\overline{s})$. The local graph $G_{P \cap C}(x)$ does not depend on x and can be computed from $G_P(s)$ and $G_C(y)$, where y is an arbitrary point in *rel int*(R), in constant time (since $G_P(s)$ and $G_C(y)$ both have constant size). Consider next a point x in the boundary of $\overline{s}$. If x is either a vertex of P or belongs to an edge of P then $G_{P \cap C}(x)$ is already known. Otherwise, $x \in s \cap$ *bd* R and $G_{P \cap C}(x)$ can be computed from $G_C(x)$ in time $O(|G_C(x)|)$.

We still need to argue that we stay within our target time. We distinguish cases according to whether R is an edge of C or a facet of C or a cross section of C and sum the cost of the three cases separately.

If R is an edge of C then $|\overline{s}| = O(|s|)$. The total cost of computing all the $\overline{s}$'s is therefore $O(|P| \cdot \log N)$. The computation of the local graphs of the relative interiors of all the $\overline{s}$'s takes time $O(|P|)$. For the computation of the local graphs of the boundary points observe that $x \in s \cap$ *bd* R implies that x is a vertex of C and that every vertex of C is contained in at most one facet of P. The total cost for computing the local graphs of boundary points is therefore $O(|C|)$.

If R is a facet of C then $R \subseteq$ *aff*(s) for at most 6 facets s of P [8] and $|\overline{s}| = O(|s| + |R|)$ [9]. The total cost of computing all the $\overline{s}$'s is therefore $O((|P| + |C|) \cdot \log N)$. The computation of the local graphs of the relative interiors of all the $\overline{s}$'s takes time $O(|P|)$. For the computation of the local graphs of the boundary points observe first that $x \in s \cap$ *bd* R implies that x is either a vertex of C or belongs to an edge of C. For the vertices we argue as in the previous paragraph. For a point x on an edge of C the local graph $G_{P \cap C}(x)$ can be computed in constant time and hence time $O(|\overline{s}|)$ suffices for all edges of C contributing to the boundary of $\overline{s}$. The total cost computing the local graphs of boundary points is therefore $O(|P| + |C|)$.

If R is a cross section of C then all edges and vertices of *bd* $\overline{s}$ are also edges

---

[8] Note that $R \subseteq$ *aff*$(s) \cap$ *aff*(s') implies *aff*$(s) =$ *aff*(s').

[9] since R is convex

and vertices of $P \cap C$. The total cost of computing all the $\bar{s}$'s and their local graphs is therefore $O((|P| + |C| + |P \cap C|) \cdot \log N)$. For the local graphs of the boundary points $x \in s \cap bd\ R$ observe that $G_{P \cap C}(x)$ can be computed in constant time if x is not a vertex of C and in time $O(|G_C(x)|)$ if x is a vertex of C, and that any vertex of C can be contained in at most one facet of P. The total cost of computing local graphs of boundary points is thus $O(|P| + |C| + |P \cap C|)$.

3.5 Full–dimensional Faces

We show how to compute the intersections between *int*(P) (the other three–dimensional face of P is not important) and the faces of C. We already know *bd* $P \cap$ *bd* C at this point and also the local graph for each point in *bd* $P \cap$ *bd* C.

Superimpose *bd* $P \cap$ *bd* C on *bd* C and obtain a planar graph embedded into the boundary of C. A traversal of this graph determines all points in *int* $P \cap$ *bd* C. The local graphs of these points are just their local graphs with respect to C.

3.6 Putting It All Together

At this point we have computed all intersection between faces of P and C involving at least one low–dimensional face and hence know *bd*$(P \cap C)$. We also know the local graphs of all points on the boundary. We now have to build the low–dimensional faces of $P \cap C$. According to Lemma 11 each face of $P \cap C$ is the union of intersections of faces of P and faces of C. We still have to perform the unions.

Let us consider the vertices first. For some vertices of P it may be the case that the new local graph indicates that the vertex now belongs to an edge or facet. If the vertex now belongs to an edge we join the two edge segments incident to the vertex[10]. If the vertex now belongs to a facet then we remove it from the description of the facet.

Consider the edges next. Edges that become part of a facet are removed from the description of the facet. We also have to determine whether several edges have to be merged into one[11]. Describe each edge by its affine hull and by a suitable encoding of its local graph (e.g., a list consisting of the coordinates of the vertices followed by a list of the edges) and determine all edges with the same description (e.g., by building a trie [Meh84, section III.1.1] for the descriptions. The nodes of this trie are dictionaries and therefore it takes time $O(l \log N)$ to insert a description of length l into the trie). Then unite all edges with the same description.

Finally, consider the facets. Determine all facets with the same affine hull and the same local graph (use a dictionary) and unite these facets (e.g., by sweeping them).

The local–graphs–data–structure of $P \cap C$ is now available. The assembly phase stays within our target time since it essentially boils down to a constant number of dictionary operations for each feature of P, C, and $P \cap C$.

10 These segments may belong to the same edge or to different edges.
11 These edges do not necessarily share an endpoint.

References

[BN88] H. Bieri and W. Nef. Elementary set operations with *d*-dimensional polyhedra. In *Computational Geometry and its Applications*, volume 333 of *Lecture Notes in Computer Science*, pages 97–112. Springer-Verlag, 1988.

[DK82] D. P. Dobkin and D. G. Kirkpatrick. Fast detection of polyhedral intersection. In *Proc. 9th Internat. Colloq. Automata Lang. Program.*, volume 140 of *Lecture Notes in Computer Science*, pages 154–165. Springer-Verlag, 1982.

[Dob90] K. Dobrindt. Algorithmen für Polyeder. Master's thesis, Fachbereich Informatik, Universität des Saarlandes, June 1990.

[EM85] H. Edelsbrunner and H. Maurer. Finding extreme points in three dimensions and solving the post office problem in the plane. *Information Processing Letters*, 21:39–47, 1985.

[EM90] H. Edelsbrunner and E.P. Mücke. Simulation of simplicity: A technique to cope with degenerate cases in geometric algorithms. *ACM Transactions on Graphics*, 9:66 – 104, 1990.

[Hof89] C.H. Hoffmann. *Geometric and Solid Modeling.* Morgan Kaufmann, San Mateo, Calif., 1989.

[Män88] M. Mäntylä. *An Introduction to Solid Modeling.* Computer Science Press, Rockville, Md., 1988.

[Meh84] K. Mehlhorn. *Data Structures and Efficient Algorithms.* Springer Verlag, 1984.

[MS85] K. Mehlhorn and K. Simon. Intersecting two polyhedra one of which is convex. In *Proc. Found. Comput. Theory*, volume 199 of *Lecture Notes in Computer Science*, pages 534–542. Springer-Verlag, 1985.

[Nef78] W. Nef. *Beiträge zur Theorie der Polyeder.* Herbert Lang Bern, 1978.

[PS85] F. Preparata and M.I. Shamos. *Computational Geometry: An Introduction.* Springer, New York Berlin Heidelberg Tokyo, 1985.

[Req80] A.A.G. Requicha. Representations for rigid solids: Theory, methods, and systems. *ACM Computing Surveys*, 12:437 – 464, 1980.

A Appendix

Let M a subset of $\mathbb{R}^d$. We will denote the complement of M by $cpl(M)$.

Definition 12. The *affine hull* $aff(M)$ of M is the intersection of all flats $N \subseteq \mathbb{R}^d$, which contain M. The *closure* $clos(M)$ of M is the intersection of all closed supersets of M. The *interior* $int(M)$ of M is the union of all open subsets of M. The *exterior* $ext(M)$ of M is defined as $int(cpl(M))$. The *relative interior* $rel\ int(M)$ of a point set M is the union of all relatively open subsets of M, i.e., open with respect to the affine hull of M.

Definition 13. M is *regular*, if $M = clos(int(M))$ and if for each element $x \in M$, there is a neighborhood U of x in $\mathbb{R}^d$ such that $int(U \cap M)$ and $ext(U \cap M)$ are connected.

In other words, a set M in $\mathbb{R}^d$ is regular, if M is closed, does not contain dangling or isolated parts of dimension $< d$, and the interior and the exterior of M are connected in a neighborhood of each element of M.

B Figures

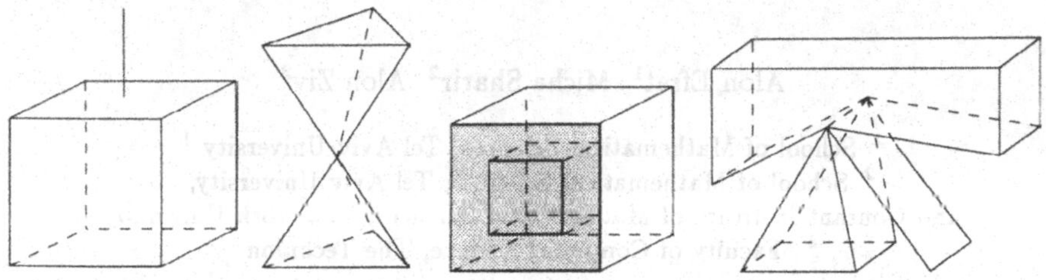

Fig. 1. Examples of general polyhedra

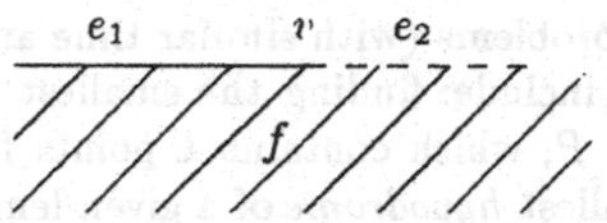

Fig. 2. Example in $\mathbb{R}^2$: the e_1 belongs to the polyhedron and e_2 does not.

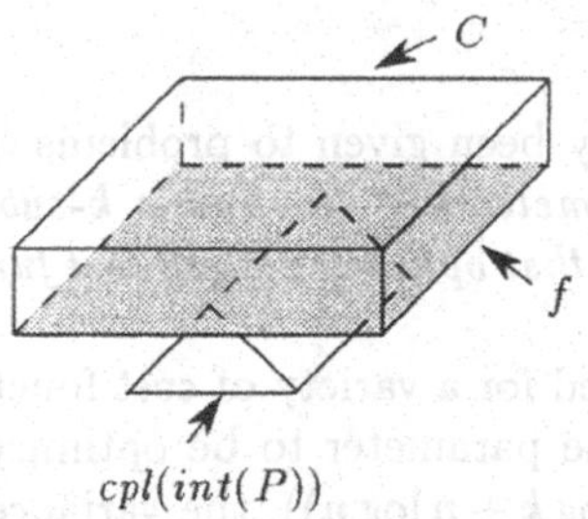

Fig. 3. Illustration of Lemma 11: the facet f of $P \cap C$ is the union of the intersection of a facet of C with $int(P)$, an edge of P and two vertices of P

Computing the Smallest k-Enclosing Circle and Related Problems*

Alon Efrat[1] Micha Sharir[2] Alon Ziv[3]

[1] School of Mathematical Sciences, Tel Aviv University †
[2] School of Mathematical Sciences, Tel Aviv University, and Courant Institute of Mathematical Sciences, New York University
[3] Faculty of Computer Science, The Technion

Abstract. We present an efficient algorithm for solving the *"smallest k-enclosing circle"* (*kSC*) problem: Given a set of n points in the plane and an integer $k \leq n$, find the smallest disk containing k of the points. We present several algorithms that run in $O(nk \log^c n)$ time, where the constant c depends on the storage that the algorithm is allowed. When using $O(nk)$ storage, the problem can be solved in time $O(nk \log^2 n)$. When only $O(n \log n)$ storage is allowed, the running time is $O(nk \log^2 n \log \frac{n}{k})$. The method we describe can be easily extended to obtain efficient solutions of several related problems (with similar time and storage bounds). These related problems include: finding the smallest homothetic copy of a given convex polygon P, which contains k points from a given planar set, and finding the smallest *hypodrome* of a given length and orientation (formally defined in Section 4) containing k points from a given planar set.

1 Introduction

Much attention has recently been given to problems of the form: *"Given a set S of n objects and a parameter $k \leq n$, find a k-subset (namely, a subset of cardinality k) of the objects that optimizes some cost function, among all possible k-subsets."*

This problem was studied for a variety of cost functions. Aggarwal et. al. [4] solve this problem when the parameter to be optimized is the diameter of the k-subset (in time $O(k^{2.5} n \log k + n \log n)$), the variance of the k-subset (in time $O(k^2 n + n \log n)$), the size of an axis-parallel enclosing square, or the perimeter of an axis-parallel enclosing rectangle (both in time $O(nk^2 \log n)$; the solution to the first problem has recently been improved to $O(n \log^2 n)$; see [6]). In [9],

† email: alone@math.tau.ac.il

* Work on this paper by the second author has been supported by NSF Grant CCR-91-22103, and by grants from the U.S.-Israeli Binational Science Foundation, the G.I.F., the German-Israeli Foundation for Scientific Research and Development, and the Fund for Basic Research administered by the Israeli Academy of Sciences.

Eppstein finds the minimum area k-gon whose vertices are points of S (in time $O(n^2 \log n + 2^{6k}n^2)$), and the minimum area convex polygon containing k points of S (in time $O(k^3n^2 + n^2 \log n)$).

In this paper we give a fast algorithm for the (apparently simpler) *"smallest k-enclosing circle" (kSC) problem: Given a set S of n points in the plane and an integer $k \leq n$, find the smallest disk containing k points of S.*

The problem arises in several applications. For example, suppose we want to partition the given set S into *clusters* of k points each. We can find the smallest k-enclosing circle of S, remove from S the k-subset that lies within the circle (this will be the first cluster), and repeat the procedure for the remaining set. This is a reasonable heuristic for clustering; it has the advantage that we can stop the process as soon as the smallest k-enclosing circle becomes too large, concluding that the remaining set can no longer be decomposed into 'sufficiently-dense' clusters of size k. We also note that the case $k = n$ gives us the well-known problem of finding the smallest enclosing disk of S, which can be solved in linear time [15].

We present an $O(nk \log^c n)$-time algorithm for solving the problem, where the constant c depends on the storage that the algorithm is allowed. With only $O(n \log n)$ storage, the algorithm solves the kSC problem in time $O(nk \log^2 n \log \frac{n}{k})$. When we allow $O(nk)$ storage, the running time slightly improves to $O(nk \log^2 n)$.

We can also apply our technique to obtain efficient solutions for other problems of this kind. Specifically:

- Given a set S of n points in the plane, a parameter $k \leq n$ and a convex polygon P with a constant number of edges, we can compute the smallest homothetic copy of P that contains k points of S, in time $O(nk \log^c n)$, where c is a constant depending as above on the storage allowed for the algorithm.
- Given a set S of n points, a parameter $k \leq n$ and a segment e, we can find a translation of e which minimizes the k-th smallest distance between e and the points of S, in time $O(nk \log^c n)$, where c is as above.

Our algorithms make use of the parametric search technique, introduced by Megiddo in [14]. We assume some familiarity of the reader with it; see [2] for an explanation of the technique and for other geometric applications of it.

The kSC problem for a planar set S is strongly related to higher-order Voronoi diagrams of S. Indeed, It is easy to verify that the center of the smallest circle containing a k-set must be on an edge of the $(k-1)$-order Voronoi diagram, and all candidate circles can be found during the construction of that diagram, The technique presented in this paper has several advantages over the higher-order Voronoi diagram technique: (i) it is simpler; (ii) it admits a storage-efficient solution; (iii) it can be easily extended to solve the related problems listed above; and (iv) it is somewhat more efficient—the best known algorithm for computing k-th order Voronoi diagrams runs in time $O(n^{1+\varepsilon}k)$ [3].

The paper is organized as follows: In Section 2 we present a simple $O(n^2 \log^2 n)$-time algorithm for solving the kSC problem; this algorithm is improved in Section 3 to get an $O(nk \log^c n)$-time algorithm, where c is as defined above. Section 4 extends these results to the computation of smallest (homothetic) k-enclosing polygons and 'hypodromes' (expansions of a segment by a disk).

2 A Naive Approach

In this section we present a simple algorithm that solves the kSC problem in $O(n^2 \log^2 n)$ time and $O(n)$ space. The time bound will be improved later (at the cost of some increase in the storage requirement).

We first introduce some terminology. We assume that we are given a planar set $S = \{p_1, \ldots, p_n\}$ of n points. We also assume that the points of S are in ***general position***, meaning in particular that no four points of S are co-circular. This assumption is not essential, but is made to simplify the description of our algorithms. Given a point p and a radius $r > 0$, we denote the circle of radius r centered at p by $C_r(p)$, and the closed disk bounded by that circle by $B_r(p)$. For any real number $r > 0$, we denote the arrangement of circles $C_r(p)$, $p \in S$, by $\mathcal{A}(S, r)$. Given an arrangement of circles $\mathcal{A}$ and a point x, the ***depth*** of x in $\mathcal{A}$, denoted by $depth_{\mathcal{A}}(x)$, is the number of circles in $\mathcal{A}$ containing x (within the closed disks that they bound). We also define $depth(\mathcal{A}) = \max\{depth_{\mathcal{A}}(x) \mid x \in \mathbb{R}^2\}$. It is obvious that for any point set S and and a positive number r,

$$\max\left\{|S \cap B_r(x)| \,:\, x \in \mathbb{R}^2\right\} \;=\; depth(\mathcal{A}(S, r))\,.$$

Hence, we can rephrase our goal as "find the smallest value of r for which $depth(\mathcal{A}(S, r)) = k$". Let r^* denote that radius. (As will follow from the analysis given below, our algorithms will also be able to produce a point having depth k in $\mathcal{A}(S, r^*)$, thus solving the original kSC problem.) This observation leads to a rather straightforward algorithm (described below), based on the parametric searching paradigm, for computing r^*. The first step in the design of such an algorithm is to produce an 'oracle' for determining whether any given radius r is too big or too small, as compared to r^*.

2.1 Oracle for the naive approach

Our algorithm, which is based on the parametric search technique of Megiddo [14], performs an implicit binary search on certain 'critical' values of r to find r^* among them. In order to do this, the algorithm uses an 'oracle' procedure which is able to answer queries of the form "given a value r, is r^* less than, equal to, or greater than r?"

As we have observed, what we need to do is to determine whether $depth(\mathcal{A}(S, r))$ is less than, equal to, or greater than k. In the first case, it is clear that $r < r^*$; similarly, in the third case we have $r > r^*$ (this follows from our general position assumption). For the second case (when the depth is exactly k), we only know that $r \geq r^*$. With some care, one can also differentiate between these two possibilities, by determining whether $depth(\mathcal{A}(S, r))$ becomes strictly smaller than k when r is decreased by an arbitrarily small amount.

Our oracle is based on the observation that if there is a point of depth k in $\mathcal{A}(S,r)$, then there must exist a point of depth $\geq k$ on one of the circles; moreover, if at least two of the circles in $\mathcal{A}(S,r)$ intersect, one of these intersection points must also be of depth $\geq k$. Hence it suffices to construct the vertices of the arrangement $\mathcal{A}(S,r)$, and compute the maximum depth of a vertex.

In order to reduce the amount of storage required by the algorithm, and also for the purpose of obtaining a parallelizable algorithm, which is required by the parametric searching technique, we implement the oracle by applying the following procedure to each of the circles $C_r(p_i)$, for $i = 1, \ldots, n$. The procedure computes the maximum depth of points that lie on its input circle; the maximum of all these output depths is the required *depth*($\mathcal{A}(S,r)$).

Procedure *Calc-Depth*(r, i)

1. Find the circles $C_r(p_j)$ that intersect $C_r(p_i)$. If no such circle exists, return 1 and exit.
2. Find the depth of the leftmost point of $C_r(p_i)$.
3. Sort the intersection points between $C_r(p_i)$ and the other circles by their (clockwise) order along $C_r(p_i)$, starting at the leftmost point of that circle.
4. Scan the sorted list of intersection points, and compute the depth of each of these points, observing that the depth of a point along $C_r(p_i)$ changes by ± 1 as we cross any point in the list, depending on whether we enter or exit a disk at that point.
5. Return the maximum depth computed at the preceding step.

It is easy to see that the time required by a call to *Calc-Depth* is $O(n \log n)$, so the total time required by a call to our oracle is $O(n^2 \log n)$. Since each circle is processed separately, the storage required is $O(n)$. However, in what follows we will use the oracle in a different, more fragmented manner, as described in the following subsection.

2.2 The generic algorithm

Fix an index $i = 1, \ldots, n$. Denote the smallest radius r for which there exists a point on $C_r(p_i)$ with depth k by r_i^* (note that r_i^* is well defined for any k). Following the parametric searching paradigm, we attempt, for each $i = 1, \ldots, n$, to perform *Calc-Depth*(r_i^*, i) in a generic manner, without knowing the value of r_i^*. For this, we need to resolve the comparisons that depend on r_i^* which are executed by the procedure. Each of these resolutions will further restrict the range in which r_i^* can lie, so that at the end of the generic execution, the control flow of the procedure is the same for all values of r in this range, which easily implies that r_i^* is the smallest point in the final range. We apply this procedure to all indices $i = 1, \ldots, n$ and compute $r^* = \min_i r_i^*$.

In more detail, this is done as follows. The first step of *Calc-Depth*(r_i^*, i) is to find the circles that intersect $C_{r_i^*}(p_i)$. This can be achieved in a constant number of parallel steps using $O(n)$ processors (the j-th processor has to determine whether $r \geq r_{ij} \equiv \frac{1}{2}|p_i - p_j|$). Similarly, the second step can also be performed in constant time using $O(n)$ processors.

In the third step of *Calc-Depth* we need to sort the intersection points of $C_{r_i^*}(p_i)$ with the other circles of that radius. Here we use (a serial simulation of) a parallel sorting scheme that uses $O(n)$ processors and $O(\log n)$ parallel steps, such as the scheme of [5]. Each parallel step of the sorting attempts to perform $O(n)$ comparisons, each asking for the relative order along $C_{r_i^*}(p_i)$ of two of its intersections with, say $C_{r_i^*}(p_j)$ and $C_{r_i^*}(p_k)$. Again, each comparison can change only at one critical value of r. Hence we obtain, for each parallel step of the algorithm, $O(n)$ critical values r_{ijk} for the radius and we proceed to search among them as above.

Performing a generic simulation of the fourth stage of *Calc-Depth* is unnecessary, as it gives us no further restriction on the interval in which r^* must lie.

Concerning the running time of this generic simulation, the first two stages require time $O(n \log^2 n)$ (this time is dominated by $O(\log n)$ calls to *Calc-Depth* with specific radii). The third step takes $O(n \log^3 n)$ time, but using a standard trick due to Cole [7], this can be reduced to $O(n \log^2 n)$. Repeating this procedure for each p_i, we obtain r^* as the minimum of all the r_i^*'s, at a total cost of $O(n^2 \log^2 n)$ (with $O(n)$ storage). We will improve this bound in subsequent sections of the paper.

3 The Improved Algorithm

The main bottleneck of the previous algorithm is that it has to check for intersections between every pair of circles. To avoid this, we first claim that there exists some initial radius r_{init}, for which $depth(\mathcal{A}(S, r_{init}))$ is at least k, but the number of intersections between the circles in this arrangement is only $O(nk)$. The existence of such a radius is a consequence of the following lemma.

Lemma 1 (Pach; see [18]). *If an arrangement $\mathcal{A}$ of n circles in the plane contains at least $9nk$ intersecting pairs of circles, then $depth(\mathcal{A}) \geq k$.*

The lemma allows us to replace, temporarily, the original problem with the following, simpler problem: "Given S and k as above, find the smallest radius r_{init} for which the number of pairs of intersecting circles in $\mathcal{A}(S, r_{init})$ is at least $9nk$". (By assuming appropriate general position, the number of intersecting pairs of circles in $\mathcal{A}(S, r_{init})$ is exactly $9nk$.) Lemma 1 implies that $r^* \leq r_{init}$. Our solution to this problem will also yield, for each circle $C_{r_{init}}(p)$, a list $L(p)$ of all circles intersecting $C_{r_{init}}(p)$ in $\mathcal{A}(S, r_{init})$; the total length of these lists is $O(nk)$.

After computing r_{init}, we will execute a variant of the algorithm described in the previous section, with the following two modifications: (i) we allow oracle calls only for $r \leq r_{init}$; and (ii) we use the lists $L(p)$ to find all vertices of $\mathcal{A}(S, r)$ along each of the circles $C_r(p)$ (this is done both in the generic algorithm and in the oracle calls). This is easily seen to reduce the running time of the previous algorithm to $O(nk \log^2 n)$, excluding the initial stage that computes r_{init}.

We now describe in detail this initial stage. It is also based on the parametric seacrhing technique. As above, we first describe the oracle used in the binary searches and then describe the generic simulation at the unknown radius r_{init}.

Before going into the analysis, we remark that r_{init} can be computed in a straightforward manner using the recent algorithm of Lenhof and Smid [12], which produces the t closest pairs in a given planar set of n points, in (optimal) time $O(t + n \log n)$. If we apply this algorithm with $t = 9nk$, and let r_{init} be half the t-th smallest distance, we obtain the required r_{init}, in time $O(n(k + \log n))$. However, we develop here an alernative technique because: (i) it can be tuned so that it uses only $O(n \log n)$ storage; (ii) it can be easily generalized to shapes other than circles, as is discussed in the next section. Also, the saving in time in this stage does not change the overall asymptotic running time of the algorithm, which is dominated by the $O(nk \log^2 n)$ cost of the subsequent phase.

3.1 The oracle: a simple version

The oracle has to answer queries of the form: "given S, k and r as above, determine whether the number of pairs of intersecting circles in $\mathcal{A}(S, r)$ is less than, equal to, or larger than $9nk$". Depending on the answer to the query, we have respectively $r < r_{init}$, $r = r_{init}$, and $r > r_{init}$. The oracle is performed using a straightforward line-sweep algorithm that counts the number of intersections between the circles. If we execute the oracle for a value of r that is greater than r_{init}, the number of intersections can be too big, and we therefore stop the oracle as soon as it encounters more than $9nk$ pairs of intersecting circles, concluding that $r > r_{init}$ in this case. Hence the running time of the oracle is $O(nk \log n)$, and it uses $O(n)$ storage.

3.2 The oracle: an improved parallelizable version

We next describe another algorithm that determines whether the number of pairs of intersecting circles in the arrangement is less than, equal to, or larger than $9nk$, which is easier to parallelize. The algorithm uses only $O(nk)$ processors, and runs in $O(\log^c n)$ parallel steps (where the constant c depends on the storage we have at hand).

The algorithm requires a data structure $DS(X)$ for answering efficiently queries of the following form, for a given planar set of points X: *"Given a query point p and a parameter r, determine whether $X \cap B_r(p)$ is empty."* To this end we use the Voronoi diagram of X, augmented with an efficient point location structure. To find whether $X \cap B_r(p) = \emptyset$, we simply find the point $q \in X$ nearest to p, and check whether the distance $d(p, q)$ is larger than r.

We build a fully balanced binary tree $\mathcal{T}$. The leaves of $\mathcal{T}$ represent the points of S, and each node v of $\mathcal{T}$ represents the subset S_v of S consisting of all points stored at the leaves of the subtree rooted at v. With each node v of $\mathcal{T}$ we also store $DS(S_v)$.

We note that $\mathcal{T}$ can be constructed in time $O(n \log n)$, as follows. We first sort the points of S by their x-coordinates, and put them in the leaves of $\mathcal{T}$ in this order. We then construct $\mathcal{T}$ bottom-up; at each new node v we merge $DS(S_{v_1})$ and $DS(S_{v_2})$, where v_1 and v_2 are the children of v. The respective Voronoi diagrams can be merged in linear time, since the sets S_{v_1} and S_{v_2} are separated [16]. Preparing the resulting diagram for efficient point location can also be performed in linear time, using the technique of [8] or of [17]. Hence the total time needed for constructing $\mathcal{T}$ is $O(n \log n)$. Note also that $\mathcal{T}$ requires $\Theta(n \log n)$ storage.

Our goal now is to find, for each $p \in S$, the set of points q such that $C_r(p)$ intersects $C_r(q)$ (or rather to count the size of this set). This is exactly the set of points for which $d(p,q) \le 2r$. We want to add up the sizes of these sets, and stop the oracle as soon as the sum first exceeds $9nk$. The algorithm uses a total of (at most) $18nk + n$ processors. Initially, it allocates a single processor to each point $p_i \in S$, and starts the procedure *Count* (described below) with the root of $\mathcal{T}$ and the point p_i as parameters.

The procedure *Count* gets a node v of $\mathcal{T}$ and a point $p \in S$ as parameters; it is also allocated a processor which executes it. The procedure runs as follows:

Procedure *Count*(v, p)

1. If v is a leaf, check whether the point p_v stored at v is different from p, and if its distance from p is $\le 2r$. Add in this case 1 to a global count N.
2. Otherwise, let v_1 and v_2 be the two children of v. Use $DS(S_{v_1})$ and $DS(S_{v_2})$ to determine whether $S_{v_1} \cap B_{2r}(p)$ and $S_{v_2} \cap B_{2r}(p)$ are nonempty. Call *Count* recursively with v_1 (resp. v_2) and p, if the first (resp. second) intersection is nonempty. Allocate a new processor if both calls are needed (so that one call uses the old processor allocated at v and the other call uses the new processor).

It can be readily verified that the value N reported by the algorithm (as accumulated from all the calls to *Count*) is twice the number of pairs of intersecting circles in $\mathcal{A}(S, r)$. Also, this number is at least as large as the number of processors at each level (except for the root) minus n; indeed, we perform a recursive call with a point p at a node v only if we are guaranteed that $B_{2r}(p)$ contains at least one point q of S_v. If $p \ne q$, this pair of points will generate two processors at each level, and will eventually cause N to be increased by 2. If $p = q$, this pair generates a single processor at each level, without affecting N, for a total of n 'spurious' processors per level. Therefore, if at the beginning of the processing of a level the total number of processors allocated is $> 18nk + n$, the algorithm can stop immediately and report that the number of pairs of intersecting circles in $\mathcal{A}(S, r)$ is larger than $9nk$.

3.3 The generic algorithm

We next describe the generic algorithm that simulates the procedure *Count* at the unknown value r_{init}. We first describe a version that is allowed to use $O(nk)$ storage, and then explain how to modify it to reduce the storage to $O(n \log n)$, at the cost of a slight increase in the running time.

The procedure processes each of the $O(\log n)$ levels of $\mathcal{T}$ separately. At an intermediate level, each of the $O(nk)$ processors locates a point p_i in the Voronoi diagram of some subset S_v of S. These locations are independent of r_{init}, so they can be performed explicitly, in $O(\log n)$ time per operation. Then (also at the leaf level) the distance from p_i to its nearest neighbor in S_v is compared with $2r$. We thus need to run a binary search to locate r_{init} among these $O(nk)$ distances, using $O(\log nk) = O(\log n)$ calls to the simple sweep-based version of the oracle, each of which takes time $O(nk \log n)$. Thus the total cost of the generic simulation is $O(nk \log^3 n)$. (Note that, since the generic simulation follows the execution of the oracle at r_{init}, we are guaranteed that the generic execution does not require more than $18nk + n$ processors.)

This can be improved to $O(nk \log^2 n)$ using Cole's trick [7]. This amounts to executing only a constant number of binary search steps at each stage of the algorithm, thereby resolving only some fixed large fraction of the number of comparisons. Nodes whose comparisons were resolved can proceed to the next level while the other nodes are stuck and have to participate again in the next round of binary search. It is easily verified that Cole's technique is indeed applicable here. In particular, one needs to observe that, at any stage of the revised algorithm, where processors can now reside at different levels of the tree, it is still true that if the number of processors exceeds $18nk + n$ then the number of pairs of intersecting circles in $\mathcal{A}(S, r)$ exceeds $9nk$. Hence, arguing as above, the generic execution does not require more than $18nk + n$ processors.

As above, at the end of the generic simulation we obtain an interval where r_{init} can lie, and the smallest endpoint of that interval is r_{init}.

In summary, we obtain a first version of the improved algorithm, which requires $O(nk \log^2 n)$ time and $O(nk)$ storage.

3.4 A space-efficient version

In the version just presented, we have to know explicitly which pairs (p, v) of a point $p \in S$ and a node v of $\mathcal{T}$ are active at any given time, and therefore $\Theta(nk)$ storage is required. The storage requirement can be reduced to $O(n \log n)$, at the cost of a slight increase in running time, as follows. For each $p \in S$, denote by *live*(p) the set of nodes of $\mathcal{T}$ that the algorithm currently searches with point p (at any execution instance). Since we can no longer maintain explicitly all the sets *live*(p), we will reconstruct each of these sets from scratch, whenever such a set is needed (which, fortunately, does not occur too often).

There are some technical difficulties in this approach. To see the issues that can arise, consider first the following initial attempt at the problem. When we process the j-th level of $\mathcal{T}$, we iterate over the points $p \in S$, and, for each point

p, we reconstruct *live*(p) by executing *Count* with p, starting at the root of $\mathcal{T}$ and proceeding until level j is reached. We then collect all critical values of r that the comparisons at level j generate, and compute their median $r_m(p)$. We repeat this process for each $p \in S$, obtaining n median values $r_m(p_1), \ldots, r_m(p_n)$. We compute the median r_m of these n medians, and call the (simple sweep-based) oracle at r_m. This resolves half of the comparisons at *live*(p), for half of the points p of S. (We would, of course, prefer to maintain all the critical values of r for all points p, but this requires too much storage.)

The problem with this approach is that now we have to repeat this procedure, but only with the critical values of r that were not yet resolved. If we still do it one level at a time, we will be spending too much time to resolve all comparisons at each level. To make the technique more efficient, we note that the generic algorithm maintains at all times an interval I where r_{init} is known to lie. Comparisons whose critical values of r lie outside I can be resolved immediately, while comparisons having a critical value inside I can be resolved only by further calls to the oracle. Our revised strategy is thus to proceed, with each $p \in S$ in turn, down the tree $\mathcal{T}$ as deep as possible, until we encounter comparisons that cannot yet be resolved. We now denote by *live*(p) the resulting set of nodes where the comparisons involving p get 'stuck'. To make this method efficient, we use (an appropriate modification of) the weighing technique of Cole [7]. That is, we give each stuck comparison at level j a weight of 4^{-j} (so that comparisons stuck higher in the tree are more important since they need to be resolved more urgently). For each point p, and each stage of the algorithm, we compute the weighted median $r_w(p)$ of the critical values of r at the nodes of *live*(p); we give $r_w(p)$ a weight equal to the sum of the weights of the nodes of *live*(p). We then compute the weighted median r_w of these medians, and call the oracle with r_w, thus resolving comparisons whose total weight is 1/4 of the overall weight of all comparisons. It can be shown, arguing as in [7], that the total weight of stuck comparions is reduced by a constant factor at each stage of the algorithm, so that, after at most $O(\log n)$ stages, we are guaranteed that no comparisons will be stuck, so the generic execution can be completed. Clearly, the storage used by the algorithm is dominated by the size of $\mathcal{T}$, and is thus $O(n \log n)$. (Note that, arguing as in the preceding version of the algorithm, we are guaranteed that the total size of all the sets *live*(p), at any given moment, does not exceed $O(nk)$.)

The running time of the procedure is estimated as follows. The total cost of all oracle calls, as just argued, is $O(nk \log^2 n)$. After each oracle call, we need to re-calculate the sets *live*(p), for each $p \in S$. This cost is $\sum_{p \in S} O(|\mathcal{T}_p| \log n)$, where $\mathcal{T}_p$ is the subtree of $\mathcal{T}$ whose root is the root of $\mathcal{T}$ and whose leaves are the nodes of *live*(p). Thus the total overhead of these re-calculations is $O(\log^2 n \cdot \sum_{p \in S} |\mathcal{T}_p|)$. To obtain a bound on this sum, we note that if a subtree of $\mathcal{T}$ has d leaves then the number of its internal nodes is $O(d \log \frac{n}{d})$. Thus, if *live*(p_i) has d_i nodes, then the corresponding sum is $O(\sum_{i=1}^{n} d_i \log \frac{n}{d_i})$, and we are also given that $\sum_{i=1}^{n} d_i \leq 18nk + n$. Simple calculation shows that the maximum value of the sum is $O(nk \log \frac{n}{k})$. Thus the total overhead of the re-calculations of the sets *live*(p) is $O(nk \log^2 n \log \frac{n}{k})$, which is easily seen to dominate the overall running time of this portion of the algorithm (including the time used by the initial construction of $\mathcal{T}$).

Recall that we have so far been discussing only the initial phase of the algorithm, which computes r_{init}. The rest of the algorithm continues by using the naive oracle and its generic simulation. However, every (generic or explicit) call to *Calc-Depth* with a specific point p requires availability of the corresponding list $L(p)$, and in this version of the algorithm we cannot afford to maintain all these lists simultaneously. We overcome this difficulty as follows: When we start a generic simulation of *Calc-Depth* with a specific $p \in S$, we first go down the tree $\mathcal{T}$ and compute the set *live*(p) at the value $r = r_{init}$; note that this set now consists only of leaves of $\mathcal{T}$ and thus gives the set of all circles $C_{r_{init}}(q)$ that intersect $C_{r_{init}}(p)$. We then proceed with the generic execution of *Calc-Depth* using only the circles given by *live*(p). The extra cost of this final calculation of the sets *live*(p), and the generic execution of the calls to *Calc-Depth* are clearly dominated by the cost of the initial stage that calculates r_{init}. We thus conclude:

Theorem 2. *Given a set S of n points in the plane and an integer $k \leq n$, one can compute the smallest circle containing k points of S in time $O(nk \log^2 n)$ and space $O(nk)$, or in time $O(nk \log^2 n \log \frac{n}{k})$ and space $O(n \log n)$.*

4 Extending the Algorithm to Other Shapes

Let P be a given convex polygon with d sides, where we consider d to be a constant. We wish to extend the technique presented above to solve the following problem: *"Given a set S of n points in the plane and a parameter $k \leq n$, find the smallest homothetic copy of P that contains k points of S."*

We regard P as given in a fixed position in which c lies at the origin. We denote any homothetic copy of P as $\alpha P + v$, where $\alpha > 0$ is the *scaling factor* of P and v is the location of the center c of this copy. It is easily verified that there exists a homothetic copy $\alpha P + v$ of P which contains k points of S if and only if there exists a point that lies in k of the n polygons $\alpha P - p_i$, for $p_i \in S$. We denote by $\mathcal{A}_P(S, \alpha)$ the arrangement formed by these polygons, and assume that S and P are in general position, as above. Then this condition is equivalent to the existence of a vertex of $\mathcal{A}_P(S, \alpha)$ at depth k (defined as above).

We can thus apply the same machinery developed in the preceding sections, with a few necessary modifications. That is, we need an oracle which, given α and k, determines whether there exists a vertex of $\mathcal{A}_P(S, \alpha)$ at depth k. We first note that, like circles, isothetic (and even homothetic) copies of a convex polygon have the property that, under an appropriate general position assumption, each pair of their boundaries intersect in at most 2 points (cf. [11]). Thus the naive algorithm can proceed in much the same way as above; the cost of the primitive operations that it performs may now depend on d, but since we assume d to be a constant, they still take constant time each.

In order to extend the improved algorithm of Section 3, we need to extend Pach's lemma (Lemma 1) to shapes other than circles. This is indeed possible:

Lemma 3 (Sharir [18]). *Let $\mathcal{A}$ be an arrangement of n homothetic copies of a closed convex set P. If the number of intersecting pairs of copies of P is larger than cnk, for some absolute positive constant c, then depth($\mathcal{A}$) is more than k.*

Another modification that the new procedure requires is in the preliminary stage that computes r_{init} (or, in our new notation, α_{init}, the smallest scaling factor for which the arrangement $\mathcal{A}_P$ has at least cnk intersecting pairs of polygons). The data structure that was used in Section 3 is based on standard Voronoi diagrams of subsets of S. Here we need to use the following generalized structure. The basic operation that we want our data structure to support is: Given a fixed subset $X \subseteq S$, a query point q and a query scaling factor α, determine whether there is $x \in X$ such that $(\alpha P + x) \cap (\alpha P + q) \neq \emptyset$. This is equivalent, as is easily checked, to asking whether $X \cap [q + \alpha(P - P)] \neq \emptyset$. Thus we can compute the generalized Voronoi diagram of X under the norm induced by $P - P$ as a unit ball (see [13, 10] for details), and preprocess it for efficient point location. The above query is answered by locating q in the diagram, and by determining whether the $(P - P)$-distance from q to its nearest neighbor in X is $\leq \alpha$.

It is now easy to see that all the algorithms described in the previous sections can be applied for the case of a convex polygon with a constant number of sides, or, for that matter, for any convex region P of sufficiently simple shape so that each of the basic operations performed by the algorithm on copies of P takes constant time.

Another case to which our (appropriately modified) algorithms can apply is the *smallest k-enclosing hypodrome problem: "Given a set S of n points in the plane, a parameter $k \leq n$, and a line segment e, find the smallest value of r for which there exists a translation v of e, such that there exist k points of S at distance no more than r from v + e."* We use the term 'hypodrome' to denote the Minkowski sum of a segment and a disk, which is the enclosing shape in this problem.

In this case we need again to replace the standard Voronoi diagrams in the data structure $DS(X)$ of Section 3.2 by a different kind of diagrams. Specifically, we need to answer queries of the form: Given a point v, find the point of S nearest to the segment $e+v$. Let S_e be the collection of parallel segments $\{p-e \,:\, p \in S\}$. Then the above query is equivalent to finding the segment of S_e nearest to the query point v. For this, we use the standard Voronoi diagram of S_e (see e.g. [19]). All other ingredients of the algorithm remain the same (including the extended version of Pach's Lemma), as is easily verified.

In summary, we have thus shown:

Theorem 4. *Given a set S of n points in the plane, an integer $k \leq n$, and a convex d-gon (where d is a constant), we can find the smallest homothetic copy of the polygon containing k points of S, in time time $O(nk \log^2 n)$ and space $O(nk)$, or in time $O(nk \log^2 n \log \frac{n}{k})$ and space $O(n \log n)$.*

Remark: If the given polygon is a square, the problem can be solved in time $O(n \log^2 n)$ (see [6]).

Theorem 5. *Given a set S of n points in the plane, an integer $k \leq n$, and a segment e, the smallest k-enclosing hypodrome for e can be found in time $O(nk\log^2 n)$ and space $O(nk)$, or in time $O(nk\log^2 n \log \frac{n}{k})$ and space $O(n\log n)$.*

References

[1] P. Agarwal, M. Sharir, and S. Toledo. New applications of parametric searching in computational geometry. to appear in *J. Algorithms*, 1993.

[2] P. K. Agarwal and J. Matoušek. Dynamic half-space range reporting and its applications. manuscript, 1992.

[3] A. Aggarwal, H. Imai, N. Katoh, and S. Suri. Finding k points with minimum diameter and related problems. In *Proc. 5th Annu. ACM Sympos. Comput. Geom.*, pages 283–291, 1989.

[4] M. Ajtai, J. Komlós, and E. Szemerédi. Sorting in $c\log n$ parallel steps. *Combinatorica*, 3:1–19, 1983.

[5] L. P. Chew and K. Kedem. Improvements on geometric pattern matching problems. In *Proc. 3rd Scand. Workshop Algorithm Theory*, volume 621 of *Lecture Notes in Computer Science*, pages 318–325. Springer-Verlag, 1992.

[6] R. Cole. Slowing down sorting networks to obtain faster sorting algorithms. *J. ACM* 31:200–208, 1984.

[7] H. Edelsbrunner, L. J. Guibas, and J. Stolfi. Optimal point location in a monotone subdivision. *SIAM J. Comput.*, 15:317–340, 1986.

[8] D. Eppstein. New algorithms for minimum area k-gons. In *Proc. 3rd ACM-SIAM Sympos. Discrete Algorithms*, pages 83–88, 1992.

[9] T. C. Kao and D. M. Mount. An algorithm for computing compacted Voronoi diagrams defined by convex distance functions. In *Proc. 3rd Canad. Conf. Comput. Geom.*, pages 104–109, 1991.

[10] K. Kedem, R. Livne, J. Pach, and M. Sharir. On the union of Jordan regions and collision-free translational motion amidst polygonal obstacles. *Discrete Comput. Geom.*, 1:59–71, 1986.

[11] H.-P. Lenhof and M. Smid. Enumerating the k closest pairs optimally. In *Proc. 33rd Annu. IEEE Sympos. Found. Comput. Sci.*, pages 380–386, 1992.

[12] D. Leven and M. Sharir. Planning a purely translational motion for a convex object in two-dimensional space using generalized Voronoi diagrams. *Discrete Comput. Geom.*, 2:9–31, 1987.

[13] N. Megiddo. Applying parallel computation algorithms in the design of serial algorithms. *J. ACM*, 30:852–865, 1983.

[14] N. Megiddo. Linear-time algorithms for linear programming in R^3 and related problems. *SIAM J. Comput.*, 12:759–776, 1983.

[15] F. P. Preparata and M. I. Shamos. *Computational Geometry: an Introduction.* Springer-Verlag, New York, NY, 1985.

[16] R. Seidel. A simple and fast incremental randomized algorithm for computing trapezoidal decompositions and for triangulating polygons. *Comput. Geom. Theory Appl.*, 1:51–64, 1991.

[17] M. Sharir. On k-sets in arrangements of curves and surfaces. *Discrete Comput. Geom.*, 6:593–613, 1991.

[18] C. K. Yap. An $O(n\log n)$ algorithm for the Voronoi diagram of a set of simple curve segments. *Discrete Comput. Geom.*, 2:365–393, 1987.

An Index Data Structure For Matrices, with Applications to Fast Two-Dimensional Pattern Matching

Raffaele Giancarlo*

AT&T Bell Laboratories, Murray Hill, NJ 07974, U.S.A.

(Extended Abstract)

Abstract

We describe a new data structure, the *submatrix trees* (*s-trees* for short). It is a forest of compacted patricia trees representing all submatrices of a given matrix $TEXT$. It can be efficiently built and used to answer quickly a wide variety of queries about $TEXT$. It can be thought of as a generalization to matrices of the suffix tree of a string [14]. We also introduce the notion of an abstract index data structure for matrices, of which the *s-trees* is an incarnation, and investigate the inherent space and time needed to build it. We show that the *s-trees* are optimal for space and within a log factor optimal for time.

1 Introduction

In recent years, there has been growing interest in two-dimensional pattern matching, due to its relevance to low-level image processing [15] and to the advent of visual databases in multimedia systems [11]. In its simplest form, the problem consists of finding all occurrences of a pattern matrix PAT into a text matrix $TEXT$. All matrices have entries defined in a totally ordered alphabet Σ. Two-dimensional pattern matching algorithms are roughly in two complementary classes (in analogy with the pattern matching algorithms for strings), each class satisfying complementary performance criteria and requirements arising in applications.

Dictionary Matching Algorithms: They preprocess a dictionary D of pattern matrices and then one can look for the occurrence of patterns of the dictionary into a text matrix $TEXT$. We discuss first the case in which the dictionary is composed of one pattern PAT of dimension $m_1 \times h_1$. Let the text be an $n \times w$ matrix. The first such algorithms, independently discovered by Baker [4] and Bird [5], need $O((m_1 \times h_1 + n \times w) \log |\Sigma|)$ time. Amir, Benson and Farach [2]

*On Leave from University of Palermo, ITALY

made a first substantial step toward alphabet independence for this problem by devising an $O((m_1 \times h_1) \log |\Sigma| + n \times w)$ time algorithm. Recently, Galil and Park [7] closed the gap by improving the time bound to $O(m_1 \times h_1 + n \times w)$. All mentioned algorithms have the drawback that they can preprocess only one pattern at a time or a dictionary of patterns that can have different widths but the same height (an $n \times m$ matrix A has heigth n and width m). Algorithms for the special case in which the patterns in the dictionary are squares, each of arbitrary heigth=width, have been obtained independently and with different approaches by Amir et al. [3] and Giancarlo [8]. Finally, Idury and Schaffer [10] have recently devised data structures and algorithms that preprocess a dictionary of pattern matrices $D = \{PAT_1, \cdots, PAT_r\}$, each of dimension $m_i \times h_i$, $1 \leq i \leq r$, and then the patterns can be searched for in a text of dimension $n \times w$. The preprocessing step takes $O((\Sigma_{i=1}^{r} m_i \times h_i)\text{polylog}(\Sigma_{i=1}^{r} m_i \times h_i))$ time and the search step takes $O((n \times w)\text{polylog}(\Sigma_{i=1}^{r} m_i \times h_i))$ time.

Library Management Algorithms: They preprocess the text matrix to build an index data structure that represents all submatrices of the text. Such data structure is the analog of the suffix tree for a string [14]. The index must support a wide variety of queries, some of which arise in two-dimensional data compression schemes [16, 17]. The most basic one is *occurrence(PAT)*, i.e., report the $oc \geq 0$ occurrences of the pattern matrix PAT in the text matrix. For a long time only straightforward data structures and algorithms were known. Indeed, using the notion of spiral strings, Gonnet [9] claims that it is possible to build a patricia tree (referred to as PAT-Tree [9]) representing all submatrices of an $n \times w$ $TEXT$ matrix in $O(n \times w \log \max(n, w))$ *expected time*. However, in the worst case, that data structure can be built in $O(\max(n, m)^2 \min(n, m)^3)$ time. Then, using the PAT-Tree, Gonnet [9] claims that one can find all occurrences of an $m_1 \times h_1$ matrix PAT in $O(\log \max(n, w))$ *expected time* independent of the number of occurrences. Again, in the worst case, the time bound to find all such occurrences is $O(m_1 \times h_1 \log |\Sigma| + oc)$. As it will be clear from the results reported in the next section, our data structure and algorithms represent a two orders magnitude improvement over the worst case time bounds obtainable by constructing the data structure proposed in [9]. Here we will not be dealing with the *expected time* performance of our data structures and algorithms.

Recently, progress has been made towards the design of efficient data structures for the special case of square matrices. Giancarlo has devised a data structure, the Lsuffix tree, that represents all square submatrices of an $n \times n$ $TEXT$ [8]. It can be built in $O(n^2 \log n)$ time. *occurrence(PAT)* takes $O((m^2 \log |\Sigma|) + oc)$ time and PAT must be a *square* matrix of width m. The only queries that the Lsuffix tree can support are limited to square matrices, so it has a limited range of applications.

2 Our Results

As implied by the above discussion, there was no efficient algorithm known that builds a two-dimensional analog of the suffix tree for general text matrices.

One of the main contributions of this paper is to provide efficient algorithms that build, dynamically change and query such data structure, which we call *s-trees*. Based on the *s-trees*, we obtain efficient algorithms for the static and dynamic versions of the following *Two-Dimensional Library Management* problem (defined below), which models various applications in low-level image processing [15], visual information management systems [11] and two-dimensional data compression schemes [16, 17].

Two-Dimensional Library Management: We have a library of texts $S = \{TEXT^1, \cdots, TEXT^r\}$, where $TEXT^i$ is an $n_i \times w_i$ matrix, $1 \leq i \leq r$. In the static version, we may preprocess the library. Then, we can ask any number of five types of queries, four of which involve an $m_1 \times h_1$ pattern matrix PAT. *occurrence*(PAT): report all occurrences of PAT in $TEXT$, for all $TEXT \in S$. *frequency*(PAT): report the number of occurrences of PAT in $TEXT$, for all $TEXT \in S$. *slp*(PAT): find the set of largest prefixes of PAT, not prefixes of each other, that occur in $TEXT$, for all $TEXT \in S$. *lp*(PAT, d): find the longest prefix of PAT of width d that occurs in $TEXT$, for all $TEXT \in S$ of height larger than width. *lrm*(d): find the matrix of width at least d and of maximum area that occurs at least twice as submatrix of the matrices in S having height larger than width.

In the dynamic version, we can update the library S by inserting in it or deleting from it some $TEXT$. The queries are as in the static version but are intermixed with update operations. Let $t(S) = \Sigma_{i=1}^{r} \max(n_i, w_i) \times \min(n_i, w_i)^2$. We assume a RAM model of computation with uniform cost criterion [1] and show the dependence of the time bounds on the alphabet size. All time bounds are worst case.

(i) *Preprocessing Static Version*. It builds the *s-trees* for the matrices in S. It takes $O(t(S)(\log|\Sigma| + \log t(S)))$ time and $O(t(S))$ space.

(ii) *Update Dynamic Version*. Let S_f be the library after the f-th update operation and let $t(S_f)$ be defined for S_f as $t(S)$ is defined for S. The *s-trees* can be updated in $O(\max(n, w) \times \min(n, w)^2(\log^2 t(S_f) + \log|\Sigma|))$ time, when an $n \times w$ matrix $TEXT$ is inserted or deleted from S_f.

(iii) *Query Answering:* The time bounds are given by the following table. Let oc denote the number of occurrences of PAT in $TEXT$, for all $TEXT \in S$, and let $\hat{w}$ be the maximum width of matrices in S (for the static bounds) or S_f (for the dynamic bounds) having height larger than width. Moreover, let $e = \min(m_1, h_1) \times \max(m_1, h_1)^2$.

	Static	Dynamic
occurrence(*PAT*)	$O(oc + m_1 h_1 \log \lvert\Sigma\rvert)$	$O(m_1 h_1 \log t(S_f) + oc)$
frequency(*PAT*)	$O(m_1 h_1 \log \lvert\Sigma\rvert)$	$O(m_1 h_1 \log t(S_f) + oc)$
slp(*PAT*)	$O(\min(e \log \lvert\Sigma\rvert,$ $m_1 h_1 \log t(S)))$	$O(\min(e \log t(S_f),$ $m_1 h_1 \log^2 t(S_f)))$
lp(*PAT*, *d*)	$O(m_1 d \log \lvert\Sigma\rvert)$	$O(m_1 d \log t(S_f))$
lmr(*d*)	$O(1)$	$O(\log\log \hat{w})$

The *s-trees* is the first data structure that efficiently support such a variety of queries, some of which have applications to information retrieval [11] while others may be used as primitive operations for two-dimensional data compression schemes [16, 17]. The query procedures are very simple and practical. We also point out that, based on the *s-trees* and in the same time bound, we can precompute a set of tables which are a space-economical representation of the *s-trees* taking only a total of $5t(S)$ memory locations. Such tables are a generalization to matrices of Manber and Myers [13] suffix arrays for strings. Many of the queries defined for *s-trees* can be efficiently supported by the tables.

Lower Bound: We also introduce the notion of an abstract index data structure, like the *s-trees*, that represents all submatrices of a given matrix and investigate the inherent space and time needed to build it. Let a *prefix* of matrix A be any submatrix of A with the same upper left corner as A. Let $S = \{TEXT^1, \cdots, TEXT^r\}$. We define an *index* for S as a tree that represents all submatrices of matrices in S and such that submatrices of matrices in S having a prefix in common share a path in the tree. Let $SPACE(I)$ and $TIME(I)$ be the minimal amount of space and time needed to build an index I for a set of matrices S. We show that $TIME(I) \geq SPACE(I) \geq ct(S)$, for some constant $c > 0$.

The lower bound establishes that our algorithm for the construction of the *s-trees* is optimal up to a log factor for time and optimal for space. It also shows that the general problem of building an index representing all submatrices of a set of matrices S is harder than the special problem of building an index representing all square submatrices of a set of square matrices. Indeed, let $n_i = w_i$, i.e., all matrices in S are squares. The general problem requires at least $t(S) = \Sigma_{i=1}^r n_i^3$ time, while there exist solutions to the special problem that require at most $O(\Sigma_{i=1}^r n_i^2 \log(\Sigma_{i=1}^r n_i^2))$ time [8].

3 Our Techniques

We have general techniques that allow us to insert (delete, resp.) a string x into (from, resp.) a patricia tree T in $O(\log|T|)$ worst case time, independent of the length of x and given some lexicographic information about x. This is a sharp departure from the way in which those operations are handled in the literature [1, 12] as our approach is similar in spirit to the way in which insertions and deletions are done in balanced binary search trees, e.g., 2-3 trees [1]. Our technique boosts the following new general trick, an instance of which allows us to gain nearly an order of magnitude time with respect to more standard ways of building our data structures. Roughly speaking, it can be described as follows. Assume that we have to build and maintain a patricia tree T for strings $x_1, \cdots, x_s$, $x_i \in \Sigma^*$, and that those strings can be cheaply compressed into strings $X_1, \cdots, X_s$ over a new alphabet Λ such that the compression preserves the lexicographic order, i.e., x_i is lexicographically smaller than x_j iff X_i is lexicographically smaller than X_j. Let T' be a patricia tree containing the compressed versions of the strings in T. When we have to insert (delete, resp.) x into (from, resp.) T, we insert (delete, resp.) its compressed version X into T' via the usual procedures [1, 12]. That will take $O(insert(X, T'))$ ($O(delete(X, T'))$, resp.) time and we also get the needed lexicographic information about X. Using our new insertion (deletion, resp.) procedures, we can then insert (delete, resp.) x into (from, resp.) T in $O(\log|T|)$ time. So, the total cost of the operation is $O(insert(X, T') + \log|T|)$ ($O(delete(X, T') + \log|T|)$, resp.) time which can be substantially better than $O(|x|\log|\Sigma|)$ time, i.e., the cost of inserting (deleting, resp.) x into (from, resp.) T in the usual way [1, 12].

4 An Abstract Index Data Structure for a Matrix

4.1 Definition

In order to define the abstract notion of an index for a matrix A, we need to be able to express any matrix in terms of one of its prefixes and an elementary notion of "character".

Consider an $n_1 \times n_2$ matrix C. There are three "immediate" extensions of C into a matrix D that has C as prefix. They are: appending a row of length n_2 to C or a column of length n_1 to C or a row of length n_2+1 and a column of length n_1 to C (see Fig. 1). So, if we want to express D in terms of its prefix C and an elementary notion of character, we need to define the following characters: $(-, n_2, x)$, a row of length n_2 given by $x \in \Sigma^*$; $(|, n_1, x)$ a column of length n_1 given by $x \in \Sigma^*$; $(\rfloor, n_2+1, n_1, x)$, a row given by the first n_2+1 characters of x and a column given by the last n_1 characters of $x \in \Sigma^*$, $|x| = n_1 + n_2 + 1$. We refer to such characters as *shapes* (see Fig. 1). Notice that we can define shapes for any length of row and column. Moreover, we can express any matrix in terms of concatenation of shapes by using the following *concatenation rule*:

Given an $n_1 \times n_2$ matrix C, we can extend it into an $(n_1+1) \times n_2$ or $n_1 \times (n_2+1)$ or $(n_1+1) \times (n_2+1)$ matrix by appending to it shape $(-, n_2, x)$ (which encodes a row) or $(|, n_1, x)$ (which encodes a column) or $(\rfloor, n_2+1, n_1, x)$ (which encodes a row and a column), $x \in \Sigma^*$, respectively. Entry $(1,1)$ of any matrix can be generated by the shape $(\rfloor, 1, 0, a)$, $a \in \Sigma$.

Let an *uncompacted submatrix index* for matrix A be a rooted tree with edges labeled by shapes and that satisfies the following constraints (see Fig. 2):

1. For each submatrix B of A, there is at least one node u such that the concatenation of the labels (shapes) on the edges of the path p from the root to u gives B. We denote by $L(u)$ the matrix obtained by concatenating the labels on p.

2. For each node u and v, the following must hold. u is ancestor of v if and only if $L(u)$ is prefix of $L(v)$.

Notice the analogy between the definition of uncompacted submatrix index and an uncompacted trie for strings [12].

In order to define an index for a matrix from an uncompacted one, we need to specify how chains of unary nodes are compacted. Consider such chain p and let u be its first node and v the last. We compact p into the edge (u, v) and assign to that edge a "generalized shape" obtained by deleting $L(u)$ from $L(v)$ (by (2.) of definition of uncompacted index, $L(u)$ is prefix of $L(v)$) (see Fig. 3). We remark, without proof, that such generalized shape can be represented in constant space (through the coordinates of its endpoints in A). Finally, we define an *index* for a matrix A as an uncompacted submatrix index in which chains of unary nodes have been compacted (see Fig. 2).

4.2 Lower Bound

We prove the lower bound for the case in which the matrix has dimension $n \times m$ and $n \geq m$. By symmetric arguments, it will hold for the case $n < m$. It can be generalized to a set of arbitrary matrices.

The *size* of an index representing a matrix A is the number of edges and nodes in it. For a given $n \times m$ matrix A, let $SIZE(A)$ be the minimum size required for an index to represent A. Let $SPACE(I)$ be the maximum $SIZE(A)$ taken over all $n \times m$ matrices. For any given matrix A, let $once(A)$ be the number of submatrices of A that occur only once in A. We can show:

Lemma 1 *Let A be an arbitrary $n \times m$ matrix, $n \geq m$. $SPACE(I) \geq c \times once(A)/n$, for some constant $c > 0$.*

Theorem 1 *Let Σ be an alphabet of at least two symbols. There exist infinitely many values of n such that for all m, $0 < m \leq n$, $SPACE(I) \geq cnm^2$, for some constant $c > 0$.*

Proof Sketch: We build a matrix A for which $once(A) \geq \Omega(n^2m^2)$. Pick $n = 2^k$, $k > 0$ and such that $n \geq 64$ and $n > 2k$. Pick any m, $0 < m \leq n$. Consider a de Bruijn sequence y of length n^2 constructed on two symbols of Σ [6]. By definition of de Bruijn sequence, each substring of y of length $2k = \log n^2$ is distinct, i.e., it occurs only once in y. We transform y into an $n \times m$ matrix A by assigning the first $n \times m$ symbols of y to A in column major order. Counting arguments show that $once(A) \geq cn^2m^2$. The theorem follows from Lemma 1. □

5 The Suffix Tree in a Nutshell

From now on, we will be working with two types of alphabets: the alphabet of atomic characters Σ and the alphabet Σ^d, d fixed, i.e., each character in Σ^d is a string over the atomic alphabet. Σ is atomic in the sense that comparison of two characters of Σ is a primitive operation that takes constant time. We refer to the "meta-characters" of Σ^d as *blocks of width* d and to the strings built out of them as *block strings of width* d. The definition of prefix and suffix of a block string is the usual one for string. We refer to a prefix or suffix of a block string as block prefix or suffix, respectively.

Consider an alphabet Λ (it may be atomic or not). Let x be a string over Λ and let \$ be an endmarker that does not match any character of Λ. We need to recall the definition of suffix tree T_x of a string $x\$$ [14]. It is a compacted trie such that, for each suffix $x[i, |x|]\$$, there is a path from the root to a leaf that "spells out" $x[i, |x|]\$$. Since \$ does not match any character of Λ, all suffixes of $x\$$ are distinct and there is a one to one correspondence between the leaves of T_x and the suffixes of $x\$$. T_x takes $O(|x|)$ space and, provided that two characters of Λ can be compared in constant time, it can be built in $O(|x| \log \min(|\Lambda|, |x|))$ time [14]. We need to know the following about the algorithm that builds the suffix tree (*MC* for short).

(a) It inserts the suffixes of x one at the time, from longest to shortest, into a tree of initially one node. Let T_{i-1} be the tree after $i-1$ insertions and $|T_{i-1}|$ be its size. When the algorithm inserts $x[i, |x|]\$$, it creates a new leaf that is associated with $x[i, |x|]\$$. The parent of that leaf is identified during insertion and may be the only new internal node created (by splitting an edge). Each insertion takes $O(\log \min(|\Lambda|, |T_{i-1}|))$ comparisons of characters and $O(\log \min(|\Lambda|, |T_{i-1}|))$ additional time, amortized over all insertions.

(b) Suffixes of x can be deleted from the suffix tree from shortest to longest. In going from T_i to T_{i-1}, the leaf associated with $x[i, |x|]\$$ is removed. If the parent of that leaf in T_i becomes a node of outdegree one, that node is also removed by merging two edges. Each deletion takes $O(1) \leq O(\log \min(|\Lambda|, |T_{i-1}|))$ time.

(c) Procedures in (a) and (b) generalize easily to a set of strings $D = \{x_1\$, \cdots, x_s\$\}$, $x_i \in \Lambda^*$. Let T be the suffix tree for the strings in D. It is of size $|T| = O(\Sigma_{i=1}^{s} |x_i|)$. Assume that we want to insert (delete, resp.) a string x and its suffixes into (from, resp.) T. We use the same insertion (deletion, resp.) order and procedures of (a) ((b), resp.). The insertion (deletion, resp.) time bound is as above with $|T_{i-1}|$ substituted by $|T|$.

6 The s-trees

We now define the *s-trees* for a matrix $A[1:n,1:m]$, $n \geq m$. By symmetry, the definition extends to the case $n < m$. When we have a set of matrices, we partition the set into two: the "tall" matrices, i.e., height larger then width, and the "fat" matrices, i.e., height smaller than width. Each set will have a distinct *s-trees* data structure representing it.

Let us restrict the types of shapes defined in Section 4 to one: $(-,d,x)$, $x \in \Sigma^d$, d fixed. That is, all submatrices of a matrix A of width d can only be obtained through concatenations of shape characters $(-,d,x)$, $x \in \Sigma^d$. Since we are adopting only one kind of shape, we can drop the encoding $-$, and for fixed d, $(-,d,x)$ becomes the block character x. So, each submatrix of A of width d is a block string of width d.

The *s-trees* is a forest of m pairs of trees. The pair $BLT_d, REFT_d$ represents all submatrices of A of width d. BLT_d represents such submatrices as block strings over the alphabet $\Lambda = \Sigma^d$.

$REFT_d$ represents the block strings of BLT_d as strings over Σ and therefore it represents the same submatrices as $REFT_d$ but as strings over Σ. Intuitively, BLT_d captures the two-dimensionality of the submatrices it represents at a "high level", while $REFT_d$ gives a "low level" description of the same data that lends itself to fast query answering (we will elaborate on this at the end of this section). We could build $REFT_d$ directly, but the advantage of having two trees will be clear in the next section.

We define only one pair BLT_d and $REFT_d$, for a fixed d, $1 \leq d \leq m$. Let $B_{j,d}$ be the block string of length n that represents $A[1:n, j:j+d-1]$, $1 \leq j \leq m-d+1$. That is, $B_{j,d}[i] = A[i, j:j+d-1]$. Notice that every submatrix $A[i:k, j:j+d-1]$ corresponds to some block prefix of some block suffix of $B_{j,d}$. That is, $B_{j,d}[i,k] = A[i:k, j:j+d-1]$. So, the patricia tree T, over the alphabet of blocks Σ^d, that represents all block suffixes of $B_{j,d}$, for $1 \leq j \leq m-d+1$, must represent all submatrices of A of width d. That is, for each submatrix C of width d of A, there is a node $u \in T$ such that the concatenation of the labels (which are blocks) gives submatrix C. Moreover, all submatrices of width d of which C is prefix are in the subtree rooted at u. Therefore, T satisfies the definition of uncompacted index for A when restricted to submatrices of width d. But a compacted version of T is the suffix tree for the block strings $B_{1,d}, B_{2,d}, \cdots, B_{m-d+1,d}$. BLT_d is that suffix tree and it satisfies the definition of index for A when restricted to submatrices of width d. It has size $O(n(m-d+1))$ because each $B_{j,d}$ is a block string of length n (set $\Lambda = \Sigma^d$ and recall from (c) of Section 5 the size of a suffix tree for a set of strings).

Notice that the outdegree of each node in BLT_d may be as large as $O(\min(\Sigma^d, n(m-d+1)))$. That is a serious problem if we try to use BLT_d for pattern matching, since "selecting which edge to cross next" may not take constant time. We overcome this problem with $REFT_d$. Indeed, $REFT_d$ is a compacted trie that represents all the block suffixes of $B_{1,d}, B_{2,d}, \cdots, B_{m-d+1,d}$ as strings over Σ. That is, it represents all the block strings represented by BLT_d as strings over Σ. The outdegree of each node in $REFT_d$ is now bounded by $|\Sigma|$. Such a reduction in outdegree does not cost us a lot in terms of space. Indeed, we can show that $REFT_d$ has $O(n(m-d+1))$ nodes.

Notice the usefulness of $REFT_d$ for pattern matching: Given a $m_1 \times d$ pattern PAT, we transform it into a string *pat* of length $m_1 d$ by concatenating its rows in row major order. Now, we can look for *pat* into $REFT_d$ in $O((m_1 \times d) \log |\Sigma|)$ time by a nearly standard search procedure in a compacted trie. In that way, we find out whether PAT is a submatrix of A.

As already stated, the *s-trees* is the forest of trees $BLT_1, REFT_1; \cdots; BLT_m, REFT_m$. Since each BLT_d is an index for A when we restrict the definition to submatrices of width d, the *s-trees* is an index for A. It takes $O(\Sigma_{d=1}^{m}(n(m-d+1)) = O(nm^2)$ space.

7 The Algorithms

We concentrate on the algorithms for the construction and dynamic maintenance of one pair BLT_d and $REFT_d$ of trees. We have two situations, which correspond to the static and dynamic versions of the Library Management Problem defined in the Introduction. We treat both situations in a unified way and model them as follows.

Assume that BLT_d and $REFT_d$ represent all submatrices of width d of a set of "tall" matrices $S = \{C_1, \cdots, C_s\}$. Refer to those trees as BT_d and RT_d, respectively. Let $B_{1,d}, \cdots, B_{m-d+1,d}$ be the block string representing all submatrices of width d of a "tall" matrix A. Now, we want to insert (delete, resp.) all block suffixes of block strings $B_{1,d}\$, B_{2,d}\$, \cdots, B_{m-d+1,d}\$$ into (from, resp.) BT_d and the strings of Σ^* corresponding to them into (from, resp.) RT_d. In the end, BT_d and RT_d will be BLT_d and REF_d for the set $S \cup \{A\}$ ($S - \{A\}$, resp.). Obviously, in the static case, S will be empty and only insertions will make sense.

The algorithms that insert or delete strings from RT_d "take a ride" on the ones that insert or delete block suffixes from BT_d. That is, we find ways so that the total time to transform RT_d into $REFT_d$ for the set $S \cup \{A\}$ ($S - \{A\}$, resp.) will be bounded by the total time to transform BT_d into BLT_d for the same set. That will result in roughly an order of magnitude saving in time with respect to the naive approach of building and maintaining $REFT_d$ only (that is why we need two trees). Here we use an instance of the techniques outlined in Section 3. The compressed strings are all the block suffixes of $B_{1,d}\$, \cdots, B_{m-d+1,d}\$$. The idea is the following.

- Block suffixes of $B_{1,d}\$, \cdots, B_{m-d+1,d}\$$, are inserted into (deleted from, resp.) BT_d according to the order established by MC: one $B_{j,d}$ at a time, $1 \le j \le m-d+1$, and its block suffixes from longest (shortest, resp.) to shortest (longest, resp.) (recall (c) of Section 5).

- For each block suffix inserted (deleted, resp.), the corresponding string is inserted in (deleted from, resp.) RT_d. We derive combinatorial properties of patricia trees that allow us to use the work of MC so that such insertion into (deletion from, resp.) RT_d will have a worst case cost of $O(\log \min(\Sigma^d, |BT_d|))$

time. Recalling from (c) of Section 5 the costs of insertion and deletion into a suffix tree for a set of strings, we will have that the cost of insertion (deletion, resp.) of strings into (from, resp.) RT_d will be bounded by the cost of insertion (deletion, resp.) into (from, resp.) BT_d.

References

[1] A.V. Aho, J.E. Hopcroft, and J.D. Ullman. *The Design and Analysis of Computer Algorithms.* Addison-Wesley, Reading, MA., 1974.

[2] A. Amir, G. Benson, and M. Farach. Alphabet independent two dimensional matching. In *Proc. 24th Symposium on Theory of Computing*, pages 59–68. ACM, 1992.

[3] A. Amir, M. Farach, R. Idury, J. La Poutre, and A. Schaffer. Improved dynamic dictionary matching. In *Proc. Fourth Symposium on Discrete Algorithms*, pages 392–401. ACM-SIAM, 1993.

[4] T.J. Baker. A technique for extending rapid exact match string matching to arrays of more than one dimension. *SIAM J. on Computing*, 7:533–541, 1978.

[5] R.S. Bird. Two dimensional pattern matching. *Information Processing Letters*, 6:168–170, 1978.

[6] N. G. de Bruijn. A combinatorial problem. *Nederl. Akad. Wetensch. Proc.*, 49:758–764, 1946.

[7] Z. Galil and K. Park. A truly alphabet independent two-dimensional pattern matching algorithm. In *Proc. 33th Symposium on Foundations of Computer Science*, pages 247–256. IEEE, 1992.

[8] R. Giancarlo. The suffix tree of a square matrix, with applications. In *Proc. Fourth Symposium on Discrete Algorithms*, pages 402–411. ACM-SIAM, 1993.

[9] G.H. Gonnet. Efficient searching of text and pictures- Extended Abstract. Technical report, University Of Waterloo- OED-88-02, 1988.

[10] I. Idury and A. Schaffer. Multiple matching of rectangular patterns. In *Proc. 25th Symposium on Theory of Computing - to appear.* ACM, 1993.

[11] R. Jain. Workshop report on visual information systems. Technical report, National Science Foundation, 1992.

[12] D.E. Knuth. *The Art of Computer Programming, VOL. 3: Sorting and Searching.* Addison-Wesley, Reading, MA., 1973.

[13] U. Manber and E. Myers. Suffix arrays: A new method for on-line string searches. In *Proc. First Symposium on Discrete Algorithms*, pages 319–327. ACM-SIAM, 1990.

[14] E.M. McCreight. A space economical suffix tree construction algorithm. *J. of ACM*, 23:262–272, 1976.

[15] A. Rosenfeld and A.C. Kak. *Digital Picture Processing*. Academic Press, 1982.

[16] D. Sheinwald, A. Lempel, and J. Ziv. Compression of pictures by finite state encoders. In R.M. Capocelli, editor, *Sequences: Combinatorics, Compression, Security and Transmission*, pages 326–347, Berlin, 1990. Springer-Verlag.

[17] J. A. Storer. Lossy on-line dynamic data compression. In R.M. Capocelli, editor, *Sequences: Combinatorics, Compression, Security and Transmission*, pages 348–357, Berlin, 1990. Springer-Verlag.

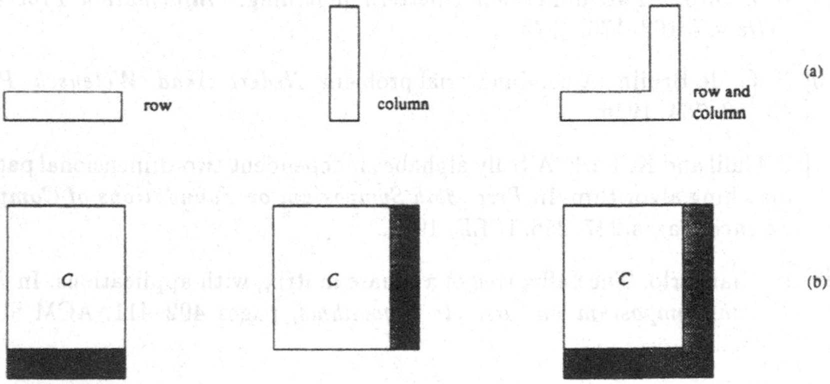

Figure 1: (a) The three possible shapes. (b) The three extensions of C into a matrix D that has it as a prefix. D is obtained from C by appending a shape to it.

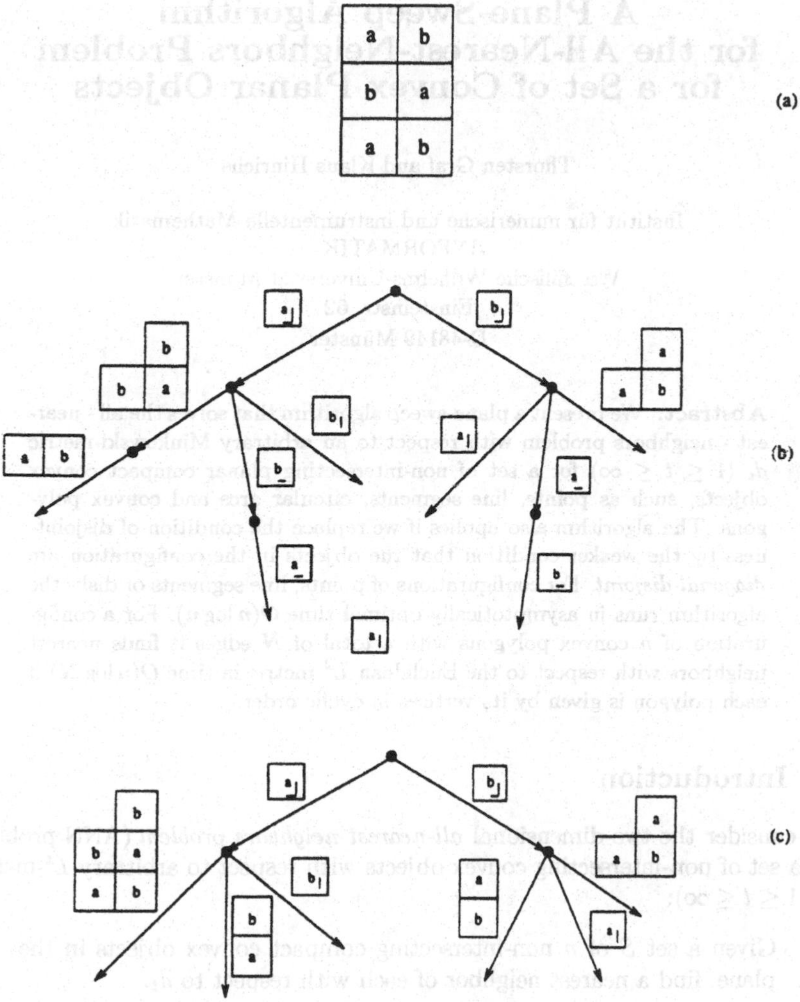

Figure 2: (b) The uncompacted submatrix index for matrix in (a). The symbols in single boxes denote whether the box is a row (−), column (|) or both (⌋). (c) An index (obtained from (b)) of matrix in (a). Notice generalized shapes on the edges.

A Plane-Sweep Algorithm for the All-Nearest-Neighbors Problem for a Set of Convex Planar Objects

Thorsten Graf and Klaus Hinrichs

Institut für numerische und instrumentelle Mathematik
-INFORMATIK-
Westfälische Wilhelms-Universität Münster
Einsteinstr. 62
D-48149 Münster

Abstract. We present a plane-sweep algorithm that solves the all - nearest - neighbors problem with respect to an arbitrary Minkowski-metric d_t $(1 \leq t \leq \infty)$ for a set of non-intersecting planar compact convex objects, such as points, line segments, circular arcs and convex polygons. The algorithm also applies if we replace the condition of disjointness by the weaker condition that the objects in the configuration are *diagonal-disjoint*. For configurations of points, line segments or disks the algorithm runs in asymptotically optimal time $O(n \log n)$. For a configuration of n convex polygons with a total of N edges it finds nearest neighbors with respect to the Euclidean L^2-metric in time $O(n \log N)$ if each polygon is given by its vertices in cyclic order.

1 Introduction

We consider the two-dimensional *all-nearest-neighbors problem* (ANN problem) for a set of non-intersecting convex objects with respect to arbitrary L^t-metrics d_t $(1 \leq t \leq \infty)$:

> Given a set S of n non-intersecting compact convex objects in the plane, find a nearest neighbor of each with respect to d_t.

The Minkowski-distance d_t of two objects $s_1, s_2 \in S$ is defined as follows:

$$\begin{aligned} d_t(s_1, s_2) &:= \min\{(|p.x - q.x|^t + |p.y - q.y|^t)^{\frac{1}{t}} : p \in s_1, q \in s_2\}, \quad 1 \leq t < \infty \\ d_\infty(s_1, s_2) &:= \min\{\max\{|p.x - q.x|, |p.y - q.y|\} : p \in s_1, q \in s_2\} \end{aligned}$$

If S consists of n points $\Omega(n \log n)$ is a well known tight lower bound for this problem in the algebraic decision tree model of computation. Optimal algorithms for this simple version of the problem are given in [9, 11, 12]; a survey is given in [8]. However, all these algorithms do not apply to sets of more complex objects, e.g. convex polygons.

For sets of line segments and circular arcs in the plane a nearest neighbor with respect to the Euclidean metric can be found in optimal time $O(n \log n)$

by first computing the Voronoi diagram ([7, 13]) and then extracting nearest neighbors from the Voronoi diagram in linear time. For a set of n polygons with a total of N edges this technique costs $\Theta(N \log N)$ time which is worse than the $O(n \log N)$ time complexity of the algorithm presented in this paper. Furthermore, the Voronoi diagram approach cannot be applied to solve the all-nearest-neighbors problem for general convex objects with respect to an arbitrary Minkowski metric.

[2] presents a plane-sweep algorithm to solve the closest-pair problem for a set of convex planar objects with respect to an arbitrary Minkowski-metric. This algorithm tests whether a new object encountered during the left-to-right sweep forms a new closest pair with any of those objects seen so far. Among these objects only those intersecting the δ-slice to the left of the sweep-line have to be considered, where δ denotes the minimal object distance found so far. A total ordering on these *active* objects is determined by their intersections with the δ-slice. During the sweep the algorithm maintains the set of active objects and the minimal distance δ between any pair of active objects which become neighbors with respect to this total ordering. New neighbor pairs are obtained either by encountering a new object or by deactivating an object which does not intersect the δ-slice any more. It seems to be surprising that this algorithm finds the correct result by just testing pairs of active objects which become neighbors during the sweep. [2] proves that this is sufficient for both an intersection free configuration and a configuration with intersecting objects, a more intuitive proof is given in [1]. The technique used in the above algorithm cannot be applied to the ANN problem for convex objects. Since we search for a nearest neighbor of each object in our configuration, we do not have a global δ for all objects but an individual δ for each object. This makes the technique used in [2] unsuitable for the ANN problem since the deactivation events do not occur in a predefined order and therefore have to be rearranged dynamically. Even worse, the individual δ-values increase the local nature of the problem, and therefore it is not necessarily true that at any time during the left-to-right sweep or the right-to-left sweep an object and any of its nearest neighbors become neighbors in the y-table, i.e. with respect to the total ordering defined in [2]. It is not difficult to construct a set of points for which this approach does not work with respect to the L^2-metric.

The algorithm PSANN (=Plane Sweep All Nearest Neighbors) presented in this paper uses four sweeps, from left to right, from right to left, from top to bottom and from bottom to top. The algorithm also works correctly if we weaken the condition of disjointness to the conditon that the objects in our configuration are ***diagonal-disjoint***, which means that for any object the x-diagonal connecting certain x-extremal and the y-diagonal connecting certain y-extremal points do not intersect any other object of S. For configurations consisting of n line segments, circular disks or convex polygons whose number of edges is treated as a constant, PSANN runs in asymptotically optimal time $O(n \log n)$. Let S be a set of n convex polygons with a total of N edges. If each polygon is given by its vertices in cyclic order then PSANN finds nearest neighbors with respect to the Euclidean metric in time $O(n \log N)$. This runtime is achieved by employing an optimal $O(\log m)$ algorithm [4] for detecting whether two convex polygons with at most m edges intersect, and an optimal $O(\log m)$ algorithm ([5, 6]) for computing the minimal Euclidean distance between two such non-intersecting convex polygons.

2 Plane-sweep applied to the ANN problem

Our algorithm PSANN is based on the well known plane-sweep method. Plane-sweep's name is derived from the image of sweeping the plane from left to right with a vertical line (front, or cross section), stopping at every transition point (event) of a geometric configuration to update the cross section, i.e. to maintain the problem-dependent *sweep invariants* which have to hold for the objects being encountered so far. All processing is done at this moving front, without any backtracking, with a look-ahead of only one object. The sweep invariants determine the events contained in the x-queue (*event queue*) and the status of the sweep which is maintained by the y-table (*sweep line structure*). In the slice between two events the properties of the geometric configuration detected so far do not change, and therefore no invariant has to be maintained and the y-table does not have to be updated. The skeleton of a plane-sweep algorithm is as follows:

```
initialize x-queue;
initialize y-table;
while not empty(x-queue) do
    { e := next(x-queue); transition(e)}
```

The procedure 'transition' is the advancing mechanism of the plane-sweep. It embodies all the work to be done when a new event is encountered; it moves the front from the slice to the left of an event e to the slice immediately to the right of e.

In this paper we apply the plane-sweep principle to solve the all-nearest-neighbors problem in a set S consisting of n convex compact objects with respect to any L^t-metric d_t $(1 \le t \le \infty)$. For each object $s \in S$ let $s[L]$ denote the smallest and $s[R]$ the largest point in s according to the following lexicographic order:

$$\forall p, q \in E^2, p \le^x q :\Longleftrightarrow (p.x < q.x) \vee [(p.x = q.x) \wedge (p.y \le q.y)].$$

Throughout this paper we write $p \le_x q$ and $p \le_y q$ instead of $p.x \le q.x$ and $p.y \le q.y$, respectively. These notations are extended to point sets as follows:

$$P \le_x Q :\Longleftrightarrow \forall p \in P\ \forall q \in Q : p \le_x q$$

For a point $p \in E^2$ the two diagonal lines (with slopes ± 1) through p subdivide the plane into four quadrants. Let us denote these quadrants as follows:

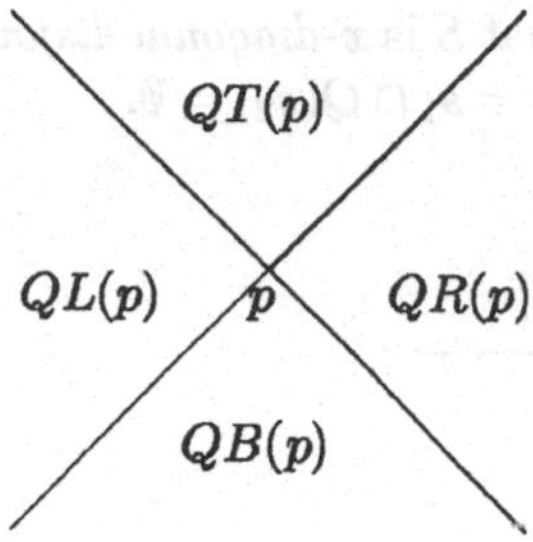

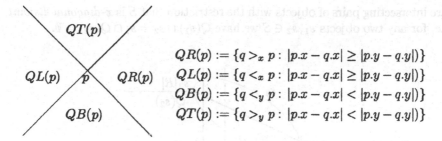

$$QR(p) := \{q >_x p : |p.x - q.x| \geq |p.y - q.y|)\}$$
$$QL(p) := \{q <_x p : |p.x - q.x| \geq |p.y - q.y|)\}$$
$$QB(p) := \{q <_y p : |p.x - q.x| < |p.y - q.y|)\}$$
$$QT(p) := \{q >_y p : |p.x - q.x| < |p.y - q.y|)\}$$

For an object $s \in S$ denote by $nn(s)$ a nearest neighbor found so far and by $\delta(s)$ the distance between s and $nn(s)$. Furthermore let $NN(s)$ be the set of all nearest neighbors of s in S and $\Delta(s)$ their distance to s.

Our algorithm PSANN uses four sweeps: from left to right, from right to left, from top to bottom and from bottom to top. We only describe the left-to-right sweep, the other sweeps work similarly. In the left-to-right sweep we find a nearest neighbor $r \in NN(s)$ for all those objects $s \in S$ for which there exist points $p \in s$ and $q \in r \cap QR(p)$ such that $d_t(p,q) = d_t(s,r)$. In the three remaining sweeps we solve the partial nearest neighbors problems with the corresponding points $p \in s$ and $q \in r \cap QL(p)$, $q \in r \cap QB(p)$ and $q \in r \cap QT(p)$. During the sweep PSANN maintains for each object $s \in S$ the smallest distance $\delta(s)$ to another object detected so far. The y-table stores the *active* objects $s \in S$ which have been encountered by the sweep-line SL and for which $s[R].x + \delta(s)$ is to the right of SL. These are exactly those objects of S encountered so far by SL which can have a nearest neighbor closer than $\delta(s)$ among the objects of S lying completely to the right of SL.

The x-queue is initialized with two sets of events: *insertion events* and *deletion events.* Insertion events are given by the left end points: an object s is inserted into the y-table when the sweep-line encounters its left end point $s[L]$. Deletion events are determined by the right end points and the smallest distance to another object detected so far: an object s is removed from the y-table as soon as the position of the sweep-line is at or to the right of $s[R].x + \delta(s)$. This may happen either if the sweep-line proceeds to the right or if $\delta(s)$ becomes smaller. Since $\delta(s)$ can be different for different active objects, the deactivation events cannot be processed in a predefined order as in [2]. Hence we have to deal with a dynamic processing of deactivation events. Repeated shrinking of $\delta(s)$ for an active object s requires left shifts of its deactivation event. We only know that the sweep-line is at or to the right of $s[R].x + \delta(s)$ when a deletion event for s is executed.

In order to allow an efficient processing of the objects contained in the y-table it is necessary to find a canonical total order on these objects. Such a total order will be defined in the following, it is the same as in [2]. For each object $s \in S$ let (Fig. 1)

$$R(s) := \{q : q \geq_x s[R], q =_y s[R]\}, \quad Q(s) := \overline{s[L], s[R]}, \quad \overline{Q}(s) := Q(s) \cup R(s)$$

For the moment we assume that there are no intersecting pairs of objects in S; we will show later that the left-to-right sweep works correctly even if there are intersecting pairs of objects with the restriction that S is *x-diagonal disjoint*, i.e. for any two objects $s_1, s_2 \in S$ we have $Q(s_1) \cap s_2 = s_1 \cap Q(s_2) = \emptyset$.

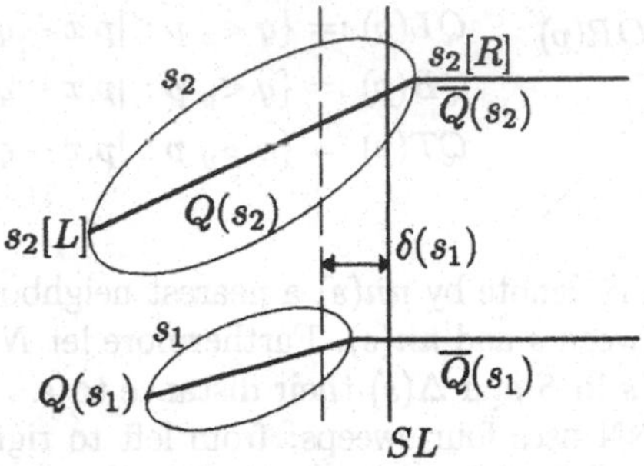

Fig. 1. Q and $\overline{Q}$ for two objects $s_1 \sqsubseteq s_2$

Let $s[SL] := \overline{Q}(s) \cap SL$ denote the *representation point* of an active object $s \in S$ with respect to a sweep-line SL. For objects $s_1, s_2 \in S$ and a sweep-line SL we define

$$s_1 \sqsubseteq s_2 :\Longleftrightarrow s_1[SL] <_y s_2[SL]$$

During the sweep the active objects are stored in the y-table with respect to $\sqsubseteq$. It seems to be necessary to exchange objects in the y-table when the sweep-line reaches an intersection point of $R(s_1)$ and $Q(s_2)$ for two active objects $s_1, s_2 \in S$ (Fig. 2).

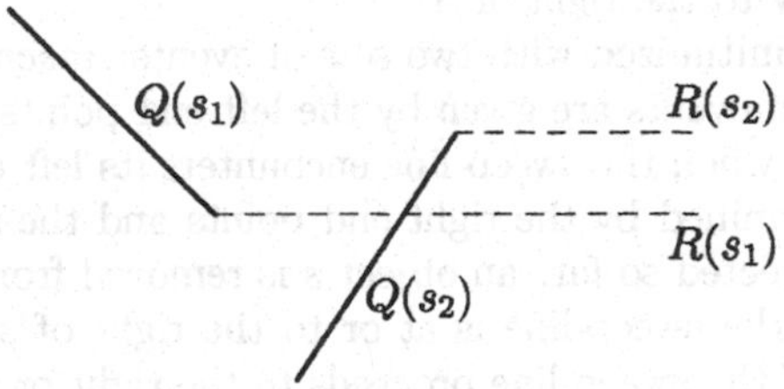

Fig. 2. Do we have to exchange s_1 and s_2?

However, this never happens since s_1 has been deactivated when the intersection point of $R(s_1)$ and $Q(s_2)$ is reached by SL: If $\tilde{p} := R(s_1) \cap Q(s_2)$ is such an intersection point with smallest x-coordinate and s_1 and s_2 are both active when $\tilde{p}$ is reached by SL then obviously s_1 and s_2 are neighbors in the y-table with respect to $\sqsubseteq$. Now s_1 has been deactivated since $\delta(s_1) \leq d_t(s_1, s_2) \leq \tilde{p}.x - s_1[R].x$. The same argument applies to the next such intersection point if we consider the set $S \setminus s_1$ instead of S, and similarly to all further intersection points. Hence active objects do not change their relative order with respect to $\sqsubseteq$ while SL is moving to the right. Since furthermore $s_1[SL] \neq_y s_2[SL]$ for two active objects $s_1 \neq s_2$ the relation $\sqsubseteq$ is a total ordering on the objects in the y-table.

During the sweep from left to right we maintain the following *sweep invariants*:

1) For each object s in the y-table $s[R].x+\delta(s)$ is to the right of the sweep-line SL.

2) For s_1, s_2 neighbored in the y-table with respect to $\sqsubseteq$ we have $d(s_1, s_2) \geq \delta(s_1)$ and $d(s_1, s_2) \geq \delta(s_2)$.

The first invariant is maintained by removing objects s from the y-table for which $s[R].x+\delta(s)$ is at or to the left of the sweep-line SL. In order to maintain the second invariant we have to compute distances of objects which become neighbors with respect to $\sqsubseteq$ in the y-table and update their δ-values, if necessary. We obtain such new neighbor pairs after inserting a new object into the y-table or after removing an object from the y-table.
It seems to be surprising that PSANN finds the correct result by just testing pairs of objects which become neighbors in the y-table with respect to $\sqsubseteq$.

3 Correctness

In this section we will answer the question why PSANN finds a nearest neighbor for each object, both in an intersection-free and in a diagonal-disjoint configuration S. As before d_t denotes an L^t-metric ($1 \leq t \leq \infty$).

Theorem 1. *For an intersection-free configuration S of n convex planar objects the algorithm PSANN finds a nearest neighbor for each object.*
Proof: Consider two objects $s_1 \in S$ and $s_2 \in S \cap NN(s_1)$ and two points $p_1 \in s_1$, $p_2 \in s_2$ such that $d_t(p_1, p_2) = d_t(s_1, s_2)$. We will show that for $p_2 \in QR(p_1)$ the objects s_1 and s_2 are neighbors with respect to $\sqsubseteq$ in the y-table when p_2 is reached by the sweep-line SL in the left-to-right sweep. This implies the existence of an event to the left of p_2 at which s_1 and s_2 become neighbors. If p_2 is contained in any of the three other quadrants $QL(p_1)$, $QB(p_1)$ or $QT(p_1)$ then s_1 and s_2 will become neighbors during the right-to-left, top-to-bottom or bottom-to-top sweep, respectively.

Now assume $p_2 \in QR(p_1)$. Consider the sweep-line SL at $p_2.x$. Obviously we have $\delta(s_1) \geq d_t(p_1, p_2)$ since $s_2 \in NN(s_1)$. If $\delta(s_1) = d_t(p_1, p_2)$ a nearest neighbor of s_1 has already been found and nothing remains to be shown. Therefore in the following we assume $\delta(s_1) > d_t(p_1, p_2)$ which implies that object s_1 is still active since $s_1[R].x+\delta(s_1) \geq p_1.x+\delta(s_1) > p_2.x$ is to the right of SL. Object s_2 is still active since it is intersected by SL. We further assume $p_2 \geq_y p_1$, the case $p_2 \leq_y p_1$ is treated analogously. Consider the closed rectangle $R_{1,2}$ spanned by the points p_1 and p_2, possibly degenerated to a line segment (Fig. 3). It is easy to see that besides p_1 and p_2 no other point of $s_1 \cup s_2$ can be contained in $R_{1,2}$: Let $\tilde{p}$ be a point in $(s_1 \cup s_2) \cap R_{1,2}$ with $\tilde{p} \notin \{p_1, p_2\}$. The conditions $p_1 \leq_y \tilde{p} \leq_y p_2$ and $p_1 \leq_x \tilde{p} \leq_x p_2$ imply that the d_t-distance of $\tilde{p}$ and the unique point in $\{p_1, p_2\}$ not contained in the same object as $\tilde{p}$ is less than $d_t(p_1, p_2)$. This is a contradiction to our choice of p_1 and p_2. In particular, the convexity of the objects implies that $Q(s_1)$, $Q(s_2)$, $T(s_1) := \overline{p_1, s_1[R]}$ and $T(s_2) := \overline{p_2, s_2[L]}$ are completely contained in the corresponding objects and therefore cannot intersect the rectangle $R_{1,2}$. This obviously implies $s_1[SL] \leq_y p_1$ and $s_2[SL] \geq_y p_2$, and by definition of $\sqsubseteq$ we get $s_1 \sqsubseteq s_2$. The identities $s_1[SL] =_y p_1$ and $s_2[SL] =_y p_2$ are equivalent to $s_1[R] = p_1$ and $s_2[L] = p_2$, respectively.

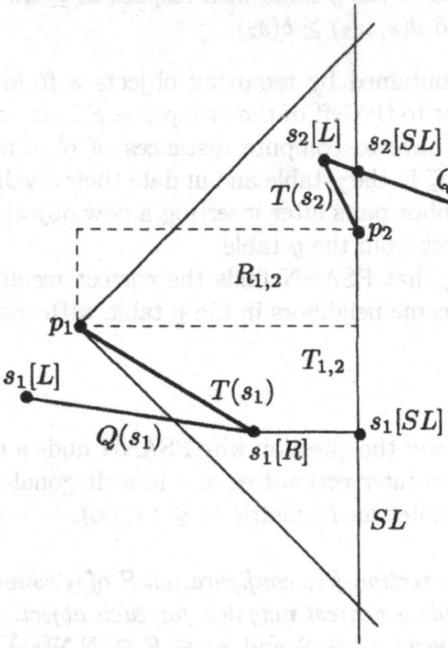

Fig. 3. Two objects with $s_2 \in NN(s_1)$, $p_2 \in QR(p_1)$ and $s_1 \sqsubseteq s_2$

Assume that there are *active* objects s^i, $1 \leq i \leq m$, with $s_1 \sqsubseteq s^1 \sqsubseteq \ldots \sqsubseteq s^m \sqsubseteq s_2$ separating s_1 and s_2 from each other when SL reaches p_2. The definition of $\sqsubseteq$ implies that

$$s_1[SL] <_y s^1[SL] <_y \ldots <_y s^m[SL] <_y s_2[SL] \quad (1)$$

We now show that $s^m[SL] \leq_y p_2$. The assumption $s^m[SL] >_y p_2$ together with (1) implies $s_2[SL] >_y p_2$ and therefore $s_2[L] >_y p_2$ since $T(s_2)$ does not intersect $R_{1,2}$. By definition of $s_2[L]$ we obviously have $s_2[L] \leq_x p_2$. Since our configuration S does not contain any intersection points, $Q(s^m)$ cannot intersect the non-degenerated triangle spanned by $T(s_2)$ and $s_2[SL]$. Together with (1) this implies an intersection point $\tilde{p}$ of $R(s^m)$ with $T(s_2)$ or $Q(s_2)$. Since s^m and s_2 are neighbors their distance has been calculated. Hence we have

$$\delta(s^m) \leq d_t(s^m, s_2) \leq d_t(s^m[R], \tilde{p}) \leq \tilde{p}.x - s^m[R].x \quad (2)$$

$\tilde{p} \leq_x p_2$ together with (2) implies $s^m[R].x + \delta(s^m) \leq p_2.x$; therefore s^m has already been deactivated in contradiction to our assumption that s^m is still active. Hence $s^m[SL] \leq_y p_2$.

Next we show that $s^1[SL] \geq_y p_1$. The assumption $s^1[SL] <_y p_1$ together with (1) implies $s_1[SL] <_y p_1$ and therefore $s_1[R] <_y p_1$. $s^1[R]$ cannot lie on the left of segment $T(s_1)$ since this would imply an intersection point of $R(s^1)$ with $T(s_1)$, and therefore s^1 would have been deactivated before. Hence there exists a point $\tilde{p} \in s^1$ contained in the trapezoid $T_{1,2}$ bounded by $\{q : p_2 \geq_x q \geq_x p_1 \wedge q =_y p_1\}$, $\{q : p_2 \geq_x q \geq_x s_1[R] \wedge q =_y s_1[R]\}$, $\{q : q =_x p_2 \wedge p_1 \geq_y q \geq_y s_1[R]\}$ and $T(s_1)$. Since the slope of $T(s_1)$ is non-positive it is easy to see that for all points $p \in T_{1,2}$ the unique point $u(p) \in T(s_1)$ with $u(p) =_y p$ is closer to p with respect to d_t than $d_t(p_1, p_2)$ and $d_t(u(\tilde{p}), \tilde{p}) < d_t(p_1, p_2)$, in contradiction to the choice of p_1 and p_2. This implies $s^1[SL] \geq_y p_1$.

Now (1), $s^m[SL] \leq_y p_2$ and $s^1[SL] \geq_y p_1$ imply $p_1 \leq_y s^1[SL] <_y \ldots <_y s^m[SL] \leq_y p_2$. In the following we show that s^1 has already been deactivated when SL reaches p_2, i.e. $s^1[R].x + \delta(s^1) \leq p_2.x$. It is easy to see that the rectangle $\tilde{R}_{1,2} := \{(x, y) : |x - p_1.x| < |p_1.x - p_2.x| \wedge p_1.y < y < p_2.y\}$ is completely contained in the L^t-disk $K^t_{\Delta(s_1)}(p_1)$ with radius $\Delta(s_1) = d_t(p_1, p_2)$ and center p_1, which shows that no point of s^1 can be contained in $\tilde{R}_{1,2}$. Together with $p_1 <_y s^1[SL] < p_2$ this implies $s^1[R].x < p_1.x - |p_2.x - p_1.x|$ and therefore $s^1[R] \in QL(p_1)$ taking into account that $p_1 \leq_y s^1[R] \leq_y p_2$. Since we have $\delta(s^1) \leq d_t(s^1, s_1) \leq d_t(s^1[R], p_1)$ it suffices to show that

$$s^1[R].x + d_t(s^1[R], p_1) \leq p_2.x \tag{3}$$

This inequality will be proved in Theorem 3. Assuming its correctness (3) implies that s^1 must have been deactivated when SL reaches p_2. This is a contradiction to our assumption that $s^1, \ldots, s^m$ are active objects separating s_1 and s_2. Hence s_1 and s_2 become neighbors before the sweep-line reaches p_2. □

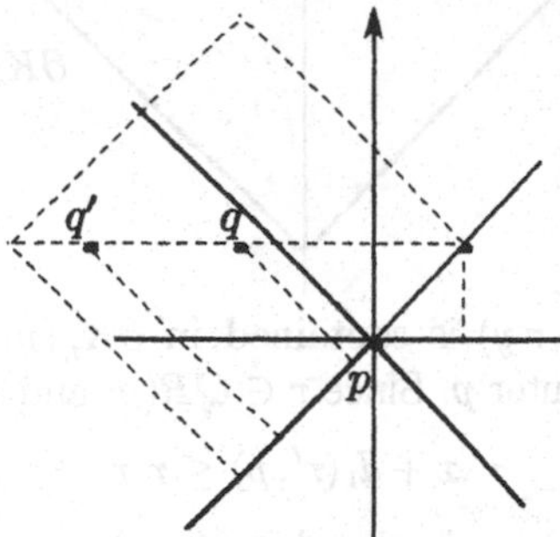

Fig. 4. Similar behavior of q and q'

To complete the proof of Theorem 1, we have to prove inequality (3). First we show that points in $QL(p)$ having the same y-coordinate behave similarly with respect to the sum of their x-coordinate and the L^t-distance to p (see Fig. 4):

Lemma 2. *For points $q, q' \in QL(p)$ with $q' =_y q$ we have*

$$q'.x + d_t(q', p) = q.x + d_t(q, p)$$

Proof: Let q be in $QL(p)$ and let us assume w.l.o.g. that $q \geq_y p$. We prove that $q.x + d_1(p,q)$ is independent of $q.x$ for fixed $q.y$. The condition $q \in QL(p)$ implies that $q \leq_x p$. Then $q.x + d_1(q,p) = q.x + |q.x - p.x| + |q.y - p.y| = q.y + p.x - p.y$ which is independent of $q.x$. □

We now prove the correctness of inequality (3). It is easy to verify that the points p_1, p_2 and $s^1[R]$ in the proof of Theorem 1 satisfy the conditions (i) - (iii) of the points p, r and l, respectively, in Theorem 3.

Theorem 3. *Given $p, l, r \in E^2$ with the conditions*

$$(i)\ r \in QR(p), l <_x r \quad (ii)\ r \geq_y l \geq_y p \text{ or } p \geq_y l \geq_y r \quad (iii)\ d_t(p,l) > d_t(p,r)$$

Then we have

$$l.x + d_t(p,l) \leq r.x.$$

Proof: W.l.o.g. we assume $p = (0,0)$ and $r \geq_y l \geq_y p$. For two arbitrary points $p \neq q \in E^2$ the following inequality holds:

$$d_1(p,q) \geq d_t(p,q) \geq d_\infty(p,q) \quad \forall t \in (1,\infty) \tag{4}$$

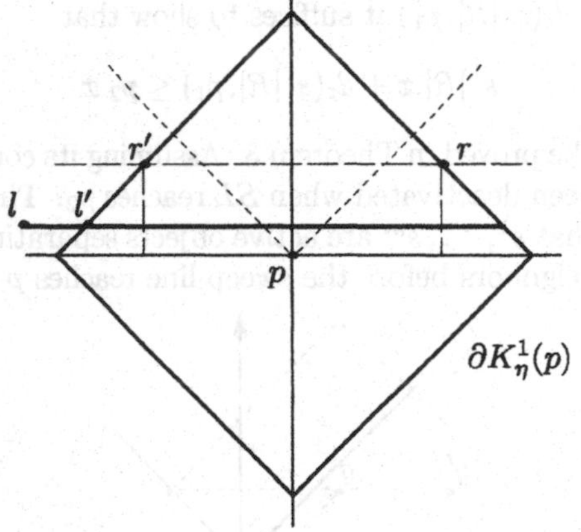

The point $r' := (-r.x, r.y)$ is contained in $\partial K^1_\eta(p)$, i.e. the L^1-circle with radius $\eta := d_1(p,r)$ and center p. Since $r \in QR(p)$ and $r' \in QL(p)$ we have

$$r'.x + d_1(r',p) \leq r.x \tag{5}$$

Conditions (i), (ii) and (iii) imply that $l \in \{(x,y) : x \leq r'.x \wedge p.y \leq y \leq r'.y\}$ and therefore $l \in QL(p)$. Denote by l' the unique point with $l =_y l'$ and $l' \in \partial K^1_\eta(p)$. By construction we have $l' \leq_x r'$. Since r', l' are in $\partial K^1_\eta(p)$ we have $d_1(p,r') = d_1(p,l')$. Furthermore by Lemma 2 we obtain

$$l.x + d_1(l,p) = l'.x + d_1(l',p) \tag{6}$$

Combining (4), (5) and (6) together with $l' \leq_x r'$ yields

$$l.x + d_t(p,l) \overset{(4)}{\leq} l.x + d_1(p,l) \overset{(6)}{=} l'.x + d_1(p,l') \leq r'.x + d_1(p,r') \overset{(5)}{\leq} r.x$$

which is the desired result. □

Theorem 4. *For a diagonal-disjoint configuration S of n convex planar objects the algorithm PSANN finds a nearest neighbor for each object.*
Proof: Using the same notations as in the proof of Theorem 1 we again restrict ourselves to the left-to-right sweep.

Assume $s_1 \in S$ intersects at least one other object in S, and let $\tilde{p} \in s_1$ be the intersection point of s_1 with minimal x-coordinate. Let s_2 be an object with $\tilde{p} \in s_1 \cap s_2$, such that no other object which contains $\tilde{p}$ lies between s_1 and s_2 with respect to $\sqsubseteq$ when $\tilde{p}$ is reached by the sweep-line. W.l.o.g. we may assume that $\tilde{p}$ is above $Q(s_1)$. Let the sweep-line SL intersect $\tilde{p}$. Since S is x-diagonal-disjoint, the segment $Q(s_2)$ cannot intersect s_1, and we may assume that $\tilde{p}$ lies on the boundary of s_1. This implies $s_1 \sqsubseteq s_2$. Note that by convexity the segments $T_1 := \overline{s_1[L],\tilde{p}}$ and $T_2 := \overline{s_2[L],\tilde{p}}$ lie completely in the objects s_1 and s_2, respectively.
Assume that when reaching the point $\tilde{p}$ with SL there are active objects

$$s_1 \sqsubseteq s^1 \sqsubseteq \ldots \sqsubseteq s^m \sqsubseteq s_2 \tag{7}$$

separating s_1 and s_2 from each other. An argument similar to the arguments used in the proof of Theorem 1 now shows that $m = 0$, i.e. s_1 and s_2 are neighbors with respect to $\sqsubseteq$ when the sweep-line reaches $\tilde{p}$.

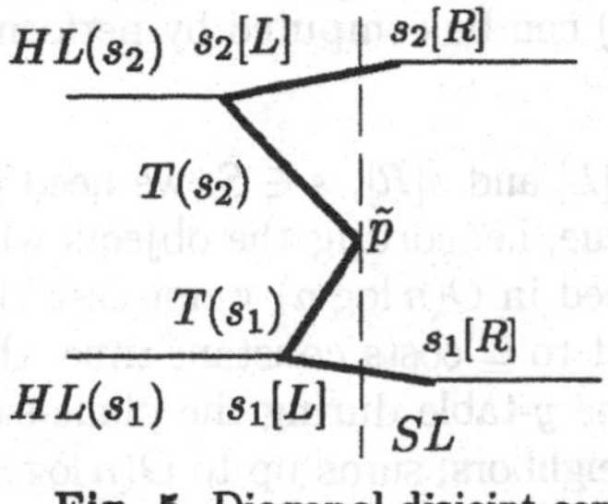

Fig. 5. Diagonal disjoint configuration

By the diagonal disjointness of S, the segments $Q(s^i)$ $(i = 1, \ldots, m)$ cannot intersect $T(s_1)$ or $T(s_2)$. If $s^i[R] \geq_x \tilde{p}$ then (7) implies that $s^i[L]$ is contained in one of the triangles spanned by $s_2[L]$, $\tilde{p}$ and $s_2[R]$ or by $s_1[L]$, $\tilde{p}$ and $s_1[R]$; these triangles may be degenerated to segments. But this contradicts the diagonal disjointness of S.
Therefore we only need to consider the case $s^i[R] <_x \tilde{p}$. Let $HL(s_1) := \{(x,y) : x \leq s_1[L].x, y = s_1[L].y\}$ and $HL(s_2)$ be defined analogously. Then $s^i[R]$ cannot be above $\overline{Q}(s_2) \cup HL(s_2)$ or below $\overline{Q}(s_1) \cup HL(s_1)$ since either this would lead to a contradiction to (7) or s^i would have been deactivated.

Hence all points $s^i[R]$ are contained in the 'wedge' bounded by $HL(s_1)$, $HL(s_2)$ and the segments $T(s_1)$ and $T(s_2)$. Since s_1 and s^1 are neighbors in the y-table, their distance has been checked and therefore we have $\delta(s_1) \leq d_t(s_1, s^1)$. The same is true for s^m and s_2. Our assumption that all objects s^i are active now implies that $s^1[R] >_y \tilde{p} \wedge s^m[R] <_y \tilde{p}$ in contradiction with $s^1 \sqsubseteq s^m$. Therefore no such active objects s^i $(i = 1, \ldots, m)$ can exist. s_1 and s_2 are neighbors in the y-table when SL reaches $\tilde{p}$ and therefore a nearest neighbor for s_1 and s_2 (with distance 0) will be found. □

4 Analysis

Let S be a set of n convex planar diagonal-disjoint objects. Let $M(s)$ denote the storage needed by algorithm PSANN for an object $s \in S$. Then obviously PSANN needs a total of $M(S) := \sum_{s \in S} M(s)$ storage, which is in $O(n)$ if each object uses $O(1)$ storage as it is true for configurations consisting of points, line segments or disks, and $O(N)$ for a configuration of n convex polygons with a total of N edges. Let the input configuration S be subset of a specific class C of compact, convex point sets of a certain type in the plane, e.g. the class of points, line segments, disks or the class of convex polygons. We assume that there exist functions g_C and h_C which depend on the underlying class C of objects such that:

- For each $s \in C$ the points $s[L]$ and $s[R]$ can be computed by at most $O(g_C)$ operations.

- For any $s_1, s_2 \in C$ their minimal distance $d_t(s_1, s_2)$ with respect to any L^t-metric d_t $(1 \leq t \leq \infty)$ can be computed by performing at most $O(h_C)$ operations.

For computing all points $s[L]$ and $s[R]$, $s \in S$, we need $O(n\, g_C)$ operations. The initialization of the x-queue, i.e. sorting the objects with respect to $s[L].x$ and $s[R].x$, can be accomplished in $O(n \log n)$ worst-case time. Each comparison of two objects with respect to $\sqsubseteq$ costs constant time; therefore the cost for all operations performed on the y-table during the plane-sweeps, i.e. inserting, deleting objects and finding neighbors, sums up to $O(n \log n)$ time in the worst case. PSANN computes the distance of at most $3(n-2)+1 = 3n-5$ pairs of objects $(n \geq 3)$, implying that all these distance computations take $O(n\, h_C)$ time. Each distance computation may result in shifting of two deactivation events which costs $O(\log n)$ and therefore in total $O(n \log n)$. Hence the plane-sweep algorithm PSANN solves the all-nearest-neighbors problem for a configuration S as defined above in time $O(n \log n + n(g_C + h_C))$. It is obvious that PSANN solves the all-nearest-neighbors problem for a configuration S of n objects in time $O(n \log n)$ and storage $O(n)$ if the underlying class C is the class of points, the class of line segments, the class of disks or another class for which we have $g_C, h_C \in O(1)$.

By using the same technique as in the Bentley-Ottman algorithm [3] for detecting line segment intersections PSANN can be modified to solve the all-nearest-neighbors problem for a configuration of n possibly intersecting line segments in time $O((n+k) \log n)$ where k is the number of intersecting line segment pairs.

Let the input configuration S consist of n convex polygons with a total of N edges each of which is represented by an array containing its vertices given in cyclic order. Then PSANN finds nearest neighbors with respect to the Euclidean metric in time $O(n \log N)$ and storage $O(N)$: For a convex polygon s given by its vertices in cyclic order the extreme points with respect to the lexicographic order can be determined in time $O(\log N)$ by Fibonacci search [4]. We can detect in optimal time $O(\log N)$ whether two of the convex polygons intersect [4]. The minimal Euclidean distance between two non-intersecting convex polygons can be computed in optimal time $O(\log N)$ ([5, 6]). Thus we have $g_C \in O(\log N)$ and $h_C \in O(\log N)$. Each of the N edges is stored exactly once, the representation of the objects in the y-table and the x-queue needs $O(n)$ storage which sums up to a storage complexity of $O(N + n) = O(N)$.

References

1. F. Bartling, Th. Graf, K. Hinrichs: A plane sweep algorithm for finding a closest pair among convex planar objects, *Preprints Angewandte Mathematik und Informatik*, Universität Münster, Bericht Nr. 1/93-I.
2. F. Bartling, K. Hinrichs: A plane-sweep algorithm for finding a closest pair among convex planar objects, A. Finkel, M. Jantzen (eds.), *STACS 92, 9th Annual Symposium on Theoretical Aspects of Computer Science*, Lecture Notes in Computer Science 577, 221-232, Springer-Verlag, Berlin, 1992.
3. J.L. Bentley and T. Ottmann, Algorithms for reporting and counting intersections, *IEEE Transactions on Computers C28*, 643-647 (1979).
4. B. Chazelle, D. P. Dobkin: Intersection of convex objects in two and three dimensions, *Journal of the ACM*, 34(1), 1-27 (1987).
5. F. Chin, C. A. Wang: Optimal algorithms for the intersection and the minimum distance problems between planar polygons, *IEEE Trans. Comput.* 32(12), 1203-1207 (1983).
6. H. Edelsbrunner: Computing the extreme distances between two convex polygons, *Journal of Algorithms* 6, 213-224 (1985).
7. S. Fortune: A sweepline algorithm for Voronoi diagrams, *Algorithmica* 2, 153-174 (1987).
8. P.-O. Fjällström. J.Katajainen, J. Petersson: Algorithms for the all-nearest-neighbors problem, *Report 92/2*, Dept. of Computer Science, University of Copenhagen, Denmark, 1992.
9. K.Hinrichs, J.Nievergelt, P.Schorn: An all-round algorithm for 2-dimensional nearest-neighbor problems, *Acta Informatica*, 29(4), 383-394 (1992).
10. P. Schorn: Robust algorithms in a program library for geometric computation, PhD Dissertation No. 9519, ETH Zürich, Switzerland, 1991
11. M.Shamos, D.Hoey: Closest-point problems, *Proceedings of the 16th Annual IEEE Symposium on Foundations of Computer Science*, 151-162 (1975).
12. P.Vaidya: An $O(n \log n)$ algorithm for the all-nearest-neighbours-problem, *Discrete & Computational Geometry*, 4, 101-115 (1989).
13. C.K. Yap: An $O(n \log n)$ algorithm for the Voronoi diagram of a set of simple curve segments, *Discrete Comput. Geometry* 2, 365-393 (1987).

Further Results on Generalized Intersection Searching Problems: Counting, Reporting, and Dynamization

Prosenjit Gupta[1] Ravi Janardan[1] Michiel Smid[2]

Abstract

In a generalized intersection searching problem, a set, S, of colored geometric objects is to be preprocessed so that given some query object, q, the distinct colors of the objects intersected by q can be reported or counted efficiently. In the dynamic setting, colored objects can be inserted into or deleted from S. These problems generalize the well-studied standard intersection searching problems and are rich in applications. Unfortunately, the techniques known for the standard problems do not yield efficient solutions for the generalized problems. Moreover, previous work on generalized problems applies only to the reporting problems and that too mainly to the static case. In this paper, a uniform framework is presented to solve efficiently the counting/reporting/dynamic versions of a variety of generalized intersection searching problems, including: 1-, 2-, and 3-dimensional range searching, quadrant searching, and 2-dimensional point enclosure searching. Several other related results are also mentioned.

Keywords: Computational geometry, data structures, dynamization, intersection searching, persistence.

[1]Department of Computer Science, University of Minnesota, Minneapolis, MN 55455, U.S.A. Email: {pgupta, janardan}@cs.umn.edu. The research of these authors was supported in part by NSF grant CCR–92–00270.

[2]Max-Planck-Institut für Informatik, W-6600 Saarbrücken, Germany. Email: michiel@mpi-sb.mpg.de. This author was supported by the ESPRIT Basic Research Actions Program, under contract No. 7141 (project ALCOM II).

1 Introduction

1.1 Intersection Searching Problems

Problems arising in diverse areas, such as computer graphics, robotics, VLSI layout design, and databases can often be formulated as ***intersection searching problems***. In a generic instance of such a problem, a set S of geometric objects is to be preprocessed into a suitable data structure so that given a query object q we can answer efficiently questions regarding the intersection of q with the objects in S. The problem comes in four versions, depending on whether we want to ***report*** the intersected objects or simply ***count*** their number, and whether S is ***static*** or ***dynamic***, i.e., changes through insertion and deletion of objects. We call these problems ***standard*** intersection searching problems to distinguish them from their ***generalized*** counterparts, which we investigate in this paper.

Typical examples of (S, q)-combinations include: (a) (points in $\mathbb{R}^d$, d-range): ***d-dimensional range searching***; (b) (d-ranges, point in $\mathbb{R}^d$): ***d-dimensional point enclosure searching***; (c) (points in $\mathbb{R}^2$, quadrant): ***Quadrant searching***; (d) (intervals on the real line, interval): ***Interval intersection searching***; (e) (horizontal line segments in $\mathbb{R}^2$, vertical line segment): ***Orthogonal segment intersection searching***. Due to their numerous applications, standard intersection searching problems have been the subject of much study and space- and query-efficient algorithms have been devised for many of them. See for example [Cha86, PS88].

1.2 Generalized Intersection Searching Problems

In a generalized intersection searching problem, the objects in S come aggregated in disjoint groups and of interest are questions regarding the intersection of q with the groups rather than with the objects. (q intersects a group if and only if it intersects some object in the group.) If we associate with each group a different color and color all the objects in the group with that color, then our goal is to report or count the distinct colors intersected by q. In the dynamic setting, an object of some (possibly new) color is inserted into S or an existing object is deleted. Note that the generalized problem reduces to the standard one when each color class has cardinality 1.

We give just one application of such generalized problems: VLSI designs often consist of several layers, each comprised of a large number (typically, thousands) of iso-oriented rectangles. Often it is necessary to wire certain subsets of these layers using vertical channels called ***vias***. Given a candidate position for a via (a point), the layout designer is faced with the problem of identifying the layers that get electrically connected, i.e., those layers that have at least one rectangle containing the via. This can be solved by assigning each layer a different color and solving an instance of the generalized 2-dimensional point enclosure searching problem.

1.2.1 Potential Approaches and Pitfalls

One approach to a generalized reporting problem is to solve the corresponding standard problem and then read off the distinct colors. However, the query time can be very high since q could intersect $\Omega(n)$ objects but only $O(1)$ distinct colors. We seek query times that are sensitive to the number, i, of distinct colors intersected, typically of the form $O(f(n)+i)$ or $O(f(n)+i \cdot g(n))$, where f and g are polylogarithmic. For a generalized counting problem, the situation is worse; it is not even clear how one can extract the answer from the answer (a mere count) to the corresponding standard problem. Thus it is clear that different techniques are needed.

Generalized intersection searching problems were first considered in [JL93], where efficient, output-sensitive solutions were presented for the generalized versions of several of the problems listed in Section 1.1. The key idea there was to first generate for each color class a sparse representation with the property that for certain special query objects q' (e.g., rays, grounded rectangles), q' intersected the sparse representation if and only if it intersected an object of the same color, and, furthermore, q' intersected only $O(1)$ objects in the sparse representation. Given q, it was decomposed into two special queries q' and q'' and a standard intersection query was answered with them. The answers to q' and q'' were then combined by eliminating duplicate colors. The time for doing this elimination could be charged to the output size.

The above approach does not yield efficient solutions to the counting and dynamic problems. For the counting problem, one could try to generate a sparse representation such that q' intersects *exactly* one (not just $O(1)$) objects in the representation and then use a standard counting structure. However, care must be taken in combining the counts returned by q' and q'', since if there is a color that is included in both counts then it should be counted only once for q; unfortunately, there seems to be no way to deduce this from the counts of q' and q''. As for the dynamic problem, the techniques of [JL93] do not lend themselves to efficient dynamization since even a single update can result in considerable change in a sparse representation.

1.3 Overview of Results and Techniques

In this paper, we present a uniform and general framework to solve efficiently the counting/reporting/dynamic versions of several generalized intersection problems, as summarized in Table 1. No solutions were known previously for any of the problems in Table 1, with the exception of the 2-dimensional point enclosure reporting problem; for this problem, an $O(n^{1.5})$-space, $O(\log n + i)$-query time solution was given in [JL93].[3] Note that in addition to the generalized counting problem described in Section 1.2 (which we call a ***type-1 counting problem***), we

[3] In recent related work, Agarwal and van Kreveld [AvK], consider the problem of reporting the simple polygons or the connected components of a set of line segments in $\mathbb{R}^2$ that are intersected by a query segment. They too cast the problem as a colored intersection searching problem.

also consider a second kind of counting problem, called a *type-2 counting problem*: Here we wish to report for each intersected color, the number of intersected objects of that color.

Generalized Problem	*Space*	*Query Time*	*Update Time*
1-D RANGE SEARCH			
Dynamic reporting	n	$\log n + i$	$\log n$
Static counting	$n \log n$	$\log n$	
	$n \log n / \log\log n$	$\log^2 n / \log\log n$	
Dynamic counting	$n \log n$	$\log^2 n$	$\log^2 n$
Static counting (type–2)	$n \log n$	$\log n + i$	
QUADRANT SEARCH			
Static reporting	n	$\log n + i$	
Static counting	$n \log n$	$\log n$	
Dynamic reporting	$n \log n$	$\log^2 n + i \log n$	$\log^2 n$
Dyn. reptg. (insertions only)	n	$\log^2 n + i$	$\log n$ (amort.)
INTERVAL INTERSECTION			
Static counting	$n \log n$	$\log n$	
Dynamic reporting	$n \log n$	$\log^2 n + i \log n$	$\log^2 n$
Dyn. reptg. (insertions only)	n	$\log^2 n + i$	$\log n$ (amort.)
2-D RANGE SEARCH			
Static counting	$n^2 \log^2 n$	$\log^2 n$	
	$n^3 \log n$	$\log n$	
Dynamic reporting	$n \log^3 n$	$\log^2 n + i \log n$	$\log^4 n$
Dyn. reptg. (insertions only)	$n \log^2 n$	$\log^2 n + i$	$\log^3 n$ (amort.)
3-D RANGE SEARCH			
Static reporting	$n \log^4 n$	$\log^2 n + i$	
1-D POINT ENCLOSURE			
Static counting	n	$\log n$	
Static counting (type–2)	n	$\log n + i$	
Dynamic reporting	n	$(i+1) \log n$	$\log n$ (amort.)
2-D POINT ENCLOSURE			
Static reporting	$n \log n$	$\log^2 n + i$	
Static counting	$n \log n$	$\log n$	
ORTH. SEGMENT INTERSECTION			
Static counting	$n \log^2 n$	$\log^2 n$	
	$n^2 \log n$	$\log n$	

Table 1: *Summary of main results. All bounds given are "big–oh" and, unless stated otherwise, are worst-case. Here $i \geq 0$ denotes the output size, i.e., the number of distinct colors intersected.*

Our results for the static counting and reporting problems are based on two main techniques: (i) persistent data structures and (ii) a simple geometric transformation. Roughly speaking, we use persistence as follows: To solve a given generalized problem we first identify a different but simpler generalized problem and devise a data structure for it that also supports updates (usually just insertions). We then make this structure partially persistent [DSST89] and query this persistent structure in an appropriate way to solve the original problem. On the other hand, the transformation method works by converting a generalized problem to a different *standard* problem and solving the latter. Although this method works (at least at this time) only for the generalized 1-dimensional range searching problem, it is crucial to our work since it also yields a dynamic data structure for the counting version of this problem, which is then used to build persistent structures for other generalized problems.

Our solutions for the dynamic problems are based on augmented data structures. Here again the transformation approach is crucial because it gives a dynamic data structure for the 1-dimensional range reporting problem, which is then used to build an augmented data structure for the dynamic quadrant reporting problem. Notice that for some dynamic problems we obtain improved solutions in the insertions-only case; these semi-dynamic solutions are vital since they are the building blocks of an efficient solution to the generalized 3-dimensional range reporting problem.

Due to space limitations, we describe here only a small subset of our results and omit all proofs and details. The full paper appears as [GJS92].

2 Generalized 1-Dimensional Range Searching

Let S be a set of n colored points on the x-axis. We wish to preprocess S so that for any query interval, q, we can report/count efficiently the distinct colors of the points in q and also update S efficiently.

In [JL93], the static reporting version of the problem was considered, and an $O(n)$-space and $O(\log n + i)$-query time algorithm was given. Using a result of [Lop91], the dynamic version of the problem can be solved in $O(n \log n)$ (resp., $O(n)$) space, $O(\log n + i)$ (resp., $O(\log^2 n + i)$) query time, and $O(\log^2 n)$ (resp., $O(\log n)$) amortized update time. Our dynamic algorithm is more efficient and much simpler. No results were known before for any of the counting problems.

We solve the problem using a transformation-based approach, as follows: For each color c, we sort the distinct points of that color by increasing x-coordinate. For each point p of color c, let $pred(p)$ be its predecessor in the sorted order; for the leftmost point of color c, we take the predecessor to be the point $-\infty$. We then map p to the point $p' = (p, pred(p))$ in the plane and associate with it the color c. Let S' be the resulting set of points. Given a query interval $q = [l, r]$, we map it to the grounded rectangle $q' = [l, r] \times (-\infty, l)$.

Lemma 2.1 *There is a point of color c in $q = [l, r]$ if and only if there is a point of color c in $q' = [l, r] \times (-\infty, l)$. Moreover, if there is a point of color c in q', then this point is unique.* □

Lemma 2.1 implies that we can solve the generalized 1-dimensional range reporting (resp., counting) problem by simply reporting the points in q' (resp., counting the number of points in q'), without regard to colors. In other words, we have transformed the generalized reporting (resp., counting) problem to the standard grounded range reporting (resp., counting) problem in two dimensions! The reporting problem can be solved easily using a priority search tree [McC85].

Theorem 2.1 *Let S be a set of n colored points on the real line. S can be preprocessed into a data structure of size $O(n)$ such that the i distinct colors of the points of S that are intersected by any query interval can be reported in $O(\log n + i)$ time and points can be inserted and deleted online in S in $O(\log n)$ time.* □

The standard counting problem can be solved using an array-augmented range tree [PS88], with fractional cascading on the arrays. For the dynamic counting problem we again use an augmented range tree, but with no fractional cascading.

Theorem 2.2 *Let S be a set of n colored points on the real line. S can be preprocessed into a data structure of size $O(n \log n)$ such that the number of distinctly-colored points of S that are intersected by any query interval can be counted in $O(\log n)$ time. Moreover, the dynamic version of the problem can be solved in $O(n \log n)$ space and $O(\log^2 n)$ query and update time, all worst case.* □

We remark that the static counting problem is also solvable in $O(n \log n)$ space and $O(\log n)$ query time using a persistence-based approach. We omit this here since the next section illustrates the persistence approach for a different problem.

3 Generalized Quadrant Searching

Let S be a set of n colored points in the plane. For any query point $q = (a, b)$, the *northeast quadrant* of q, denoted by $NE(q)$, is the set of all points (x, y) in the plane such that $x \geq a$ and $y \geq b$. We show how to preprocess S so that for any $NE(q)$ we can solve the static reporting and counting problems and the dynamic and semi-dynamic (insertions-only) reporting problem efficiently. We present an application of these results in Section 4.2.

3.1 The Static Reporting and Counting Problems

Consider the reporting problem. We first solve a related problem: Preprocess a set of colored points on the real line so that the i distinct colors intersected by a query interval $[a, \infty)$ can be reported efficiently and, moreover, points can be inserted efficiently.

For each color, we determine the rightmost point of that color and store all these points in sorted order at the leaves of a balanced search tree T and thread the leaves into a doubly-linked list L. Moreover, we store all the colors in a balanced search tree CT. With each color, we store the rightmost point of that color. A query can be answered by simply scanning L from right to left, stopping when a point to the left of a is encountered. The query time is $\Theta(i)$ and the space is $O(n)$. To insert a point p of color c we use CT to determine if p is to the right of the c-colored point (if any) currently in L and if so then we delete that point from T and insert p. This takes $O(\log n)$ time and causes $O(1)$ memory modifications in the list L.

To solve the quadrant reporting problem, we consider the points of S in non-increasing y-order and insert their x-coordinates into a partially persistent version of the above list L. (The trees T and CT are needed only to do updates efficiently in the current list. They are not used for queries and so need not be made persistent.) Given $NE(q)$, where $q = (a, b)$, we determine the smallest y-coordinate in S that is greater than or equal to b and query the corresponding version of L with $[a, \infty)$.

Theorem 3.1 *A set S of n colored points in the plane can be preprocessed in $O(n \log n)$ time into a data structure of size $O(n)$ such that for any query point q, the i distinct colors of the points lying in the northeast quadrant of q can be reported in $O(\log n + i)$ time.* □

For the static quadrant counting problem, the 1-dimensional problem that we solve is counting the number of distinct colors in any query interval $[a, \infty)$. For this we use the tree T above (but not L). Each internal node now stores a count of the number of leaves in its subtree. Given q, we identify $O(\log n)$ canonical nodes v such that q spans the range of v but not that of v's parent and sum the counts at all these nodes v. As above, the structure supports updates in $O(\log n)$ time but the number of memory modifications now is $O(\log n)$. To solve the quadrant counting problem, we make this structure partially persistent.

Theorem 3.2 *A set S of n colored points in the plane can be preprocessed in $O(n \log n)$ time into a data structure of size $O(n \log n)$ such that the number of distinctly-colored points lying in the northeast quadrant of any query point q can be reported in $O(\log n)$ time.* □

3.2 Dynamic Quadrant Reporting

We store the points of S in nondecreasing y-order from left to right at the leaves of a $BB(\alpha)$ tree T. For any node v of T, let $Y(v)$ be the y-coordinate stored at the leftmost leaf in $T(v)$. At v we store an auxiliary structure, $D(v)$, which is an instance of the structure of Theorem 2.1 for dynamic generalized 1-dimensional range searching. $D(v)$ is built on the x-coordinates of the points in $T(v)$.

Given a query $NE(q)$, where $q = (a, b)$, we proceed as follows: Let v be the current node in the search; initially, v is the root of T. If v is ***nil*** then we return. If $b \leq Y(v)$ then we query $D(v)$ with $[a, \infty)$ and return. If $b > Y(\mathit{right}(v))$ then we recursively search $T(\mathit{right}(v))$; otherwise ($b \leq Y(\mathit{right}(v))$), we query $D(\mathit{right}(v))$ with $[a, \infty)$ and then recursively search $T(\mathit{left}(v))$.

Insertion/deletion of a point is done using the worst-case updating strategy for $BB(\alpha)$ trees [WL85].

Theorem 3.3 *Let S be a set of n colored points in the plane. S can be stored in a data structure of size $O(n \log n)$ such that for any query point q, the i distinct colors of the points lying in q's northeast quadrant can be reported in $O(\log^2 n + i \log n)$ time and points can be inserted and deleted in $O(\log^2 n)$ time.* □

3.3 Semi-dynamic Quadrant Reporting

The results of Section 3.2 can be improved substantially if only insertions are allowed. Our scheme uses $O(n)$ space, has a query time of $O(\log^2 n + i)$, and an amortized insertion time of $O(\log n)$. More importantly, this result is one of the building blocks of an efficient scheme for generalized 3-dimensional range searching which we describe in Section 5.

For each color c, we determine the c-maximal points. (A point p is called *c-maximal* if it has color c and if there are no points of color c in p's northeast quadrant.) We discard all points of color c that are not c-maximal. In the resulting set, let the predecessor, *pred*(p), of a c-colored point p be the c-colored point that lies immediately to the left of p. (For the leftmost point of color c, the predecessor is the point $(-\infty, \infty)$.) With each point $p = (a, b)$, we associate the horizontal segment with endpoints (a', b) and (a, b), where a' is the x-coordinate of *pred*(p). This segment gets the same color as p. Let S_c be the set of such segments of color c. (Note that the segments in S_c form a staircase.) The data structure consists of the following:

1. A structure $\mathcal{T}$ storing the segments in the sets S_c, where c runs over all colors. $\mathcal{T}$ supports the following query: given a point q in the plane, report the segments that are intersected by the upward-vertical ray starting at q. Moreover, it allows segments to be inserted and deleted. We implement $\mathcal{T}$ as the structure given in [CJ90]. This structure uses $O(n)$ space, supports insertions and deletions in $O(\log n)$ time, and has a query time of $O(\log^2 n + l)$, where l is the number of segments intersected.
2. A balanced search tree, *CT*, storing all colors. For each color c, we maintain a balanced search tree, T_c, storing the segments of S_c by increasing y-coordinate. This structure supports the following operation: Given a c-colored point p in the plane, find all segments in S_c that intersect the southwest quadrant, $SW(p)$, of p. Moreover, after the segments have been found, p must be inserted into T_c.

Let $s_1, \ldots, s_k$ be the segments, in increasing y-order, that intersect $SW(p)$. These segments can be found by binary search in T_c. To insert p, we do the following on T_c. We delete $s_2, \ldots, s_k$ from T_c, insert the horizontal segment that starts at p and stretches left to the x-coordinate of the left endpoint of s_k, and keep only that part of s_1 that stretches to the right of p's x-coordinate. The entire operation can be done in $O((k+1)\log n)$ time (or even $O(\log n + k)$ time).

To answer a quadrant query, $NE(q)$, we query $\mathcal{T}$ with the upward-vertical ray from q and report the colors of the l segments intersected. Since the query can intersect at most two segments in any S_c, we have $l \leq 2i$, and so the query time is $O(\log^2 n + i)$.

Let p be a c-colored point that is to be inserted into S. If c is not in *CT*, then we insert it into *CT* and insert the horizontal, leftward-directed ray emanating from p. If c is present already, we find the segments $s_1, \ldots, s_k$ as just described. We update *CT* and then perform the same updates on $\mathcal{T}$. Hence, an insertion takes $O((k + 1)\log n)$ time. By a charging argument it can be shown that the amortized insertion time is $O(\log n)$.

Theorem 3.4 *Let S be a set of n colored points in the plane. There exists a data structure of size $O(n)$ such that for any query point q, we can report the i distinct colors of the points that are contained in the northeast quadrant of q in $O(\log^2 n + i)$ time. Moreover, if we do n insertions into an initially-empty set then the amortized insertion time is $O(\log n)$.* □

4 Generalized 2-Dimensional Range Searching

We show how to preprocess a set S of n colored points in the plane so that for any axes-parallel query rectangle $q = [a, b] \times [c, d]$, we can solve the static counting, dynamic reporting, and semi-dynamic reporting problems efficiently.

No results were known for these problems. The static reporting version had been solved in [JL93] in $O(n \log^2 n)$ (resp., $O(n \log n)$) space and $O(\log n + i)$ (resp., $O(\log^2 n + i)$) query time.

4.1 The Static Counting Problem

We first solve the problem for $q' = [a, \infty) \times [c, d]$. We sweep over the points of S by nonincreasing x-coordinate and insert their y-coordinates into a partially persistent version of the structure of Theorem 2.2 for dynamic 1-dimensional range counting. Given q', we access the version corresponding to the smallest x-coordinate x_0 such that $x_0 \geq a$ and query it with $[c, d]$.

To solve the problem for $q = [a, b] \times [c, d]$, we build for each distinct x-coordinate $\hat{x}$ in S an instance of the above structure for those points whose x-coordinate is less than or equal to $\hat{x}$. Given q, we access the structure corresponding to the greatest x-coordinate x_1 in S such that $x_1 \leq b$ and query it with q'.

Theorem 4.1 *A set S of n colored points in the plane can be preprocessed in $O(n^2 \log^2 n)$ time into a data structure of size $O(n^2 \log^2 n)$ such that for any axes-parallel query rectangle $q = [a, b] \times [c, d]$, the number of distinctly-colored points in q can be reported in $O(\log^2 n)$ time.* □

4.2 The Dynamic and Semi-dynamic Reporting Problems

Our solution is based on the dynamic quadrant reporting structures of Sections 3.2 and 3.3 We first solve the problem for $q' = [a, b] \times [c, \infty)$.

We store the points of S in sorted order by x-coordinate at the leaves of a $BB(\alpha)$ tree T'. At each internal node v, we store an instance of the structure of Theorem 3.3 for *NE*-queries (resp., *NW*-queries) built on the points in v's left (resp., right) subtree. To answer a query q', we do a binary search down T', using $[a, b]$, until we reach the first node v (if any) such that a and b are in different subtrees of v. We query the structures at v using the *NE*-quadrant and the *NW*-quadrant derived from q' and then combine the answers. Updates on T' are performed as in [WL85].

Lemma 4.1 *A set S of n colored points in the plane can be preprocessed into a structure of size $O(n \log^2 n)$ such that the i distinct colors of the points lying inside a grounded axes-parallel query rectangle $q = [a, b] \times [c, \infty)$ can be reported in $O(\log^2 n + i \log n)$ time. The structure supports updates in $O(\log^3 n)$ time.* □

We can now solve the problem for $q = [a, b] \times [c, d]$ as follows: We store the points of S by sorted y-order at the leaves of a $BB(\alpha)$ tree T. At each internal node v, we build an instance of the structure of Lemma 4.1 for points in v's left (resp., right) subtree to answer queries with upward-grounded (resp., downward-grounded) rectangles. To answer a query, we search in T with $[c, d]$ and decompose q at some node v of T into two grounded rectangles and query v's auxiliary structures with these. Updates are done as in [WL85].

Theorem 4.2 *A set S of n colored points in the plane can be preprocessed into a structure of size $O(n \log^3 n)$ such that the i distinct colors of the points lying inside an axes-parallel query rectangle can be reported in $O(\log^2 n + i \log n)$ time. The structure supports updates in $O(\log^4 n)$ time.* □

Similarly, we can solve the semidynamic problem. (The only difference is that the structure of Theorem 3.3 is replaced by the one of Theorem 3.4.)

Theorem 4.3 *Let S be a set of n colored points in the plane. There exists a data structure of size $O(n \log^2 n)$ such that for any query rectangle $[a, b] \times [c, d]$, we can report the i distinct colors of the points that are contained in it in $O(\log^2 n + i)$ time. Moreover, if we do n insertions into an initially-empty set then the amortized insertion time is $O(\log^3 n)$.* □

5 Generalized 3-Dimensional Range Searching

The semidynamic structure of Theorem 4.3 coupled with the idea of persistence allows us to go up one dimension and solve the following problem: Preprocess a set S of n colored points in 3-space, so that for any query box $q = [a, b] \times [c, d] \times [e, f]$ the i distinct colors of the points inside q can be reported efficiently.

First consider queries $q' = [a, b] \times [c, d] \times [e, \infty)$. We sort the points by non-increasing z-coordinates, and insert them in this order into a partially persistent version of the structure of Theorem 4.3, taking only the first two coordinates into account. To answer q', we access the version corresponding to the smallest z-coordinate greater than or equal to e and query it with $[a, b] \times [c, d]$.

Lemma 5.1 *Let S be a set of n colored points in 3-space. There exists a data structure of size $O(n \log^3 n)$ that can be built in $O(n \log^3 n)$ time, such that for any query box $[a, b] \times [c, d] \times [e, \infty)$, we can report the i distinct colors of the points that are contained in it in $O(\log^2 n + i)$ time.* □

From Lemma 5.1 we immediately get the following for any query box q:

Theorem 5.1 *Let S be a set of n colored points in 3-space. There exists a data structure of size $O(n \log^4 n)$ that can be built in $O(n \log^4 n)$ time, such that for any query box $[a,b] \times [c,d] \times [e,f]$, we can report the i distinct colors of the points that are contained in it in $O(\log^2 n + i)$ time.* □

6 Generalized 2-Dimensional Point Enclosure Searching

We show how to preprocess a set S of n colored, axes-parallel rectangles in the plane so that given a query point $q = (a, b)$, we can solve efficiently the reporting and counting problems.

We create a *segment tree*, T, on the distinct x-coordinates of the vertical sides of the rectangles in S. Each node of T will contain a data structure for the generalized 1-dimensional point enclosure problem for an appropriate set of y-intervals, which we now define: Let $r = [x_1, x_2] \times [y_1, y_2]$ be a rectangle of S. Let v be a node of T such that the range of v is contained in $[x_1, x_2]$, but the range of v's parent is not. Then, rectangle r—or more precisely, the interval $[y_1, y_2]$—is *associated* with v. (However, we do not necessarily store $[y_1, y_2]$ at v.)

Now we define the set of y-intervals that are *allocated* to v—these are the intervals on which the auxiliary structure of v is built. If v is the root of T then the intervals allocated to it are just the intervals associated with it. If v is a non-root node of T, then let $v_1, v_2, \ldots, v_m = v$ be the nodes on the path from the root, v_1, to v. For each color c, let $s_1, \ldots, s_k$ be the pairwise disjoint y-intervals that form the union of all c-colored y-intervals that are allocated to $v_1, \ldots, v_{m-1}$. Moreover, let $t_1, \ldots, t_l$ be the pairwise disjoint y-intervals that form the union of all c-colored y-intervals that are associated with v. Then, we allocate to v the pairwise disjoint c-colored y-intervals that span $\bigcup_{i=1}^{l} t_i \setminus \bigcup_{i=1}^{k} s_i$.

We take the intervals of all colors allocated to v and store them in the data structure for generalized 1-dimensional point enclosure reporting given in [JL93]. For a set of m intervals, this structure uses $O(m)$ space and has a query time of $O(\log m + i)$. By a careful counting argument (see [GJS92]), it can be shown that the total space is $O(n \log n)$.

Given a query point $q = (a, b)$, we proceed as follows: Assume that a is in the range of the root of T; otherwise, we can stop. We do a binary search in T for a and query with b the auxiliary structure of each node v visited.

Theorem 6.1 *A set S of n colored, axes-parallel rectangles in the plane can be preprocessed into a data structure of size $O(n \log n)$ so that for any query point q, the i distinct colors of the rectangles containing q can be reported in $O(\log^2 n + i)$ time.* □

Similarly, we can solve the counting problem by using a structure for generalized 1-dimensional point enclosure counting (see Table 1.). In this case, the auxiliary structures (which are arrays) admit fractional cascading and so the query time is $O(\log n)$.

Theorem 6.2 *A set S of n colored, axes-parallel rectangles in the plane can be preprocessed into a data structure of size $O(n \log n)$ such that the number of distinctly-colored rectangles that contain a query point can be determined in $O(\log n)$ time.* □

References

[AvK] P.K. Agarwal and M. van Kreveld. Connected component and simple polygon intersection searching. This Proceedings.

[Cha86] B.M. Chazelle. Filtering search: a new approach to query-answering. *SIAM Journal on Computing*, 15:703–724, 1986.

[CJ90] S.W. Cheng and R. Janardan. Efficient dynamic algorithms for some geometric intersection problems. *Information Processing Letters*, 36:251–258, 1990.

[DSST89] J.R. Driscoll, N. Sarnak, D.D. Sleator, and R.E. Tarjan. Making data structures persistent. *Journal of Computer and System Sciences*, 38:86–124, 1989.

[GJS92] P. Gupta, R. Janardan, and M. Smid. Further results on generalized intersection searching problems: counting, reporting, and dynamization. Technical Report TR–92–72, Dept. of Computer Science, University of Minnesota, 1992. Submitted.

[JL93] R. Janardan and M. Lopez. Generalized intersection searching problems. *International Journal of Computational Geometry & Applications*, 3:39–69, 1993.

[Lop91] M. Lopez. *Algorithms for composite geometric objects*. PhD thesis, Department of Computer Science, University of Minnesota, Minneapolis, Minnesota, 1991.

[McC85] E.M. McCreight. Priority search trees. *SIAM Journal on Computing*, 14:257–276, 1985.

[PS88] F.P. Preparata and M.I. Shamos. *Computational Geometry – An Introduction*. Springer–Verlag, 1988.

[WL85] D.E. Willard and G.S. Lueker. Adding range restriction capability to dynamic data structures. *Journal of the ACM*, 32:597–617, 1985.

Generalized Approximate Algorithms for Point Set Congruence

Paul J. Heffernan *
Dept. of Mathematical Sciences
Memphis State University, Memphis, Tennessee 38152

Abstract

We address the question of determining if two point sets are congruent. This task consists of defining the concept of point congruence, and developing algorithms that test for it. We introduce the (ε,k)-*map*, a device which pictorially represents the degree of congruence between two point sets. Point set congruence has been studied by previous researchers, but we feel that our definitions offer a more general and powerful approach to the problem. By using the paradigm of *approximate algorithms*, we are able to construct the (ε,k)-map efficiently.

1 Introduction

In this paper we address the following question: given two point sets in $\mathbb{R}^d$, what do we mean when we say that they are "congruent," and how do we determine congruency? To answer this question, we must define congruency, and provide efficient algorithms that test for it. The practical motivation behind our study comes from computer vision, where an observed image is compared to a hypothesized model. While previous researchers have formalized and studied this problem, we feel that the formulations often have been too restrictive. In particular, an elegant formulation has been introduced by [AMWW] (also [Ba]) and studied further by [HS]. It is this formulation, which we call ε-congruence, which we generalize here, in an attempt to lessen its restrictions while preserving its strengths.

Roughly speaking, the ε-congruence formulation is as follows. The input is assumed to be two equal cardinality, planar point sets, with a given metric (usually L_2 or L_∞) and a given family of isometries (an isometry is a mapping from $\mathbb{R}^d$ to itself that preserves distances). There are two problem versions, decision and optimization. In the decision problem, one asks for a given value $\varepsilon > 0$ whether an isometry from the family can be applied to one point set and a matching found between the sets such that the points in each matched pair

*Supported in part by the U.S. Army Research Office, Grant No. DAAL03-92-G-0378

are within distance ε of each other. In the optimization problem, one asks for the smallest value ε that returns an affirmative answer in the decision problem.

Formulation of the model as a point set is appropriate in applications where the model consists of a number of small pieces, such as locations on a map, or the model is a figure that can be represented by its key boundary points, such as letters of an alphabet. The ε-congruence formulation has both advantages and drawbacks. One of its main strengths is that it allows for the real-world possibility of noisy input—that is, imprecisions in recording the image points—by considering a pair of points to be matched if they are merely within a tolerance distance ε of each other. Every point of both the model and image is important, since each point must be matched to a nearby partner. This allows us to distinguish between models that vary in only a few points. On the other hand, if even one image point encounters an amount of noise exceeding the tolerance value, we may falsely answer that an image does not resemble a model. Another concern with the strict definition of [AMWW] is the requirement that the two point sets have equal cardinality. Among the imperfections of image registrations are missing points, which fail to appear on the image, and spurious points, which are erroneously introduced into the image. The possibility of such events renders the equal cardinality assumption impractical. Additionally, in [AMWW] attention is limited to 2-dimensional point sets.

Our goal in this paper is to abandon the rigidity of the traditional definition of ε-congruency. In lieu of the decision problem, we propose the following generalized version: for a given $\varepsilon > 0$, find the largest number of point pairs that can be placed within distance ε of each other, under some matching and isometry from the family. The generalized optimization problem asks, for a given value k, to find the smallest ε for which the decision problem returns a value of at least k. We introduce and show how to compute a structure called the *(ε,k)-map*, which stores the complete generalized solution. We drop the assumption that the point sets must be of equal cardinality, and we expand our methods to higher dimensional point sets. We consider a variety of metrics and families of isometries. We even consider the case of *projection congruency*, in which a 2-dimensional image is compared to a 3-dimensional model. As a result of these efforts, we build a more powerful framework for congruency of point sets.

While Alt, et al. [AMWW] give algorithms for the ε-congruence problem, most suffer from high worst-case run-times. A way to obtain efficient algorithms while preserving the formulation is through the use of *approximate algorithms* for ε-congruence, first introduced by Schirra [Sc1, Sc2]. Faster approximate algorithms have subsequently been developed by Heffernan and Schirra [HS]. In order to develop algorithms for our generalized definition of point set congruence, we draw on the techniques of [HS], and develop a body of approximate algorithms.

2 Generalized approximate congruence

In the point set congruence problem, we are given sets A and B. We wish to match points of A to points of B so that as many pairs as possible are as close as possible. In this sense we have a bicriteria problem with a trade-off between the two goals. In order to obtain workable formulations, we can fix one goal while optimizing the other. Under this approach, one problem consists of matching the points in order to maximize the number of "close" pairs, where "close" is determined by a fixed tolerance value. The corresponding problem fixes the number of close pairs and asks for the minimum possible closeness parameter. In either problem, we are allowed to impose an isometry from a given family on one of the point sets.

We formalize these notions as follows. We are given two point sets A and B in $\mathbb{R}^d$, with $|A| = n$, $|B| = m$, $n \leq m$, a family of isometries $\mathcal{I}$, and a metric $dist(\cdot,\cdot)$ on $\mathbb{R}^d$. We define two functions,

$$k_{max} : \mathbb{R}^+ \rightarrow \{2,\ldots,n\} \text{ and } \varepsilon_{opt} : \{2,\ldots,n\} \rightarrow \mathbb{R}^+,$$

which describe the congruence between A and B. The definitions are as follows.

- We say that A and B are *(ε,k)-congruent* if there exist $I \in \mathcal{I}$ and an injection (1-to-1 function) $\ell : A \rightarrow B$ such that $dist(\ell(a), I(a)) \leq \varepsilon$ for at least k points $a \in A$.
- For $\varepsilon > 0$, $k_{max}(\varepsilon)$ (or $k_{max}(\varepsilon, A, B)$) is the maximum value k such that A and B are (ε,k)-congruent.
- For $k \in \{2,\ldots,n\}$, $\varepsilon_{opt}(k)$ (or $\varepsilon_{opt}(k, A, B)$) is the smallest value $\varepsilon > 0$ such that A and B are (ε,k)-congruent.

Combining the notions behind the functions k_{max} and ε_{opt} leads naturally to a structure that we call the *(ε,k)-map*, an example of which is pictured in Figure 1. The (ε,k)-map plots k against ε, and it consists of both a YES region and a NO region of ordered pairs (ε,k), such that A and B are (ε,k)-congruent if and only if (ε,k) is in the YES region. The (ε,k)-map allows one to see the interaction between the two goals of obtaining a small closeness parameter and a large matching. One wishes for a point in the upper-left corner of the map, while the YES region is anchored in the lower-right corner. The boundary between the two regions is a step-wise curve from lower-left to upper-right (the peculiarity of a discontinuous curve serving as a boundary occurs because k assumes discrete values). In essense, the (ε,k)-map is a picture of the congruence between A and B.

The (ε,k)-map offers some clear advantages over the traditional definition of ε-congruence. The latter deals only with matchings of size n, and therefore limits its perspective to the top line of the (ε,k)-map. Much additional information is contained in the full (ε,k)-map: two pairs of point sets may have the same value of $\varepsilon_{opt}(n)$ but a sharp difference at $\varepsilon_{opt}(n-i)$ for a small integer i. Indeed, large recording errors for a few image points can increase $\varepsilon_{opt}(n)$ greatly, while

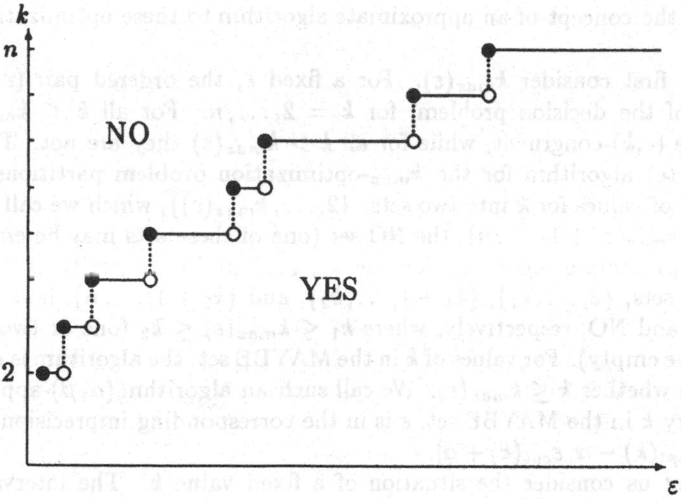

Figure 1: The (ε,k)-map

$\varepsilon_{opt}(n-i)$, where i equals the number of such "outliers," gives a much truer picture. We should also note that the (ε,k)-map drops the assumption of equal cardinality point sets. The relaxation allows the formulation to encompass the case of lost or spurious points in the image. Once such points exist, the idea of a "perfect matching" between the two point sets becomes less meaningful, and the concept of the (ε,k)-map more appropriate. For these reasons we feel that the (ε,k)-map is the true manner in which to think about approximate congruence.

3 Generalized approximate algorithms

Approximate decision algorithms were introduced in [Sc1, Sc2], and expanded in [HS]. Here we present a generalized description of approximate algorithms suitable for (ε,k)-congruence. For fixed ε and k, the decision problem asks whether input point sets A and B are (ε,k)-congruent. It is natural to believe that this question is most difficult when ε is near $\varepsilon_{opt}(k)$, so we will allow an approximate algorithm not to answer in such a situation. An *(α,β)-approximate decision algorithm* for (ε,k)-congruence of A and B is one which (1) correctly determines whether A and B are (ε,k)-congruent whenever $\varepsilon \notin [\varepsilon_{opt}(k)-\alpha, \varepsilon_{opt}(k)+\beta]$, and (2) either correctly determines (ε,k)-congruence or chooses not to answer when $\varepsilon \in [\varepsilon_{opt}(k)-\alpha, \varepsilon_{opt}(k)+\beta]$. We call $[\varepsilon_{opt}(k)-\alpha, \varepsilon_{opt}(k)+\beta]$ the *imprecision interval*. A major strength of approximate decision algorithms is that they never return an incorrect answer; if a particular query (ε,k) is too difficult, the algorithm simply says that it is unsure. Previous approximate decision algorithms have been defined only for $k=n$.

Earlier, we stated two optimization problems: maximizing k for fixed ε (i.e. computing $k_{max}(\varepsilon)$) and minimizing ε for fixed k (computing $\varepsilon_{opt}(k)$). We wish to extend the concept of an approximate algorithm to these optimization problems.

Let us first consider $k_{max}(\varepsilon)$. For a fixed ε, the ordered pair (ε,k) is an instance of the decision problem, for $k = 2,\ldots,n$. For all $k \leq k_{max}(\varepsilon)$, A and B are (ε,k)-congruent, while for all $k > k_{max}(\varepsilon)$ they are not. Therefore, a (complete) algorithm for the k_{max}-optimization problem partitions the set $\{2,\ldots,n\}$ of values for k into two sets: $\{2,\ldots,k_{max}(\varepsilon)\}$, which we call the YES set, and $\{k_{max}(\varepsilon)+1,\ldots,n\}$, the NO set (one of these sets may be empty).

An approximate algorithm for the $k_{max}(\varepsilon)$ problem partitions $\{2,\ldots,n\}$ into three sets, $\{2,\ldots,k_1\}$, $\{k_1+1,\ldots,k_2\}$, and $\{k_2+1,\ldots,n\}$, labelled YES, MAYBE, and NO, respectively, where $k_1 \leq k_{max}(\varepsilon) \leq k_2$ (one or two of these sets may be empty). For values of k in the MAYBE set, the algorithm is unable to determine whether $k \leq k_{max}(\varepsilon)$. We call such an algorithm (α,β)-approximate if, for every k in the MAYBE set, ε is in the corresponding imprecision interval, i.e. $\varepsilon \in [\varepsilon_{opt}(k)-\alpha, \varepsilon_{opt}(k)+\beta]$.

Now let us consider the situation of a fixed value k. The interval $\mathbb{R}^+$ of possible values for ε is divided into the NO subinterval $(0,\varepsilon_{opt}(k))$ and the YES subinterval $[\varepsilon_{opt}(k),\infty)$; these intervals are returned by a (complete) algorithm for the $\varepsilon_{opt}(k)$ problem when it is given input k. An approximate algorithm for the $\varepsilon_{opt}(k)$ problem, when given k, returns intervals $(0,\varepsilon_1)$, $[\varepsilon_1,\varepsilon_2)$, and $[\varepsilon_2,\infty)$, labelled NO, MAYBE, and YES, respectively, where $\varepsilon_1 \leq \varepsilon_{opt}(k) \leq \varepsilon_2$. As above, the MAYBE set represents those values of ε for which the algorithm is not able to determine (ε,k)-congruence. Such an algorithm is (α,β)-approximate if, for every ε in the MAYBE set, ε is in the corresponding imprecision interval; that is, $\varepsilon_1,\varepsilon_2 \in [\varepsilon_{opt}(k)-\alpha, \varepsilon_{opt}(k)+\beta]$.

4 Building the (ε,k,ν)-approximate map

In this section we show how to use a (γ,γ)-approximate algorithm for the $k_{max}(\varepsilon)$ problem in order to build a (ε,k,ν)-*approximate map* (see Figure 2). A (ε,k,ν)-approximate map is similar to an (ε,k)-map, except that in the approximate map, $\varepsilon_{opt}(k)$ for a given k is represented, not as a single point, but as an interval of length no more than ν that contains $\varepsilon_{opt}(k)$. The (ε,k)-congruent region is separated from the non-congruent region not by a step-wise curve, but by a region, which we call the imprecision region (or, the MAYBE region).

Our approach consists of a two-part search. The first part, which we call Phase I, finds for each $k \in \{2,\ldots,n\}$ an interval $(c,2c)$ that contains $\varepsilon_{opt}(k)$. In Phase II, a binary search is performed on each interval $(c,2c)$ until the interval size is at most ν.

Since a (γ,γ)-approximate algorithm is more precise for small values of γ, we expect its run-time to vary inversely with γ. Actually, we will see that the run-times for our approximate algorithms for the $k_{max}(\varepsilon)$ problem vary directly with ε/γ, so we attempt to keep this value small when constructing our (ε,k)-map algorithm. We will see that the time-complexity of Phase II dominates that of Phase I. We will describe the general (ε,k)-map estimation method now. Later

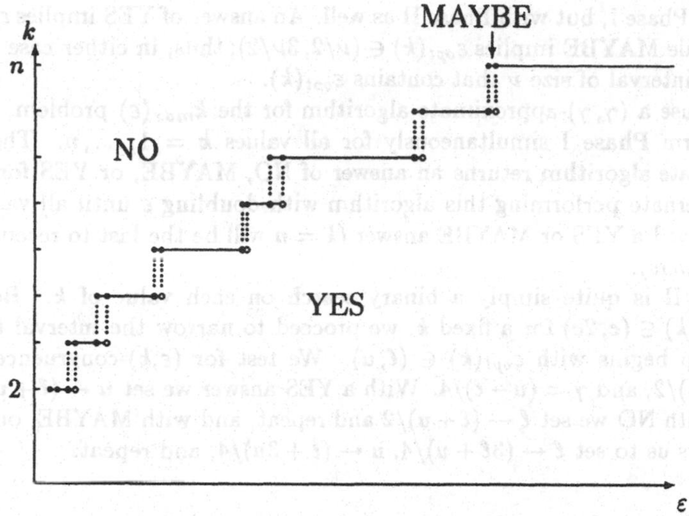

Figure 2: The (ε, k, ν)-approximate map

we will give (γ,γ)-approximate algorithms for the $k_{max}(\varepsilon)$ problem in various dimensions $\mathbb{R}^d$ and under various families of isometries $\mathcal{I}$, and after analyzing each such algorithm we will compute the time-complexity of the corresponding (ε, k)-map estimation.

The idea behind Phase I is simple: compute $k_{max}(\varepsilon)$, initially for a small value ε, then repeatedly double ε and recompute $k_{max}(\varepsilon)$, until ε is as large as $\varepsilon_{opt}(k)$. At each step we use a (γ,γ)-approximate algorithm, where we set $\gamma = \varepsilon/2$. Let us first describe how Phase I works for a single value k, and then generalize the approach for all k.

We begin by setting $\varepsilon = \nu$, and then alternating the steps of testing for (ε, k)-congruence by using a $(\varepsilon/2, \varepsilon/2)$-approximate algorithm, and doubling ε, until a test returns an answer of YES or MAYBE. (Note that an answer of YES (NO) to an (ε, k)-congruence test implies that $\varepsilon \geq \varepsilon_{opt}(k)$ $(\varepsilon < \varepsilon_{opt}(k))$.) At this point, we suspect that $\varepsilon_{opt}(k)$ is roughly within a factor of 2 of ε, since the test for $(\varepsilon/2, k)$-congruence returned a NO answer while the one for (ε, k)-congruence returned YES or MAYBE. If the answer for (ε, k)-congruence is YES, then we realize that $\varepsilon_{opt}(k) \in (\varepsilon/2, \varepsilon)$, which is an interval of the form $(c, 2c)$. If the answer is MAYBE, we see that $\varepsilon_{opt}(k) \in (\varepsilon/2, 3\varepsilon/2)$, since we used a $(\varepsilon/2, \varepsilon/2)$-approximate algorithm to test for (ε, k)-congruence. This interval is not of the form $(c, 2c)$, but we can trim it to this form through one additional test for (ε, k)-congruence with $\gamma = \varepsilon/4$: an answer of YES, MAYBE, or NO, respectively, yields an interval $(\varepsilon/2, \varepsilon)$, $(3\varepsilon/4, 5\varepsilon/4)$, or $(\varepsilon, 3\varepsilon/2)$.

There is one special case, obtaining a YES or MAYBE answer on the initial test, when $\varepsilon = \nu$. Either answer implies that we are immediately finished not only with Phase I, but with Phase II as well. An answer of YES implies $\varepsilon_{opt}(k) \in (0, \nu)$, while MAYBE implies $\varepsilon_{opt}(k) \in (\nu/2, 3\nu/2)$; thus, in either case we have found an interval of size ν that contains $\varepsilon_{opt}(k)$.

If we use a (γ, γ)-approximate algorithm for the $k_{max}(\varepsilon)$ problem, then we can perform Phase I simultaneously for all values $k = 1, \ldots, n$. The (γ, γ)-approximate algorithm returns an answer of NO, MAYBE, or YES for each k, so we alternate performing this algorithm with doubling ε until all values of k have received a YES or MAYBE answer ($k = n$ will be the last to receive one of these answers).

Phase II is quite simply a binary search on each value of k. Beginning with $\varepsilon_{opt}(k) \in (c, 2c)$ for a fixed k, we proceed to narrow the interval to width ν. A step begins with $\varepsilon_{opt}(k) \in (\ell, u)$. We test for (ε,k)-congruence, where $\varepsilon = (\ell + u)/2$, and $\gamma = (u - \ell)/4$. With a YES answer we set $u \leftarrow (\ell + u)/2$ and repeat, with NO we set $\ell \leftarrow (\ell + u)/2$ and repeat, and with MAYBE, our choice of γ allows us to set $\ell \leftarrow (3\ell + u)/4$, $u \leftarrow (\ell + 3u)/4$, and repeat.

5 Algorithms for $k_{max}(\varepsilon)$

In this section we describe (γ, γ)-approximate algorithms for $k_{max}(\varepsilon)$, under various isometry families. We will use the Euclidean metric, although our algorithms work for any L_p metric. Our algorithms are based largely on the (γ, γ)-approximate decision algorithms for ε-congruence of [HS].

All approximate algorithms employ a similar strategy; they (1) discretize the set of isometries, $\mathcal{I}$, into a finite set of candidates, $\mathcal{C}$, and (2) test the point sets for congruence under each candidate. The methods of [HS] employ an additional tactic of perturbing each point in A and B in order to impose structure; the resulting multisets are denoted $A^\#$ and $B^\#$. This method consists of laying a lattice onto $\mathbb{R}^d$, with adjacent lattice points distance λ apart, where λ is proportional to γ. $A^\#$ and $B^\#$ are obtained by moving each point of A and B to the nearest lattice point. Note that no point is perturbed more than $\sqrt{d}\lambda/2$. We will see later how working with $A^\#$ and $B^\#$ instead of A and B allows an improvement in time-complexity.

Each algorithm requires that the following conditions hold for some fixed γ_1, γ_2.

Condition 1: For each $I \in \mathcal{I}$ that admits (ε,k)-congruence for some $k \in \{2, \ldots, n\}$, there exists a candidate isometry $C \in \mathcal{C}$ such that $dist(C(a^\#), I(a^\#)) \leq \gamma_1$ for all $a^\# \in A^\#$.

Condition 2: For each $k \leq n$, if $A^\#$ and $B^\#$ are $(\varepsilon - \gamma_2, k)$-congruent for a candidate isometry $C \in \mathcal{C}$, then A and B are (ε,k)-congruent.

Condition 3: For each $k \leq n$, if $A^\#$ and $B^\#$ are $(\varepsilon + \gamma_1 + \gamma_2, k)$-congruent for *no* candidate isometry $C \in \mathcal{C}$, then A and B are not (ε,k)-congruent.

In the discussion that follows, we let $k_{max}(\mu, A^\#, B^\#; C)$ represent the size of the largest matching between $A^\#$ and $B^\#$ with tolerance μ under isometry

C. Table 1 gives an approximate algorithm for $k_{max}(\varepsilon, A, B)$, assuming we have constructed $\mathcal{C}$ and have a procedure that computes $k_{max}(\mu, A^{\#}, B^{\#}; C)$.

[1] **Compute set of candidates $\mathcal{C}$**
[2] **$k_{YES} \leftarrow 1, k_{MAYBE} \leftarrow 1$**
[3] **for all $C \in \mathcal{C}$ do**
[4] **if $k_{max}(\varepsilon + \gamma_1 + \gamma_2, A^{\#}, B^{\#}; C) > k_{MAYBE}$**
then $k_{MAYBE} \leftarrow k_{max}(\varepsilon + \gamma_1 + \gamma_2, A^{\#}, B^{\#}; C)$ fi
[5] **if $k_{max}(\varepsilon - \gamma_2, A^{\#}, B^{\#}; C) > k_{YES}$**
then $k_{YES} \leftarrow k_{max}(\varepsilon - \gamma_2, A^{\#}, B^{\#}; C)$ fi od;
[6] **YES $\leftarrow \{2, \ldots, k_{YES}\}$**
[7] **MAYBE $\leftarrow \{k_{YES} + 1, \ldots, k_{MAYBE}\}$**
[8] **NO $\leftarrow \{k_{MAYBE} + 1, \ldots, n\}$**

Table 1: (γ, γ)-approximate algorithm for $k_{max}(\varepsilon, A, B)$

Theorem 1 ***If Conditions 2 and 3 are satisfied, then the algorithm of Table 1 is a (γ, γ)-approximate algorithm for $k_{max}(\varepsilon, A, B)$, where $\gamma = \gamma_1 + 2\gamma_2$.***

In our algorithms, $\gamma_2 = 2\sqrt{d}\lambda/2$, since no point of A or B is perturbed more than $\sqrt{d}\lambda/2$. We will also choose γ_1 proportional to λ for each algorithm, so $\gamma = \Theta(\lambda)$.

Implementation of an algorithm of this form requires (1) a method for constructing $\mathcal{C}$, and (2) a method for computing $k_{max}(\mu, A^{\#}, B^{\#}; C)$. The first question depends on the family of isometries $\mathcal{I}$, and will be explored below. The second question leads us a graph theoretic formulation and solution.

The matching problem between $A^{\#}$ and $B^{\#}$ can be formulated as a max-flow problem. To represent graph networks, we will use the notation (V, E, c), where V and E are the vertex and edge sets, respectively, and $c : E \rightarrow \mathbb{N}$ gives the edge capacities. Consider the network

$$G(C, \mu, A^{\#}, B^{\#}) = (\{s, t\} \cup U \cup V, E, c)$$

where C is a candidate isometry, $U = \{u_1, \ldots, u_n\}$ represents the points in $A^{\#}$ and $V = \{v_1, \ldots, v_m\}$ the points in $B^{\#}$,

$$E = \{(s, u_i) \mid 1 \le i \le n\} \cup \{(v_j, t) \mid 1 \le j \le m\} \cup \{(u_i, v_j) \mid dist(I(a_i^{\#}), b_j^{\#}) \le \mu\},$$

and $c(e) = 1$ for all $e \in E$. The max-flow on G from source s to sink t is equal to $k_{max}(\mu, A^{\#}, B^{\#}; C)$.

At this point, we exploit the special structure of the perturbed sets $A^{\#}$ and $B^{\#}$, in order to efficiently compute a max-flow. Feder and Motwani [FM] have presented a technique for improving graph algorithms called the *compression graph*, which is well suited to $A^{\#}$ and $B^{\#}$. Let $U_k^{\circ} = \{u_i \mid a_i$ is moved onto $a_k^{\circ}\} \subset U$ and $V_l^{\circ} = \{v_j \mid b_j$ is moved onto $b_l^{\circ}\} \subset V$. If C moves gridpoint a_k° into the μ-neighborhood of b_l°, this generates a complete bipartite

subgraph with node sets U_k° and V_l°. For every such bipartite clique, we remove the edges in $U_k^\circ \times V_l^\circ$, add a new node w_{kl}, and add edges $\{(u_i, w_{kl}) \mid u_i \in U_k^\circ\}$ and $\{(w_{kl}, v_j) \mid v_j \in V_l^\circ\}$ with capacity 1 each. Thereby we replace $|U_k^\circ| \cdot |V_l^\circ|$ old edges by $|U_k^\circ| + |V_l^\circ|$ new edges. We call the resulting graph $G^\#_{\text{comp}}(C, \mu, A^\#, B^\#)$.

A max-flow in $G^\#_{\text{comp}}(C, \mu, A^\#, B^\#)$ corresponds to a max-flow in $G(C, \mu, A^\#, B^\#)$. We can show that Dinic's algorithm computes a max-flow on $G^\#_{\text{comp}}(C, \mu, A^\#, B^\#)$ in $O(\sqrt{n+m})$ phases, each in time proportional to the new number of edges. The reason for this is that Dinic's algorithm takes $O(\sqrt{\text{number of nodes}})$ phases in a simple 0-1 network. A node is called simple if it has indegree 1 or outdegree 1, and a network is called simple if all nodes are simple. Note that the nodes w_{kl} are not simple in the compression of a graph; however, in a manner analogous to the proof on the number of phases of Dinic's algorithm for simple 0-1 networks in [Me, page 81], it can be shown that Dinic's algorithm takes $O(\sqrt{\text{number of simple nodes}})$ phases in a 0-1 network, if no edge connects non-simple nodes.

The μ-neighborhood of any point contains $O((\mu/\gamma)^d)$ grid points (since $\gamma = \Theta(\lambda)$). Hence each point of $A^\# \cup B^\#$ contributes to $O((\mu/\gamma)^d)$ bipartite cliques. Thus the number of edges in graph $G^\#_{\text{comp}}(C, \mu, A^\#, B^\#)$ is $O((n+m)(\mu/\gamma)^d)$, which yields a total time of $O((n+m)^{1.5}(\mu/\gamma)^d)$ to compute $k_{max}(\mu, A^\#, B^\#; C)$. The graph $G^\#_{\text{comp}}$ can be constructed in time $O((n+m)\log^{d-1}(n+m))$ using standard range-searching techniques [PS].

Note that the original graph G is a simple 0-1 network, for which a max-flow can be computed in $O((n+m)^{1/2}nm)$ time. Since $m = \Theta(n)$ for a typical problem instance, we see that the use of the compression graph $G^\#_{\text{comp}}$ that is enabled by the structure of the perturbed sets $A^\#$ and $B^\#$ reduces the run-time in terms of n.

We now discuss the construction of the set of candidates $\mathcal{C}$ for various isometry groups. Throughout this section, we let $\delta = \max\{\text{diam}(A), \text{diam}(B)\}$ be the larger of the diameters of the sets A and B. For each isometry group $\mathcal{I}$ we must find a set of candidates $\mathcal{C}$ that satisfies:

Condition 1: For each $I \in \mathcal{I}$ that admits (ε,k)-congruence for some $k \in \{2, \ldots, n\}$, there exists a candidate isometry $C \in \mathcal{C}$ such that $dist(C(a^\#), I(a^\#)) \leq \gamma_1$ for all $a^\# \in A^\#$.

If $\mathcal{I}$ equals the set of translations on $\mathbb{R}^d$, then let T_{xy} represent the translation that maps point x to point y. By fixing point x and choosing as candidates all translations T_{xy} where y is a lattice point, we satisfy the above condition, with $\gamma_1 = \sqrt{d}\lambda/2$ (recall that each point of $\mathbb{R}^d$ is no more than distance $\sqrt{d}\lambda/2$ from the nearest lattice point). In order to obtain a finite set of candidates, we note that if a candidate C admits (ε,k)-congruence for some $k \geq 2$, then C maps the convex hulls of $A^\#$ and $B^\#$ within a distance ε of each other. Thus, a collection $\mathcal{C}$ of $|\mathcal{C}| = O(((\varepsilon+\delta)/\lambda)^d) = O((\delta/\gamma)^d)$ candidates suffices.

Let $\mathcal{I}$ be the set of isometries on $\mathbb{R}^d$. Since an isometry is an affine mapping, it can be described by its action on $d+1$ affinely independent points. We will construct a collection $\mathcal{C}$ that satisfies Condition 1, where each candidate isometry is a set of $d+1$ affinely independent points. Fixing the first point is equivalent to choosing a translation. Once the first point is fixed, each remaining point is free to move on a d-dimensional hypersphere centered at the first fixed point, and after $d' \leq d-1$ affinely independent points have been fixed, each point not in the affine subspace of the fixed points is free to move on a $(d-d'+1)$-dimensional hypersphere. Once d affinely independent points have been fixed, each point not in the affine subspace can assume one of two possible values, since only a reflection remains to be determined.

We construct the candidates by choosing fixed points one at a time. For the first point we choose $O((\delta/\gamma)^d)$ points in such a manner that every translation that causes the convex hulls of $A^{\#}$ and $B^{\#}$ to intersect has its fixed point within distance $O(\gamma)$ of the fixed point of a candidate. When $d' \leq d-1$ points have been fixed, the set of potential choices for the next fixed point is represented by all points on a hypersphere. Since no free point moves on a hypersphere of radius greater than δ, we choose $O((\delta/\gamma)^{(d-d'+1)})$ candidate points on the hypersphere of radius δ centered at the first fixed point, such that no point on this hypersphere is more than distance $O(\gamma)$ from a candidate point. The last fixed point represents selecting a reflection and therefore presents only two choices. If the dimension d is a constant, then by choosing λ and the constant in the $O(\gamma)$ term above sufficiently small, Condition 1 is satisfied. The total number of candidates is $O((\delta/\gamma)^{d(d+1)/2})$.

6 Time complexity

The run-time of a (γ,γ)-approximate algorithm for $k_{max}(\varepsilon)$ is equal to $O(T \cdot |C|)$, where $|C|$ equals the number of candidates, and T the time to test each candidate of C. We saw above that $T = O((n+m)^{1.5}(\varepsilon/\gamma)^d)$. The value $|C|$ varies with the isometry family, and with the above observations we obtain the following time-complexities for a (γ,γ)-approximate algorithm for $k_{max}(\varepsilon)$: $O((n+m)^{1.5}(\varepsilon/\gamma)^d(\delta/\gamma)^d)$ if $\mathcal{I}$ is the set of translations in $\mathbb{R}^d$ for $A, B \in \mathbb{R}^d$, and $O((n+m)^{1.5}(\varepsilon/\gamma)^d(\delta/\gamma)^{d(d+1)/2})$ if $\mathcal{I}$ is the set of isometries in $\mathbb{R}^d$ for $A, B \in \mathbb{R}^d$.

Let us examine the total run-time to construct a (ε,k,ν)-approximate map. To compute the number of Phase I tests for a given k, consider that we test for (ε,k)-congruence for all values $\varepsilon = 2^i\nu$, $i = 0,\ldots,j$, where $\varepsilon_{opt}(k) \in (2^{j-1}\nu, 3\cdot 2^{j-1}\nu]$. This implies $j = \log_2(\varepsilon_{opt}(n)/\nu)$ tests until we receive a YES or MAYBE answer for all k, since $k = n$ will be the last to receive such an answer. Also, at each value $\varepsilon = 2^i\nu$, one additional test may have to be performed because certain values of k receive for the first time there an answer of MAYBE, but one test suffices even if several values of k receive a MAYBE at the same value $\varepsilon = 2^i\nu$. Thus, the total number of values of ε tested in Phase I is $O(\log(\varepsilon_{opt}(n)/\nu))$; since $\gamma = \varepsilon/2$ or $\gamma = \varepsilon/4$ at all times during Phase I, the total time is $O((n+m)^{1.5}(\delta/\nu)^x \log(\varepsilon_{opt}(n)/\nu))$, where the exponent x is determined by the family of isometries $\mathcal{I}$ and the dimension d.

Phase II, for a given k, is a binary search on an interval $(c, 2c]$ that contains $\varepsilon_{opt}(k)$. Each search query is a test for (ε,k)-congruence, where ε is in $(c, 2c]$

and thus approximately equal to $\varepsilon_{opt}(k)$, while γ is cut in half after each query. Therefore, since $T = O((n+m)^{1.5}(\varepsilon/\gamma)^d)$ and $|C|$ varies inversely with γ for isometry families that we considered, the sum of all $k_{max}(\varepsilon)$ tests always will be dominated by the last one, for which γ is approximately the same size as ν. If we perform Phase II over all n values of k, we get a total time of $O(n(n+m)^{1.5}(\delta/\nu)^x(\varepsilon_{opt}(n)/\nu)^d)$, with x as above, since $\varepsilon_{opt}(n) \geq \varepsilon$ for every test value ε. We see that Phase II dominates Phase I.

In summary, implementing Phase I and Phase II together builds a (ε,k,ν)-approximate map. Using the (γ,γ)-approximate algorithms for $k_{max}(\varepsilon)$ that we have described yields the following time-bounds for constructing (ε,k,ν)-approximate maps:

$O(n(n+m)^{1.5}(\varepsilon_{opt}(n)/\nu)^d(\delta/\nu)^d)$ if $\mathcal{I}$ is the set of translations in $\mathbb{R}^d$, for $A, B \in \mathbb{R}^d$;

$O(n(n+m)^{1.5}(\varepsilon_{opt}(n)/\nu)^d(\delta/\nu)^{d(d+1)/2})$ if $\mathcal{I}$ is all isometries in $\mathbb{R}^d$, for $A, B \in \mathbb{R}^d$.

7 Projective congruence

In ***projective congruence***, we compare a 2-dimensional point set B to a 3-dimensional set A. Rather than choosing just an isometry and a matching (between the elements of the point sets), finding a congruence between the points sets now consists of first choosing an orthogonal projection of the 3-dimensional set A to $\mathbb{R}^2$, and then choosing an isometry on the resulting 2-dimensional set, and finally a matching. We develop an approximate algorithm for this problem by approaching it in a similar manner as before. The Phase I-Phase II approach motivates a version of the $k_{max}(\varepsilon)$ problem, in which the set of all possible compositions of projection and isometry on A must be discretized into candidates that closely approximate all reasonable choices. Discretizing the candidates resembles the same task for isometries in $\mathbb{R}^3$, except that the projection of the point set to an affine space of one lesser degree results in the loss of one degree of freedom; in fixing the first point (which determines the translation), we realize that we have two dimensions of freedom and not three, since motion parallel to the direction of projection is irrelevant. The total number of candidates on the last test of Phase II is $O((\delta/\nu)^5)$, and the max-flow problems here are between point sets in $\mathbb{R}^2$, so the total run-time is $O(n(n+m)^{1.5}(\varepsilon_{opt}(n)/\nu)^2(\delta/\nu)^5)$.

8 Conclusion

We have generalized the concept of ε-congruence in order to encompass a greater number of situations; our generalized definition is called (ε,k)-congruence, and it is represented pictorially by the (ε,k)-map. Because we desired small run-times, we chose to develop approximate algorithms that test for (ε,k)-congruence, using many of the techniques of [HS]. While we did not discuss complete algorithms for (ε,k)-congruence, such algorithms would extend naturally from some of the complete algorithms for ε-congruence given by [AMWW].

While we have achieved run-times with small exponents for n and m, it is troubling that the diameter δ appears. This suggests that our approximate algorithms are best suited for relatively dense point sets. Ironically, for ε-congruence, complete algorithms perform best on point sets that are *ε-separated* [AMWW, AKMSW] (i.e. no two points of the same set are within distance ε of each other). One open question is to explore which types of point sets are best suited to our approximate algorithms and which to complete algorithms. As a final open question, we ask if there are ways to reduce $|C|$ for the various isometry families we studied, since any reduction leads automatically to improved algorithms for $k_{max}(\varepsilon)$ and the (ε,k,ν)-approximate map.

References

[AMWW] H. Alt, K. Mehlhorn, H. Wagener, and E. Welzl, "Congruence, similarity, and symmetries of geometric objects," *Discrete and Computational Geometry*, 3 (1988), pp. 237-256.

[AKMSW] E.M. Arkin, K. Kedem, J.S.B. Mitchell, J. Sprinzak, M. Werman, "Matching points into noise regions: combinatorial bounds and algorithms," *ORSA J. on Computing*, 4 (1992), pp. 375-386.

[Ba] H.S. Baird, *Model-Based Image Matching Using Location*, MIT Press, 1984.

[FM] T. Feder and R. Motwani, "Clique partitions, graph compression and speeding-up algorithms," in *Proc. of the 23rd ACM Symp. on Theory of Computing*, 1991.

[HS] P.J. Heffernan and S. Schirra, "Approximate decision algorithms for point set congruence," in *Proc. of 8th ACM Symp. on Computational Geometry* (1992), pp. 93-101; to appear in *Computational Geometry: Theory and Applications.*

[Ma] G.E. Martin, *Transformation Geometry*, Springer Verlag, 1982.

[Me] K. Mehlhorn, *Data Structures and Algorithms 2: Graph Algorithms and NP-completeness*, Springer-Verlag, 1984.

[PS] F.P. Preparata and M.I. Shamos, *Computational Geometry — An Introduction*, Springer-Verlag, 1985.

[Sc1] S. Schirra, *Über die Bitkomplexität der ε-Kongruenz*, Diplomarbeit, Universität des Saarlandes, Saarbrücken, Germany, 1988.

[Sc2] S. Schirra, "Approximate decision algorithms for approximate congruence," *Information Processing Letters*, 43 (1992), pp. 29-34.

Approximating Shortest Superstrings with Constraints

(Extended Abstract)

Tao Jiang* and Ming Li**

Abstract. Various versions of the shortest common superstring problem play important roles in data compression and DNA sequencing. Only recently, the open problem of how to approximate a shortest superstring given a set of strings was solved in [1, 9]. [1] shows that several greedy algorithms produce a superstring of length $O(n)$, where n is the optimal length. However, a major problem remains open: Can we still linearly approximate a superstring in polynomial time when the superstring is required to be consistent with some given negative strings, *i.e.*, it must not contain any negative string. The best previous algorithm, Group-Merge given in [6, 9], produces a consistent superstring of length $\theta(n \log n)$. The negative strings make the problem much more difficult and, as we will show, a greedy-style algorithm cannot achieve linear approximation for this problem.

We present polynomial-time approximation algorithms that produce consistent superstrings of length $O(n)$, for two important special cases: (a) When no negative strings contain positive strings as substrings; (b) When there are only a constant number of negative strings. The algorithms are obtained by making an essential use of the Hungarian algorithm which can find an optimal cycle cover on weighted graphs.

The other main objective of this paper is to analyze the performance of some greedy-style algorithms for this problem. Due to their time efficiency and simplicity, greedy algorithms are of practical importance. We introduce a new analysis showing that when no negative strings contain positive strings, a greedy algorithm achieves $O(n^{4/3})$ and $O(n)$ if the number of negative examples is further bounded by some constant.

1 Introduction

Given a finite set of strings, the *shortest common superstring* problem is to find a shortest possible string s such that every string in the set is a substring of s. In a more general setting, given a finite set of positive strings and a finite set of negative strings, the *shortest consistent superstring* problem is to find the shortest possible string s such that every positive string is a substring of s and no negative string is a substring of s.

* Supported in part by NSERC Operating Grant OGP0046613. Address: Department of Computer Science, McMaster University, Hamilton, Ont. L8S 4K1, Canada. E-mail: jiang@maccs.mcmaster.ca

** Supported in part by the NSERC Operating Grant OGP0046506. Address: Department of Computer Science, University of Waterloo, Waterloo, Ont. N3L 3G1, Canada. E-mail: mli@math.waterloo.edu

A solution of the latter problem implies a solution to the former problem. Both problems have applications in data compression practice and DNA sequencing procedures [8, 9, 12, 13]. The former problem is well-known. The latter problem occurs when merging certain strings is prohibited. For example, in a shot-gun DNA sequencing procedure, certain combinations of fragments make no biological sense and are thus excluded. This problem is especially important in the recently developed DNA sequencing by hybridization (SBH) technique [2, 10]. In an SBH procedure, a biochemist first tests the membership of a large number of oligonucleotide probes (*i.e.*, short strings of nucleotides) in the target sequence and then tries to infer the target sequence. The second step is essentially to construct a superstring consistent with the observed membership of the oligonucleotides.

The latter problem can also be formulated as a learning problem: Given a set of positive examples, as substrings of a string to be learned, and a set of negative examples, as strings that are not substrings of the string to be learned, finding a short consistent superstring implies efficient learning in Valiant's PAC learning model [6, 9, 16].

Both problems are NP-hard [3, 4]. Only recently, the open problem of how to approximate a shortest common superstring with constant factors has been partially solved in [6, 9] and completely solved in [1]. [1] shows several greedy algorithms produce a superstring of length $O(n)$ for a given set of strings, where n is the optimal length. But it is not known how such algorithms would perform in the presence of negative strings. The Group-Merge algorithm in [6, 9] actually works even in the presence of negative strings. But that algorithm only achieves an $O(n \log n)$ approximation.

It turns out that negative strings are very hard to deal with. We can show that none of the above algorithms achieve linear approximation. In particular, we will give an $\Omega(n^{1.5})$ lower bound for the greedy algorithms. An $\Omega(n \log n)$ lower bound for Group-Merge is shown in [6]. Thus we have to develop new algorithms and new proofs to solve the question. Remember a linear approximation algorithm for the shortest consistent superstring problem is also a linear approximation algorithm for the shortest common superstring problem.

We give polynomial-time approximation algorithms that produce a consistent superstring of length $O(n)$ for several important special cases, *viz.*, (a) when no negative strings contains positive strings and (b) when there are only a constant number of negative strings. This implies a $\log n$ multiplicative factor improvement on the sample complexity for string learning [6], when the negative examples satisfy one of these special properties. Case (a) is interesting in practice since it corresponds to the situation when restrictions are imposed on the merging of a pair of input strings. Also the assumption seems to hold in an SBH procedure since the strings involved (*i.e.*, oligonucleotides) usually have roughly the same length [10]. Case (b) arises in a shot-gun DNA sequencing procedure where usually only a limited number of combinations are disallowed. The algorithms all rely on the Hungarian algorithm to find optimal cycle covers on some weighted graphs derived from the input strings in their first stages.

Our other main goal is to study greedy algorithms, because of their practical importance. Although we have obtained a polynomial-time algorithm that linearly approximates the shortest consistent superstring for several important special cases, the polynomial power is too large to be practical. For example, it is not uncommon that we are given 10^6 strings of 100 characters to compress. Simple greedy algorithms running in $O(ml_{max} \log m)$ time [14, 15], where m and l_{max} are the number and maximum length of input strings, would require about 10^8 steps. On a supercomputer this, say, takes one second. However Group-Merge would take a million seconds on the same computer, and our new polynomial algorithms would take billions of seconds to finish the task. Remember also, due to their simplicity and efficiency, greedy algorithms are actually implemented and are used routinely by computers or human hands in biochemistry labs and elsewhere. It is thus of great interests to study the performance of greedy algorithms in the presence of negative strings. In Section 4, we will introduce a new method for analyzing greedy algorithms and show that when no negative string contains positive strings as substrings, a greedy algorithm achieves $O(n^{4/3})$ approximation and, moreover, when there are only a constant number of such negative strings it achieves $O(n)$ approximation. We also conjecture that the greedy algorithm actually achieves $O(n)$ when no negative string contains positive strings and hope that some of the analysis techniques developed in this paper will be useful in proving such a linear bound.

2 Some Basic Definitions and Facts

Let $P = \{s_1, \ldots, s_m\}$ be a set of *positive* strings and $N = \{t_1, \ldots, t_k\}$ a set of *negative* strings, over some alphabet Σ. Without loss of generality, we assume that sets P and N are "substring free", *i.e.*, no string s_i (or t_i) is contained in any other string s_j (or t_j, respectively). Moreover, we assume that no negative string t_i is a substring of any positive string s_j. A *consistent superstring* for (P, N) is a string s such that each s_i is a substring of s and no t_i is a substring of s. In this paper, we will use n and $OPT(P, N)$ interchangeably for the length of a *shortest* consistent superstring for (P, N). Our goal is to find a consistent superstring for (P, N) whose length is as close to $OPT(P, N)$ as possible. Since it is NP-hard to decide if (P, N) has a consistent superstring if we require a superstring s to be a string over the same alphabet Σ [7], in this paper we will allow s to be a string over $\Sigma \bigcup \{\#\}$, where $\# \notin \Sigma$ is a *delimiter* symbol, so that a trivial consistent superstring (*i.e.*, $s_1\#s_2\#\cdots\#s_m$) always exists. Note that this assumption is actually consistent with practice. *E.g.*, when compressing a set of strings into a single superstring, we can always introduce new delimiters if necessary.

Most of the following definitions are introduced in [1]. For two distinct strings s and t, let v be the longest string such that $s = uv$ and $t = vw$. We call $|v|$ the (amount of) *overlap* between s and t, and denote it as $ov(s,t)$. Furthermore, u is called the *prefix* of s with respect to t, and is denoted $pref(s,t)$. We call $|pref(s,t)| = |u|$ the *distance* from s to t, and denote it as $d(s,t)$. So, the string $uvw = pref(s,t)t$, of length $d(s,t)+|t| = |s|+|t|-ov(s,t)$ is the shortest superstring of s and t in which s appears (strictly) before t, and is also called the *merge* of s and t and denoted $m(s,t)$. It is useful to extend the above definitions to also include the case $s = t$. The *self-overlap* of a string s, denoted $ov(s,s)$, is the length of the longest string v such that $s = uv = vw$ for some *non-emtpy* strings u and w. The extension of the other definitions is straightforward. Given a list of strings $s_{i_1}, s_{i_2}, \ldots, s_{i_r}$, we define the superstring

$s = \langle s_{i_1}, \ldots, s_{i_r} \rangle$ to be the string $pref(s_{i_1}, s_{i_2})pref(s_{i_2}, s_{i_3}) \cdots pref(s_{i_{r-1}}, s_{i_r})s_{i_r}$. That is, s is obtained by maximally overlapping $s_{i_1}, s_{i_2}, \ldots, s_{i_r}$ *in order*. Define $first(s) = s_{i_1}$ and $last(s) = s_{i_r}$. Note, for two strings s and t obtained by merging strings in P, $ov(s,t)$ in fact equals $ov(last(s), first(t))$, and as a result, the merge of s and t is $\langle first(s), \ldots, last(s), first(t), \ldots, last(t) \rangle$. Observe that since P is substring-free, a shortest common superstring for P must be $\langle s_{i_1}, \ldots, s_{i_m} \rangle$ for some permutation $\langle i_1, \ldots, i_m \rangle$. However, this is not true when there are negative strings, *i.e.*, a shortest consistent superstring for (P, N) may not be $\langle s_{i_1}, \ldots, s_{i_m} \rangle$ for any permutation $\langle i_1, \ldots, i_m \rangle$, since adjacent strings in the superstring may or may not be maximally overlapped.

For any weighted digraph G, a *cycle cover* (or *path cover*) of G is a set of vertex-disjoint cycles (or paths, resp.) covering all nodes of G. (A cycle cover is called an *assignment* in [1], due to the fact that its linear programming form is the same as the well-known assignment in operations research.) The cover is said to be optimal if it has the smallest weight. Denote the weight of an optimal cycle cover of graph G as $CYC(G)$. We will consider cycle covers on a weighted complete digraph G_P derived from the positive strings. Digraph $G_P = (V, E, d)$ has m vertices $V = \{1, \ldots, m\}$, and m^2 edges $E = \{(i,j) : 1 \le i, j \le m\}$. Here we take as weight function the distance $d(,)$: edge (i,j) has weight $w(i,j) = d(s_i, s_j)$, to obtain the *distance graph*. The string represented by a path $i_1, \ldots, i_r$ is $\langle s_{i_1}, \ldots, s_{i_r} \rangle$. As observed in [1], we have

$$CYC(G_P) \le OPT((P, \emptyset))$$

Denote by $w(c)$ the *weight* of a cycle c. For convenience, define the length of a cycle c, denoted $l(c)$, to be its weight and the length of a path $p = i_1, \ldots, i_r$, denoted $l(p)$, to be $|\langle s_{i_1}, \ldots, s_{i_r} \rangle| = d(s_{i_1}, s_{i_2}) + \ldots + d(s_{i_{r-1}}, s_{i_r}) + |s_{i_r}|$.

The *factor* of a string s, denoted $factor(s)$ is the shortest string u such that $s = u^i v$ for some positive integer i and prefix v of u (v may be null). The *period* of s, denoted by $period(s)$, is $|factor(s)|$. A string s is said to be *i-periodic* if $i \le \frac{|s|}{period(s)} < i + 1$. A string is *fully periodic* if it is at least 4-periodic. A string s is *prefix-periodic* (or *suffix-periodic*) if s is not fully periodic and s has a fully periodic prefix (or suffix, respectively) of length at least $3|s|/4$. (The reason for choosing the specific numbers 4 and 3/4 here can be seen from the proof of Theorem 2.) Call a string *periodic* if it is either fully periodic or prefix-periodic or suffix-periodic. Suppose s is a prefix-periodic string and $s = uv$, where u is the longest fully periodic prefix of s. Then u is called the *periodic prefix* of s and v is the *non-periodic suffix* of s. Similarly, if s is a suffix-periodic string and $s = uv$, where v is the longest periodic suffix of s, then v is called the *periodic suffix* of s and u is the *non-periodic prefix* of s.

For example, s_0 = ababababab is a fully periodic string with $factor(s_0)$ = ab and $period(s_0) = 2$, s_1 = ababababacc is a prefix-periodic string with periodic prefix u_1 = ababababa and non-periodic suffix v_1 = cc, and s_2 = bbbbabababab is a suffix-periodic string with non-periodic prefix u_2 = bbb and periodic suffix v_2 = babababab.

Two strings are *equivalent* if they are cyclic shifts of each other. Two fully periodic strings are *compatible* if they have equivalent factors. Two prefix- (or suffix-) periodic strings are *compatible* if one of their periodic prefixes (suffixes, resp.) is a suffix (prefix, resp.) of the other, and one of their non-periodic suffixes (prefixes, resp.) is a prefix (suffix, resp.) of the other. Informally speaking, two periodic strings have a "large" overlap if and only if they are compatible. Note that compatibility is an equivalence relation. *E.g.*, string s_1 = abababababc is compatible with s_2 = bababababcd, but not with s_3 = ababababababac and s_4 = bababababdd.

3 Linear Approximation: Two Special Cases

In this section we present polynomial time algorithms which produce a consistent superstring of length $O(n)$, where n is the optimal length, for two special cases: (a) when no positive string is a substring of any negative strings; and (b) when there are only constant number of negative strings.

The best previous algorithm for shortest consistent superstrings is Group-Merge introduced in [6, 9], which achieves $\theta(n \log n)$ approximation [6]. The lower bound in fact holds for $N = \emptyset$. Thus Group-Merge does not work well for the special cases that we are interested in. Although we suspect that greedy algorithms may achieve linear approximation in these special cases, so far we can only prove an upper bound $O(n^{4/3})$ (to be presented in next section). Thus we must search for new algorithms. Our departure point is the algorithm Concat-Cycles which is used to find a common superstring of length $O(n)$ in [1], when there are no negative strings. The essence of Concat-Cycles is the Hungarian algorithm [11] which can find an optimal cycle cover for any given weighted digraph. But, [1] did not need the full power of the Hungarian algorithm. We will fully utilize its power.

3.1 When No Negative Strings Contain Positive Ones

In this subsection, we show that when no negative strings contain positive strings as substrings, an algorithm can achieve linear approximation. This special case is natural, it corresponds to the restrictions that we impose on the merges of input strings. In some practice, we may want to forbid some 'bad merges' to happen.

As mentioned above, our algorithm works in a way similar to Concat-Cycles [1] and uses the Hungarian algorithm to find an optimal cycle cover on the distance graph derived from the input strings. It is shown in [1] that if we have an optimal cycle cover of the distance graph, then opening each cycle into a path arbitrarily and simply concatenating the strings associated with the paths yields a superstring of length $O(n)$.

We informally describe the construction of our algorithm. Let $P = \{s_1, \ldots, s_m\}$ and $N = \{t_1, \ldots, t_k\}$. First construct the distance graph G_P for the positive strings as defined in Section 2, except that for each pair i, j such that $m(s_i, s_j)$ contains a negative string we remove the edge (i, j) from G_P. Note that, sicne no negative strings contain any positive strings, each path in the new G_P corresponds to a string consistent with N.

Several problems need to be solved: (i) G_P may not have a cycle cover; (ii) Even if G_P has a cycle cover, it is not clear if $CYC(G_P) = O(n)$, which is essential to achieving an $O(n)$ bound on the length of superstring produced.

(i) is easy to solve. We can just add a sufficient number of the *delimiter nodes* (called # nodes) to G_P. Each such node represents a delimiter #. We set the weights as follows: $w(\#, \#) = 0$ and for each node i, $w(i, \#) = |s_i|$ and $w(\#, i) = 1$. Call the resulting graph $G_{P\#}$. Clearly $G_{P\#}$ always has a cycle cover. Note that the use of delimiter nodes is consistent with our definition of a superstring. Also observe that $CYC(G_{P\#}) \leq CYC(G_P)$ if the latter exists, as we can always let the delimiter nodes form a cycle with zero weight.

But it is still not obvious that $CYC(G_{P\#}) = O(n)$. The reason is that in a shortest consistent superstring, two adjacent strings may or may not be maximally overlapped and maximally overlapping two strings sometimes prevents better arrangement, because of the presence of negative strings. So (ii) is resolved by considering a special form of consistent superstrings. First observe that for any pair of strings s and t, if s is not suffix-periodic and t is not prefix-periodic, then there is at most one way of overlapping s and t with s in front to achieve a large amount of overlap (*i.e.*, $\geq 3\max\{|s|, |t|\}/4$). Thus, if the overlap between s and t is large, then the overlap must equal $ov(s,t)$. Now define a *normal* superstring for (P, N) to be a superstring of the form: $u_1\#u_2\#\ldots\#u_r$, where each u_i is either s_i or $m(s_i, s_j)$ for some $i \neq j$. Denote the length of a shortest normal consistent superstring for (P, N) by $OPT1(P, N)$.

Lemma 1. *If P does not contain any periodic strings, then $OPT1(P, N) = O(n)$.*

Proof. (Sketch) Let s be a shortest consistent superstring for (P, N). Order strings $s_1, \ldots, s_m$ according to their first occurrences in s. Suppose the sequence is $s_{i_1}, \ldots, s_{i_m}$. Cut the sequence into segments *maximally* from left to right such that each segment satisfies (i) it contains a single string, or (ii) the first and last string in it overlap by at least 3/4 of their maximum length. Let $(a_1, b_1), \ldots, (a_r, b_r)$ be the pairs of the first and last strings of the segments. For each pair (a_j, b_j) with $a_j \neq b_j$, since, a_j, b_j are non-periodic and their overlap in s is at least $3\max\{|a_j|, |b_j|\}/4$, they overlap by precisely $ov(a_j, b_j)$ in s. Hence the string $m(a_j, b_j)$ is a consistent superstring covering all the strings in segment j. Let $u_j = m(a_j, b_j)$ if $a_j \neq b_j$ or a_j otherwise. Then $u = u_1\#\ldots\#u_r$ is a consistent superstring for (P, N). Note that $\sum_{j=1}^{r} |u_j| \leq \sum_{j=1}^{r} |a_j| + |b_j|$. By some careful calculation we can show that $\sum_{j=1}^{r} |a_j| < 8|s| = 8n$ and $\sum_{j=1}^{r} |b_j| < 8n$. □

Observe that in the proof of above lemma, for each segment $j = a_j, \ldots, b_j$, $\langle a_j, \ldots, b_j\rangle = m(a_j, b_j)$, since P is substring free and the strings in the segment are overlapped by a large amount. Thus the constructed consistent superstring u in fact corresponds to a Hamiltonian path on $G_{P\#}$ and the lemma implies that if P does not contain any periodic strings, then

$$CYC(G_{P\#}) \leq OPT1(P, N) = O(n)$$

Although we can actually show that the above result holds for any set P of strings, we do not need the stronger result here since we will process the periodic strings separately anyway, for some other reason.

Before we formally present our linear approximation algorithm, we need to describe a simple greedy algorithm Greedy 1, which is a straightforward extension of the algorithm Greedy discussed in [1, 14, 15].

Algorithm Greedy 1

1. Choose two (different) strings s and t from P such that $m(s,t)$ does not contain any string in N and $ov(s,t)$ is maximized. Remove s and t from P and replace them with the merged string $m(s,t,)$. Repeat step 1. If such s and t could not be found, go to step 2.
2. Concatenate the strings in P, inserting delimiters # if necessary.

Our approximation algorithm combines Greedy 1 and the Hungarian algorithm:

1. Put the fully periodic strings in P into set X_1, the prefix-periodic strings into set X_2, the suffix-periodic strings into set X_3, and other strings into set Y.
2. Divide X_1, X_2, and X_3 further into groups of compatible strings. Run Greedy 1 on each group separately.
3. Construct the graph $G_{Y\#}$ as described above. Find an optimal cycle cover of $G_{Y\#}$. Open each cycle into a path and thus a string.
4. Concatenate the strings obtained in steps 2 and 3, inserting #'s if necessary.

Theorem 2. *Given (P,N), where no string in N contains a string in P, the above algorithm produces a consistent superstring for (P,N) of length $O(n)$.*

Proof. (Sketch) We know from the above discussion that the optimal cycle cover found in step 3 has weight $CYC(G_{Y\#}) = O(OPT1(Y,N)) = O(OPT(Y,N)) = O(n)$. Since the strings in Y are non-periodic, it is easy to show that their merges are at most 4-periodic. The strings that are at most 4-periodic do not have large self-overlap. More precisely, $ov(s,s) < 4|s|/5$ for any s that is at most 5-periodic. Thus opening a cycle into a path can at most increase its length by a factor of 5. This shows the strings obtained in step 3 have a total length at most $5 \cdot CYC(G_{Y\#}) = O(n)$.

Now we consider the strings produced in step 2. Let $U_1, \ldots, U_r$ be the compatible groups for X_2. (The proof for X_1 and X_3 are similar.) It follows from Lemma 9 in [1] that for any two fully periodic strings x and y, if x and y are incompatible, then $ov(x,y) < period(x) + period(y)$. By our definition of periodicity, for any $u_i \in U_i$, $u_j \in U_j$, $i \neq j$, $ov(u_i,u_j) < (|u_i| + |u_j|)/4 + \max\{|u_i|,|u_j|\}/4 < 3\max\{|u_i|,|u_j|\}/4$. Thus, informally speaking, strings belonging to different groups do not have much overlap with each other. It can be shown by a calculation as in the proof of Lemma 1 that we can afford losing such "small overlaps" in constructing an $O(OPT(X_2,N))$ long consistent superstring for (X_2,N), since replacing each such overlap with a plain concatenation in a shortest consistent superstring for (X_2,N) will at most increase its length by a factor of 8. Hence we have the following lemma:

Lemma 3. $\sum_{i=1}^{r} OPT(U_1,N) = O(OPT(X_2,N)) = O(n)$.

To complete the proof, it suffices to prove that Greedy 1 produces a consistent superstring of length $O(OPT(U_i,N))$ for each group U_i. A key observation in this proof is that because the strings in U_i are all compatible with each other, the large overlaps are unique in the following sense: for any $s,t \in U_i$, if $m(s,t)$ does not contain any negative examples, then $ov(s,t)$ must occur in the construction of a shortest consistent superstring. Thus, Greedy 1 can actually identify all the *correct* (*i.e.*, used in the construction of a shortest consistent superstring) large overlaps and perform the corresponding merges. Greedy 1 will ignore all the small overlaps (including the correct ones) and replaces them with concatenation. But this is fine as observed before. □

Remark. Since the Hungarian algorithm runs in $O(m^3)$ time on a graph of m nodes, our algorithm has time complexity $O(m^3 l_{max})$, where l_{max} is the maximum length of the input strings.

0.2 With Constant Number of Negative Strings

In this section, we consider the case $|N| \leq c$, for some constant c. However here we allow any kind of negative strings. We present a linear approximation algorithm for this special case. A sketch of the construction of our algorithm is given below.

Again, given input (P, N) with $|N| \leq c$, we remove the periodic strings from P and process them separately to obtain a consistent superstring of length $O(n)$ for these periodic strings, as shown in the previous subsection. So from now on assume that P does not contain any periodic strings.

Let G_P be the distance graph for P as defined in Section 2. Again obtain the graph $G_{P\#}$ by adding a sufficient number of delimiter nodes to G_P. Now we have to find an optimal cycle cover of $G_{P\#}$ that is consistent with N, *i.e.*, each cycle should not contain a path whose associated string violates the negative strings in N. Let $CYC(G_{P\#}, N)$ denote the weight of an optimal consistent cycle cover of graph $G_{P\#}$. By the proof of Lemma 1, we have

$$CYC(G_{P\#}, N) \leq OPT1(P, N) = O(n)$$

So our construction has to utilize the fact that $|N| \leq c$. Our main idea is to make many (but polynomial number) copies of $G_{P\#}$; each is slightly modified (some edges deleted) according to the negative strings. The graphs are constructed so that the inconsistent cycle covers are prevented. In other words, in these graphs, every cycle cover is consistent with N. We also want these graphs to give all possible consistent cycle covers of $G_{P\#}$ collectively. Thus by running the Hungarian algorithm on each of them we can find an optimal consistent cycle cover of $G_{P\#}$.

We describe how to construct these graphs. Consider a negative string t. Observe that there may be many paths in $G_{P\#}$ violating t. We have to prevent them all. Let $i_1, \ldots, i_r$ be the path (not necessarily simple) with maximum number of nodes in $G_{P\#}$ such that $\langle s_{i_1}, \ldots, s_{i_r} \rangle$ is contained in t. Observe that the path is unique. Let P_1 include the strings in P such that merging any string in P_1 to the left of $\langle s_{i_1}, \ldots, s_{i_r} \rangle$ would cover a prefix of t and, let P_2 include the strings in P such that merging any string in P_2 to the right of $\langle s_{i_1}, \ldots, s_{i_r} \rangle$ would cover a suffix of t. For convenience, let s_{i_0} denote a string in P_1 and $s_{i_{r+1}}$ denote a string in P_2 generically. The following lemma essentially says that in each consistent cycle cover of $G_{P\#}$, the path $i_0, i_1, \ldots, i_r, i_{r+1}$ has to be completely broken.

Lemma 4. *Each consistent cycle cover of $G_{P\#}$ can be transformed into one without increasing the weight such that there is an index $0 \leq j \leq r$ such that none of the edges of the form (i_a, i_b), where $a \leq j < b$ and $m(s_{i_a}, s_{i_b}) = \langle s_{i_a}, \ldots, s_{i_b} \rangle$, are used in the cover.*

Thus, for the negative string t, we construct $r+1 \leq m = |P|$ graphs $G_t^0, G_t^1, \ldots, G_t^r$ from $G_{P\#}$ by deleting some edges. We make sure that in each G_t^j, all the edges (i_a, i_b) satisfying the condition in the above lemma are broken. Thus each cycle cover of G_t^j is consistent with the string t. By the lemma, there exists some j such that an optimal consistent cycle cover of G_t^j is also an optimal consistent cycle cover of $G_{P\#}$. Then starting again from each G_t^j, we repeat the above procedure for another negative string, constructing more graphs. This eventually gives us at most m^c graphs. One of such graphs must contain an optimal consistent cycle cover of $G_{P\#}$.

So the last step of our algorithm is to run the Hungarian algorithm and obtain an optimal cycle cover for each of these graphs, and choose a cycle cover with the minimum weight. By the above discussion, the chosen cycle cover is an optimal consistent cycle cover of $G_{P\#}$. Then we simply open the cycles into paths and concatenate the corresponding strings. Since the strings are non-periodic, this results in a superstring of length at most $5CYC(G_{P\#}, N) \leq 5n$.

Theorem 5. *Given (P, N) with $|N| \leq c$ for some constant c, the above algorithm outputs a consistent superstring of length $O(n)$.*

Remark. It is possible to give an alternative proof of the above theorem using overlap properties of negative strings when opening cycles, but this analysis would give a higher approximation factor.

4 Greedy Solutions

Our main objective is to analyze the greedy-style algorithms in the presence of negative strings. This, we believe, will also better our understanding about the original shortest common superstring problem. Because of their simplicity, time-efficiency, and appeal to common sense, greedy algorithms are routinely used in computer programs and by human hands. For example, greedy algorithms usually run in $O(ml_{max} \log m)$ time [14, 15], where m and l_{max} are the number and maximum length of input strings, whereas our new algorithms and Group-Merge would require at least $O(m^3 l_{max})$ time. In practice, it is possible to have a large number of input strings, and in such cases, it is infeasible to use algorithms with time complexity $\Omega(m^2 l_{max})$. This makes greedy algorithms the only practical algorithms known so far.

Although greedy algorithms can achieve $O(n)$ approximation when there are no negative strings, our the next theorem shows that generally they do not work well when negative strings are present. For simplicity, we just prove a lower bound for the algorithm Greedy 1 described in the previous section.

Theorem 6. *For some input (P, N), Greedy 1 produces a superstring of length $\Omega(n^{1.5})$.*

Proof. We will construct two sets of strings P, N to force Greedy 1 to output a superstring of length $\theta(n^{1.5})$. The true shortest superstring in our mind is: $b^n a^k \# a^k b^n$ where $k = \theta(\sqrt{n})$. P contains $a\#a$, and the pairs of positive strings:

$b^{n+1-i(i+1)/2}a^i, a^ib^{n+1-i(i+1)/2}$, $1 \le i \le \sqrt{n}$. If b^ia^j, a^jb^i is in P, then N contains $a^{j+1}b^ia^j$ and $a^jb^ia^{j+1}$.

Thus the first pair of strings in P will merge (wrongly) to ab^na, Negative strings a^2b^na and ab^na^2 will prevent further merge to ab^na. Then the second pair in P will merge to $a^2b^{n-2}a^2$, and so on. Hence we will end up with $\sqrt{n}$ strings of form $a^ib^ja^i$ with total length $\Omega(n^{1.5})$. Because of negative strings in N, they must be concatenated to a final string of length $\Omega(n^{1.5})$. □

If we inspect the construction in the above proof carefully, we observe that some positive strings are substrings of negative strings. Such positive strings trick Greedy 1 into bad traps. If we forbid such things to happen, can a greedy algorithm do better? The answer turns out to be positive. Our result will show that negative strings which contain positive strings are essentially responsible for the bad cases. In the following, we present a greedy algorithm which produces a consistent superstring of length $O(n^{4/3})$, when no negative string contains positive strings. (Recall that we have given an $O(n)$ approximation algorithm for this special case in section 3, with time complexity $O(m^3 l_{max})$.) The algorithm combines Greedy 1 with another algorithm Mgreedy 1, which is a straightforward extension of the algorithm Mgreedy in [1]. We first describe Mgreedy 1.

Algorithm Mgreedy 1

1. Let (P, N) be the input and T be empty.
2. While P is non-empty, do the following: Choose $s, t \in P$ (not necessarily distinct) such that $m(s,t)$ does not contain any string in N and $ov(s,t)$ is maximized. If $s \ne t$, then remove s and t from P and replace them with the merged string $m(s,t,)$. If $s = t$, then just move s from P to T. If such s and t could not be found, move all strings in P to T.
3. Concatenate the strings in T, inserting delimiters # if necessary.

It is not easy to prove a nontrivial upper bound on the performance of Greedy 1, nor is it easy for Mgreedy 1. The trouble maker again is the periodic strings. So we will consider an algorithm which processes the periodic and non-periodic strings separately:

1. Put the fully periodic strings in P into set X_1, the prefix-periodic strings into set X_2, the suffix-periodic strings into set X_3, and other strings into set Y.
2. Divide X_1, X_2, and X_3 further into groups of compatible strings. Run Greedy 1 on each group separately.
3. Run Mgreedy 1 on set Y.
4. Concatenate the strings obtained in steps 2 and 3, inserting #'s if necessary.

Theorem 7. *Given (P, N), where no string in P is a substring of a string in N, the above algorithm returns a consistent superstring of length $O(n^{4/3})$.*

Proof. (Outline) The proof is actually quite involved and needs 8 lemmas. So here we only give a very rough sketch. The complete proof is given in the full version of this paper. By the proof of Theorem 2, the strings produced in step 2 have total length $O(n)$. So it remains to analyze step 3. The proof of the $4n$ bound for Mgreedy in [1] essentially uses the fact that Mgreedy actually selects the edges (representing merges) following a Monge sequence [5] on the distance graph and finds an optimal cycle cover. However, with the presence of negative strings, a distance graph may or may not have a Monge sequence. (The negative strings lengthen some edges.) Thus we have to use a different strategy.

Our analysis can be roughly stated as follows. Consider the distance graph G_Y and view Mgreedy 1 as choosing edges in the graph G_Y: When Mgreedy 1 merges strings s and t, it chooses the edge $(last(s), first(t))$. Initially, we fix a path cover C on G_Y such that the total length of the paths in C is $O(n)$. We analyze Mgreedy 1 on Y with respect to the initial cover C. As Mgreedy 1 merges strings, we update the cover by possibly breaking a path into two or joining two paths into one or turning a path into a cycle. The merges performed by Mgreedy 1 are divided into several classes. A merge is *correct* if it chooses an edge in some current path or cycle. Otherwise the merge is *incorrect*. An incorrect merge is a *jump merge* if it breaks two potential correct merges simultaneously. Suppose in a jump merge Mgreedy 1 chooses an edge (x, y). Let x' be the current successor of x and y' the current predecessor of y, in their respective paths/cycles. That is, the choice of edge (x, y) prevents us from choosing the edges (x, x') and (y', y) in the future. Then the merge is *good* if $m(y', x')$ does not contain any negative string. Otherwise the merge is *bad*. Clearly the type of a merge performed by Mgreedy 1 depends the initial cover C and how we update paths and cycles.

We choose the initial cover C and the updating rule such that (i) Strings in each initial path overlap "a lot"; (ii) Only bad jump merges will increase the total length of the current paths and cycles. Then we can prove an upper bound $O(|C|^{3/2})$ on the total number of bad jump merges, implying an upper bound $O(n^{4/3})$ on the length of the superstring produced by Mgreedy 1. A key fact used in this proof is that the strings in Y do not have a long periodic prefix or suffix and thus, for any two strings there is a *unique* way of overlapping them to achieve a large amount of overlap. □

If the number of negative strings is bounded by some constant, we can show that our algorithm in fact achieves linear approximation.

Corollary 8. *Given (P, N), where no string in P is a substring of a string in N and $|N| \leq c$, c is a constant, then the above algorithm returns a consistent superstring of length $O(n)$.*

Proof. (Idea) Observe that using each negative string in N, a path can only interfere some other path just once. Thus each negative string can cause no more than $O(n)$ extra cost. Hence the resulting superstring is of length $O(n)$ following the proof of Theorem 7. □

5 Concluding Remarks

We have given polynomial-time linear approximation algorithms for two special cases of the shortest consistent superstring problem. It still remains open if a polynomial-time linear approximation algorithm exists for the general case. We suspect that our $O(n^{4/3})$ upper bound on the performance of a greedy algorithm, in the special case when no negative strings contain positive strings, can be improved to $O(n)$.

6 Acknowledgments

We have enjoyed and benefited a lot from working with A. Blum, J. Tromp, and M. Yannakakis on the shortest common superstring problem.

References

1. A. Blum, T. Jiang, M. Li, J. Tromp, M. Yannakakis. Linear approximation of shortest superstrings. *Proc. 23rd ACM Symp. on Theory of Computing*, 1991, 328-336; also to appear in *J. ACM*.
2. R. Drmanac and C. Crkvenjakov. Sequencing by hybridization (SBH) with oligonucleotide probes as an integral appraoch for the analysis of complex genomes. *International Journal of Genomic Research* 1-1, 1992, 59-79.
3. J. Gallant, D. Maier, J. Storer. On finding minimal length superstring. *Journal of Computer and System Sciences 20*, 1980, 50-58.
4. M. Garey and D. Johnson. *Computers and Intractability*. Freeman, New York, 1979.
5. A. Hoffman. On simple linear programming problems. *Convexity: Proc. of Symposia in Pure Mathematics, Vol. 7* (V. Klee, Ed.). American Mathematical Society, 1963.
6. T. Jiang and M. Li. Towards a DNA sequencing theory (revised version). Submitted to *Information and Computation*, 1991.
7. T. Jiang and M. Li. On the complexity of learning strings and sequences. *Proc. 4th Workshop on Computational Learning*, 1991, 367-371; also to appear in *Theoretical Computer Science* 119, 1993.
8. A. Lesk (Edited). *Computational Molecular Biology, Sources and Methods for Sequence Analysis*. Oxford University Press, 1988.
9. M. Li. Towards a DNA sequencing theory. *Proc. 31st IEEE Symp. on Foundations of Computer Science*, 1990, 125-134.
10. P. Pevzner and R. Lipshutz. Towards DNA sequencing by hybridization. Manuscript, 1993.
11. C. Papadimitriou and K. Steiglitz. *Combinatorial Optimization: Algorithms and Complexity*. Prentice-Hall, 1982.
12. H. Peltola, H. Soderlund, J. Tarhio, and E. Ukkonen. Algorithms for some string matching problems arising in molecular genetics. *Information Processing 83 (Proc. IFIP Congress)*, 1983, 53-64.
13. J. Storer. *Data compression: methods and theory*. Computer Science Press, 1988.
14. J. Tarhio and E. Ukkonen. A greedy approximation algorithm for constructing shortest common superstrings. *Theoretical Computer Science 57*, 1988, 131-145.
15. J. Turner. Approximation algorithms for the shortest common superstring problem. *Information and Computation 83*, 1989, 1-20.
16. L. G. Valiant. A theory of the learnable. *Comm. ACM 27(11)*, 1984, 1134-1142.

Tree Reconstruction from Partial Orders

Sampath Kannan[1]* and Tandy Warnow[2]**

[1] University of Arizona, Tucson, AZ.
[2] Sandia National Laboratories, Albuquerque, NM.

Abstract. The problem of constructing trees given a matrix of interleaf distances is motivated by applications in computational evolutionary biology and linguistics. The general problem is to find an edge-weighted tree which most closely approximates (under some norm) the distance matrix. Although the construction problem is easy when the tree exactly fits the distance matrix, optimization problems under all popular criteria are either known or conjectured to be *NP*-complete. In this paper we consider the related problem where we are given a partial order on the pairwise distances, and wish to construct (if possible) an edge-weighted tree realizing the partial order. In particular we are interested in partial orders which arise from *experiments* on triples of species. We will show that the consistency problem is *NP-hard* in general, but that for certain special cases the construction problem can be solved in polynomial time.

1 Introduction

Constructing edge-weighted trees from distance data is a classical problem motivated by applications in molecular biology, computational linguistics, and other areas. Here we are given an n-by-n distance matrix M and asked to find a tree T with leaves $1, 2, \ldots, n$ such that the path distance d_{ij}^T in the tree closely approximates the matrix M. When $d_{ij}^T = M_{ij}$ the matrix is said to be *additive* and efficient algorithms exist for constructing trees from additive distance data (see [14, 8, 2] and others). Various optimization criteria for the problem were proposed, and many *NP*-hardness results were published ([3, 4] and others). In fact, in [4] it is shown that for one of the standard optimization criteria (finding a minimum sized tree T such that $d_{ij}^T \geq M_{ij}$, where the size of the tree is the sum of the edge weights in the tree), there is a constant $\epsilon > 0$ such that no polynomial time algorithm can approximate the optimal solution within a ratio of n^ϵ unless *P=NP*. Thus, constructing trees from distance data is a hard problem and the usual heuristic approaches are unlikely to lead to reasonable solutions.

However, for many applications, the actual numeric data is quite unreliable (see [4, 6] for discussions of how interspecies distances are derived in computational molecular biology and why the data is unreliable). One way of handling

* Supported in part by NSF Grant CCR9108969
** This work began when this author was visiting DIMACS, and was supported in part by DOE contract number DE-AC04-76DP00789.

this unreliability is to assume that distances are given with error bars. This approach was taken by Farach, Kannan and Warnow in [4]. In this paper, we will take a different approach, and assume that we have confidence only in relative information, so that our input will be given in the form of a partial order on the pairwise distances. We are particularly interested in the problem where the partial order is constructed from *experiments* on triples of species, where these experiments are of one of two types:

Total Order Model (TOM) ; A *TOM* experiment on i, j, k determines the total order of the three pairwise distances $d(i,j), d(j,k), d(i,k)$, with equality or strict inequality indicated.

Partial Order Model (POM) : A *POM* experiment on i, j, k determines the minimum elements of $d(i,j), d(j,k), d(i,k)$.

These models are inspired by the model of Kannan, Lawler, and Warnow [10], in which rooted trees are constructed using experiments which determine the *rooted* topology for any three species.

We study the following problems under these models:

Consistency: Given a partial order on a set of distances, does there exist an edge-weighted tree which realizes this partial order?

Construction: Given the ability to perform an experiment, how quickly can we construct an edge-weighted tree realizing the experiments (here we assume the experiments *are* realizable).

The essential difference between the consistency problem and the construction problem is that we are *not* allowed to perform additional experiments to determine the consistency of a given set of experiments. This makes a substantial difference in complexity, with the consequence that determining consistency is NP-Complete for some of the models below where construction can be done efficiently.

We present the following results:

1. The problem of determining whether a set of POM or TOM experiments is consistent with a tree is *NP*-complete, for weighted trees as well as for unweighted trees without degree two nodes. This result is described in Section 2.
2. We can construct trees which are unweighted and without nodes of degree two in $O(n^3)$ time from TOM experiments, and in $O(n^4)$ time from POM experiments. These results are described in Sections 3 and 4, respectively.

Constructing unweighted trees without nodes of degree two is motivated by the work of Winkler[15], who considered the related *Discrete Metric Realization* Problem, in which one is given an n-by-n distance matrix M and an integer k, and the task is to create a graph G with n distinguished vertices and at most k edges, such that the shortest path in the graph G between x_i and x_j is exactly equal to M_{ij}. Winkler showed that this problem is *strongly* NP-complete for general graphs and for unweighted graphs without nodes of degree two.

2 Consistency of TOM or POM Experiments

In this section we show that determining whether a set of either TOM or POM experiments is consistent with a tree is *NP*-Complete, for weighted trees as well as for unweighted trees without nodes of degree two. We begin by considering a related problem of constructing trees using unrooted *quartets*, where a quartet is an unrooted tree on four leaves, i, j, k, l. Each quartet q is constrained to contain an edge e so that $q - \{e\}$ describes a partition of the four leaves into two sets of two leaves each. We indicate this by writing $q = (ij, kl)$. Thus, q indicates that the topology on leaves i, j, k, l is as follows:

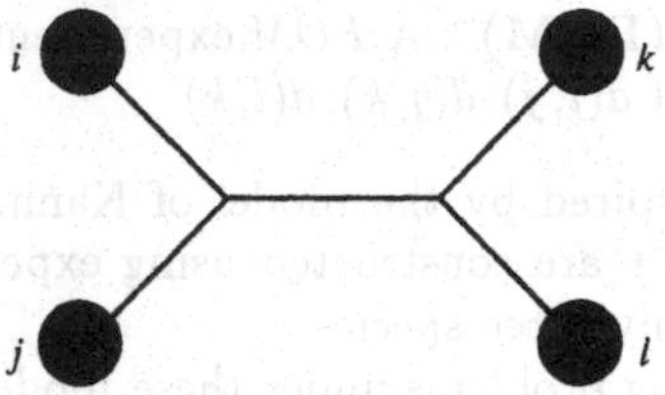

The Unrooted Quartet Consistency (UQC) Problem is as follows:

Problem: Unrooted Quartet Consistency
Input: A set Q of quartets on the species set $S = \{s_1, s_2, \ldots, s_n\}$.
Question: Does there exist a tree T with leaves labeled by the species of S such that if $q = (ij, kl) \in Q$, then there is an edge e in T such that i, j are on one side of e and k, l are on the other side.

The UQC problem was shown *NP-Complete* by Steel in [13], and is polynomially equivalent to the Perfect Phylogeny Problem and Triangulating Colored Graphs Problem, which were shown *NP-Complete* by Bodlaender, Fellows and Warnow in [1].

We can now prove that determining consistency of TOM experiments is *NP*-Complete.

Theorem 1. *TOM Consistency is* NP-*Complete.*

Proof. The reduction is from the *UQC* problem. Let I be an instance of the UQC problem and let (ab, cd) be one of the topology constraints.

We create two new leaves x and y and write down total order constraints for the triples $(x, y, a), (x, y, b), (x, y, c)$ and (x, y, d) as follows:

$d(x, a) < d(y, a) < d(x, y)$
$d(x, b) < d(y, b) < d(x, y)$
$d(y, c) < d(x, c) < d(x, y)$
$d(y, d) < d(x, d) < d(x, y)$

It is not hard to see that any edge-weighted tree satisfying each of the four constraints above also satisfies the topology constraint imposed by the quartet. Thus any tree that is consistent with the constraints of the TOM problem is consistent with the constraints of the UQC problem.

Conversely suppose there is a tree that is consistent with the constraints of the UQC problem. We augment this tree with the addition of the newly defined leaves such as x and y as follows. For each quartet (ab, cd), let $e = (u, v)$ be an edge in the tree separating ab from cd, with u on the a, b side, and v on the c, d side. In the TOM problem we introduced dummy species x and y such that a, b are closer to x than to y, and c, d are closer to y than to x. Attach these leaves x and y to u and v respectively and set $w(x, u) = w(y, v) = n^2$. All edges in the original tree are left at unit weight. It is then clear that the augmented tree satisfies all the constraints of the derived instance. Hence we have a valid reduction showing the NP-Completeness of the consistency problem for TOM experiments.

□

It is easy to see that this proof implies that the consistency of POM experiments is also an *NP*-Complete problem, and for both cases, the weights on the edges are integers between 1 and n^2. What is not quite as obvious is that consistency of TOM and POM experiments is still *NP*-Complete when we restrict ourselves to trees of unit weight and without degree two nodes. We now prove this for TOM experiments, since the proof for POM experiments follows along the same lines.

Theorem 2. *Determining Consistency of TOM experiments with unit weight trees is* NP-*complete.*

Proof. That the problem is in *NP* is trivial; thus we need only show that it is *NP*-hard. By the above proof, determining whether a set of TOM experiments is consistent with a tree in which every edge has integer weight bounded by n^2 is *NP*-Complete. We will show that this problem (bounded integer weight TOM experiment consistency) reduces to unit weight TOM experiment consistency.

Let $S = \{s_1, s_2, \cdots, s_n\}$ be a set of leaves and $\mathcal{E}$ a set of k TOM experiments. Let $S' = S \cup \{x_1, x_2, \cdots, x_{2n^3}\}$ be another set of leaves. We will show that $S, \mathcal{E}$ is consistent with a weighted tree T if and only if $S', \mathcal{E}$ is consistent with an unweighted tree T' without degree two nodes. It is clear that we can assume that T contains no nodes of degree two. So suppose T realizes the TOM experiments. By our construction, we can assume that T has only integer weights on the edges, and that these integer weights are bounded by n^2. For each edge e in T with weight $w(e) > 1$, we introduce $w(e) - 1$ additional internal nodes to that edge and $w(e) - 1$ new leaves hanging off those new internal nodes. We then set the weight of the newly created edges to be 1. It is clear that this new tree T' still satisfies the original set of experiments $\mathcal{E}$, and that it contains at most $2n^3$ additional leaves. Setting $S'=$ leaves(T'), we note that $S', \mathcal{E}$ is realized by T'.

For the converse, if the experiments in E can be realized by a unit-weighted tree T on leaf set S and at most $2n^2$ additional leaves, since E does not contain any constraints involving the additional leaves, E is realized by the tree T with the additional leaves stripped off.

□

As a consequence, we have the following corollary:

Corollary 3. *Let $\leq_P$ be a partial order defined on the set of pairwise distances, d_{ij}. Then determining whether $\leq_P$ is compatible with an edge-weighted (or unweighted) tree is* NP*-Complete.*

3 An $O(n^3)$ algorithm for constructing unweighted trees from TOM experiments

We examine the problem of constructing a tree T whose edges are unweighted (equivalently, $w(e) = 1$ for all edges $e \in E(T)$) and which has no nodes of degree two. We will assume here the ability to perform an experiment on any triple of species i, j, k, where the output of the experiment on the triple i, j, k will be a linear ordering on the three pairwise distances, with either equality or strict inequality indicated. We use $O(i, j, k)$ to indicate this linear ordering.

Sibling Sets Since our trees are not constrained to be binary, the sibling relation may have more than two siblings in each set of siblings. The general structure of the algorithm is to discover a set of siblings and connect them. The result will be a *supernode*, which will in fact represent subtree induced by an edge-deletion of T. Note that any tree where each internal node has degree three must have at least two sets of sibling nodes. Once we locate a sibling set $i_1, i_2, \ldots, i_r$, we 'collapse' the nodes representing $i_1, i_2, \ldots, i_r$ into a supernode. Recursively, we may collapse several supernodes together if they are discovered to constitute a sibling set. We will determine sibling sets by computing the transitive closure of the pairwise relationship of siblinghood (i.e. determining when a pair of (super)nodes are siblings). We therefore need to show how to determine siblinghood of pairs of supernodes, since from that relation we can determine sibling sets.

We will always maintain the following invariants: we will always know the exact tree that any particular supernode represents and each supernode will represent a portion of the overall tree which is connected to the rest of the tree by exactly one edge. We will abuse the notation slightly and use 'supernode' to represent not only a collection of the original vertices, but also the tree structure that we have computed on them. The root of a supernode V is defined to be the node $r(V)$ in V at which the rest of the tree is attached to V. For each supernode V we choose a 'representative' v which is a leaf of the original tree that is contained in V and which is nearest to the root of V. We denote the representative of a supernode V by $rep(V)$. We define $d(V) = d(rep(V), r(V))$. Initially, the supernodes are singleton sets and $d(V) = 0$ for all supernodes V. We will also say that V_1 and V_2 are siblings if $d(r(V_1), r(V_2)) = 2$.

We will clarify each step of the algorithm in what follows. In particular it is important to note that a 'collapse' of a set of nodes does not mean throwing away any of the individual nodes; in fact, even after we collapse a set of nodes we will need to refer to experiments involving individual nodes in the set that has been collapsed. Also note that the recursive application of the algorithm is not straightforward since after a single collapse we are not allowed to perform experiments on the new leaf that is created.

3.1 Determining Sibling Pairs

We have three techniques for determining whether V_1 and V_2 are siblings, depending on how large $|d(V_1) - d(V_2)|$ is.

Given two supernodes V_1 and V_2 with representatives i_1 and i_2, and given any leaf $j \notin V_1 \cup V_2$, we can determine whether j is closer to i_1 or i_2, and use this information to determine whether V_1 and V_2 are siblings. We therefore make the following definitions:

LT: $(V_1, V_2) \in LT$ if and only if there exists a leaf $j \in T - (V_1 \cup V_2)$ such that $d(j, i_1) < d(j, i_2)$.

GT: $(V_1, V_2) \in GT$ if and only if there exists leaf $j \in T - (V_1 \cup V_2)$ such that $d(i_2, j) < d(j, i_1)$.

EQ (*"Equals"*.) $(V_1, V_2) \in EQ$ if and only if there exists leaf $j \in T - (V_1 \cup V_2)$ such that $d(i_1, j) = d(i_2, j)$.

We now describe the details of how we determine whether two supernodes are siblings. We begin with the case where $d(V_1) = d(V_2)$.

Suppose that $d(V_1) = d(V_2)$ and i_1 and i_2 are the representatives of V_1 and V_2 respectively. If V_1 and V_2 are siblings, then for all $j \notin V_1 \cup V_2$, $d(i_1, j) = d(i_2, j)$. Thus, we will find that $(V_1, V_2) \in EQ$ and $(V_1, V_2) \notin LT \cup GT$. On the other hand, if V_1 and V_2 are not siblings, then there is an internal node v closer to the root of V_1 than to V_2, and if j is a leaf in the subtree rooted at v on the path from V_1 to V_2, then $d(i_1, j) < d(i_2, j)$. Thus, $(V_1, V_2) \in LT$. We therefore can characterize the pairs of supernodes V_1, V_2 such that $d(V_1) = d(V_2)$ which *must* be siblings as follows:

Lemma 4. *Let V_1 and V_2 be supernodes with $d(V_1) = d(V_2)$. Then V_1 and V_2 are siblings* if and only if $(V_1, V_2) \notin LT \cup GT$. *This can be determined in $O(n)$ time.*

Equally simple is the case where $d(V_1) = d(V_2) + 1$. For this case we have the following lemma whose proof is omitted.

Lemma 5. *Let V_1 and V_2 be supernodes such that $d(V_1) = d(V_2) + 1$. Then V_1 and V_2 are siblings* if and only if $(V_1, V_2) \notin EQ \cup LT$. *This can be determined in $O(n)$ time.*

In the case where $d(V_1) \geq d(V_2) + 2$, once again let i_1 and i_2 be the representatives of V_1 and V_2 respectively. If $(V_1, V_2) \in EQ \cup LT$ we can conclude immediately that V_1 and V_2 are not sibling supernodes. However, the converse does not hold since if the roots of V_1 and V_2 are connected by a path of length three we would still have $(V_1, V_2) \notin EQ \cup LT$. We need a different technique to handle this case.

We assume that V_1 and V_2 are siblings and derive either a contradiction or confirmation of this assumption. Since we know the entire structure of V_1 and V_2, under our assumption above, we know the length of the path P from i_1 to i_2. We have two cases. If P is of even length, let x be the midpoint of P. Because of our assumption, we can identify the above node x which lies within V_1. Since x has degree greater than 2 we can find a leaf y which lies in a branch of x other than the ones containing i_1 or i_2 (y lies within V_1 as well and we can identify such a y quickly). Then V_1 and V_2 are siblings if and only if $d(i_1, y) = d(i_2, y)$. If P is of odd length, the proof is similar — we identify the two middle nodes w and x in P and check that for leaves y_1 and y_2 hanging off third branches of w and x respectively, $d(y_1, i_1) < d(y_1, i_2)$ and $d(y_2, i_1) > d(y_2, i_2)$. Again these conditions will hold if and only if V_1 and V_2 are actually siblings. We summarize the above in the following lemma.

Lemma 6. *Let V_1 and V_2 be supernodes with $d(V_1) \geq d(V_2) + 2$. We can determine whether V_1 and V_2 are siblings in $O(n)$ time.*

Remark 1: It is possible using clever data structures for maintaining the structure of supernodes to reduce the time to check siblinghood in this case to sublinear. However, we don't do this, since in the other cases we do need linear time to check siblinghood.

3.2 Implementation and Running Time

It is clear that the invariants will be maintained after the detection of sibling sets and the merger of the constituent supernodes into a bigger supernode. When we have only three supernodes left we stop and construct the tree. We then have to check that this tree is in fact consistent with all of the experiments that have been performed.

In order to see that the running time is $O(n^3)$, note that we perform $O(n^2)$ tests of siblinghood initially (between every pair of given species). Since each test of siblinghood takes $O(n)$ time, this initial step costs $O(n^3)$. Now suppose we determine that two supernodes I and J are siblings. We then need to look at the remaining supernodes to see if any of these are also siblings to I (and hence to J). Hence, to determine the sibling set containing I will cost us $O(n)$ sibling tests, for a total cost of $O(n^2)$. This cost is only incurred by a new sibling pair, and since there are at most $O(n)$ sibling sets, this only occurs $O(n)$ times. Hence, the overall cost is $O(n^3)$.

Note that the algorithm above is optimal since we have to check that the tree is correct with respect to $\Omega(n^3)$ experiments. However, we might consider a related promise problem where the task is to construct a tree under the promise that such a tree exists. In this case, the lower bound does not hold and the problem of whether there is a more efficient algorithm is open.

4 An $O(n^4)$ algorithm for constructing unweighted trees from POM experiments

The algorithm we use for constructing unweighted trees from POM experiments uses some of the same techniques as in the previous section. Here, an experiment on i, j, k only returns pair(s) which have minimum distance, and does not totally order that set. We indicate this experimental outcome by $O(i, j, k)$ as before, but understand this to be a set of pairs.

As in the previous algorithm, we work with supernodes and determine siblinghood pairs, from which we will derive sibling sets.

4.1 Determining Sibling Pairs

We again have techniques for determining siblinghood of supernodes V_1 and V_2, which depend upon $|d(V_1) - d(V_2)|$. Since we do not obtain a total order on distances, we use different definitions of the sets *EQ*, *LT*, and *GT*, in which we may combine information from different experiments in order to infer the existence of internal nodes v for which we can determine the relative proximity of v to the leaves i_1 and i_2.

LT: $(V_1, V_2) \in LT$ if and only if there exists leaf $j \in T - (V_1 \cup V_2)$ such that $(i_1, j) \in O(i_1, i_2, j)$ but $(i_2, j) \notin O(i_1, i_2, j)$, *OR* there exist leaves j, k in $T - (V_1 \cup V_2)$ such that $(i_1, j) \in O(i_1, j, k)$ and $(i_2, j) \notin O(i_2, j, k)$. *Membership in this set indicates the existence of an internal node v so that $d(i_1, v) < d(i_2, v)$.*

GT: $(V_1, V_2) \in GT$ if and only if there exists leaf $j \in T - (V_1 \cup V_2)$ such that $(i_2, j) \in O(i_1, i_2, j)$ but $(i_1, j) \notin O(i_1, i_2, j)$, *OR* there exist leaves j, k in $T - (V_1 \cup V_2)$ such that $(i_2, j) \in O(i_2, j, k)$ and $(i_1, j) \notin O(i_1, j, k)$. *Membership in this set indicates the existence of an internal node v so that $d(i_1, v) > d(i_2, v)$.*

EQ (*"Equals".*) $(V_1, V_2) \in EQ$ if and only if there exists leaf $j \in T - (V_1 \cup V_2)$ such that $\{(i_1, j), (i_2, j)\} \subset O(i_1, i_2, j)$. *Membership in this set indicates the existence of an internal node v so that $d(i_1, v) = d(i_2, v)$.*

In order to be able to use membership or non-membership in these sets EQ, GT, and LT, we need two more lemmas. First we make the following definition.

Definition: Let i be a leaf, v an internal node, and c a variable or a constant whose value is a positive integer.

Suppose it is known that $d(i, v) \geq c$. An experiment on i, r, s is said to be (c, i, v)-*critical* if from the outcome of the experiment on (i, r, s) and previous experiments, it is possible to deduce whether $d(i, v) = c$.

In other words, an experiment i, r, s is (c, i, v)-*critical* if its outcome can be predicted from the assumption that $d(i, v) = c$ and if the predicted outcome will equal the true outcome if and only if $d(i, v) = c$.

We can now prove the following lemma:

Lemma 7. *Let v be an internal node separating T into three subtrees, T_1, T_2, and T_3. Let i be a node in T_1 with $d(i, v) \geq k$ where k is a constant or a variable whose value is a positive integer and $min_{x \in T_2} d(x, v) = t \geq k + 2$. Then there exists leaves $r, s \in T_2$ such that i, r, s is a (k, i, v)-critical experiment.*

Proof. Define the distance function $d^*(i, r) = d(v, r) + k$. Trivially, $d(i, r) \geq d^*(i, r)$ for all $r \in T_2$ with equality if and only if $d(i, v) = k$. Let x be the leaf in T_2 of shortest d^*-distance from i, and let $d^*(i,x) = L$. By construction, $L \geq 2k+2$. In the path from i to x let c be the node which is at d^*-distance $\lfloor \frac{L}{2} \rfloor$ from i. Since $L \geq 2k+2$, $c \in T_2$. As before, the removal of c from the tree splits the leaves into three sets, $S_1(c), S_2(c)$, and $S_3(c)$. Let $i \in S_1(c)$ and $x \in S_2(c)$, and let y be the node closest to i in $S_3(c)$. The d^*-distance from y to c is at least equal to $\lceil \frac{L}{2} \rceil$ because of the way that x was chosen. If the d^*-distance from y to c is less than or equal to $\lceil \frac{L+1}{2} \rceil$, then the experiment i, x, y is a (k, i, v)-critical experiment. Suppose, on the other hand that the d^*-distance from y to c is greater than $\lceil \frac{L+1}{2} \rceil$. Then let the path from i to y be of length L' and let c' be the node along this path of d^*-distance $\lfloor \frac{L'}{2} \rfloor$ from i. Then c lies strictly between i and c' and we can repeat the argument by considering i and y and the third branch out of c'. Ultimately this procedure has to terminate since the tree is finite. At this point we will have found suitable x and y so that (i, x, y) is a (k, i, v)-critical experiment.

□

Note that in general it is not possible to make the above proof constructive in the obvious manner since we do not know the structure of the overall tree. Thus finding suitable x and y could take $O(n^2)$ time in the worst case.

If V_1 and V_2 were siblings, since we already know the structure of V_1 and V_2, we can predict the outcome of every experiment involving leaves only in $V_1 \cup V_2$. It is thus clear that a necessary condition for V_1 and V_2 to be siblings is that the *predicted* outcome $\mathcal{PO}(i,j,k)$ must equal the actual outcome of the experiment, $O(i,j,k)$, for every $\{i,j,k\} \subset V_1 \cup V_2$.

We also need the following lemma:

Lemma 8. *Let V_1 and V_2 be supernodes with representatives i_1 and i_2, and let $v \notin V_1 \cup V_2$ be a node on the path between i_1 and i_2. Let v separate T into the subtrees $S_t(v)$, $t = 1,2,3$ with $i_1 \in S_1(v), i_2 \in S_2(v)$, and let j be the leaf in $S_3(v)$ closest to v. Then there exist leaves $k, l \in T - (V_1 \cup V_2)$ such that the experiments on i_1, i_2, k and l indicate that*

$$(V_1, V_2) \in LT \text{ if } d(i_1, v) < d(i_2, v),$$
$$(V_1, V_2) \in GT \text{ if } d(i_1, v) > d(i_2, v).$$

Proof. If $d(i_1,v) < d(i_2,v)$, then $O(i_1,i_2,j) = \{(i_1,i_2\}$ or else $(i_1,j) \in O(i_1,i_2,j)$. If $(i_1,j) \in O(i_1,i_2,j)$ then $(i_2,j) \notin O(i_1,i_2,j)$ so that $(V_1,V_2) \in LT$. On the other hand, if $(i_1,j) \notin O(i_1,i_2,j)$ then we can deduce that $d(j,v) > d(i_2,v) > d(i_1,v)$ so that $d(j,v) \geq d(i_1,v)+2$. We can therefore apply Lemma 7 and deduce the existence of leaves r, s in $T - (V_1 \cup V_2)$ so that the experiment on (i_1, r, s) is $(d(i_1,v), i_1, v)$-critical. It is then easy to see that the experiments on i_1, i_2, r and s determine that $d(i_2,v) > d(i_1,v)$ so that $(V_1,V_2) \in LT$. A similar analysis shows that $(V_1,V_2) \in GT$ if $d(i_1,v) > d(i_2,v)$.

□

Determining Siblinghood when $d(V_1) = d(V_2)$

We can now prove the following:

Lemma 9. *Let V_1 and V_2 be supernodes with $d(V_1) = d(V_2)$. Then V_1 and V_2 are siblings if and only if the ordered pair $(V_1, V_2) \notin GT \cup LT$.*

Proof. The necessity is obvious. For the sufficiency, apply Lemma 8.

□

Determining Siblinghood when $|d(V_1) - d(V_2)| = 1$

Lemma 10. *Let V_1 and V_2 be supernodes with $d(V_1) = d(V_2)+1$, with representatives i_1 and i_2 respectively, and assume that the path P between the roots of the supernodes has length at least four. Then there exists experiments indicating that $(V_1, V_2) \in LT$.*

Proof. Let v be the node on P adjacent to the root of V_1. Then $d(v, i_1) = d(V_1)+1 = d(V_2)+2 < d(i_2, v)$. By Lemma 8, there exists leaves $a, b \in T-(V_1 \cup V_2)$ such that the experiments on a, b, i_1, i_2 indicate that $(V_1, V_2) \in LT$.

□

However, we still need to be able to determine that V_1 and V_2 are not siblings when the distance between their roots is exactly three. For this case we have a different technique.

As noted before, if V_1 and V_2 are siblings, then $d(v, i_1) = d(v, i_2) + 1$ for all nodes $v \in T - (V_1 \cup V_2)$. In particular, the parity of the distances $d(v, i_1)$ and $d(v, i_2)$ will be different for every node $v \in T - (V_1 \cup V_2)$. Let P be the path from the root r_1 of V_1 to the root r_2 of V_2. Suppose that P has length exactly three. Let v be the node adjacent to r_1 in P, and let x be a leaf in the subtree below v. Then $d(x, i_1) = d(x, i_2)$, so that the parities of $d(x, i_1)$ and $d(x, i_2)$ are identical. We will show that we can determine the existence of a leaf x so that $d(x, i_1)$ and $d(x, i_2)$ have the same parity, when V_1 and V_2 have distance exactly three apart.

Lemma 11. *Let V_1 and V_2 be supernodes of distance three apart and with $d(V_1) = d(V_2) + 1$. Then there exists an experiment indicating that $(V_1, V_2) \in EQ$ or there exists experiments indicating that parity(i_1, x) = parity(i_2, x), for a leaf $x \in T - (V_1 \cup V_2)$.*

Proof. Let v be the node closest to r_1 in the path from V_1 to V_2. As indicated before, every leaf x in the subtree Q rooted at v is equidistant to i_1 and i_2. Suppose for some $x \in Q$, $(i_1, x) \in O(i_1, i_2, x)$. Then $\{(i_1, x), (i_2, x)\} \subset O(i_1, i_2, x)$ so that $(V_1, V_2) \in EQ$. Otherwise, for all $x \in Q$, $(i_1, i_2) = O(i_1, i_2, x)$, so that $d(i_1, i_2) < d(i_1, x) = d(i_2, x)$. Let y be the node of longest distance from i_1 within Q. We will show that we can determine the parity of the distance from i_1 to y.

If the distance $d(i_1, y)$ is even, then the node z halfway along the path from i_1 to y is an element of Q. For any leaf t in the subtree Q' rooted at z not containing either i_1 or y, $O(i_1, y, t) = \{(i_1, t), (y, t)\}$. This indicates that $d(i_1, y)$ is even. On the other hand, if the distance $d(i_1, y)$ is odd, then for all t, $\{(i_1, t), (y, t)\} \not\subset O(i_1, y, t)$. Thus, we can determine the parity of $d(i_1, y)$ for y the leaf furthest from i_1 in Q. However, by construction, y is also the leaf of longest distance to i_2 in Q, so that the parity of $d(i_2, y)$ can also be determined. By our construction, $d(i_2, y) = d(i_1, y)$, so that the parity is the same.

□

We summarize our findings in the following theorem.

Theorem 12. *Let V_1, V_2 be supernodes with $d(V_1) = d(V_2) + 1$. Then we can in $O(n^2)$ time determine whether V_1 and V_2 are siblings.*

Proof. We first check whether $(V_1, V_2) \in LT \cup EQ$ by examining $O(n^2)$ experiments involving at least one of i_1 and i_2. If $(V_1, V_2) \in LT \cup EQ$ then we know immediately that V_1 and V_2 cannot be siblings. However, if $(V_1, V_2) \notin LT \cup EQ$, it is still possible that they are not siblings but have a path of distance exactly three between them. By the previous lemma, if the distance between V_1 and V_2 were exactly three, we would be able to determine that $(V_1, V_2) \in EQ$ or else that $d(x, i_1)$ and $d(x, i_2)$ have the same parity, for some $x \in T - (V_1 \cup V_2)$, by examining $O(n^2)$ experiments. Should we find some such x we know that V_1 and V_2 are not siblings. Otherwise the only possibility is that they are in fact siblings.

□

Determining Siblinghood when $|d(V_1) - d(V_2)| \geq 2$

Determining siblinghood in this case is a straight-forward application of lemma 7. More precisely, suppose, wlog that $d(V_1) \geq d(V_2) + 2$. Then we know that $d(r(V_1), i_2) \geq d(V_2) + 2$ with equality if and only if V_1 and V_2 are siblings. Thus we can find x and y (within V_1 in this case) which confirm or refute the assumption of siblinghood of V_1 and V_2.

Theorem 13. *Let V_1 and V_2 be supernodes such that $d(V_2) - d(V_1) \geq 2$, and let $i_1 = rep(V_1)$. Then V_1 and V_2 are siblings if and only if for all leaves $j, k \in V_2$, $\mathcal{PO}(i_1, j, k) = O(i_1, j, k)$. Furthermore, we can determine whether $\mathcal{PO}(i_1, j, k) = O(i_1, j, k)$ for all $j, k \in V_2$ in $O(|V_2|)$ time, only knowing the structures of V_1 and V_2.*

The proof follows as for the case where the experiments are TOM experiments.

4.2 Implementation and Running Time

At the top level the construction algorithm for the POM model is identical to the algorithm for the TOM model. The difference is only in the procedure for testing siblinghood of two supernodes. In the POM model this takes $O(n^2)$ time leading to an overall running time of $O(n^4)$.

5 Conclusions and Open Problems

The models we have presented in this paper strictly generalizes any distance-based models of tree reconstruction since we can infer order information given actual distances. It may be possible to even incorporate some tolerance in the distance values by considering $d(i,j)$ to be less than $d(k,l)$ only if $d(k,l) - d(i,j) > B$ for some tolerance parameter B.

There are obvious optimization questions related to the construction questions we have considered. for a given set $\mathcal{E}$ of experiments (TOM or POM), what is the maximum cardinality subset of $\mathcal{E}$ which is consistent with a tree? Since consistency of TOM and POM experiments is an *NP*-Complete problem, whether the tree is weighted or unweighted, these problems are *NP*-hard.

The major open question is whether there are polynomial time algorithms to reconstruct *weighted* trees in the TOM and POM models.

References

1. H. Bodlaender, M. Fellows, and T. Warnow, *Two strikes against perfect phylogeny*, Proceedings, ICALP, Vienna, Austria, July 1992.
2. J. Culberson and P. Rudnicki, *A fast algorithm for constructing trees from distance matrices*, Information Processing Letters, 30 (1989), pp. 215-220.
3. W.H.E. Day, *Computational complexity of inferring phylogenies from dissimilarity matrices*, Bulletin of Mathematical Biology, Vol. 49, No. 4, pp. 461-467, 1987.
4. M. Farach, S. Kannan, and T. Warnow, *A robust model for finding optimal evolutionary trees*, to appear, Algorithmica, Special issue on Computational Biology, also to appear, Proceedings of the Symposium on the Theory of Computing (STOC), San Diego, CA, 1993.
5. J.S. Farris, *Estimating phylogenetic trees from distance matrices*, Am. Nat., 106, pp. 645-668, 1972.
6. J. Felsenstein, *Numerical methods for inferring evolutionary trees*, The Quarterly Review of Biology, Vol. 57, No. 4, Dec. 1982.
7. W.M. Fitch and E. Margoliash, *The construction of phylogenetic trees*, Science 155:29-94, 1976.
8. J. Hein, *An optimal algorithm to reconstruct trees from additive distance matrices*, Bulletin of Mathematical Biology, Vol. 51, No. 5, pp. 597-603, 1989.
9. J. Hein, *A tree reconstruction method that is economical in the number of pairwise comparisons used*, Mol. Biol. Evol. 6(6), pp. 669-684, 1989.
10. S. Kannan, E. Lawler, and T. Warnow, *Determining the evolutionary tree*, Proc. First Annual ACM-SIAM Symp. on Discrete Algorithms, San Francisco, Jan. 1990, also, to appear, J. of Algorithms.
11. W.-H. Li, *Simple method for constructing phylogenetic trees from distance matrices*, Proc. Natl. Acad. Sci. USA, 78:1085-89, 1981.
12. N. Saitou and M. Nei, *The neighbor-joining method: a new method for reconstructing phylogenetic trees*, Mol. Biol. Evol. 4:406-25, 1987.
13. M.A. Steel, *The complexity of reconstructing trees from qualitative characters and subtrees*, Journal of Classification, Vol. 9, 1992.
14. M.S. Waterman, T.F. Smith, M. Singh, and W.A. Beyer, *Additive evolutionary trees*, J. Theor. Biol., 64, pp. 199-213, 1977.
15. P. Winkler, *The complexity of metric realization*, SIAM J. Discrete Math, Vol. 1, No. 4, 1988.

Improved Parallel Depth-First Search in Undirected Planar Graphs

Ming-Yang Kao* Shang-Hua Teng† Kentaro Toyama‡

Abstract

We present an improved parallel algorithm for constructing a depth-first search tree in a connected undirected planar graph. The algorithm runs in $O(\log^2 n)$ time with $n/\log n$ processors for an n-vertex graph. It hinges on the use of a new optimal algorithm for computing a cycle separator of an embedded planar graph in $O(\log n)$ time with $n/\log n$ processors. The best previous algorithms for computing depth-first search trees and cycle separators achieved the same time complexities, but with n processors. Our algorithms run on a parallel random access machine that permits concurrent reads and concurrent writes in its shared memory and allows an arbitrary processor to succeed in case of a write conflict.

1 Introduction

Depth-first search is one of the most useful techniques for solving a wide variety of graph problems [4, 11, 35]. Starting from a vertex, depth-first search constructs a tree, called a *depth-first search tree*, during the course of searching the graph. The *depth-first search problem* is defined as follows: given a graph and a distinguished vertex, construct a tree that corresponds to performing depth-first search in the graph starting from the given vertex.

A depth-first search tree can be computed sequentially in optimal linear time for any given graph [4, 11, 35]. The problem of performing depth-first search in parallel has been studied by numerous authors. For ordered depth-first search, the problem seems to be highly sequential in nature [33]. For unordered depth-first search, while randomized NC algorithms have been found for general graphs [2, 3], deterministic algorithms are known only for certain classes of graphs, including chordal graphs [26], series parallel graphs [19], undirected planar graphs [16, 20, 21, 34], and directed planar graphs [22, 23, 24].

In this paper, we study the depth-first search problem for connected undirected planar graphs. Our main objective is to reduce the total work of parallel

---

*Department of Computer Science, Duke University, Durham, NC 27706. Supported in part by NSF Grant CCR-9101385.

†Department of Mathematics, Massachusetts Institute of Technology, Cambridge, MA 02139.

‡Department of Computer Science, Yale University, New Haven, CT 06520.

depth-first search while achieving the same time complexity as the best previous algorithms. Our algorithm hinges on the use of a new optimal algorithm for computing cycle separators. A *cycle separator* of an undirected graph is a vertex-simple cycle whose removal divides the graph into connected components of size at most a constant fraction of the size of the original graph.

The step of computing a cycle separator is at the core of almost all previous parallel depth-first search algorithms for undirected planar graphs [20, 21, 34]. Known algorithms for this step include the first NC algorithm given by Smith [34] that runs in $O(\log^2 n)$ time on $n^2/\log^2 n$ processors, and the more efficient algorithm of He and Yesha [20] that runs in $O(\log n)$ time with n processors.

These cycle separator algorithms all lead to depth-first search tree algorithms. Smith's depth-first search algorithm [34] runs in $O(\log^3 n)$ time on n^4 processors. Ja'Ja and Kosaraju [21] found an algorithm that reduces the number of processors required to n while maintaining the same time complexity. He and Yesha [20] gave algorithms that take $O(\log^2 n)$ time and n processors.

In this paper, we show that for a connected undirected embedded planar graph with n vertices, a cycle separator can be found in $O(\log n)$ time with $n/\log n$ processors. As a consequence, a depth-first search tree rooted at a specified vertex can be computed in $O(\log^2 n)$ time with $n/\log n$ processors. These algorithms run on a parallel random access machine that allows concurrent reads and concurrent writes in its shared memory, and in case of a write conflict, permits an arbitrary processor to succeed (arbitrary-CRCW PRAM [25]).

Section 2 reviews important concepts and facts about planar graphs. Sections 3 and 4 present our algorithms for computing cycle separators and depth-first search trees, respectively. Section 5 concludes with open problems.

2 Basics of undirected planar graphs

From this point onward, all graphs in this paper are assumed to be undirected with no multiple edges or self loops, unless otherwise indicated.

There are several essentially equivalent definitions for an embedding of a graph in a plane [7, 8, 12, 18, 36, 37]. An intuitive one is that of a *topological embedding* [37], where each vertex of a graph is mapped to a point in the plane, and each edge is associated with a curve segment between the points of its two end vertices.

A topological embedding of a graph is called *planar* if no two curve segments intersect except at a common endpoint, in which case the two corresponding edges have a common end vertex. A graph is called *planar* if it has a planar topological embedding. An *embedded* planar graph is one accompanies by a planar topological embedding.

After the curve segments of an embedded planar graph are removed from the plane, the plane is divided into regions. Each region is called a *face* of the embedded planar graph. Exactly one of the faces is unbounded; it is called the *external* face. The *boundary* of a face is the set of vertices and edges surrounding that face. The vertices on the boundary of the external face are called the *external* vertices.

Theorem 2.1 (Euler) *An embedded planar graph with n vertices, m edges, and f faces satisfies* $n - m + f = 2$.

Corollary 2.2 *An n-vertex planar graph with* $n \geq 3$ *has at most* $3n - 6$ *edges.*

2.1 Planar combinatorial embeddings and data structures

This paper employs an algorithmically compact representation for a planar topological embedding of a connected graph, namely, a planar combinatorial embedding [12], [30].

Let G be a connected embedded planar graph. Each edge of G is replaced by two directed edges with the opposite directions. These directed edges are called the *darts*.

A *planar combinatorial embedding* of G is a permutation ϕ of the darts of G satisfying the following conditions:

- For each dart e, the darts e and $\phi(e)$ point from the same vertex.
- For each vertex v, the darts pointing from v form a cyclic permutation in ϕ to specify the clockwise cyclic ordering of the darts that radiate from v in G.

Let α be the permutation of the darts such that for each dart e the dart $\alpha(e)$ is in the opposite direction of e. Then the orbits of the permutation $\psi = \phi \cdot \alpha$ form the boundaries of the faces of G. To be precise, let f be a face of G and let e be a counterclockwise dart on the boundary of f. Then the sequence of darts $e, \psi(e), \psi^2(e), \ldots$ trace the boundary of f in the counterclockwise direction.

Theorem 2.3 ([32]) *A planar combinatorial embedding of a connected planar graph with n vertices can be computed deterministically in* $O(\log n)$ *time with* $n \log\log n / \log n$ *processors and probablistically in* $O(\log n)$ *time with* $n/\log n$ *processors.*

2.2 Connected components and spanning trees

Theorem 2.4 ([15]) *The connected components of a planar graph with n vertices can be computed in* $O(\log n)$ *time with* $n/\log n$ *processors.*

Theorem 2.5 ([15]) *A spanning tree of a connected planar graph with n vertices can be computed in* $O(\log n)$ *time with* $n/\log n$ *processors.*

3 Computing cycle separators for planar graphs

A key subroutine in most parallel algorithms for depth-first search is to find graph separators with certain connectivity structures [2, 3, 20, 21, 22, 23, 24, 34].

In this paper, we employ path and cycle separators defined as follows.

Let G be a graph. A vertex subset of G is 2/3-*heavy* for G if it contains more than 2/3 of the vertices in G.

Let n be the number of vertices in G. A *separator* of G is a set S of vertices such that each of the connected components in $G - S$ contains at most $2n/3$ vertices, i.e., none of the connected components in $G - S$ is 2/3-heavy for G.

A *path* (respectively, *cycle*) separator of G is a vertex-simple path (respectively, cycle) whose vertices form a separator. A single vertex is considered a cycle of length zero. Thus, if the removal of a vertex separates a graph, the vertex is a cycle separator.

Section 3.1 explains the procedure for computing a path separator of a connected embedded planar graph. Section 3.2 describes how to convert a path separator into a cycle separator for a connected embedded planar graph.

3.1 Computing a path separator

Let G be a connected embedded planar graph with n vertices. Our idea for computing a path separator of G is to first convert it into an embedded outer planar graph G'. (An embedded *outer* planar graph is one whose edges can be partioned into two sets C and E such that C forms a vertex-simple cycle and every edge in E is placed outside C and connected between two vertices of C.) We will then find a separator of G' that consists of two vertices and use this separator to construct a path separator of G.

Theorem 3.1 *Given a connected embedded planar graph with n vertices, a path separator can be computed in $O(\log n)$ time with $n/\log n$ processors.*

To prove this theorem, we detail our algorithm for computing a path separator of G as follows:

Remark. Steps 1–3 convert G into an outer planar graph G'.

1. Compute an arbitrary spanning tree T of G and the set D of the edges not in T.

2. Find an Euler tour U of T with the properties below.

   The edges incident at each vertex in T are traversed in the clockwise cyclic ordering in the embedding of G. Each edge of T is traversed twice and is replicated twice in U. If a vertex is of degree d in T, it is replicated d times in U. Note that with this duplication U is a vertex-simple cycle.

   Remark. To intuitively visualize the effects of this and the next step, we can imagine that T is a balloon without air whose surface is formed by the duplicates of the tree edges. In this step, we blow air into T to form U. Correspondingly, G is pushed from the inside of T out to form a new planar graph whose edges are either on or outside the balloon, i.e., G becomes an outer planar graph. This effect is formally described in the next step.

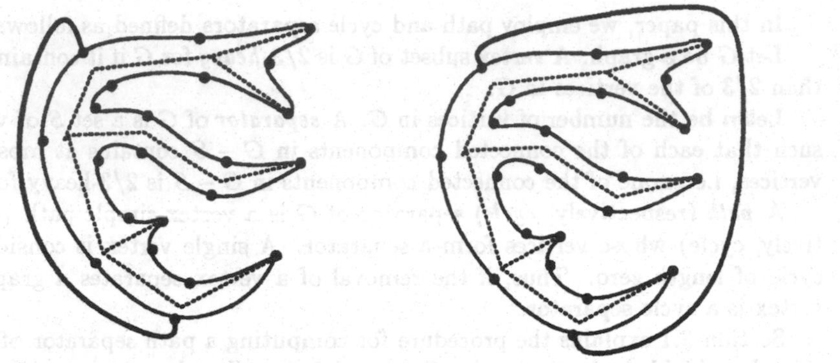

Figure 1: The dotted lines represent T. The solid lines represent U with the points being vertices on the tour. The dark line is an edge e of D. The two figures demonstrate the two positions that e can assume under the flip operation.

3. For each edge $e = \{u, v\}$ in D, reconnect e between the copies u' and v' of u and v in U defined below. Let G' be the graph formed by U and the reconnected edges.

   Let e_1 and e_2 be the two edges of T incident at u such that e_1, e, e_2 are in the clockwise cyclic ordering around u in the embedding of G and that e_1 and e_2 are adjacent in T in the cyclic ordering around u. Let e_1' and e_2' be the copies of e_1 and e_2 such that e_2' is next to e_1' in the clockwise direction of U. Let u' be the copy of u that is a common endpoint of e_1' and e_2'. The vertex v' is similarly defined.

   Remark. It can be proved that G' is an embedded outer planar graph with no edges inside U. Steps 4–9 compute a two-vertex separator $\{x, y\}$ of G'.

4. For each vertex v in U, assign weight $1/d$ to v where d is the degree of v in T.

   Remark. This step is based on the fact that if a vertex is of degree d in T, then it is replicated d times in U. The total weight of U is n.

5. Compute the prefix sums of U starting from an arbitrary vertex.

   Remark. As a result of this step, the total weight of any subpath of U can be computed in constant time with one processor.

6. Reposition each edge $e \in D$ in the embedding of G' as follows.

   The two endpoints of e partition U into two subpaths P_1 and P_2. Let C_1 and C_2 be the cycles formed by e and P_1 and by e and P_2. Let P_1' and P_2' be P_1 and P_2 without the endpoints of e. Then either P_2' is outside C_1 or P_1' is outside C_2. If P_2' is outside C_1, we say that e *covers* P_1'; otherwise, e *covers* P_2'. At least one of these two positions results in e covering a path with a total weight at most $n/2$. Wherever necessary, we flip e from one position to the other (see Figure 1) to ensure that e covers a path with a total weight at most $n/2$. Consequently, the total weight of U strictly between two adjacent external vertices of G' is at most $n/2$. G' remains an embedded planar graph with no edges inside U.

7. Find an external vertex z_0 of G' and the subpath U' of U starting from z_0 running clockwise to the vertex right before z_0 in U.

8. Find the last vertex $y \in U'$ such that y is an external vertex of G' and the total weight of U' from z_0 to y is at most $2n/3$.

9. Find the external vertex z_1 of G' that is right after y in U'. If z_1 exists and the total weight of U' from y to z_1 is more than $n/3$, then let $x = z_1$. Otherwise, let $x = z_0$.

   Remark. It can be proved that $\{x, y\}$ is a separator of G'. The next step uses this separator to compute a path separator of G.

10. Return the tree path of T between x and y.

We now prove the correctness of the above algorithm and show that it achieves the time and processor complexity stated in Theorem 3.1. The next two lemmas establish the correctness of the algorithm.

Lemma 3.2 *Let R_1' be the subpath of U' between x and y. Let R_1 be R_1' without x and y. Let R_2 be the path $U - R_1'$. Then the total weights of R_1 and R_2 are at most $2n/3$ each.*

Proof. There are three cases depending on the outcome of Step 9.

Case 1: z_1 exists and the total weight of U' from y to z_1 is more than $n/3$. As a result of Step 6, because no external vertex of G' exists between x and y in U, the total weight of R_1 is at most $n/2$. Next, because the total weight of R_1' is more than $n/3$, the total weight of R_2 is less than $2n/3$.

Case 2: z_1 does not exist. By the choice of y, the total weight of R_1 is at most $2n/3$. As for R_2, because the total weight of U' is n, there must be an edge in D between z_0 and y that covers R_2. Thus, as a result of Step 6, the total weight of R_2 is at most $n/2$.

Case 3: z_1 exists and the total weight of U' from y to z_1 is at most $n/3$. By the choice of y, the total weight of R_1 is at most $2n/3$. As for R_2, the total weight of R_1' must be more than $n/3$; otherwise either z_1 or some external vertex after z_1 should have been chosen for y. Thus, the total weight of R_2 is less than $2n/3$. □

Lemma 3.3 *The tree path of T between x and y forms a separator of G.*

Proof. Let R_1 and R_2 be the two paths defined in Lemma 3.2. Let V_1 and V_2 be the sets of vertices in G whose duplicates appear in R_1 and R_2. Let S be the tree path of T between x and y.

Because G' has no edges inside the cycle U and because x and y are external vertices of G', the paths R_1 and R_2 are disconnected in G' after x and y are removed. Because the traversal of T by U and the reconnecting of the nontree edges are consistent with the embedding of G, V_1 and V_2 can only share vertices from S. Therefore, $V_1 - S$ and $V_2 - S$ are disconnected in $G - S$. By the weight assignment in Step 4, the numbers of vertices in $V_1 - S$ and $V_2 - S$ are bounded by the weights of R_1 and R_2, respectively. Thus by Lemma 3.2, S is a path separator of G. □

The next lemma finishes the proof of Theorem 3.1.

Lemma 3.4 *The path-separator algorithm runs in $O(\log n)$ time with $n/\log n$ processors.*

Proof. It suffices to show that each of the steps can be carried out in $O(\log n)$ time with $n/\log n$ processors as follows. All the steps not mentioned in this paragraph can be done in a straightforward manner. Step 1 is done by means of Theorem 2.5. Step 2 uses known optimal algorithms for Euler tour [6]. Step 5 uses optimal algorithms for prefix computation [28, 29] and list ranking [5, 10, 17]. Step 6 is discussed in detail below. Step 10 can be done with tree contraction.

In Step 6, let D' be the edges in D that need to be flipped. Each edge in D can be tested for membership in D' with $O(1)$ time and one processor using the prefix sums computed in Step 5. Therefore, D' can be found in $O(\log n)$ time with $n/\log n$ processors.

Because the total weight of U is n and each edge in D' covers a path with a total weight more than $n/2$, the subpaths of U covered by the edges in D' are nested. Let P' be the innermost of these subpaths. Let e be the edge in D' that covers P'. Let P be the subpath of U that extends P' to include the endpoints of e. Let C be the cycle formed by e and P.

The edges in D' can be flipped by making C the new external face for G'. Because e is the edge in D' such that the total weight of the path covered by it is the smallest, e and C can easily be computed in $O(\log n)$ time with $n/\log n$ processors. Thus Step 6 can be done within the desired complexity. □

3.2 Computing a cycle separator from a path separator

Let G be a connected embedded planar graph with n vertices. Let P be a path separator of G. To compute a cycle separator of G from P, we will delete from the two ends of P as many vertices as possible such that P remains a separator. We will then show that there is a path from one end of P through some vertices in $G - P$ to the other end of P. Because P is a separator, this path and P form a cycle separator.

Theorem 3.5 *Given a connected embedded planar graph with n vertices, a cycle separator can be computed in $O(\log n)$ time with $n/\log n$ processors.*

Proof. Our algorithm for computing a cycle separator of G is as follows:

1. Compute a path separator $P = u_1, \cdots, u_p$ of G.

2. Compute as follows the largest index s such that some connected component Z_s of $G-\{u_1,\ldots,u_{s-1}\}$ is 2/3-heavy for G:

 2-1. Compute the connected components G_i of $G-\{u_1,\cdots,u_p\}$.

 2-2. Compute a spanning tree T_i for each G_i.

 2-3. For each T_i, find an edge e_i that connects T_i to the vertex in P with the largest possible index.

 2-4. Compute the spanning tree T of G formed by the trees T_i and the edges e_i.

 2-5. For each u_j, compute the number n_j of vertices in the subtree W_j of T rooted at u_j.

 2-6. Compute the largest index s with $n_s > 2n/3$.

 2-7. Compute the subgraph Z_s of G induced by the vertices in W_s.

 Remark. $u_s \in Z_s$.

3. Compute the path P' formed by $u_1,\ldots,u_s$.

 Remark. P' remains a path separator.

4. Find the smallest index t such that some connected component Z_t of $G-\{u_{t+1},\ldots,u_s\}$ is 2/3-heavy for G.

 Remark. This step is similar to Step 2 with the indexing of P reversed.

5. Compute the path P'' formed by $u_t,\ldots,u_s$.

 Remark. P'' remains a path separator.

6. Compute a vertex-simple path Q in $Z_s \cup Z_t$ between u_s and u_t.

 Remark. $Z_s \cup Z_t$ is connected since Z_s and Z_t are both 2/3-heavy for G.

7. Return the cycle formed by P'' and Q.

The above algorithm is analyzed as follows. Step 2 is the most subtle step of the algorithm. Its goal is to shorten the u_p-end of P as much as possible while maintaining its separator property. A naive way of computing the index s is to apply binary search on P. This would take more than the desired optimal complexity. The algorithm instead computes s based on the following observations. For each j, let P_j be the path $u_j,\cdots,u_p$. After P_j is discarded from P, a new connected component is formed in $G-(P-P_j)$ that combines P_j and some connected components G_i in $G-P$. By the maximality of the index of the endpoint of e_i in P, the component G_i is connected to P_j in $G-(P-P_j)$ if and only if e_i connects to P at some u_k with $k \geq j$. Therefore, n_j is indeed the number of vertices in the new connected component in $G-(P-P_j)$. The trees computed in Steps 2-2 and 2-4 allow the numbers n_i to be efficiently computed by tree contraction. Because P is a separator of G, by the maximality of s, the path P' remains a separator. Step 4 is symmetric to Step 2. Thus, P'' remains a separator.

By the choice of W_s, the vertex u_s is in Z_s. By symmetry, u_t is in Z_t. Because both Z_s and Z_t are 2/3-heavy for G, they intersect and $Z_s \cup Z_t$ is connected. Thus, Q exists and $P'' \cup Q$ forms a cycle. Because Z_s is formed by the subtree of T at u_s, it contains no vertices from $u_1, \cdots, u_{s-1}$. Similarly, Z_t contains no vertices from $u_{t+1}, \cdots, u_s$. Therefore, Q contains no vertices from $u_{t+1}, \cdots, u_{s-1}$ and the cycle formed by Q and P'' is vertex-simple. Because P'' is a separator of G, this cycle is a cycle separator. This proves the correctness of the cycle-separator algorithm.

To prove the time complexity, we show how to carry out each step of the algorithm in $O(\log n)$ time on $n/\log n$ processors. All the steps not mentioned below can be done in a straightforward manner. Step 1 is done by means of Theorem 3.1. Steps 2-1 and 2-2 are done by means of Theorems 2.4 and 2.5. Step 2-3 uses an optimal algorithm for computing maxima. Step 2-5 uses tree contraction [1, 9, 13, 14, 27, 31]. Steps 4 and 5 are symmetric to Steps 2 and 3. Step 6 is done by means of Theorem 2.5. □

4 Parallel depth-first search in a planar graph

Theorem 4.1 *Let G be a connected planar graph with n vertices. Let r be a vertex in G. A depth-first search tree of G rooted at r can be computed in $O(\log^2 n)$ time with $n/\log n$ processors.*

Proof. The proof is by means of a known reduction from cycle separators to depth-first search trees [20, 21, 34]. For convenience of analysis, this reduction is included below:

1. If G has only one vertex, then return the singleton tree formed by r.

2. Compute a cycle separator C of G.

3. Find a path P from r to C such that P and C intersect at only one vertex u.

4. Compute the path S formed by P and C with only one of the two edges adjacent to u in C.

5. Compute the connected components G_i of $G - S$.

6. For each G_i, perform the following steps:

   6-1. Find an edge $e_i \in G$ that connects G_i at the farthest possible vertex from r in S.

   6-2. Find the endpoint r_i of e_i in G_i.

   6-3. Recursively compute a depth-first search tree T_i of G_i rooted at r_i.

7. Return the tree formed by the path S, the trees T_i, and the edges e_i.

The correctness of the above procedure is known [20, 21, 34]. To analyze its complexity, by Theorem 2.3 we may assume that G is embedded. Because S is a path separator of G, each recursive call in Step 6-3 processes a subgraph of G with at most 2/3 of the vertices in the graph at the previous recursive call. Thus the depth of recursion is $O(\log n)$ and the subgraphs of G processed at each level of recursion are disjoint. For the remainder of the proof, we only need to show that each level of recursion takes $O(\log n)$ time on $n/\log n$ processors. All the steps not mentioned below can be done in a straightforward manner. Step 2 is done with Theorem 3.5. Step 3 is done with Theorem 2.5 and tree contraction techniques [1, 9, 13, 14, 27, 31]. Step 5 is done with Theorem 2.4. Step 6-1 uses an optimal algorithm for finding maxima. □

5 Open problems

The total work of our depth-first search algorithm is $O(n \log n)$, which may not be optimal, since the best sequential algorithms take $O(n)$ time for planar graphs. This immediately suggests two open problems:

- Can the time complexity of parallel planar depth-first search be reduced to $O(\log n)$ on n processors?
- Is there an $O(n)$-work parallel algorithm that runs in polylogarithmic time or even $o(n)$ time for conducting depth-first search in planar graphs?

References

[1] K. Abrahamson, N. Dadoun, D. G. Kirkpatrick, and T. Przytycka, *A simple tree contraction algorithm*, Journal of Algorithms, 10 (1989), pp. 287–302.

[2] A. Aggarwal and R. J. Anderson, *A random NC algorithm for depth first search*, Combinatorica, 8 (1988), pp. 1–12.

[3] A. Aggarwal, R. J. Anderson, and M. Y. Kao, *Parallel depth-first search in general directed graphs*, SIAM J. Comput., 19 (1990), pp. 397–409.

[4] A. V. Aho, J. E. Hopcroft, and J. D. Ullman, *The Design and Analysis of Computer Algorithms*, Addison-Wesley, 1974.

[5] R. J. Anderson and G. L. Miller, *Deterministic parallel list ranking*, Algorithmica, 6 (1991), pp. 859–868.

[6] M. Atallah and U. Vishkin, *Finding euler tours in parallel*, Journal of Computer and System Sciences, 29 (1984), pp. 330–337.

[7] C. Berge, *Graphs*, North-Holland, New York, second revised ed., 1985.

[8] B. Bollobás, *Graph Theory*, Springer–Verlag, New York, 1979.

[9] R. COLE AND U. VISHKIN, *The accelerated centroid decomposition technique for optimal tree evaluation in logarithmic time*, Algorithmica, 3 (1988), pp. 329–346.

[10] ——, *Faster optimal prefix sums and list ranking*, Information and Computation, 81 (1989), pp. 334–352.

[11] T. H. CORMEN, C. L. LEISERSON, AND R. L. RIVEST, *Introduction to Algorithms*, The MIT Press, Cambridge, Massachusetts, 1991.

[12] J. EDMONDS, *A combinatorial representation for polyhedral surfaces*, American Mathematical Society Notices, 7 (1960), p. 646.

[13] H. GAZIT, G. L. MILLER, AND S. H. TENG, *Optimal tree contraction in the EREW model*, in Concurrent Computations: Algorithms, Architecture, and Technology, S. T. amd B.W. Dickinson and S. Schwartz, eds., Plenum, New York, 1988, pp. 139–156.

[14] A. M. GIBBONS AND W. RYTTER, *An optimal parallel algorithm for dynamic expression evaluation and its applications*, Information and Computation, 81 (1989), pp. 32–45.

[15] T. HAGERUP, *Optimal parallel algorithms on planar graphs*, Information and Computation, 84 (1990), pp. 71–96.

[16] ——, *Planar depth-first search in* $O(\log n)$ *parallel time*, SIAM J. Comput., 19 (1990), pp. 678–704.

[17] Y. HAN, *An optimal linked list prefix algorithm on a local memory computer*, IEEE Transactions on Computers, 40 (1991), pp. 1149–1153.

[18] F. HARARY, *Graph Theory*, Addison-Wesley, 1969.

[19] X. HE, *Efficient parallel algorithms for series parallel graphs*, Journal of Algorithms, 12 (1991), pp. 409–430.

[20] X. HE AND Y. YESHA, *A nearly optimal parallel algorithm for constructing depth first spanning trees in planar graphs*, SIAM J. Comput., 17 (1988), pp. 486–491.

[21] J. JA'JA AND S. KOSARAJU, *Parallel algorithms for planar graphs and related problems*, IEEE Trans. Circuits and Systems, 35 (1988), pp. 304–311.

[22] M. Y. KAO, *All graphs have cycle separators and planar directed depth-first search is in DNC*, in Lecture Notes in Computer Science 319: the 3rd Aegean Workshop on Computing, Springer-Verlag, 1988, pp. 53–63.

[23] ——, *Planar strong connectivity helps in parallel depth-first search*, in Proceedings of the 1992 International Computer Symposium, 1992, pp. 309–316.

[24] M. Y. Kao and P. N. Klein, *Towards overcoming the transitive-closure bottleneck: Efficient parallel algorithms for planar digraphs*, in Proceedings of the 22nd Annual ACM Symposium on Theory of Computing, 1990, pp. 181–192.

[25] R. Karp and V. Ramachandran, *A survey of parallel algorithms for shared-memory machines*, in Handbook of Theoretical Computer Science: Algorithms and Complexity, J. van Leeuwen, ed., vol. A, Elsevier Science Publishers B. V., 1990, pp. 869–941.

[26] P. N. Klein, *Efficient parallel algorithms for chordal graphs*, in Proceedings of the 29th Annual IEEE Symposium on Foundations of Computer Science, 1988, pp. 150–161.

[27] S. R. Kosaraju and A. L. Delcher, *Optimal parallel evaluation of tree-structured computations by raking*, in Lecture Notes in Computer Science 319: the 3rd Aegean Workshop on Computing, Springer-Verlag, 1988, pp. 101–110.

[28] C. P. Kruskal, L. Rudolph, and M. Snir, *The power of parallel prefix*, IEEE Transactions on Computers, C-34 (1985), pp. 965–968.

[29] R. E. Ladner and M. J. Fischer, *Parallel prefix computation*, Journal of the ACM, 27 (1980), pp. 831–838.

[30] G. L. Miller, *Finding small simple cycle separators for 2-connected planar graphs*, Journal of Computer and System Sciences, 32 (1986), pp. 265–279.

[31] G. L. Miller and J. H. Reif, *Parallel tree contraction, part 1: Fundamentals*, in Advances in Computing Research: Randomness and Computation, S. Micali, ed., vol. 5, JAI Press, Greenwich, CT, 1989, pp. 47–72.

[32] V. Ramachandran and J. H. Reif, *An optimal parallel algorithm for graph planarity*, in Proceedings of the 30th Annual IEEE Symposium on Foundations of Computer Science, 1989, pp. 282–287.

[33] J. H. Reif, *Depth-first search is inherently sequential*, Information Processing Letters, 20 (1985), pp. 229–234.

[34] J. R. Smith, *Parallel algorithms for depth-first search I. Planar graphs*, SIAM J. Comput., 15 (1986), pp. 814–830.

[35] R. Tarjan, *Depth-first search and linear graph algorithms*, SIAM J. Comput., 1 (1972), pp. 146–160.

[36] W. Tutte, *Graph Theory*, vol. 21 of Encyclopedia of Mathematics and its Applications, Addison-Wesley, 1984.

[37] A. T. White, *Graphs, Groups, and Surfaces*, North-Holland, 1973.

On Approximating the Longest Path in a Graph

(Preliminary Version)

David Karger * ** *Rajeev Motwani** *G.D.S. Ramkumar* ***

Department of Computer Science, Stanford University
Stanford, CA 94305

Abstract. We consider the problem of approximating the longest path in undirected graphs and present both positive and negative results. A simple greedy algorithm is shown to find long paths in dense graphs. We also present an algorithm for finding paths of a logarithmic length in *weakly* Hamiltonian graphs, and this result is the best possible. For sparse random graphs, we show that a relatively long path can be obtained. To explain the difficulty of obtaining better approximations, we provide some strong hardness results. For any $\epsilon < 1$, the problem of finding a path of length $n - n^\epsilon$ in an n-vertex Hamiltonian graph is shown to be **NP**-hard. We also show that no polynomial time algorithm can find a constant factor approximation to the longest path problem unless **P** = **NP**. Finally, it is shown that if any polynomial time algorithm can approximate the longest path to a ratio of $2^{O(\log^{1-\epsilon} n)}$, for any $\epsilon > 0$, then **NP** has a quasi-polynomial deterministic time simulation.

1 Introduction

The area of approximation algorithms for **NP**-hard optimization problems has received a lot of attention over the past two decades [11, 13]. It has now become apparent that even the approximate solution of a large class of **NP**-hard optimization problems remains outside the bounds of feasibility. For example, a sequence of results [9, 7, 2, 1] established the intractability of approximating the largest clique in a graph, culminating in the result of Arora, Lund, Motwani, Sudan and Szegedy [1] that for some constant $\delta > 0$, there does not exist any polynomial time algorithm which will approximate the maximum clique within a ratio of n^δ unless **P** = **NP**. They also established that unless **P** = **NP**, there do not exist polynomial time approximation schemes (PTAS) for optimization problems which are **MAX SNP**-hard [14].

In this context, a major outstanding open problem is that of determining the approximability of the *longest path* in an *undirected* graph. The optimization version of this problem is **NP**-hard since it includes the Hamiltonian path

* Supported by NSF Grant CCR-9010517, and grants from Mitsubishi and OTL.
** Supported an NSF Graduate Fellowship.
*** Supported by a grant from Toshiba Corporation.

problem as a special case. Therefore, it is natural to look for polynomial-time algorithms with a small performance ratio, where the performance ratio is defined as the ratio of the longest path in the input graph to the length of the path produced by the algorithm. Our results attempt to pin down the best possible performance ratio achievable by a polynomial-time approximation algorithms for longest paths. We provide some approximation algorithms for this problem, but unfortunately the performance ratio of these algorithms is as weak as in the case of the best-known approximation algorithms for clique [6]. We conjecture that the situation for longest paths is essentially as bad as for this problem, i.e. if there exists an approximation algorithm which has a performance ratio of n^δ, for some constant $\delta > 0$, then $\mathbf{P} = \mathbf{NP}$. We explain the difficulty of obtaining better performance guarantees for longest path approximations by providing hardness results which come fairly close to establishing this conjecture.

In Section 2, we present several polynomial time approximation algorithms for longest paths. A simple greedy algorithm is presented and it is shown that it finds long paths in dense graphs. In the case of cliques, the extreme hardness of the problem led to the study of special inputs where the optimum was guaranteed to take on an extreme value; for example, the approximation of cliques in graphs containing a linear-sized clique is studied by Boppana and Halldorsson [6]. We therefore formulate the problem of finding long paths in Hamiltonian graphs. It is easy to see that there is no essential difference between the cases where the input graph has Hamiltonian paths or Hamiltonian cycles, and we concentrate on the latter case. Our second algorithm finds paths of a logarithmic length in Hamiltonian graphs. In fact, we show that this algorithm will find such paths in a much larger class of graphs, viz. *weakly Hamiltonian* graphs, or even *1-tough* graphs. Some variants of this algorithm are also analyzed. This result is the best possible in the sense that we can demonstrate the existence of such graphs where the longest path is of logarithmic length.

The hard case appears to be that of sparse Hamiltonian graphs. In Section 3, we consider sparse random Hamiltonian graphs and show that it is possible to find paths of length $\Omega(\sqrt{n}/\log n)$. Surprisingly, this algorithm works in any graph obtained by adding *any number* of random edges to a Hamiltonian cycle. This result partially answers an open question posed by Broder, Frieze and Shamir [5]. They had considered the problem of finding Hamiltonian cycles in graphs obtained by adding a relatively large number of random edges to a Hamiltonian cycle.

In Section 4, we provide hardness results for the problem of approximating the longest path. We first consider the problem of finding long paths in Hamiltonian graphs and show that for any $\epsilon < 1$, it is impossible to find paths of length $n - n^\epsilon$ in an n-vertex Hamiltonian graph unless $\mathbf{P} = \mathbf{NP}$. The problem of finding long paths is easier for Hamiltonian graphs than for arbitrary inputs. Therefore, it is not surprising that we can prove much stronger negative results in general input graphs. We first prove a self-improvability result for the longest path problem. Combining this with the recent results on the intractability of approximation problems which are **MAX SNP**-hard, we obtain that no polynomial time algorithm can find a constant factor approximation for the longest path problem unless $\mathbf{P} = \mathbf{NP}$. We conjecture that the result can be strengthened to say that for some constant $\delta > 0$, finding an approximation of ratio n^δ is also **NP**-hard. As evidence towards this conjecture, we show that if any polynomial time algorithm can approximate the longest path to a ratio of $2^{O(\log^{1-\epsilon} n)}$, for any $\epsilon > 0$, then **NP** has a quasi-polynomial deterministic time simulation. The hardness results apply even to the special case where the input consists of *bounded degree* graphs.

Before describing our results in greater detail we review some related work. Monien [12] showed that an $O(k!\,nm)$ time algorithm finds paths of length k in a Hamiltonian graph with n vertices and m edges. Our results are an improvement on this since in polynomial time Monien's algorithm can only find paths of length $O(\log n/\log\log n)$. Furer and Raghavachari [10] present approximation algorithms for minimum-degree spanning tree which delivered absolute performance guarantees. From this we can derive a polynomial-time algorithm for finding logarithmic length paths in Hamiltonian graphs, matching our result for that case. However, note that our result even in that case is more general in that it applies to a wider class of graphs, viz. the weakly Hamiltonian graphs. No hardness results for longest paths were known earlier, although a seemingly related problem has been studied by Berman and Schnitger [7]. They show that the above conjecture is true for the problem of approximating the longest *induced* path in an undirected graph. Note that the induced path problem is strictly harder and their hardness result does not carry over to the problem under consideration here. Bellare [3] provides hardness results for the approximability of a generalization of the longest paths problem called the *longest color-respecting path* problem. Our results are strictly stronger since there are no color constraints on the paths.

2 Algorithms for Finding Long Paths

In this section, we present a polynomial time algorithm for finding paths of a logarithmic length in a 1-tough graph, and show that this is the best possible result for such graphs. But first we describe an exceedingly simple algorithm which finds long paths in dense graphs.

Consider a graph G with n vertices and m edges, and let $d = m/n$. The greedy algorithm for long paths works as follows. Repeatedly choose a vertex of degree smaller than d, and remove this vertex and all incident edges from the graph. This process terminates when the residual graph has minimum degree at least d. Clearly, the residual graph cannot be empty since at most $(d-1)n$ edges can be removed from the graph in the process of deleting small degree vertices. In a graph with minimum degree d, any maximal path has length at least d and a simple greedy algorithm finds such a path.

Theorem 1. *The greedy algorithm finds a path of length at least d in a graph with density $d = m/n$.*

We now describe the algorithm for 1-tough graphs. The notion of 1-tough and weakly Hamiltonian graphs was introduced by Chvatal [8]. The latter are defined by a necessary condition for Hamiltonicity obtained via an integer linear programming formulation. We omit the formal definition and instead provide three properties of weakly Hamiltonian graphs – the "1-2-3" properties.

Definition 2. A graph $G = (V, E)$ is said to be *1-tough* if for any set $U \subset V$, the induced subgraph $G[V - U]$ has at most $|U|$ connected components.

Definition 3. A graph $G = (V, E)$ is said to have a *2-factor* if there exists $E' \subset E$ such that in the graph $G' = (V, E')$, each vertex $v \in V$ has degree 2.

Definition 4. A graph $G = (V, E)$ is said to be *3-cyclable* if for every three vertices $u, v, w \in V$, there exists a cycle in the graph G containing u, v and w.

Theorem 5. [Chvatal] *Any weakly Hamiltonian graph $G = (V, E)$ is* 1-tough, *has a* 2-factor *and is* 3-cyclable

The proof of the following theorem is obvious.

Theorem 6. *Every Hamiltonian graph is weakly Hamiltonian.*

We now present an algorithm, Algorithm LONG-PATH, which finds a path of length $\Omega(\log n)$ in any 1-tough graph. Since weak Hamiltonicity is a more restricted property, this algorithm can be used to generate such a path in weakly Hamiltonian graphs as well. The basic idea behind the algorithm is the following. Having found a path of a certain length k, the removal of those k vertices can create at most k connected components. Each of these components has an inherited toughness property, and the end-points of the current graph must have neighbors in each component. We then extend the path into the largest such component, and repeat the entire process.

To analyze this algorithm we generalize the notion of 1-toughness. This allows us to characterize the subgraphs obtained during the execution of this algorithm. Notice that as we remove vertices from G one-by-one, the graph may not decompose into the maximum possible number of components allowed by 1-toughness. Thus, at some later stage the removal of even one additional vertex may cause more than one new component to be created. However, the total number of components created up to any stage cannot exceed the total number of vertices removed till then. To capture the notion of a bounded potential to generate many components by just one incremental vertex deletion, we make the following definition.

Definition 7. A graph $G = (V, E)$ is said to be *p-deficient* if for any set $U \subset V$, the induced subgraph $G[V - U]$ has at most $|U| + p - 1$ components.

In terms of this definition, a graph is 1-tough if and only if it is 1-deficient.

Lemma 8. *Given an input graph $G(V, E)$ of deficiency $p \geq 1$, Algorithm LONG-PATH computes a path of length at least $\frac{\log n - p}{2}$.*

Proof. The proof proceeds by induction on n. The base case of the induction is easy: for $n = 1$, the algorithm computes a path of length 1 and the value of $(\log n - p)/2$ non-positive. Now consider a graph $G = (V, E)$ with n vertices and deficiency p, At the first step, the algorithm arbitrarily chooses a vertex $v \in V$ and removes it from the graph. Suppose that the graph now decomposes into k components; clearly, $1 \leq k \leq p$, since G is of deficiency p. It must be the case that v has a neighbor in each of these k components. The algorithm now picks the largest component W in the resulting graph, chooses a neighbor of v in W as the new end-point and recurses on $G[W]$.

Clearly, the size of W is at least $\frac{n-1}{k}$. We claim that $G[W]$ has deficiency at most $p - k + 2$. This can be proved by reductio ad absurdum. If the claim were not true, there would be a set of vertices $U \subseteq W$ such that the induced subgraph $G[W - U]$ has at least $|U| + p - k + 2$ components. But then, by removing $U \cup \{v\}$ from G, the number of components we would obtain would be at least $(|U| + p - k + 2) + k - 1 = (|U| + 1) + p$. Since G is p-deficient, we have a contradiction.

By the inductive hypothesis, on input $G[W]$ the algorithm would find a path of length at least

$$\frac{\log |W| - (p - k + 2)}{2} = \frac{\log \frac{n-1}{k} - (p - k + 2)}{2}$$

Hence, in terms of the original input G, the algorithm finds a path of length at least

$$1 + \frac{\log \frac{n-1}{k} - (p - k + 2)}{2}.$$

We claim that this is always at least $(\log n - p)/2$, thereby establishing the inductive hypothesis. □

Theorem 9. *Algorithm LONG-PATH computes a path of length $\Omega(\log n)$.*

It is not very hard to see that the above algorithm can be viewed as picking a long path in a depth-first tree. This results in the following corollary.

Corollary 10. *Let G be a 1-tough graph with n vertices. Then the depth-first search tree of G has depth $\Omega(\log n)$.*

We can show that at any level of the depth-first tree, the number of nodes at that level is at most equal to the number of nodes in all the previous levels, so that the tree which is found is very close to being a binary tree (in an "average" sense). From this fact we can derive an extremely simple algorithm to find a path of length k in time $k^2!$ in any Hamiltonian graph. First build a depth-first search tree. If its depth exceeds k, then we are done. Otherwise, find some subtree of size between k^2 and $2k^2$. Such a subtree must exists by arguments similar to those above. The path from the root to this subtree, which contains at most k vertices, cuts the subtree off from the rest of the graph, so the Hamiltonian path can enter and then exit the subtree at most k times. It follows that this subtree contains a path of length k, which can be found by exhaustive search.

Theorem 11. *Let G be a Hamiltonian graph on n vertices. Then in G a path of length k can be constructed in time $O(k^2!)$.*

We now prove an upper bound of $O(\log n)$ on the longest path in 1-tough graphs. In fact, we prove that there exist graphs satisfying the 1-2-3 properties which have a longest path of length $O(\log n)$. We define a graph $G_k(V_k, E_k)$ for each non-negative integer k. The graph consists of $2^k - 1$ *normal* vertices and 2 *super* vertices. The normal vertices are arranged in a complete binary tree. The tree is augmented by an edge between each pair of siblings. We will distinguish between the *tree* edges and the *sibling* edges. The two super vertices are called s_1 and s_2, and they are adjacent to every other vertex in the graph, including each other. The size of V_k is $n = 2^k + 1$.

Theorem 12. *The graph G_k is 1-tough, has a 2-factor and is 3-cyclable.*

Lemma 13. *The longest path in the graph G_k not containing the super vertices is of length $\Theta(\log n)$*

Observe that the longest path in G_k is at most 3 times longer than the longest path not including the super vertices.

Theorem 14. *The longest path in the graph G_k is of length $\Theta(\log n)$.*

3 Finding Long Paths in Sparse Random Graphs

We now turn to the issue of random Hamiltonian graphs. It should be noted that the amplification described previously becomes ineffectual when we consider finding paths of length n^ϵ. It is therefore an interesting coincidence that on random Hamiltonian graphs, a path of length $\Omega(\sqrt{n}/\log n)$ can in fact be found. The following analysis applies to numerous distributions on graphs—in fact, to any random graph obtained by adding any number of edges to a Hamiltonian cycle, such that each non-Hamiltonian edge has an equal probability of occurring. For example, the random edges could form a $G_{n,p}$ graph, or a random regular graph [4]. For the case of random Hamiltonian graphs with average degree much larger than 3, Broder et al [5] give an algorithm which actually finds the Hamiltonian cycle in polynomial time. However, they leave open the question of what can be done for sparse graphs, e.g. graphs of degree 3.

Theorem 15. *Let G be a random 3-regular Hamiltonian graph on n vertices. There is a polynomial time algorithm which finds a path of length $\Omega(\sqrt{n}/\log n)$ in G, with high probability of success.*

A random 3-regular Hamiltonian graph can be viewed as a Hamiltonian cycle with a random perfect matching added. We will therefore refer to the edges on the Hamiltonian path as *Hamiltonian edges*, and will refer to the other edges as *matching edges.* We will call the neighbor of a vertex along a matching edge its *match.* The key property of the random edges used in this analysis is that for any vertex, its match is chosen uniformly at random from among the other vertices. Finally, let the *Hamiltonian distance* between two vertices be the length of the shortest path between them made up entirely of Hamiltonian edges.

The algorithm proceeds as a series of trials. In each trial, we start at a randomly selected vertex of the graph and perform a random walk, with the modification that the last edge traversed in not considered as a possibility for the next step, so that we do not immediately back up along the walk. Mark each vertex as it is visited. The trial ends when we visit a marked vertex, and succeeds if we visit $k = \Omega(\sqrt{n}/\log n)$ vertices before encountering a vertex which is already marked (during this or any previous trial). We claim that with high probability, we will find a path of length k within $O(\log n)$ trials.

Consider the tth trial, $t = O(\log n)$. At the end of this trial at most kt vertices can have been marked. Analyze the walk during this trial as a series of epochs. An epoch ends when the walk traverses a matching edge. Since at each step of the walk a matching edge is traversed with probability 1/2, epochs have length $O(\log n)$ with high probability. Call a vertex *safe* if its Hamiltonian distance from any marked vertex exceeds $\Omega(\log n)$. Call an epoch *safe* if it begins at a safe vertex. It follows that with high probability no marked vertex is visited during a safe epoch. Thus the path continues to extend safely so long as safe epochs occur.

Consider the end of a safe epoch. This happens when a matching edge is traversed. Since the endpoints of a matching edge are random, this means that the starting vertex of the next epoch is chosen uniformly at random from among the unvisited vertices. Since at most kt vertices are marked, there are $O(kt\log n)$ unsafe vertices; thus the next epoch is unsafe with probability $O(kt\log n/n) = O(\log n/\sqrt{n})$. It follows that with constant probability $\Omega(\sqrt{n}/\log n)$ safe epochs occur during the trial. Clearly the length of the path which is found is equal to at least the number of epochs, so we get the desired path length. Furthermore, since a trial ends when we visit an unsafe vertex, we can treat the trials as independent. Thus with high probability one of the $O(\log n)$ trials succeeds.

In fact, the algorithm itself can be made deterministic. All that is required is that an epoch (in the sense described above) ends in $O(\log n)$ steps. To ensure this we use a form of "universal non-traversal sequence." Consider assigning labels to the edges of the graph in the standard traversal sequence manner. It is simple to interpret a sequence over $\{0,1\}$ so that when one arrives at a given vertex via some edge, the next symbol in the sequence determines which of the other two edges should be traversed to leave that vertex. We can now modify traversal sequences in the following way to produce walks which do not "back up." Given a traversal sequence drawn from $\{1,2\}^*$, interpret it as follows: at each step, examine the label on the edge the walk arrived from. If i is the traversal symbol and j the label of the edge just traversed, then next traverse the edge labeled $(i+j) \bmod 3$.

Consider the particular modified traversal sequence which causes a traversal of the Hamiltonian path. This is a sequence of length n, and it therefore follows that some sequence of length $\log n + 1$ does not occur as a substring of the traversal sequence (since there are $2n$ such sequences). If we try all $2n$ such sequences, we will eventually stumble upon one satisfying the property. If we construct our walk using this sequence (repeated over and over again) then it is clear that we will traverse a matching edge at least once every $2\log n$ steps of our walk, and then the analysis given above is applicable.

The random walk algorithm can be generalized to any other random graph distribution, so long as the endpoint of a randomly chosen non-Hamiltonian edge out of v is uniformly distributed among the vertices of G. The argument goes through unchanged, except in the case that some of the vertices of G have degree two (*i.e*, only Hamiltonian edges. If such vertices are forbidden in the distribution, then we have the desired result immediately.

The degree two situation can be rectified if for any input graph we first "shortcut" any degree two vertex v. replacing (u, v) and (v, w) by (u, w). Any path we construct in the new graph yields a longer path in the original graph. If the original graph had the uniform endpoint distribution property, then so does the new graph. If the number of vertices in the shortcut graph is less than $n^{2/3}$, then there must have been a path of degree two vertices of length $n^{1/3}$ in the original graph. If not, then we find a path of length at least $n^{1/3}$ in the new graph, yielding a path of at least this length in the original graph.

4 Hardness Results

It is important to keep in mind that it may be possible to achieve better performance guarantees on graphs which are known to be Hamiltonian, than on arbitrary input graphs. We first consider the "easier" problem and prove that, for any constant $\epsilon < 1$, no polynomial time algorithm can find a path of length $n - n^{\epsilon}$ in an n-vertex Hamiltonian graph, unless $\mathbf{P} = \mathbf{NP}$. Next we demonstrate a *self improvability* result for approximating longest paths. The self-improvability result is used to show that finding constant factor approximations to the longest path problem is **NP**-hard. We also provide evidence that it is unlikely to be the case that there exists any $o(n^{\delta})$ ratio approximation algorithm for this problem. Our results extend to showing the hardness of the longest path problem even in the case of *bounded degree* graphs.

4.1 Hardness Result for Hamiltonian Graphs

The following theorem easily generalizes to the approximation of longest paths in arbitrary (non-Hamiltonian) graphs. But the subsequent results are much stronger for the more general problem.

Theorem 16. *For any $\epsilon < 1$, the problem of finding a path of length $n - n^{\epsilon}$ in a Hamiltonian graph is* **NP***-complete.*

We present only a sketch of the proof. Let $G = (V, E)$ be a Hamiltonian graph on n vertices, and define $K = n^{\epsilon/(1-\epsilon)}$. We define an *auxiliary graph* $G' = (V', E')$. Informally, G' consists of K *copies* of G "glued" together in a cycle by K "special" vertices. The K copies of G are arranged in a cycle and any two successive copies of G are separated by a special vertex which is connected to all the vertices of both the copies. In more formal terms, $V' = (V \cup \{v_s\}) \times \{1, 2, \ldots, K\}$, where v_s is a special vertex. Any two vertices (v_1, k_1) and (v_2, k_2) in V' share an edge in E' if and only if one of the following conditions hold: $k_1 = k_2$ and $(v_1, v_2) \in E$; $v_1 = v_s$ and $k_2 = (k_1 + 1) \pmod K$; or, $v_1 = v_s$ and $k_2 = (k_1 - 1) \pmod K$.

Let A be an algorithm which is guaranteed to find a path of length at least $n - n^\epsilon$ in a Hamiltonian graph on n vertices. Consider what happens when we run algorithm A on G'. We claim that A finds a Hamiltonian path in at least one of the copies of G in G'. Since there are only a polynomial number of such copies, this gives a polynomial time algorithm for finding a Hamiltonian path in G.

To see this claim, we first observe that any simple path in G' is a disjoint collection of simple paths in copies of G "stitched" together by special vertices. In this collection, there can be at most one simple path per copy of G, since there is exactly one special vertex to enter a copy of G and one special vertex to leave the copy, and these vertices cannot be visited twice in a simple path. Since $|V'| = (n + 1)K$, running A on G' gives a path P of length l, where

$$l > (n+1)K - ((n+1)K)^\epsilon > (n+1)K - n^{(1+\epsilon/(1-\epsilon))\epsilon} = (n+1)K - K.$$

In other words, P includes all but at most $K - 1$ vertices of G'. Since there are K copies of G, we can conclude that P includes *all* the vertices of at least one copy of G. Since all vertices of this copy of G occur contiguously within P, the path P includes a Hamiltonian path from that copy of G.

4.2 Self-Improvability and Hardness Results for Longest Paths

We now show that if the longest path can be approximated to some constant k in polynomial time, then we can approximate it to *any* constant k', also in polynomial time. To this end, we define the following novel definition of a graph product.

Definition 17. For a graph $G(V, E)$, its *edge square* graph $G^2(V^2, E^2)$ is obtained as follows. Replace each edge $e = (u, v)$ in G by a copy of G, call it G_e, and connect both u and v to each vertex in G_e. The vertices u and v are referred to as the *contact* vertices of G_e.

If the graph G has n vertices, then G^2 will have at most n^3 vertices. Observe that the edge square of a Hamiltonian graph need not be Hamiltonian. It is precisely for this reason that the subsequent results apply only to the more general problem of approximating the longest path in arbitrary graphs, as opposed to the possible easier problem where the input is guaranteed to be Hamiltonian. The following lemma relates the length of the longest paths in G and G^2.

Lemma 18. *Let $G = (V, E)$ be a graph whose longest simple path is of length l. Then, the longest path in G^2 is of length at least l^2, and given a path of length m in G^2, we can obtain in polynomial time a path of length $\sqrt{m} - 1$ in G.*

Theorem 19. *If the longest path problem has a polynomial time algorithm which achieves a constant factor approximation, then it has a PTAS.*

Proof. Let A_k be a polynomial time approximation algorithm with a performance ratio of $k > 1$, i.e. it obtains a path of length l/k in a graph with longest path length l. Let p be the smallest integer exceeding

$$\log \frac{2 \log k}{\log(1+\epsilon)}$$

Given an input graph G with longest path length l, suppose we run the algorithm A_k on the graph G^{2^p} obtained by squaring G repeatedly p times; this yields a path of length at least l^{2^p}/k in G^{2^p}. Using Lemma 18, a simple calculation shows that we obtain in G a path of length at least

$$\left(\frac{l^{2^p}}{k}\right)^{1/2^p} - p \geq \frac{l}{\sqrt{1+\epsilon}} - p \geq \frac{l}{1+\epsilon}$$

provided $l \geq 2p(1+\epsilon)/\epsilon$. Observe that for small l, we can compute the optimal solution by brute-force in polynomial time. Moreover, the running time of the resulting algorithm is polynomial for fixed ϵ, since the graph G^{2^p} has at most n^{3^p} vertices. This gives the desired PTAS. □

We now show the non-existence of a PTAS for longest paths. The proof is based on the recent results of Arora et al [1].

Theorem 20. *There is no PTAS for the longest path problem, unless* **P** = **NP**.

Proof. We first claim that it suffices to demonstrate the hardness result when the longest path problem is restricted to instances containing a Hamiltonian *cycle*. Clearly, if it is **NP**-hard to find a constant factor approximation to the longest path in graphs with a Hamiltonian cycle, then this is also the case for the problem of approximating the longest path in arbitrary graphs.

Let us denote by TSP(1,2) the problem of finding an optimal traveling salesman tour in a complete graph where all edge lengths are either 1 or 2. The approximation version of this problem has been shown to be **MAX SNP**-hard by Papadimitriou and Yannakakis [15]. Their reduction is from a version of the MAX 3SAT problem which is also **MAX SNP**-complete. We claim that the following statement can be obtained as a consequence of their result. Suppose that for every $\delta > 0$ there is a polynomial time algorithm which on any instance of TSP(1,2) with optimum value n returns a tour of cost at most $(1+\delta)n$, then MAX 3SAT has a PTAS. Observe that having an optimum solution of size n corresponds to having a Hamiltonian cycle using only the edges of length 1.

Suppose now that we have a PTAS for the longest path problem in Hamiltonian graphs. In particular, this implies that we can find paths of length at least $(1-\delta)n$, for any fixed $\delta > 0$, in graphs which possess a Hamiltonian path. Clearly, this works equally well with graphs which possess a Hamiltonian cycle. Given any instance of TSP(1,2) with optimum value n, the edges of weight 1 form a Hamiltonian graph. Using the PTAS for longest paths, we can construct a path of length $(1-\delta)n$ using only the edges of weight 1. This can be converted into a tour of weight at most $(1+\delta)n$ in the instance of TSP(1,2) by extending the path with the edges of length 2. Thus, the PTAS for the longest path problem can be used to obtain a PTAS for such instances of TSP(1,2).

But the results of Arora et al [1] show that if any **MAX SNP**-hard problem has a PTAS, then **P** = **NP**. By the reduction of Papadimitriou and Yannakakis, we may conclude that MAX 3SAT has a PTAS, and therefore **P** = **NP**. □

The self-improvability result implies the **NP**-hardness of constant factor approximations to longest paths. Unfortunately, this result does not hold for the case of Hamiltonian graphs since the edge product of a graph with itself does not preserve Hamiltonicity. The next corollary follows by combining Theorems 19 and 20.

Corollary 21. *There does not exist a constant factor approximation algorithm for the longest path problem, unless* **P** = **NP**.

Since finding constant factor approximations to the longest path problem is hard, the next step would be to try to find weaker approximations. The following theorem provides evidence that finding such weaker approximations may also be very difficult. The proof uses the edge product to amplify the gap in the longest path lengths. We defer the details of the proof to the final version of the paper.

Theorem 22. *If there is a poly-time algorithm for the longest path problem with a performance ratio of* $2^{O(\sqrt{\log n})}$, *then* $\mathbf{NP} \subseteq \mathbf{DTIME}\left(2^{O(\log^5 n)}\right)$.

The previous theorem can be strengthened if one can define an edge squaring operation on graphs which only squares the number of vertices, instead of cubing it as currently defined. We achieve the same effect by considering a restricted class of graphs, namely graphs of bounded degree. Since the number of edges in such graphs is linear, the number of vertices in the edge squared graph is quadratic instead of cubic. All the above hardness results can also be extended to the case where the input graphs are of bounded degree.

We briefly explain the reason here. The TSP(1,2) reduction of Papadimitriou and Yannakakis also applies to the case where the edges of length 1 induce a bounded degree graph. Recall from Lemma 18 that in the graph G^2, the only vertices that had their degrees increased were the contact vertices u and v. In the bounded degree case, we need to modify the construction of the graph G^2 so that the maximum degree of the vertices remains the same during the squaring operation, while maintaining the longest-path properties. To achieve this, we consider the problem of approximating the longest path between a specified pair of vertices, say s and t. A hardness result for this problem can be easily extended to the more general longest path problem. Now, instead of using a "fan-out" from vertices u and v to $G_{(u,v)}$, we connect vertices u and v only to vertices s and t of $G_{(u,v)}$, respectively. It can be shown that regardless of the number of squaring operations, the maximum degree in the graph increases by at most 1. The remaining argument is identical to the case of general graphs. We obtain the following theorem by using the new edge squaring operation; the proof is omitted.

Theorem 23. *For $\epsilon > 0$, if there is a polynomial time algorithm for approximating the longest path to a ratio of $2^{O(\log^{1-\epsilon} n)}$ then $\mathbf{NP} \subseteq \mathbf{DTIME}\left(2^{O(\log^{1/\epsilon} n)}\right)$. The result holds even for the special case of bounded degree graphs.*

The argument used in the proof of this theorem can be easily extended to show the following.

Theorem 24. *For any $\delta > 0$, if there is a polynomial time algorithm for approximating the longest path to a ratio of $2^{O\left(\frac{\log n}{\log\log n}\right)}$ then $\mathbf{NP} \subseteq \mathbf{DTIME}\left(2^{O(n^{\delta})}\right)$.*

References

1. S. Arora, C. Lund, R. Motwani, M. Sudan and M. Szegedy. Proof Verification and Hardness of Approximation Problems. In *Proc. 33rd FOCS*, pages 14–23, 1992.
2. S. Arora and S. Safra. Approximating Clique is NP-complete. In *Proc. 33rd FOCS*, pages 2–13, 1992.
3. M. Bellare. Interactive Proofs and Approximation. *IBM Research Report* RC 17969, April 1992.
4. B. Bollobas. *Random Graphs*. Academic Press, 1985.
5. A. Broder, A.M. Frieze and E. Shamir. Finding hidden Hamiltonian cycles. In *Proc. 23rd STOC*, pages 182–189, 1991.
6. R. B. Boppana and M. M. Halldorsson. Approximating maximum independent sets by excluding subgraphs. In *Proc. 2nd SWAT*, pages 13–25, 1990.
7. P. Berman and G. Schnitger. On the Complexity of Approximating the Independent Set Problem. *Information and Computation*, vol. 96, pages 77–94, 1992.
8. V. Chvatal. Hamiltonian cycles. in *The Traveling Salesman Problem: A Guided Tour of Combinatorial Optimization*, (ed: E.L. Lawler et al), pages 402–430, 1985.
9. U. Feige, S. Goldwasser, L. Lovász, S. Safra, and M. Szegedy. Approximating clique is almost NP-complete. In *Proc. 32nd FOCS*, pages 2–12, 1991.
10. M. Furer and B. Raghavachari. Approximating the Minimum Degree Spanning Tree to within One from the Optimal Degree. In *Proc. 3rd SODA*, pages 317–324, 1992.
11. Michael R. Garey and David S. Johnson. *Computers and Intractability: A Guide to the Theory of NP-Completeness*. W. H. Freeman, 1979.
12. B. Monien. How to find long paths efficiently. *Annals of Discrete Mathematics*, vol. 25, pages 239–254, 1984.
13. R. Motwani. Lecture Notes on Approximation Algorithms. Report No. STAN-CS-92-1435, Department of Computer Science, Stanford University (1992).
14. C. H. Papadimitriou and M. Yannakakis. Optimization, Approximation, and Complexity Classes. In *Proc. 20th STOC*, pages 229–234, 1988.
15. C. H. Papadimitriou and M. Yannakakis. The traveling salesman problem with distances one and two, *Mathematics of Operations Research*, to appear.

Designing Multi-Commodity Flow Trees

Samir Khuller * Balaji Raghavachari † Neal Young ‡

Abstract

The traditional multi-commodity flow problem assumes a given flow network in which multiple commodities are to be maximally routed in response to given demands. This paper considers the multi-commodity flow network-design problem: given a set of multi-commodity flow demands, find a network subject to certain constraints such that the commodities can be maximally routed.

This paper focuses on the case when the network is required to be a tree. The main result is an approximation algorithm for the case when the tree is required to be of constant degree. The algorithm reduces the problem to the minimum-weight balanced-separator problem; the performance guarantee of the algorithm is within a factor of 4 of the performance guarantee of the balanced-separator procedure. If Leighton and Rao's balanced-separator procedure is used, the performance guarantee is $O(\log n)$.

1 Introduction

Let a graph $G = (V, E)$ represent multicommodity flow demands: the weight of each edge $e = \{a, b\}$ represents the demand of a distinct commodity to be transported between the sites a and b. Our goal is to design a network, in which the vertices of G will be embedded, and to route the commodities in the network. The maximum capacity edge of the network should be low in comparison to the best possible in *any* network meeting the required constraints. For example, the weight of each edge could denote the expected rate of phone calls between two sites. The problem is to design a network in which calls can be routed minimizing the maximum bandwidth required; the cost of building the network increases with the required bandwidth.

We consider the case when the network is required to be a tree, called the *tree congestion problem*. Given a tree in which the vertices of G are embedded, the load on an edge e is defined as follows: delete e from T. This breaks T into two connected

*Department of Computer Science, University of Maryland, College Park, MD 20742. E-mail : `samir@cs.umd.edu`.

†Computer Science Department, Pennsylvania State University, University Park, PA 16802. E-mail : `rbk@cs.psu.edu`. Part of this work was done while this author was visiting UMIACS.

‡Institute for Advanced Computer Studies, University of Maryland, College Park, MD 20742. E-mail : `young@umiacs.umd.edu`. Research supported in part by NSF grants CCR-8906949 and CCR-9111348.

components. If S is the set of vertices from G in one of the connected components, then *load*(e) is equal to

$$W(S,\bar{S}) = \sum_{(x,y)\in E, x\in S, y\in \bar{S}} w(x,y).$$

In other words, the demand of each edge $e = \{a,b\}$ in G, maps to the unique path in T from a to b, and loads each edge on the path. The load of a single edge is the sum of the demands that load this edge.

In this paper we study two different versions of this problem.

1.1 Routing Tree Problem

The following problem was proposed and studied by Seymour and Thomas [ST].

Definition 1 *[ST] A tree T is called a* routing tree *if it satisfies the following conditions:*

- *The leaves of T correspond to vertices of G.*
- *Each internal vertex has degree 3.*

The congestion of T is the maximum load of any edge of T. The congestion of G, denoted by β_G, is defined to be the minimum congestion over all routing trees T of G.

We would like to find a routing tree T with minimum congestion (that achieves β_G).

Seymour and Thomas showed that this problem is NP-hard by showing that graph bisection can be reduced to this problem. They also showed that in the special case when G is planar, the problem can be solved optimally in polynomial time.

We provide a polynomial time approximation algorithm for the congestion problem when G is an arbitrary graph. Our algorithm computes a routing tree T whose congestion is within an $O(\log n)$ factor from the optimal congestion (Section 3). The algorithm extends to the case when the routing tree is allowed to have vertices of higher degree.

1.2 Congestion Tree Problem

We also study the case when T is required to be a spanning tree of a given feasibility graph G_F. We show that the problem is NP-complete (Section 4). In the special case when G_F is complete, we show that an optimal solution can be computed in polynomial time[1]. We conjecture that using ideas similar to the ones used to solve the routing tree problem, one can design an $O(\log n)$ approximation scheme for the congestion tree problem.

[1] We actually show that if the Gomory-Hu cut tree T_{GH} of G [GH, Gu] is a subgraph of G_F then T_{GH} is an optimal solution.

1.3 Main Ideas

Our algorithm is a simple divide-and-conquer algorithm that uses the Leighton-Rao [LR] balanced separator algorithm to split the graph. By a naive application of the LR algorithm, one obtains an $O(\log^2 n)$ approximation factor. Our main contribution is to show that by a subtle application of LR, one can actually obtain an $O(\log n)$ approximation factor. We suspect that this kind of an application of LR will actually be useful for other problems as well (in improving approximation factors by a factor of $\log n$).

2 Preliminaries

A *cut* in a graph G is a set of edges which separate G into two pieces S and $\bar{S} = V \setminus S$. A cut can be represented by the vertex set S. The *weight* of a cut S, denoted by $W(S, \bar{S})$, is the sum of the weights of those edges which have one endpoint in S and one endpoint in $\bar{S}$. We use $W(v)$ to refer to the sum of the weights of the edges incident to v. A cut S is *b-balanced* if $n \cdot b \leq |S| \leq (1-b) \cdot n$. The definition is extended to the case when vertices are weighted as follows. Let U be a non-negative weight function on the vertices and let $U(S)$ be the sum of the weights of all the vertices in S. A cut S is b-balanced if

$$b \cdot U(V) \leq U(S) \leq (1-b) \cdot U(V)$$

Definition 2 *A* λ-approximate minimum b-bisector *is a b-balanced cut whose weight is at most λ times the weight of a minimum-weight $\frac{1}{3}$-balanced cut, for some constant $b \leq \frac{1}{3}$.*

The following result was proved by Leighton and Rao ([LR], Section 1.4).

Theorem 2.1 ([LR]) *It is possible to compute an $O(\log n)$-approximate minimum $\frac{1}{4}$-bisector in polynomial time.*

The above theorem can be extended to the case when vertices are given non-negative weights [Rao, Tar].

Definition 3 *Let T be a tree and let u be a vertex of degree two in T. Let v and w be the neighbors of u. The following operation is said to* short-cut *u in T - delete u from T and add the edge $\{v, w\}$. Short-cutting T implies the deletion of all vertices of degree two by short-cutting them in arbitrary order.*

3 Routing Tree Problem

$W(v)$ corresponds to the total weight between v and other vertices and is called the load of a vertex. Note that the load of any vertex v is a lower bound on β_G, because the edge incident to the leaf corresponding to v in any routing tree has to handle this load.

Lemma 3.1 *For any vertex v,* $W(v) \leq \beta_G$.

Given a procedure to compute a λ-approximate minimum b-bisector, our algorithm finds a routing tree whose congestion is at most λ/b times the optimal congestion.

3.1 Lower Bounds

We show two ways of finding lower bounds on the weight of the optimal solution. First, we show that the weight of a minimum-weight balanced separator is a lower bound on β_G. Second, we show that the optimal solution for the problem in a subgraph G' induced by an arbitrary set of vertices $V' \subset V$ is a lower bound on the optimal solution of G. This implies that an optimal solution to a sub-problem costs no more than any feasible solution to the whole problem.

Lemma 3.2 *Let $G = (V, E)$ be a graph with non-negative weights on the edges. Suppose we are given a non-negative weight function $U(v)$ on the vertices. Let the weight of each vertex be at most one-half of the total weight of all the vertices. Let Q be the weight of a minimum-weight b-balanced separator of G for any $b \leq 1/3$. Then $Q \leq \beta_G$.*

Proof. Let T be a routing tree with congestion β_G. Each edge e of T naturally induces a cut in G as follows: delete e from T to obtain subtrees T_1 and T_2. Let S_e be the set of vertices in G that are leaves of T_1 (this yields a cut in G). Clearly, $W(S_e, \overline{S_e})$ is the congestion on edge e and hence $W(S_e, \overline{S_e}) \leq \beta_G$. Since T is a tree of degree three, and by the assumption on the weights of vertices, it contains at least one edge e' which yields a b-balanced separator. Since Q is the minimum b-balanced separator of G we have $Q \leq W(S_{e'}, \overline{S_{e'}}) \leq \beta_G$. □

Lemma 3.3 *Let $G = (V, E)$ be a graph. Let H be a subgraph of G. Then $\beta_H \leq \beta_G$.*

Proof. Let T be a routing tree with congestion β_G. We will generate a routing tree T_H for H from T such that the load of any edge in T_H is at most the load of some edge in T. We generate the tree T_H from T as follows. Let V_H be the vertex set of H. Mark the leaves of T corresponding to V_H. Repeatedly delete the unmarked leaves of T until it has no unmarked leaves. Delete all vertices of degree two by short-cutting the tree, thus yielding T_H. The tree that we generate has V_H as its leaves and all its internal vertices have degree three. Hence it is a routing tree for H. Cuts in T_H can be associated with corresponding cuts in T and hence the load on any edge in T_H is at most the load of its corresponding edge in T. □

3.2 The Routing Tree Algorithm

Our basic approach is to subdivide the graph into pieces which are smaller by a constant fraction using an approximately minimum bisector. Since computing a minimum-weight balanced separator is also NP-hard, we use approximation algorithms designed by Leighton and Rao [LR] for computing approximately minimum-weight balanced separators (or approximate minimum bisectors). The solutions for the pieces are obtained recursively. All internal vertices of the solution tree have degree three except for the root. The two trees are glued together by creating a new root and making the roots of the pieces as the children of the new root. If implemented naively, this procedure leads to an $O(\log^2 n)$ factor approximation. Using balancing techniques, we improve the performance ratio to $O(\log n)$.

Suppose S, a subset of the vertices representing a subproblem, is split into two pieces S_1 and S_2 using an approximate bisector. When the problem is solved recursively on the two pieces, the main obstacle to obtaining an $O(\log n)$ approximation is the following. In the worst case, it is possible that most of the load corresponding to $W(S, \bar{S})$ may fall on S_1 or S_2. If this happens repeatedly, an edge can be overloaded proportionally to its depth in the tree. *To avoid this, it is necessary to partition the demand from $\bar{S}$ roughly equally among the pieces S_1 and S_2.* The following idea solves the problem and leads to an $O(\log n)$ approximate solution. Suppose we define a weight $U(v)$ for each vertex v in S according to the amount of demand from v to the set $\bar{S}$. Now when we split S, we use a cut that splits the vertices of S into sets of roughly equal weights. Lemma 3.2 guarantees that the minimum value of such a cut is a lower bound on β_S, which is a lower bound on β_G by Lemma 3.3.

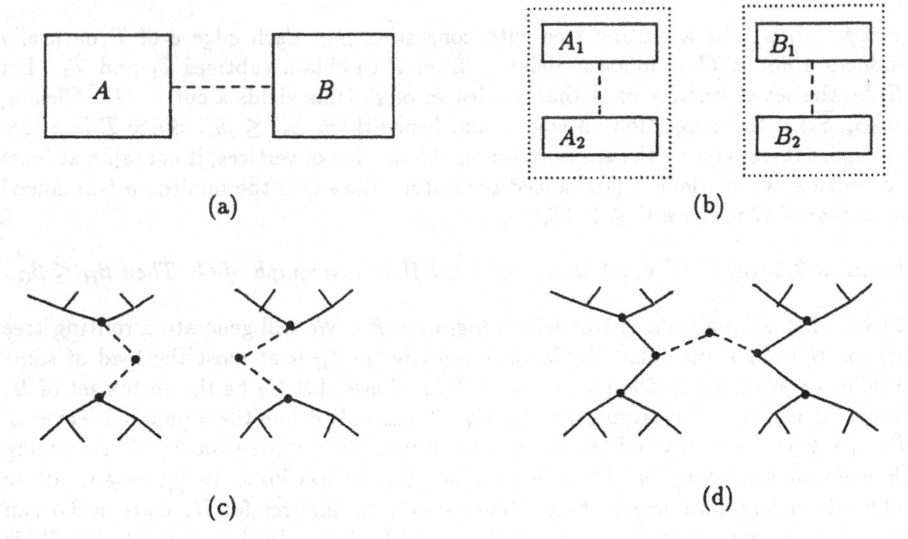

Figure 1: Example to illustrate algorithm.

We illustrate the recursive step of the algorithm by an example in Fig. 1. The algorithm first splits graph G into A, B by using an approximate bisector. Each vertex in A is assigned a weight equal to the total demand it has to vertices in $\bar{A}$. Similarly vertices in B are assigned weights corresponding to their demands from $\bar{B}$. The algorithm now recursively splits A and B by approximate bisectors. The weight of each vertex in A_1 is now increased by its demand to vertices in A_2 (similarly for sets A_2, B_1, B_2). The problem is solved recursively on each piece. These recursive calls return with respective trees as solutions for the pieces A and B as shown. By adding new edges and a new root vertex, the solution for the entire graph is obtained.

The algorithm given in Fig. 2 implements the above ideas. Given a graph G, ROUTE-TREE(V) returns a routing tree for G. To make sure that the root of the tree has degree three, we can discard the root by short-cutting it.

ROUTE-TREE(S) — *Find a routing tree for S.*
1 **If** $|S| = 1$ **then Return** S as a tree on a single vertex.
2 For each $v \in S$, fix its weight $U(v)$ to be $W(\{v\}, \bar{S})$.
Let the sum of the weights of the vertices in S be U_S.
3 **If** for any vertex v, $U(v) \geq U_S/2$ and $U_S \neq 0$ **then**
4 ROUTE-TREE($S \setminus \{v\}$)
5 Create a new tree T by attaching the above tree and v as the children of a new root r. **Return** T.
6 Find an approximate minimum-weight $\frac{1}{4}$-balanced separator for the subgraph induced by S in G (if $U_S = 0$, find an unweighted balanced separator). Let this break S into pieces S_1 and S_2.
7 ROUTE-TREE(S_1)
8 ROUTE-TREE(S_2)
9 Create a new tree T by attaching the two trees generated above as the children of a new root vertex. **Return** T.

Figure 2: Approximation Algorithm to Find a Routing Tree

Let the algorithm use a λ-approximate minimum $\frac{1}{4}$-bisector in Line 6. If Leighton and Rao's [LR] balanced separator algorithm is used, $\lambda = O(\log n)$. The following theorem shows that the load of any edge is at most 4λ times the optimal congestion. We use induction to prove that our load-balancing technique splits the load properly.

Theorem 3.4 (Performance) *The algorithm in Fig. 2 finds a routing tree T for G such that $\beta_T \leq 4\lambda\beta_G$.*

Proof. The proof proceeds by induction on the level of recursion. In the first call of ROUTE-TREE, G is split into two pieces S and $\bar{S}$ using an approximate bisector. We then find routing trees for S and $\bar{S}$ and connect the two roots with an edge e. The load on e is $W(S, \bar{S})$. By Lemma 3.2, the weight of a minimum-weight balanced separator is a lower bound on β_G. The weight of the separator the algorithm uses is guaranteed to be at most λ times the weight an optimal separator. Hence the load on edge e is at most $\lambda\beta_G$. This satisfies the induction hypothesis.

For the induction step, let us consider the case when we take a set S and split it into two pieces S_1 and S_2 (see Fig. 3). Let L be the load on the edge connecting the tree for S to its parent. Similarly, let L_i ($i = 1, 2$) be the load on the edge connecting the tree for S_i to its parent. Inductively, $L \leq 4\lambda\beta_G$. We show that each $L_i \leq 4\lambda\beta_G$.

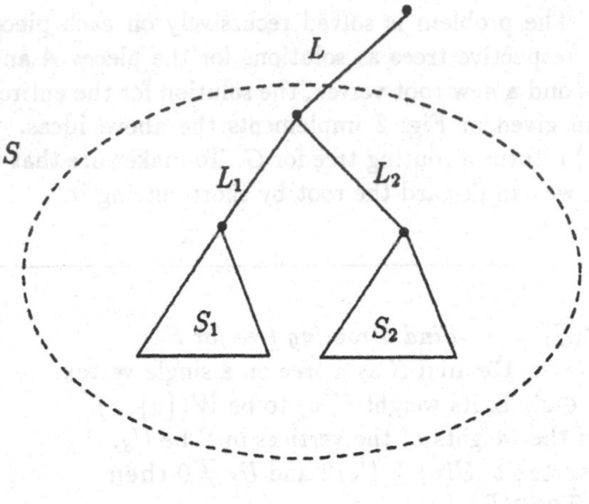

Figure 3: Inductive proof.

Let U be the weight function defined by the algorithm in this recursive call. Note that $L = U(S) = W(S, \bar{S})$ and $L_i = W(S_i, \bar{S}_i) = W(S_i, \bar{S}) + W(S_1, S_2)$. Also observe that $U(S_i) = W(S_i, \bar{S})$.

Case 1: If there is some vertex v in S whose weight $U(v)$ is more than $U(S)/2$, then we split S as $S_1 = \{v\}$ and $S_2 = S \setminus \{v\}$. Since $L_i = U(S_i) + W(S_1, S_2)$ and $U(S_1) > U(S)/2 > U(S_2)$ it follows that $L_1 > L_2$. This is because $U(S)$ is the sum of $U(S_1)$ and $U(S_2)$. It remains only to bound L_1. The demand from v, $W(v)$, is a lower bound on the congestion (by Lemma 3.1) and therefore $\beta_G \geq W(v) = L_1$. Hence both L_1 and L_2 satisfy the induction hypothesis.

Case 2: Otherwise, the algorithm distributed $U(S)$ into the weights of the vertices of S and then used a λ-approximate $\frac{1}{4}$-bisector of S. By the induction hypothesis, the edge from the subtree of S to its parent has a load L $(= U(S))$ of at most $4\lambda\beta_G$.

Since $W(S_i, \bar{S}) = U(S_i) \leq \frac{3}{4}U(S)$ and $W(S_1, S_2) \leq \lambda\beta_G$ (by Lemmas 3.2 and 3.3) we have:

$$L_i = W(S_i, \bar{S}) + W(S_1, S_2) \leq 3\lambda\beta_G + \lambda\beta_G.$$

□

Theorem 3.5 (Running Time) *The routing tree algorithm in Fig. 2 runs in polynomial time.* □

Corollary 3.6 *The algorithm in Fig. 2 finds in polynomial time a routing tree T for G such that $\beta_T = O(\log n)\beta_G$.*

Note: Our algorithm also handles the case when vertices of G are allowed to be internal vertices of the output tree. Lemmas 3.2 and 3.3 are valid in this case also. The lower bound in Lemma 3.1 weakens by a factor of 3. This lower bound is not critical to the performance ratio, so the performance ratio of the algorithm is unchanged.

Our algorithm can be generalized to find routing trees when every internal vertex may have degree up to k, for any $k \geq 3$. We obtain the same $O(\log n)$ approximation factor, independent of k. An algorithm obtaining an approximation factor of n/k is straightforward and is useful as k approaches n.

4 General Congestion Problem

4.1 NP-Completeness

In this section we show that the following problem is NP-complete. Given a graph $G = (V, E)$ representing a demand network. Each edge $e = \{a, b\}$ has a nonnegative weight $w(e)$ that represents the demand between the sites a and b. We are also given a feasibility graph G' and an integer D. The problem is to find a tree T that is a subgraph of G', such that when the demands of the edges in G are mapped to the tree T the congestion on each edge is at most D.

The reduction is done from the k Edge-Disjoint Paths Problem, known to be NP-Complete [GJ].

k Edge-Disjoint Paths Problem: Given an undirected graph $H = (V, E)$, and sets $S = \{s_1, s_2, \ldots, s_k\}$ and $T = \{t_1, t_2, \ldots, t_k\}$ are there k mutually edge-disjoint paths $P_1, P_2, \ldots, P_k$ such that P_i connects s_i with t_i ?

It is easy to see that this problem can be reduced to the general tree congestion problem. For the reduction we construct G' from H. For each vertex $u \in V$, if u has degree $d(u)$, we create a clique on $d(u)$ vertices, $u_1, u_2, \ldots, u_{d(u)}$. For each edge from v to w we introduce an edge from v_i to w_j where these are *distinct* vertices (not shared with any other edges). (Informally, each vertex is "exploded" into a clique, and the edges incident on the vertex are made incident on distinct clique vertices.) The demand graph G has edges between s_i and t_i (for all i). If there is a solution to the disjoint paths problem, clearly that yields a congestion tree with bandwidth one. The set of paths P_i can form cycles, but these cycles can be "pried" apart in G' since we replaced each vertex with a clique. These can now be connected to form a congestion tree with bandwidth one.

If there is a solution to the congestion tree problem it is clear that this yields a solution to the edge-disjoint paths problem (the demand edge from s_i to s_j gets mapped to a path in the tree and causes a load of one on each edge). Since the bandwidth is restricted to one, no other path can use the same edge (even when we go from G' to H).

4.2 Polynomially Solvable Case

In this section we show that when $T_{GH} \subseteq G_F$ (the feasibility graph contains the Gomory-Hu cut tree) we can solve the congestion problem optimally. (This is certainly the case when G_F is a complete graph.)

Given the demand graph G, we compute the Gomory-Hu cut tree T_{GH} [GH, Gu]. This is the tree that is used to route the calls. This yields an optimal solution for the following reason: consider any edge $e = \{s, t\}$ with load $L(e)$. T_{GH} has the property that $L(e)$ is the value of the s-t min cut. Clearly any s-t min cut is a lower bound on the optimal congestion.

Theorem 4.1 *T_{GH} is an optimal solution to the congestion problem.*

References

[GJ] M. R. Garey and D. S. Johnson, "Computers and Intractability: A guide to the theory of NP-completeness", *Freeman, San Francisco* (1979).

[GH] R. E. Gomory and T. C. Hu. Multi-terminal network flows. *Journal of SIAM*, 9(4): 551–570, 1961.

[Gu] D. Gusfield. Very simple methods for all pairs network flow analysis. *SIAM Journal on Computing*, 19(1): 143–155, 1990.

[LR] F. T. Leighton and S. Rao. An approximate max-flow min-cut theorem for uniform multicommodity flow problems with applications to approximation algorithms. In *Proc. 29th Annual Symp. on Foundations of Computer Science*, pages 422–431, October 1988. White Plains, NY.

[Rao] S. Rao. Personal communication.

[ST] P. Seymour and R. Thomas. Call routing and the rat catcher. *Workshop on Algorithms and Combinatorial Optimization*, March 1991. Atlanta, GA.

[Tar] É. Tardos. Personal communication.

A Fully Dynamic Approximation Scheme for All-Pairs Shortest Paths in Planar Graphs

Philip N. Klein* Sairam Subramanian**

Department of Computer Science, Brown University, Providence, R. I. 02912, USA

{pnk, ss}@cs.brown.edu

Abstract. In this paper we give a fully dynamic approximation scheme for maintaining all-pairs shortest paths in planar networks. Given an error parameter ϵ such that $0 < \epsilon \leq 1$, our algorithm maintains approximate all-pairs shortest-paths in an undirected planar graph G with nonnegative edge lengths. The approximate paths are guaranteed to be accurate to within a $(1+\epsilon)$-factor. The time bounds for both query and update for our algorithm is $O(\epsilon^{-1} n^{2/3} \log^2 n \log D)$, where n is the number of nodes in G and D is the sum of its edge lengths.

Our approximation algorithm is based upon a novel technique for approximately representing all-pairs shortest paths among a *selected* subset of the nodes by a *sparse substitute graph*.

1 Introduction

Designing dynamic algorithms for graph problems is a challenging area of research, and has attracted much interest motivated by many important applications in network optimization, VLSI layout, and distributed computing. An algorithm for a given graph problem is said to be *dynamic* if it can maintain the solution to the problem as the graph undergoes changes. These changes could be additions or deletions of edges, or a change in the cost of some edge (if applicable). In such a setting an *update* denotes an incremental change to the input, and a *query* is a request for some information about the current solution. We expect the dynamic algorithm to handle both queries and updates quickly i.e. in time that is substantially less than it would take to solve the problem from scratch every time the input changes. Unfortunately designing fully dynamic algorithms seems to be considerably harder than designing their sequential counterparts, and very few graph problems have fully dynamic solutions.

* Research supported by NSF grant CCR-9012357 and NSF PYI award CCR-9157620, together with PYI matching funds from Thinking Machines Corporation and Xerox Corporation. Additional support provided by DARPA contract N00014-91-J-4052 ARPA Order 8225.

** Research supported in part by a National Science Foundation Presidential Young Investigator Award CCR-9047466 with matching funds from IBM, by NSF research grant CCR-9007851, by Army Research Office grant DAAL03-91-G-0035, and by the Office of Naval Research and the Defense Advanced Research Projects Agency under contract N00014-91-J-4052 and ARPA order 8225.

In this paper we consider the problem of maintaining shortest-path information in planar graphs. The shortest path problem is a fundamental optimization problem since many applications can be formulated as shortest-path problems. Furthermore, a number of more complex problems can be solved by procedures which use shortest-path algorithms as subroutines.

Given a graph G with m nodes and n edges the shortest path between any two nodes can be computed efficiently by using Dijkstra's algorithm [2,7] in $O(m + n\log n)$ time. For planar graphs a faster algorithm due to Frederickson [6] runs in $O(n\sqrt{\log n})$ time. However, in the dynamic realm this problem is much less well-understood. Though there are many algorithms for the dynamic problem (see for example [1,4,8], see also [3]), none of them can simultaneously handle both updates and queries in time that is sublinear in the input size.

In this paper we present two solutions to the dynamic shortest-path problem on planar graphs. We first give a dynamic algorithm that computes exact shortest paths and is only marginally better than solving the problem from scratch. We then show that if we are willing to settle for approximate answers then substantial improvements are possible in both the query and update times. We say a path is an ϵ-approximate shortest path if its length is at most $1+\epsilon$ times the distance between its endpoints.

Theorem 1. *For any $0 < \epsilon \leq 1$, there exists a fully dynamic data structure to maintain ϵ-approximate all-pairs shortest-path information in planar undirected graphs with nonnegative edge lengths. The time per operation is $O(\epsilon^{-1} n^{\frac{2}{3}} \log^2 n \log D)$ where D is the sum of the lengths of all the edges and n is the number of nodes in the graph. The time for queries, edge-deletion and changing lengths is worst-case, while the time for adding edges is amortized.*

Our approximation algorithm is based upon a novel technique for compactly representing approximate all-pairs shortest paths among a set of k *selected* nodes by a substitute graph with the following properties:

- Each edge uv in the substitute graph corresponds to a path π from u to v in G.
- Each shortest path between selected nodes in G is approximated to within a $(1+\epsilon)$-factor by a two-edge path in the substitute graph.

The size of the substitute graph depends both on the number of selected nodes and on their distribution over the faces in G. In particular given an n-node undirected planar graph with nonnegative edge-lengths that sum to D, we have the following bounds on the size of sparse substitutes:

Theorem 2 Face-boundary substitute. *If there are k selected nodes, all on the boundaries of a constant number of faces of G then there exists a sparse substitute graph having $O(\epsilon^{-1} k \log k \log D)$ edges that approximates the all-selected-node-pair shortest paths in G to within a $(1+\epsilon)$-factor. Furthermore, the substitute graph can be constructed in $O(\epsilon^{-1} n \log^2 n \log D)$ time.*

The remainder of the paper is organized as follows: In section 2 we give a generic algorithm based on the techniques of Frederickson [5,9,10], for maintaining all-pairs shortest paths in an approximate or exact setting. We use the generic algorithm to for maintaining exact all-pairs shortest path information using $O(n)$-time per operation. We then show how to considerably speed up the dynamic algorithm at the expense of a slight loss in accuracy by using face-boundary substitutes. In section 3 we address the issue of constructing face-boundary substitutes, and in section 4 we discuss some extensions of our algorithm.

2 A generic algorithm for maintaining shortest paths

Let G be an n-node planar undirected graph with nonnegative integral edge-lengths. Let D be the sum of lengths. The *length* of a path π from u to v is simply the sum of the lengths of the edges in π. A minimum-length path from u to v is called a shortest path. The all-pairs shortest path problem is the problem of finding shortest paths between all pairs of nodes in G. Here we consider the problem of building dynamic data structures to maintain all-pairs shortest paths.

In this section we describe a simple generic algorithm for maintaining the shortest-path information in G in a fully dynamic environment. We then show how this generic algorithm can be used to derive an efficient approximation scheme by using face-boundary substitutes of Theorem 2.

Throughout this paper we assume that all the edge-additions are planarity-preserving. To see whether edge-additions preserve planarity we can run the planarity-testing algorithm from [10] in the background to prevent addition of edges that destroy planarity. Doing this only increases the time-complexity of our update operations by a constant factor. In the descriptions of our algorithms we will not explicitly mention these additional steps.

Our generic algorithm is based upon Frederickson's [5] idea of dividing the graph G into a number of clusters and using the partial solutions for different clusters to derive the solution for the entire graph. This clustering approach was first used in the context of dynamic algorithms by Galil and Italiano [9] to derive a fully dynamic data structure for maintaining two and three-vertex connectivity information in a planar graph, and later used by Galil, Italiano, and Sarnak [10] to develop a fully dynamic planarity-testing algorithm.

Galil Italiano and Sarnak [9,10] use the separator algorithm due to Lipton and Tarjan [12] to repeatedly divide the underlying planar graph into clusters. We borrow this technique and their terminology. For reasons that will be apparent soon, in this paper we will use the planar separator algorithm by Miller [13].

Given a subset B of nodes, Millers's algorithm uses a cycle of $O(\sqrt{n})$ nodes to divide a two-connected triangulated graph into two pieces each containing no more than two-thirds of the nodes of B. To find a separator in a graph that need not be two-connected or triangulated, we first add zero-weight dummy nodes and edges to the graph. There is one dummy node per face of the original graph, with dummy edges connecting the dummy node to the nodes on the boundary of the original face. The resulting graph is two-connected and triangulated. By applying Miller's algorithm to this graph, we obtain a cycle separator. The set X of non-dummy nodes of the separator need not form a cycle in the original graph. However, it does divide the graph into two pieces such that, in the subgraph H induced on one piece together with the nodes of X, the nodes of X all lie on the boundary of a single face.

Definition 3. A *k-cluster partition* of a two-connected n-node planar graph G is a partitioning of G into $O(k)$ edge-induced subgraphs $G_1, G_2, \cdots, G_r$ (where $r = ck$ for a suitable constant c) such that the following properties hold:

1. Each subgraph G_i contains $O(n/k)$ edges.
2. The number of *boundary nodes* in each G_i is $O(\sqrt{n/k})$, where a node is a boundary node if it is in more than one subgraph. Note that a node u can be a boundary node in G_i only if it is an endpoint of some edge in G_i.
3. In each subgraph G_i the boundary nodes all lie on the boundaries of a constant number of *faces*. (Note that a face of G_i need not be a face of G.)

Lemma 4 Frederickson. *Given a two-connected planar graph G, a k-cluster partition can be obtained in $O(n \log n)$ time.*

Consider a k-cluster partition of G. In such a partition we define the *parent* of an edge uv (denoted by G_{uv}) to be the subgraph G_i that contains it. Similarly, if u is a non-boundary node, we define its parent (denoted G_u) to be the subgraph G_i containing it. If u is a boundary node, we arbitrarily select one of the subgraphs G_i that contains u, and assign it to be u's parent (denoted by G_u).

To initialize, we find an k-cluster partition of G (the optimal value for k will be derived later), and precompute *substitute graphs* that represent the boundary-to-boundary shortest paths (either exactly or approximately) in each of the subgraphs $G_1, G_2, \cdots, G_r$. These substitute graphs are then unioned to form a skeletal graph S. The skeletal graph S is a compact representation for the shortest-paths among the boundary nodes and is used in the query-stage to compute the distance between the two query points.

To answer a query concerning the distance between two given nodes u and v, we form an auxiliary graph H by unioning the regions containing u and v along with the skeletal graph S. We then run a sequential shortest-path algorithm on the auxiliary graph H to compute the distance between u and v. It is easy to see that if the substitute graphs represent ϵ-approximate shortest paths, then the shortest path in H from u to v is an ϵ-approximate shortest path in the original graph.

To perform an update (whenever an edge is added, deleted, or its length is changed) we recompute the substitute graphs for all the subgraphs that are affected by the change in the input. Lemma 5 shows that only a constant number of subgraphs are affected during a single update operation.

Lemma 5. *Adding, deleting, or changing the length of an edge affects only a constant number of subgraphs in the cluster partition. Therefore, only a constant number of substitute graphs need to be recomputed to modify the skeletal graph S.*

Proof: We discuss the case of a *delete* operation. The arguments for addition and changing edge-lengths are similar. Consider deleting the edge uv. This results in a change in the subgraph G_{uv}. Furthermore, it is possible that deleting this edge could make either u (or v) a non-boundary node (this could happen if all the remaining edges incident at u (v) now lie in a single subgraph). Such a change in the boundary-status can affect only G_u and G_v. Therefore, to modify S we need only recompute G_{uv}, G_u, and G_v. □

A careful look at our update procedure shows that repeated requests for adding edges can result in an excess of boundary nodes, since every time an edge uv is added either u or v may become a boundary node. Therefore, the generic algorithm recomputes the cluster partition after the number of *add* operations exceeds the value of a preset parameter *limit*. The values for k and *limit* will be different for different versions of the generic algorithm.

We are now ready to discuss our bounds for maintaining exact and approximate shortest paths. To maintain exact shortest paths, we set k to be $\sqrt{n}$, and set *limit* to be $n^{1/4}\log n$. For each G_i we let the substitute graph $\hat{G}_i$ be the all-pairs distance-graph that has as its node-set the boundary-nodes of G_i, and has an edge uv for each pair of boundary nodes u, v. The length of uv in $\hat{G}_i$ is set to be equal to the length of the shortest u-v path that lies entirely within G_i. Since $k = \sqrt{n}$ the number of nodes in any G_i is $O(\sqrt{n})$ at all times. Also, since we recompute the cluster partition once every $n^{1/4}\log n$ *adds*, no subgraph has more than $n^{1/4}\log n$ boundary nodes.

To bound the query time we note that the number of nodes in S at any time is $O(n^{3/4})$ and the number of edges in S at any time is $O(n)$. It is easy to see that the same bounds hold for the size of the auxiliary graph H as well. Therefore, using Fibonacci heaps [7], Dijkstra's algorithm can be implemented to run on H in $O(n)$ time.

To bound the update time we note that for a set B_i of boundary nodes in G_i the boundary-to-boundary shortest paths can be found in $|B_i|\sqrt{n}\log n$ by running $|B_i|$ copies of Dijkstra's algorithm, one from each boundary node. Therefore, the substitute graph $\hat{G}_i$ can be constructed in $O(n^{3/4}\log^2 n) = O(n)$ time. Thus, the time needed to modify the data structure for any update operation is $O(n)$. We also recompute the cluster-partition and all the substitute graphs once every $n^{1/4}\log n$ *adds*. The time taken to recompute the cluster partition and to build all the substitute graphs is $\sqrt{n} \times O(n^{3/4}\log n) = O(n^{5/4}\log n)$. Amortizing this over $n^{1/4}\log n$ *adds* gives us our bound.

We now turn our attention to maintaining approximate shortest paths. It is easy to see that the reason for bad query and update times in our exact algorithm is the dense representation $\hat{G}_i$ we use to encode boundary-to-boundary shortest paths in G_i. We now show that by using the face-boundary-substitute of Theorem 2 we can get a much more efficient algorithm at the expense of some accuracy in the answers.

To get the bounds in Theorem 1 we set both k and *limit* to be $n^{1/3}$, and use the face-boundary-substitutes to represent all-boundary-pair shortest paths within any region. The algorithm for maintaining approximate shortest paths is otherwise identical to the one for exact shortest paths except for the following difference: Since the size of the face-boundary substitutes depends on the distribution of the boundary nodes we modify our updating procedure slightly. Whenever the number of *boundary faces* (faces that contain boundary nodes) in some cluster G_i exceeds three we apply Miller's separator algorithm to the graph with dummy nodes added. In order to reduce the number of boundary faces, we choose a separator X that divides up the dummy nodes corresponding to boundary faces. Each of the two resulting pieces has at most two-thirds of the four boundary faces, hence at most two old boundary faces. The new separator introduces an additional boundary face, for a total of three per piece. If the separator includes dummy nodes corresponding to old boundary faces, then in the resulting pieces these faces are merged with the new boundary face. Hence each piece ends up with at most three boundary faces.

To bound the time taken for a query we note that at any time the skeletal graph S has $O(\epsilon^{-1}n^{2/3}\log n \log D)$ edges. Initially, each of the $n^{1/3}$ subgraphs contribute $O(\epsilon^{-1}n^{1/3}\log n \log D)$ edges to S. This is because the face-boundary substitute constructed by Theorem 2 has $O(\epsilon^{-1}n^{1/3}\log n \log D)$ edges. Successive *add* operations cause new nodes to be labeled boundary nodes, increasing the size of S. If an *add* operation results in a split as described above, then $O(n^{1/3})$ nodes may become boundary nodes, resulting in an addition of $O(\epsilon^{-1}n^{1/3}\log n \log D)$ new edges to S. However, this process can continue for only $n^{1/3}$ *add* operations before we recompute the partition. Therefore the number of edges in S at any time during the computation is $O(\epsilon^{-1}n^{2/3}\log n \log D)$. Hence the auxiliary graph H also obeys the same bound on the number of edges. Thus an execution of Dijkstra's algorithm on H takes $O(\epsilon^{-1}n^{2/3}\log^2 n \log D)$ time, giving us the bound on the query time.

Arguments similar to the one for the exact version of the algorithms show that the worst-case time required for a *delete* operation and the amortized time required for an *add* operation are both $O(\epsilon^{-1}n^{2/3}\log^2 n \log D)$. We therefore get the bounds of Theorem 1.

3 Constructing a face-boundary sparse substitute

In this section we address the issue of constructing a face-boundary substitute to approximately represent the all-pairs shortest paths in G among a set N of $O(\sqrt{n})$ selected nodes (distributed over a constant number of faces).

In subsection 3.1 we describe a basic sparsification technique that is the key to the construction of our substitutes, and in subsection 3.2 we give a simple divide-and-conquer procedure that repeatedly uses the sparsification technique to construct a face-boundary substitute.

3.1 A basic sparsification technique

Let ϵ an error parameter and d a distance parameter, and let P be a path in G of length $O(d)$. In this subsection we show how to sparsely represent selected node-pair shortest paths that intersect P and are are between d and $2d$ in length. In particular, we show that there exists a substitute graph with $O(\epsilon^{-1}k)$ edges that approximates these shortest paths to within a $(1+\epsilon)$-factor. We call this substitute a *crossing substitute.*

To construct the crossing substitute we proceed as follows: We first divide P into $O(\epsilon^{-1})$ segments of length at most $\epsilon d/2$ each. The first node x_i in each segment s_i is identified as the *segment node* of that segment. We now perform single-source shortest-path computations in G from each of the $O(\epsilon^{-1})$ segment nodes. Our crossing substitute consists of a collection of stars, one for each segment. The star for segment s_i has x_i as its center and the selected nodes as its leaves. The edge between a selected node u and x_i is labeled with the weight of the shortest path from x_i to u.

We claim that these stars accurately represent (roughly) d-length shortest paths between selected nodes that intersect P. To prove our claim consider a path π

between selected nodes u and v that passes via segment s_i of some separating path. As shown in Figure 1, there is an alternate path between the endpoints u and v that detours through x_i along s_i. It is also easy to see that the additional length acquired because of the detour is at most $2\epsilon d/2 = \epsilon d$.

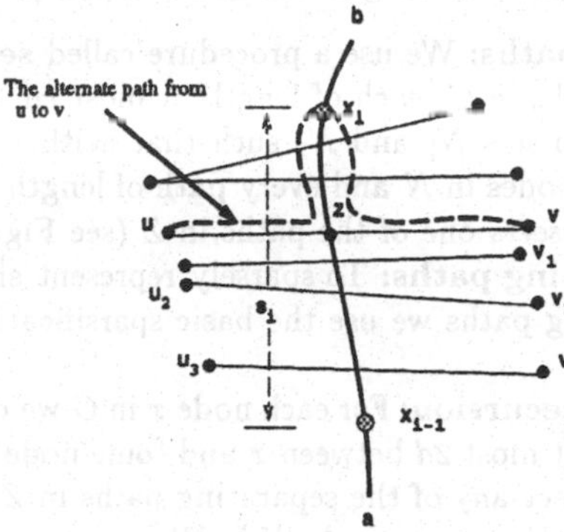

Fig. 1. Approximating a path that passes through segment s_i

In the substitute we approximate π by two edges, ux_i and x_iv. Since ux_i and x_iv represent shortest paths between their respective endpoints, the length of the two-edge path ux_i, x_iv is no more than the alternate path of Figure 1. Therefore, the two-edge path is at most ϵd longer than π. This combined with the fact that π is at least d in length implies the two-edge path ux_i, x_iv is a $(1+\epsilon)$-factor approximation of π.

To bound the number of edges in our substitute we note that it consist of $O(\epsilon^{-1})$ stars each of which has at most k edges. The bound on the size therefore follows. To bound the time required to compute the crossing substitute we note that the most expensive step is the single-source shortest path computations from the segment nodes. Since there are $O(\epsilon^{-1})$ segment nodes, the entire computation can be carried out in $O(\epsilon^{-1} n \log n)$ time.

3.2 Face-boundary substitutes

Let N be a set of $k = O(\sqrt{n})$ *selected* nodes in G, distributed over a constant number of faces and let $\epsilon > 0$ be an error parameter. In this section we show how to construct a face-boundary substitute that approximates the all-pairs shortest paths between selected nodes to within a $(1+\epsilon)$ factor.

Here we describe a procedure for sparsification when all the selected nodes lie on the boundary of a single face f. The case of multiple faces is similar. Our sparsification algorithm uses a divide-and-conquer mechanism and makes use of the basic sparsification technique from subsection 3.1. Since the basic sparsification technique works best with paths that are all roughly of the same length. To accommodate this, we borrow a *grouping* technique from [11].

We consider the selected-node-pair shortest paths in $\log D$ different groups and for each group we build a different substitute. The dth substitute approximates selected-node-pair shortest paths in the range $[d, 2d]$. These $\log D$ substitutes are unioned to get a substitute that sparsely represents paths of all lengths.

To sparsely represent selected-node-pair shortest paths with lengths in the range $[d, 2d]$ we therefore use the following divide-and-conquer procedure:

1. **Find separating paths:** We use a procedure called **separate** that gives a set Z of one or two paths in G, each of length at most rd, that divide the selected nodes in N into two sets N_1 and N_2 such that neither of them has more than three-fourth of the nodes in N and every path of length less than or equal to $4d$ from N_1 to N_2 intersects one of the paths in Z (see Figure 2).
2. **Sparsify intersecting paths:** To sparsely represent shortest paths that cross one of the separating paths we use the basic sparsification technique from subsection 3.1.
3. **Divide G for the recursion:** For each node x in G we determine whether there is a path of length at most $2d$ between x and some node in N_1 (respectively N_2) that does not intersect any of the separating paths in Z. If there is such a path then we place x in V_1 (respectively V_2). We construct G_1 and G_2 by taking node-induced subgraphs of V_1 and V_2 respectively. It is easy to see that no node is in both V_1 and V_2 at the same time because that would imply that there is a path of length at most $4d$ from some node in N_1 to a node in N_2 that does not intersect any separating path.
   To find the nodes of G that go into V_1, combine all the nodes of N_1 into a single supernode S and find the nodes of G that are within a distance $2d$ from S in $G - Z$. The set V_2 is determined similarly.
4. **Recursive computation:** Recursively compute sparse substitutes for $[d, 2d]$-shortest paths in G_1 and G_2 between the nodes in N_1 and N_2 respectively. Union the two recursive substitutes along with the crossing substitute found in step 2 to get the face-boundary substitute for all the selected nodes.

Before we analyze our procedure for sparsification we describe procedure **separate**. Consider a division of N into four subsets A_1, A_2, A_3, and A_4 as shown in Figure 2. If there is a path of length at most $4d$ from A_1 to A_3 or A_2 to A_4 then we can use it to divide N into subsets N_1 and N_2 as shown in Figure 2. Note that the nodes of N_1 are topologically separated from those of N_2 by the separator. Thus every path from N_1 to N_2 crosses the separating path.

When there is no single path of the desired length that divides N into roughly equal subsets we can use two paths as shown in Figure 2. Each of the paths has length at most $4d$, and their endpoints a_1, b_1 and a_2, b_2 satisfy the following properties: Node a_1 is in A_1 while node a_2 is in A_3; b_1 and b_2 occur after a_1 and a_2 respectively in the cyclic order around f; and the separation between a_i and b_i, $i = 1, 2$ is the maximum possible (in terms of their placement on the boundary of f) under the previous constraints. As shown in Figure 2 the exterior of face f is topologically separated into three regions R_1, R_2, and R_3. Any path between N_1 and N_2 that has one of its endpoints in R_1 or R_2 is forced to cross one of the separating paths. Also, by the maximality of the separation between a_i and b_i there is no path from N_1 to N_2 of length less than or equal to $4d$ that lies entirely in R_3.

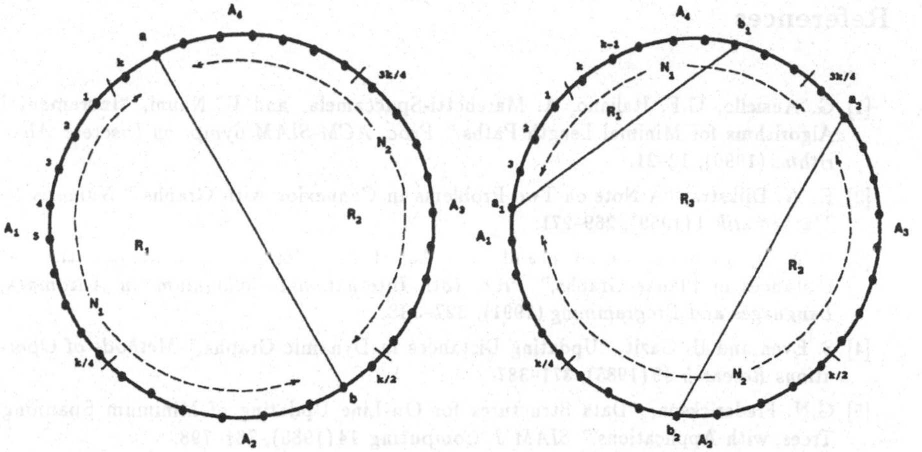

Fig. 2. Using one or two paths to cut the boundary

These paths can be found by performing a constant number of single-source shortest-path computations from appropriate supernodes that are created by temporarily merging all the nodes from a subset (A_1, A_2, A_3, or A_4) into a single node. Therefore, the time needed to construct the separating paths is $O(n \log n)$.

To bound the size of the substitute we note that the substitute for shortest paths in any range $[d, 2d]$ is constructed in $O(\log k)$ stages. Furthermore, at each level in the recursion the sum of the sizes of all crossing substitutes is $O(\epsilon^{-1}k)$. Hence the total size of the substitute for group d is $O(\epsilon^{-1}k \log k)$. This coupled with the fact there are $O(\log D)$ groups implies that the size of the substitute is $O(\epsilon^{-1}k \log k \log D)$.

To bound the running time of our sparsification procedure we note that the time required for constructing the crossing substitute in step 2 is $O(\epsilon^{-1} n \log n)$. The time needed to find separating paths and the time required for dividing G are subsumed by the time for constructing the crossing substitute. This in conjunction with the fact that the recursion depth is $O(\log k)$ and the fact that there are $\log D$ groups gives us the bounds for Theorem 2.

4 Extensions

These techniques can be extended to handle directed planar graphs as well. In particular, using a slightly different basic sparsification procedure it is possible to maintain $(1+\epsilon)$-factor approximations for shortest paths in directed planar graphs. The time bounds for queries and updates is $O(\epsilon^{-2} n^{2/3} \log^5 n \log D)$ for the directed case (see [14] for details). These techniques are also useful in solving the dynamic reachability problem in planar digraphs [15].

References

[1] G. Ausiello, G.F. Italiano, A. Marchetti-Spaccamela, and U. Nanni, "Incremental Algorithms for Minimal Length Paths," *Proc. ACM-SIAM Symp. on Discrete Algorithms* (1990), 12–21.

[2] E. W. Dijkstra, "A Note on Two Problems in Connexion with Graphs," *Numerische Mathematik* 1 (1959), 269–271.

[3] H. N. Djidev, G. E. Pantziou, and C. D. Zaroliagis, "Computing Shortest Paths and Distances in Planar Graphs," *Proc. 18th International Colloquium on Automata, Languages and Programming* (1991), 327–338.

[4] S. Even and H. Gazit, "Updating Distances in Dynamic Graphs," *Methods of Operations Research* 49 (1985), 371–387.

[5] G.N. Frederickson, "Data Structures for On-Line Updating of Minimum Spanning Trees, with Applications," *SIAM J. Computing* 14 (1985), 781–798.

[6] G.N. Frederickson, "Fast Algorithms for Shortest Paths in Planar Graphs, with Applications," *SIAM Journal on Computing* 16 (1987), 1004–1022.

[7] M.L. Fredman and R.E. Tarjan, "Fibonacci Heaps and their Uses in Improved Network Optimization Algorithms," *Journal of the Association for Computing Machinery* 34 (1987), 596–615.

[8] E. Fuerstein and A. Marchetti-Spaccamela, "Dynamic Algorithms for shortest Path Problems in Planar Graphs," *Proc. 17th International Workshop on Graph Theoretic Concepts in Computer Science* (1991), 187–197.

[9] Z. Galil and G. F. Italiano, "Maintaining biconnected components of dynamic planar graphs," *Proc. 18th Int. Colloquium on Automata, Languages, and Programming.* (1991), 339–350.

[10] Z. Galil, G.F. Italiano, and N. Sarnak, "Fully Dynamic Planarity Testing," *Proc. 24th Annual ACM Symposium on Theory of Computing* (1992), 495–506.

[11] P. N. Klein and S. Sairam, "A Parallel Randomized Approximation Scheme for Shortest Paths," *Proc. 24th ACM Symp. on Theory of Computing* (1992), 750–758.

[12] R.J. Lipton and R.E. Tarjan, "A Separator Theorem for Planar Graphs," *SIAM Journal of Applied Mathematics* 36 (1979), 177–189.

[13] G. Miller, "Finding Small Simple Cycle Separators for 2-Connected Planar Graphs," *Journal of Computer and System Sciences* 32 (1986), 265–279.

[14] Sairam Subramanian, "Parallel and Dynamic Graph Algorithms: A Combined Perspective," Department of Computer Science, Brown University, Ph.D. Thesis, 1993.

[15] Sairam Subramanian, "A Fully Dynamic Data Structure For Reachability in Planar Digraphs," Department of Computer Science, Brown University, Technical report , 1993, submitted to the European Symposium on Algorithms.

On Fat Partitioning, Fat Covering and the Union Size of Polygons*

(extended abstract)

Marc van Kreveld

Abstract

The complexity of the contour of the union of simple polygons can be $O(n^2)$ in general. In this paper, a necessary and sufficient condition is given for simple polygons which guarantees smaller union complexity. A δ-corridor in a polygon is a passage between two edges with width/length ratio δ. If a set of polygons with n vertices in total has no δ-corridors, then the union size is $O((n \log\log n)/\delta)$, which is close to optimal in the worst case. The result has many applications to basic problems in computational geometry, such as efficient hidden surface removal, motion planning, injection molding, etc. The result is based on a new method to partition a simple polygon P with n vertices into $O(n)$ convex quadrilaterals, without introducing angles smaller than $\pi/12$ radians or narrow corridors. Furthermore, a convex quadrilateral can be covered (but not partitioned) with $O(1/\delta)$ triangles without introducing small angles. The maximum overlap of the triangles at any point is two. The algorithms take $O(n \log^2 n)$ and $O(n \log^2 n + n/\delta)$ time for partitioning and covering, respectively.

1 Introduction

The primary motivation of this research is to determine for what sets of geometric objects (closed curves), the contour of the union has small complexity. When the union size is small, many geometric problems can be solved more efficiently and with simpler algorithms than in the general case. We give some examples later in this section.

Upper bounds on the union size have been found for several types of objects. Kedem et al.[6] show that the contour of the union of a set of n pseudo-discs in the plane has linear description size (a set of pseudo-discs is a set of simply connected regions of which any two boundaries intersect at most twice). It is easy to see that the contour of the union of a set of n isothetic rectangles can have $\Omega(n^2)$ connected components, and therefore quadratic description size, by placing them in a grid-like pattern. Since two isothetic rectangles intersect at

*The research of the author is supported by an NSERC international fellowship. Author's address: School of Computer Science, McGill University, 3480 University St., Montréal, Québec, Canada, H3A 2A7.

most four times, the question arises what the maximum union size is of sets of unbounded regions of which every two boundaries intersect at most three times. This case was settled by Edelsbrunner et al.[4], who show that the contour size is $O(n\alpha(n))$, and there are $O(n\alpha(n))$ connected components in the contour (where $\alpha(n)$ is the extremely slowly growing functional inverse of Ackermann's function). These bounds are tight in the worst case. Some other results on the maximum union size were obtained by Alt et al.[2].

Recently, Matoušek et al.[9] observed that for triangles, a quadratic lower bound example can only be constructed if the triangles have sharp angles. They proved that for a set of triangles of which any interior angle is at least δ, for some constant $\delta > 0$, the union determines only $O(n)$ holes, and the contour size is $O(n \log \log n)$. Notice that two such triangles can intersect six times. The constant of proportionality in these results is $O(1/\delta^3)$, whereas the best known lower bound example gives a constant $\Omega(1/\delta)$.

In this paper we extend the results from [9] to the case of simple polygons. Also, we prove that the dependency on δ is $\Theta(1/\delta)$ for fat triangles and polygons. For polygons, the fatness condition that each angle is bounded from below by a constant clearly is not good enough, because the lower bound example with rectangles still holds. To obtain a necessary and sufficient condition to bound the union size of simple polygons, we make the following definitions (see Figure 1):

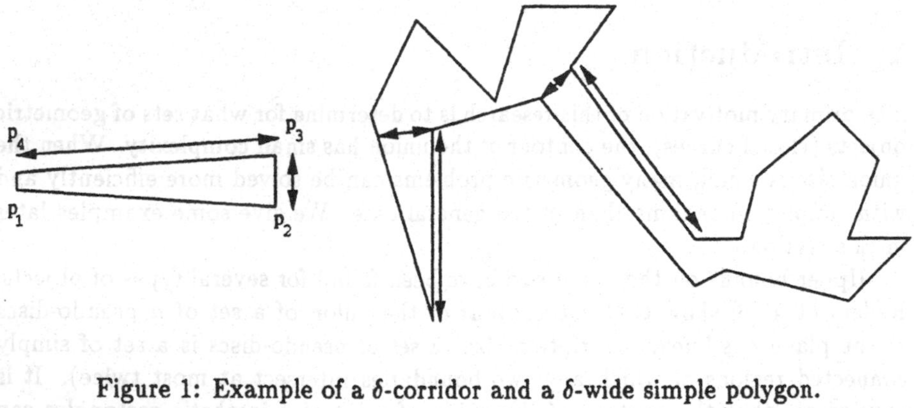

Figure 1: Example of a δ-corridor and a δ-wide simple polygon.

Definition 1 *For any* $0 < \delta \leq 1$, *a* δ-corridor *is a convex quadrilateral* Q *with vertices* p_1, p_2, p_3, p_4 *such that* $\angle p_1p_2p_3 = \angle p_2p_3p_4$ *and* $\angle p_3p_4p_1 = \angle p_4p_1p_2$, *and* $|\overline{p_1p_2}| = |\overline{p_3p_4}| = \delta \cdot \max\{|\overline{p_2p_3}|, |\overline{p_1p_4}|\}$.

For any $0 < \delta \leq 1$, *a simple polygon* P *(or any set of edges) is* δ-wide *if for any two edges* e *and* e' *of* P, *and any four points* $p_1, p_2 \in e$ *and* $p_3, p_4 \in e'$ *that are the vertices of a* γ*-corridor* Q *such that* interior$(Q) \subseteq$ interior(P), *it follows that* $\gamma \geq \delta$.

If δ is a constant, we refer to P as a *wide polygon.* If P is δ-wide, then the minimum angle at any vertex of P is at least $2\arcsin(\delta/2) \geq \delta$ radians. The main result of this paper is:

Theorem 1 *Let S be a set of δ-wide simple polygons with n vertices in total. The contour of the union of the polygons in S has complexity $O((n \log\log n)/\delta)$.*

The bound is close to optimal, because a lower bound example where the contour size is $\Omega(n/\delta + n\alpha(n))$ will be given ($\alpha(n)$ is the extremely slowly growing functional inverse of Ackermann's function). We list a couple of problems where the bound on the union size is important for the complexity of algorithms:

- *Hidden surface removal:* The simple and efficient hidden surface removal algorithm of Katz et al.[5] runs in time $O((U(n) + k)\log^2 n)$, where $U(n)$ is the maximum union size of n objects of the given type, and k is the size of the resulting visibility map. Our result implies that their algorithm can be used in a wider range of cases. The best known general hidden surface removal algorithm (for objects with a given depth order) takes $O(n^{1+\epsilon} + n^{2/3+\epsilon}k^{2/3})$ time for any fixed $\epsilon > 0$, see [1].
- *Motion planning:* Several motion planning algorithms compute the free placement space FP of the robot, that is, all placements for which the robot does not intersect any obstacle (see e.g. [7, 11]). Let the robot be a simple polygon with k vertices and the obstacles a set of simple polygons with n vertices in total. Assume that the robot can only translate. Then FP is the complement of the Minkowski difference of the robot and all obstacles, and the worst-case complexity of FP is $\Theta((nk)^2)$. If the robot and the obstacles are δ-wide, then FP has worst-case complexity $O((nk \log\log nk)/\delta)$, since the Minkowski difference of two δ-wide polygons is again δ-wide.
- *Injection molding for polyhedra:* A problem that arises in injection molding for a three-dimensional polyhedron P is to find the orientation of P for which the number of local maxima is minimum. Bose et al.[3] give an $O(n^2)$ time algorithm for polyhedra with n vertices. They also show that for certain restricted classes of polyhedra, this can be improved to $O(nk \log^{O(1)} n)$ time, where k is the minimum number of local maxima. The improvement makes use of the union size of n wide convex polygons.

Other applications include point containment queries in simple polygons, red-blue intersection detection of two sets of polygons, and ray shooting in 3-dimensional space. The full paper contains more details.

The method we use to obtain the bound on the union size is interesting in its own right. Basically, we try to partition a δ-wide simple polygon (with no small angles) into $O(n/\delta)$ fat triangles[1], and then the result follows immediately

[1] In this paper, we define a triangle or quadrilateral to be fat if it is wide. This does not conflict with existing definitions of fat triangles such as in [9].

from the work of Matoušek et al.[9] mentioned before. Unfortunately, such a partitioning does not always exist. A wide convex quadrilateral exists that cannot be partitioned into $O(1)$ fat triangles, see Figure 2 (in the full paper we show that $\Omega(\log m)$ fat triangles are needed). Therefore, we construct a *covering* of δ-wide polygons.

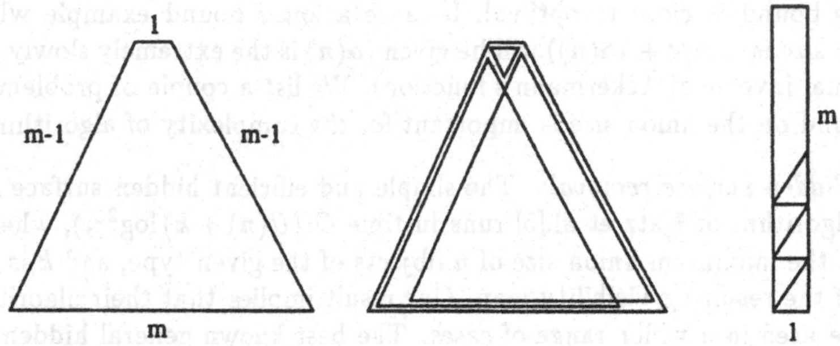

Figure 2: The leftmost wide quadrilateral cannot be partitioned into $O(1)$ fat triangles, but it can be covered by 3 fat triangles. The rightmost non-wide rectangle cannot be partitioned or covered by $O(1)$ fat triangles.

Definition 2 *A set S of quadrilaterals and triangles (only triangles) is a* weak Steiner quadrilateralization *(*weak Steiner triangulation*) of a simple polygon P if S is a partitioning of P. The set S is a* strong Steiner quadrilateralization *(*strong Steiner triangulation*) if additionally, no edge of any quadrilateral or triangle contains Steiner points in its interior. A set S of triangles is a k-*covering *of P if any point in the interior of P lies in the interior of at most k triangles of S, and the union of the triangles in S is P.*

We consider what simple polygons admit partitionings and coverings using fat quadrilaterals and triangles, using a set S of small size. It is easy to see that a rectangle with edge lengths 1 and m cannot be k-covered with a set S of constant size when m is large. The rectangle is not wide, and therefore, we consider partitionings and coverings of δ-wide polygons in this paper. We do not assume that δ is a constant.

Let P be any δ-wide simple polygon, and let $\gamma = \min\{\delta, 1 - \frac{1}{2}\sqrt{3}\}$. We show that P admits a weak Steiner quadrilateralization using $O(n)$ γ-wide quadrilaterals and triangles. Furthermore, we show that any δ-wide quadrilateral can be 2-covered using $O(1/\delta)$ γ-wide triangles. Consequently, P can be 2-covered using $O(n/\delta)$ γ-wide triangles. Thus, wide polygons do not admit a fat triangulation of linear size: bounds must depend on the ratio of edge lengths of the polygon as well. However, fat quadrilateralizations and fat triangle 2-coverings of linear size do exist for wide polygons.

In the remainder of the paper, any polygon is simple by default. Furthermore, we let a quadrilateral be a *convex* 4-gon.

In Section 2 we show that any simple polygon P with n vertices can be partitioned in $O(n \log^2 n)$ time into $O(n)$ quadrilaterals and triangles, such that no angle smaller than $\pi/12$ is created, and no δ-corridor with $\delta < 1 - \frac{1}{2}\sqrt{3}$. Consequently, a δ-wide polygon can be partitioned into $O(n)$ γ-wide quadrilaterals and triangles, where $\gamma = \min\{\delta, 1 - \frac{1}{2}\sqrt{3}\}$. The main construction tool is the *Edge Voronoi Diagram*, see e.g. Lee[8] and Yap[12]. We show in Section 3 that a δ-wide quadrilateral with all angles at least some positive constant can be 2-covered with $O(1/\delta)$ fat triangles. No angle smaller than $\pi/10$ radians is introduced. Section 4 shows that the contour size of the union of δ-wide polygons with n vertices in total is $O((n \log\log n)/\delta)$, and that the contour size can be $\Omega(n/\delta + n\alpha(n))$ (assuming $1/n \leq \delta \leq 1$). We close with the conclusions and open problems in Section 5.

2 Partitioning simple polygons preserving wideness

In this section we show how to construct a weak Steiner quadrilateralization of any wide polygon, while preserving the wideness in any resulting subpolygon. Since the subdivision is weak, we may omit all vertices with angle π from the polygon. We give some elementary properties of δ-corridors and segments that are interior to a polygon P, which are straightforward to verify.

Lemma 1 *Two segments of which the supporting lines make an angle α cannot form a δ-corridor with $\delta < 2\sin(\alpha/2)$.*

Lemma 2 *Let C be a circle whose interior lies completely inside a polygon P. Any segment $\overline{pq}$ between two points p and q on C such that $\overline{pq}$ makes an angle α with the tangents to C at p and q cannot create a δ-corridor with $\delta < \frac{1-\cos\alpha}{2\sin\alpha}$.*

Lemma 3 *Let P be a polygon, let $e = \overline{uv}$ be an edge of P and let w be a vertex of P. If the interior of Δuvw does not intersect P, then the addition of the segment that bisects $\alpha = \angle uwv$ to partition P cannot create δ-corridors with $\delta < 2\sin(\alpha/4)$ in the two subpolygons.*

2.1 Partitioning a simple polygon into 6-gons

The *Edge Voronoi Diagram* inside P of the edges of P, denoted EVD, is a subdivision of the interior of P into regions where one edge is closest. See Figure 3 for an example. The EVD is also called the *medial axis* or *internal skeleton*.

The arcs and nodes of the EVD form a tree with n leaves, one for each vertex of P. Any arc is a line segment or paraboloid arc. The arc of the EVD incident to a leaf v (a vertex of P) is the bisector of the angle of the two edges of P that are incident to v. In non-degenerate cases, no node of the EVD has degree greater than 3, and there are exactly $n-2$ nodes with degree 3. Each node is the center

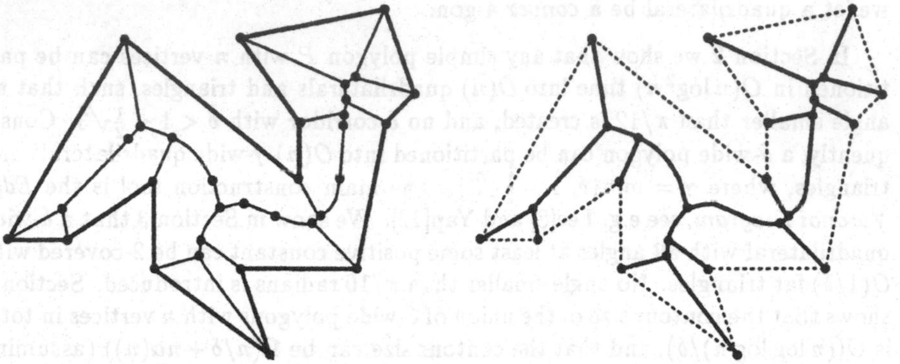

Figure 3: Left: the EVD of a simple polygon. Right: the tree T of the EVD with only nodes of degree 1 or 3.

of a circle that is at equal distance to 3 edges of P, 2 edges and 1 vertex of P, 1 edge and 1 vertex of P, or 2 vertices of P. No edge or vertex of P is nearer to that vertex of the EVD. It follows that the interior of the circle is empty.

We are only interested in the n leaves and the $n-2$ vertices of degree 3, which form a tree T when we identify the two arcs incident to each node of degree 2 (see Figure 3). T has the property that there exists a vertex v of degree 3 whose removal partitions the tree into 3 subtrees, each with at most $\lfloor n/2 \rfloor$ leaves of T. We use the largest circle with empty interior centered at v to partition polygon P into subpolygons, without creating δ-corridors with small δ.

Lemma 4 *Let P be a simple polygon and C a circle such that* interior(C) $\subset$ interior(P) *and which intersects P in 2 points. The two points can be connected with 1 or 2 segments, such that in either of the two subpolygons no δ-corridors are created with $\delta < \frac{1}{2}\sqrt{2} - \frac{1}{2}$. If C intersects P in 3 points, then these points can be connected with at most 4 segments, such that each of the 3 subpolygons that arise use at most 2 of these segments, and no δ-corridor is created with $\delta < 1 - \frac{1}{2}\sqrt{3}$. Furthermore, a γ-wide quadrilateral may be created with $\gamma > 1/2$.*

A proof is omitted. See Figure 4 for the cases.

The following algorithm is used to partition any simple polygon P with n vertices into 6-gons:

1. Compute the EVD of P. Consider it as a graph, and remove each node of degree 2 by identifying the arcs incident to it to obtain the tree T. Select a vertex v in T whose removal gives 3 subtrees which have at most $\lfloor n/2 \rfloor$ leaves each.

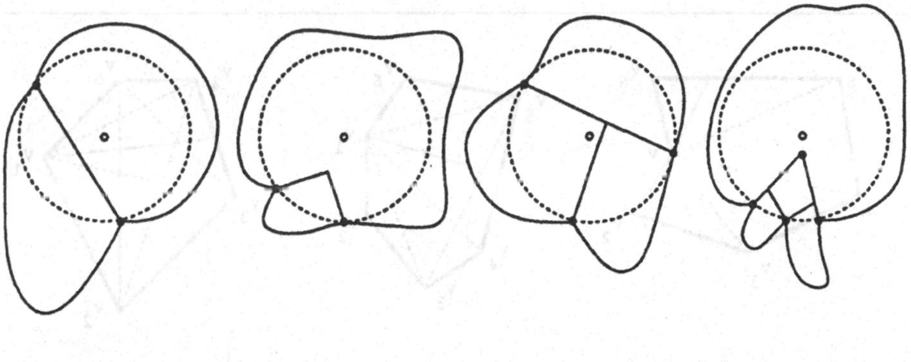

Figure 4: Splitting a simple polygon without creating narrow corridors using empty circles.

2. Compute the largest empty interior circle centered at v. It intersects P in 2 or 3 points. Add the segments as used in Lemma 4, to obtain 2 or 3 subpolygons of P (and, possibly, a wide quadrilateral).

3. For each subpolygon that has at least 7 vertices, recursively subdivide it.

Theorem 2 *Any δ-wide simple polygon P with n vertices can be partitioned in $O(n \log^2 n)$ time into $O(n)$ γ-wide polygons with at most 6 vertices and* $\gamma = \min\{\delta, 1 - \frac{1}{2}\sqrt{3}\}$.

Proof: Notice that splitting P as in Lemma 4 yields subpolygons with at most 3 new vertices. Therefore, the above algorithm gives subpolygons with at most $\lfloor n/2 \rfloor + 3$ vertices. As long as $n > 6$, a polygon is partitioned into subpolygons with fewer vertices. The wideness guarantee of the resulting polygons follows from Lemma 4. The time bound of the algorithm follows easily from the $O(n \log n)$ time algorithms for computing an EVD [8, 12]. □

2.2 Partitioning 6-gons, 5-gons and 4-gons

Let P be a non-convex 6-gon with vertices $v_1, \ldots, v_6$, and assume w.l.o.g. that v_1 is a reflex vertex (see Figure 5, left). We show how to partition P in $O(1)$ convex pieces. Since $\angle v_6v_1v_5 + \angle v_5v_1v_4 + \angle v_4v_1v_3 + \angle v_3v_1v_2 = \angle v_6v_1v_2 > \pi$, at least one of the four angles is $> \pi/4$. (If any of v_3, v_4, v_5 is not visible from v_1, it can simply be removed from consideration; the argument still holds with better constants.) We draw a segment from v_1 to bisect this angle. The two new angles in the two subpolygons at v_1 are at least $\pi/8$, and therefore the largest new angle at v_1 is at least $\pi/8$ less than $\angle v_6v_1v_2$. By Lemma 3, no δ-corridors with $\delta < 2\sin(\pi/16)$

are created. We continue this procedure until every subpolygon of P is convex. The same idea can be used for non-convex 5-gons and 4-gons.

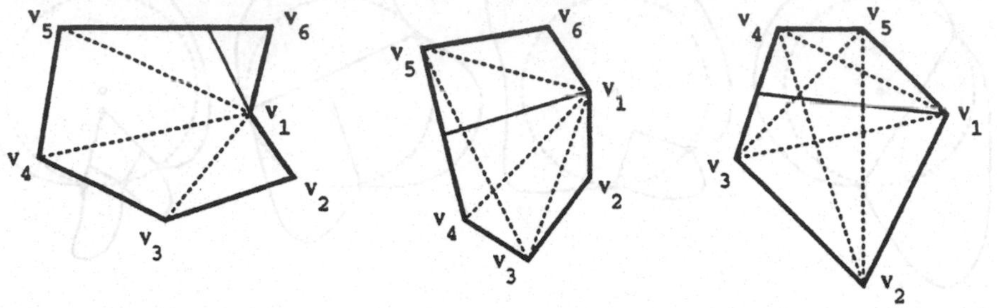

Figure 5: Left: one step of the partitioning of a 6-gon into convex pieces. Middle: partitioning a convex 6-gon. Right: partitioning a convex 5-gon.

Let P be a convex 6-gon with vertices $v_1, \ldots, v_6$. P is partitioned into a 5-gon and a 4-gon as follows. Consider the triangle $\Delta v_1 v_3 v_5$, and assume w.l.o.g. that v_1 has angle $\angle v_5 v_1 v_3 \geq \pi/3$ (see Figure 5, middle). Therefore, $\angle v_3 v_1 v_4 \geq \pi/6$ or $\angle v_4 v_1 v_5 \geq \pi/6$. If $\angle v_3 v_1 v_4 \geq \pi/6$, then we bisect this angle with a segment to obtain a 5-gon and a 4-gon without introducing angles smaller than $\pi/12$ or δ-corridors with $\delta < 2\sin(\pi/24)$ by Lemma 3. The other case is the same.

Let P be a convex 5-gon with vertices $v_1, \ldots, v_5$. P is partitioned into two 4-gons as follows. Consider the sequence $v_1 v_3 v_5 v_2 v_4$, which forms the star inscribed in P (see Figure 5, right). The sum of the five angles at the five points of the star is π, and hence, at least one the five angles is at least $\pi/5$. Assume w.l.o.g. that it is at v_1. We bisect the angle $\angle v_3 v_1 v_4$ with a segment to obtain two 4-gons without introducing angles smaller than $\pi/10$ or δ-corridors with $\delta < 2\sin(\pi/20)$ by Lemma 3.

The above partitionings lead to the following result:

Theorem 3 *A δ-wide simple polygon P with n vertices can be partitioned in $O(n \log^2 n)$ time into $O(n)$ γ-wide quadrilaterals and triangles, where* $\gamma = \min\{\delta, 1 - \frac{1}{2}\sqrt{3}\}$.

Remark: The partitioning methods of this section do not introduce angles smaller than $\pi/12$ radians.

3 Covering quadrilaterals by triangles

We show that a δ-wide quadrilateral Q can be covered by $O(1/\delta)$ fat triangles, if every angle of Q is at least some constant. Otherwise, one or two triangles of the covering will have a sharp angle equivalent to the one or two sharp angles of Q. No three triangles in the covering overlap in a positive area region.

The first stage of the covering is to separate angles that are smaller than $\pi/5$, using one triangle that contains that angle. Then we separate all angles greater than $4\pi/5$, without creating angles less than $\pi/5$ or greater than $4\pi/5$. See the full paper for the details.

Let Q be a quadrilateral with all angles at least $\pi/5$ and at most $4\pi/5$. We consider again the Edge Voronoi Diagram of the edges of Q. The EVD contains two nodes of degree 3, which are the centers of circles that touch Q in three edges. Orient Q such that the center of the larger circle is vertically above the center of the smaller circle (see Figure 6).

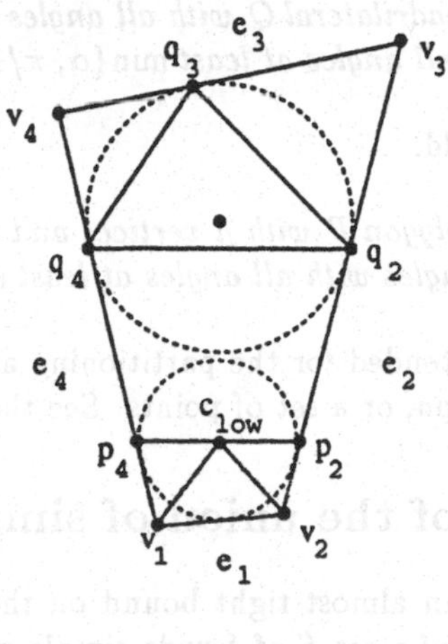

Figure 6: Quadrilateral Q: the points q_2, q_3, q_4 are the points where the larger circle touches Q, and p_2, p_4 are chosen such that the edge $\overline{p_4p_2}$ contains the center of the smaller circle and is parallel to $\overline{q_4q_2}$.

In the full paper we show that the following triangles have all angles at least $\pi/10$: $\Delta q_4q_3v_4$, $\Delta q_2q_3q_4$, $\Delta q_2v_3q_3$, $\Delta v_1c_{low}p_4$, $\Delta v_2c_{low}v_1$, $\Delta v_2p_2c_{low}$. Therefore, the problem reduces to partitioning or covering the trapezoid T with vertices p_4, p_2, q_2, q_4. This trapezoid is a corridor, as defined in the Introduction.

Let ℓ_q be the line through q_4 and q_2, and let ℓ_p be the line through p_4 and p_2. There are two cases. Firstly, it may be the case that the line ℓ_p lies above the line ℓ_q. Then we already have a 2-covering of Q. Secondly, the line ℓ_p may lie below the line ℓ_q. If their distance is smaller than $\frac{\pi}{20}\overline{p_4p_2}$, then we define a new line ℓ'_p below ℓ_p such that if $p'_4 = \ell'_p \cap e_4$ and $p'_2 = \ell'_p \cap e_2$, then we choose ℓ'_p such that the distance between ℓ_q and ℓ'_p is $\frac{\pi}{20}|p'_4p'_2|$. In the full paper we show that this is always possible with ℓ'_p intersecting both e_4 and e_2. We use the following result.

Denote by h the thickness of T, which is the smallest distance between $\overline{p_4p_2}$ and $\overline{q_4q_2}$. Denote by α the angle of the lines supporting e_2 and e_4. We have $0 \leq \alpha \leq 4\pi/5$. In the full paper we show:

Lemma 5 *Let T be a trapezoid with $h \geq \frac{\pi}{20}|\overline{p_4p_2}|$. If $h \leq |\overline{p_4p_2}|$, then T can be partitioned into 3 or 5 fat triangles. If $h \geq |\overline{p_4p_2}|$ and $\alpha \geq \pi/10$, then T can be covered by 3 fat triangles of which at most two overlap. If $h \geq |\overline{p_4p_2}|$ and $\alpha < \pi/10$, then T can be partitioned into $O(1/\delta)$ fat triangles if T is a δ-corridor. The triangles have all angles at least $\pi/10$.*

All ingredients to obtain a 2-covering of a quadrilateral by triangles with a guarantee on minimum angles have been sketched, and we conclude:

Theorem 4 *A δ-wide quadrilateral Q with all angles at least α can be 2-covered by $O(1/\delta)$ triangles with all angles at least* $\min\{\alpha, \pi/10\}$.

Theorems 3 and 4 yield:

Theorem 5 *A δ-wide polygon P with n vertices and all angles at least α can be 2-covered by $O(n/\delta)$ triangles with all angles at least* $\min\{\alpha, \pi/12\}$.

The results can be extended for the partitioning and covering of the faces of a planar straight line graph, or a set of points. See the full paper for details.

4 The contour of the union of simple polygons

In this section we show an almost tight bound on the maximum complexity of the contour of the union of a set S of δ-wide simple polygons with n vertices in total. We assume in this section that $1/n \leq \delta \leq 1$ (this restriction is only made for an easier statement of the results).

Theorem 6 *Let S be a set of δ-wide polygons with n vertices in total. The maximum complexity of the contour of the union for S is $\Omega(n/\delta + n\alpha(n))$ and $O((n \log\log n)/\delta)$, where $\alpha(n)$ is the extremely slowly growing functional inverse of Ackermann's function.*

Proof: (Sketch, see also Figure 7.) The lower bound follows from a straightforward construction and from [9]. To prove the upper bound, let P be any polygon in S with m vertices. We cut off a tiny equi-angular triangle t at every sharp angle of P, such that t either lies completely in the union of the other polygons, $\bigcup_{Q \in S\setminus\{P\}} Q$, or in the complement of the union. We obtain a polygon P' with at most $2m$ vertices and no sharp angles. We cover P' as in Theorem 5, using $O(m/\delta)$ fat triangles. This is done for every polygon in S, and the union $\bigcup_{P\in S} P'$ is the union of $O(n/\delta)$ fat triangles. Consequently, the complexity of the union is $O((n \log\log n)/\delta)$ by the result from [9]. Then we add the tiny equi-angular

Figure 7: Separating vertices with sharp angles of simple polygons.

triangles to the union, and it can be shown that they increase the complexity of the contour with $O(n)$ vertices and edges. □

In the same way we can show:

Theorem 7 *Let S be a set of δ-wide polygons with n vertices in total. The maximum complexity of the boundaries of all cells covered by at most k polygons of S is $O((nk \log\log(n/k))/\delta)$.*

5 Conclusions and open problems

In this paper we studied the complexity of the contour of the union of a set of simple polygons. The notion of δ-wide polygons was introduced, where the value of δ influences the contour size. Almost tight upper and lower bounds on the maximum union contour size were given. The main idea was to cover each polygon by a set of fat triangles, and then use the result of Matoušek et al.[9] on the union size of fat triangles. We also showed that a *partitioning* of a polygon into fat triangles cannot give the desired bounds, because too many fat triangles will needed.

The partitioning and covering algorithms presented in this paper require $O(n \log^2 n)$ and $O(n \log^2 n + n/\delta)$ time, respectively. It may be possible to improve upon this bound. We remark, however, that for most applications it is not necessary to perform the actual partitioning or covering, but instead regard the techniques as a proof that the union size is not large. To compute the actual union of a set of δ-wide polygons, a straightforward $O((n \log^2 n \log\log n)/\delta)$ algorithm exists, see e.g. Kedem et al.[6]. A slightly more efficient, but randomized, algorithm can be obtained by using the algorithm of Miller and Sharir[10], see also [9]. A further speedup may be possible. A third open problem we take from [9]: The maximum union contour size of n fat triangles is $O(n \log\log n)$ and $\Omega(n\alpha(n))$. There is a gap to be closed. More tight bounds would immediately give more tight bounds for the union contour size of δ-wide polygons.

References

[1] Agarwal, P.K., and M. Sharir, Applications of a new partitioning scheme. *Discr. & Comp. Geom.*, to appear.

[2] Alt, H., R. Fleischer, M. Kaufmann, K. Mehlhorn, S. Näher, S. Shirra, and C. Uhrig, Approximate motion planning and the complexity of the boundary of the union of simple geometric figures. *Proc. 6th ACM Symp. Comp. Geom.* (1990), pp. 281–289.

[3] Bose, J., M. van Kreveld, and G. Toussaint, *Filling polyhedral molds.* Tech. Rep. SOCS 93.1, School of Computer Science, McGill University, 1993. Extended abstract in these proceedings.

[4] Edelsbrunner, H., L. Guibas, J. Hershberger, J. Pach, R. Pollack, R. Seidel, M. Sharir, and J. Snoeyink, On arrangements of Jordan arcs with three intersections per pair. *Discr. & Comp. Geom.* 4 (1989), pp. 523–539.

[5] Katz, M.J., M.H. Overmars, and M. Sharir, Efficient hidden surface removal for objects with small union size. *Proc. 7th ACM Symp. Comp. Geom.* (1991), pp. 31–40.

[6] Kedem, K., R. Livne, J. Pach, and M. Sharir, On the union of Jordan regions and collision-free translational motion amidst polygonal obstacles. *Discr. & Comp. Geom.* 1 (1986), pp. 59–71.

[7] Latombe, J.-C., *Robot motion planning.* Kluwer Academic Publishers, Boston, 1991.

[8] Lee, D.T., Medial axis transformation of a planar shape. *IEEE Trans. Pattern Anal. Mach. Intel.* 4 (1982), pp. 363–369.

[9] Matoušek, J., N. Miller, J. Pach, M. Sharir, S. Sifrony, and E. Welzl, Fat triangles determine linearly many holes. *Proc. 32nd IEEE Symp. Found. Comp. Science* (1991), pp. 49–58.

[10] Miller, N., and M. Sharir, *Efficient randomized algorithms for constructing the union of fat triangles and of pseudodiscs.* Manuscript, 1993.

[11] Sharir, M., Efficient algorithms for planning purely translational collision-free motion in two and three dimensions. *Proc. IEEE Int. Conf. on Robotics and Automation* (1987), pp. 1326–1331.

[12] Yap, C.K., An $O(n \log n)$ algorithm for the Voronoi diagram of a set of simple curve segments. *Discr. & Comp. Geom.* 2 (1987), pp. 365–393.

A Time-Randomness Tradeoff for Selection in Parallel

Danny Krizanc *

School of Computer Science
Carleton University
Ottawa, Ontario K1S 5B6

Abstract. In this paper we study the problem finding the median on a parallel comparison tree (PCT). Results due to Valiant [11], Meggido [3] and Reishcuk [9] show a provable gap between the randomized and deterministic parallel complexity of selection. We prove a tight tradeoff between the amount of randomness used by an algorithm for this problem and its performance, measured by the time it requires to complete its computation with a given failure probability. The tradeoff provides a smooth bridge between the deterministic and randomized complexity of the problem.

1 Introduction

The use of randomness in computation is well established (see Rabin [7]). It has been applied in sequential, distributed and parallel computation with great success, in many cases providing efficient solutions where no deterministic solutions are known and in some cases where comparably efficient deterministic solutions are provably impossible. To help explain this phenomena we propose to consider the randomness an algorithm uses as a resource and study its effect on the efficiency of the algorithm. We prove a tight tradeoff between the amount of randomness used by an algorithm and its performance, measured by the time it requires to complete its computation with a given failure probability. Such a tradeoff was first shown by Krizanc et al. [2, 5] for the problem of oblivious routing on hypercubic networks. A number of other researchers have also studied the effect of limiting randomness in computation [1, 4, 8, 10].

In this paper we consider the problem of selection in parallel. Valiant [11] showed a $\Omega(\log\log n)$ deterministic lower bound for the problem of finding the median of n elements on an n processor parallel computation tree (PCT). Reischuk [9], and independently Meggido [3], proved the existence of a gap between the randomized and deterministic complexities of the problem by providing a n processor randomized PCT which selects the median of n elements in $O(1)$ time with high probability.

* Research was supported in part by the Natural Sciences and Engineering Research Council of Canada, research grant OGP0137991.

In section 3, we demonstrate how to convert Valiant's lower bound into a tradeoff between the amount of randomness required by a PCT to find the median, its run-time and its probability of failure. In section 4 we extend Reischuk's algorithm (using results due to Pippenger [6] to show the existence of randomized PCT strategies matching the bounds given in Section 3. The techniques used for both the lower and upper bounds are analogous to those used by Krizanc et al. and may be applicable to other problems.

2 Preliminaries

Valiant [11] introduced the parallel computation tree (PCT) model for studying parallelism in the classical comparison problems of maximum, median, merging and sorting. The input for each problem is a set of n elements on which a linear ordering is defined. The basic operation available to processors is the comparison of two elements. With p processors, p comparisons may be performed simultaneously in one step. Depending on which of the 2^p possible results is attained, the next set of p comparisons is chosen. The computation ends when sufficient information is discovered about the relationships of the elements to specify the solution to the given problem. The *deterministic time complexity* of a problem in this model is the number of steps required for the worst case input or the minimum depth of a tree solving the problem, as a function of the size of the input set and the number of processors used.

The model is easily extended to allow random computations. In the randomized parallel computation tree (RPCT)model, at each step we introduce a probability distribution over the choice of which p comparisons are to be performed. In this case, the *randomized time complexity* of a RPCT is the expected number of steps required on the worst case input.

We say a RPCT is *uniform* if all its random choices are generated by a fixed number of independent calls to the same random process having a finite number of outcomes (e.g., coin flips). Consider the problem, called *k-selection*, of selecting the kth largest element of an n element set. This includes the problems of finding the maximum ($k = 1$) and median ($k = \frac{n}{2}$). We

Claim: Any p processor, uniform RPCT strategy for k-selection is equivalent to a strategy $A = \langle S, D \rangle$ where

1. $S = \{s_1, s_2, \ldots\}$, a set of p processor, deterministic PCT strategies for k-selection,

2. $D = \{p_1, p_2, \ldots\}$, a probability distribution over S with $Prob(s_i) = p_i$,

3. the strategy used on input permutation π is: Choose $s \in S$ with distribution D and solve π using strategy s.

Furthermore the number of independent calls to the random process required in the worst case is the same for both strategies and therefore the number of independent random bits required in the worst case is also the same.

Proof: Let d be the maximum number of calls the given uniform RPCT makes to its underlying random process on all possible runs. Let $x_1 \cdots x_d$ be a string of d possible outcomes of the random process. Each such string defines a (not necessarily distinct) deterministic PCT strategy, $s(x_1 \cdots x_d)$, where the random choices of the uniform RPCT are governed by the outcomes of the string. Let $p(x_i)$ be the probability of x_i occurring as an outcome of the random process. Let $p(x_1 \cdots x_d) = p(x_1) \cdot \ldots \cdot p(x_d)$ be the probability of a string of d independent outcomes being $x_1 \cdots x_d$. Then the strategy having $s(x_1 \cdots x_d)$ occurring with probability $p(x_1 \cdots x_d)$ is of the required form and exactly simulates the given uniform RPCT. Furthermore the number of independent calls to the random process in the worst case is d for both strategies. □

Let $A = \langle S, D \rangle$ be a uniform RPCT strategy solving the k-selection problem. Let m_π^i be the number of steps required by the deterministic PCT, s_i, on the input π of size n. We say

$$T_q^A \leq t \Leftrightarrow \forall \pi \sum_{\{i \mid m_\pi^i \geq t\}} p_i \leq q.$$

Thus a strategy, A, has complexity $T_q^A \leq t$, if for all permutations it solves the k-selection problem within time t with probability greater than $1-q$. We denote by B^A the worst case number of independent random bits the strategy requires. The following fact is easily derived:

Fact 1 $B^A \geq \max_i \{-log\ p_i \mid p_i \neq 0\}$.□

3 A Lower Bound for k-Selection

The main result of this section is a relationship between a time-processor tradeoff for k-selection and its corresponding time-randomness tradeoff. Let $t_{\text{ksel}}(p, n)$ be the minimum over all p processor, deterministic PCT strategies for k-selection of the number of steps required by a PCT on its worst case input of size n. We prove

Theorem 1 *Let $0 < q < 1$ and let $r \geq 1$. Let $A = \langle S, D \rangle$ be a p processor, uniform RPCT strategy for k-selection with inputs of size n. Say $t_{\text{ksel}}(rp, n) \geq t$. If $T_q^A < t$ then $B^A \geq \log r - log q$.*

Proof: From any subset, S', of S of size r, we can form an rp processor, deterministic PCT strategy for k-selection by running in parallel each of the strategies in S' and terminating whenever one of the strategies terminates. Since $t_{\text{ksel}}(rp, n) \geq t$, there exists a permutation taking time greater than t for this strategy. Therefore each of the strategies in S' must take at least time t on this permutation. This implies any subset of S of size r or less must have probability totaling less than q. Therefore there must be at least $\frac{r}{q}$ strategies in S with positive probability. One of these strategies must have probability less than $\frac{q}{r}$. The result now follows from fact 1. □

Corollary 1 *Let $0 < q < 1$, let $r \geq 1$. Let $A = \langle S, D \rangle$ be a p processor, uniform RPCT strategy for finding the kth largest element of n elements. For some $c = \Theta(1)$ we have*

1. If $1 < pr < n$ *then if* $T_q^A < \frac{n}{pr} + \log\log pr - c$ *then* $B^A \geq \log r - \log q$*;*
2. If $4 \leq 2n \leq pr \leq \frac{n(n-1)}{2}$ *then if* $T_q^A < \log\log n - \log\log\frac{pr}{n} - c$ *then* $B^A \geq \log r - \log q$*.*

Proof: The results follow immediately from the bounds on finding the maximum of n elements with pr processors, $t_{1sel}(pr, n)$, given in [11] and the fact that finding the maximum of n elements is reducible to finding the kth of $n + k$ elements. □

We are particularly interested in the case of PCT's using n processors:

Corollary 2 *Let* $0 < q < 1$ *and let* $\frac{1}{\log n} \leq \epsilon \leq 1$*. Let* $A = \langle S, D \rangle$ *be an* n *processor, uniform RPCT strategy for finding the* k*th largest of* n *elements. For some* $c = \Theta(1)$ *we have, if* $T_q^A < \log(\frac{1}{\epsilon}) - c$ *then* $B^A \geq \epsilon \log n - \log q$*.* □

An examination of the proofs of theorem 1 and the lower bound in [2, 5] suggests that they both follow from the same general principle: one can trade randomness for processors. This is especially evident in theorem 1. If we keep the error probability constant, by adding one bit of randomness we get the same time lower bound for half the number of processors.

4 Upper Bounds for k-Selection

In this section, combining results of Pippenger on deterministic selection [6] and Reischuk on randomized selection [9] we prove the existence of uniform RPCTs matching the bounds given by corollary 2.

The following fact is due to Pippenger [6]:

Fact 2 *Let* $\frac{1}{\log n} \leq \epsilon \leq 1$*. There exists a deterministic PCT strategy using* $n^{1+\epsilon}$ *processors solving the* k*-selection problem for inputs of size* n *in time* $O(\log(\frac{1}{\epsilon}) + c)$ *where* $c = \Theta(1)$*.* □

Below we present a modified deterministic version of Reischuk's n processor selection algorithm which we call (k, ϵ)-SELECT. Let $X = \{x_1, \ldots, x_n\}$ be an n element set. Let k be the rank of the element of X to be selected and let $\frac{1}{\log n} \leq \epsilon \leq 1$.

Procedure (k, ϵ)-SELECT

1. Let $X' = \{x_1, \ldots, x_{n^{1-\frac{\epsilon}{6}}}\}$;
2. Define

$$t_1 = \max\{0, \lfloor k\frac{n^{1-\frac{\epsilon}{6}} + 1}{n+1} - n^{1-\frac{\epsilon}{3}}\rfloor\},$$

$$t_2 = \min\{n+1, \lceil k\frac{n^{1-\frac{\epsilon}{6}} + 1}{n+1} + n^{1-\frac{\epsilon}{3}}\rceil\};$$

3. Select s_1 equal to the element of rank t_1 of X';
4. Select s_2 equal to the element of rank t_2 of X';
5. Compare every element of X with s_1;

6. Compare every element of X with s_2;
7. Define

$$A = \{x \in X | x < s_1\},$$
$$B = \{x \in X | s_1 \leq x \leq s_2\},$$
$$C = \{x \in X | s_2 < x\};$$

8. (a) If $|A| < k$ and $|A| + |B| \geq k$ then let $Y = B$ and $k' = k - |A|$,
   (b) If $|A| \geq k$ then let $Y = A$ and $k' = k$,
   (c) If $|A| + |B| < k$ then let $Y = C$ and $k' = k - |A| - |B|$;
9. Select x equal to the k'th element of Y;
10. Output x;

End Procedure.

The correctness of the procedure is easily verified. Using Reischuk's probabilistic analysis we are able to prove

Lemma 1 *Let $\frac{1}{\log n} \leq \epsilon \leq 1$. On an input of size n, chosen uniformly at random, the algorithm (k,ϵ)-SELECT runs in time $O(\log(\frac{1}{\epsilon}) + c)$ with probability greater than $1 - \exp\{-\frac{1}{4}n^{1-\frac{\epsilon}{2}} + O(\log n)\}$, where $c = \Theta(1)$.*

Proof: Note that as far as the analysis is concerned, choosing a sample of size $n^{1-\frac{\epsilon}{6}}$ at random is equivalent to choosing the first $n^{1-\frac{\epsilon}{6}}$ elements of a randomly chosen set.

Using the algorithm of fact 2, steps 3 and 4 take $O(\log(\frac{1}{\epsilon}) + c)$ time. Steps 5 and 6 take constant time with n processors. Following Reischuk's analysis [9] we have the probability that $|Y| > n^{1-\frac{\epsilon}{6}}$ is less than $\exp\{-\frac{1}{4}n^{1-\frac{\epsilon}{2}} + O(\log n)\}$. Therefore, with probability greater than $1 - \exp\{-\frac{1}{4}n^{1-\frac{\epsilon}{2}} + O(\log n)\}$, step 9 requires $O(\log(\frac{1}{\epsilon}) + c)$ time. □

Theorem 2 *Let $2\frac{\log(5\log n)}{\log n} \leq \epsilon \leq 1$ and $\frac{1}{n} \leq q \leq \frac{1}{2}$. For the problem of selecting the kth of n elements there exists a uniform RPCT strategy, $A = \langle S, U\rangle$, with $B^A = \epsilon \log n - \log q$ for which $T_q^A = O(\log(\frac{1}{\epsilon}) + c)$, where $c = \Theta(1)$ and U is the uniform distribution assigning equal probability to each strategy in S.*

Proof: For each permutation, π, define an n processor deterministic PCT strategy s_π as follows: Perform permutation π on the input and then run (k,ϵ)-SELECT on the resulting permutation. We show that a randomly chosen set of such strategies works with non-zero probability and therefore a set, S, with the desired property exists.

We say a permutation π_1 *succeeds for* an input π_0 *in time t* if s_{π_1} runs in time t on π_0. Lemma 1 is equivalent to the statement: For a fixed permutation, π_0, a randomly chosen permutation, π_1, succeeds for π_0 in time $O(\log(\frac{1}{\epsilon}) + c)$ with probability greater than $1 - \exp\{-\frac{1}{4}n^{1-\frac{\epsilon}{2}} + O(\log n)\}$.

Claim: Let $\Pi = \{\pi_1, \ldots, \pi_{\frac{n^\epsilon}{q}}\}$ be a random set of $\frac{n^\epsilon}{q}$ deterministic PCT strategies. Let E_Π be the event, for all permutations π,

$$|\{s_{\pi_i} \in \Pi | \pi_i \text{ fails (does not succeed) for } \pi \text{ in } O(\log(\tfrac{1}{\epsilon}) + c)\}| < q|\Pi|.$$

Then $Prob(E_\Pi) > 0$.

Proof : We bound the probability of the event, $\bar{E}_\Pi$, there exists a permutation π, such that

$$|\{s_{\pi_i} \in \Pi | \pi_i \text{ fails for } \pi \text{ in time } O(\log(\tfrac{1}{\epsilon}) + c)\}| \geq q|\Pi|.$$

For a given permutation π, the probability that more than $q|\Pi|$ of the random permutations in Π fail for π is less than

$$\left(\frac{1}{e^{\frac{1}{4}n^{1-\frac{\epsilon}{2}} - O(\log n)}}\right)^{n^\epsilon} \cdot \binom{\frac{n^\epsilon}{q}}{n^\epsilon} < \frac{2^{\frac{n^\epsilon}{q}}}{e^{\frac{1}{4}n^{1+\frac{\epsilon}{2}} - n^\epsilon O(\log n)}} < \frac{1}{n!}$$

whenever

$$q \geq n^{-1}, \epsilon \geq 2\frac{\log(5\log n)}{\log n}.$$

Thus the probability there exists some permutation, among the $n!$ possible permutations, for which more than $q|\Pi|$ of the strategies in Π fail in time $O(\log(\frac{1}{\epsilon}) + c)$ is strictly less than $n! \cdot \frac{1}{n!} = 1$. That is, $\text{Prob}(E_\Pi) = 1 - \text{Prob}(\bar{E}_\Pi) > 0$.□

From the claim we may conclude that there exists a Π for which E_Π holds. This Π may be used to define A satisfying the conditions of the theorem. □

5 Conclusions

Comparing theorem 2 with corollary 2 we see that there exist uniform RPCT strategies using an optimal number of random bits and running in time optimal to within an additive constant. However, the upper bound presented here is nonconstructive. A result of Karloff and Raghavan [1] implies a constructive scheme for k-selection, A, with $B^A = O(\log n)$ and $T^A_{n^{-c}} = O(1)$. Their result (actually dealing with sorting on a PRAM) uses simple pseudo-random number generators. It does not immediately answer the problem of constructing a uniform RPCT strategy for k-selection matching the bound given in theorem 2.

We commented earlier that the lower bound in this paper and that in [2] follow from the same general principle: one can trade processors for randomness. Similarly the (nonconstructive) upper bounds in each case take the same form: reduce the amount of randomness by choosing from a subset of all objects rather than the full set. It would be interesting to find other examples of time-randomness tradeoffs in parallel computation and/or find general conditions under which such tradeoffs exist.

Acknowledgements. The author would like to thank Les Valiant, Eli Upfal, David Peleg, Allan Borodin, Mihaly Gereb and Thanasis Tsantilas for helpful discussions.

References

1. H. J. Karloff and P. Raghavan, *Randomized Algorithms and Pseudorandom Numbers*, Proc. of 20th ACM Symp. on Theory of Computing, 1988, pp. 310-321.
2. D. Krizanc, D. Peleg and E. Upfal, *A Time-Randomness Tradeoff for Oblivious Routing*, Proc. of 20th ACM Symp. on Theory of Computing, 1988, pp. 93-102.
3. N. Meggido, *Parallel Algorithms for Finding the Maximum and the Median Almost Surely in Constant-time*, Carnegie-Mellon University, Oct. 1982.
4. K. Mulmuley, *Randomized Geometric Algorithms and Pseudo-Random Generators*, Proc. of 33rd Symp. on Foundations of Computer Science, 1992, pp. 90-100.
5. D. Peleg and E. Upfal, *A Time-Randomness Tradeoff for Oblivious Routing*, SIAM J. of Computing **20**, (1989), pp. 396-409.
6. N. Pippenger, *Sorting and Selecting in Rounds*, SIAM J. of Computing **16**, (1987), pp. 1032-1038.
7. M. O. Rabin, *Probabilistic Algorithms*, in Algorithms and Complexity, Recent Results and New Directions, (J. F. Traub, Ed.), Academic Press, New York, (1976), pp. 21-40.
8. A. G. Ranade, *Constrained Randomization for Parallel Communication*, Yale Univeristy Technical Report TR-511, 1987.
9. R. Reischuk, *Probabilistic Parallel Algorithms for Sorting and Selection*, SIAM J. of Computing **14**, (1985), pp. 396-409.
10. J. Schmidt, A. Seigel and A. Srinivasan, *Chernoff-Hoeffding Bounds for Applications with Limited Independence*, Proc. of 4th Symp. on Discrete Algorithms, 1993, pp. 331-340.
11. L. G. Valiant, *Parallelism in Comparison Problems*, SIAM J. of Computing **4**, (1975), pp. 348-355.

Detecting Race Conditions in Parallel Programs that Use One Semaphore

Hsueh-I Lu   Philip N. Klein*   Robert H. B. Netzer

Department of Computer Science, Brown University, Providence, RI 02912, USA**

Abstract. We address a problem arising in debugging parallel programs, detecting race conditions in programs using a single semaphore for synchronization. It is NP-complete to detect races in programs that use many semaphores. For the case of a single semaphore, we give an algorithm that takes $O(n^{1.5}p)$ time, where p is the number of processors and n is the total number of semaphore operations executed. Our algorithm constructs a representation from which one can determine in constant time whether a race exists between two given events.

1 Introduction

Race condition detection is a crucial aspect of developing and debugging shared-memory parallel programs. Explicit synchronization is usually added to such programs to coordinate access to shared data, and *race conditions* result when this synchronization does not force concurrent processes to access data in the expected order. One way to dynamically detect races in a program is to trace its execution and analyze the traces afterward. A central part of dynamic race detection is to compute from the trace the order in which shared-memory accesses were *guaranteed* by the execution's synchronization to have executed. Accesses to the same location not guaranteed to execute in some particular order are considered a race. When programs use semaphore operations for synchronization,[1] some operations (belonging to different processes) could have potentially executed in an order different than what was traced. In this paper, we present fast algorithms for computing the order in which an execution's semaphore operations *could have* executed for the only case when it is tractable: programs that use a single semaphore. Our algorithms can be used to exactly detect race conditions in executions of such programs. Past work has shown that exactly detecting races in programs that use multiple semaphores is NP-complete [9], and has developed exact algorithms for other cases where the problem is efficiently solvable (programs that use types of synchronization weaker than semaphores) [6,8], and heuristics for the multiple semaphore case [4,7]. The complexity for the case of a single semaphore has been an open question.

Our main results are two fast algorithms, and are the first known polynomial-time algorithms that allow exact race detection in programs that use semaphores. Our goal was to solve this problem as efficiently as possible, since parallel programs are typically long-running, and the

* Research supported by NSF grant CCR-9012357 and NSF PYI award CCR-9157620, together with PYI matching funds from Thinking Machines Corporation and Xerox Corporation. Additional support provided by DARPA contract N00014-91-J-4052 ARPA Order No. 8225.

** Email addresses: {hil,pnk,rn}@cs.brown.edu.

[1] When using a semaphore, a V operation increments the semaphore, and a P operation waits until the semaphore is greater than zero and then decrements the semaphore. P operations are typically used to wait (synchronize) until some condition is true (such as a shared buffer becoming non-empty), and V operations typically signal that some condition is now true.

resulting large traces must be analyzed quickly by our algorithms. Our first algorithm determines whether any two given semaphore operations could have executed in a different order than during execution (and can be used to detect whether a race exists between any two particular events) and runs in time and space linear in the total number of semaphore operations. Our second algorithm answers this question for all pairs of operations (and can detect all races in the execution) and runs in $O(n^{1.5}p)$ time, where n is the number of operations and p is the number of processes in the execution.

Computing the order in which an execution's semaphore operations could have executed requires solving a scheduling problem. To determine whether any two operations, a and b where guaranteed to have executed in a fixed order, we must determine if some *valid* schedule of the execution's operations exists in which b precedes a. A valid schedule is an interleaving of the p processes' semaphore operations that honors the semantics of semaphore-style synchronization; i.e., a linear ordering of the operations such that at each point in the ordering, the number of V operations is never exceeded by the number of P operations (meaning that the semaphore is always nonnegative). Then, if the trace indicates that a preceded b in the actual execution, but a valid schedule exists in which b precedes a, then a and b could have executed in either order.

This scheduling problem can be reduced to a special case of *sequencing to minimize maximum cumulative cost (SMMCC)*. Given an acyclic directed graph G with costs on the nodes, a schedule is a topological ordering of the nodes; i.e. an ordering of the nodes consistent with the arcs. The cumulative cost of the first i nodes of such a schedule is just the sum of the cost of these nodes. Thus minimizing the maximum cumulative cost is an attempt to make sure that the cumulative cost stays low throughout the schedule. The SMMCC problem is NP-complete in general [1,5], but Abdel-Wahab and Kameda present an $O(n^2)$-time algorithm for the special case where G is a series-parallel graph [2] (the time bound was later improved to $O(n \log n)$ [3]). As part of this solution, they give an $O(n \log p)$-time algorithm applicable when G is a chain graph, a graph consisting of a union of p disjoint directed paths.

The problem we address can be reduced to the SMMCC problem in a chain graph augmented with one inter-chain edge. In particular, suppose we want to determine whether there is a valid schedule in which b precedes a. We add an edge from b to a, assign costs to the nodes (-1 if the node is a P-operation, $+1$ if a V-operation), and compute the minimum maximum cumulative cost. The cost is zero if there is a valid schedule, and positive otherwise. The augmented chain graph is not series-parallel, so the algorithm of Abdel-Wahab and Kameda is not applicable. We show that the SMMCC problem can nevertheless be solved in polynomial time. In fact, for the special case of interest, that in which the costs are ± 1, we give a linear-time algorithm.

Theorem 1. *Suppose G is a graph consisting of p disjoint chains comprising n nodes, where each node represents either a P-operation or a V-operation. For any two nodes a and b of G, one can determine in $O(n)$ time whether there is a valid schedule in which b precedes a.*

In the application, parallel debugging, it is important to exactly detect all races. Hence we need to determine the above for all pairs of nodes a and b. Fortunately, there is a *compact representation* of this information. The relation "a precedes b in all valid schedules" (written $a \prec b$) is a partial order consistent with the original chain graph. To represent this information, it is sufficient that we indicate, for each node a, and for each chain C not containing a, the first node b in C such that a precedes b in all valid schedules. This representation has size $O(np)$, where n is the number of nodes and p is the number of chains.

The representation can be used to determine in constant time whether there is a race between two given operations a and b. A race exists if either operation can precede the other. To determine whether b can precede a, we determine the first node in b's chain that is preceded by a in all valid

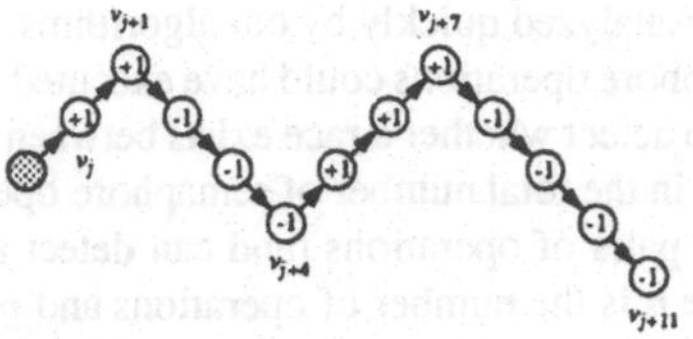

Fig. 1. A hump H of 12 nodes: $v_j, \ldots, v_{j+11}$. The cost of each node is in the circle. By definition $s(H) = -2$, $h(H) = 2$, and $\bar{h}(H) = 4$. Both of v_{j+1} and v_{j+7} are peaks of H, but only v_{j+1} is useful.

schedules. If this first node is numbered later than b, then b can precede a. If not, b cannot precede a.

We therefore consider the complexity of constructing such a representation. Clearly it can be constructed by a sequence of calls to the algorithm of Theorem 1. We show how to do much better; in fact the time required by our algorithm is only $O(\sqrt{n})$ times the time required simply to write down the output.

Theorem 2. *Suppose G is as in Theorem 1. The compact representation of the relation "a precedes b in all valid schedules" can be constructed in $O(n^{1.5}p)$ time.*

2 Definitions and Lemmas

For the remainder of the paper, we shall use G' to denote an acyclic graph, and G to denote a chain graph whose node costs are ± 1. Let $\mu(G')$ denote the lowest maximum cumulative cost of a schedule of G'. Thus we shall be interested in computing $\mu(G \cup \{ab\})$ for pairs a, b of nodes of G.

Now we introduce some terminology having to do with schedules. Much of this terminology is adapted from that in [2]. A *segment* of a schedule is a consecutive subsequence. Let $H = v_j v_{j+1} \cdots v_k$ be a sequence consisting of some of the nodes of G. The *cost* of H, denoted $s(H)$, is the sum of the costs of its nodes. The *height* of a node v_ℓ in H is defined to be the sum of the costs of the nodes v_1 through v_ℓ. The *height* of H, denoted $h(H)$, is the maximum height of any node in H. A node of maximum height in H is called a *peak*. The *reverse height* of H, denote $\bar{h}(H)$, is the height of H minus the cost of H. Note that height and reverse height are nonnegative.

A *chain* is a connected graph with all indegrees and outdegrees at most one. We are concerned primarily with graphs G consisting of the disjoint union of chains. For convenience, we assume in such a graph the existence of an *initial pseudonode*, preceding all nodes, and a *terminal pseudonode*, following all nodes, each of cost zero.

Suppose H is a chain in G' containing a peak v_ℓ such that (1) every node of H preceding v_ℓ has nonnegative height, and (2) every node of H following v_ℓ has height at least the cost of H. In this case, we call H a *hump* in G', and we say v_ℓ is a *useful peak* of H. This definition is illustrated in Figure 1.

We say a hump is an N-*hump* if its cost is negative, a P-*hump* if its cost is positive, and an L-*hump* if its cost is zero. As part of their scheduling algorithm for series-parallel graphs, Abdel-Wahab and Kameda show that in linear time a sequence of nodes can be decomposed into a series of humps. It can be proved that the humps have the following properties.

Hump decomposition properties:

- The series consists of N-humps, followed by at most one L-hump, followed by P-humps.

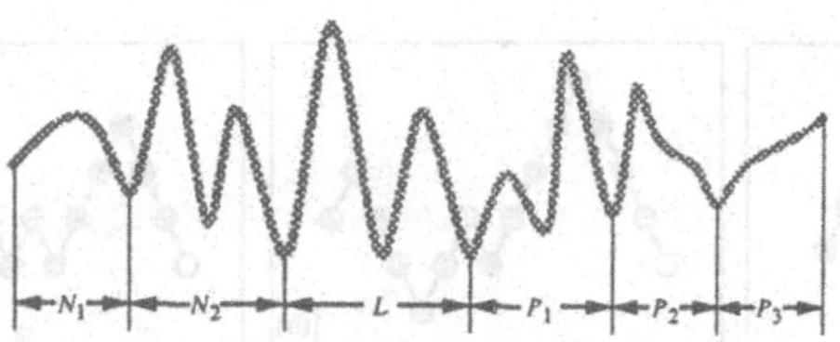

Fig. 2. A chain which is decomposed into two N-humps, one L-hump, and three P-humps.

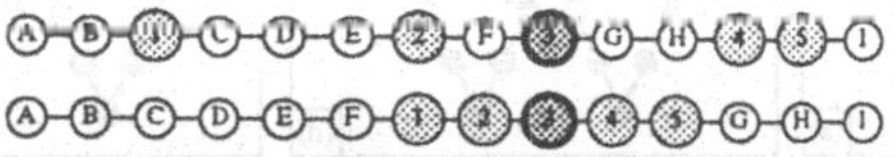

Fig. 3. The second sequence of nodes is obtained from the first one by clustering the numbered nodes to node 3.

- The N-humps are in increasing order of height, and the P-humps are in decreasing order of reverse height. Furthermore, for any two N-humps in the series, the reverse height of the one ahead is less than the height of the one behind. For any two P-humps in the series, the reverse height of the one ahead is greater than the height of the one behind.
- Every peak of the sequence occurs at or before the first P-hump.

For example, the chain in Figure 2 is decomposed into two N-humps, one L-hump, and three P-humps. When we refer to *the* humps of a chain or a graph, we mean the humps obtained by this hump decomposition process. We refer the process of carrying out this decomposition on a sequence as *constructing the humps* of the sequence. A node is called a *hump boundary* of the sequence if it is the last node of some hump of the sequence. Let $\text{hb}(S)$ denote the set of hump boundaries of some sequence S. One can prove the following property.

Hump boundary property: For sequences S_1 and S_2, we have $\text{hb}(S_1S_2) \subseteq \text{hb}(S_1) \cup \text{hb}(S_2)$. Furthermore, if $\text{hb}(S_1S_2)$ contains the last node of S_1, then $\text{hb}(S_2) \subseteq \text{hb}(S_1S_2)$.

It will turn out that once we decompose a sequence of nodes into humps, we need not be concerned with the internal structure of these humps. We need only store the heights, costs, and boundaries of the humps. Thus a sequence consisting of ℓ humps can be represented by a length-ℓ sequence of triples (height, cost, boundary). We call this sequence the *hump representation* of the sequence. Using the hump boundary properties, one can straightforwardly derive the hump representation of S_1S_2 from the hump representation of S_1 and that of S_1, and can derive the hump representation of S_2 from that of S_1S_2.

Three lemmas are useful to our results. They are all generalizations of lemmas in [2]. The first lemma concerns an operation on a schedule called *clustering* the nodes of a hump. Suppose H is a hump in G, and let v be a useful peak of H. Let S be a schedule of G. If all the nodes of H are consecutive in S, we say H is *clustered in* S. Consider the humps obtained in applying hump decomposition to the chains of G; if each of these humps is clustered in S, we say the schedule S is *clustered*. If a hump is not clustered in a schedule, we can modify the schedule to make it so. To *cluster the nodes of* H *to* v is to change the positions of nodes of H other than v so that all the nodes of H are consecutive, and the order among nodes of H is unchanged. An example is shown in Figure 3.

Lemma 3. *Let G' be an acyclic graph with costs on nodes and H be a hump in G'. Suppose S is a schedule of G'. If T is obtained from S by clustering all nodes in H to a useful peak of H, then T is a schedule of G' and $h(T) \leq h(S)$.*

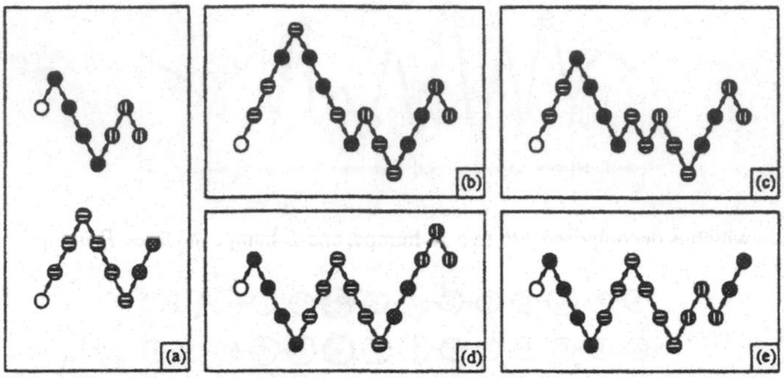

Fig. 4. (a) The graph G is composed of two chains. The first chain is composed of an N-hump followed by a P-hump. The second chain is composed of an L-hump followed by a P-hump. (b) A schedule for G of height four. (c) The schedule obtained from the previous one by clustering the N-hump to its useful peak. (d) A clustered schedule of G of height two. This one is obtained from the previous schedule by clustering every hump. (e) A clustered schedule of G of minimum height.

The proof is a modification of the proof in [2] of a similar lemma that requires G' to be a chain graph. An example is shown in Figure 4. The height of the schedule in (c) is smaller than that of the schedule in (b). It follows from Lemma 3 that there is always a clustered optimal schedule of G. Two clustered schedules of the graph in Figure 4-(a) are shown in Figure 4-(d) and (e). The latter is optimal.

Consider a series $S_1 \dots S_m$ of subsequences. It is in *standard form* if it satisfies the following properties:

Standard form properties:

- The series consists of S_i's with negative costs, followed by S_j's with zero costs, followed by S_k's with positive costs;
- The S_i's with negative costs are in nondecreasing order of height, and the S_k's with positive costs are in nonincreasing order of reverse height.

Lemma 4. *Let G' be an acyclic graph with node costs. Let A, B, S_1, S_2, and S_3 be subsequences of nodes in G'. Suppose $S = S_1AS_2BS_3$, $T_1 = S_1BAS_2S_3$, and $T_2 = S_1S_2BAS_3$. If the series of BA is in standard form then either $h(S) \geq h(T_1)$ or $h(S) \geq h(T_2)$.*

The proof of Lemma 4 is a modification of a similar lemma in [2] that requires A and B to be humps and requires S, T_1, and T_2 to be schedules of G'.

The following lemma is immediate from Lemma 4; we simply let S_2 be empty.

Lemma 5. *Let G' be an acyclic graph with node costs. Let A, B, and S_2 be subsequences of nodes in G'. Suppose $S = S_1ABS_2$ and $T = S_1BAS_2$. If the series of BA is in standard form then $h(S) \geq h(T)$.*

A schedule of G is in *standard form* if it is clustered and its series of humps in G is in standard form. Let T be any schedule of G in standard form. Recall that by Lemma 3 there is always an optimal schedule S of G which is clustered. The humps of G, while clustered both in T and in S, may not be in the same order. However, any two humps of the same chain of G must be in the same order in T and in S, else either T or S is not a schedule. Take two consecutive humps in S that are from different chains and that are not in the same order as in T, and exchange their positions. By Lemma 5, the resulting ordering has height no more than S. By a series of such exchanges, we eventually obtain T from S. It follows that T's height is no more than that of S, and hence that T is optimal. This argument shows that every schedule in standard form is an optimal schedule of G. The algorithm of Abdel-Wahab and Kameda for chain graphs G consists in decomposing the chains into humps and then merging the orderings of humps into a series of humps in standard form; they show this can be done in $O(n \log p)$ time. Let us call this algorithm the *hump-merging algorithm.*

3 The Linear-time Algorithm for a Single Pair

In this section we shall present an algorithm of computing $\mu(G \cup \{e\})$, where $e = ba$, and a and b are nodes of G. The algorithm consists of two steps. The first step merges $G \cup \{e\}$ into three chains with an extra arc. The second step computes the minimal maximum cumulative cost of the merged graph.

3.1 Merging $p - 2$ Chains into One Chain

We use $C(a)$ and $C(b)$ to denote the chains containing a and b, respectively. We assume $C(a) \neq C(b)$, for otherwise either no schedule exists or $\mu(G \cup \{e\}) = \mu(G)$, depending on the relative order of a and b. In the first step of our algorithm we use the hump-merging algorithm to find an optimal schedule C^* for the other $p - 2$ chains. Let $G^* = C(a) \cup C(b) \cup C^* \cup \{e\}$. The following theorem implies that the minimal maximum cumulative cost of $G \cup \{e\}$ is the same as that of G^*.

Theorem 6. *Let G' be an acyclic graph with node costs, and let C_1 and C_2 be two parallel chains in G'. Suppose G'' is the graph obtained from G' by merging C_1 and C_2 using the hump-merging algorithm. The minimal maximum cumulative cost of G' is equal to that of G''.*

Proof. Since G'' is more restrictive than G', clearly the cost of a schedule for G'' is at least that for G'. To prove the reverse inequality, we first use Lemma 3 to show that, without loss of generality, an optimal schedule for G' can be assumed to consist of the humps of G' in standard form. By Lemmas 4 and 5, we can assume moreover that the humps occur in the same order as in the merge of C_1 and C_2. The resulting schedule is also a schedule for G'', proving the theorem. □

The following observation is the basis for our linear-time algorithm for merging.

Observation 1 *Let C be a chain, and let y be the nth node of C. The hump decomposition algorithm applied to C yields $O(\sqrt{n})$ humps containing nodes occurring before y. Moreover, the hump decomposition applied to p chains comprising n nodes yields $O(\sqrt{np})$ humps.*

Proof. Recall that the costs of nodes are either $+1$ or -1. Hence a hump of height ℓ contains at least ℓ nodes. By the second hump decomposition property, the heights of a series of N-humps must all be different. Thus any sequence of k N-humps comprises $\Omega(k^2)$ nodes. The same property holds for P-humps, and there is at most one L-hump. This proves the first statement.

Suppose there are ℓ_i nodes in the ith chain, where $\ell_1 + \cdots + \ell_p = n$. The total number of humps is $\sum_i O(\sqrt{\ell_i})$, which can be shown to be $O(np)$. □

We now describe the algorithm for merging $p - 2$ chains. To initialize, we apply the hump decomposition algorithm to obtain hump representations for each of the chains. This takes $O(n)$

time. The remainder of the algorithm consists of $O(\log p)$ iterations. In each iteration the chains are paired up and each pair is merged into one chain. At the beginning of the ith iteration, we have the hump representation for each of $p/2^{i-1}$ chains. By Observation 1, the total number of humps is $O(\sqrt{np/2^{i-1}})$. Given a pair of chains C' and C'' having respectively n_1 and n_2 humps, we can merge the pair into a single chain and find the humps of this new chain in time $O(n_1+n_2)$. Hence the time for the ith iteration is $O(\sqrt{np/2^{i-1}})$. Therefore the time complexity of all $O(\log p)$ iterations is $\sum_i O(\sqrt{np/2^{i-1}}) = O(\sqrt{np})$, which is $O(n)$.

3.2 Finding an Optimal Schedule of Three Chains with One Arc

Let (V_1, V_2) denote an ordered partition of $V - \{a\}$ that is consistent with G (i.e. no edge of G goes from a node in V_2 to a node in V_1) and such that all nodes of $C(a)$ preceding a are in V_1 and all nodes of $C(a)$ preceded by a are in V_2. We call such an ordered partition *proper*. Note that a proper partition is determined by the last node of $C(b)$ in V_1 and the last node of C^* in V_1. We call the boundaries right after these last nodes *breakpoints*. For the case that all nodes of C^* are in V_2, the breakpoint in C^* is the boundary right before the first node of C^*. Because there are at most n possible breakpoints in $C(b)$ and at most n possible breakpoints in C^*, the number of proper (V_1, V_2)-partitions is $O(n^2)$.

We say a schedule is *consistent* with (V_1, V_2) if it has the form $S_1 a S_2$, where S_1 is an ordering of V_1 and S_2 is an ordering of V_2. Let $h(V_1, V_2)$ be the minimum height of a schedule consistent with (V_1, V_2). Clearly the minimum height of a schedule for G^* in which b precedes a is given by

$$\mu(G^*) = \min_{(V_1,V_2)} h(V_1, V_2), \tag{1}$$

where the minimum is taken over all proper partitions (V_1, V_2) of $V - \{a\}$ such that $b \in V_1$.

To find $h(V_1, V_2)$, let $G^*(V_1)$ and $G^*(V_2)$ be the subgraphs of G^* induced by V_1 and V_2, respectively. Each consists of three parallel chains. We apply the hump-merging algorithm to find optimal schedules S_1^* and S_2^* of these graphs. Then $S_1^* a S_2^*$ is a schedule of G^* consistent with (V_1, V_2), and its height is

$$\max\{\mu(G^*(V_1)),\ \ \mu(G^*(V_2)) + \sum_{v \in V_1 \cup \{a\}} s(v)\}. \tag{2}$$

Any consistent schedule of lower height would induce better schedules for $G_1^*(V_1)$ and $G^*(V_2)$, a contradiction. Thus $S_1^* a S_2^*$ is optimal, and $h(V_1, V_2)$ is given by (2).

To compute $\mu(G^*)$ according to (1), it is sufficient to consider each of the $O(n^2)$ proper partitions, and compute (2) for each. In order to achieve linear time, however, we must restrict the set of breakpoints considered. The following observation is a step in that direction.

Observation 2 *We need only consider breakpoints in $C(b)$ and C^* that are hump boundaries.*

Proof. Let S be an optimal schedule for G^*. Let us construct humps for C^* and for the subchain of $C(b)$ which is composed of nodes after b in $C(b)$. By Observation 1, each has $O(\sqrt{n})$ humps. Let T be obtained from S by clustering these humps. By Lemma 3, T is also an optimal schedule. By definition of clustering, each of these humps is consecutive in T, so none is separated by a. Thus in the schedule T, the element a must occur between two of these humps. It follows that in the equation (1), we may restrict the proper partitions to those where the breakpoints in $C(b)$ and C^* lie between these humps. □

By careful examination of these hump boundaries, we can further restrict the search for an optimal schedule. Namely, for any breakpoint in $C(b)$, we show below how to quickly determine just two breakpoints in C^* such that there is an optimal schedule using one of these two. We exploit this property in a procedure to compute the value of (1). Let $C_1(a)$ be the subchain of $C(a)$ up to but not including a, and let $C_2(b)$ be the subchain beginning after a. Let $C_1(b)$ be the subchain of $C(b)$ up to and including b, and let $C_2(b)$ be the remainder of $C(b)$. In the procedure below we assume that hump decompositions are given for C^*, $C_1(a)$, $C_2(a)$, $C_1(b)$, and $C_2(b)$. All these hump decompositions can be found in $O(n)$ time in a preprocessing step.

For the sake of our all-pairs algorithm, we present a procedure MINSCHED that takes an additional parameter. Given a node y of $C(b)$, MINSCHED computes the value of (1) where the minimum is taken over all proper partitions (V_1, V_2) where $b \in V_1$ and $y \in V_2$. Assuming the hump decompositions described above are provided, the time required is $O(\sqrt{n\ell})$, where ℓ is the number of nodes in $C(b)$ between b and x. When y is the terminal pseudonode, the value computed is the minimum height of a schedule for G^* in which b precedes a. In this case the time required is $O(\sqrt{n|C(b)|})$.

The procedure MINSCHED is as follows. Consider the humps of $C_2(b)$. By Observation 1, there are $O(\sqrt{\ell})$ hump boundaries occurring before y. We will try each of these hump boundaries as a breakpoint for $C(b)$. For each, we determine in $O(\sqrt{|C^*|})$ time two candidate breakpoints in C^*, each a hump boundary. Combining a breakpoint for $C(b)$ with a breakpoint for C^* yields a proper partition (V_1, V_2). It follows by hump boundary property that from the given hump decompositions we can obtain the humps of $C(b) \cap V_1$ and $C(b) \cap V_2$ in $O(\sqrt{|C(b)|})$ time (the hump boundaries are a subset of those appearing in the given hump decompositions). We similarly obtain the humps of $C^* \cap V_1$ and $C^* \cap V_2$. Using these humps we compute $\mu(G^*(V_1))$ and $\mu(G^*(V_2))$ in $O(\sqrt{n})$ time, and thus determine the value of (2). The total time required by the procedure is thus $O(\sqrt{\ell n})$.

We commence the proof that for any breakpoint in $C(b)$, we can determine two breakpoints in C^* that are sufficient to examine. First, we know from Observation 2 that we need only consider boundaries between humps of C^*. Every such boundary is either next to an N-hump (i.e. between two N-humps or just before the first N-hump or just after the last) or next to a P-hump, or both. We show that by scanning all the boundaries in the first category, one can be determined which is at least as good a breakpoint as all the others in that category, and similarly for the second category.

Lemma 7. *For any fixed breakpoint in $C(b)$, a breakpoint next to an N-hump can be found in $O(\sqrt{n})$ time that is just as good as all the other such breakpoints.*

The analogous lemma holds for the breakpoints next to P-humps, and is proved symmetrically. The proof of Lemma 7 relies on the following observation, which follows directly from Lemma 5.

Observation 3 *Let G' be an acyclic graph with node costs. Let S_1, S_2, A, and B be subsequences of nodes in G', where $s(B) \leq 0$. Suppose $S = S_1ABS_2$ and $T = S_1BAS_2$. If either (1) $h(A) \geq h(B)$, or (2) $s(A) \geq 0$, then $h(S) \geq h(T)$.*

Let $N_1, N_2, \ldots, N_q$ be the series of N-humps in C^*, which is in standard form. We must select one best breakpoint from among the $q+1$ occurring between these humps. For $i = 0, 1, \ldots, q$, let (W_i, W_i') be the proper partition with $N_1, \ldots, N_i$ in W_i and $N_{i+1}, \ldots, N_q$ in W_i'. We define $T_i = S^*(W_i, W_i')$. Note that $N_1, \ldots, N_q$ are in the same order in $T_0, \ldots, T_q$. Without loss of generality, we may assume that all humps other than $N_1, \ldots, N_q$ are also in the same order in $T_0, \ldots, T_q$ as well. Suppose $T_0 = H_1H_2 \cdots H_\ell a H_{\ell+1} \cdots H_m$, where each H_j, $1 \leq j \leq m$, is a hump belonging to $C_1(a)$, $C_2(a)$, $C_1(b)$, $C_2(b)$, or C^*. By definition of T_0, every hump N_i occurs after a in T_0. For $j = 0, 1, \ldots, \ell$, let A_j denote the series of humps in T_0 from H_j to H_ℓ, i.e. $A_j = H_jH_{j+1} \cdots H_\ell$.

For $k = 0, 1, \ldots, q$, let B'_k denote the portion of T_0 starting at a, and ending just before the occurrence of N_k. Let B_k be B'_k but with all humps N_i omitted. Thus, if N_i is the j_i^{th} hump of T_0 for $i = 1, \ldots, q$, then

$$B_k = aH_{\ell+1} \cdots H_{j_1-1}H_{j_1+1} \cdots H_{j_2-1}H_{j_2+1} \cdots H_{j_k-1}.$$

Note that B_k is a segment of each of $T_{k-1}, T_k, \ldots, T_q$. The following four claims help us relate the height of T_k to the heights of T_{k-1} and T_{k+1}.

Claim 1 *If there exists* $j \leq \ell$ *such that* $h(H_j) \geq h(N_k)$, *then* $h(T_{k-1}) \geq h(T_k)$.

Proof. Since $h(A_j B_k) \geq h(H_j) \geq h(N_k)$, by the first condition of Observation 3

$$h(A_j B_k N_k) \geq h(N_k A_j B_k)$$

. Because $h(H_j) \geq h(N_k) > h(N_{k-1}) > \cdots > h(N_1)$, all of $N_1, \ldots, N_{k-1}$ must be scheduled before H_j in T_{k-1}. It follows that $A_j B_k N_k$ is a segment of T_{k-1}. Note that $N_k A_j B_k$ is a segment of T'_k where the (V_1, V_2)-partition of T'_k is the same as that of T_k. Hence $h(T_{k-1}) \geq h(T'_k) \geq h(T_k)$. □

Claim 2 *If there exists* $j \leq \ell$ *such that* $s(A_j B_k) \geq 0$, *then* $h(T_{k-1}) \geq h(T_k)$.

Proof. Since $s(A_j) + s(B_k) = s(A_j B_k) \geq 0$ and $s(B_k) < 0$, we have $s(A_j) > 0$. This implies that at least one P-hump exists in A_j. Let $H_{j'}$ be the first P-hump in A_j. We know that $s(A_{j'}) \geq s(A_j)$. Hence $s(A_{j'} B_k) \geq 0$. By the second condition of Observation 3, we have $h(A_{j'} B_k N_k) \geq h(N_k A_{j'} B_k)$. Because $H_{j'}$ is a P-hump, all of $N_1, \ldots, N_{k-1}$ must be scheduled before $H_{j'}$ in T_{k-1}. It follows that $A_{j'} B_k N_k$ is a segment of T_{k-1}. Note that $N_k A_{j'} B_k$ is a segment of T'_k where the (V_1, V_2)-partition of T'_k is the same as that of T_k. Hence $h(T_{k-1}) \geq h(T'_k) \geq h(T_k)$. □

Claim 3 *If for every* $1 \leq j \leq \ell$, $h(H_j) < h(N_k)$ *and* $s(A_j B_k) < 0$, *then* $h(T_{k-1}) \leq h(T_k)$.

Proof. There is a $1 \leq j' \leq \ell$ such that $A_{j'} B_k N_k$ is a segment of T_{k-1} and $N_k A_{j'} B_k$ is a segment of T_k. Since $s(A_{j'} B_k) < 0$, if we can prove that $h(N_k) > h(A_{j'} B_k)$, then $h(T_{k-1}) \leq h(T_k)$ by the first condition of Observation 3. Note that $h(A_{j'} B_k) = \max\{h(A_{j'}), s(A_{j'}) + h(B_k)\}$. Let us show that $h(N_k)$ is actually greater than both elements of the maximum.

Suppose H_{j^*} is the hump containing the last peak of $A_{j'}$. Namely $h(A_{j'}) = s(H_{j'}) + s(H_{j'+1}) + \cdots + s(H_{j^*-1}) + h(H_{j^*})$. By the third hump decomposition property, H_{j^*} cannot be later than the first P-hump. This implies that none of $H_{j'}, H_{j'+1}, \ldots, H_{j^*-1}$ is a P-hump. Thus $s(H_{j'}) + s(H_{j'+1}) + \cdots + s(H_{j^*-1}) \leq 0$. It follows that $h(N_k) > h(H_{j^*}) \geq h(A_{j'})$.

By the second hump decomposition property and the fact that B_k is composed of N-humps, we have $h(N_k) > \bar{h}(B_k) = h(B_k) - s(B_k)$. Since $s(A_{j'}) + s(B_k) = s(A_{j'} B_k) < 0$, it follows that $h(N_k) > h(B_k) + s(A_{j'})$. □

Claim 4 *If for every* $1 \leq j \leq \ell$, $h(H_j) < h(N_k)$ *and* $s(A_j B_k) < 0$, *then for every* $1 \leq j \leq \ell$, *we have* $h(H_j) < h(N_{k+1})$ *and* $s(A_j B_{k+1}) < 0$.

Proof. Since N_{k+1} is a N-hump, $h(H_j) < h(N_k) < h(N_{k+1})$. Note that

$$B_{k+1} = B_k H_{j_k+1} H_{j_k+2} \cdots H_{j_{k+1}-1},$$

where $H_{j_k+1}, H_{j_k+2}, \ldots, H_{j_{k+1}-1}$ are all N-humps. It follows that $s(B_{k+1}) \leq s(B_k)$. Hence $s(A_j B_{k+1}) \leq s(A_j B_k) < 0$. □

Input: $C_1, C_2, \ldots, C_p$.
Output: For every C_i and C_j, for every node v in C_i, output the arc vw where w is the first node in C_j such that $v \prec w$.

Step 1 Construct humps for every possible pair of subchains.
Step 2 For every $1 \leq i \leq p$ do
Step 3 Build a merge tree T_i^* from all chains except C_i.
Step 4 For every $1 \leq j \leq p$ do
Step 5 From T_i^* compute $C_{i,j}^*$, the chain merged from all chains except C_i and C_j.
Step 6 For every node v in C_i do
Step 7 Find the first node w^* in C_j such that $v \prec w^*$.
Step 8 Output the arc vw^*.

Fig. 5. The $O(n^{1.5}p)$-time algorithm of computing the $\prec$ relation among every pair of nodes.

Now we can prove Lemma 7. Let $1 \leq k \leq q$ be the least number such that for every $1 \leq j \leq \ell$, $h(H_j) < h(N_k)$ and $s(A_j B_k) < 0$. By Claim 4, for every $k < i \leq q$, $h(H_j) < h(N_i)$ and $s(A_j B_i) < 0$. By Claim 3 we know

$$h(T_{k-1}) \leq h(T_k) \leq h(T_{k+1}) \leq \cdots \leq h(T_q). \tag{3}$$

Note that for every $0 \leq i \leq k-1$, either $h(H_j) \geq h(N_i)$ or $s(A_j B_i) \geq 0$. By Claim 1 and 2,

$$h(T_0) \geq h(T_1) \geq \cdots \geq h(T_{k-2}) \geq h(T_{k-1}). \tag{4}$$

It follows from (3) and (4) that the boundary between N_{k-1} and N_k is one of the best breakpoints among N-humps of C^*. Note that if such a k does not exist, then the boundary right after N_q is as least as good as all the other boundaries among N-humps.

Let $H_{j'}$ be the highest hump among all H_j's, $1 \leq j \leq \ell$. In particular, if $h(H_{j'}) < h(N_k)$ then $h(H_j) < h(N_k)$ for all $1 \leq j \leq \ell$. Let $H_{j''}$ be the hump that $s(H_1 H_2 \cdots H_{j''-1})$ is the smallest. Since $s(H_1 H_2 \cdots H_\ell)$ is fixed, $s(A_{j''})$ is the largest among all A_j's. This implies that if $s(A_{j''} B_k) < 0$ then $s(A_j B_k) < 0$ for all $1 \leq j \leq \ell$. In order to locate the best breakpoint between among N-humps of C^*, we just iteratively scan the N-humps. For each N-hump N_k in turn, we test if $h(H_{j'}) < h(N_k)$ and $s(A_{j''} B_k) < 0$ are both true. Since only $q+1$ iterations are required, we need only $O(q)$ time. The lemma follows by using Observation 1 to show that $q = O(\sqrt{n})$.

4 The Algorithm for All Pairs

Recall that G is composed of p chains of n nodes. In this section we shall show how to decide the $\prec$ relation for all pairs of nodes in G. The linear-time algorithm for a single pair of nodes, applied to all $O(n^2)$ pairs, takes $O(n^3)$ time. We present a faster algorithm, one that runs in time $O(n^{1.5}p)$. The algorithm is outlined in Figure 5. We elaborate it in the following subsections.

Step 1: Let ℓ_i be the number of nodes in C_i. For every chain C_i, we construct humps for every possible pair of subchains C_i' and C_i'' where $C_i = C_i' C_i''$. There are two passes, one for C_i' and the other for C_i''. The first pass is as follows. To initialize, let C_i' be the hump composed of the first node of C_i. We then execute a loop of $\ell_i - 1$ iterations. In the kth iteration, we add the

$(k+1)^{st}$ node of C_i to C_i', and then reconstruct humps for C_i'. The second pass is very similar to the first one. Initially let C_i'' be the hump composed of the last node of C_i. In the kth iteration, we add the $(k+1)^{st}$ last node of C_i to C_i'', and then reconstruct humps for C_i''.

Since the lengths of C_i' and C_i'' are always less than or equal to ℓ_i, the number of humps constructed in every iteration is always bounded by $O(\sqrt{\ell_i})$. Whenever a new node is added to C_i', it takes only $O(\sqrt{\ell_i})$ time to reconstruct humps for the lengthened subchain. It follows that the time complexity of Step 1 is $\sum_{1\le i\le p} O(\ell_i\sqrt{\ell_i}) = O(n^{1.5})$.

Step 3: In this step the algorithm constructs a rooted binary tree of chains to help implement step 5. The leaves of the tree are all the chains except the ith (in hump representation). Every nonleaf vertex is (the hump representation of) the chain resulting from merging the vertex's children. Thus the root is the chain resulting from merging all the original p chains except the ith.

By the same analysis as in Subsection 3.1, building a merge tree takes $O(\sqrt{np})$ time. Note that the humps of every chain in the merge tree are constructed at the same time. The hump representation uses $O(1)$ space for every constructed hump. Since there are $O(\sqrt{np})$ humps in this merge tree, the space used for the tree is only linear.

Step 5: For a given i and j, we must compute C_{ij}^*. It is easy to see that there is a set of $O(\log p)$ vertices of the binary tree constructed in step 2 such that the leaves below these vertices comprise all chains except the ith and the jth. We compute the amortized complexity of Step 5. Fix a value of i. In step 3 the algorithm constructed a merge tree whose leaves were all the chains except the ith. We use this merge tree in executing step 5, namely computing C_{ij}^*, for $j = 1, \ldots, p$. We show that the total time required is $O(p\sqrt{n})$. Hence the amortized complexity of step 5 is $O(\sqrt{n})$.

For each value of j, we recompute the vertices of the merge tree whose chains include the jth chain, working from the leaf to the root. Once the root has been recomputed, we obtain the chain merged from all chains but the ith and the jth, i.e. C_{ij}^*. Thus the root gets recomputed a total of p times, each of its children gets recomputed $p/2$ times, and so on. In general, each vertex at level k gets recomputed $p/2^k$ times.

Now we evaluate the sum of times required for recomputing the vertices at level k. The total time required is proportional to the number of humps in chains belonging to vertices at level $k+1$. The number of chains at level $k+1$ is 2^{k+1}, so the number of humps is $O(\sqrt{n2^{k+1}})$. This, then, is the time required for recomputing all the vertices at level k, once each. Since each vertex at level k gets recomputed $p/2^k$ times, the total time invested in recomputing vertices at level k is $O((p/2^k)\sqrt{n2^{k+1}})$, which is $O(p\sqrt{n/2^k})$. Finally, we sum over all levels k, obtaining $O(p\sqrt{n})$, as required.

Steps 6–8: Our goal is to determine for each v in C_i the first node w^* in C_j such that $v \prec w$. For the recursion, we address a more complicated goal, in which we are given an interval in which to seek w^*. Namely, given a node v in C_i, and given two nodes x and y of C_j such that w^* is guaranteed to be between them, find w^*.

First we give a recursive procedure BINSEARCH(v, x, y) based on binary search. Let w be the node of C_j located midway between x and y. We use the procedure MINSCHED of Section 3 to determine the value of (1) where the minimum is taken over all proper partitions (V_1, V_2) where $w \in V_1$ and $y \in V_2$. If the value is zero, we conclude that there is a valid schedule in which w precedes v, and hence $v \not\prec w$. It follows that w^* is located between w and y. We therefore recurse on this interval.

If the value is positive, we claim that there is no valid schedule in which w precedes v, i.e. no schedule in which w precedes v whose maximum cumulative cost is zero. For suppose there were such a schedule S. By our assumption that w^* lies between x and y, we have $v \prec y$, so there is no valid schedule in which y precedes v. In particular, v precedes y in S. Let V_1 be the set of nodes preceding v in S, and let V_2 be the set of nodes following v. Then (V_1, V_2) is a proper partition where $w \in V_1$ and $y \in V_2$, and we have that $h(V_1, V_2)$ is no more than the cost of S, which is zero. This contradiction proves our claim. Hence $v \prec w$, and we can recurse on the interval between x and w.

The time required by MINSCHED is $O(\sqrt{n\ell})$, where ℓ is the number of nodes in the interval from x to y. The length of that interval is halved in each recursion, so the time required by BINSEARCH is a geometric series and hence is also $O(\sqrt{n\ell})$.

If Step 6 were just a simple FOR-loop, then the time complexity of Step 6–8 would be $O(\ell_i\sqrt{n\ell_j})$. By using divide-and-conquer for Step 6–8, however, we achieve time $O(\sqrt{n\ell_i\ell_j})$ as follows. For the recursion, we are given an interval $[p, q]$ of C_i and an interval $[x, y]$ of C_j. For every v in the first interval, we need to find the corresponding w^*. We are given that for every such v, the corresponding w^* lies in the interval $[x, y]$ of C_j. The procedure is as follows. Let v be the node of C_j midway between p and q, and call BINSEARCH(v, x, y) to find the corresponding w^*. It then follows that for every v' between p and v, the corresponding w'^* lies between x and w^*, and for every v' between v and q, the corresponding w'^* lies between w^* and y. We therefore recurse with these intervals. Let ℓ_i be the number of nodes in the interval $[p, q]$, and let ℓ_j be the number of nodes in the interval $[x, y]$. The recurrence for the time required by the procedure is

$$T(1, \ell_j) \leq c\sqrt{n\ell_j}$$
$$T(\ell_i, \ell_j) \leq c\sqrt{n\ell_j} + \max_{0\leq\ell\leq\ell_j} \{T(\ell_i/2, \ell) + T(\ell_i/2, \ell_j - \ell)\}.$$

It can be proved that therefore the time complexity is $O(\sqrt{n\ell_i\ell_j})$.

Overall Complexity: The complexity of the algorithm is dominated by the time for steps 6–8, which is $\sum_{i,j} O(\sqrt{n\ell_i\ell_j})$. This can be shown to be $O(n^{1.5}p)$, proving Theorem 2.

References

[1] Abdel-Wahab, H. M., "Scheduling with Application to Register Allocation and Deadlock Problems," University of Waterloo, PhD Thesis, 1976.

[2] Abdel-Wahab, H. M. & Kameda, T., "Scheduling to Minimize Maximum Cumulative Cost Subject to Series-parallel Precedence Constraints," *Operations Research* 26 (1978), 141–158.

[3] ———, "On Strictly Optimal Schedules for the Cumulative Cost-Optimal Scheduling Problem," *Computing* 24 (1980), 61–86.

[4] Emrath, P. A., Ghosh, S. & Padua, D. A., "Event Synchronization Analysis for Debugging Parallel Programs," *Supercomputing '89* (November 1989), 580–588.

[5] Garey, M. R. & Johnson, D. S., *Computers and Intractability—A Guide to the Theory of NP-Completeness*, W. H. Freeman and Company, 1979.

[6] Helmbold, D. P. & McDowell, C. E., "A Class of Synchronization Operations that Permit Efficient Race Detection," *University of California at Santa Cruz Technical Report* (January 1993).

[7] Helmbold, D. P., McDowell, C. E. & Wang, J-Z., "Analyzing Traces with Anonymous Synchronization," *International Conference on Parallel Processing* (August 1990), II70–II77.

[8] Netzer, R. H. B. & Ghosh, S., "Efficient Race Condition Detection for Shared-Memory Programs with Post/Wait Synchronization," *International Conference on Parallel Processing* (August 1992), II242–II246.

[9] Netzer, R. H. B. & Miller, B. P., "On the Complexity of Event Ordering for Shared-Memory Parallel Program Executions," *International Conference on Parallel Processing* (August 1990), II93–II97.

An Algorithm for Finding Predecessors in Integer Sets

Bruce Maggs[1] and Monika Rauch[2]

[1] NEC Research Institute, 4 Independence Way, Princeton, NJ 08540.
[2] Siemens ZFE BT SE 14, D-81730 München, Germany.

Abstract. This paper presents a data structure that supports the operations of inserting and deleting elements drawn from a universe $U = \{0 \ldots u-1\}$ into a set S, and for finding the predecessors of elements of U in S. We consider both random inputs and worst-case inputs. In the case of random inputs, we show that each operation can be performed in constant time, with high probability, provided that the total number of operations is at most polynomial in u. In Yao's cell probe model, the algorithm uses $O(n)$ space, where n is the maximum size that S achieves during the sequence of operations. The algorithm can be implemented with the same performance in the RAM model with word size $O(\log u)$ at the cost of performing u^ϵ preprocessing operations, and using u^ϵ additional space, for any fixed $\epsilon > 0$. The algorithm can be modified so that even in the worst case no operation takes more than $O(\log \log u)$ time, at the cost of using $O(u)$ additional space.

1 Introduction

This paper presents an algorithm for implementing the following operations on a set S drawn from a universe $U = \{0, 1, \ldots, u-1\}$:

1. *insert*(x): insert x into S, if $x \notin S$,
2. *delete*(x): delete x from S, if $x \in S$,
3. *pred*(x): find the largest $y \in S$, such that $y \leq x$, or indicate that no such y exists.

Several algorithms for implementing these operations with good worst-case running time are already known. Van Emde Boas [5] gave an algorithm with $O(\log \log u)$ worst-case time per operation using $O(u)$ space. We call this data structure the *VEB data structure*. Also, using dynamic perfect hashing a randomized algorithm can be constructed with expected time $O(\log \log u)$ per operation using $O(n)$ space, where n is the maximum size achieved by S [4]. Ajtai [1] proved

This research was conducted while the second author was at the Department of Computer Science, Princeton University, and at NEC Research Institute.

a non-constant lower bound on the query time of any algorithm that works for arbitrary size n and uses space polynomial in n in Yao's cell probe model [6]. Neither the VEB data structure nor the data structure based on dynamic perfect hashing have optimal performance in the case of random inputs.

In the case of random inputs, it is relatively straightforward to design a data structure for which the expected time to perform each insert and delete operation is constant. The simplest of these have poor worst-case performance, however, and even in the case of random inputs some operations are likely to take more than constant time. In Section 3 we describe one of these algorithms.

The algorithm in this paper performs well on both worst-case and random inputs. In the case of random inputs, every operation takes only constant time, with high probability, provided that the total number of operations is at most polynomial in u. In the case of worst-case inputs, the time per operation is $O(\log \log u)$. The algorithm uses as subroutines an algorithm due to Ajtai, Fredman and Komlós [2] (AFK) to handle random inputs and the VEB algorithm to handle worst-case inputs. We describe the AFK algorithm in Section 2.

1.1 The Worst-Case and Random Input Models

In this paper we analyze the behavior of the algorithm on both worst-case and randomly-chosen inputs. For worst-case inputs, an adversary is allowed to perform any set of m insert, delete, or predecessor operations. For randomly-chosen inputs, the adversary prepares a sequence of length m indicating which type of operation (insert, delete, or predecessor query) is to be performed at each step, *before any of these operations are actually performed.* The adversary does not specify which elements are involved in insert and delete operations; they are chosen at random when the operations are executed. An insert operation chooses an element at random from $U - S$ and inserts it into S. A delete operation chooses an element at random from S and deletes it from S. When a predecessor query operation is executed, the adversary can examine the data structure and then choose the element for the query. The goal of the adversary is to maximize the time required to perform the slowest operation.

1.2 Outline

The remainder of this paper is organized as follows. In Section 2 we describe the AFK algorithm. In Section 3 we describe an algorithm with good performance in the case of random inputs. Although this algorithm has poor worst-case behavior, and some operations are likely to take more than constant time, it introduces some of the ideas used later in the paper. In Section 4 we describe an algorithm with good worst-case and random-case behavior. We conclude in Section 5 with some comments on possible extensions to the results in this paper.

2 AFK Encodings

The AFK algorithm implements the insert, delete, and predecessor query operations in constant time in Yao's cell probe model provided that $n = O(\log u / \log\log u)$. It represents the set S by a table that contains the elements of S and a $\log u$-bit number that identifies a compressed trie. We refer to this representation of a set as the *AFK encoding* of the set. A query is answered by searching the trie. Each leaf in the trie has a pointer to an element in the table. Upon reaching a leaf, the pointer from that leaf is followed to the table. In Yao's cell probe model, only two probes are needed to implement a query: one to read the identifier of the trie, and the other to access the element in the table. No probes are required to search the trie in this model. An element is inserted by first adding it to the end of the table, and then computing a new trie that contains the element as a leaf with a pointer to the element's position in the table. A deletion removes an element from the table, then moves the last element in the table into the empty slot. It then computes a new trie in which the pointer to the element that has been moved is updated, and in which the deleted element does not appear. The number of different tries is at most $n!\,4^n(\log u)^{n-1}$ [2], so that if $n = O(\log u / \log\log u)$, each trie can be identified with only $\log u$ bits.

Although Ajtai, Fredman, and Komlós described their algorithm in the cell probe model, it can also be implemented in the RAM model. To implement their algorithm on a RAM with word size $O(\log u)$, we first precompute all possible operations on S and store them in a *look-up table*. The inputs to each operation are a trie and an element. The output of each operation is either another trie or an element. Since there may be as many as u tries and u different elements, the look-up table must have u^2 entries. Each entry consists of an element or a trie, depending on the operation. The time to precompute the look-up table is $O(u^2 \log u / \log\log u)$, since the time to compute the outcome of any operation is at most the size of S, $O(\log u / \log\log u)$.

Both the space and the precomputation time can be reduced to u^ϵ for any fixed $\epsilon > 0$ using the following trick. The universe is split into $\sqrt{u}$ subuniverses, each of which contains $\sqrt{u}$ consecutive elements. At the top level, we maintain a table whose entries are pointers to non-empty subuniverses. The table also contains $\frac{1}{2}\log u$ bits that identify the top-level trie. At the bottom level, there is a table for each non-empty subuniverse. An operation is implemented by examining the high-order $\frac{1}{2}\log u$ bits of an element and searching the top-level trie to determine which subuniverse the element or its predecessor belongs to, and then completing the operation in the table for the subuniverse. Notice that at each level only $\frac{1}{2}\log u$ bits are necessary to specify an element, and only $\frac{1}{2}\log u$ bits are necessary to specify a trie. Thus, precomputing all possible operations on tries and elements takes only $O(u \log u / \log\log u)$ time and requires only $O(u)$ space for the look-up table. Applying this trick increases the time required to implement each operation by a constant factor, increases the space required at run-time by at most a constant factor, and decreases the size of the largest set that can be stored by a constant factor. However, the time remains constant, the space remains $O(n)$, and the size of the set remains $c'(\log u/\log\log u)$, for some constant c'. By applying this trick a constant number of times, the precomputation time and the space needed for the look-up table can be reduced to u^ϵ for any fixed $\epsilon > 0$.

It is also not difficult to extend the AFK algorithm (in the RAM model) to the case $n = (\log u)^c$ for any constant c using a similar trick [3]. For example, by breaking the n keys in S into $\sqrt{n}$ groups of $\sqrt{n}$ keys each and using a two-level scheme to represent S, it is possible to store a set of size $c''(\log u/\log\log u)^2$ for some constant c''.

It is also possible to implement the operation of splitting a set S into two equal-sized sets in constant time using precomputation and table look-up, provided that S contains at most $(\log u)^c$ elements.

The performance of the AFK encoding is summarized by the following theorem.

Theorem 1. *For any fixed constants $c > 0$ and $\epsilon > 0$, there exists an AFK encoding that can implement the insert, delete, and predecessor query operations in constant time provided that the maximum size of the set S does not exceed $(\log u)^c$. The data structure uses u^ϵ space and requires u^ϵ time for preprocessing.*

3 A Simple Algorithm with Good Random-Case Behavior

In this section we describe a simple algorithm with good expected performance in the case of random inputs for arbitrarily large sets S.

The algorithm periodically rebuilds the data structure. Suppose that S contains k elements at the time that the data structure is rebuilt. The backbone of the data structure is an array a containing k pointers. With each element $a[i]$ in the array, we associate a sequence of u/k consecutive elements in U. We call such a sequence an *interval*. An insert, delete, or predecessor operation begins by examining the corresponding array pointer.

The ith element in the array points to a (possibly empty) linked list containing any elements of S that lie in the corresponding interval. An insert operation is implemented by inserting the new element at the front of the linked list. A delete operation is performed by searching the linked list. A predecessor operation is performed by first searching the linked list for a predecessor, and then, if necessary, searching the array for an element with a smaller index that points to a non-empty linked list, and searching that list. The algorithm rebuilds the data structure whenever the number of elements in S rises to $2k$ or drops to $k/2$.

Each time the array is rebuilt, the time spent rebuilding is at most proportional to the number of operations since the last rebuild.

It is not difficult to show that the expected time to perform each randomly-chosen insert or delete operation is constant (not counting the cost of periodically rebuilding the data structure) and the expected time to perform m insert and delete operations is $O(m)$ (including the cost of rebuilding the data structure). The algorithm also uses $O(n)$ space, where n is the maximum size achieved by S. In the worst case, however, m operations may take as much as $\Omega(mn)$ time, and even in the case of random inputs, it is likely that the adversary can choose queries that will take more than constant time. Using an idea described in Section 4.3, it is possible to move the rebuilding operations "to the background" so that at most a constant amount of time is spent rebuilding between operations. Thus, the expected time per insert or delete operation can be made constant, including the cost of rebuilding.

4 The Algorithm

This section presents an algorithm for implementing insert, delete, and predecessor query operations with good performance in the case of both random and worst-case inputs. We begin in Section 4.1 by describing an algorithm (called Algorithm A) in which the cost of each random insert, delete, or query operation is constant, with high probability. In Algorithm A, however, the data structure must periodically be rebuilt, and rebuilding takes more than constant time. We analyze Algorithm A in Section 4.2. In Section 4.3 we present a refinement of Algorithm A (called Algorithm B) in which rebuilds are performed in the background, so that no operation takes more than constant time, with high probability. Both Algorithm A and Algorithm B use $O(n + u^{\epsilon})$ space. In Section 4.4, we show how to modify Algorithm B so that no operation takes more than $O(\log\log u)$ time in the worst-case. The resulting algorithm is called Algorithm C. Algorithm C uses $O(u)$ additional space. In the case of random inputs, Algorithm C has the same performance as Algorithm B.

4.1 Algorithm A

As in Section 3 we build a data structure that is rebuilt periodically. To be more precise let k be the number of elements in the data structure during the last rebuild. The data structure is rebuilt if $k \leq 10$ or after $k/10$ operations since the last rebuild. Algorithm A maintains an array a of $k/(\gamma \log u)$ pointers, where γ is a constant that will be determined later. Associated with each array entry is an *interval* of $\frac{u}{k}\gamma \log u$ consecutive elements of U. The array entry points to a *bucket* that stores any elements of S from the corresponding interval. Note that the expected number of elements in each bucket is $\gamma \log u$.

Whenever the data structure is rebuilt, we create a bucket for each interval and make each pointer point to the bucket containing the elements of S in its interval. The subset s of S stored in a bucket is represented by an AFK encoding if its size is at most $20\gamma \log u$. If the size of s is larger than $20\gamma \log u$, then it is represented by a linked list. To build either representation takes time linear in the size of the set. Thus, the time for the rebuild is proportional to the number of elements in S during the rebuild.

To answer a *pred(x)* operation, we first look up the entry $a[\lfloor x/\lceil \frac{u}{k}\gamma \log u\rceil \rfloor]$, i.e., $a[i]$ can be accessed by the items from the interval $i\lceil \frac{u}{k}\gamma \log u\rceil, \ldots, (i+1)\lceil \frac{u}{k}\gamma \log u\rceil - 1$, called *interval(i)* for $0 \leq i \leq \frac{k}{\gamma \log u} - 1$. Let b be the bucket that is pointed to by $a[i]$. If b is not empty, we look for the predecessor of x in the subset s stored in b (either by using the look-up table for the AFK encoding or by examining the list for a linked list). If x is smaller than any element in s or if b is empty, then we search the array for a pointer to a non-empty bucket that precedes b and then search that bucket.

An *insert(x)* operation inserts x into the bucket b that contains the elements of S that lie in the same interval as x. In case of an AFK encoding this can be done by a table look-up in constant time. In case of a linked list the element is added to the beginning of the list. During an insertion, we switch from an AFK encoding to a list representation if the subset s stored in b becomes larger than $20\gamma \log u$. To determine the elements stored in the AFK encoding we execute $O(\log u)$ predecessor and delete operations on the AFK encoding. Then we construct a linked list of elements. This takes time $O(\log u)$. All further operations use the linked list. In Section 4.2 we show that it is very unlikely that a bucket is ever represented by linked list.

A *delete(x)* operation is similar to an insertion. First we determine the bucket b to which the interval of x is mapped. Then we remove x from b either by table look-up for an AFK encoding or by examination of the linked list. Note that a delete operation never changes the representation of a bucket.

4.2 Analysis of Random Inputs to Algorithm A

In this section we show that in the case of random inputs, every operation can be implemented in constant time (not counting the time spent rebuilding the data structure), with high probability, provided that the number of operations is at most polynomial in u. In the case of random inputs, the adversary presents in advance a list of length m indicating what type of operation is to be performed at each step, where $m = u^b$ for some constant b. If the operation is an insertion or deletion then the element inserted or deleted is not known until the operation is actually executed, at which point it is chosen at random. In the case of a query, the adversary can choose the element when the operation is executed.

The following lemma shows that at any time all sets S of the appropriate size are equally likely to be chosen.

Lemma 2. *After i insertions and d deletions each set of $i-d$ elements of U is equally likely to be stored in S.*

Proof. By symmetry. □

Let $X_{i,t}$ denote the number of elements in the ith bucket after t operations. The following lemma bounds the probability that $X_{i,t}$ is large.

Lemma 3. *For all i and t, $P(X_{i,t} \geq c) \leq 2^{-c}$ for $c \geq \frac{11}{5}e^2\gamma \log u$.*

Proof. Let Y_i denote the number of elements in S at the last rebuild before operation i. Then

$$\begin{aligned} P(X_{i,t} \geq c \mid Y_i = k) &\leq \frac{\binom{\frac{u}{k}\gamma \log u}{c}\binom{u-c}{\frac{11}{10}k-c}}{\binom{u}{\frac{11}{10}k}} \\ &= \frac{\binom{\frac{u}{k}\gamma \log u}{c}\binom{\frac{11}{10}k}{c}}{\binom{u}{c}} \\ &\leq \left(\frac{\frac{11}{10}e^2\gamma \log u}{c}\right)^c \\ &= 2^{-c}, \end{aligned}$$

for $c \geq \frac{11}{5}e^2\gamma \log u$. The first inequality is derived by observing that if S contained k elements at the last rebuild before operation i, then it can contain at most $11k/10$ elements after i operations, and by Lemma 2 all distinct sets of $11k/10$ elements are equally likely to be stored in S. The second line is derived using the equation $\binom{a}{b}\binom{b}{c} = \binom{a-c}{b-c}\binom{a}{c}$. The third line is derived using the inequalities $\left(\frac{a}{b}\right)^b \leq \binom{a}{b} \leq \left(\frac{ae}{b}\right)^b$. □

The following lemma bounds the probability that the ith bucket is empty after t operations.

Lemma 4. $P(X_{i,t} = 0) \leq e^{-\frac{9}{10}\gamma \log u}$.

Proof. Let Y_i denote the number of elements in S at the last rebuild before operation i. Then

$$\begin{aligned} P(X_{i,t} = 0 \mid Y_i = k) &\leq \frac{\binom{u-\frac{u}{k}\gamma \log u}{\frac{9}{10}k}}{\binom{u}{\frac{9}{10}k}} \\ &= \frac{(u - \frac{u}{k}\gamma \log u) \times \cdots \times (u - \frac{u}{k}\gamma \log u - \frac{9}{10}k + 1)}{u \times \cdots \times (u - \frac{9}{10}k + 1)} \end{aligned}$$

$$\leq \left(1 - \frac{\gamma \log u}{k}\right)^{\frac{9}{10}k}$$
$$\leq e^{-\frac{9}{10}\gamma \log u}.$$

The first inequality is derived by observing that if S contained k elements at the last rebuild before operation i, then it contains at least $9k/10$ elements after i operations, and by Lemma 2 all distinct sets of $9k/10$ elements are equally likely to be stored in S. □

Lemma 5. *For any fixed constants $\delta > 0$, $b > 0$, and $\epsilon > 0$, there is a constant $\gamma > 0$ such that Algorithm A uses $O(n + u^\epsilon)$ space and the probability that any of the first u^b operations takes more than constant time is at most $u^{-\delta}$ (not counting the cost of periodically rebuilding the data structure).*

Proof. An operation takes more than constant time only if some bucket grows to contain more than $20\gamma \log u$ elements or if a bucket becomes empty. By Lemmas 3 and 4, the probability that the ith bucket is either too big or empty after t operations is at most $2^{-20\gamma \log u} + e^{-\frac{9}{10}\gamma \log u}$. By assumption, the total number of operations is u^b for some constant b, and there are a total of at most u buckets at any step. Thus, by making γ large enough we can ensure that the probability that any bucket is too large or empty at any time step is at most $u^{-\delta}$ for any fixed constant $\delta > 0$. □

4.3 Algorithm B

One of the disadvantages of Algorithm A is that it must occasionally stop and rebuild the data structure, which takes more than constant time. Algorithm B will build a new data structure "in the background" so that no operation will ever take more than constant time, with high probability.

The basic idea is that the algorithm maintains three data structures (called *tables*) simultaneously. One table is used for answering queries and is called the *current* table. The other two tables are kept in case the number of elements becomes either too large or too small for the current table. One is called the *small* table and the other is called the *large* table. Let k be the number of elements in the set when the current table came into use. Then $k/10$ operations will be performed using the current table. While these operations are being performed, the large and small tables are constructed. The large table is designed to accomodate $11k/10$ elements (i.e., it has $\frac{11}{10}k/\gamma \log u$ buckets), and the small table is designed to accomodate $9k/10$ elements. For the first $k/20$ operations, a constant amount of work is performed after each operation to reclaim the space used by the two previously constructed tables that are not the current table. For the next $k/20$ operations, a constant amount of work is performed constructing the new large and small tables. Since the tables take $O(k)$ time to construct, they will be ready after these $k/20$ operations are performed. Also, during these $k/20$ operations, every insert or delete operation is performed simultaneously on all three tables. If after $k/10$ operations the set S contains fewer than k elements, then the small table becomes the current table. If S contains k or more elements then the large table becomes the current table.

Theorem 6. *For any fixed constants $\delta > 0$, $b > 0$, and $\epsilon > 0$, there is a constant $\gamma > 0$ such that Algorithm B uses $O(n + u^\epsilon)$ space and the probability that any of the first u^b operations takes more than constant time is at most $u^{-\delta}$.*

Proof. The proof is essentially identical to the proof of Lemma 5. One minor difference is that although the small table is designed to hold $9k/10$ elements, the set S may contain as many as $k - 1$ elements when it becomes the current table. Similarly, although the large table is designed to hold $11k/10$ elements, S may contain only k elements when it comes the current table. However, this overloading or underloading makes only small differences in the constants in the proofs of Lemma 3, Lemma 4, and Lemma 5. □

4.4 Algorithm C

In this section we describe an extension to Algorithm B that guarantees that no operation takes more than $O(\log\log u)$ time in the worst case. The new algorithm, called Algorithm C, uses a VEB data structure as an auxiliary data structure. As a consequence, it requires $O(u)$ additional space.

The main difference between Algorithm B and Algorithm C is in the way overflowing buckets and empty buckets are handled. In Algorithm C, when the number of elements in a bucket exceeds $20\gamma \log u$, the bucket is split into two buckets of size $10\gamma \log u$, and the smallest element in each of these buckets is inserted into the VEB data structure. A pointer is stored from each of these two elements to the AFK encodings of the corresponding buckets. The array entry for the corresponding interval is marked so that all future operations on the interval take place in the VEB data structure. Furthermore, once an element has been inserted into the VEB data structure, all rebuilds of the data structure stop, and only the current table is used for the duration of the algorithm.

If a bucket becomes empty, then the corresponding array entry is marked, and, if the bucket for the preceding interval has not already been marked, then this bucket is inserted into the VEB data structure and its array entry is marked. This guarantees that if a bucket is empty, then the predecessor for the elements in the corresponding interval has been inserted into the VEB data structure.

A *pred*(x) operation is implemented as in Algorithm B, except that it searches the VEB data structure whenever it encounters a marked array entry. Upon reaching a leaf of the VEB data structure, the algorithm searches the AFK encoding of the bucket stored at that leaf.

An *insert*(x) operation is implemented as in Algorithm B, with the following exceptions. If the array entry for the interval containing x is marked, then the VEB data structure is searched, and the element is inserted into the AFK encoding of the bucket that is found at that leaf. If the new element is smaller than the smallest previously stored element in the interval, then the old smallest element is deleted from the VEB data structure, and the new smallest element is inserted. A pointer is stored from the new smallest element to the AFK encoding of the bucket. If the number of elements in a bucket exceeds $20\gamma \log u$, then the smallest element in the bucket is deleted from the VEB data structure (if necessary) and the bucket is split into two equal-sized buckets whose smallest elements are both inserted into the VEB data structure.

A *delete*(x) operation is handled as follows. If the array entry is marked, then the VEB data structure is searched, and the element is deleted from the bucket that is found at the leaf. If x is the smallest element in the bucket, then it is deleted from the VEB data structure, and the new smallest element is inserted (unless the bucket is now empty). A pointer is stored from the new smallest element to the AFK encoding of the bucket. On the other hand, if the array entry is not marked, then the deletion is handled as in Algorithm B. If the bucket becomes empty, then we insert the bucket of the preceding interval into the VEB data structure, as described above.

The following theorem summarizes the performance of Algorithm C.

Theorem 7. *Algorithm C uses $O(u)$ space. In the case of random inputs, for any fixed constants $\delta > 0$ and $b > 0$, there is a constant $\gamma > 0$ such that the probability that Algorithm C uses more than constant time to perform any of the first u^b operations is at most $u^{-\delta}$. In the worst case, Algorithm C completes every operation in $O(\log\log u)$ time.*

Proof. In the case of randomly-chosen inputs, Algorithm C has the same performance as Algorithm B, provided that no bucket becomes empty or overflows. Thus, the random-case performance of Algorithm C is given by Theorem 6. In the worst case, since each operation performs only a constant number of operations in the VEB data structure, the time for each operation is $O(\log\log u)$. □

5 Extensions

This section describes some possible extensions to the results in this paper.

In this paper, the adversary generating random inputs is a little weak. Before any of the operations are executed, the adversary must present a list of length m indicating what type of operation is to be performed at each step. The insert and delete operations are then executed without further input from the adversary. One way to strengthen the adversary is to allow the adversary to choose the type of each operation after seeing the results of the previous operations. As before, once the type of operation is determined, the actual element inserted or deleted is chosen at random. With this additional power, the adversary can bias the probability that a particular set S is created. Nevertheless, we conjecture that the time bounds given for the algorithm of Section 4 hold even in this stronger model if the number of operations is polynomial in u.

Algorithm C uses the VEB algorithm as a subroutine, but does not exploit any special properties of this data structure. In fact, the performance of Algorithm C can be made to match the performance of the data structure with the best worst-case performance and space utilization by using that data structure instead.

6 Acknowledgements

The authors would like to thank Arne Andersson for pointing out a simple data structure with good performance in the case of random inputs and for finding several problems in an earlier version of this paper. We would also like to thank Bernard Chazelle, Brandon Dixon, Sandy Irani, and Bob Tarjan for many fruitful comments.

References

1. M. Ajtai. A lower bound for finding predecessors in Yao's cell probe model. *Combinatorica*, 8(3):235–247, 1988.
2. M. Ajtai, M. Fredman, and J. Komlós. Hash Functions for Priority Queues. *Information and Control*, 63:217–225, 1984.
3. B. Chazelle. CS-593C Lecture Notes. Department of Computer Science, Princeton University. Spring 1989.
4. K. Mehlhorn and St. Näher. Bounded Ordered Dictionaries in $O(\log\log N)$ time and $O(n)$ Space. *Info. Proc. Lett.*, 35(4):183–189, 1990.
5. P. van Emde Boas. Preserving order in a forest in less than logarithmic time and linear space. *Info. Proc. Lett.*, 6(3):80–82, 1977.
6. A. Yao. Should tables be sorted. *J. Assoc. Comput. Mach.*, 28(3):615–628, 1981.

The Exhaustion of Shared Memory: Stochastic Results

Robert S. Maier*[1] and René Schott[2]

[1] University of Arizona, Dept. of Mathematics, Tucson, Arizona 85721, U.S.A.
[2] Université de Nancy 1, C.R.I.N. and INRIA-Lorraine, F-54506 Vandœuvre-lès-Nancy, France

Abstract. We analyse a model of exhaustion of shared memory. The memory usage of a finite number of dynamic data structures is modelled as a Markov chain, and the asymptotics of the expected time until memory exhaustion are worked out, in the limit when memory availability and memory needs scale proportionately, and are taken to infinity. This stochastic model subsumes the model of colliding stacks previously treated by the authors, and gives rise to difficult mathematical problems. However, analytic results can be obtained in the limit. Our analysis uses a technique of matched asymptotic expansions introduced by Naeh et al. [11]. The technique is applicable to other stochastically modelled discrete algorithms.

1 Introduction

Central to the probabilistic analysis of dynamic data structures, and storage allocation generally, is the following question. What is the probability that over a specified time interval, the size of a data structure exceeds some specified bound? In real-world applications of dynamic data structures this question arises in the context of possible memory exhaustion. The *expected time until memory exhaustion* may be used to quantify the performance of a stochastically modelled on-line algorithm.

In models of dynamic data structures it is sometimes possible to investigate the expected memory exhaustion time combinatorially, using generating function techniques [2]. Recently, Kenyon-Mathieu and Vitter [4] and Louchard, Kenyon and Schott [7] have investigated maximum data structure size using probabilistic methods. There has however been little work on the exhaustion of *shared* memory, or on 'multidimensional' exhaustion, where one of a number of inequivalent resources becomes exhausted.

In this paper we go beyond previous work by studying the exhaustion of shared storage: we consider the interaction of q independent processes $P_1, \ldots, P_q$, each with its own memory needs. We allow the processes to allocate and deallocate r different, non-substitutable resources (types of memory): $R_1, \ldots, R_r$.

* Supported in part by the U.S. National Science Foundation under grant NCR-90-16211.

In the context of a very general stochastic model, we employ formal asymptotic expansions [5, 11] to estimate the time until memory exhaustion.

We model resource limitations, and define memory exhaustion, as follows. At any time s, process P_i is assumed to have allocated some quantity $y_i^j(s)$ of resource R_j. (Both time and resource usage are taken to be discrete, so that $s \in \mathbb{N}$ and $y_i^j(s) \in \mathbb{N}$.) Process P_i is assumed to have some maximum need m_{ij} of resource R_j, so that

$$0 \leq y_i^j(s) \leq m_{ij} \tag{1}$$

for all s. m_{ij} may be infinite; if finite, it is a hard limit which the process P_i never attempts to exceed. The resources R_j are limited, so that

$$\sum_{i=1}^{q} y_i^j(s) < m_j \tag{2}$$

for $m_j - 1$ the total amount of resource R_j available for allocation. Resource exhaustion occurs when some process P_i issues an unfulfillable request for a quantity of some resource R_j. Here 'unfulfillable' means that fulfilling the request would violate one of the inequalities (2).

The state space Q of the memory allocation system is the subset of $\mathbb{N}^{qr}$ determined by (1) and (2). This polyhedral state space is familiar: it is used in the banker's algorithm for deadlock avoidance. However most treatments of deadlocks (see, e.g., Habermann [3]) assume that processes request and release resources in a mechanical way: a process P_i requests increasing amounts of each resource R_j until the corresponding goal m_{ij} is reached, then releases resource units until $y_i^j = 0$, and repeats. (The r different goals of the process need not be reached simultaneously, of course.) This is a powerful assumption: it facilitates a classification of system states into 'safe' and 'unsafe' states, the latter being those which can lead to deadlock. It is however an idealization, especially in the context of dynamic data structures.

In this paper we shall analyse a *stochastic* model for transitions within Q. We assume that regardless of the system state, each process P_i with $0 < y_i^j < m_{ij}$ can issue either an allocation or deallocation request for resource R_j. The probabilities of the different sorts of request may depend on the current state vector (y_i^j). In other words we take the state of the storage allocation system as a function of time to be a multidimensional Markov chain; this is an alternative approach which goes at least as far back as Ellis [1]. In such a model states may not be classified unequivocally as safe or unsafe.

Our goal is the estimation of the memory exhaustion time τ, if initially the r types of resources are completely unallocated: $y_i^j = 0$ for all i, j. We are particularly interested in *asymptotics*: the consequences of expanding the resource limits $m_j - 1$ and the per-process maximum needs m_{ij} (if finite) on the expected time to exhaustion. For this we must specify the transition probabilities of the Markov chain. First we make the realistic assumption that at each time $s \in \mathbb{N}$ and for each pair (i, j), the change $\xi_i^j \stackrel{\text{def}}{=} \Delta y_i^j$ in the usage of resource R_j by process P_i due to allocations or deallocations has a *negative expectation*, provided of course that $y_i^j > 0$. (The somewhat less realistic case of positive expectation has been considered by Yao [13].) Under this assumption the Markov chain manifests a sort of stability: the initial state is an equilibrium state, although large fluctuations away from it occasionally occur, and eventually end in memory exhaustion. We also assume that the q processes behave identically, and that their needs for the r different resources are equal. These last assumptions make the mathematics symmetric: for all i, j we have $m_i = m'$ and $m_{ij} = m''$ for some m', m''.

At each time $s \in \mathbb{N}$ the state vector y_i^j will change by ξ_i^j, an independent instance of a $\mathbb{Z}^{qr}$-valued random increment vector. We assume that at each time s, exactly one request is issued. There is probability $(qr)^{-1}$ of any process $P_{i'}$ issuing a request concerning any resource type $R_{j'}$, on account of symmetry. If this occurs, the random vector ξ_i^j will equal $\xi\delta_{i,i'}\delta_{j,j'}$, in which ξ is a $\mathbb{Z}$-valued random variable signifying the requested change in resource allocation. We allow the distribution of ξ to depend on the current state as follows: it is some specified function of $y_{i'}^{j'}/m'$, the amount of resource $R_{j'}$ currently allocated to process $P_{i'}$ as a fraction of the total amount available. This yields a very general but reasonable model: each process evolves independently, and its behavior *vis-à-vis* some resource (the probabilities of its issuing allocation and deallocation requests for the resource) depends only on that fraction of the resource which it has currently allocated.

Actually the preceding description of the distribution of ξ_i^j can hold only in the interior of Q; near the boundary the transition probabilities must differ. For example if $y_i^j = m''$, then ξ_i^j cannot be positive. In effect there must be 'reflecting boundary conditions' on the faces of Q determined by (1), just as there are final states, or 'absorbing boundary conditions' on the exhaustion faces determined by (2).

We shall answer the following 'scaling up' question: if for certain positive α' and α'' we take $m' = \lfloor N\alpha' \rfloor$, $m'' = \lfloor N\alpha'' \rfloor$, what are the large-$N$ asymptotics of $\mathsf{E}\tau$, the expected time until exhaustion of one of the resources occurs? Scaling up signifies increasing both resource limits and (if present) per-process maximum resource needs, but keeping process *dynamics* fixed: the distribution of ξ, the random variable signifying a requested change in resource usage (either positive or negative), does not depend directly on N.

A very special case of our model would set

$$\mathsf{P}\{\xi = k\} = \begin{cases} p, & k = 1; \\ 1-p, & k = -1 \end{cases} \tag{3}$$

for some fixed $p < \frac{1}{2}$. With this choice, p would be the probability of an allocation request (for one resource unit) and $1-p$ the probability of a deallocation. The model defined by (3), with two processes and one resource, has a long history. Knuth ([6], Ex. 2.2.13) proposed a model very similar to it: a model of *colliding stacks* in which two stacks (heights y_1^1 and y_2^1) are allowed to grow from opposite sides of a fixed block of memory, and memory exhaustion occurs when the stacks collide. In his model m'' was effectively infinite, so the state space Q was a triangle in $\mathbb{N}^2$, with two reflecting edges ($y_1^1 = 0$, $y_2^1 = 0$) and a memory exhaustion edge ($y_1^1 + y_2^1 = m'$, the total memory available).

The colliding stacks model has been treated by Louchard and Schott [8] and by Maier [9]; the treatment of Maier covers the case in which the transition probabilities are nontrivially state-dependent. In addition, Louchard and Schott have treated the case of finite m''. If $m'/2 < m'' < m'$ then Q becomes a pentagon, with sides $y_1^1 = 0$, $y_2^1 = 0$, $y_1^1 = m''$, $y_2^1 = m''$, $y_1^1 + y_2^1 = m'$. When ξ is distributed according to (3) they showed that $\mathsf{E}\tau$ grows to leading order *exponentially* as $N \to \infty$, and worked out the growth rate; they obtained asymptotics of higher order as well. They also showed that in the large-N limit, the distribution of the system state across the exhaustion edge of the pentagon, when memory exhaustion finally occurs, is *uniform*. (Flajolet [2] had earlier derived this combinatorially for the Knuth model.)

That $\mathsf{E}\tau$ grows exponentially in N is intuitively obvious, but the uniform distribution over exhaustion states is counterintuitive. Maier [9] showed that this distribution is an artifact: if for each process the probability p of issuing an allocation request is allowed to depend on the fraction of memory currently allocated, very different behavior may obtain. If p is a *decreasing* function of this fraction (so that the model is 'increasingly contractive', with large fluctuations away from the initial state strongly suppressed), then the final state will be concentrated close to the center of the exhaustion face of Q.

In Section 2 we review the asymptotic techniques of Matkowsky, Schuss and coworkers [5, 11], and in Sections 3 and 4 apply them to the general model: q and r arbitrary, and the distribution of the increment ξ_i^j, for any pair (i, j), allowed to depend in some specified way on y_i^j/m''; equivalently, on y_i^j/N. Our results are summarized in the sequence of theorems beginning with Theorem 2. We obtain precise large-N asymptotics of $\mathsf{E}\tau$ if transition probabilities are constant, and somewhat less precise asymptotics for increasingly contractive models. The technique introduced in Ref. [11] allows, in principle, a *complete* determination of the asymptotics of the mean exhaustion time $\mathsf{E}\tau$: not merely the rate of exponential growth with N, but also the pre-exponential factor. We also comment on the limiting distribution over exhaustion states.

The problem of general q and r was left open in Ref. [8], and is now largely solved. Conclusions, and a brief discussion of the applicability of our approach to other stochastically modelled algorithms, appear in Section 5.

2 Absorption Time Theory

We summarize the ideas of Matkowsky, Schuss *et al.*, adapted to the case of Markov chains. (Their earlier paper Ref. [5] concentrated on Markov chains; their recent key paper Ref. [11] deals with the exit problem for continuous-time Markov processes.) Their technique of 'matched asymptotic expansions' is precisely what is needed to obtain the large-N asymptotics of $\mathsf{E}\tau$.

Let $\mathcal{Q} \subset \mathbb{R}^d$ be a *normalized state space*: we assume that $\mathcal{Q}$ is closed, connected and bounded with a sufficiently well-behaved boundary, and that $0 \in \mathcal{Q}$. Define $Q \subset \mathbb{Z}^d$, an N-dependent state space, to be the set of all d-tuples of integers $(y_1, \ldots, y_d)$ such that $N^{-1}(y_1, \ldots, y_d) \in \mathcal{Q}$. We introduce a $\mathbb{Z}^d$-valued increment random variable ξ, whose distribution is parametrized by the current normalized state $x \in \mathcal{Q}$. A Markov chain $y(0), y(1), \ldots$ on Q is defined by specifying that any time s, $\Delta y(s) \stackrel{\text{def}}{=} y(s+1) - y(s)$ has the same distribution as $\xi(y/N)$. Since we want the state 0 to be a point of stable equilibrium, we require that the 'mean drift' field $\mathsf{E}\xi(\cdot)$ on $\mathcal{Q}$ have 0 as a stable fixed point.

We must specify appropriate boundary conditions for this Markov chain. Some parts of the boundary of $\mathcal{Q}$, which we denote $\partial\mathcal{Q}$, are taken to be 'absorbing.' The corresponding states in Q are viewed as final states; whenever the system enters such a state, its evolution terminates. Also, some portions of the boundary may be 'reflecting': the transition probabilities of the chain, i.e., the distribution of ξ, are modified there so as to prevent departure from $\mathcal{Q}$.

For any N, Q is a finite set. The transition matrix of the Markov chain on Q, $\mathbf{T} = (T_{yz}) \in \mathbb{R}_+^{Q \times Q}$, is a substochastic matrix whose elements follow from the distribution of ξ. It is strictly substochastic (i.e., $\sum_{z \in Q} T_{yz} \le 1$ for all y rather than $\sum_{z \in Q} T_{yz} = 1$ for all y) on account of the absorption on the boundary. By standard Perron-Frobenius theory [10], $\mathbf{T}$ has a left eigenvalue $\lambda_1 > 0$ of maximum modulus, and a corresponding positive left eigenvector $\rho = (\rho_y) \in \mathbb{R}_+^Q$. If $\mathbf{T}$ were stochastic, this eigenvector would be interpreted as the (unnormalized) stationary density of the chain, and the eigenvalue would equal 1. But since it is not, $\lambda_1 < 1$ and ρ is interpreted as a 'quasi-stationary' density: the distribution of the system state over Q, if one conditions on the event that after a very long time, absorption on the boundary has failed to occur. (One expects [11] that ρ is strongly concentrated near the point of stable equilibrium 0, much like a stationary density.) $1 - \lambda_1$ is a *limiting* (or *quasi-stationary*) *absorption rate*: we shall see that it falls to zero exponentially in N. Other eigenvalues of $\mathbf{T}$, it turns out, are well separated from λ_1 as $N \to \infty$, so we are justified in approximating the mean absorption time $\mathsf{E}\tau$ by $(1-\lambda_1)^{-1}$ in the large-N limit. The technique of Ref. [11] is really a technique for computing the large-N asymptotics of $1 - \lambda_1$.

The idea is to approximate the quasi-stationary eigenvector ρ_y, when N is large, in three different regions: (1) near $y = 0$, where most of the probability is concentrated, (2) along certain trajectories between 0 and the absorbing boundary, and (3) in a neighborhood of the absorbing boundary. By matching these three expressions together a consistent set of approximations to ρ is obtained. Surprisingly, to leading order in N (as $N \to \infty$) it is possible to do all this without knowing the exponentially small quantity $1 - \lambda_1$: in effect, λ_1 may be taken equal to unity throughout. After ρ is sufficiently well approximated, the large-N asymptotics of $1 - \lambda_1$ are computed as an asymptotic absorption rate:

$$1 - \lambda_1 \sim \frac{\sum_{y \in Q}(1 - \sum_{z \in Q} T_{yz})\rho_z}{\sum_{y \in Q} \rho_y} . \tag{4}$$

Here $(1 - \sum_{z \in Q} T_{yz})\rho_y$ is the amount of probability absorbed at state y per time step, on account of the substochasticity; it is nonzero only for y sufficiently near the absorbing boundary.

The numerator in Eq. (4) will typically be much smaller than the denominator (which is simply a normalization factor) because ρ_y, for y near the absorbing boundary, will when N is large be exponentially smaller than ρ_y, for y near the point of stable equilibrium. In fact the denominator may be computed from the approximation to ρ_y in Region 1 alone. The numerator is computed from the approximation to ρ_y in Region 3, and the approximation in Region 2 merely serves to ensure consistency between the approximations in Regions 1 and 3.

3 The $q = 1, r = 1$ Model

We first apply the approach of the last section to a one-dimensional model: the $q = 1$, $r = 1$ case of the memory exhaustion model of Section 1. This is a model of a single process P_1 and a single resource R_1 available in limited quantities. The process will allocate and deallocate resource units singly, with the respective probabilities allowed to depend on the fraction of the resource which is currently allocated.

To model this, consider a normalized state space $\mathcal{Q} = [0, \alpha]$. The corresponding N-dependent state space will be $Q = \{0, \ldots, \lfloor N\alpha \rfloor\}$. (There is clearly no need to distinguish between α' and α'' here.) Let the random increment variable $\xi = \Delta y$, representing a change in memory utilization, have discrete density given by (3), but with p allowed to be a function of the normalized state $x \overset{\text{def}}{=} y/N$. We assume $0 < p(x) < \frac{1}{2}$ for all $x \in \mathcal{Q}$, so the state 0 (the initial state) is stable. The state $\lfloor N\alpha \rfloor$ is the only final state.

The transition probabilities T_{yz} of the Markov chain on Q must be modified when $y = 0$ or $y = \lfloor N\alpha \rfloor - 1$. The initial state 0 is reflecting; we take $T_{00} = 1 - p_0$, $T_{01} = p_0$, for some parameter $p_0 \in (0, 1]$ which specifies how 'hard' the reflection is. (In the computing context p_0 is the probability that the process P_1, if the resource has been completely deallocated, will immediately issue an allocation request.) And since the state $\lfloor N\alpha \rfloor$ is final and probability is 'lost' at state $\lfloor N\alpha \rfloor - 1$, we set $T_{\lfloor N\alpha \rfloor - 1, \lfloor N\alpha \rfloor} = 0$. With these choices the model becomes identical to the one-stack model of Maier [9].

Consider the quasi-stationary density ρ, approximated as the solution of $\rho\mathbf{T} = \rho$, in Region 1. In this region $y = o(N)$, i.e., $x = o(1)$. To leading order $\mathsf{P}\{\Delta y = +1\} = p(0)$ and $\mathsf{P}\{\Delta y = -1\} = 1 - p(0)$; it is clear that

$$\rho_y = \begin{cases} [p(0)/(1 - p(0))]^y, & y > 0; \\ p(0)/p_0, & y = 0 \end{cases} \tag{5}$$

satisfies $\rho\mathbf{T} = \rho$. The parameter p_0 affects ρ_0 only.

The behavior of ρ_y in Region 3 is similar. Here $\lfloor N\alpha \rfloor - y = o(N)$, i.e., $\alpha - x = o(1)$. As in Region 1, transition probabilities are essentially constant: $\mathsf{P}\{\Delta y = +1\} = p(\alpha)$ and $\mathsf{P}\{\Delta y = -1\} = 1 - p(\alpha)$. We set $\rho_{\lfloor N\alpha \rfloor} = 0$ since probability, once absorbed in the final state, never returns. The approximation

$$\rho_y = C_3 \left([p(\alpha)/(1 - p(\alpha))]^y - [p(\alpha)/(1 - p(\alpha))]^{\lfloor N\alpha \rfloor} \right) \tag{6}$$

will satisfy $\rho\mathbf{T} = \rho$. C_3 will be chosen to ensure consistency between (5) and (6).

To ensure consistency between the approximations to ρ in Regions 1 and 3, we use a special sort of approximation in the intermediate Region 2. We take

$$\rho_y = K(y/N) \exp\left(-NW(y/N)\right) . \tag{7}$$

This sort of approximate, but asymptotically correct, solution to an eigenvector (or eigenfunction) equation is traditionally called a 'WKB solution' by applied mathematicians ([11]; see Section 3 of Ref. [5] for the Markov chain case). The two functions $K(\cdot)$ and $W(\cdot)$ on Q are determined by the x-dependence of the density of ξ. With the choice (3), it suffices to take

$$W(x) = \int_0^x \log[(1-p(z))/p(z)]\, dz \tag{8}$$

$$K(x) = C_2[p(x)(1-p(x))]^{-1/2} \tag{9}$$

with C_2 some constant. It is an elementary, though tedious, exercise to verify that with these choices for $W(\cdot)$ and $K(\cdot)$ we have $\rho\mathbf{T} = \rho$, i.e.,

$$\rho_{y-1}p((y-1)/N) + \rho_{y+1}(1-p((y+1)/N)) = \rho_y \tag{10}$$

to leading order in N as $N \to \infty$.

(For a general $d = 1$ model, with ξ not restricted to ± 1 values, the formulæ for $W(x)$ and $K(x)$ would differ; in general $W(x) = \int_0^x p^*(z)\, dz$, with $p^* = p^*(x)$ defined as the positive solution to the implicit equation $\mathsf{E}e^{p^*\xi(x)} = 1$. See Maier [9] for more.)

The WKB solution (7) is of exactly the form needed to interpolate between (5) and (6). It has asymptotics

$$\rho_y \sim C_2[p(0)(1-p(0))]^{-1/2}[p(0)/(1-p(0))]^y, \quad y = o(N) \tag{11}$$

$$\rho_y \sim C_2[p(\alpha)(1-p(\alpha))]^{-1/2}e^{-NW(\alpha)}[p(\alpha)/(1-p(\alpha))]^{y-\lfloor N\alpha\rfloor} \tag{12}$$

with (12) holding when $y = \lfloor N\alpha \rfloor - o(N)$. Equating coefficients in (5) and (11) gives $C_2 = [p(0)(1-p(0))]^{1/2}$. Comparison between (6) and (12) shows that the WKB solution matches up with the *first* term in (6); the second (constant) term is important only near the boundary. And by equating coefficients in (6) and (12) we get

$$C_3 = [K(\alpha)/K(0)]e^{-NW(\alpha)}[p(\alpha)/(1-p(\alpha))]^{-\lfloor N\alpha\rfloor} . \tag{13}$$

The approximations (5) and (6) may now be substituted into the denominator and numerator of (4) respectively. (The numerator comprises only one term, since $1-\sum_{z\in Q} T_{yz}$ is nonzero only if $y = \lfloor N\alpha\rfloor - 1$, in which case it equals $p(\alpha)$.) A bit of algebra yields a simple expression for $1-\lambda_1$, and using the fact that $\mathsf{E}\tau \sim (1-\lambda)^{-1}$ in the large-N limit we get the following. (We assume that $N\alpha$ is an integer; otherwise trivial changes are necessary.)

Theorem 1. *In the one-process, one-resource (i.e., $q = r = 1$) memory model, when memory units are allocated and deallocated singly the mean time until exhaustion has asymptotics*

$$\mathsf{E}\tau \sim \sqrt{\frac{p(0)p(\alpha)(1-p(\alpha))}{1-p(0)}}(1-2p(\alpha))^{-1}\left(\frac{1}{1-2p(0)}+\frac{1}{p_0}\right)\exp\left(NW(\alpha)\right) \quad (14)$$

as $N \to \infty$. The exponential growth rate $W(\alpha)$ may be computed from (8).

So the exponential growth rate is $W(\alpha)$, and to compute it one must know the extent to which the allocation probability depends on normalized state x. (If the constraint of single allocations and deallocations is dropped, the growth rate generalizes to $\int_0^\alpha p^*(z)\,dz$.) The pre-exponential factor is however more complicated, and depends on the hardness parameter p_0 of the model.

The formula (14), in slightly different notation, appears as Theorem 8.1 of Maier [9]. But the present derivation is much shorter and cleaner.

4 Analysis of the General Model

By building on the results of the preceding section, we can analyse the Markov shared memory exhaustion model with an arbitrary number of processes q and resources (i.e., memory types) r. We shall see that the technique of Section 2 yields the complete asymptotics of $\mathsf{E}\tau$. It can in fact be viewed as an extension of the Wentzell-Freidlin approach used by Maier [9], which yields only the exponential growth rate.

First consider the case of a single resource. The state space and normalized state space are polyhedra; respectively

$$Q = \{\, y \in \mathbb{N}^q : 0 \le y_i \le \lfloor N\alpha'' \rfloor,\ \sum_{i=1}^{q} y_i \le \lfloor N\alpha' \rfloor \,\}, \quad (15)$$

$$\mathcal{Q} = \{\, x \in \mathbb{R}^q : 0 \le x_i \le \alpha'',\ \sum_{i=1}^{q} x_i \le \alpha' \,\}. \quad (16)$$

(We assume $q\alpha'' > \alpha'$, so the exhaustion face of $\mathcal{Q}$ is nonempty.) The system state evolves as follows. A process P_i is selected equiprobably from the set $\{P_1, \ldots, P_q\}$, and the corresponding y_i is incremented by ξ_i. As in the $q = 1$ model, the distribution of ξ_i may depend on $x_i \stackrel{\text{def}}{=} y_i/N$.

We first consider the case when ξ_i has density given by (3), with no dependence at all on x_i; $p < \frac{1}{2}$, so the state 0 is stable. This is the case of single allocations and deallocations, with their respective probabilities taken independent of the number of resource units currently allocated. We must modify the transition probabilities on the reflecting faces of Q; in particular, the faces where one or more of the y_i equals zero. If the process P_i has no quantity of the resource allocated, we take $\xi_i = \Delta y_i$ equal to $+1$ with probability p_0, and to 0 with probability $1-p_0$. Here $p_0 \in (0,1]$ is a 'hardness' parameter, as in the $q=1$ model: it specifies the probability that complete deallocations are immediately followed by allocations.

This choice of reflecting boundary conditions makes the random processes $y_i(s)$ as independent as possible of one another near the equilibrium state 0. The quasi-stationary density ρ in Region 1, approximated as the solution of $\rho\mathbf{T}=\rho$, will be

$$\rho_y = (p/p_0)^{|y|_0}[p/(1-p)]^{\sum_{i=1}^{q} y_i} \tag{17}$$

which generalizes the constant-p case of (5). Here $|y|_0$ is the number of zero components of the vector $y=(y_1,\ldots,y_q)$, and the parameter p_0 affects ρ_y only if $|y|_0>0$.

In the intermediate Region 2, which is the interior of Q, a WKB solution is easily constructed. First note that if $p(x)$ is independent of x, the $q=1$ WKB solution of (7), (8) and (9) specializes to

$$\rho_y = C_2'[p/(1-p)]^y \tag{18}$$

for C_2' some constant. (This solution clearly satisfies (10) in the interior of Q.)

$$\rho_y = C_2'[p/(1-p)]^{\sum_{i=1}^{q} y_i} \tag{19}$$

is the generalization to the $q>1$ case.

As in the $q=1$ case, on account of absorption of probability on the exhaustion face we set $\rho_y=0$ for all y satisfying $\sum_{i=1}^{q} y_i = \lfloor N\alpha' \rfloor$. Moreover we modify the transition matrix elements T_{yz} of the Markov chain when y is one unit away from the boundary. If $\sum_{i=1}^{q} y_i = \lfloor N\alpha' \rfloor - 1$ we take

$$T_{yz} = \begin{cases} q^{-1}(1-p), & \text{if } z_i = y_i - 1 \text{ for exactly one } i \\ 0, & \text{if } z_i = y_i + 1 \text{ for exactly one } i \end{cases} \tag{20}$$

where the zero probability assigned to each of the q possible allocations would normally be $q^{-1}p$. This alteration makes the transition matrix substochastic. With these choices, the approximation

$$\rho_y = C_3\left([p/(1-p)]^{\sum_{i=1}^{q} y_i} - [p/(1-p)]^{\lfloor N\alpha' \rfloor}\right) \tag{21}$$

will satisfy $\rho\mathbf{T}=\rho$ in Region 3.

The WKB solution (19) interpolates between the approximations (17) and (21) of Regions 1 and 3. Comparison between (17) and (19) gives $C_2'=1$. Comparison between (19) and the first term of (21) yields $C_3=1$. The second term is important, in a relative sense, only near the boundary.

Now that $C_3=1$ is known, the approximations (17) and (21) may be substituted into the denominator and numerator of (4). The summands in the numerator are nonzero only if y is one unit away from the boundary, in which case the factor $1-\sum_{z\in Q} T_{yz}$ equals p. The number of such nonzero summands (they are all equal to each other) is to leading order the number of states on the exhaustion face of Q. This is N^{q-1} times a combinatorial factor, essentially the area of a truncated simplex in $\mathbb{R}^q$. Let us denote this factor $A(\alpha', \alpha'', q)$. After simplifying the expression (4), and using the fact that $\mathsf{E}\tau \sim (1-\lambda_1)^{-1}$, we get the following. (We assume $N\alpha'$ is an integer.)

Theorem 2. *In the one-resource (i.e., $r = 1$, q arbitrary) shared memory model, when memory units are allocated and deallocated singly, with probabilities that are independent of current allocations, the mean time until exhaustion has asymptotics*

$$\mathsf{E}\tau \sim (1-2p)^{-1}p^q\left(\frac{1}{1-2p}+\frac{1}{p_0}\right)^q A(\alpha', \alpha'', q)^{-1}N^{-(q-1)}((1-p)/p)^{N\alpha'} \quad (22)$$

as $N \to \infty$. The areal factor $A(\alpha', \alpha'', q)$ is computed as stated above.

The $q = 2$ case of this theorem was previously obtained as Theorem 2.13 of Louchard and Schott [8] (in the special case when α'' is effectively infinite, and $p_0 = p$). But the $q > 2$ case is new. We see that if $q > 1$ the large-N asymptotics of $\mathsf{E}\tau$ will necessarily include a power law factor proportional to $N^{-(q-1)}$, as well as the leading (dominant) exponential growth.

It follows from our derivation that in the $N \to \infty$ limit, the system state at exhaustion time will be uniformly distributed across the exhaustion face of Q. This is because the rate of absorption of probability is, by the equality of the nonzero summands in the numerator in (4), independent of position on the face. So when p is constant, the limiting uniform distribution originally discovered by Flajolet [2] occurs irrespective of the number of processes.

Theorem 2 may be generalized in several directions; for example, to the case of batch allocations and deallocations of memory units. If the q processes can allocate and deallocate units in multiples rather than singly, $\xi_i = \Delta y_i$ will take on values other than ± 1, and formula (22) must be modified. The asymptotics will however be qualitatively similar: an exponential growth of $\mathsf{E}\tau$ with N, and a pre-exponential factor proportional to $N^{-(q-1)}$. The details may appear elsewhere.

It is more interesting to consider the generalization of Theorem 2 to the case when the probability of an allocation by a process P_i is allowed to depend on the currently allocated resource fraction $x_i = y_i/N$. In this case the transition probabilities T_{xy} will vary over the interior of Q. A WKB approximation $K(x)\exp(-NW(x))$ to the quasi-stationary density in Region 2 may still be constructed, but if $q > 1$ its behavior will be quite different from that of the WKB solution (19). We discuss this matter briefly.

The function $W(x)$ necessarily satisfies a so-called eikonal equation, a nonlinear partial differential equation on Q. (See Ref. [5]; also Section 3 of Ref. [11] for the continuous-time case.) This equation is the same as the Hamilton-Jacobi equation solved by Maier [9], the solution of which is the 'quasi-potential' or 'classical action function' of Wentzell-Freidlin theory. Maier showed (in the $q = 2$ colliding stacks case, but the proof generalizes) that if the density of each ξ_i is given by (3), but with p taken to be a *decreasing* function of the allocation fraction x_i, then the behavior of $W(x)$ will be as follows. It will not depend on $\sum_{i=1}^q x_i$ alone, so it will not be constant over the exhaustion face. It will attain a quadratic minimum at the center of the face, where $x = q^{-1}\alpha'(1,1,\ldots)$. The WKB solution, restricted to the exhaustion face, will accordingly display a *Gaussian falloff* around this point; this Gaussian will have a $\Theta(N^{1/2})$ standard deviation.

The approximation in Region 3, with which the WKB solution must match, will exhibit a similar transverse falloff. So the sum in the numerator of (4), used in computing $1-\lambda_1$, will be of magnitude $N^{(q-1)/2}$ rather than of magnitude N^{q-1}. The following is a consequence.

Theorem 3. *In the one-resource (i.e., $r = 1$, q arbitrary) shared memory model, when memory units are allocated and deallocated singly, with the probability of an allocation by each process assumed to be a* decreasing *function of the fraction of memory units allocated to it (the 'increasingly contractive' case), the mean time until exhaustion has $N \to \infty$ asymptotics*

$$\mathsf{E}\tau \sim C(\alpha', \alpha'', q) N^{-(q-1)/2}((1-p)/p)^{N\alpha'}. \tag{23}$$

It is not a simple matter to compute the constant $C(\alpha', \alpha'', q)$ in (23) when $q > 1$. (This requires solving the 'transport equation' [5] satisfied by the pre-exponential factor $K(x)$ in the WKB solution.) However, it follows from our derivation that the system state at exhaustion time will be within $O(N^{1/2})$ units of the state $q^{-1}\alpha' N(1,1,\ldots)$. To within this accuracy, the processes $P_1, \ldots, P_q$ will have *equal* quantities of the resource R_1 on hand when exhaustion occurs. We emphasize that this is special to the 'increasingly contractive' case, in which $p(x)$ is a decreasing function of x.

Due to space constraints we must state without proof the generalization of Theorems 2 and 3 to the case of an arbitrary number of resources r.

Theorem 4. *In the general shared memory model with r types of memory, when memory units are allocated and deallocated singly, the $N \to \infty$ asymptotics of the mean time until exhaustion will be given by formula (22) (in the case when allocation probabilities are independent of the number of resource units allocated) or by formula (23) (in the case when they are a decreasing function of the fraction allocated). In both cases the formula for $\mathsf{E}\tau$ must be multiplied by a factor r^{-1}.*

5 Conclusions

We have seen that in the limit as the scaling parameter N is taken to infinity, it is possible to obtain analytic results on the expected time until shared memory exhaustion. Our Theorems 2 through 4 are the strongest obtained on this.

Since our stochastic model is very general, our results have wide applicability. They apply in particular to Markov models of dynamic data structures whose transition probabilities are nontrivially state-dependent. This is an advance beyond the work of Kenyon-Mathieu and Vitter [4] and Louchard, Kenyon and Schott [7]. Permitting state-dependence allows the modelling of processes implementing *exhaustion avoidance strategies*, by restricting memory allocations.

It would be interesting to compare our results with simulations. Ellis [1] performed some simulations of the one-resource model. But since he took $p \equiv \frac{1}{2}$ and used a slightly different definition of 'final state,' our results are not directly comparable. In the $q = 2$, $r = 1$ colliding stacks model Flajolet [2] confirmed that the asymptotically uniform distribution over final states mentioned after Theorem 2 is indeed seen in simulations. To date there have been no simulations of the radically different case when the allocation and deallocation probabilities vary as functions of the number of allocated resource units. Simulation of 'increasingly contractive' models, in particular, remains to be done.

We expect that asymptotic techniques similar to those of this paper will prove useful in analysing the behavior of stochastically modelled on-line algorithms whenever (1) the algorithm state space is naturally viewed as a subset of some finite-dimensional Euclidean space, and (2) there is some scaling parameter N, governing the size of this set, which is being taken to infinity. This includes the analysis of buddy systems and related storage allocators. Many years ago Purdom and Stigler [12] began the stochastic analysis of binary buddy systems. Our approach may well permit the computation, in the case of an arbitrary number of block sizes, of the asymptotics of the expected time until memory exhaustion. Work on this is now under way.

References

1. C. A. Ellis. Probabilistic models of computer deadlock. *Inform. Sci.* **12** (1977), 43–60.
2. P. Flajolet. The evolution of two stacks in bounded space and random walks in a triangle. In *Proc. MFCS '86*, LNCS #233, pp. 325–340. Springer-Verlag, 1986.
3. A. N. Habermann. System Deadlocks. In *Current Trends in Programming Methodology*, edited by K. M. Chandy and R. T. Yeh, volume 3. Prentice-Hall, 1978.
4. C. M. Kenyon-Mathieu and J. S. Vitter. The maximum size of dynamic data structures. *SIAM J. Computing* **20** (1991), 807–823.
5. C. Knessl, B. J. Matkowsky, Z. Schuss, and C. Tier. An asymptotic theory of large deviations for Markov jump processes. *SIAM J. Appl. Math.* **46** (1985), 1006–1028.
6. D. E. Knuth. *Fundamental Algorithms*, volume 1 of *The Art of Computer Programming*. Addison-Wesley, second edition, 1981.
7. G. Louchard, C. Kenyon, and R. Schott. Data structures maxima. In *Proc. FCT '91*, LNCS #529, pp. 339–349. Springer-Verlag, 1991.
8. G. Louchard and R. Schott. Probabilistic analysis of some distributed algorithms. *Random Structures and Algorithms* **2** (1991), 151–186.
9. R. S. Maier. Colliding stacks: a large deviations analysis. *Random Structures and Algorithms* **2** (1991), 379–420.
10. H. Minc. *Nonnegative Matrices*. Wiley, New York, 1988.
11. T. Naeh, M. M. Kłosek, B. J. Matkowsky, and Z. Schuss. A direct approach to the exit problem. *SIAM J. Appl. Math.* **50** (1990), 595–627.
12. P. W. Purdom, Jr. and S. M. Stigler. Statistical properties of the buddy system. *J. ACM* **17** (1970), 683–697.
13. A. C. Yao. An analysis of a memory allocation scheme for implementing stacks. *SIAM J. Computing* **10** (1981), 398–403.

Minimum Weight Euclidean Matching and Weighted Relative Neighborhood Graphs

Andy Mirzaian *

1 Introduction

A matching in a graph is a subset of the edges, no two of which share a vertex. The weight of a matching is the sum of its edge weights. The *minimum weight matching* problem is to find a maximum cardinality matching with the least possible weight in a given weighted graph. The first polynomial time algorithm for this problem was proposed by the pioneering work of Edmonds [3,4]. Lovász and Plummer [13] provide a comprehensive study of graph matching. Also, [12,17] are general sources for the subject. Galil [9] provides a lucid survey of the area up to 1986. Let n and m, respectively, denote the number of vertices and edges of the given graph. The best known time bound for the minimum weight matching problem is $O(n^3)$ for dense graphs [7,12], and $O(n(m + n \log n))$ for sparse graphs [6,8].

This paper is concerned with the *Euclidean Minimum Weight Matching* (EMWM) problem: given $2n$ point sites in the plane that form the vertices of an underlying complete graph with Euclidean interpoint distances as edge weights, find a minimum weight perfect matching (ie, cardinality n) of the point sites. Currently, the best known algorithm is by Vaidya [20] which requires $O(n^{2.5}(\log n)^4)$ time and $O(n \log n)$ space.

The aim of this paper is to work towards a more geometric solution of the EMWM problem. We start with Edmonds' primal-dual method. One result of this paper is a proposed $O((n^2 + \mathcal{F}) \log n)$ time, $O(n)$ space, algorithm for MWEM based on the Weighted Voronoi Diagram of the $2n$ sites, where the weights are dynamically changing and are related to the linear programming dual variables. The $\mathcal{F}$ term is the number of *edge-flips* in the Weighted Voronoi Diagram during the matching algorithm. We conjecture that $\mathcal{F}$ is close to $O(n^2)$. The main contribution of this paper includes further exploration of the geometric properties of EMWM. More specifically, we show the following:

- The vertex dual variables in EMWM remain nonnegative. Hence the constraints corresponding to trivial and nontrivial blossoms become the same.

*Address: Dept. of Computer Science, York University, Toronto, Canada, M3J 1P3. Andy@CS.YorkU.Ca. Author's research was partly supported by an NSERC grant.

This allows us to associate circular disks centered at the point sites whose (nonnegative) radii are related to vertex and blossom dual variables. These radii are considered as the site weights.

- We generalize the Relative Neighborhood Graphs and Gabriel Graphs to their weighted versions. (For the unweighted versions see [1,11,14,19].) The Weighted Relative Neighborhood Graph (WRNG) is a subgraph of the Weighted Gabriel Graph (WGG), which is a subgraph of the Weighted Delaunay Diagram (WDD). Both WRNG and WGG are planar straight-line graphs, and when the weights are according to EMWM, they are connected graphs and span all the sites.

- The (straight-line) *admissible edges* form a subgraph of the WRNG of the sites. Therefore, there are $O(n)$ admissible edges at a given time and they are noncrossing. (The matching edges are a subset of the admissible edges.) This enables us to search the sparse WDD (or WGG or WRNG) edges, rather than the underlying complete graph, in order to maintain the admissible edges. However, we now have to pay the overhead for maintaining the WDD (or WGG or WRNG) edges, since they change (through edge-flips) as the weights change.

The rest of the paper is organized as follows. Section 2 gives a brief overview of Edmonds' algorithm. Section 3 develops the new geometric results listed above. Section 4 presents the proposed new algorithm. Section 5 offers further discussion and some concluding remarks. Some proofs and implementation details are omitted from this extended abstract and can be found in the full version of the paper [16].

2 The LP and Edmonds' Algorithm

Here we will consider Edmonds' linear programming formulation as adapted by Lovász and Plummer [13]. Let $G = (V, E)$ be a given weighted graph with edge weights $d_e = d_{vu} = d(v, u)$ for $e = (v, u) \in E$. We assume $|V| = 2n$ and G contains a perfect matching. A *blossom* is an odd cardinality subset of V. A blossom is called *trivial* if it is a singleton (a vertex); otherwise it is called *nontrivial*. Let $\mathcal{B}$ denote the set of all blossoms of G. Let x be a real vector with an entry for each edge of G (with the interpretation that $x_e = 1$ if e is a matched edge and $x_e = 0$ otherwise). Similarly, let d denote the vector of edge weights. We say an edge e is incident to a blossom B if exactly one endpoint of e is in B. We let ∇B denote the set of all edges incident to blossom B. Let $x(B) = \Sigma\{x_e | e \in \nabla B\}$. The linear programming formulation of the problem is

$$\begin{array}{lll} \textit{minimize} & d^T \cdot x & \\ \textit{subject to:} & x_e \geq 0 & (\textit{for each edge } e \in E) \\ & x(B) = 1 & (\textit{for each trivial blossom } B) \\ & x(B) \geq 1 & (\textit{for each nontrivial blossom } B)\ . \end{array}$$

We use a dual variable α_B for each blossom $B \in \mathcal{B}$. Let $\alpha(e) = \alpha(u,v) = \Sigma\{\alpha_B | e \in \nabla B\}$ for any edge $e = (u,v) \in E$. The dual program consists of the following objective and constraints:

$$\begin{array}{lll} \textit{maximize} & \Sigma_{B\in\mathcal{B}}\, \alpha_B & \\ \textit{subject to}: & \alpha_B \geq 0 & (\textit{for each nontrivial blossom } B) \\ & \alpha(e) \leq d_e & (\textit{for each edge } e \in E)\,. \end{array}$$

In Section 3 we will show that these linear programs can be further simplified by removing the distinction between trivial and nontrivial blossoms if the underlying graph is complete and the edge weights form a distance metric (such as the Euclidean case).

Edmonds' algorithm is a primal-dual method applied to the above linear programs. The algorithm maintains dual feasibility (starting with $\alpha(B) = 0$ for every blossom $B \in \mathcal{B}$). It also maintains an integral primal solution (starting with $x_e = 0$ for each edge $e \in E$) that satisfies all the primal constraints except that for some blossoms B it may have $x(B) = 0$. Any such solution is a matching of G though not necessarily a perfect matching. The algorithm proceeds through n phases. In each phase, it increases the number of matched edges by one and resolves one of the violated primal constraints. We let M denote the set of matched edges during the execution of the algorithm. To ensure optimality, the algorithm maintains the complementary slackness conditions:

$$\begin{array}{l} \textit{if } x(B) = 0 \textit{ then } \alpha(B) = 0, \textit{ for each blossom } B \in \mathcal{B} \\ \textit{if } \alpha(e) < d_e \textit{ then } x_e = 0, \textit{ for each edge } e \in E. \end{array}$$

The *slack* for an edge $e = (u,v) \in E$ is the quantity $\textit{slack}(e) = \textit{slack}(u,v) = d_e - \alpha(e) \geq 0$. An edge $e \in E$ is called *admissible* if $\alpha(e) = d_e$, that is, with a zero slack. Dual feasibility and the complementary slackness conditions imply that the edges in the (optimum) matching are admissible. We call a blossom B *active* if it is either trivial or $\alpha(B) > 0$. A biproduct of Edmonds' algorithm is that the set of active blossoms are nested. That is any two active blossoms are either disjoint or one includes the other. This allows us to represent the active blossoms by the *blossom structure forest*. The trivial blossoms form the leaves of the structure. A blossom B_1 is a child of blossom B_2 in the structure, if B_2 is the smallest active blossom that properly contains B_1. The roots of the blossom structure forest will be called *maximal blossoms*. A blossom (or vertex) B is called *exposed* if $x(B) = 0$, otherwise it is called *matched*. A pair of vertices/blossoms incident to a matched edge are called mates. To help construct an augmenting path during a phase, the algorithm maintains an *alternating forest*, using only admissible edges, in the (conceptually) shrunken graph where blossoms are shrunk to single nodes. The nodes of the forest are the maximal blossoms labeled S or T. The remaining maximal blossoms are labeled F. The roots of the forest are exactly the exposed maximal blossoms and are labeled S. The path from a root to any leaf in this forest is an alternating path and the nodes alternate between S-blossoms and T-blossoms. A T-blossom has exactly one child — its S-blossom mate. The label of a vertex is the label of the maximal blossom that contains it. Now imagine increasing (decreasing) $\alpha(B)$ for each maximal S- (T-) blossom, all by the same amount θ, until either a new edge becomes admissible, or $\alpha(B)$ becomes 0 for some maximal T-blossom B. The quantity θ is obtained from the following equations:

$$\begin{aligned}
\theta_{SS} &= \min \{slack(u,v)/2 \mid (u,v) \in E, u \text{ and } v \text{ in distinct maximal } S-blossoms\} \\
\theta_{FS} &= \min \{slack(u,v) \mid (u,v) \in E, u \text{ is an } F-vertex,\ v \text{ is an } S-vertex\} \\
\theta_{T} &= \min \{\alpha(B) \mid B \text{ is a maximal } T-blossom\} \\
\theta &= \min \{\theta_{SS},\ \theta_{FS},\ \theta_{T}\}\ .
\end{aligned}$$

Each such θ calculation corresponds to a *stage*. Each of the n phases of the algorithm consists of a number of stages of the following types: (1) a dual variable change stage, (2) a tree growing stage, (3) a blossom deactivation (or expansion) stage, (4) a blossom activation (or shrinking) stage, and (5) an augmentation stage. The following lemma gives a bound on the number of stages.

Fact 2.1 ([10,20]) *Within each of the n phases of the algorithm, the following quantities are $O(n)$: the number of alternating tree growings, blossom deactivations, blossom activations, dual variable changes, the total number of different maximal blossoms, and the number of times θ_{SS}, θ_{FS}, θ_T, θ are computed.*

Let W be any subset of V. We define $\rho(W) = \Sigma\{\ \alpha(B) \mid W \subseteq B \in \mathcal{B}\}$. Let $lca(u,v)$ denote the blossom that is the lowest common ancestor of vertices u and v in the blossom structure forest. Assume the lowest common ancestor is V if u and v are not in the same tree of the forest. (Note that $\rho(V) = 0$.) Then, $\rho(u,v) = \rho(lca(u,v))$. The following lemma is needed later.

Lemma 2.2 *Consider an arbitrary edge $e = (u,v) \in E$. Then, $\alpha(e) = \rho(u) + \rho(v) - 2\rho(u,v)$. Thus, if u and v are not in the same active blossom, then $\alpha(e) = \rho(u) + \rho(v)$.*

We summarize this section by the following theorem (See [10,16] for details.)

Theorem 2.3 *Excluding the maintenance of θ_{FS} and θ_{SS}, Edmonds' algorithm can be implemented to run in $O(n \log n)$ time per phase, $O(n^2 \log n)$ time total, and $O(n)$ space.*

3 New Geometric Results

The remaining issue is how to maintain θ_{SS} and θ_{FS} efficiently. The rest of the paper concentrates on this issue for the Euclidean case. So, now G is a complete graph on $2n$ point sites in the plane as its vertices, and $d(u,v)$ is the Euclidean distance between sites u and v.

Theorem 3.1 *Consider Edmonds' algorithm applied to the minimum weight matching problem on a complete weighted graph where the edge weights form a distance metric (such as the EMWM problem). Then the dual variables corresponding to the vertices (that is, trivial blossoms) remain nonnegative.*

Proof: This can be proved by a primal argument. Here we use an argument based on the dual variables. Suppose to the contrary that for some vertex v, $\alpha(v)$ becomes negative in Edmonds' algorithm. Consider the first time that this occurs. Just prior to that time, v must have been a maximal T-blossom. At this point $\rho(v) = \alpha(v) > 0$, and v is incident to two maximal active S-blossoms B_1 and B_2 via two admissible edges. These are the parent and the unique child of v in its alternating tree. Suppose these two admissible edges are (v, u) and (v, w), with $u \in B_1$ and $w \in B_2$. At this point we have

$$d(v,u) = \rho(v) + \rho(u) \ , d(v,w) = \rho(v) + \rho(w) \ , \ d(u,w) \geq \rho(u) + \rho(w) \ .$$

These follow from Lemma 2.2, the dual feasibility, and the fact that edges (v, u) and (v, w) are admissible. Also, v is a T-vertex and $\rho(v)$ is decreasing, while u and w are S-vertices and hence $\rho(u)$ and $\rho(w)$ are increasing. This progress will stop when the third inequality above becomes equality, before $\rho(v)$ becomes negative. That is, at this point we have

$$\begin{aligned} \rho(v) &= (d(v,u) + d(v,w) - d(u,w))/2 \ , \\ \rho(u) &= (d(u,v) + d(u,w) - d(v,w))/2 \ , \\ \rho(w) &= (d(w,v) + d(w,u) - d(v,u))/2 \ . \end{aligned}$$

These are nonnegative since the edge distances satisfy the triangle inequality. Also, at this point $d(u, w) = \rho(u) + \rho(w)$. Thus, the edge (u, w) between two distinct S-blossoms B_1 and B_2 has become admissible. So, v, B_1 and B_2 will be shrunk into a new S-blossom, and $\alpha(v) = \rho(v) \geq 0$ will be fixed. A contradiction. □

Thus, without loss of generality, we can add the constraints $\alpha(B) \geq 0$, for each trivial blossom B, to the dual linear program for EMWM and obtain the following:

$$\begin{aligned} &\textit{minimize} && d^T \cdot x \\ &\textit{subject to}: && x_{uv} \geq 0 && (\textit{for each pair of sites } u, v) \\ & && x(B) \geq 1 && (\textit{for each blossom } B)\ . \end{aligned}$$

$$\begin{aligned} &\textit{maximize} && \Sigma_{B \in \mathcal{B}}\, \alpha_B \\ &\textit{subject to}: && \alpha_B \geq 0 && (\textit{for each blossom } B) \\ & && \alpha(u,v) \leq d(u,v) && (\textit{for each pair of sites } u, v)\ . \end{aligned}$$

Now, dual feasibility implies $\rho(v) \geq \alpha(v) \geq 0$ for each vertex v. Let $disk(v)$ denote the circular disk centered at v with the nonnegative radius $\rho(v)$. For each active blossom B associate the planar region $Region(B) = \bigcup_{v \in B} disk(v)$.

Lemma 3.2 *Suppose Edmonds' algorithm is applied to EMWM. Then, for each active blossom B (ie, $\alpha(B) > 0$), the interior of Region(B) is connected.*

Lemma 3.3 *Consider any two distinct maximal blossoms B_1 and B_2. Then, the interiors of Region(B_1) and Region(B_2) are disjoint.*

Figure 1: Blossom clusters in EMWM.

Corollary 3.4 *Two vertices are in the same maximal active blossom if and only if they are in the same connected component of the interior of* $\bigcup_{v \in V}$ *disk*(v).

In what follows, we will make repeated use of the following crucial lemma:

Lemma 3.5 *During Edmonds' algorithm applied to EMWM, there is no pair of distinct vertices u and v such that disk*(u) *is included in the interior of disk*(v). *That is,* $|\rho(u) - \rho(v)| \le d(u,v)$, *for each pair of sites u and v.*

Proof: By dual feasibility and Lemma 2.2, at all times we have $d(u,v) \ge \rho(u) + \rho(v) - 2\rho(u,v)$. Let us define $r_u = \rho(u) - \rho(u,v)$ and $r_v = \rho(v) - \rho(u,v)$. Thus, $d(u,v) \ge r_u + r_v$. We note that $r_u = \Sigma\{\ \alpha(B) \mid u \in B \subset lca(u,v)\ \} \ge 0$. Similarly, $r_v \ge 0$. Therefore, the two disks centered at u and v with radii, respectively, r_u and r_v have disjoint interiors. Hence, $|r_u - r_v| \le d(u,v)$. By adding $\rho(u,v)$ to these two radii, we obtain $|\rho(u) - \rho(v)| \le d(u,v)$. □

Figure 1 shows the clustering structure of blossoms and subblossoms in an example. The admissible edges are shown in straight line-segments, the bold ones are matched edges.

3.1 The WVD, WDD, WGG and WRNG

Weighted Voronoi Diagram: Let $\rho(s)$ be the weight of each site s. For any point x in the plane, the *weighted distance* of x from site s is defined as $\delta_s(x) = d(x,s) - \rho(s)$. Consider the location of the points that have smaller or equal weighted distance from site u than v. This region is empty if $\rho(u) < \rho(v) - d(u,v)$. Otherwise, it includes u and is bounded by the bisector of u and v, denoted by $H(u,v)$, which is one branch of a hyperbola with foci u and v. This bisector is bending towards the site with smaller weight (it is a line if both sites have equal weight and it is a half line if $|\rho(u) - \rho(v)| = d(u,v)$). This region (if nonempty) is star shaped and u is one of its kernel points. The Weighted Voronoi cell (or Voronoi region) of site s, $Vor(s)$, is the set of points in the plane that are at least as close (in the sense of weighted distance) to s than to any other site. A Weighted Voronoi region $Vor(s)$ is star shaped with s as one of its kernel points, and its boundary consists of a chain of hyperbolas. The subdivision of the plane by Weighted Voronoi regions of the sites is their Weighted Voronoi Diagram (WVD). The vertices and edges of the subdivision are called Voronoi vertices and Voronoi edges, respectively. In the absence of degeneracy, Voronoi vertices have degree three. For more details on the general properties of WVD see for example [2,5,18].

Remark: *In this paper the weights are not entirely general due to Lemma 3.5. Some of the properties that we prove here do not hold for general weights.*

Weighted Delaunay Diagram: The WDD is the topological dual of the Weighted Voronoi diagram. The sites are the vertices of WDD, and there is an edge between a pair of sites u and v, if $Vor(u)$ and $Vor(v)$ share a common boundary edge. In the unweighted case WDD is a straight-line triangulation of the sites and is called Delaunay Triangulation. In the weighted case, WDD is possibly a multi-graph and can be drawn quasi-straight-line as follows. A non-empty Voronoi region $Vor(u)$ is star-shaped with site u one of its kernel points. First suppose $Vor(u)$ has a non-empty interior. Draw a line-segment between u and each Voronoi vertex on the boundary of $Vor(u)$. These segments partition $Vor(u)$ into sectors. We can associate each sector with the boundary edge of $Vor(u)$ it contains. This boundary edge is called the *base* of the sector. Now, if $Vor(u)$ and $Vor(v)$ share a common boundary edge, then consider the sectors of $Vor(u)$ and $Vor(v)$ with the common base. If the common base does not intersect the line segment (u, v), then draw the edge between u and v as two connected straight line segments within the two sectors with the connection point on the common base. Draw the edge as straight-line, if the segment (u, v) intersects the common base. The common boundary between a pair of Voronoi regions may consist of several bases. In that case we will have the corresponding multiple edges in the WDD. In the absence of degeneracy, and if none of the Voronoi regions are empty, this drawing gives a quasi-straight-line topological triangulation (that is, each bounded face is incident to three sites). A site u is on the boundary of the unbounded (exterior) face, if and only if $disk(u)$ touches the boundary of the convex-hull of $\bigcup_{v \in V} disk(v)$. In the presense of degeneracy, a Voronoi region might be a half-line or a line segment, and some internal faces may not be triangles. These cases can be resolved by a slight perturbation of the weights: in the first case thicken the half-line or the line segment slightly to endow it an interior; in the second case triangulate these non-triangular faces by adding to them a maximal number of non-crossing chords.

Weighted Gabriel Graph: The WGG is the straight-line graph whose vertices are the given sites, and the line segment between sites u and v is an edge of the graph if and only if the (closed) line segment (u, v) intersects the bisector $H(u, v)$ and this intersection point c_{uv} is on the common boundary of $Vor(u)$

and $Vor(v)$. In other words, let $c_{uv} = (u,v) \cap H(u,v)$ (if it exists). Then (u,v) is an edge of the WGG, if and only if c_{uv} is closer (in terms of weighted distance) to u and v than to any other site. Point c_{uv} is called the *Weighted Gabriel Center* of sites u and v. The following is now obvious.

Theorem 3.6 *The Weighted Gabriel Graph is a straight-line subgraph of the Weighted Delaunay Diagram.*

Below, we will prove some additional properties of Weighted Gabriel Graphs (and later for Weighted Relative Neighborhood Graphs) when the weights satisfy Lemma 3.5.

Lemma 3.7 *The Weighted Gabriel Center c_{uv} of each pair of sites u and v exists and is on the line segment (u,v).*

Lemma 3.8 *For each site u, $Vor(u)$ is a nonempty star shaped region and u is one of its kernel points.*

Theorem 3.9 *The Weighted Gabriel Graph is a connected straight-line planar graph and spans all the sites.*

Proof: The fact that WGG is straight-line planar follows from the quasi-linear drawing and planarity of WDD and Theorem 3.6. Now assume to the contrary that WGG is not connected. Consider a pair of sites u and v that are disconnected in WGG. Select u and v such that $\delta_u(c_{uv}) = (d(u,v) - \rho(u) - \rho(v))/2$ is minimum possible. If the choice is not unique, select a pair (u,v) among them such that $d(u,v)$ is minimum. If the choice is still not unique, select a pair (u,v) among the latter such that the minimum x-coordinate of u or v is as small as possible (break the tie arbitrarily). If there is no other site w, such that $\delta_w(c_{uv}) \leq \delta_u(c_{uv})$, then by definition, the line segment (u,v) is an edge in WGG, a contradiction. Now, assume there is a site w such that $\delta_w(c_{uv}) \leq \delta_u(c_{uv})$. Then $d(u,w) < d(u,c_{uv}) + d(w,c_{uv})$. Thus, $2\delta_u(c_{uw}) = d(u,w) - \rho(u) - \rho(w) < \delta_u(c_{uv}) + \delta_w(c_{uv}) \leq 2\delta_u(c_{uv})$. Thus, $\delta_u(c_{uw}) < \delta_u(c_{uv})$. Similarly, $\delta_v(c_{vw}) < \delta_v(c_{uv})$. Then, by the selection criterion of (u,v), we conclude that WGG contains a path between u and w and a path between v and w. Thus, u and v are connected, a contradiction. □

Weighted Relative Neighborhood Graph: The WRNG is also a straight-line graph with the given sites as vertices, and is defined as follows. Let $\delta(a,b) = d(a,b) - \rho(a) - \rho(b)$ be the (symmetric) relative distance between sites a and b. The line segment (a,b) is an edge of WRNG if and only if $\delta(a,b) \leq max\{\delta(a,c), \delta(b,c)\}$, for any other site c.

Theorem 3.10 *The Weighted Relative Neighborhood Graph is a connected spanning subgraph of the Weighted Gabriel Graph.*

Proof: Suppose (a,b) is an edge of WRNG. Then, for any other site c we have $\delta(a,b) \leq max\{\delta(a,c),\delta(b,c)\}$. Thus, $\delta_c(c_{ab}) = d(c,c_{ab}) - \rho(c) \geq d(a,c) - d(a,c_{ab}) - \rho(c) = \delta(a,c) - \delta(a,b)/2$. Similarly, $\delta_c(c_{ab}) \geq \delta(b,c) - \delta(a,b)/2$. Thus, $\delta_c(c_{ab}) \geq max\{\delta(a,c),\delta(b,c)\} - \delta(a,b)/2 \geq \delta(a,b)/2 = \delta_a(c_{ab})$. This implies (a,b) is an edge in WGG. The fact that WRNG is also a spanning connected graph can be proved similar to that of the previous theorem. □.

3.2 The Admissible Edges

In this subsection we establish some connection between the admissible edges and the structures discussed in the previous subsection, such as Weighted Relative Neighborhood Graphs. Recall that $\rho(a,b) = \rho(lca(a,b)) = \Sigma\{\rho(B) \mid a \in B, b \in B\}$. Furthermore, by Lemma 2.2 and dual feasibility, we have $d(a,b) \geq \alpha(a,b) = \rho(a) + \rho(b) - 2\rho(a,b)$. So, for an admissible edge (a,b) we have $\delta_a(c_{ab}) = -\rho(a,b) = (d(a,b) - \rho(a) - \rho(b))/2$. Consider applying Edmonds' algorithm to EMWM. We have the following:

Lemma 3.11 *For any three sites a,b,c, we have $\rho(a,b) \geq min\{\rho(a,c),\rho(b,c)\}$.*

Proof: The proof is based on the nested structure of the blossoms. Details omitted from this extended abstract. □

Theorem 3.12 *Admissible edges form a subgraph of the Weighted Relative Neighborhood Graph.*

Proof: Let (a,b) be an admissible edge. Let c be any other site. By Lemma 3.11 we have $\delta(a,b) = d(a,b) - \rho(a) - \rho(b) = -2\rho(a,b) \leq max\{-2\rho(a,c), -2\rho(b,c)\} \leq max\{d(a,c) - \rho(a) - \rho(c), d(b,c) - \rho(b) - \rho(c)\} = max\{\delta(a,c),\delta(b,c)\}$. This implies (a,b) is an edge of WRNG. □

Corollary 3.13 *At any time during the algorithm there are only $O(n)$ admissible edges and they are non-crossing.*

3.3 The Edge-Flip Criterion

In this subsection we show how to compute the θ change needed before an edge-flip in WDD occurs. There are two kinds of flips possible and are explained below. Let the triple (x_i, y_i, r_i) denote a circle C_i with radius r_i whose center is at Cartesian coordinates (x_i, y_i). We say three circles C_i, $i = 1,2,3$, are *collinear*, if there is a line tangent to the three given circles, and all three circles are on the same side of the line. Also, we say four circles C_i, $i = 1,2,3,4$, are *cocircular*, if there is a circle C tangent to the four given circles, and all four circles are on the same side of C, that is, all external tangents, or all internal tangents to C. We need the following two lemmas.

Lemma 3.14 *Three circles $C_i = (x_i, y_i, r_i)$, $i = 1,2,3$, are collinear, only if* $A^2 = B^2 + C^2$, where $A = \begin{vmatrix} x_1 & y_1 & 1 \\ x_2 & y_2 & 1 \\ x_3 & y_3 & 1 \end{vmatrix}$, $B = \begin{vmatrix} r_1 & y_1 & 1 \\ r_2 & y_2 & 1 \\ r_3 & y_3 & 1 \end{vmatrix}$, $C = \begin{vmatrix} r_1 & x_1 & 1 \\ r_2 & x_2 & 1 \\ r_3 & x_3 & 1 \end{vmatrix}$.

Lemma 3.15 *Four circles $C_i = (x_i, y_i, r_i)$, $i = 1,2,3,4$, are cocircular, only if $4AB + C^2 + D^2 - E^2 = 0$, where*

$$A = \begin{vmatrix} 1 & x_1 & y_1 & r_1 \\ 1 & x_2 & y_2 & r_2 \\ 1 & x_3 & y_3 & r_3 \\ 1 & x_4 & y_4 & r_4 \end{vmatrix}, \quad B = \begin{vmatrix} z_1 & x_1 & y_1 & r_1 \\ z_2 & x_2 & y_2 & r_2 \\ z_3 & x_3 & y_3 & r_3 \\ z_4 & x_4 & y_4 & r_4 \end{vmatrix}, \quad C = \begin{vmatrix} 1 & z_1 & y_1 & r_1 \\ 1 & z_2 & y_2 & r_2 \\ 1 & z_3 & y_3 & r_3 \\ 1 & z_4 & y_4 & r_4 \end{vmatrix},$$

$$D = \begin{vmatrix} 1 & z_1 & x_1 & r_1 \\ 1 & z_2 & x_2 & r_2 \\ 1 & z_3 & x_3 & r_3 \\ 1 & z_4 & x_4 & r_4 \end{vmatrix}, \quad E = \begin{vmatrix} 1 & z_1 & x_1 & y_1 \\ 1 & z_2 & x_2 & y_2 \\ 1 & z_3 & x_3 & y_3 \\ 1 & z_4 & x_4 & y_4 \end{vmatrix}, \text{ and } z_i = x_i^2 + y_i^2 - r_i^2 .$$

As the weights of the sites change, the Weighted Delaunay Diagram changes by edge-flips. There are two kinds of edge-flips (corresponding to Lemmas 3.14 and 3.15). An edge-flip of the first kind occurs when an unbounded Voronoi edge (and the corresponding dual edge in WDD) appears or disappears. This happens when the Voronoi vertex incident to the unbounded Voronoi edge moves to (or from) infinity. This occurs when the disks of the three sites that share that Voronoi vertex on their common boundary become collinear. In Lemma 3.14 the three circles $C_i = (x_i, y_i, r_i)$ are the associated three sites, where r_i is $\rho(i)$ if site i is an F-vertex, $\rho(i) + \theta$ if i is an S-vertex, and it is $\rho(i) - \theta$ if i is a T-vertex. We substitute these values in the stated condition of the lemma and compute the smallest positive value of θ. The condition is a second degree polynomial in θ. An edge-flip of the second kind occurs when a bounded Voronoi edge shrinks to zero length, that is, its two incident Voronoi vertices coincide. Consider an edge (a, b) of WDD that is incident to two triangles (a, b, c) and (a, b, d). Flipping edge (a, b) means replacing it with edge (c, d). The corresponding condition is given in Lemma 3.15. Similar to the above, this involves solving the stated condition of Lemma 3.15 for the unknown θ. In general, the condition becomes a 6-th degree polynomial in θ. We will assume that this equation can be solved for the smallest positive root θ in $O(1)$ time.

4 The New Algorithm

The proposed new algorithm for EMWM is the following modification of Edmonds' algorithm. During the initialization we also construct, in $O(n \log n)$ time, the Weighted Delaunay Diagram with zero weights. As the weights change, the WDD changes when an edge-flip occurs as discussed in the previous subsection. Suppose $\theta(a, b) \geq 0$ is the minimal change to θ needed after which edge (a, b) of

WDD should be flipped. Let us modify the definition of θ as follows:

$$\theta_V = min\{\ \theta(a,b)\ \mid (a,b)\ is\ an\ edge\ of\ WDD\ \}$$
$$\theta = min\{\ \theta_V, \theta_{SS}, \theta_{FS}, \theta_T\ \}.$$

We can maintain θ_V and execute an edge-flip event in $O(\log n)$ time, using a priority queue of WDD edges, with $\theta(a,b)$ as the priority of edge (a,b). When an edge-flip occurs, the neighboring four edges in WDD must also be checked and their priorities be updated. When a site incident to a WDD edge changes label (S, F, or T), we need to update its priority.

We maintain θ_{FS} and θ_{SS} essentially similar to [10], but applied to the dynamically changing WDD. (For details see [16].) Suppose $\mathcal{F}_i$ edge flips occur in phase i, for a total of $\mathcal{F} = \Sigma_{i=1}^{n} \mathcal{F}_i$. The total number of edges involved in phase i is $E_i = O(n + \mathcal{F}_i)$. In the full version of the paper we show that the total time for phase i is $O(E_i \log n)$. We conclude:

Theorem 4.1 *The proposed new algorithm solves the MWEM problem in $O((n^2 + \mathcal{F})\log n)$ time and $O(n)$ space, where $\mathcal{F}$ is the total number of edge-flips during the algorithm.*

5 Discussion

The aim of this paper has been to devise a more geometric solution of the weighted Euclidean matching problem. A number of open questions regarding weighted relative neighborhood graphs and their relation with the EMWM problem remain. On a related note, let us also mention that the fractional version of EMWM gives rise to a circle packing problem which admits a more efficient solution [15].

Another remaining problem is to find a tight upper bound on $\mathcal{F}$. A very crude estimate of $\mathcal{F}$ can be obtained as follows. From [5] we know that the WVD is the projection of the lower envelope of vertical circular cones. These are identical circular cones in one-to-one correspondence with the sites, such that each of their axes is perpendicular to the xy–plane and intersects it at the corresponding site, and the radius of the circular intersection of the cone with the xy–plane is the weight of the site. Consider the lower envelope of only the S-cones, corresponding to S-vertices. Call this lower envelope, the S-surface. Define T-surface and F-surface similarly. During a stage, the vertex labels do not change, the S-surface is moving down at a constant rate, the T-surface is moving up at the same constant rate, and the F-surface is stationary. As these three surfaces interact with each other, their lower envelope goes through at most $O(n^2)$ combinatorial changes. Since there are a total of $O(n^2)$ stages, we conclude the total number of combinatorial changes, that is $\mathcal{F}$, is $O(n^4)$. This analysis is of course very crude, since it considers the worst for each stage and does not take into account the correlations inherent in the algorithm. We conjecture that $\mathcal{F}$ is close to $O(n^2)$.

References

[1] P.J. Agarwal and J. Matoušek. Relative neighborhood graphs in three dimensions. In *Proc. 3rd Annual ACM-SIAM Symp. on Discrete Algorithms*, pages 58–65, 1992.

[2] F. Aurenhammer. Voronoi diagrams – a survey of a fundamental geometric data structure. Technical Report B 90-09, Freie Universität, 1990.

[3] J. Edmonds. Maximum matching and a polyhedron with 0,1-vertices. *J. of Research of National Bureau of Standards*, 69B:125–130, 1965.

[4] J. Edmonds. Paths, trees, and flowers. *Canadian J. Math.*, 17:449–467, 1965.

[5] S. Fortune. A sweepline algorithm for Voronoi diagrams. *Algorithmica*, 2:153–174, 1987.

[6] M.L. Fredman and R.E. Tarjan. Fibonacci heaps and their uses in improved network optimization algorithms. *J. ACM*, 34:596–615, 1987.

[7] H.N. Gabow. *Implementations of algorithms for maximum matching on nonbipartite graphs.* PhD thesis, Comp. Sci. Dept., Stanford Univ., 1973.

[8] H.N. Gabow. Data structures for weighted matching and nearest common ancestors with linking. In *Proc. 1st Annual ACM-SIAM Symp. on Discrete Algorithms*, pages 434–443, 1990.

[9] Z. Galil. Efficient algorithms for finding maximum matching in graphs. *ACM Comp. Surveys*, 18:23–38, 1986.

[10] Z. Galil, S. Micali, and H.N. Gabow. An $O(EVlogV)$ algorithm for finding a maximal weighted matching in general graphs. *SIAM J. Computing*, 15:120–130, 1986.

[11] J. Jarmoczyk, M. Kowaluk, and F.F. Yao. An optimal algorithm for constructing β-skeletons in L_p metric. *SIAM J. Computing*, to appear.

[12] E.L. Lawler. *Combinatrotial Optimization: Networks and Matroids.* Holt, Reinhart and Winston, New York, 1976.

[13] L. Lovász and M.D. Plummer. *Matching Theory.* Annals of Discrete Math 29, North-Holland, 1986.

[14] D.W. Matula and R.R. Sokal. Properties of Gabriel graphs relevant to geographic variation search and the clustering of points in the plane. *Geogr. Analysis*, pages 205–222, 1980.

[15] A. Mirzaian. Optimum circle packing with specified centers. in preparation.

[16] A. Mirzaian. Minimum weight Euclidean matching and weighted relative neighborhood graphs. Technical Report CS-92-08, Dept. of Computer Science, York University, Canada, Sept. 1992.

[17] C.H. Papadimitriou and K. Steiglitz. *Combinatorial Optimization : Algorithms and Complexity.* Prentice-Hall, Englewood Cliffs, NJ, 1982.

[18] M. Sharir. Intersection and closest-pair problems for a set of planar discs. *SIAM J. Computing*, 14:448–468, 1985.

[19] G.T. Toussaint. The relative neighborhood graph of a finite planar set. *Pattern Recognition*, 12:261–268, 1980.

[20] P.M. Vaidya. Geometry helps in matching. *SIAM J. Computing*, 18:1201–1225, 1989.

Efficient Approximate Shortest-Path Queries Among Isothetic Rectangular Obstacles

Pinaki Mitra and Binay Bhattacharya

School of Computing Science
Simon Fraser University
Burnaby, B.C, Canada V5A 1S6.

Abstract. In this paper we consider the problem of approximate rectilinear shortest-path query between two arbitrary points in the presence of n isothetic and disjoint rectangular obstacles. We present an algorithm which reports a path whose length is at most three times the optimal path length between two arbitrary corner points and at most seven times the optimal path length between two arbitrary points. Our algorithm takes $O(n \log^3 n)$ preprocessing time, $O(n \log^2 n)$ space and $O(\log^2 n)$ query time for the distance problem. The actual path can be reported in $O(\log^2 n + k)$ where k is the number of segments in the reported path. Thus we exhibit a tradeoff between a previous result in [6] in which an exact solution of this query problem is given at the expense of $O(n\sqrt{n})$ preprocessing and $O(\sqrt{n} + k)$ query time.

1 Introduction

The shortest-path problem received considerable attention in computational geometry literatures. The shortest path inside a simple polygon from a given source point can be computed in linear time [8]. Subsequently in [7] an algorithm has been given to answer the shortest-path query inside a simple polygon between two arbitrary query points. Similar research has been done for the shortest-path problem in free space. The rectilinear shortest-path problem in free space was studied in [3], [4], [10], [11], [14] and several other papers. This problem has applications in areas like VLSI routing and motion planning. In this paper we consider the rectilinear shortest-path problem in the presence of $\mathcal{R} = \{R_1, R_2, \ldots, R_n\}$, a set of n isothetic and disjoint rectangular obstacles. From a given source the rectilinear shortest-path problem avoiding a set of isothetic disjoint rectangles was studied in [4]. In [6] the condition, of the source being given as the input, was relaxed. In that paper an exact solution was given to answer the shortest-distance query between two arbitrary query points in $O(\sqrt{n})$ time and the shortest-path query in $O(\sqrt{n}+k)$ time (k is the number of segments in the path) with $O(n\sqrt{n})$ preprocessing. The approach was based on reducing the original problem to the shortest-path problem of a directed acyclic planar graph, called carrier graphs. Subsequently that graph was preprocessed using planar separators for the shortest-path query. In [1] a parallel algorithm is given to solve this problem with $O(n^2)$ processors and $O(\log^2 n)$ time in the

CREW PRAM model of computation. Their approach was based on recursive decomposition of the set of rectangles in $\mathcal{R}$ using staircase separators.

Many times in motion planning problems instead of obtaining the optimal path researchers have investigated for near optimal paths which can be obtained by spending much less time. This issue was addressed in [2]. A survey on different such variations of the shortest-path problems is presented in [12]. In this paper we present an algorithm to solve the query problem for the approximate shortest path avoiding the rectangles in $\mathcal{R}$ between two arbitrary points. The algorithm reports a path whose length is at most three times the optimal path length between two arbitrary corner points and at most seven times the optimal path length between two arbitrary points. For this we spend $O(n \log^3 n)$ preprocessing to answer the approximate shortest-distance query in $O(\log^2 n)$ time. The actual path can be reported in $O(\log^2 n + k)$ time where k is the size or the total number of segments in the reported path. Our approach for solving this problem uses the staircase separator of [1] which can produce a balanced splitting of the set of rectangles in R, and Voronoi diagram computation on the sparse visibility graph, introduced in [3], to solve the rectilinear shortest-path problem in the presence of obstacles.

Our overall organization of the paper is as follows. In **section 2** we introduce certain definitions. In **section 3** we make a brief review of some existing results on the rectilinear shortest-path problem, which we shall be using later. In **section 4.1** we present an algorithm to answer the approximate shortest-distance query between two arbitrary corner points of $\mathcal{R}$ in $O(\log^2 n)$ time using $O(n \log^3 n)$ preprocessing and $O(n \log^2 n)$ space. Subsequently in **section 4.2** we relax the corner point restriction. Lastly in **section 5** we summarize our contributions and present some problems for further research.

2 Definitions and Notations

Let us introduce a few terms which will be used later in the paper.

Let $\mathcal{H} = \{h_1, h_2, \ldots, h_{4n}\}$ denote the set of corner points of all rectangles in $\mathcal{R}$.

For any point p, p_x and p_y denote its x and y coordinates respectively.

$SD(s,t)$ will denote the shortest distance between s and t in presence of $\mathcal{R}$. Similarly $SP(s,t)$ will denote the shortest path between s and t in presence of $\mathcal{R}$.

$ASD(s,t)$ will denote the approximate shortest distance between s and t in presence of $\mathcal{R}$.

Given a point s and the set of orthogonal disjoint rectangles $\mathcal{R}$, $+X+Y$ path from s is defined as follows. Start from s, move in $+x$ direction until an obstacle is encountered. Then start moving in $+y$ direction till the motion towards $+x$ direction can be resumed. This procedure is repeated to build up $+X+Y$ path from s. In a similar way $+Y+X$, $+Y-X$, $-X+Y$, $-X-Y$, $-Y-X$, $-Y+X$, $+X-Y$ paths are defined.

3 A Brief Review of Some Existing Results

For the sake of completeness in this section we shall briefly introduce the following two structures from existing literature.

- **Staircase Separator :** The notion of the staircase separator was introduced in [1]. The staircase separator S of $\mathcal{R}$ is an orthogonal chain avoiding the set of obstacles in $\mathcal{R}$. S has $O(n)$ segments and is monotone along both x and y directions. S splits $\mathcal{R}$ to two subsets R_1 and R_2 in such a way that both $|R_1| \leq 7n/8$ and $|R_2| \leq 7n/8$. Also S can be constructed in $O(n \log n)$ time. The existence of S greatly facilitates in the design of divide-and-conquer type algorithms for problems on rectangles since we can obtain a balanced decomposition.
- **Sparse Visibility Graphs :** This graph was introduced in [3]. Given $\mathcal{R}$ we can maintain the shortest-path information between any two corner points $h_i, h_j \in \mathcal{H}$ in this graph. This graph $G = (V, E)$ is defined as follows :
  - From each corner point of every $R_i \in \mathcal{R}$ we extend both of the obstacle edges incident on it until they hit other obstacles. We call these two new points shadow points [Fig. 1(a)]. In the illustration p_m and p_n are shadow points of h_k. Let us denote the set of all shadow points by $\mathcal{P}$. There is an edge between each point and its corresponding shadow points. Let us denote these set of edges by E_1. For each obstacle edge $e_i = (h_i, h_j)$ if projected shadows are $p_l, p_{l+1}, \ldots, p_m \in \mathcal{P}$ in order then following edges $(h_i, p_l), (p_l, p_{l+1}), \ldots, (p_m, h_j)$ are included in E. Here we note that if there is no shadow point on an edge e_i then we include that edge in the graph. Let us denote the set of all edges of this type by E_2.
  - We take a vertical line which passes through a point which has the median value x_m with respect to x coordinates of points belonging to $\mathcal{H}$. We join an edge between each point $h_i \in \mathcal{H}$ to the point $h'_i = (x_m, h_{i_y})$ if it is visible from h_i. Let $h'_l, \ldots, h'_m$ be those Steiner points in sorted order along y directions. Then we join every two adjacent points in that order by an edge if they are mutually visible. Please refer to Fig. 1(b) for this construction. We apply this construction recursively for all points with the value of x coordinate less than x_m and separately for all points with x coordinate values greater than x_m. The same construction is repeated by horizontal splittings. Let E_3 denote the set of edges added by this recursive construction both in vertical and horizontal directions. Also let $\mathcal{Q}$ denote the set of Steiner points introduced by this construction. Here $|\mathcal{Q}| \in O(n \log n)$.

  So in $G = (V, E)$ we have
  - $V = \mathcal{H} \cup \mathcal{P} \cup \mathcal{Q}$.
  - $E = E_1 \cup E_2 \cup E_3$.

  Each edge of the graph is associated with an weight which is L_1 distance between its two endpoints. This graph can be computed in $O(n \log^2 n)$ time. The algorithm is described in [3]. Between any pair of corner points $h_i, h_j \in \mathcal{H}$ the shortest path will be maintained in G, i.e., a subgraph of G.

Here we would like to note that the carrier graph of [6] is more sparse compared to the sparse visibility graph of [3]. But the shortest path between every pair of h_i and h_j is always maintained in the sparse visibility graph. But that is not so in case of carrier graphs. For this reason we shall use the sparse visibility graph.

4 Query Between Two Arbitrary Points

Let s and t denote the two query points. Without loss of generality we shall assume that t lies in the first quadrant of s, i.e., in other words $s_x < t_x$ and $s_y < t_y$. For the first part of our discussion we shall concentrate on the approximate shortest-path query between two arbitrary corner points of $\mathcal{R}$. Subsequently we shall relax this restriction. This is done using the technique used in [1], [6] by reducing the problem of query between two arbitrary points to a query constrained between corner vertices of rectangles. The exact procedure is described in more detail in the later part of the section.

4.1 Query Between Two Arbitrary Corner Points

Now let R_1 and R_2 denote the two subsets of $\mathcal{R}$ after being split by the staircase separator S. Let h_i and h_j be two arbitrary corner points belonging to R_1 and R_2 respectively. The following property which holds for the shortest path between h_i and h_j have been used in [1]. For the sake of completeness we give its proof here.

Lemma 1. *There exists a shortest path between h_i and h_j whose intersection with the staircase S must contain at least one of the horizontal or vertical projection of the corner points of $R_1 \cup R_2$ on S.*

Proof : Let P be the shortest path computed between h_i and h_j using the algorithm of [4]. Now we consider the part of the shortest path which connects the last vertex along P in R_1, namely h_k to the first vertex along P in R_2 namely h_l. Then from the result in [4] we know that h_k and h_l must be visible, and, if we consider the rectangular region $\mathbf{X}$ with the diagonal $h_k h_l$, must be free of any other obstacle. Please refer to Fig. 2 for illustration. In this proof we assume the separating staircase S is of $+Y + X$ type. For other cases exactly similar argument will hold. Now the intersection of $\mathbf{X}$ with S must occur at two points which are either horizontal or vertical projections of h_k or h_l on S. For example in our illustration of Fig. 2 p and q are those points which are horizontal and vertical projections of h_k on S. But in any case we can force the portion of the shortest-path P which connects h_k and h_l to move through those projected points without affecting the length of the path. □

Preprocessing Phase of the Algorithm The preprocessing algorithm consists of building a search tree whose every node maintains the approximate shortest-path information for corner points of $\mathcal{R}$. We shall name this tree the approximate shortest route search tree, or **ASRST**. Now we give a high-level description of the preprocessing phase of our algorithm.

Algorithm Preprocess :

Comments : The results of the computation of the first three steps are stored at the current node of **ASRST**. Both subtrees, built up recursively in the last step, are maintained as children of the current node of **ASRST**.

- **Step I.** Compute the staircase separator S, which gives a balanced decomposition of $\mathcal{R}$ into two subsets R_1 and R_2 as described in [1].
- **Step II.** Project each corner point of R_1 and R_2, both horizontally and vertically onto S, if the line joining the projected point and the corresponding corner point doesn't intersect any other obstacle. Let $\mathcal{T} = \{t_1, t_2, \ldots, t_p\}$ denote the set of all such projected points. This set can be computed by the trapezoidation algorithm.
- **Step III.** Compute the Geodesic Voronoi Diagram of $\mathcal{T}$ on $\mathcal{H}$, the set of corner points in $\mathcal{R}$. More formally we partition $\mathcal{H}$ into subsets $H_1, H_2, \ldots, H_p$ such that $\bigcup_{i=1}^{p} H_i = \mathcal{H}$ and for each corner point $h_j \in H_i$, t_i is geodesically nearer to it compared to $t_j \neq t_i$. The detailed algorithm to compute this efficiently will be given later in this section.
- **Step IV.** Recursively build up **ASRST** on both R_1 and R_2. □

Now let us consider two arbitrary corner points of $\mathcal{R}$, say h_i and h_j, which lie on two opposite sides of the staircase S. Let t_k and t_l belonging to $\mathcal{T}$ be the geodesically nearest point of h_i and h_j respectively. Now we consider the following path :

$Q = SP(h_i, t_k) \cup SP(t_k, t_l) \cup SP(t_l, h_j)$

Here we note that $SP(t_k, t_l)$ is taken along the staircase S. Then the following property holds on the path Q.

Theorem 2. *The length of the path Q between h_i and h_j is at most three times the length of the shortest path $SP(h_i, h_j)$.*

Proof : Please refer to Fig. 3(a) and 3(b) for illustrations. From **Lemma 1** we know that there always exists a shortest path whose intersection with $\mathcal{T}$ is nonempty. Let P be the shortest path having this property. Let $t_i \in \mathcal{T}$ be one of the point in $P \cap \mathcal{T}$. There are a few cases to be considered here. In one case t_k is above t_l and in another case t_k is below t_l. But since the same argument holds in both cases we shall elaborate on the first case only. Again three possible cases can arise when the first condition holds.

- **Case I.** t_i is beteween t_k and t_l.
- **Case II.** t_i is above t_k.

- **Case III.** t_i **is below** t_l.

Since Case III and Case II are symmetrical we shall argue for the first two cases only.

- **Case I:** Here let us consider the ratio between $SD(h_i, t_k) + SD(t_k, t_i)$ to $SD(h_i, t_i)$. To design the worst case situation we have to make $SD(t_k, t_i)$ as large as possible. Now we claim that $SD(t_k, t_i) \leq 2 \times SD(h_i, t_i)$. This follows from subsequent arguments.

  From the triangle inequality of L_1 metric we have :-
  $SD(t_k, t_i) = L_1(t_k, t_i) \leq L_1(h_i, t_k) + L_1(h_i, t_i)$
  $\leq SD(h_i, t_k) + SD(h_i, t_i) \leq SD(h_i, t_i) + SD(h_i, t_i)$
  The last inequality holds because t_k is the geodesic Voronoi neighbour of h_i in the set T and therefore $SD(h_i, t_k) \leq SD(h_i, t_i)$.

  Thus we have $SD(t_k, t_i) \leq 2 \times SD(h_i, t_i)$.
  Therefore the total distance traversed is equal to :-

  $$SD(h_i, t_k) + SD(t_k, t_i) \leq SD(h_i, t_i) + 2 \times SD(h_i, t_i) = 3 \times SD(h_i, t_i) \quad (1)$$

  Following exactly similar argument we obtain

  $$SD(h_j, t_l) + SD(t_l, t_i) \leq 3 \times SD(h_j, t_i) \quad (2)$$

  Therefore adding (1) and (2) we obtain :-

  $SD(h_i, t_k) + SD(t_k, t_l) + SD(t_l, h_j) \leq 3 \times (SD(h_i, t_i) + SD(t_i, h_j)) = 3 \times SD(h_i, h_j)$

- **Case II:** Following the same argument of Case I we claim that

  $SD(h_j, t_l) + SD(t_l, t_i) \leq 3 \times SD(h_j, t_i)$

  $$\Rightarrow SD(h_j, t_l) + SD(t_l, t_k) \leq 3 \times SD(h_j, t_i) \quad (3)$$

  since $SD(t_l, t_k) \leq SD(t_l, t_i)$.

  Again t_k is the geodesic Voronoi neighbour of h_i. Therefore

  $$SD(h_i, t_k) \leq SD(h_i, t_i) \quad (4)$$

  Combining (3) and (4) we have

  $SD(h_i, t_k) + SD(t_l, t_k) + SD(h_j, t_l) \leq SD(h_i, t_i) + 3 \times SD(h_j, t_i) \leq 3 \times (SD(h_j, t_i) + SD(t_i, h_i)) = 3 \times SD(h_i, h_j)$ □

So from the above theorem we can see that Q is a near optimal path between h_i and h_j. Now we shall describe how to compute the **Step III** of the algorithm **Preprocess** efficiently. For that purpose we shall make use of the result in [13] which computes the Voronoi diagram of a subset of vertices in a graph. More formally given a weighted graph $G = (V, E)$ and $V' \subset V$ where $|V'| = k$, we want to partition V into subsets $V_1, V_2, \ldots, V_k$ such that $\bigcup_{i=1}^{k} V_i = V$ and $v_i \in V'$ is nearer, to all vertices of V_i in G, compared to other $v_j \in V'$, $j \neq i$. In all cases we shall assume edge weights to be non-negative. Here we briefly describe the method of [13] to compute this partition.

Lemma 3. *Given a weighted graph $G = (V, E)$ and $V' \subset V$ we can compute the above mentioned Voronoi partition of V in $O(|E| \log |V|)$ time.*

Proof: We add a new vertex s and join it to each vertex $v_i \in V'$ with edges of weight 0. Now we run the single-source shortest-path algorithm from s in this transformed graph. Now in the shortest-path tree of s, all vertices hanging in the subtree rooted at $v_i \in V'$ will be put into the set V_i. From the construction it directly follows that any vertex in V_i is nearer to v_i than any other $v_j \in V'$ where $j \neq i$. The desired complexity can be achieved using the heap implementation of [5] for the single source shortest-path algorithm. □

Lemma 4. *The* **step III** *of the algorithm* **Preprocess** *can be computed in $O(n \log^2 n)$ time at the first level of recursion.*

Proof : The Voronoi diagram of $\mathcal{T}$ on $\mathcal{H}$ can be computed in the following way :-

- **Step I.** Compute the sparse visibility graph G_1 for $\mathcal{T} \cup R_1$ and G_2 for $\mathcal{T} \cup R_2$.
- **Step II.** Apply the algorithm of [13] described in **Lemma 3** to both G_1 and G_2 respectively with all vertices of $\mathcal{T}$ as the set V'.

The computation of **step I** requires $O(n \log^2 n)$ time [3]. Since $|V| = O(n \log n)$ and $|E| = O(n \log n)$ for the sparse visibility graph G **step II** can be carried out in $O(n \log^2 n)$ time from **Lemma 3**. □

Theorem 5. *The algorithm* **Preprocess** *requires $O(n \log^3 n)$ time and $O(n \log^2 n)$ space.*

Proof : The time required to compute the staircase separator in **step I** is $O(n \log n)$ [1]. To compute $\mathcal{T}$ using trapezoidation will take $O(n \log n)$ time. Therefore using **Lemma 4** we obtain the following recurrence relation for the overall time complexity $T(n)$, for the algorithm **Preprocess** as :-

$$T(n) \leq T(\alpha n) + T((1 - \alpha)n) + O(n \log^2 n),\ 1/8 \leq \alpha \leq 7/8$$

Therefore, $T(n) \in O(n \log^3 n)$. At each level of recursion to store the complete approximate shortest-path information the space required is $O(n \log n)$. Therefore the total space required is $O(n \log^2 n)$. □

Algorithm to answer the query In this section we shall describe how to answer the approximate shortest-distance query between two arbitrary corner points. Let h_i and h_j be the two query points. We use the following procedure to answer the approximate shortest-distance query between h_i and h_j.

Algorithm Query :
Input : h_i, h_j; **Output :** $ASD(h_i, h_j)$;

- **Step I.** Using binary search over the staircase S stored in root of **ASRST**, determine if h_i and h_j lie on the opposite or the same side of S or on S.
- **Step II.** If h_i and h_j lie on opposite sides of S or if at least one of h_i or h_j lie on S compute the geodesic Voronoi neighbours of h_i and h_j from the **ASRST** built during the preprocessing step. Let t_k and t_l respectively be the geodesic Voronoi neighbours of h_i and h_j. We report $SD(h_i, t_k) + SD(t_k, t_l) + SD(t_l, h_j)$ as the approximate shortest distance between h_i and h_j, i.e., $ASD(h_i, h_j)$ and **stop**. Otherwise if h_i and h_j lie on the same side of S then go to **step III**.
- **Step III.** Go to the corresponding child node in the **ASRST** tree, depending on which side of S h_i and h_j are, and repeat from **step I**.

Lemma 6. *Algorithm* **Query** *reports a distance which is at most three times the optimal path length between h_i and h_j in $O(\log^2 n)$ time.*

Proof : The proof of the fact that the distance reported by the algorithm **Query** is at most three times the optimal path length directly follows from **Theorem 2**.

To carry out a binary search in **step I** requires $O(\log n)$ time. **Step II** can be computed in $O(1)$ time. Therefore the worst case running time $Q(n)$ of the algorithm **Query** follows from the following recurrence :-

$$Q(n) \leq Q(7n/8) + O(\log n)$$

Therefore $Q(n) \in \log^2 n$. □

Thus we have established the following theorem.

Theorem 7. *Between two arbitrary corner points belonging to $\mathcal{H}$, an approximate orthogonal shortest distance through a path avoiding $\mathcal{R}$ whose length is at most three times the optimal length can be obtained in $O(\log^2 n)$ query time using $O(n \log^3 n)$ preprocessing and $O(n \log^2 n)$ space.*

4.2 Removal of the Corner Point Restriction

In this section we shall relax the restriction of the two query points being two corner points in the set $\mathcal{H}$. The approach here is very similar to the methods of [1], [6]. Let those two query points be s and t respectively. Using the algorithm of [9] we preprocess the planar subdivision formed by the trapezoidation of $\mathcal{R}$

for ray shooting queries along both horizontal and vertical directions. In the subsequent discussion we shall use $Dist(a,b)$ to denote L_1 distance between two points a and b. Then we use the following steps to determine the approximate shortest distance between s and t.

Algorithm General-Query :

Comments : Please refer to Fig. 4 for the illustration.

- **Step I.** Shoot a horizontal ray towards right from s. Let h_1 and h_2 be two vertices on the edge of the rectangle hit by the ray. From t we shoot another horizontal ray towards left and let h_3 and h_4 be two vertices on the edge of the rectangle hit by the ray. If $h_{1_x} > h_{3_x}$ then $ASD_x(s,t) = +\infty$ else compute $ASD_x(s,t)$ as

  $$min\{Dist(s,h_i)+ASD(h_i,h_j)+Dist(h_j,t)\} \text{ where } i \in \{1,2\} \text{ and } j \in \{3,4\}.$$

  In the above expression $ASD(h_i,h_j)$, where $i \in \{1,2\}$ and $j \in \{3,4\}$, is computed using the algorithm **Query**.
- **Step II.** Shoot a vertical ray up from s. Let h'_1 and h'_2 be two vertices on the edge of the rectangle hit by the ray. Shoot another vertical ray downwards from t. Let h'_3 and h'_4 denote the corresponding corner points. If $h'_{1_y} > h'_{3_y}$ then $ASD_y(s,t) = +\infty$, else compute $ASD_y(s,t)$ as

  $$min\{Dist(s,h'_i)+ASD(h'_i,h'_j)+Dist(h'_j,t)\} \text{ where } i \in \{1,2\} \text{ and } j \in \{3,4\}.$$

  In the above expression $ASD(h'_i,h'_j)$, where $i \in \{1,2\}$ and $j \in \{3,4\}$, is computed using the algorithm **Query**.
- **Step III.** If $h'_{2_y} \leq h_{4_y}$ and $h'_{2_x} \leq h_{4_x}$ then compute $ASD_{xy}(s,t) = Dist(s,h'_2)+ ASD(h'_2,h_4) + Dist(h_4,t)$ else $ASD_{xy}(s,t) = +\infty$. If $h_{1_x} \leq h'_{3_x}$ and $h_{1_y} \leq h'_{3_y}$ then compute $ASD_{yx}(s,t) = Dist(s,h_1) + ASD(h_1,h'_3) + Dist(h'_3,t)$ else $ASD_{yx}(s,t) = +\infty$. If all $ASD_x(s,t) = ASD_y(s,t) = ASD_{xy}(s,t) = ASD_{yx}(s,t) = +\infty$ then $SD(s,t)$ is equal to $|t_x - s_x| + |t_y - s_y|$ else $ASD(s,t) = min\{ASD_x(s,t), ASD_y(s,t), ASD_{xy}(s,t), ASD_{yx}(s,t)\}$.

Lemma 8. *The algorithm* **General-Query** *correctly computes the approximate shortest distance between s and t, which is at most seven times the optimal distance, and the time complexity of the algorithm is $O(\log^2 n)$.*

Proof : Since t lies in the first quadrant of s the shortest path between them can be of three types from the result of [4] :

- **Case I.** The shortest path is monotone only in $+x$ direction.
- **Case II.** The shortest path is monotone only in $+y$ direction.
- **Case III.** The shortest path is monotone in both $+x$ and $+y$ directions.

In **Case I** from the result of [4] we know that the shortest path will start from s and will either go through h_1 or h_2 and will enter t through either h_3 or h_4. So **Step I** of the algorithm **General-Query** exhaustively computes all four types of paths. Also the minimum of all four types of paths computed in **Step I** must be at most three times the shortest path between s and t in **Case I**.

In a similar way we observe that **Step II** of the algorithm **General-Query** computes a path which is at most three times the optimal in **Case II**.

Thus if **Case I** or **Case II** holds after the completion of **Step I** and **Step II**, i.e., in **Step III**, by finding the minimum of $ASD_x(s,t)$ and $ASD_y(s,t)$, we are guaranteed to get the approximate shortest distance between s and t through a path which is at most three times the optimal path.

Otherwise if all $ASD_x(s,t) = ASD_y(s,t) = ASD_{xy}(s,t) = ASD_{yx}(s,t) = +\infty$ as is the case between s and t' in Fig. 4, then from the result of [4] we know that **Case III** definitely holds. So in that case we get the exact shortest distance between s and t in **Step III**. If **Case III** holds and $ASD_{xy}(s,t) \neq +\infty$ then we bound the path length in the following way. If the shortest-path between h_2' and h_4 is monotone in both directions then the fact that $ASD(s,t) \leq 3 \times SD(s,t)$ follows directly from **Theorem 2**. Otherwise as in the situation between s and t_1 in Fig. 4 we have $ASD(t_1,s) \leq Dist(t_1,q) + ASD(q,h_2') + Dist(h_2',s)$. Now $SD(q,h_2') \leq SD(q,x) + SD(x,h_2') \leq 2 \times SD(x,h_2')$, since $h'_{2_x} < q_x$ and $h'_{2_y} < q_y$ implies $SD(q,x) = Dist(q,x) \leq SD(x,h_2')$. Therefore $ASD(t_1,s) \leq Dist(x,q) + Dist(t_1,x) + 6 \times SD(x,h_2') + Dist(h_2',s) \leq Dist(x,q) + 6 \times SD(t_1,s) \leq 7 \times SD(t_1,s)$. In other situations when **Case III** holds following similar arguments we can show $ASD(s,t) \leq 7 \times SD(s,t)$.

The time required by the algorithm **General-Query** is $O(\log n)$ time for the point location query and $O(\log^2 n)$ time for each approximate shortest distance computation between corner points using the algorithm **Query** in **Step I** and **Step II**. Therefore the overall time complexity of the algorithm **General-Query** is $O(\log^2 n)$. □

Here we note that the actual path between s and t can be reported in $O(\log^2 n + k)$ time where k is the number of segments in the reported path. This follows from the fact that if at least one among $ASD_x(s,t)$ or $ASD_y(s,t)$ is not equal to infinity then the actual path can be reported from the approximate shortest-path information maintained in the preprocessing phase in **ASRST**. Otherwise if both $ASD_x(s,t) = ASD_y(s,t) = +\infty$ then we already know that the shortest path between between s and t is monotone in both $+x$ and $+y$ directions. In that case the exact shortest path can be reported by growing two monotone staircases, one $+Y+X$ type staircase from s and another $-X-Y$ type staircase from t. The existence of intersection between these two types of paths follows from [4] because the shortest path between s and t is monotone in both $+x$ and $+y$ directions. Also these two staircase paths are readily available in the sparse visibility graph computed in the preprocessing phase.

Thus we have established the following theorem.

Theorem 9. *Between two arbitrary query points an approximate orthogonal shortest distance through a path whose length is at most seven times the optimal* *length can be obtained in $O(\log^2 n)$ query time using $O(n \log^3 n)$ preprocessing and $O(n \log^2 n)$ space. The actual path can be reported in $O(\log^2 n + k)$ time where k is the size of the output path.*

5 Conclusion and Open Problems

In this paper we have studied approximation algorithms for the rectilinear shortest-path query problem in the presence of a set of isothetic disjoint rectangles. The next step should be to design efficient approximation algorithms for the shortest-path query problem in the presence of other types of polygonal obstacles.

References

1. M. J. Atallah and D. Z. Chen, Parallel rectilinear shortest paths with rectangular obstacles, *Proc. 2nd Annual ACM Symposium on Parallel Algorithms and Architecture*, 1990, pp. 270-279.
2. K. Clarkson, Approximation algorithms for shortest path motion planning, *Proc. 19th Annual ACM Symposium on Theory of Computing*, 1987, pp. 56-65.
3. K. L. Clarkson, S. Kapoor and P. M. Vaidya, Rectilinear shortest paths through polygonal obstacles, *Proc. 3rd Annual Conf. Computational Geometry*, 1987, pp. 251-257.
4. P. J. de Rezende, D. T. Lee and Y. F. Wu, Rectilinear shortest paths in the presence of rectangular obstacles, *Discrete Comput. Geom. 4*, 1989, pp. 41-53.
5. E. W. Dijkstra, A note on two problems in connexion with graphs, *Numer. Math. 1*, 1959, pp. 269-271.
6. H. Elgindy and P. Mitra, Orthogonal shortest route queries among axes parallel rectangular obstacles, *Int. J. of Comput. Geom. and Applications*, to appear.
7. L. J. Guibas and J. Hershberger, Optimal shortest path queries in a simple polygon, *Proc. 3rd Annual Symposium on Computational Geometry*, 1987, pp. 50-63.
8. L. Guibas, J. Hershberger, D. Leven, M. Sharir and R. Tarjan, Linear time algorithms for visibility and shortest path problems inside simple polygons, *Proc. 2nd Annual Conf. Computational Geometry* , 1986, pp. 1-13.
9. D. G. Kirkpatrick, Optimal search in planar subdivisions, *SIAM J. Comp. 12*, 1983, pp. 28-35.
10. D. T. Lee and F. P. Preparata, Eucledian shortest paths among rectilinear barriers, *Networks, 11*, pp. 393-410.
11. J. S. B. Mitchell, L_1 shortest paths among polygonal obstacles in the plane, *Algorithmica, 8(1)*, 1992, pp. 55-88.
12. J. S. B. Mitchell, Algorithmic approaches to optimal route planning, *Technical Report No. 937, School of Operations Research and Industrial Engineering, Cornell University*, 1990.
13. K. Mehlhorn, A faster approximation algorithm for the Steiner problem in graphs, *Information Processing Letters 27*, 1988, pp. 125-128.
14. Y. F. Wu, P. Widmayer, M. D. F. Schlag and C. K. Wong, Rectilinear shortest paths and minimum spanning trees in the presence of rectilinear obstacles, *IEEE Transactions on Computers, 36(3)*, 1987, pp. 321-331.

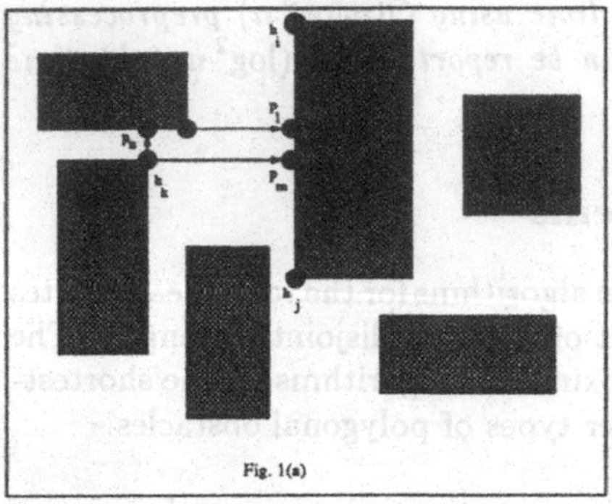

Fig. 1(a)

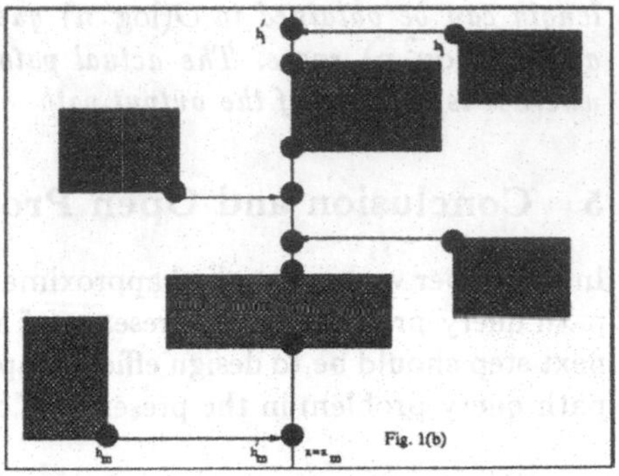

Fig. 1(b)

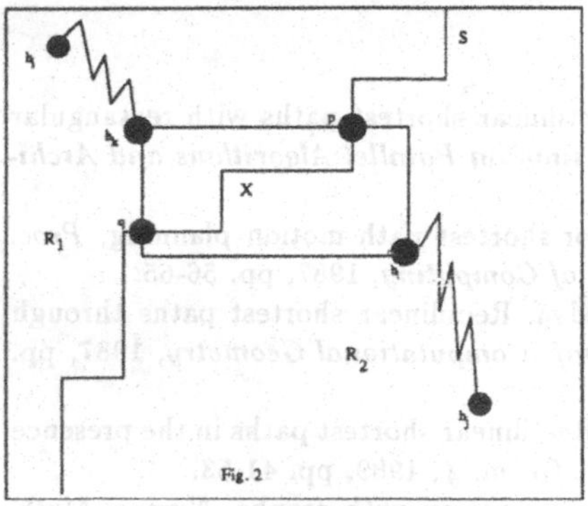

Fig. 2

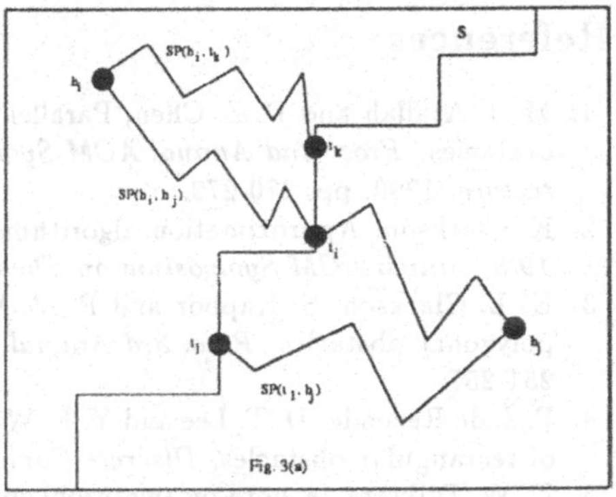

Fig. 3(a)

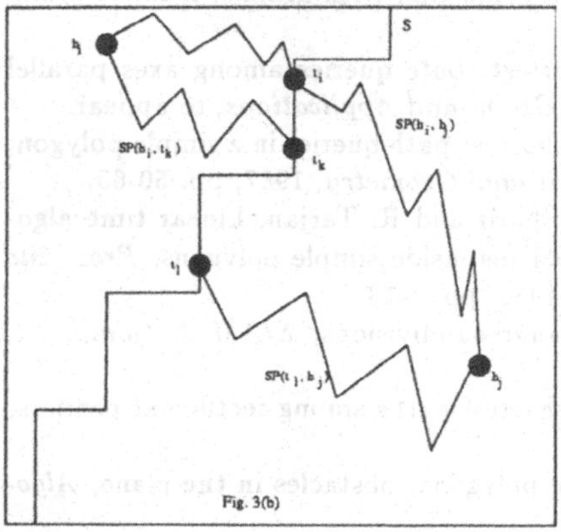

Fig. 3(b)

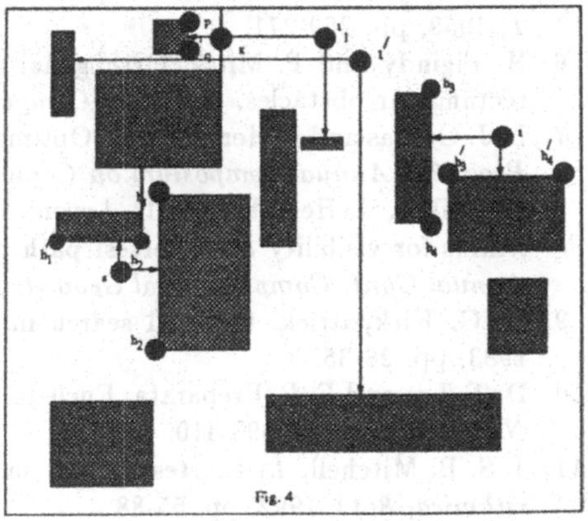

Fig. 4

Counting and Reporting Red/Blue Segment Intersections

Larry Palazzi Jack Snoeyink*
Department of Computer Science
University of British Columbia

Abstract

We simplify the red/blue segment intersection algorithm of Chazelle et al: Given sets of n disjoint red and n disjoint blue segments, we count red/blue intersections in $O(n \log n)$ time using $O(n)$ space or report them in additional time proportional to their number. Our algorithm uses a plane sweep to presort the segments; then it operates on a list of slabs that efficiently stores a single level of a segment tree. With no dynamic memory allocation, low pointer overhead, and mostly sequential memory reference, our algorithm performs well even with inadequate physical memory.

1 Introduction

Geographic information systems frequently organize map data into various layers. Users can make custom maps by overlaying roads, political boundaries, soil types, or whatever features are of interest to them. The ARC/INFO system [7] is organized around this model; even a relatively inexpensive database like the Digital Chart of the World [8] contains seventeen layers, several with sublayers. An algorithm for map overlay must be able to handle large amounts of data and compute the overlay quickly for good user response performance.

We consider a geometric abstraction of the map overlay problem. Suppose $\mathcal{R}$ is a set of red line segments in the plane and $\mathcal{B}$ is a set of blue segments such that no interiors of segments of the same color intersect. The ***red/blue segment intersection problem*** asks for an efficient algorithm to count or report the red/blue intersections.

Chazelle et al. [3] give output-sensitive solutions for this problem, meaning that the running time of their algorithms depends on the amount of output. They outlined relatively simple algorithms that run in $O(n \log^2 n + K)$ time and use $O(n \log n)$ space, where K is the number of intersections for the reporting problem and $K = 1$ for the intersection counting problem. We describe their method in section 2. They also state that the space can be reduced to linear by streaming [6] and the time to $O(n \log n + K)$ by a dynamic form of fractional cascading [4, 5], which they admit is complicated. This paper presents an alternate way to reduce space that yields a much simpler approach to reducing the time.

*Supported in part by an NSERC Research Grant

The red/blue intersection problem was first considered while researchers were searching for general output-sensitive line segment intersection routines. Shamos and Hoey [12] gave a plane-sweep algorithm to detect an intersection in $\Theta(n \log n)$ time. Bently and Ottman [1] turned their algorithm into a general intersection reporting procedure that runs in $O((n+K)\log n)$ time and uses linear space. Mairson and Stolfi [10] applied plane-sweep to the red/blue intersection problems and obtained $O(n)$ space algorithms for reporting in $O(n \log n + K)$ time and for counting in $O(n \log n + K^{1/2})$ time. Chazelle and Edelsbrunner [2] finally gave an output-sensitive algorithm for the general intersection problem: their algorithm runs in $O(n \log n + K)$ and uses $O(n + K)$ space. One can, of course, use these general routines to report red/blue intersections if one is willing to pay the time penalty of Bently-Ottman or the space penalty of Chazelle and Edelsbrunner. They do not, however, adapt to efficiently solve the intersection-counting problem.

2 Preliminaries

The conceptual structure of the red/blue intersection procedure of Chazelle et al. [3] is easy to state once we define the ***hereditary segment tree*** data structure to store the set $S = \mathcal{R} \cup \mathcal{B}$ of red and blue segments. As the name implies, this is a modification of the ***segment tree*** [11]. Our definition of a segment tree is slightly non-standard—it uses midpoints instead of endpoints to define vertical slabs.

Let $\{x_1, x_2, ..., x_k\}$ be the set of distinct x-coordinates of segment endpoints in increasing order. Form the set of midpoints $M = \{m_1, m_2, ..., m_{k+1}\}$, where $m_1 = -\infty$, $m_{k+1} = \infty$ and $m_i = (x_{i-1} + x_i)/2$, for $1 < i \le k$. Then form a balanced binary tree on k leaves such that the ith leaf node ν_i is associated with the ***leaf slab*** $s(\nu_i)$ of all points whose x-coordinates lie in the halfopen interval $[m_i, m_{i+1})$. Notice that the leaf slab $s(\nu_i)$ contains at most one endpoint of a red or blue segment. Each internal node ν' is associated with the slab $s(\nu')$ that is the union of the leaf slabs in the subtree rooted at ν'.

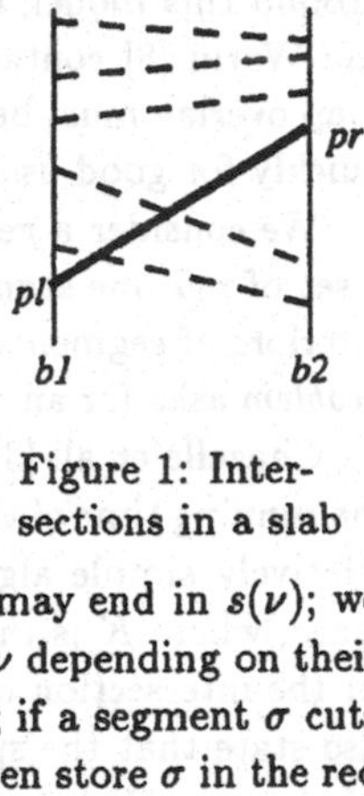

Figure 1: Intersections in a slab

Now, let us look at the relation of the red and blue segments to the slab $s(\nu)$ of an internal or leaf node ν. Some segments may end in $s(\nu)$; we call them ***short*** in ν and store them in red or blue short lists in ν depending on their color. Others, which we call ***long***, cut completely through $s(\nu)$; if a segment σ cuts through $s(\nu)$ and not through the parent's slab $s(parent(\nu))$, then store σ in the red or blue ***long*** list for ν.

Lemma 2.1 *On each level of the tree, a segment is stored in at most two short lists and two long lists.*

Proof: The slabs stored at any given level of the tree are disjoint. A segment, σ, is stored *short*, therefore, in the at most two slabs that contain its endpoints. σ is stored *long* in at most one child of each node where σ is stored *short*. ∎

We can efficiently compute the intersections between long red segments and blue segments in the slab $s(\nu)$. To begin, sort the long red segments vertically within the slab $s(\nu)$. Then clip each blue segment to the slab and locate the endpoints of each clipped blue segment in the red long list by binary search. The red segments that a clipped blue segment intersects are exactly those between the blue endpoints—one can report them in time proportional to their number or count them in constant time by subtracting the ranks of the segments above the blue endpoints (see figure 1). Similarly, one can report the intersections of short red segments with long blue segments that appear in the slab.

If we perform this procedure for every tree node ν—reporting the intersections between the long red and long and short blue segments and the long blue and short red segments—then we can show that every intersection is reported exactly once.

Lemma 2.2 *Every intersection point is the intersection of a long segment and another segment in exactly one slab.*

Proof: Consider an intersection point of a red segment r and a blue segment b, namely $p = r \cap b$. In the leaf slab that contains p, there is at most one short segment, so either r or b must be long. Assume that b is long, and if r is also long assume that r is not stored higher than b.

Let ν be the node that stores b as *long*. By the assumptions, r is stored either as long or short at ν, so the intersection point p will be reported at ν. Since the portion of b containing p is stored long only at ν and the portion of r is not stored long above ν, the point p is reported only at ν. ∎

How much space and time is taken by this procedure, excluding the amount used to report output? If we construct the entire segment tree, each segment is stored in at most four slabs per level by lemma 2.1, so the total space is $O(n \log n)$. In each slab we sort long segments and locate long and short segments; both can be done in $O(\log n)$ time per segment. This gives a total of $O(n \log^2 n)$ time.

This algorithm and analysis is contained in Chazelle et al. [3]. They also state that one can remove the logarithmic factor from the space by the technique of *streaming* [6]: rather than building the entire segment tree, one builds the succession of root to leaf paths, starting with the path to the leftmost leaf and ending with the path to the rightmost leaf. In moving from one path to the next only the nodes that change need to be recomputed. They also state that a logarithmic factor can be removed from the time bound by using a dynamic form of fractional cascading [4]: because each endpoint will be located in $O(\log n)$ lists, sharing elements between the lists allow repeated searches to be performed more efficiently.

In the next section we develop an alternate way of reducing the space to linear that gives an easier way to reduce the time to $O(n \log n)$. It has the advantage that it eliminates the segment tree data structure and replaces it by a list of slabs. This is an advantage because the overhead of a segment tree is not always negligible in practice. Also, the capability of preprocessing data into sorted segments allows us to perform the sort on a data set only once and store it in its sorted format for future reference. Another major advantage is due to the sequential nature of our algorithm. This results in localized memory references which reduces memory swapping considerably and allows running of large numbers of segments.

3 The improved algorithm

We have already made a key observation that allows us to reduce the space to linear. Since by lemma 2.1 each segment appears in at most four lists in each level, we can store a complete level of the segment tree in linear space. Thus, rather than streaming, we will compute the tree level by level.

An extra logarithmic factor enters the time complexity of Chazelle et al. [3] in two ways: sorting the segments in the long lists within each slab and locating in these long lists the endpoints of segments that are clipped to the slab. We can do most of the sorting and point location in advance: Define the *aboveness* relation on sets in the plane: $A \succ B$ if there are points $(x, y_A) \in A$ and $(x, y_B) \in B$ with $y_A > y_B$ (see figure 2). When applied to disjoint convex sets, the aboveness relation is acyclic [11].

Lemma 3.1 *The aboveness relation for disjoint convex sets in the plane is acyclic.*

Proof: Suppose that $A_1 \succ A_2 \succ \cdots \succ A_k \succ A_1$ is a cycle of minimum length. We will derive a contradiction. (Note that $k > 2$.)

Let $x_0 = \min_{\forall j} \max \{ x \mid (x, y) \in A_j \}$ and let A_i be a set that attains x_0 as in figure 2. The line $x = x_0$ must intersect A_{i-1} above A_i above A_{i+1} since A_{i-1} and A_{i+1} are comparable to A_i. Omitting A_i, therefore, gives a smaller cycle, which contradicts minimality. ∎

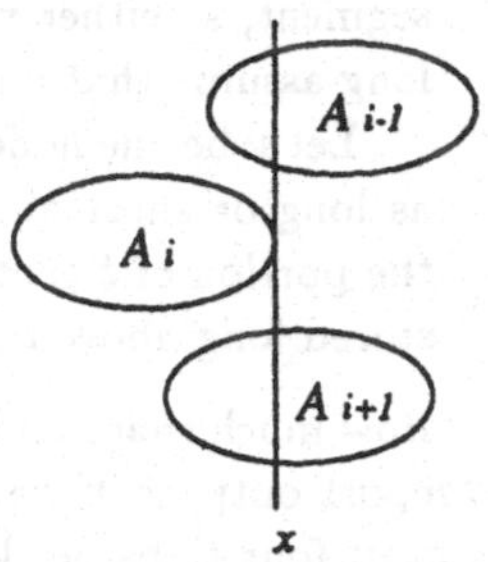

Figure 2: $A_{i-1} \succ A_i \succ A_{i+1}$ and $A_{i-1} \succ A_{i+1}$.

On the set of red segments and blue endpoints, extend the partial order defined by $\succ$ to a total order. In section 3.2 we describe how to do this efficiently using a simple sweep algorithm. If we add the red segments and blue endpoints to a level of the tree according to this order, two things happen automatically: 1) in each slab, the long red segments are inserted in sorted order, and 2) when a blue endpoint appears in a slab then the segment immediately above it was the last to be added to the slab. Thus, the sorting of long red segments can be omitted and (original) endpoint location is a simple matter of looking at the last long red segment added to the slab containing the blue endpoint.

The only task that remains is locating the clipped ends of blue segments. If we total-order the set of blue segments and red endpoints and treat them analogously, then we obtain the clipped blue ends in sorted order along the slab boundaries. Merging this with the long red list gives us the ranks of all clipped blue endpoints, and takes time proportional to the number of long segments and endpoints.

In section 3.1, we describe the data structure requirements for our algorithm. Section 3.2 outlines the sweep algorithm for pre-sorting the segments and points. Section 3.3 outlines the intersection algorithm. Section 3.4 discusses how to handle degenerate cases (in sections 3.2 and 3.3 we will assume that no degeneracies exist in our data).

3.1 Data Structure Requirements

We define a global structure for storing information on each colour. We need one structure for *red* information and one for *blue* information. In each structure, we store the following: the number of points (twice the number of segments), the list of segments stored as point pairs, the list of endpoints sorted by x-coordinate (used by the sorting phase) and the sorted list of segments and endpoints (created by the sorting phase and passed to the intersection phase). To store point information, we define a structure that holds the (x, y) coordinate of each point p, the index to the current slab containing p, a count of the number of long segments above p and a pointer to be used in a linked list of point structures.

During the sorting phase of the algorithm, we use two tree structures. The first tree, which we call the *search tree*, maintains the segments that currently intersect the sweep line in sorted order. The segments are maintained such that segments in the left subtree of any segment, S, are below S and segments in the right subtree are above S. We use this tree for finding the predecessor (segment directly above) of the current point being swept. The second tree, which we call the *sweep tree*, is built during the sweep by making each segment a child of the segment immediately above its right endpoint. When the sweep is complete, the sweep tree holds the set of segments and endpoints so that an inorder traversal gives a total ordering consistent with the aboveness relation.

Figure 3 illustrates a sweep tree. We modify the standard trick of using child and sibling pointers to represent a higher degree tree as a binary tree [9, p. 333], and use pointers to the rightmost child and left sibling. The segment d is the child of c (because its right endpoint is below c), segments b and c are siblings, and c (along with b) is the child of a. The data structures for both trees are the standard structures for binary trees. Each node has a left (sibling) pointer and a right (child) pointer. The search tree and sweep tree structures are needed during the sorting phase only. A segment in the search tree also has a pointer to its location in the sweep tree.

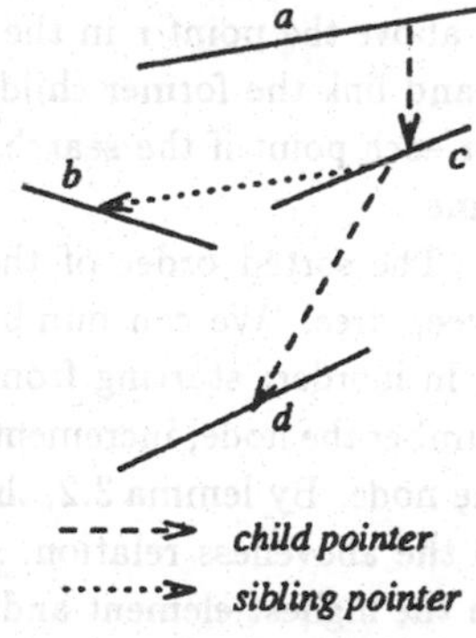

Figure 3: Sweep Tree

For the intersection phase, we define a structure to store the slab information. For each slab, we define two head pointers to linked lists storing the long segments on the left and right boundaries of the slab. We also store two counts for the number of long segments above points on the left and on the right boundaries of the slab. A list of these structures represents the list of slabs at the current level of the segment tree. Our convention is to number the slabs starting from zero so that each *even/odd* pair of slabs represents the two nodes in the segment tree that will be merged together in the next level of the segment tree. To form the acutal slab boundaries we need the list of midpoints—the points between x-coordinates of endpoints as described in section 2.

3.2 The Sweep Algorithm For Pre-Sorting

We initialize the sweep tree to a node H containing a horizontal line from $(-\infty, \infty)$ to (∞, ∞) so that all segments and endpoints will be below this line. Before the sweep begins, a node for each segment and endpoint is pre-allocated. As the sweep proceeds, these nodes are linked together forming a forest of trees. Eventually, all of these subtrees will be linked to H, forming the final sweep tree. The head node, H, will then have the entire list of segments and endpoints as its child subtree and will not have any siblings. The total sorted order of the segments and points can then be recovered by traversing this tree in inorder.

Next, the sets $\mathcal{R}$ and $\mathcal{B}$ are each sorted individually by the smaller x-coordinate of each segment. The endpoints in these lists will be swept from left to right by increasing x-coordinate. We then call this sorting routine once for the red segments and blue endpoints, and once for the blue segments and red endpoints, creating two sorted lists. We describe the proceedure only for red segments and blue endpoints.

The sweep begins with the line $x = -\infty$ that intersects only the dummy red segment from $(-\infty, \infty)$ to (∞, ∞). When the first endpoint, p, of a red segment is encountered, we insert the segment into the search tree. When the second red endpoint, q, is encountered, we delete the red segment $\overline{pq}$ from the search tree and, in the sweep tree, link $\overline{pq}$ as the child of the segment above q and make the former child the sibling of $\overline{pq}$. When a blue point r is encountered, we find the red segment, s, above the point r in the search tree and, in the sweep tree, link r as the child of s and link the former child as the sibling of r. This process takes logarithmic time for each point if the search tree is kept balanced—all other operations are constant time.

The sorted order of the segments and points can now be recovered from the sweep tree. We can number the nodes of the tree from 1 to the highest number, n, in inorder: starting from the root, we recursively number the siblings of a node, number the node, increment the counter, and then recursively number the children of the node. By lemma 3.2, this gives us a list of segments and points sorted according to the aboveness relation. The first element (segment or endpoint) in the list will be the highest element and the last element in the list will be the lowest element.

Lemma 3.2 *Ordering the segments and endpoints stored in the nodes in increasing order of node numbers gives a total order that is consistent with the aboveness relation.*

Proof: Define the *rightward path* for a segment (or point) s to begin at the left endpoint of s. It then repeatedly extends to the right endpoint of the segment it lies on and extends vertically to the segment above the right endpoint. The rightward paths for two segments (or points) s and t cannot cross: when they meet, they must meet along a segment where they will join.

Look at the segment u where the rightward paths for s and t join. (Recall that a dummy infinite segment is stored at the root of the sweep tree.) if s is above t, then either $u = s$ or s and t are in subtrees of the sweep tree that are rooted at children of u. In the former case, s is the parent of t and an inorder traversal of the sweep tree numbers children of s after s. In the latter case, the root of the subtree containing s is a sibling to the left of the root of the subtree containing t; again, an inorder traversal numbers s before t. ∎

3.3 The Red/Blue Intersection Algorithm

This algorithm takes two topologically ordered lists of segments and points, assigns each segment and point to its slabs and computes the number of intersections in each slab. Figure 4 outlines our intersection algorithm. We pass the head pointers to the sorted lists of red segments and blue endpoints, and blue segments and red endpoints to this routine. When all of the red/blue intersections are found, we return the total number of intersections.

We create a list of slab boundaries (by `prepare_first_slabs()`) so that each slab contains one segment endpoint as described in section 2. This routine also stores with each point p the index to the slab containing p. Now, we count the number of intersections found at each level in the segment tree and return the sum of these totals.

For each level in the segment tree, we must assign each long segment to the proper slabs. To do this the routine `make_longs()` traverses the list of segments and endpoints and inserts the long segments into the slabs in sorted order from highest to lowest. When a segment, S, is encountered, we examine the left and right slab indices, l and r, already stored with each endpoint of S. If l and r are adjacent slabs, or the same slab, then S is not stored long anywhere (see segment S_3 in figure 5). If l is an even slab, then S is stored long in slab $l+1$ (see segments S_2 and S_4 in figure 5). If r is an odd slab, then S is stored long in slab $r-1$ (see segment S_1 and S_4 in figure 5). When an endpoint, p, is encountered, the number of long segments already in the slab is recorded with p as the count of long segments (of opposite colour) above p.

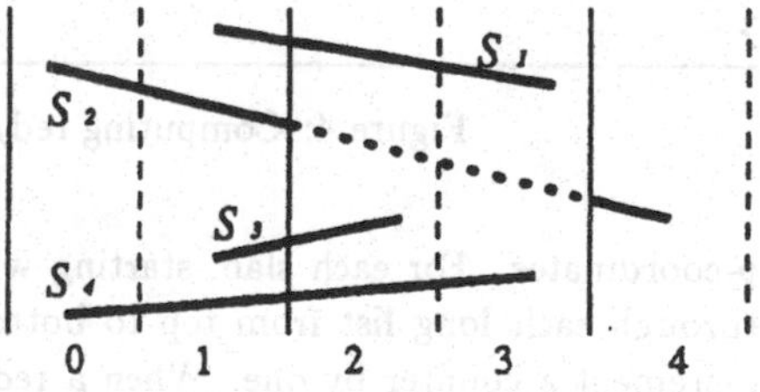

Figure 5: Inserting long segments into slabs

Next, we find the intersections between long red segments and long blue segments (by the routine `total_long_long()`). For each long red segment, we count the number of long blue segments above each of its endpoints. To do this, we *merge* the long red list with the long blue list along both boundaries of each slab by comparing

```
long int intersection_count(red, blue)
    color_data *red, *blue;
{                       /* count intersections given sorted orders      */
  register int i,j;

  nslabs = prepare_first_slabs(red, blue); /* initialize slabs,           */
                          /* storing the slab index for each point */
  do        /* for each level in the segment tree                */
    { clear_slabs(nslabs, red); /* initialize slabs                 */
      clear_slabs(nslabs, blue);
      make_longs(red);              /* put long segments in place      */
      make_longs(blue);

      total_long_long(nslabs, red, blue); /* find intersections        */
      total_long_short(nslabs, red, blue);
      total_long_short(nslabs, blue, red);

      fix_inslab(red);  /* fix indices and halve number of slabs */
      fix_inslab(blue);
      for (i = 1, j = 2; j < nslabs; i++, j +=2)
        mid[i] = mid[j];
      nslabs =  (nslabs + 1) >>1;
      mid[nslabs] = COORDMAX; /* set last midpoint on the right      */
    }
  while (nslabs > 1);    /* we are done when one slab remains        */
  return(total);         /* return number of intersections found     */
}
```

Figure 4: Computing red/blue segment intersections

y-coordinates. For each slab, starting with the left boundary of the slab, we step through each long list from top to bottom. When a blue segment is crossed, we increment a counter by one. When a red segment is reached, we store the current counter with the left red endpoint as its *above* count. (Note that we do not actually create a final merged list since we set the red endpoint counts *during* the merging process.) Similarly, we perform the merge on the right slab boundary. Then by subtracting the left and right counts for each long red segment, we obtain the number of blue segments that cross (intersect) the long red segment. For example, back in figure 1 the number of long blue segments above the left endpoint, *pl*, of the long red segment is 5, and the number for the right endpoint, *pr*, is 3. So the number of intersections along the long red segment is 2. We add the absolute value of the difference between the left and right endpoint counts to the total intersection count.

Finally, we count the intersections between long blue segments and short red segments, remembering that we have already stored the number of long blue segments above each red endpoint with each original red endpoint (in the routine `make_longs()`). We traverse the *original* list of red segments in sorted order. For each red segment we know the slabs containing each endpoint of the segment. If both endpoints are inside the same slab, we simply add the absolute difference of the counts for each endpoint to the total number of intersections. If the slabs are different, we must count the number of long blue segments above each point on the slab boundaries. Starting with the first short red segment, we clip the segment to the slab boundary. As before, we use the y-value of this intersection point to find the number of long blue segments above it on the slab boundary. The absolute difference between this count and the count stored with the starting endpoint is added to the total number of intersections. We do the same for the second short segment.

We use the same procedure to find the intersections between long blue segments and short red segments. Once this is completed for all slabs, we proceed to the next level in the segment tree by throwing out every other slab boundary and merging pairs of adjacent slabs. Then, we update the slab indices stored with each endpoint (by the routine `fix_inslab()`). When all levels of the segment tree have been processed, we return the total number of intersections found.

3.4 Special Cases

So far, we have assumed that no degenerate cases exist in our data. If such situations do arise, we must ensure that they are handled properly.

One possible degeneracy occurs when a red segment, $\overline{rs}$, intersects a blue endpoint, b_1 (b_1 lies somewhere on $\overline{rs}$). This can be handled in the sorting phase of the algorithm. When we reach b_1, we must decide whether b_1 is above or below $\overline{rs}$. To determine this, we must examine the second blue endpoint b_2. If we make one endpoint above $\overline{rs}$ and one below $\overline{rs}$, then we can ensure that the intersection point will be detected during the intersection algorithm. So, if b_2 is above $\overline{rs}$, then we choose b_1 to be below $\overline{rs}$. If b_2 is below $\overline{rs}$, then we choose b_1 to be above $\overline{rs}$. This also handles the situation when a red and blue segment intersect at an endpoint.

If the point b_2 is also on $\overline{rs}$, then the two segments are colinear. In this case, we must test to see if the two segments intersect only at an endpoint, in which case we arbitrarily make one endpoint above $\overline{rs}$ and one below $\overline{rs}$. If there are an infinite number of intersection points, then we report this to the user (in addition to the final intersection count).

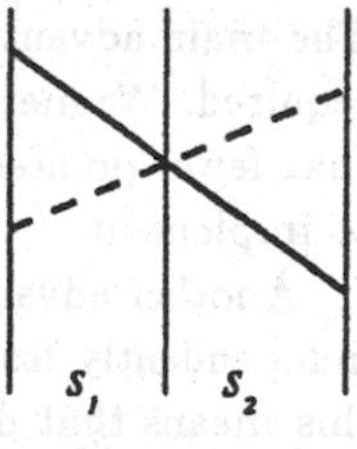

Figure 6: Intersection on boundary

A more interesting situation occurs when a red and blue segment intersect at the boundary dividing two slabs, s_1 and s_2 (left and right respectively), as in figure 6. We want the intersection point to be detected in only one slab, either s_1 or s_2. By our definition of slabs in section 2, only the left slab boundaries are stored with each slab, not the right boundaries. This means that the intersection should be detected in s_2, not in s_1. When we merge the y-values on slab boundaries for finding the number of long segments above a point, if we detect equal y values then we can count the intersection point only if we are on the *left* slab boundary.

Data Set	Number of Segments	No. that Intersect	Sun 4/75 (secs) total	after	SGI Crimson total	after
Complete	400 × 400	160 000	0.42	0.27	0.14	0.09
Grid	4 000 × 4 000	16 000 000	5.93	4.00	1.77	1.12
	40 000 × 40 000	1.6×10^9	79.40	52.07	29.81	19.95
	200 000 × 200 000	4×10^{10}	—	—	181.88	123.79
Horizontal	4 000 × 4 000	15297	7.19	5.30	1.89	1.34
& Slanted	4 000 × 4 000	290 876	10.70	7.18	2.88	1.79
	40 000 × 40 000	1 523 785	118.67	79.93	35.55	24.32
	40 000 × 40 000	29 249 076	181.72	99.52	61.23	36.94
roads/survey	11 074 × 239	536	8.87	5.33	2.70	1.51
roads/vegitat	11 074 × 5 562	202	14.12	8.98	4.130	2.44
forest/compart	116 359 × 8 053	4 637	128.82	80.97	40.22	24.39
biogeo/compart	235 635 × 8 053	3 548	—	—	81.93	49.73
biogeo/forest	235 635 × 116 359	50 045	—	—	136.66	92.53

Table 1: Total and "after-sorting" execution times on a 16 Meg Sun 4/75 (spark 2) and a 64 Meg SGI Crimson

4 Results of implementation

We have implemented this algorithm in about 750 lines of C, excluding the I/O and debugging code. Total execution times and time after the initial topological sorting are reported in table 1. Synthetic data sets and GIS data from Littleton, Colorado, and the UBC research forest were used on a Sun 4/75 and a Silicon Graphics Crimson. By way of comparison, the direct implementation (checking all pairs of segments) on the Crimson takes 0.5 seconds for 400 segments of each colour, 50.88 seconds for 4 000, and over 80 minutes for 40 000.

5 Conclusions

The main advantage of this algorithm is that a segment tree data structure is not required. We merely store one level of the segment tree as a list of slabs. This means that fewer pointers are needed, less memory is required and the algorithm is easier to implement.

Another advantage is that the sorting of the segments and endpoints is done first, independently from the intersection calculation phase. In GIS overlay applications, this means that data can be pre-sorted just once prior to storage. Future accesses to this data need not sort again. This would save considerable time with little or no additional memory costs.

Acknowledgements

We thank Otfried Schwarzkopf for discussions on red/blue intersection problems, Jerry Maedel for data from the UBC research forest, and Scott Andrews for converting the program to read GIS data.

References

[1] J. L. Bentley and T. A. Ottmann. Algorithms for reporting and counting geometric intersections. *IEEE Transactions on Computers*, C–28(9):643–647, 1979.

[2] B. Chazelle and H. Edelsbrunner. An optimal algorithm for intersecting line segments in the plane. In *Proceedings of the 29th IEEE Symposium on Foundations of Computer Science*, pages 590–600, 1988.

[3] B. Chazelle, H. Edelsbrunner, L. Guibas, and M. Sharir. Algorithms for bichromatic line segment problems and polyhedral terrains. Technical Report UIUC DCS–R–90–1578, Dept. Comp. Sci., Univ. Ill. Urbana, 1990.

[4] B. Chazelle and L. J. Guibas. Fractional cascading: I. A data structuring technique. *Algorithmica*, 1:133–162, 1986.

[5] B. Chazelle and L. J. Guibas. Fractional cascading: II. Applications. *Algorithmica*, 1:163–191, 1986.

[6] H. Edelsbrunner and M. H. Overmars. Batched dynamic solutions to decomposable searching problems. *Journal of Algorithms*, 6:515–542, 1985.

[7] ESRI White Paper Series. Environmental Systems Research Institute, Inc. *ARC/INFO: GIS Today and Tomorrow*, Mar. 1992.

[8] Federal Geomatics Bulletin, 4(1), 1992. GIS Division, Energy, Mines and Resources. Ottawa, Canada.

[9] D. E. Knuth. *Fundamental Algorithms*, volume 1 of *The Art of Computer Programming*. Addison-Wesley, second edition, 1973.

[10] H. G. Mairson and J. Stolfi. Reporting line segment intersections. In R. Earnshaw, editor, *Theoretical Foundations of Computer Graphics and CAD*, number F40 in NATO ASI Series, pages 307–326. Springer-Verlag, 1988.

[11] F. P. Preparata and M. I. Shamos. *Computational Geometry—An Introduction*. Springer-Verlag, New York, 1985.

[12] M. I. Shamos and D. Hoey. Geometric intersection problems. In *Proceedings of the 17th IEEE Symposium on Foundations of Computer Science*, pages 208–215, 1976.

Repetitive Hidden-Surface-Removal for Polyhedral Scenes

Marco Pellegrini
Dept. of Computer Science, King's College,
The Strand, London WC2R 2LS U.K. marco@dcs.kcl.ac.uk

Abstract

The repetitive hidden-surface-removal problem can be rephrased as the problem of finding the most compact representation of all views of a polyhedral scene that allows efficient on-line retrieval of a single view. In this paper we present a novel approach to the hidden surface removal problem in repetitive mode. We assume that a polyhedral scene in 3-space is given in advance and is preprocessed off-line into a data structure *independently from any view-point*. Afterwards, the data structure is accessed repeatedly with view-points given on-line and the portions of the polyhedra visible from each view-point are produced on-line. This mode of operation is close to that of real interactive display systems.

Let n be the number total of edges, vertices and faces of the polyhedral objects and let k be the number of vertices and edges of the image. The main result of this paper is a collection of data structures answering hidden-surface-removal queries in almost-optimal output-sensitive time $O(k \log^2 n)$ using $O(n^{2+\epsilon})$ storage for the case of axis oriented polyhedra, or more generally for c-oriented polyhedra. If only linear storage is allowed, hidden-surface-removal queries are answered in time $O(k \log n + \min\{n \log n, kn^{1/2+\epsilon}\})$. A continuous trade off between storage and query time is given and the data structures are made dynamic under insertion and deletion of polyhedral objects. The polyhedra may intersect and may have cycles in the dominance relation.

For general polyhedral scenes we answer queries in time $O(k \log^2 n)$, using $O(n^{4+\epsilon})$ storage.

1 Introduction

1.1 The problem

The Hidden Surface Removal problem (HSR for short) for polyhedral scenes in 3-space is important for graphical applications and has attracted much interest in the research community in recent years (e.g. [11],[5],[29],[3],[17], [23],[13] and [22]). A good survey of old and recent results can be found in [10].

Given a set P of objects bounded by polygonal faces which we assume to be opaque, a view-point v, and a view-plane V, the problem consists in computing the visible portion of the scene as viewed from v and projected on V. We consider the "object space" variant of this problem where the output of the algorithm is a planar graph representing the subdivision of V into maximal connected regions in each of which a single face of an object (or the background) is seen. This subdivision is called *visibility map* of P from v, denoted by $M(v, P)$. Other solutions are classified as "image space" since the aim is to compute for each pixel of the screen the object visible at that pixel [31].

We assume in this paper that the polyhedra are given in advance and are preprocessed into a data structure *independently from any view-point*. Afterwards the data structure is accessed repeatedly with query view-points and the visibility map for each viewpoint is produced on-line (*repetitive mode*). This scenario should be compared with the *single-shot mode*, in which *both* the polyhedral scene *and* a view-point are processed together to produce the output visibility map. No data structure is saved and if a new viewpoint is given the process is repeated from scratch. In the real use of display systems the repetitive mode of operation is more natural than the single-shot one.

Display of views are often done in real-time and it is therefore important to optimize the query time over other cost components. Our first objective in this paper is to achieve optimal output sensitive query time (up to polylogarithmic factors) and minimize the worst case storage requirement. We also consider query-time/storage trade offs.

1.2 The aspect graph approach

Traditionally, the problem of efficiently computing the visibility map of a polygonal scene has been tackled using the *aspect graph* approach [18,19,26,15]. For a survey of results see [15,16]. The aspect graph of a polyhedral scene is a graph in which each node is associated with a region of R^3 where the same combinatorial visibility map is visible. Provided that we can store efficiently the visibility maps at the nodes of the graph and that, given a view-point we can find efficiently the corresponding node in the graph, the storage depends mainly on the number of nodes and edges in the aspect graph.

In [15] Gigus, Canny and Seidel give an algorithm for computing the aspect graph of *orthographic views* (i.e with viewpoint at infinity) of a general polyhedral scene. They use $O(n^4 \log n + |G| \log |G|)$ time to produce the aspect graph G. This graph is stored in a data structure of size $O(|G|)$ and orthographic HSR-queries can be solved in time $O(\log |G| + k)$, where k is the size of the output visibility map. The method in [15] does not seem to generalize immediately to computing also the aspect graph of *perspective views* (i.e. with a proper point as view-point). For perspective views a method in [27] computes the aspect graph in time $O(n^5 + n^2|G|)$ and $O(n^2|G|)$ working storage.

In [27] it is shown that for n non-intersecting triangles the number of orthographic views is $\Theta(n^6)$ in the worst case and the number of perspective views is $\Theta(n^9)$. In [30] it is shown that for axis oriented polyhedra with n edges the number of orthographic views can be $\Theta(n^6)$. In [12] it is shown that for a polyhedral terrain with n edges there can be $\Omega(n^5\alpha(n))$ orthographic views and $\Omega(n^8\alpha(n))$ perspective views.

In this paper we show that $O(n^{4+\epsilon})$ preprocessing time and storage are always sufficient to answer efficiently *both orthographic and perspective HSR-queries* for general polyhedral scenes. For axis oriented polyhedra $O(n^{2+\epsilon})$ pre-processing and storage are always sufficient.

1.3 Other previous results on repetitive HSR

In [22], Mulmuley builds a spatial decomposition $D(P)$ for a general non-intersecting polyhedral scene P. The size of $D(P)$ is $\Theta(n^2)$ in the worst case, and ranges from linear to quadratic depending on the input. Given a query view-point v the visibility map from v can be computed on-line by traversing this decomposition. With $\mathcal{V}(v, P)$ we denote the set of all point in R^3 visible from v. Note that the visibility map $M(v, P)$ is isomorphic to the boundary of $\mathcal{V}(v, P)$. The time bound for the construction of $M(v, P)$ is proportional to the number of features of $\mathcal{V}(v, P) \cap D(P)$. It is not difficult to devise an example P, v, in which the size of $M(v, P)$ is constant but but the size of $\mathcal{V}(v, P) \cap D(P)$ is quadratic. Mulmuley leaves open the problem of producing $M(v, P)$ in time proportional to its actual size. In this paper we solve this problem.

A variation of the HSR problem that received some attention in the research community is obtained when we constrain the view-point to be on a given line L. This problem has applications in flight simulation. The approach used by [6] and [22] is to precompute all topologically different visibility maps for viewpoints on L using a sweeping approach. Such results are thus close to the aspect graph approach. Let us define by K the number of topological changes in the visibility map $M(v, P)$ for v moving on L. With K_t we denote the transparent topological changes, that is the number of changes in the projection of edges of P as v moves on L. With K_s we denote the semi-opaque topological changes (i.e. changes on the projection of edges in P, such that the segments involved can see each other locally). In general $1 \leq K \leq K_s \leq K_t \leq n^3/3$. The algorithm in [22] computes all K changes in time $O((K_s + n^2\alpha(n)) \log n)$. The algorithm in [6] computes all K changes in time $O((n^2 + K_t) \log n)$. Since since we can build examples in which K is $O(1)$ and $K_s, K_t = \Omega(n^2)$, these solutions do not exhibit a guaranteed output-sensitive behaviour. The methods in [22,6] do not seem to be able to make use of a restriction of the input to axis-oriented polyhedra.

1.4 Previous results on one-shot HSR for axis-oriented polyhedra

A polyhedron p_1 *dominates* polyhedron p_2 if p_1 partially obstructs the view of p_2 from v. We can classify output sensitive HSR algorithms in two categories according to their reliance on the existence of an acyclic dominance relation for the objects as seen from the viewpoint.

The algorithm proposed by de Berg and Overmars in [13] for HSR on axis-oriented polyhedra works in *single-shot mode*, with a time complexity $O((n+k) \log n)$, without requiring acyclicity of the dominance relation.

Since our algorithm departs from the one in [13] it is worth to discuss it in more detail. The method in [13] is based on using as primitive operations (i) ray-shooting queries on "curtains" and (ii) vertex-visibility queries. Each component of $M(v,P)$ is traced using these two primitives. The $O(n \log n)$ overhead term accounts for the construction of the data structures and for performing a vertex-visibility query for each vertex in P. Trying to use this method for repetitive HSR we run into the following problems. The data structures for ray-shooting on curtains and for vertex-visibility depend on the viewpoint v and cannot be updated in sublinear time for a new viewpoint v'. Moreover, in order to obtain optimal output sensitive query time we cannot afford to consider independently each vertex for visibility.

In [9] there is a method for maintaining the view of axis-oriented polyhedra from a *fixed point of view* when we are allowed to insert and delete polyhedra.

Preparata et al. [28] compute $M(v,P)$ for axis-oriented polyhedra in time $O((n+k) \log n \log\log n)$ provided that the dominance relation is acyclic.

1.5 Previous results on one-shot HSR for general polyhedra

When the dominance relation on general non-intersecting polyhedra is acyclic, Overmars and Sharir [23] give two algorithms running in time $O(n\sqrt{k} \log n)$ and $O(n^{4/3} \log^{2/3} n + k^{3/5} n^{4/5+\epsilon})$. Later this second algorithm has been improved in [3] to run in time $O(n^{2/3-\epsilon} k^{2/3+\epsilon} + n^{1+\epsilon} + k^{1+\epsilon})$.

For objects with the additional property that the union of their projection has small complexity Katz et al. [17] give an algorithm running in time $O((U(n) + k) \log^2 n)$ where $U(n)$ is the maximum size of the union of the projection of n such objects. This methods applies to disks and balls ($U(n) = O(n)$), fat triangles ($U(n) = O(n \log\log n)$) and terrains ($U(n) = O(n\alpha(n))$). For terrains Reif and Sen [29] obtain an $O((n+k) \log n \log\log n)$ algorithm.

We can check the acyclicity of a set of polygons in 3-space in time $O(n^{4/3+\epsilon})$ using an elegant algorithm in [14]. This algorithm is highly dynamic and depends heavily on the knowledge of the viewpoint v. If we discover a cycle in the dominance relation we can eliminate it by splitting an object in two. It is not known how to produce such splits optimally (partial results are in [7]).

The first output-sensitive HSR algorithm that works well even in presence of cycles is in [11] and runs in time $O(n^{1+\epsilon}\sqrt{k})$. We depart from this algorithm too, therefore it is worth to give some details. The basic idea is to use a sweeping line approach to the construction of $M(v,P)$. The event queue is initialized with the vertices of P and the computation proceeds using primitives for ray-shooting on curtains and for ray-shooting from the viewpoint v. Trying to adapt this method for repetitive HSR we run into the following problems. We cannot afford to initialize the queue with n vertices most of which are potentially invisible and again the data structures depend heavily on v and cannot be updated efficiently for a new view-point.

The method in [11] is improved in [2] to run in time $O(n^{2/3+\epsilon}k^{2/3} + n^{1+\epsilon} + k)$.

All the methods used for one-shot hidden surface removal on general polyhedral scenes are off the optimal time $O(k)$ by a polynomial factor unless $k = \Omega(n^2)$. Thus if we use one-shot methods for the repetitive HSR problem we do not achieve an optimal query time. This discrepancy is more evident for output visibility maps of small size.

1.6 Summary of results

For a set of axis-oriented polyhedra with n vertices, edges and faces we can compute hidden-surface-removal queries from any point $v \in R^3$ in time $O(k \log^2 n)$ using $O(n^{2+\epsilon})$ storage (Theorem 4). The query time is output-sensitive and optimal up to polylogarithmic factors. If only linear storage is allowed, we can answer HSR-queries in time $O(k \log n + \min\{n \log n, kn^{1/2+\epsilon}\})$ (Corollary 1). This bound improves the best previous result in [13] for small values of k (i.e. $k < \sqrt{n}$). We also have a continuous trade off between storage and query time: with m storage we can answer HSR queries in time $O(k \log n + \min\{n \log n, kn^{1+\epsilon}/m^{1/2}\})$ (Theorem 5).

We consider the dynamization of our data structure under insertion and deletion of axis oriented polyhedral objects form the scene. We obtain an update time bound which depends only on the number of objects inserted or deleted and not on the (potentially much larger) number of objects present in the scene (Section 4.2).

For a general set of polyhedra with n vertices, edges and faces we can compute HSR-queries from any point $v \in R^3$ in time $O(k \log^2 n)$ using $O(n^{4+\epsilon})$ storage (Theorem 7). This result shows that for complex general scenes $O(n^{4+\epsilon})$ storage is always sufficient to store all views. Previous results based on the aspect graph approach require much higher storage in the worst case.

Note that we place no other restriction on the polygonal scene. In particular, we do not require disjointness of the input polyhedra. As a consequence the polygonal scene can be specified as the union of polyhedral objects with a potential saving in the number of vertices, edges and faces required as input.

We do not require acyclicity in the dominance relation. Computing or checking the acyclicity of the dominance relation is a quite complex and time-consuming task. Moreover, the existence of cycles depends heavily on the viewpoint v and thus some re-checking is required at each query. Our algorithms avoid all these intricated issues.

1.7 The method

In order to obtain our result on repetitive HSR we mix and modify the approaches in [13] and [11].

We use a sweeping line approach, but we avoid initializing the queue with all the vertices. Instead, we discover the visible vertices on the fly. Vertices that are not visible are never put in the queue.

The second challenge is to make our data structure independent of v. We pay this freedom with an initial increase in storage from linear to quadratic. Then we can return to linear storage by trading off storage with query time

In order to achieve these two goals we must support primitive operations more powerful than those in [13] and [11]. In particular, besides supporting ray-shooting queries we must be able to count efficiently how many edges of the polyhedral scene meet query *triangles* and query *pyramids*. Moreover, we apply Megiddo's parametric search technique [21] in order to obtain even more powerful operations. Intuitively, we need to detect efficiently the first moment in which a growing triangle or pyramid intersects a feature of P. We implement all these operations for axis-oriented polyhedra by a reduction to half-plane range queries.

The algorithm in Section 2 is quite general and works also for general polyhedral scenes when we provide an implementation for the primitive operations. The implementation of the primitive operations for general polyhedra is based on results in [24] on collision-free simplices.

The paper is organized as follows: In Section 2 we describe an high-level generic algorithm for solving hidden-surface removal queries which we will denote as the HSRA algorithm. In Section 3 we discuss the implementation of the primitive operations for a set of axis-oriented polyhedra. In Section 4 we discuss query-time/storage trade-offs and dynamization of the data structures under insertions and deletions. In Section 5 we discuss the case of general polyhedral scenes.

2 The HSRA Algorithm

In this section we give a generic algorithm for solving HSR-queries, which we will specialize later for particular types of polyhedral scenes. The general structure is reminiscent of Bentley and Ottman line-sweep algorithm for reporting intersections of segments in the plane [4]. We take a fixed plane at pre-processing time, independently of any viewpoint, as our view-plane V onto which the visibility map is projected. We require that $v \notin V$ but it is easy to overcome this constraint. We also fix a direction on V which we call vertical.

The visibility map $M(v, P)$ will be constructed by sweeping a vertical line $l(x)$ on V. As in [4] we need to keep two data structures: a dictionary *Line_status*(x) which stores the intersection of $M(v, P)$ with the current line $l(x)$, and the priority queue *Event_queue*(x) which stores the list of "events" in sorted order along the x-axis.

As a matter of convention we will denote with e an edge in 3-space and with e' its projection of V. Let e' be an edge stored in *Line_status*(x) which is the projection of edge e in 3-space. While e' is traced ahead (left-to-right) we detect the *first* of the following events.

(i) The right end-point of e. At this event the edge e' should be deleted from *Line_status*.

(ii) Edge e_1 from face t_1 obscures e, which is therefore no longer visible. We delete e' from *Line_status* and insert e'_1 if not already present.

(iii) Edge e_1 from face t_1 is partially obscured by e. We insert e'_1 in *Line_status* if not already present.

(iv) Edge e intersects face t_1. If t is the face containing e, then the intersection of t and t_1 is a new visible edge, which is inserted in *Line_status*.

(v) The intersection of three facets of polyhedra in P. All new visible edges from this intersection point are inserted in *Line_status*.

There is a further event that cannot be discovered by tracing a single edge present in *Line_status*(x):

(vi) A visible vertex q of a face in P projects on $l(x)$ from v. In this case all the edges incident to v are inserted in the data structure.

Event (vi) is quite important because it allows us to find on-line all the connected components of the visibility map.

By a discussion in [13,10] these are all and only the interesting events. Since each edge is traced until visible and discarded afterwards until it might be visible again, at each instant *Line_status*(x) stores a faithful representation of $M(v, P) \cap l(x)$. Since each event corresponds to a feature of the visibility map it can be charged to the output size k. Each event is not detected more then a constant number of times. An edge can be inserted and deleted several times but each operation corresponds to a distinct feature of $M(v, P)$. The algorithm proceeds by executing the update corresponding to the next event in *Event_queue* and by updating *Event_queue* itself with the new events induced by the edges which have been manipulated. All the operations on *Line_status* and *Event_queue* take $O(\log n)$ time using standard data structures. The primitive operations needed to discover events are the following:

1. *Ray-shooting queries.* Given a point p and a ray ρ from p, determine the first face of a polyhedron in P intersected by ρ, and the intersection point.

2. *Triangle-emptiness queries*: Given a triangle t in 3-space determine whether t intersects any face or edge or vertex in P.

3. ***Minimal-empty triangle queries***: Given a family of triangles $t(s)$ depending on a positive real value s such that $s_1 < s_2$ implies $t(s_1) \subset t(s_2)$, report the minimum value of s for which $t(s)$ intersects a feature of P and the feature of P for which this happens.

4. ***Minimal-empty pyramid queries***: Given a family $\Gamma(s)$ of pyramids with quadrangular base, depending on a positive real value s such that $s_1 < s_2$ implies $\Gamma(s_1) \subset \Gamma(s_2)$, report the minimum value of s for which $\Gamma(s)$ intersects a feature (vertex, edge, face) of P and the feature of P for which this happens.

Setting up Line_status. Initially we set up *Line_status*(0) in the following way. We take a point a on $l(0)$ and we ***shoot a ray*** from v towards a. We determine the point q on the first face f hit by the ray. By definition q is visible. Then we take on the face f a the line through q which projects from v onto $l(0)$. We parametrize this line the generic point $p(s)$ where $p(0) = q$ and s is a positive real parameter. Let $\bar{s} > 0$ be the value of s for which $p(s)$ is on the boundary of f.

We consider the triangle $t(s)$ of vertices $v, q, p(s)$. We make a triangle emptiness query with $t(\bar{s})$, if $t(\bar{s})$ is empty we have discovered the first point on a visible edge of $M(v, P)$. By shooting a ray from v towards $p(\bar{s})$ we find a point on the first visible face behind f (or the background) and we repeat the tracing procedure upwards on this new face (or background).

If $t(\bar{s})$ is not empty, we use the minimal-empty triangle query on the family of triangles $t(s)$ and we find a point on a visible edge of $M(v, P)$. This point belongs to an edge of a face f' different from f. We repeat this tracing procedure upwards on f'. Repeating the tracing procedure from q downwards, we complete the construction of the initial data structure *Line_status*(0).

Computing events. Let e' be an edge in *Line_status*, e the edge in 3-space generating e', q a visible point of e projecting from v onto $l(x)$, and f a visible face containing e (if both faces incident to e are visible, choose one arbitrarily). We parametrize the edge e by a generic point $p(s)$ where $p(0) = q$ and s is a positive real value. Let $\bar{s} > 0$ be the value of s for which $p(s)$ is the right endpoint of e.

1) If the triangle $v, q, p(\bar{s})$ is empty then we have an event (i). Otherwise, by asking a a minimum-empty triangle query on $v, q, p(s)$ we detect the first event of type (ii), (iv) or (v).

2) By shooting a ray from v towards q we find a point q' on the first visible face f' different from f. By projecting e onto f' and repeating the steps at 1) we can detect an event of type (iii). Note that we deal implicitly also with the case when both faces incident to e are visible. In this case f' is one of these faces.

3) In order to detect events of type (vi) we proceed in the following way. We consider a pair of consecutive edges e'_1 and e'_2 in *Line_status*. Let f be the face of P visible between e_1 and e_2 at $l(x)$. We project both e_1 and e_2 onto this face obtaining e''_1 and e''_2. We consider a parametrization of e''_1 and e''_2 with generic points $p_1(s)$ and $p_2(s)$ such that the segment $p_1(s)p_2(s)$ projects on V into a segment parallel to $l(x)$. We ask minimum pyramid emptiness queries with a pyramid of vertices $v, p_1(0), p_2(0), p_1(s), p_2(s)$. The value of s returned corresponds to a new visible vertex of P or to some other event (i)-(v) relative to the edges e'_1 and e'_2. In order to detect vertices and edges lying on the face f we must consider the query pyramid as a closed set.

Initially we compute events for the edges and pairs of consecutive edges in *Line_status*(0). On the fly, at each update of *Line_status*, we insert new events in *Event_queue* relative to the edges just updated and their neighbours in the linear order stored in *Line_status*.

When a new visible vertex q is discovered (event (vi)), we insert in *Line_status* the new edges incident to q. For axis-oriented polyhedra in general position the number of edges incident to a vertex is bounded by a constant, therefore we can afford to check every edge for its visibility. The case of general polyhedral scenes is more complex since the degree of a vertex is not bounded.

If q projects from v on the sweeping line at $l(\bar{x})$, we consider the perturbed position $l(\bar{x}+\delta)$ for an infinitesimal $\delta > 0$. We choose a point a to the right of q that projects on $l(\bar{x}+\delta)$ and we use the procedure for setting up *Line_status* in the neighborhood of q (i.e. at $l(\bar{x}+\delta)$ instead of $l(0)$). Using a as the initial target for ray-shooting. We stop the procedure when we find a visible edge not incident to q above and below a. Any such edge must be already present in *Line_status*$(\bar{x})$, so it is easy to check this halting condition. All the visible adges discovered during this scan are incident to q. Conversely, any visible edge incident to q is discovered. The number of primitive queries is proportional to the number of visible edges incident to q.

3 Primitive operations for axis-oriented polyhedra

In this section we discuss the implementation of the four primitive operations used by algorithm HSRA when the input is restricted to axis-oriented polyhedra.

3.1 Ray-shooting

From results in [25] we can answer ray-shooting queries on axis-parallel polyhedra in time $O(\log n)$ using $O(n^{2+\epsilon})$ storage. Using this primitive we can check whether v is interior to any polyhedron in P. We send a ray from v in an arbitrary direction. If the ray meets a face of P coming from the interior of the polyhedron containing this face we conclude that v is an interior point of P. From now on we consider v to be external to P.

3.2 Empty triangle queries

Given a set of axis-oriented polyhedra P and a triangle t we want to check whether t meets any polyhedron in P. In our application we use triangles t such that at least one vertex in external to any polyhedron in P. The following lemma holds:

Lemma 1 *A triangle t with a vertex disjoint from P intersects P of and only if an edge of P meets t or an edge of t meets a face in P.*

Let $l(s)$ be the line spanning the segment s and *aff*(t) be the plane spanning the triangle t. We have the following lemma:

Lemma 2 *For any segment s and triangle t not co-planar:*

$$s \cap t \neq \emptyset \iff s \cap \text{aff}(t) \neq \emptyset \wedge l(s) \cap t \neq \emptyset$$

From Lemma 1 and Lemma 2 we derive the following theorem:

Theorem 1 *Given a set of axis-parallel polyhedra P of complexity n, we can answer empty triangle queries of algorithm HSRA in time $O(\log n)$ using data structures of total size $O(n^{2+\epsilon})$.*

Proof. We can easily check whether an edge of t meets a face of P using three ray-shooting queries. In order to check edges of P against t we partion those edges into three sets of parallel edges and we test each set independently.

Let S be a set of parallel segments in 3-space and t our query triangle. We build a multilevel data structure on the set S in order to answer the queries. The first level is queried with the plane *aff*(t) spanning t.

It is crucial for the analysis of the storage required by this data structure to recall that the planes *aff*(t) used in by the HSRA algorithm in Section 2 are of a special kind. Every such plane must contain an edge of P, or or must be parallel to the sweeping line. We have thus four sub-families of planes, where planes in each sub-family must all meet a fixed point at infinity. We project the segments in S from such point at infinity onto the auxiliary plane obtaining a planar set of segments S'. We project the plane *aff*(t) from such point at infinity onto the auxiliary plane obtaining a line l_t. The query at the first layer is reduced to detecting the segments in S' intersected by the line l_t. This in turns is resolved by applications of halfplane range searching.

More precisely, we dualize the problem and we build the arrangement of lines $D(V(S'))$ over the set of vertices of S'. The query is the point $D(l_t)$. By locating $D(l_t)$ in the arrangement of planes $D(V(S'))$ we can retrieve efficiently the subset of segments in S that meet $aff(t)$ as a union of canonical sets prestored in the arrangement.

The second level of the multilayer data structure is build on the canonical sets of the first layer. For each such canonical set S_i we project the segments onto a plane orthogonal to the common direction of segments in S_i, obtaining a set of points S_i'. Also we project triangle t onto this plane obtaining a triangle t'. Now we can use the planar triangle range searching techniques in [8,20] in order to determine whether there is any point of S_i' in t'. If this is the case we have detected a collision between t and a segment of S.

The overall storage is $O(n^{2+\epsilon})$ and the query time is $O(\log n)$, by using standard partition tree approaches to the construction of the data structure. ∎

3.3 Minimum empty triangle queries

Let us consider a family of planar triangles in 3-space $t(s)$ indexed by a positive real parameter s, forming an *inclusion chain* (i.e. $s_1 < s_2 \Rightarrow t(s_1) \subset t(s_2)$). It is easy to see that the property of intersecting P is monotone for such set of triangles. There is a value s^* such that for all $s < s^*$ $t(s)$ does not meet P and for all $s > s^*$ t meets P. Therefore we can use Megiddo's parametric search technique to transform the algorithm of Theorem 1 into an algorithm for finding s^* [21]. The query time of the transformed algorithm is the square of the time of the original one.

Theorem 2 *Given a set of axis-parallel polyhedra P of complexity n, we can answer minimum empty triangle queries of algorithm HSRA in time $O(\log^2 n)$ using data structures of total size $O(n^{2+\epsilon})$.*

3.4 Minimal empty pyramid queries

Let us consider a family of pyramids in 3-space $\Gamma(s)$ indexed by a positive real parameter s, forming an *inclusion chain* (i.e. $s_1 < s_2 \Rightarrow \Gamma(s_1) \subset \Gamma(s_2)$). Again, the property of $\Gamma(s)$ to intersect P is monotone in s: there is a value s^* such that for all $s < s^*$ $\Gamma(s)$ does not meet P and for all $s > s^*$ $\Gamma(s)$ meets P. Therefore we can use Megiddo's parametric search technique to transform an algorithm for detecting empty pyramids into an algorithm for finding s^*.

Lemma 3 *A pyramid Γ with a vertex disjoint from P intersects P of and only if an edge of P meets a face of Γ or an edge of Γ meets a face in P or a vertex of P is contained in Γ.*

Theorem 3 *Given a set of axis-parallel polyhedra P of complexity n, we can answer minimum empty pyramid queries of algorithm HSRA in time $O(\log^2 n)$ using data structures of total size $O(n^{2+\epsilon})$.*

Proof. From the above discussion is is sufficient to set up a test for pyramid emptiness and apply the parametric search transformation. From Lemma 3 we have three cases to check. We can check whether an edge of Γ meets a face in P using the ray-shooting data structure. We can test whether a face of Γ (triangle or quadrangle) meets an edge of P using the result of Theorem 1 (though we need a slight modification to deal with the quadrangular face of Γ within the stated storage bound). The last case we need to test is an instance of simplex range searching on the set of vertices of P. The pyramid has faces belonging to four subfamilies of planes. Each subfamily is constrained to contain a fixed point at infinity. Using a multilayer data structure we can split this query into several planar halfplane range queries using the same trick as in the proof of Theorem 1. The total storage is $O(n^{2+\epsilon})$ and the query time is $O(\log n)$. The total query time for minimum empty triangle queries becomes $O(\log^2 n)$ after the application of Megiddo's parametric search technique. ∎

3.5 Putting the pieces together

The general HSRA algorithm of Section 2 and the implementation of the primitive operations in Section 3 prove the following theorem:

Theorem 4 *Given an axis-parallel polyhedral scene P with n vertices, edges and faces, we can build a data structure $\mathcal{D}_1(P)$ of size $O(n^{2+\epsilon})$ such that for a query point v the visibility map $M(v,P)$ can be computed in time $O(k\log^2 n)$, where k is the output size.*

4 Extensions

4.1 Storage-query time trade offs

The primitive operations of algorithm HSRA are implemented in Section 3 by a reduction to (several) halfplane range queries on planar set of points. Such queries can be solved using the techniques in [8,20] in $n^{1+\epsilon} \leq m \leq n^{2+\epsilon}$ storage and $T = O(n^{1+\epsilon}/m^{1/2})$ query time. When we apply the parametric search technique to superlogarithmic algorithms (as it is the case for the query time in the trade-off case) we use a more sophisticated form of parametric search that needs a parallel version of the halfplane range searching algorithm [21]. The data structures in [8,20] are based on a partition-tree approach and the query time depends on the number of nodes in the partition tree visited during the query. We can allocate dynamically processors to the nodes visited during the search, thus we need $p = O(n^{1+\epsilon}/m^{1/2})$ processors. The parallel query time T' is given by the depth of the tree which in $O(\log n)$. After the parametric-search transformation we have a total query time $O(T'p + TT'\log n) = O(n^{1+\epsilon}/m^{1/2})$. Also, we can run in parallel the algorithm in [13], which has better performances when k is large.

The above discussion leads to the following theorem:

Theorem 5 *Given set P of axis-parallel polyhedra with n vertices, edges and faces, we can build a data structure $\mathcal{D}_2(P,m)$ of size $n^{1+\epsilon} \leq m \leq n^{2+\epsilon}$ such that for a query point v the visibility map $M(v,P)$ can be computed in time $O(k\log n + \min\{n\log n, kn^{1+\epsilon}/m^{1/2}\})$, where k is the output size.*

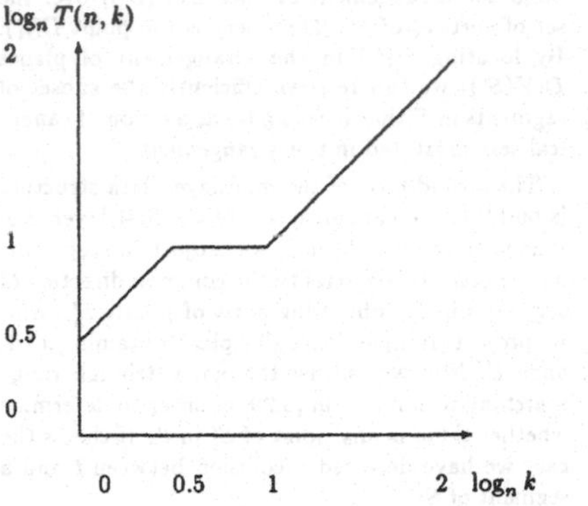

Figure 1: Running time in logarithmic scale as function of k (linear storage).

If we allow only linear storage we obtain a query time roughly order of $k\sqrt{n}$ which is asymptotically better than the one-shot algorithm of [13] for $k \leq \sqrt{n}$. If we run the algorithm in [13] and the HSRA algorithm in parallel we can attain the following result:

Corollary 1 *Given a set P of axis-parallel polyhedra with n vertices, edges and faces, we can build a data structure $\mathcal{D}_3(P)$ of size $O(n)$ such that for a query point v the visibility map $M(v,P)$ can be computed in time $O(k\log n + \min\{n\log n, kn^{1/2+\epsilon}\})$, where k is the output size.*

The result of Corollary 1 is significant for those scenes in which only a small part of the total number of polyhedra is visible at any given time. A graph of the running time $T(n,k)$ as a function of k in logarithmic scale is in Figure 1.

4.2 Dynamization

Since all data structures are multilayer data structures based on the CSW scheme [8,3] or on Matoušek 's scheme [20] we can use a dynamization result in [3]:

Theorem 6 ([3]) *Given a set of* n *points in* R^d *and a parameter* $n^{1+\epsilon} \le m \le n^{d+\epsilon}$ *one can maintain the CSW-partitioning structure in* $O(m/n^{1-\epsilon})$ *amortized time as we insert or delete a point, and can answer half-space range queries in time* $O(n^{1+\epsilon}/m^{1/d})$.

Thus the amortized update time for $\mathcal{D}_2(P,m)$ is $O(m/n^{1-\epsilon})$.

Actually we can do better than this in practice. Notice that all of the data structures for answering primitive queries of the HSRA algorithm can be modified to *count* the number of features (vertices, edges, faces) intersecting the query objects. Let P be the initial set of axis-oriented polyhedra, P_I the set of axis-oriented polyhedra inserted up to the present moment and P_D the set of axis oriented polyhedra deleted form P and P_I up to the present moment. Let $C_g(S,q)$ be the number of features of type g (vertex, edge, face) of a set of polyhedra S intersecting the query object q (ray, triangle, pyramid). Clearly:

$$C_g((P \cup P_I)/P_D, q) = C_g(P,q) + C_g(P_I,q) - C_g(P_D,q) \quad (1)$$

Thus we need to modify the $\mathcal{D}_2(P,m)$ only to accommodate insertion. Moreover, the amortized time for an insertion is only $O(|\mathcal{D}_2(P_I,m)|/|P_I|^{1-\epsilon})$. Since in real applications we expect that over a session of use of the system $|P_I| << |P|$, we obtain a very small slowdown of the system due to reconstructions. A similar argument holds for P_D. Extensive reconstruction of the data structure in order to maintain the overall performance can be done off-line between sessions.

5 Primitive operations for general polyhedra

In this section we discuss the implementation for the several primitive operations used in the HSRA algorithm in the case of general polyhedral scenes.

1. *Ray-shooting queries.* From results in [25] we can answer ray-shooting queries in time $O(\log n)$ using $O(n^{4+\epsilon})$ storage.

2. *Empty triangle queries.* From results in [21] we can answer triangle emptiness queries in time $O(\log n)$ using $O(n^{4+\epsilon})$ storage.

3. *Minimal-empty triangle queries.* We use the data structure in [24] to answer triangle emptiness queries. We modify the query method by applying Megiddo's parametric search [21], thus obtaining a method to determine the smallest empty triangle in a family of triangles forming a chain of inclusions. The query time becomes $O(\log^2 n)$.

4. *Minimal-empty pyramid queries.* A data structure described in [24] answers simplex emptiness queries in time $O(\log n)$ using $O(n^{4+\epsilon})$ storage. This method can be easily adapted to answer emptiness queries with any convex polyhedron with a constant number of facets (e.g. a pyramid with quadrangular base). Similarly to the previous case we modify the query algorithm by applying Megiddo's parametric search [21]. Minimal-empty pyramid queries can be solved in time $O(\log^2 n)$ using $O(n^{4+\epsilon})$ storage.

The HSRA Algorithm together with the above primitives leads to the following result:

Theorem 7 *Given a set* P *of polyhedra with* n *vertices, edges and faces, we can build a data structure* $\mathcal{D}_4(P)$ *of size* $O(n^{4+\epsilon})$ *such that for a query point* v *the visibility map* $M(v,P)$ *can be computed in time* $O(k \log^2 n)$, *where* k *is the output size.*

Most of the argument used for axis-oriented polyhedra hold for general scenes with somewhat degraded performances. In particular using results in [1] we have the following corollary:

Corollary 2 *Given a set P of polyhedra with n vertices, edges and faces, we can build a data structure $\mathcal{D}_5(P,m)$ of size $n^{1+\epsilon} \leq m \leq n^{4+\epsilon}$ such that for a query point v the visibility map $M(v,P)$ can be computed in time $O(kn^{1+\epsilon}/m^{1/4})$, where k is the output size.*

The method of Corollary 2 compares favourably with the one-shot methods for small values of k.

6 Conclusions

In this paper we have shown several results on the repetitive hidden-surface removal problem for polyhedral scenes. We considered both axis-oriented polyhedra, for which better bounds are attained, and general polyhedral scenes. We reduce the HSR problem to combinations of half-space and half-plane range searching, in conjunction with a general sweeping line algorithmic skeleton. For several visibility problems in 3-space we improve on the previously known best bounds.

References

[1] P. Agarwal and J. Matoušek. Range searching with semialgebraic sets. In *Proc. of the 17th Symp. on Mathematical Foundations of Computer Science*, number 629 in Lecture Notes in Computer Science, pages 1–13, 1992.

[2] P. Agarwal and J. Matoušek. Ray shooting and parametric search. In *Proceedings of the 24th Annual ACM Symposium on Theory of Computing*, pages 517–526, 1992.

[3] P. K. Agarwal and M. Sharir. Applications of a new space partitioning technique. In *Proceedings of the 1991 Workshop on Algorithms and Data Structures*, number 519 in Lecture Notes in Computer Science, pages 379–391. Springer Verlag, 1991.

[4] J. Bentley and T. Ottman. Algorithms for reporting and counting geometric intersections. *IEEE Trans. on Computers*, C-28:643–647, 1979.

[5] M. Bern. Hidden surface removal for rectangles. In *Proceedings of the 4th ACM Symposium on Computational Geometry*, pages 183–192, 1988.

[6] M. Bern, D. Dobkin, D. Eppstein, and R. Grossman. Visibility with a moving point. In *Proceedings of the 1st ACM-SIAM Symposium on Discrete Algorithms*, pages 107–117, 1990.

[7] B. Chazelle, H. Edelsbrunner, L. Guibas, R. Pollack, R. Seidel, M. Sharir, and J. Snoeyink. Counting and cutting circles of lines and rods in space. In *Proceedings of the 31st Annual Symposium on Foundations of Computer Science*, 1990.

[8] B. Chazelle, M. Sharir, and E. Welzl. Quasi-optimal upper bounds for simplex range searching and new zone theorems. In *Proceedings of the 6th ACM Symposium on Computational Geometry*, pages 23–33, 1990.

[9] M. de Berg. Dynamic output-sensitive hidden surface removal for c-oriented polyhedra. *Computational Geometry: Theory and Applications*, 2:119–140, 1992.

[10] M. de Berg. *Efficient Algorithms for ray-shooting and hidden surface removal.* PhD thesis, Utrecht University, Dept. of Comp. Sci., 1992.

[11] M. de Berg, D. Halperlin, M. Overmars, J. Snoeyink, and M. van Kreveld. Efficient ray-shooting and hidden surface removal. In *Proceedings of the 7th ACM Symposium on Computational Geometry*, pages 21–30, 1991.

[12] M. de Berg, D. Halperlin, M. Overmars, and M. van Kreveld. Sparse arrangements and the number of views of polyhedral scenes. Manuscript, June 1992.

[13] M. de Berg and M. Overmars. Hidden surface removal for axis-parallel polyhedra. In *Proceedings of the 31st IEEE Symposium on Foundations of Computer Science*, pages 252–261, 1990.

[14] M. de Berg, M. Overmars, and O. Schwarzkopf. Computing and verifying depth orders. In *Proceedings of the 8th ACM Symposium on Computational Geometry*, pages 138–145, 1992.

[15] Z. Gigus, J. Canny, and R. Seidel. Efficiently computing and representing aspect graphs of polyhedral objects. *IEEE Transactions on pattern Analysis and Machine Intelligence*, 13:542–551, 1991.

[16] Z. Gigus and J. Malik. Computing the aspect graphs for line drawings of of polyhedral objects. *IEEE Transactions on pattern Analysis and Machine Intelligence*, 12:113–122, 1990.

[17] M. Katz, M. Overmars, and M. Sharir. Efficient hidden surface removal for objects with small union size. In *Proceedings of the 7th ACM Symposium on Computational Geometry*, pages 31–40, 1991.

[18] J. Koenderlink and J. van Doorn. The singularities of visual mapping. *Biological Cybernetics*, 24:51–59, 1976.

[19] J. Koenderlink and J. van Doorn. The internal representation of solid shape with respect to vision. *Biological Cybernetics*, 32:211–216, 1979.

[20] J. Matoušek. Efficient partition trees. In *Proceedings of the 7th ACM Symposium on Computational Geometry*, pages 1–9, 1991.

[21] N. Megiddo. Applying parallel computation algorithms in the design of sequential algorithms. *J. of ACM*, 30:852–865, 1983.

[22] K. Mulmuley. Hidden surface removal with respect to a moving view point. In *Proceedings of the 23th Annual ACM Symposium on Theory of Computing*, pages 512–522, 1991.

[23] M. Overmars and M. Sharir. Output-sensitive hidden surface removal. In *Proceedings of the 30th IEEE Symposium on Foundations of Computer Science*, pages 598–603, 1989.

[24] M. Pellegrini. On collision-free placements of simplices and the closest pair of lines in 3-space. To appear in *SIAM J. on Computing*. Preliminary version in the 8th ACM Symp. on Comp. Geom. with the title 'Incidence and nearest-neighbor problems for lines in 3-space' pp. 130-137, 1992.

[25] M. Pellegrini. Ray shooting on triangles in 3-space. *Algorithmica*, 9:471–494, 1993.

[26] W. Plantiga and C. Dyer. An algorithm for constructing the aspect graph. In *Proceedings of the 27th IEEE Symposium on Foundations of Computer Science*, pages 123–131, 1986.

[27] W. Plantiga and C. Dyer. Visibility, occlusion and the aspect graph. *Int. J. of Computer Vision*, 5:137–160, 1990.

[28] F. Preparata, J. Vitter, and M. Yvinec. Output sensitive generation of the perspective view of isothetic parallelepipeds. In *Proceedings of the 2nd Scandinavian Workshop on Algorithm Theory*, number 447 in Lecture Notes in Computer Science, pages 71–84. Springer Verlag, 1990.

[29] J. Reif and S. Sen. An efficient output-sensitive hidden-surface removal algorithm and its parallelization. In *Proceedings of the 4th ACM Symposium on Computational Geometry*, pages 193–200, 1988.

[30] J. Snoeyink. The number of views of axis-parallel objects. *Algorithmic Review*, 2:27–32, 1991.

[31] I. Sutherland, R. Sproul, and R. Schumaker. A characterization of ten hidden-surface algorithms. *Computing Surveys*, 6(1):1–55, 1974.

On reconfigurability of VLSI linear arrays

Roberto De Prisco[1] and Angelo Monti[2]

[1] Department of Computer Science, Columbia University, New York, N.Y. 10027
[2] Dipartimento di Informatica, Universitá di Pisa,
Corso Italia, 40 - 56125 PISA, Italy

Abstract. Fault tolerance through the incorporation of redundancy and reconfiguration is quite common. In a redundant linear array of processing elements, k redundant links of fixed lengths are provided to each element of the array in addition to the regular links connecting neighboring processors. The redundant links may be activated to bypass faulty elements. The number and the distribution of faults can have severe impact on the effectiveness of such a method. In this paper we study the problem of deciding whether a pattern of n blocks of faults is catastrophic for a redundant array. We prove that, for arrays provided of bidirectional links, the problem requires time $O(kn)$. In the unidirectional case we propose an algorithm whose complexity is $O(n)$ when the array has only one redundant link, and $O(kn \log k)$ otherwise. When the pattern is not catastrophic we are interested to obtain a *reconfiguration set* and, in particular, an *optimal reconfiguration set* (i.e. one with maximal number of working elements). We prove that such a problem is NP-hard, when the links are bidirectional. For the unidirectional case, instead, we present an algorithm of complexity $O(kng)$, where g is the length of the longest link.

1 Introduction

In a linear array of N processing elements one faulty element in any location is sufficient to stop the flow of information from one side to the other. Without the provision of fault tolerance capabilities the yield of very large VLSI chips for such architecture would be so poor to make its production unacceptable.

In a redundant linear array, besides the regular links connecting neighboring processing elements, extra links (called *bypass links*) are included in a regular fashion. These redundant links can be activated in a reconfiguration phase to bypass faulty processors.

This approach has some inherent limits. Making the realistic assumption that the length of the longest link is small with respect to the number of processing elements, regardless of any amount of redundancy, there are sets of faults occurring at strategic positions which affect the chip in an unrepairable way, see [4]. Such sets of faults, called *Catastrophic Fault Patterns*, have been extensively studied in [1,6,7] but only for the specific case in which the number of faults in the pattern is exactly g, that is the length of the longest bypass link. For a fault pattern, having at least g faults, is a necessary condition to be catastrophic, see [4].

In [6] an $O(g^2)$ algorithm is presented recognizing Catastrophic Fault Patterns consisting of g faults, for redundant arrays with a single bypass link of length g. An $O(kg)$ algorithm that solves the problem in the more general case of $k \geq 1$ bypass links, has been devised. The problem representation and the solutions techniques presented in the previous works are not easily adaptable for the general case of any number m, $g \leq m \leq N$, of faults.

In this paper, following a completely different approach, we consider the general problem of testing whether a fault pattern consisting of m faults (subdivided in $n \leq m$ blocks of consecutive faulty processors) is catastrophic. Moreover, when the fault pattern is not catastrophic, we consider the problem of finding an optimal reconfiguration set, that is a set of bypass links to activate so that connection between the two ends of the array is recovered utilizing the maximum number of working processors.

Our results are the following.

The problem of testing whether a set of n blocks of faulty processing elements is catastrophic for a redundant array with k bypass links is solved in time $O(kn)$ when the links in the array are bidirectional. When the links are unidirectional we present a solution requiring time $O(nk \log k)$ for $k > 1$, and $O(n)$ for $k = 1$.

Finally we show that the problem of finding an optimal reconfiguration set when the links are bidirectional is NP-hard, while, in the unidirectional case, such a problem may be solved in time $O(kng)$, where g is the length of the longest bypass link.

The remaining of this paper is organized as follows. Basic concepts and a formal definition of the problem are introduced in Section 2. A test algorithm for arrays with bidirectional links is given in Section 3. In Section 4 a test algorithm for arrays with unidirectional links is provided. Section 5 contains results on the optimal reconfiguration sets.

2 Preliminaries.

The basic components of a redundant linear array are the ***processing elements***, or simply processors, and ***links***. There are two kinds of links: ***regular*** or ***bypass***. Regular links exist between neighboring processors, while the bypass links connect non-neighbors processors. The bypass links are used strictly for reconfiguration purposes when faulty processors are detected.

More precisely, let $A = \{p_1, ..., p_N\}$ denote a linear array of identical processing elements connected by regular links (p_i, p_{i+1}), $1 \leq i < N$. Let $G = \{g_1, ..., g_k\}$ be an ordered set of integers. We say that A has *redundancy* G if, for each g_t, $1 \leq t \leq k$, there is a bypass link (p_i, p_{i+g_t}), $1 \leq i \leq N - g_t$. Notice that the set G does not contains the regular link. We will denote by g the length of the longest bypass link, i.e. $g = g_k$.

At the ends of A two special processors, called I (for Input) and O (for Output), are responsible for the I/O functions of the system. We assume that I is connected to $p_1, ..., p_g$ while O is connected to $p_{N-g+1}, ..., p_N$ so that bottlenecks at the borders of the array are avoided.

Example 1. The following picture shows a linear array of 20 processing elements with redundancy $G = \{4\}$.

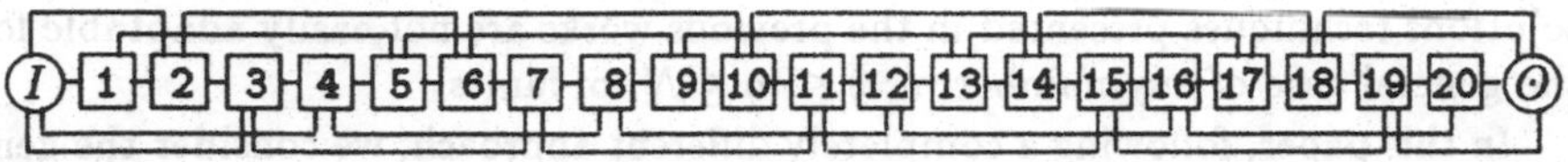

We refer to this structure as a *redundant array* or simply as an *array*. The array is called *bidirectional* or *unidirectional* according to the nature of the links. We will admit faults occurring in the processors only (i.e. both I and O and the links always operate correctly). Moreover we will refer to a processor p_i also as processor i or simply as p_i.

Definition 1 *For a redundant linear array A, a fault pattern F is a set of pairs of positive integers* $F = \{(f_1, l_1), (f_2, l_2), ..., (f_n, l_n)\}$, *where* $f_i + l_i < f_{i+1} \leq N - l_n + 1, 1 \leq i < n$.

Each pair (f_i, l_i) identifies the block of faulty processors $p_{f_i}, p_{f_i+1}, ..., p_{f_i+l_i-1}$. Hence a *faulty processor* p_z is such that $f_i \leq z < f_i + l_i$ for some $i, 1 \leq i \leq n$. Not faulty processors are *working* processors.

A *path* from a working processor i_0 to an eventually faulty processor i_s is a sequence of processors $i_0, i_1, ..., i_{s-1}, i_s$ such that, for each $j = 0, 1, ..., s-1$, processor i_j is a working processor connected by a link to processor i_{j+1} and $i_j = i_z$ if and only if $j = z$, $0 \leq j, z \leq s$. The length of the path is s. An *escape path* is a path from I to O.

In order to formally handle paths we give an efficient representation of them. Usually the flow of the computation pass from processor i to processor $i+1$, hence to represent a path we can indicate only those processors for which the computation does not pass to the next processor. Formally we give the following definition of a path.

Definition 2 *Given a fault pattern F, a path P is represented as a triple consisting of a starting processor* p_u, *an ending processor* p_v *and a set of pair of integers* $\{(e_1, a_1), (e_2, a_2), ..., (e_q, a_q)\}$, *where* $1 \leq e_i \leq N$, $e_i \neq e_j$ *if* $i \neq j$ *and* $-k \leq a_i \leq k$, *for each* $i = 1, 2, ...q$.

According to the above definition, processor e_i, $i = 1, 2, ..., q$, has active a link that is not the regular one. The active link of processor e_i is defined according to a_i, namely:

⊙ If $a_i = 0$ the active link is $(p_{e_i}, p_{e_i - 1})$

⊙ If $a_i < 0$ the active link is $(p_{e_i}, p_{e_i - g_{(-a_i)}})$

⊙ If $a_i > 0$ the active link is $(p_{e_i}, p_{e_i + g_{a_i}})$

All other processors have active the regular link. The path represented by P is the sequence of processors obtained starting from p_u and following the active links. It is obvious that this sequence must not contain a faulty processor, except, eventually, the ending processor p_v, and that it must ends in p_v. An escape path is a path P for which $p_u = I$ and $p_v = O$. In representing escape paths we will omit the processors I and O. Since by activing the link a_i of the processor e_i, for $i = 1, 2, ..., q$, we reconfigure the system (or achieve a path from p_u to p_v), we call this set *reconfiguration set.*

Definition 3 *Given a redundant array A, a fault pattern F is catastrophic for A if and only if no escape path exists.*

Given a fault pattern F for a redundant array A, we will focalize our attention on that part of A beginning at processor p_{f_1-g+1} and ending at processor $p_{f_n+l_n+g-1}$. We will call *fault zone* this part of the array. Moreover, since all the processors are indistinguishable, without loss of generality, we will assume that the fault zone begins at processor p_1, i.e. $f_1 = g$.
A block of maximal length of working processors in the fault zone will be called *chunk*. More formally we give the following definition:

Definition 4 *Given an array A with redundancy G and a fault pattern F, $chunk_i$, $1 \leq i \leq n-1$, is the block of processors between $f_i + l_i - 1$ and f_{i+1}, i.e. the block $p_{f_i+l_i}, ..., p_{f_{i+1}-1}$. In particular $chunk_0$ is the block of processors $p_{f_1-g+1}, ..., p_{f_1-1}$, i.e. the first $g-1$ processors of the fault zone, and $chunk_n$ is the block of processors $p_{f_n+l_n}, ..., p_{f_n+l_n+g-1}$, i.e. the last $g-1$ processors of the fault zone.*

3 A Testing algorithm for bidirectional arrays.

Given a bidirectional array A of N processors and redundancy $G = \{g_1, \ldots, g_k\}$ and a fault pattern of m faults grouped in $n \leq m$ blocks, consider the graph associated to the array after the removal of the faulty elements and their incident links. The problem of testing if the fault pattern is catastrophic for A is equivalent to the one of testing if, in the graph, the vertices I and O are connected. By applying a standard algorithm for the test of connectivity we would obtain an $O((N-m)k)$ time complexity. However in the case of redundant arrays the usual assumption is that N is much greater than m. We propose a better algorithm running in time $O(nk)$.

To represent the problem we construct a different graph $H = (V, E)$ as follows. Let F be the fault pattern. The set V of vertices is $\{C_0, C_1, \cdots, C_n\}$, where C_i's represent the chunks of F. We will write $p_x \in C_i$ to means that the processor p_x belong to the $chunk_i$ which is represented by the vertex C_i. For each i and j, $0 \leq i, j \leq n$, $i \neq j$, the edge (C_i, C_j) belong to E if and only if there are two processors, $p_x \in C_i$ and $p_y \in C_j$, and an integer t, $1 \leq t \leq k$, such that $| y - x |= g_t$, that is the two processors are connected by a bypass link.

Lemma 1 *F is not catastrophic for A, if and only if C_0 is connected with C_n in graph H.*

Proof. We prove both implications.

Suppose that C_0 and C_n are connected in H. Let $C_0 = C_{i_1} \ldots C_{i_s} = C_n$, $2 \leq s \leq n$ be a path from C_0 to C_n. Let (x_j, y_{j+1}), $1 \leq j < s$ be a pair of indices such that $p_{x_j} \in C_{i_j}$, $p_{y_{j+1}} \in C_{i_{j+1}}$ and $\mid x_j - y_{j+1} \mid = g_t$ for some $t \in \{1 \ldots k\}$ (note that such a pair must exist since $(C_{i_j}, C_{i_{j+1}}) \in E$). Construct the sequences of working processors S_j, $1 \leq j < s-1$, where $S_j = (p_{y_j} \ldots p_{x_{j+1}})$ if $x_{j+1} \geq y_j$, and $S_j = (p_{x_{j+1}}, p_{x_{j+1}-1} \ldots p_{y_j})$ otherwise. It is easy to see that the sequence of processors given by $\{I \ldots p_{x_1}\} S_1 \ldots S_{s-2} \{p_{y_s} \ldots O\}$ is an escape path. Hence F is not catastrophic for A.

Suppose that F is not catastrophic for A. Let P be an escape path and $p_x \in P$ be the last processor such that $x < f_1$ and $p_y \in P$ the first processor such that $y > f_n + l_n$.
Clearly $p_x \in C_0$ and $p_y \in C_n$. Now, following the path from p_x to p_y and recording the chunks traversed, we obtain a sequence $C_{i_1} \ldots C_{i_t}$, $t \geq 2$. By construction $(C_{i_j}, C_{i_{j+1}})$, $1 \leq j < t$, belongs to E, so there is a path in H from C_0 to C_n. □

In Figure 1 is given a procedure, called GRAPH constructing the graph H. Inputs to GRAPH are the fault pattern F and the redundancy G. The output is the graph H represented by its adjacency list. An analysis of the procedure shows that, some redundant information may be present in the adjacency list (i.e. an edge may appear more than once in the same list) however this is not a problem from a complexity point of view.

```
GRAPH(F,G)
  x0 = f1 - g + 1, y0 = f1 - 1
  L(C0) = ∅
  for i = 1 to n - 1 do
        xq = fi + li
        yq = fi+1 - 1
        L(Cq) = ∅
  endfor
  xn = fn + ln, yn = fn + ln + g - 1
  L(Cn) = ∅
  for t = 1 to k do
      ADJACENCY(gt)
  endfor
return
```

```
ADJACENCY(h)
   j = 1
   for i = 0 to n - 1 do
       while xi + h > yj do
           j = j + 1
       endwhile
       while j ≤ n and xj ≤ yi + h do
           L(Ci) ← Cj
           L(Cj) ← Ci
           j = j + 1
       endwhile
       if yj-1 ≥ xi+1 + h then j = j - 1
   endfor
return
```

Fig. 1. Algorithm GRAPH

Lemma 2 *Algorithm* GRAPH *constructs the graph H.*

Proof. Algorithm GRAPH constructs the adjacency list of graph H. The adjacency list of node C_i is $L(C_i)$, for $i = 0, 1, ..., n$. Chunk C_i begins at processor x_i and ends at processor y_i. The values of x_i's and y_i's are computed in the first **for** of GRAPH. To prove the lemma we have to show that C_s is inserted into $L(C_i)$ and C_i is inserted into $L(C_s)$ if and only if (C_i, C_s) is an edge of H.

Suppose that (C_i, C_s) is an edge of H. Without loss of generality assume $i < s$. By definition of H there exist two integers, z and t, such that $1 \leq t \leq k$, $x_i \leq z - g_t \leq y_i$, and $x_s \leq z \leq y_s$. Hence we have that $x_i + g_t \leq y_s$ and $x_s \leq y_i + g_t$.
Consider the i-th iteration of the **for** in ADJACENCY(g_t) . At the beginning of this iteration at least one among $x_i + g_t \leq y_{j-1}$ and $j = i + 1$ holds. So j will be incremented until $x_i + g_t \leq y_j$ and for all the j' such that $x_i + g_t \leq y_{j'}$ and $x_{j'} \leq y_i + g_t$, $C_{j'}$ will be inserted into $L(C_i)$ and C_i will be inserted into $L(C_{j'})$. Hence also C_s will be inserted into $L(C_i)$ and C_i will be inserted into $L(C_s)$.

Suppose that C_s is inserted into $L(C_i)$ and C_i is inserted into $L(C_s)$. Assume $i < s$. It must exists a g_t such that $x_i + g_t \leq y_s$ and $x_s \leq y_i + g_t$. This implies there exists an integer z such that $x_i + g_t \leq y_i + g_t$ and $x_s \leq z \leq y_s$. Hence $p_{z-g_t} \in C_i$ and $p_z \in C_s$ so, by definition of H, $(C_i, C_s) \in E$. □

Theorem 1 *The problem to test whether a fault pattern of n blocks is a CFP for a bidirectional redundant array with k bypass links is solvable in time $O(kn)$.*

Proof. By Lemmas 1 and 2 it is sufficient to show that the algorithm GRAPH requires time $O(kn)$ and the thesis will follow noting that connectivity in a graph $H = (V, E)$ may be tested in time $O(|V| + |E|)$, see for example [3].
Since the first **for** requires clearly $O(n)$ time, we have simply to prove that a call of the routine ADJACENCY requires time $O(n)$.
Let l_i be the increment of the variable j at the $i - th$ iteration of the **for** in ADJACENCY. Clearly the $i - th$ iteration of the **for** will cost $O(l_i)$ and $\sum_{i=0}^{n-1} l_i = n$. Hence we conclude that ADJACENCY has complexity $O(n)$. □

4 A Testing Algorithm for unidirectional arrays.

In this section we study the problem of testing whether a fault pattern is catastrophic for a redundant array with unidirectional links. In this case the technique used for the bidirectional case does not work.

To cope with this problem we use a different approach and we present a solution requiring time $O(n)$ when the redundant array has only one bypass link, and time $O(nk \log k)$ when k is greater than 1.

We call a *block* a set of consecutive (fault or working) processors , for which each processor has a path that does not involve other processors of the block, i.e. each processor is reachable from I.

In the following we will denote a chunk (or block) by the pair (x, y) where p_x and p_y are the first and the last processor in the chunk (or block), respectively. We will say that a pair (x, y) is *minimal* in a set X of pairs if, for each (u, v) in X, it holds $x \leq u$, we will say that it is *maximal* if, for each (u, v) in X, it results $u \leq x$.

In Figure 2 is shown the algorithm TEST, that, given a fault pattern F and the link redundancy G of the array, test if F is catastrophic or not.

```
TEST(F,G)
begin
   S = {(f1 − g + 1, f1 − 1)}; B = {(f1 − g + 1, f1 − 1)}
   for i = 1,...,n−1 do
         insert (fi + li, fi+1 − 1) into S
   endfor
   while S ≠ ∅ and B ≠ ∅ do
         let (x', y') be a minimal element of B
         let (x, y) be a minimal element of S
         case
               y' < x     : delete (x', y') from B
               y < x'     : delete (x , y ) from S
               otherwise :
                          x̄ := max(x, x')
                          for i = 1,...,k do
                                insert (x̄ + gi, y + gi) into B
                          endfor
                          delete (x, y) from S
         endcase
   endwhile
   if in B there is a pair (x', y'), y' + g > fn + ln then return False
                                                      else return True
end
```

Fig. 2. The algorithm TEST

Lemma 3 *Given a fault pattern F for a linear array with link redundancy G, algorithm* TEST *return True if and only if F is catastrophic.*

Proof: Let (x_i, y_i) denote the i-th chunk inserted into S. It is easy to show by induction on i, that if and only if there is a path from p_{f_1-g+1} to p_z, $x_i \leq z \leq y_i$, blocks (x'_j, y'_j) such that $x'_j \leq z + g_t \leq y'_j$, $1 \leq j \leq k$, are inserted into B.

Suppose that F is not catastrophic. Let P be an escape path. We can assume, without loss of generality that P pass through processor $f_1 - g + 1$. Since P is an escape path, it must bypass the fault zone. Let p_u and p_v two consecutive processors in the path P (i.e. there is a link g_t between them), such that $u < f_n$ and $v > f_n + l_n$. Clearly there is a path from processor p_{f_1-g+1} to processor p_u. Hence a block (x', y'), that contains v, i.e. $x' \leq v \leq y'$, will be inserted into B. Such block can not be deleted anymore. Therefore the **if** will return *False*.

Suppose that TEST returns *False*. Then a block (x', y') with $y' > f_n + l_n$ has been inserted into B. This implies the existence of a path from p_{f_1-g+1} to a processor p_z, with $f_n + l_n < z \leq y'$. Such a path, can be easily extended to an escape path. □

Theorem 2 *The problem of testing if a fault pattern of n blocks is a CFP for a unidirectional array is solvable in time $O(n)$ when the array has only one bypass link and time $O(nk \log k)$ when there are $k > 1$ bypass links.*

Proof. By Lemma 3 it is sufficient to show that the algorithm requires time $O(n)$ when $k = 1$ and $O(nk \log k)$ otherwise. We start proving the following facts.

⊙ The number of iterations performed in the **while** is $O(kn)$. Indeed, during each iteration there must be a deletion from S or B, and k insertions into B occur always with a deletion from S. Recalling that initially S contains n elements and B is a singleton, we have that after at most $O(nk)$ iterations one among S and B is empty.

⊙ Since chunks in S are inserted in increasing order, we can organize S as a queue, so that insertions and deletions in S and picking the minimal element of S take constant time. Hence the first **for** requires $O(n)$ time.

Next, we prove the bounds on the running time of TEST.

Case $k = 1$. In this case also the set B, like the set S, can be organized as a queue: indeed new blocks to insert into B are produced in increasing order. Hence insertions and deletions in B and picking the minimal and maximal elements of B take constant time. Since $k = 1$ the internal **for** inside the **while** take no time, hence the complexity of the **while** is $O(n)$. Finally observe that the **if** test can be performed in constant time by picking the maximal element of B.

Case $k > 1$. In this case the new blocks to be inserted into B are not produced in increasing order, so that we have to spend a bit to efficiently handle set B. We organize B in k subsets $B_i, 1 \leq i \leq k$, each containing blocks produced using the link g_i. When a new block is generated using the link g_i, we insert it in the corresponding B_i. The inserted block is maximal in B_i, hence each B_i can be organized as a standard queue. Moreover, we organize the k "heads" of the queues B_i, that contain the minimal elements of each B_i, as an heap which provides the minimal element of B. With this data structure, we can insert and delete in B and we can pick the minimal element of B in $O(\log k)$ time.

Each iteration of the **while** may require time $O(\log k)$, $O(1)$ or $(k \log k)$ depending on the **case** inside the **while**. Since the number of **while** iterations is $O(kn)$, we have only to show that the number of **while** iterations that require time $(k \log k)$ is $O(n)$. This follows easily, noting that each of these iterations requires a deletion from S and the number of elements in S at the beginning is

n. Finally the **if** test may be done in time $O(k)$ by picking the maximal elements of each B_i. □

5 Finding optimal escape paths

When a fault pattern is not catastrophic we are interested in finding escape paths. The algorithms for test if a fault pattern is catastrophic provided in Sections 3 and 4 can be easily modified, without modifying the behavior of their asymptotic time complexity, in order to obtain, when the fault pattern is not catastrophic, an escape path. The escape paths so obtained can be not the best ones. In this section we are interested in escape paths that avoid the fault processors bypassing the minimum number of working processors, i.e. the escape paths with maximal length. We call such escape paths *optimal escape paths* and we call *optimal reconfiguration set* a reconfiguration set that achieve an optimal escape path.

We prove that finding an optimal escape path for a bidirectional redundant array is a NP-hard problem. We provide instead an algorithm that find an optimal escape path for a unidirectional redundant array in time $O(kng)$ where g is the length of the longest bypass link in the array.

Consider the following *Maximum Reconfiguration Length* (MRL for short) problem.

Definition 5 *The MRL problem is: given a bidirectional redundant array A consisting of N processors, with link redundancy G, a fault pattern F and a fixed positive integer K, is there an escape path of length at least K?*

The following lemma holds.

Lemma 4 *The MRL problem is NP-complete.*

Proof. We reduce the problem of testing whether there exists an hamiltonian path between two given vertices of a graph (HP for short), that, notoriously is NP-complete (see [2]), to the MRL problem. Since it is easy to give a non deterministic polynomial time algorithm that solves the MRL problem we conclude that MRL is NP-complete.
Let $H = (V, E)$ be the input graph for the HP problem. Without loss of generality, assume that $V = \{1, 2, \ldots n\}$ and that 1 and n are the vertices to be tested. Consider the following MRL instance: the array A consists of $N = 5 \cdot 2^n - 5$ processors, $G = \{2^n\} \cup \{| 2^j - 2^i | /(i, j) \in E\}$, $F = \{(2^n, 2^n - 1)\} \cup \{(2^{n+1} + 2^i - 2, 2^i - 1)/1 \leq i \leq n\}$ and $K = 2^{n+1} + n - 2$.
Building this MRL instance requires, clearly, polynomial time.
Observe that F consists of $n+1$ blocks of faults. Hence there are $n+2$ chunks. The longest bypass link has length 2^n. Thus $chunk_0$ and $chunk_{n+1}$ consist of $2^n - 1$ processors, whereas it is easy to see that $chunk_i$, $i = 1, 2, ..., n$ consist of one

processor. The blocks of faults between $chunk_0$ and $chunk_1$ and between $chunk_n$ and $chunk_{n+1}$ both consists of $2^n - 1$ processors. This implies the following fact.

Fact 1. Any escape path must pass through $chunk_1$ and $chunk_n$, and that if $chunk_i$, $1 < i < n$ is traversed by the path it must be traversed after $chunk_1$ and before $chunk_n$.

Block of faults between $chunk_i$ and $chunk_{i+1}$, $i = 1, 2, ... n-1$ consists of $2^i - 1$ processors. This implies that the distance between $chunk_i$ and $chunk_j$ is $|2^i - 2^j|$. We will denote the distance between $chunk_i$ and $chunk_j$ by $d_{i,j} = 2^j - 2^i$, supposing that $i < j$ (since the distance is symmetric this is not a loss of generality). It is easy to see that for $1 \leq i < j, u < v \leq n$ we have that $d_{i,j} = d_{u,v}$ if and only if $i = u$ and $j = v$.

Fact 2. Graph H is isomorphic to the graph consisting of $chunk_i$, $i = 1, 2, ..., n$ and their incident links.

This fact follow immediately from the definition of G. Indeed $chunk_i$ and $chunk_j$ are connected by a bypass link of length $d_{i,j}$ if and only if in H vertexes i and j are connected by an edge. Moreover, since no other two chunks are at distance $d_{i,j}$, this bypass link connects only $chunk_i$ and $chunk_j$.

Now we can prove that there is an escape path of length at least K if and only if there is an hamiltonian path between vertices 1 and n in the graph H.

Suppose that there is an escape path of length K. Since there are exactly K working processors each processor is involved in the escape path. Since all the chunks are traversed by Facts 1 and 2, we conclude that exists an hamiltonian path between vertices 1 and n in H (recall that by definition of escape path each processor can be traversed at most once).

Conversely, given an hamiltonian path between vertices 1 and n in H, by Fact 2 it correspond to a path from $chunk_1$ to $chunk_n$. By Fact 1 this path traverses once all $chunk_i$, $i = 2, 3, ..., n-1$ so it can be easily extended to an escape path of length K. □

The NP-completeness of MRL clearly implies the NP-hardness of the problem of finding an optimal escape path for a bidirectional redundant array.

When the array is unidirectional, the problem of finding an optimal escape path may be solved in time $O(kng)$ where g is the length of the longest bypass link. To obtain this result we need two preliminary lemmas.

In figure 3 is defined a procedure, called OPTIMUM_SET, that, given the redundancy G of a unidirectional array A and a fault pattern F not catastrophic for A, constructs an optimal reconfiguration set for A.

Lemma 5 OPTIMUM_SET *is correct and constructs an optimal reconfiguration set in time $O(\ell k)$ where ℓ is the number of processors in the fault zone.*

Proof. Let us define the set $B[s] = \{i | (i = s - g_t, t = 1, 2, ... k \text{ or } i = s - 1), p_i \text{ is not faulty}\}$. Observe that since the array is unidirectional we can reach p_s only from one of $\{p_i | i \in B[s]\}$.

Fix an integer z, $f_1 - g + 1 < z \leq f_n + l_n + g - 1$. It is easy to see by induction that the following invariant holds: at the iteration of the **while** for which $i = z$,

LENGTH[z] is the length of a longest path from processor $f_1 - g + 1$ to processor z, and REC_SET[z] is a reconfiguration set that achieve a path (from $p_{f_1 - g+1}$ to p_z) of such length.
Thus LENGTH[$f_n + l_n + g - 1$] is the length of the longest path from processor $f_1 - g + 1$ to processor $f_n + l_n + g - 1$, and REC_SET[$f_n + l_n + g - 1$] is a reconfiguration set that achieves a path of such length. Moreover, since the array is unidirectional, any optimal escape path must pass through $f_1 - g + 1$ and $f_n + l_n + g - 1$. This means that REC_SET[$f_n + l_n + g - 1$] is an optimal reconfiguration set.
The complexity of OPTIMUM_SET is easily computed: the first **for** take $O(\ell)$ time. The **while** with the nested **for** takes $O(\ell k)$ time. Hence the algorithm runs in time $O(\ell k)$. □

```
OPTIMUM_SET(F, G, P)
  for i = f1 − g + 1 to fn + ln + g − 1 do
      LENGTH[i]=0
      REC_SET[i]= ∅
  endfor
  i = f1 − g + 1
  while i < f1 − g + 1 do
    if pi is a working processor then
      for h ∈ {1} ∪ G do
            if LENGTH[i]+1 >LENGTH[i + h] then
               LENGTH[i + h]=LENGTH[i]+1
               if h > 1 then
                  Let t be the integer s.t. h = gt
                  REC_SET[i + h]=REC_SET[i] ∪ {(i, t)}
               endif
            endif
      endfor
    endif
      i = i + 1
  endwhile
  P=REC_SET[fn + g − 1]
return
```

Fig. 3. The algorithm OPTIMUM_SET

Lemma 6 *Let $F = \{(f_1, l_1) \ldots (f_n, l_n)\}$, be a fault pattern for an unidirectional array with redundancy G. If there exists an ordered set of integers $S = \{j_1, j_2, ..., j_s\}$, $j_s < n$, such that, $chunk_{j_i}$, $1 \le i \le s$, has more than $2g - 4$ processors then F is CFP if and only if at least one among the fault patterns*

$F_1 = \{(f_1, l_1) \ldots (f_{j_1}, l_{j_1})\}$, $F_2 = \{(f_{j_1+1} + l_1, l_{j_1+1}) \ldots (f_{j_2}, l_{j_2})\}$, $\ldots, F_s = \{(f_{j_{s-1}+1}, l_{j_{s-1}+1}) \ldots (f_n, l_n)\}$ *is a CFP.*

Proof. It is clear that if at least one among $F_1, F_2, \ldots, F_s$ is catastrophic then F is also catastrophic. Hence, to prove the lemma we have to show that if F is catastrophic then at least one among $F_1, F_2, \ldots, F_s$ is catastrophic. For the sake of contradiction, suppose that F is catastrophic and none $F_1, F_2, \ldots, F_s$ is catastrophic. Since $chunk_{j_i}$ has more than $2g-4$ processors any escape path for F_i ends before the beginning of any escape path for F_{i+1}. Then concatenating the escape paths of $F_1, F_2, \ldots, F_s$ we can construct an escape path for F, contradicting the hypothesis that F is catastrophic. □

Theorem 3 *Given an unidirectional array A with k redundant links and whose longest link has length g, the problem of finding a reconfiguration set for a non catastrophic fault pattern F of n blocks of faults is solvable in time $O(kng)$.*

Proof. By Lemma 6, split the fault pattern F in s fault patterns, $F_1, \ldots F_s$, $1 \le i \le s$. An optimal reconfiguration set for F is given by the union of the s reconfiguration sets obtained applying the OPTIMUM_SET procedure to $F_1, \ldots F_s$, $1 \le i \le s$.
Let n_i be the number of blocks in F_i, $1 \le i \le s$. Clearly $\sum_{j=1}^{s} n_i = n$. Since the number of elements in the fault zone for F_i is less than $n_i \cdot g + (n-1)(2g-3) + 2g - 2$ (remember that each chunk has less than $2g-4$ processor and that F_i is not catastrophic for A, hence each block has less than g elements), by Lemma 5 constructing an optimal reconfiguration set for F_i takes $O(n_i g)$. Hence, constructing an optimal reconfiguration set for F takes $O(\sum_{j=1}^{s} n_j gk)$ that is $O(kng)$. □

References

1. R. De Prisco and A. De Santis "Catastrophic faults in reconfigurable VLSI linear arrays", Preprint 1993.
2. M. Garey, and D. Johnson, "Computers and Intractability", *Freeman, New York, 1979.*
3. E. Horowitz, and S. Sahni, "Fundamentals of computer algorithms", *Computer Science Press, 1978.*
4. A. Nayak, "On reconfigurability of some regular architectures", Ph.D Thesis, Dept. System and Computer Engineering, Carleton University, Ottawa, Canada, 1991.
5. A. Nayak and N. Santoro, "Bounds on performance of VLSI processor arrays", in *5th Int'l Parallel Processing Symposium*, Anaheim, California, May 1991.
6. A. Nayak, N. Santoro, and R. Tan, "Fault-intolerance of reconfigurable systolic arrays", In *Proc. 20th Int. Symp. on Fault Tolerant Computing, FTCS'20*, pp. 202–209, 1990.
7. L. Pagli and G. Pucci, "Reliability analysis of redundant VLSI arrays", Preprint 1992.

Reconstructing Strings from Substrings (Extended Abstract)

Steven S. Skiena* **Gopalakrishnan Sundaram**

Department of Computer Science, State University of New York
Stony Brook, NY 11794-4400

1 Introduction

Suppose that there was an unknown string S, over a known alphabet Σ, containing the secret to life. We are permitted to ask questions of the form "is s a substring of S?", where s is a specific query string over Σ. We are not told where s occurs in S, nor how many times it occurs, just whether or not s a substring of S. Our goal is to determine the exact contents of S using as few queries as possible.

Although this tale is perhaps over-dramatic, it is not completely inaccurate. The problem arises in sequencing by hybridization (SBH) [2, 3, 5, 10, 14], a new and promising approach to DNA sequencing which offers the potential of reduced cost and higher throughput over traditional gel-based approaches.

The basic sequencing by hybridization procedure attaches a set of single-stranded fragments to a substrate, forming a *sequencing chip*. A solution of radiolabeled single-stranded target DNA fragments are exposed to the chip. These fragments hybridize with complementary fragments on the chip, and the hybridized fragments can be identified using a nuclear detector. The target DNA can now be sequenced based on the constraints of which strings are and are not substrings of the target. Pevzner and Lipshutz [14] give an excellent survey of the current state of the art in sequencing by hybridization, both technologically and algorithmically.

In this paper, we consider a variety of problems with application to sequencing by hybridization. First, we develop a theory of interactive sequencing by hybridization, based on reconstructing strings from substrings. On receiving a sample to sequence, the system will decide which substring it would like to query, and synthesize or extract the corresponding primer. Based on whether

***This work was partially supported by NSF Research Initiation Award CCR-9109289, New York State Science and Technology Foundation Grant RDG-90171, and an ONR Young Investigator Award.**

or not this primer hybridizes, it will propose a new query, and the process will repeat until the sample is completely sequenced. This approach is applicable to Crkvenjakov and Drmanac's target down approach to sequencing by hybridization [3, 14], although the problem is also of independent interest.

In this paper:

- We provide tight bounds on the complexity of reconstructing unknown strings from substring queries. Specifically, we show that $(\alpha-1)n+\Theta(\alpha\sqrt{n})$ queries are sufficent to reconstruct an unknown string, where α is the alphabet size and n the length of the string. This matches the information-theoretic lower bound for binary strings. Further, we show that $\sim \alpha n/4$ queries are necessary, which is within a factor of 4 of our upper bound for larger alphabets. This lower bound holds even for a stronger model which returns the number of occurrences of s as a substring of S, instead of simply whether it occurs.
- We demonstrate that subsequence queries are significantly more powerful than substring queries. Specifically, $O(n \lg \alpha + \alpha \lg n)$ subsequence queries suffice for reconstructing, matching the information theoretic lower bound.
- In certain applications, it may already be known that the unknown string is one of a small set of possibilities, and hence can be determined faster than an arbitrary string. We show that building an optimal decision tree is NP-complete, then give an approximation algorithm which gives trees within a constant multiplicative factor of optimal, with the constant depending upon α.

The conventional approach to sequencing by hybridization uses prefabricated chips instead of interactive queries. In the classical sequencing chip $C(m)$, all 4^m single-stranded oligonucleotides of length m are attached to the surface of a substrate. For example, in $C(8)$ all $4^8 = 65,536$ octamers are used.

Pevzner's algorithm [15] for reconstruction using classical sequencing chips interprets the results of a sequencing experiment as a subgraph of the de Bruijn graph, such that any Eulerian path corresponds to a possible sequence. Thus the reconstruction is not unique unless the subgraph consists entirely of a directed induced path. The strength of this requirement means that large sequencing chips are needed to reconstruct relatively short strands of DNA. For example, the classical chip $C(8)$ suffices to reconstruct 200 nucleotide long sequences in only 94 of 100 cases [13], even in error-free experiments.

However, additional information about the sequence is often available, in particular its length. We show that length can be used to help disambiguate the sequence. For example, observe that the digraph in Figure 1 has a unique postman walk of length $w + x + y + z + 6a + b$ even though it contains a three cycles. The postman walk of length $w+x+y+z+9a+b$ is not unique, however. Although the problem of constructing a postman walk of length l in a weighted graph is NP-complete, we present an algorithm which tests the uniqueness of a postman walk of given length l in polynomial-time for any unweighted digraph.

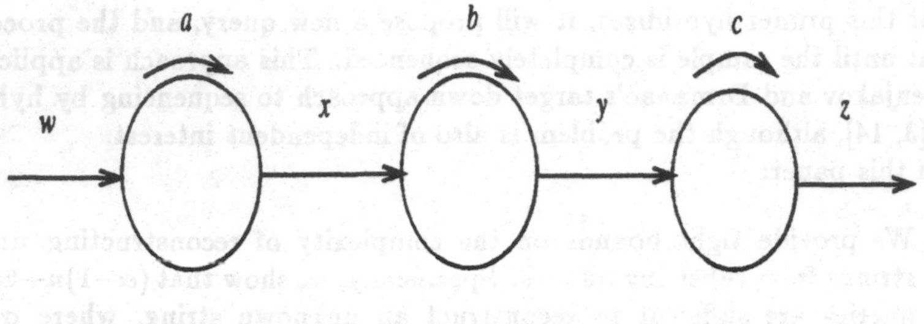

Figure 1: A non-trivial graph with a unique postman tour of length $w + x + y + z + 6a + b$.

This new algorithm, based on a restricted-size variant of the knapsack problem, significantly increases the resolving power of classical sequencing chips.

2 Reconstructing Unknown Strings

In this section, we consider the problem of reconstructing strings from substring queries. Any interactive strategy for determining strings can be specified by a *decision tree* [11]. A decision tree is a rooted binary tree, where each internal node is labeled by a substring query and each leaf by a candidate string. For each node, all leaf nodes of the left subtree contain the given query substring, while none of the leaf nodes of the right subtree do. Figure 2 gives an optimal

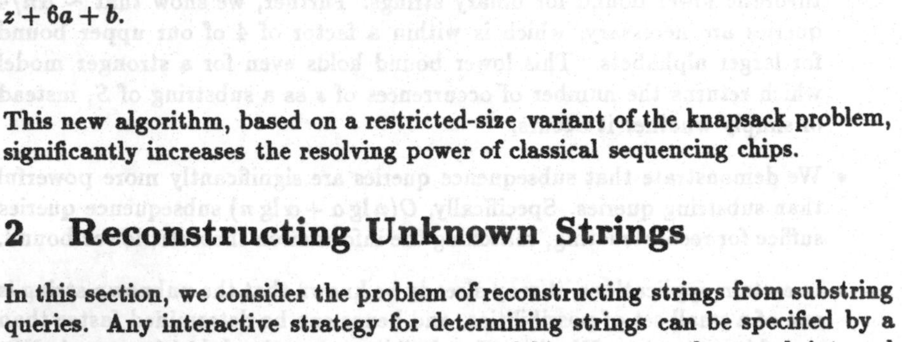

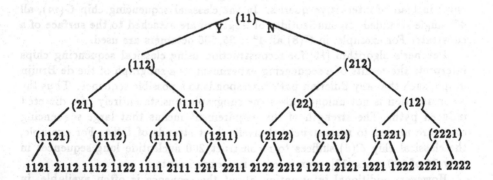

Figure 2: An optimal decision tree for four-character binary strings.

decision tree for four-character binary strings. This tree is of height 4, which is clearly optimal for any binary tree with 16 leaves. Such perfect binary trees are not always possible - consider the set of all two-character binary strings,

{11, 12, 21, 22}, or all binary strings of length six. For each node in the tree, we seek a substring query which partitions the set of candidate strings as evenly as possible. Guibas and Odlyzko [7] and Wilf [16] consider the problem of counting the number of strings with a given set of substrings and forbidden substrings. However, the resulting formulae are far too cumbersome apply to constructing large decision trees.

We use the following notation for strings. The *length* of a string S, $|S|$, is the number of characters it contains. We use α to denote the length of the alphabet. The string representing the concatenation of strings a and b will be denoted ab. If $S = ab$, then a is a prefix of S and b a suffix of S – note that a string of length n has n distinct non-empty prefixes and n non-empty suffixes. If $S = ab$, then $S - b = a$. The string formed by concatenating i copies of s will be denoted s^i.

2.1 Unrestricted Candidate Strings

In this section, we consider interactive strategies for determining unknown, unrestricted strings. Suppose the string is known to be of length n. Any such strategy can be represented by a decision tree with α^n leaves. This exponential growth prohibits the explicit construction of decision trees for even modest-length strings, and so our strategies are presented as algorithms which generates the next query in response to the results of previous queries. We can use lower bounds on the height of decision trees to assess how good the worst-case complexity of our strategies are.

Lemma 1 *$n \lg \alpha$ substring queries are necessary to determine an unknown string S on alphabet Σ, where $|S| = n$ and $|\Sigma| = \alpha$.*

Proof: By an information theoretic argument, $\lg(\alpha^n)$ bits are requires to specify S. The result follows since the result of each substring query provides at most one bit of information about S. ∎

Theorem 2 *An unknown string S of known length n can be reconstructed in $(\alpha - 1)n + 2\lg n + O(\alpha)$ substring queries. If n is unknown, $(\alpha - 1)n + \Theta(\alpha\sqrt{n})$ queries suffice.*

Proof: First we show a slightly weaker result, that any unknown string of known length can be reconstructed in $\alpha(n + 1)$ queries. Begin by making substring queries of single character substrings, so after at most α queries we know a character of S. Let s be a known substring of S and $\Sigma = \{\sigma_1, \sigma_2, \ldots, \sigma_\alpha\}$. In general, we can increase the length of this known substring by one character by querying on the strings $s\sigma_i$, for $1 \leq i \leq \alpha$. At least one of these query strings must be a substring of S, unless s is a suffix of S. When s can no longer be extended, s is a suffix of S and we can continue the process by prepending each character to the known substring, until it is of length n and S is determined.

Now we show how to reduce the multiplicative constant by one. By performing queries emulating a binary search, in $\lg(n + 1)$ queries we can determine largest l such that ${\sigma_\alpha}^l$ is a substring of S. If $l > 0$, initialize s to ${\sigma_\alpha}^l$, otherwise use another $\alpha - 1$ queries to determine a single character of S.

Now suppose that s is a known substring of S, but $s\sigma_i$ is not, for all $1 \leq i < \alpha$. Therefore, either s is the suffix of S or $S\sigma_\alpha$ is a substring of S. We will assume the latter and continue with the procedure. If this suffix of σ_α's reaches length i+1, we must have extended past the right end of S, so $s = s'{\sigma_\alpha}^j$ where $0 \leq j \leq i$ and s' is a string whose last character is not σ_α. If i is sufficiently small, we can find j and thus the right end of S using a linear search with $j+1$ queries. If i is sufficiently large, it is more efficient to performing a binary search, determining the longest such substring s of S in $\lg(i+1)$ queries. We continue this phase until s is a suffix of S. With the suffix and n known, we repeatedly extend s by one prepended character (using $\alpha - 1$ queries) until S is completely determined.

To analyze the complexity of this strategy, observe that each character of s is determined using at most $\alpha - 1$ queries. An additional $(l - j + 1) \cdot (\alpha - 1)$ queries may have been "wasted" determining the non-existant suffix of S, but if so we may charge this to our savings in determining the first l characters of S in the first $\log(n + 1)$ queries, leaving a excess of only $O(\alpha)$ queries.

If n is unknown, we will use a different strategy to identify the ends of S. After finding a single character of S in α queries, we will extend the known substring s by one character to the right in $\alpha - 1$ queries as before, provided s does not end with an unverified suffix of ${\sigma_\alpha}^{\sqrt{m}}$, where $m = |s|$. If it does, we will issue a query to verify s. If $s \in S$ the procedure continues, but if not it means that the string of unverified σ_α's is terminated by the right end of S. The correct suffix can be determined using $\sqrt{m}$ queries of the form $s - {\sigma_\alpha}^i$ for $1 \leq i \leq \sqrt{m}$. We can then repeat the previous procedure with suffix replaced by prefix to determine the rest of S.

In general, $\alpha - 1$ queries are used to determine each character of S. At most $2\alpha\sqrt{n}$ queries are wasted building the non-existant prefix and suffix, independent of the T_n verification queries, where $T_n = T(n - \lceil\sqrt{n}\rceil) + 1$ and $T_1 = 1$. We observe that each query for $|s| > n/2$ verifies at least $\sqrt{n/2}$ characters, and so at most $(n/2)/\sqrt{n/2} = \sqrt{n/2}$ queries are necessary to verify the final $n/2$ characters of S. Therefore,

$$T_n \leq \sum_{i=1}^{\lg n} \sqrt{n/2^i} = \sqrt{n} \sum_{i=1}^{\lg n} 1/2^{i/2} \leq 2\sqrt{n} \sum_{i=1}^{(\lg n)/2} 1/2^i \leq 4\sqrt{n}$$

so $(\alpha - 1)n + \Theta(\alpha\sqrt{n})$ total queries suffice to determine S. ∎

We note that the bounds in Lemma 1 and Theorem 2 are tight for binary strings and when $|S| = 1$.

2.2 A Lower Bound for Reconstruction

In this section, we give an adversarial argument to show that $n(\alpha-3)/4$ substring queries are necessary, approaching our upper bound within a factor of 4. Further, $(n\alpha/4)(1 - o(1))$ queries remain necessary even under a stronger model, which returns the number of times a query string occurs in S.

For ease of presentation, let us initially assume that S is a *circular* string of length $n = \alpha^m$, for some m. To prove the lower bound, we formulate the problem of determining S as a two-person query game. Holmes and Moriarty play a game in which Holmes (the great detective) seeks to determine the unknown string S, where Moriarty chooses S and acts as an oracle, replying either 'yes' or 'no' to each of Holmes' substring queries. The game is over once Holmes finds out S. Of course, Moriarty is Holmes' adversary in Conan Doyle's famous mystery stories.

We assume that both Holmes and Moriarty have infinite computing power. Moriarty does not have to choose S at the start of the game, but may instead maintain the set Q of strings consistent with the results of all previous queries, i.e., he plays an adversary strategy. Provided Q is non-empty, he can always choose any sequence from Q for S.

Suppose Moriarty tells Holmes at the beginning of the game that S is a de Bruijn sequence $d(\alpha, m)$. A de Bruijn sequence $d(\alpha, m)$ of span m is a circular sequence of length α^m, which contains all the strings of length m exactly once. For example, 0000100110101110 and 0000101001101110 are two distinct de Bruijn sequences of span 4 on the binary alphabet $\{0, 1\}$.

Good [6] demonstrated how to construct the sequences of $D(\alpha, m)$, by building a directed graph $G(\alpha, m)$ where the vertex set of $G(\alpha, m)$ represents each string of length $m-1$ on α. Create a directed edge between $x = x_1 \ldots x_{m-1}$ and $y = y_1 \ldots y_{m-1}$ iff $x_i = y_{i-1}$ for $i = 2, \ldots, m-1$. Therefore, each edge is labeled by a string of length m, namely xy_{m-1}. In summary, G contains $\alpha^{m-1}(= n/\alpha)$ vertices and $\alpha^m(= n)$ edges, with each vertex of indegree and outdegree α. There exists a 1-1 correspondence between $D(\alpha, m)$ and the set of all Eulerian circuits of G. Hence, finding an unknown de Bruijn sequence in $D(\alpha, m)$ is equivalent to finding an unknown Eulerian circuit in $G(\alpha, m)$.

Let f_v be a 1-1 function, $f_v : \Sigma \rightarrow \Sigma$. An Eulerian circuit c of $G(\alpha, m)$ and a vertex v defines such a 1-1 function where $f_v(x)$ is mapped to y when we enter the vertex $v = x_1 \ldots x_{m-1}$ in c by the edge xv and leave v by the edge vy. Thus c defines a set of $|V|$ 1-1 functions $F = \{f'_1, \ldots, f'_{n/\alpha}\}$. Note that the converse is not true, i.e., an arbitrary set of functions may not give an Eulerian circuit because the implied circuit may not be connected.

We can represent f_v as an $\alpha \times \alpha$ permutation matrix. To keep track of the information Moriarty has revealed to Holmes, he maintains the state of F using a set of n/α $\alpha \times \alpha$ matrices $f_1, \ldots, f_{n/\alpha}$. At any moment in time, each bit of each f_v in F is either 'yes', 'no', or 'unknown'. If $f_v[x, y] =$ '*unknown*', then Holmes does not know whether xvy is a substring of S. If $f_v[x, y] =$ '*no*', then S does not contain xvy as a substring. If the $f_v[x, y] =$ '*yes*', then xvy is a substring of S. There can be exactly one 'yes' in each row and column. Thus each matrix element determines whether a specific $m+1$ length string is contained in s.

We now we propose a simple strategy for Moriarty to answer queries. Let q be Holmes' query string. For a query q with $|q| \leq m$, Moriarty will always say 'YES', because any de Bruijn sequence in $D(\alpha, m)$ contains all substrings of length m. Therefore, from Holmes's perspective it never pays to ask queries of length $\leq m$. Suppose $|q| \geq m+1$. Let $q_1, \ldots, q_{|q|-m}$ be the $(m+1)$-length substrings of q such that q_i appears before q_j in q, for $i < j$. Note that these

q_i's must be distinct, since S is a de Bruijn sequence. Let z be the smallest index such $f_v[x, y] =$ '*unknown*', where $q_z = xvy$ and $v = x_1 \ldots x_{m-1}$, i.e., z is smallest index such that Holmes does not know whether S contains q_z or not. Moriarty will reply 'YES' if all the matrix elements corresponding to all substrings in q are 'yes' and respond 'NO' if there is one entry corresponding to a substring q is 'no'.

When Moriarty replies 'NO' and z exists, he changes the state of F by:

1. Removing all strings from Q which contains q_z as a substring.
2. Setting $f_v[x, y] =$ '*no*'.
3. Maintaining a row-column constraint. If there is a partition of Σ into B_1, B_2 such that the all entries of the induced submatrix $B_1 B_2$ of f_v are 'no', then all entries in the induced submatrix $B_2 B_1$ of f_v are set 'no'.
4. A mapping $y = f_v(x)$ is said to be *forbidden* if all circuits in Q do not contain xvy as a substring, where $v = x_1 \ldots x_{m-1}$. Change all forbidden mappings from 'unknown' to 'no'.
5. Recur on the changes until the state of the matrix reaches a fixed point.

Therefore, Moriarty interprets Holmes's question as a single query of length $m+1$, and answers accordingly. (Therefore, from Holmes's point of view it always suffices to ask queries of length $m+1$.)

Lemma 3 *Moriarty's strategy always maintains a non-empty set Q of consistent Eulerian cycles.*

How many queries are asked by Holmes before all entries in the matrices of F are filled with 'yes' and 'no's? Here, we outline the analysis, with details to appear in the complete paper.

To characterize the forbidden mappings, let us consider the graph G which evolves dynamically as the game proceeds. Initially, $G = G(\alpha, m)$. Whenever there is a mapping $y = f_v(x)$ implying xv is mapped to vy, then xv and vy are removed from G and replaced by one edge from x to y, labeled xvy. Whenever Moriarty replies 'NO' to a query 'Is xv mapped to vy?', the edges xv and vy cannot be replaced by an edge xvy in G. Isolated vertices are discarded from G as they are created, and by the end of the game G consists a single vertex with a self loop labeled S. With the exception of isolated vertices, we do not permit G to become disconnected, because that would leave S as two disconnected Eulerian circuits and hence not a de Bruijn sequence.

Under what condition can G possibly become disconnected? After a series of queries, articulation vertices may appear in G. Suppose there exists an articulation vertex v such that the components of $G - v$ can be partitioned into two sets, S_1 and S_2, such that all edges in $S_1 \cup \{v\}$ which are incident on v can only be mapped to edges within $S_1 \cup \{v\}$. In this case it is clearly impossible to traverse all the edges in G exactly once. Therefore, if an articulation vertex is created in G during the game, Moriarty must ensure that this situation does

not arise. For example, suppose only one remaining edge xv in $S_1 \cup \{v\}$ can be mapped into an edge in $S_2 \cup \{v\}$. Then, Moriarty must forbid all assignments of the form $y = f_v(x)$, for all edges yv in $S_1 \cup \{v\}$. We claim that all forbidden mappings are mappings of this form.

Lemma 4 ***Forbidden mappings are created only when there exists an articulation vertex v in G such that the components of $G - v$ can be partitioned into two sets, S_1 and S_2, such that there exists a vertex $x \in S_1$ such that for all $v_1 \in S_1 - x$ and all $v_2 \in S_2$, $f_v(v_1) \neq v_2$. Further, these forbidden mappings are exactly $f_v(x) \neq v_1$ for all $v_1 \in S_1$.***

Lemma 5 ***Let S_1, S_2 and T_1, T_2 represent the partitions of $G - v$ creating forbidden mappings in f_v, where (1) the forbidden mappings associated with S were created prior to those of T and (2) the vertex x defining the forbidden mapping belongs to T_1, not T_2. Then either $T_1 \subset S_1$ or $T_1 \subset S_2$.***

Hence, each time a set of forbidden mappings are created, the number of forbidden mappings is at most the number of blue bits gained plus one. There can be at most α sets of forbidden mappings created as each vertex has degree of α.

Corollary 1 ***Reconstructing a circular string of length n on Σ requires at least $n(\alpha - 3)/4$ substring queries.***

Theorem 6 ***$\alpha n/4 - O(n)$ substring queries are necessary to reconstruct a string of length n, even if the frequency of the query substring is also returned.***

Proof: In the previous discussion we assumed that the unknown string S was a circular string. We may relax this assumption by defining, for any Eulerian circuit c, an associated linear string S_c of length $|c| + m$. Let σ be a character in Σ. S_c begins and ends with σ^m, corresponding to breaking c at a specific self-loop in G. The $m + 1$ substrings of S_c are identical to that of S.

Now suppose that each substring query returns a frequency count, instead of 'yes' or 'no'. Observe that a query string q occurs at most once in $G(\alpha, m)$ if $|q| > m$, and exactly $\alpha^{m-|q|}$ times if $|q| \leq m$. In S_c, a string of length q occurs at most once if $|q| > m$, and occurs $\alpha^{m-|q|}$ unless it is of the form σ_1^l, which occurs exactly $\alpha^{m-|q|} + (m - l + 1)$ times. Thus returning the frequency count returns no additional information about S_c, and the lower bound result holds for the stronger model. ∎

2.3 Restricted Candidate Strings

In certain applications, we may have much more knowledge about an unknown string S than just its alphabet and length. For example, there may be a previously defined set of forbidden substrings, none of which can be in S. More generally, we can consider the problem where we are given as input a set of candidates C, and limit the problem to identifying which member of C is S.

Such *model-based* recognition problems occur frequently in testing and classification procedures. In general, we are given a finite set of candidates $C = \{C_1, \ldots, C_n\}$ and a finite set of tests $T = \{T_1, \ldots, T_m\}$, where $T_i \in C$. Each test distinguishes the candidates in T_i from those in $C - T_i$.

Hyafil and Rivest [8] proved that the problem of constructing a minimum height or minimum path-length decision tree is NP-complete. Despite this result, there is some hope for being able to construct optimal decision trees for special types of models and queries. For example, optimal decision trees for non-degenerate polygonal models all sharing a common point can be efficiently constructed [1], although the problem becomes hard if either the common point or degeneracy assumption is removed.

Theorem 7 *Building a minimum height decision tree for substring queries is NP-complete.*

Proof: Our proof is based on the reduction of Hyafil and Rivest [8], who transform exact-cover-by-three-sets to minimum height decision tree. In exact-cover-by-three-sets, we are given as input a universal set U and a set V of three element subsets of U. We seek a subset of $|U|/3$ elements of V whose union is U.

In the reduction of [8], the set of candidates $C = U \cup \{a, b, c\}$, where a, b, c are three elements not in U. The test set T consists of union of V and the set of singleton sets for C. They show that this problem has a decision tree of height $|U|/3 + 2$ iff there exists an exact cover for the original instance. We will construct a set of $n = |C|$ strings on an alphabet of $|V| + 3$, such that the only non-trivial substring queries emulate members of T.

The characters of the ith string, for $1 \leq i \leq |U|$ correspond to the three-element subsets of V containing U_i, so exactly 3 strings contain any such character. The last three strings are each only one character long, consisting of the only occurence of each of the remaining three characters, representing a, b, and c.

Observe that the only substring queries contained in at least three strings are the single character strings coorresponding to tests. Further, these characters are in exactly three strings. By the analysis of [8], there is a decision tree of height $|U|/3 + 2$ iff there is an exact-cover-by-three-sets. ∎

Since the problem is NP-complete, we seek heuristics which can deliver provably good although non-optimal height trees. The most natural strategy is the greedy heuristic, which for each internal node selects the substring which partitions the candidate set as evenly as possible. It has been shown [1] that the greedy tree has height at most $\lg m$ times that of the optimal tree, where m is the number of objects to be distinguished, and that in general this bound is tight. However, for the special case of substring queries, we give a constant factor approximation algorithm, with the constant a function of α.

Lemma 8 *Let M be a set of strings on alphabet Σ. There exists a string s which is contained in at least $|M|/\alpha$ strings from M.*

Lemma 9 *Let M be a set of strings with a common substring s. Then there exists a string longer than s common to at least $|M|/(2\alpha)$ of the strings.*

Lemma 10 *For any set of strings M on alphabet Σ, there exists a substring s which is contained in at least $|M|/(2\alpha+1)$ and at most $2\alpha|M|/(2\alpha+1)$ strings.*

Theorem 11 *For any set of strings M on an alphabet Σ, there exists a decision tree of height at most $\log_{(2\alpha+1)/(2\alpha)}|M|$. Further, such a tree can be constructed in $O(N \lg |M|)$ time, where N is the total number of characters in M.*

Proof: By constructing s according to Lemma 10, we obtain a substring which partitions at least $2\alpha/(2\alpha+1)$th of the largest subset of strings at each level of the tree. A tree with such a partition at each node has the specified height. ∎

We note that the strategy of Theorem 11 can be applied to the set of all α^n strings, providing an alternate solution to the problem of Section 2.1. However, the resulting constants are inferior to those of Theorem 2.

2.4 Reconstruction via Subsequence Queries

Given two sequences $a = a_1 \ldots a_n$ and $b = b_1 \ldots b_m$, with $m \leq n$, we say b is a *subsequence* of a if for some $1 \leq i_1 < \ldots < i_m \leq n$, we have $a_{i_h} = b_h$, for all h, $1 \leq h \leq m$. In this section, we consider the problem of reconstructing an unknown string S over Σ using *subsequence* queries, ie. asking whether q is a subsequence of S. Our main result is that $O(n \lg \alpha + \alpha \lg n)$ subsequence queries suffice to reconstruct an unknown string of length n. The matches the information-theoretic lower bound of Lemma 1 and proves that subsequence queries are more powerful than substring queries for reconstruction.

Our basic strategy is first to determine the character composition of S, and then interleave the resulting subsequences of S into longer subsequences. By repeatedly interleaving the resulting subsequences, we reconstruct S.

For each character x in Σ, we can use subsequence queries to perform a binary search to count how many times x occurs in S, for the problem is equivalent to asking for the largest i such that x^i is a subsequence of S. Even if n is unknown, we can perform one-sided binary searches, so $2\alpha \lg n$ subsequence queries suffice to determine the character composition of S. Then, we can use repeated one-sided binary search to merge the sequences:

Lemma 12 *Let A and B be character-disjoint subsequences of S, where $a = |A|$, $b = |B|$, and $a \leq b$. We can interleave A and B to obtain a length $a+b$ subsequence of S using at most $2.18a + b + 5$ subsequence queries.*

Theorem 13 $2\alpha \lg n + 1.59n \lg \alpha + 5\alpha$ *subsequence queries suffice to reconstruct a unknown string S of unknown length n.*

3 Verifying Strings with Substrings

Strictly speaking, a sequence S is uniquely determined by a classical sequencing chip $C(m)$ iff the implied subgraph of the de Bruijn graph $D(\alpha, m)$ is exactly a path. However, often we have additional information about the sequence which we can exploit to show that $C(m)$ implies S. In this section, we show how specifying the length of S can be used to increase the resolving power of classical sequencing chips.

A special application of these results is in using sequencing by hybridization to verify the validity of a DNA sequence derived from a different experimental procedure. Let S' be an experimentally determined sequence of S. We need to verify whether $S' = S$ using substring queries. In general, our objective is to minimize the cost in deciding whether $S = S'$, where the cost of a query is somehow a function of its length. Certain strings are expensive to verify under such a model, for example, verifying $a^l b a^l$ requires asking at least one query of length $l+1$. Minimizing the length of the longest query is equivalent to asking what is the smallest classical sequencing chip $C(k)$ such that $C(k)$ verifies S, given that we know $|S| = n$.

This problem can be stated as testing whether an unweighted digraph has a unique *postman walk* of length n. In a digraph $G = (V, E)$, a walk W from x to y is a sequence of vertices $x, v_1, \ldots, v_k, y$ such that $(v_i, v_{i+1}) \in E$. A walk W is called a *postman walk* if all edges in E appear in W at least once. A postman walk W is said to be a *minimum postman walk* if there does not exist a postman walk W' such that $|W'| < |W|$.

Let A be the set of all k-substrings contained in S. As in the de Bruijn graph construction of Section 2.2, for each k-string $x_1, \ldots, x_k$ in A, we create two vertices $x = x_1, \ldots, x_{k-1}$ and $y = x_2, \ldots, x_k$ and directed edge from x to y. It can be easily seen that there exists a unique string of length l, containing all k-strings in A if and only if there exists a unique postman walk of length $l = n - k + 1$ in G.

Finding the minimum postman walk is known as the *Chinese postman problem* [9]. Polynomial algorithms based on bipartite matching exist for directed and undirected graphs [4], although the problem is NP-complete for mixed graphs [12].

Lemma 14 *Let G be a directed graph containing a vertex of indegree or outdegree greater than 2. Then for any positive integer k, a postman walk W of length k in G cannot be unique.*

Lemma 15 *In any postman walk W, where $|W| > |P|$, there is a vertex of G which appears at least three times in W.*

Lemma 16 *If a postman walk W of length k is unique, then (1) the minimum postman walk P is unique and (2) W can be constructed by adding redundant self walks to P.*

Lemma 16 gives a necessary condition for a unique postman walk W, with $|W| \geq |P|$, i.e., W is constructed by adding redundant self walks to P. For a walk W, we define the operation *add self walk, ASW(w,i)*, as inserting the self walk $w = (x, z_1, \ldots, x)$ at x^i. Also, two walks W and W' are equivalent if W' can be obtained from W by moving the self walk $x^i, \ldots, x^{i+1}$ to $x^j, \ldots, x^{j+1}$ in

W, for some $j \neq i$. We say W' is obtained from *W* by applying the *MOVE(x,i,j)* operation to *W*.

Lemma 17 *In G, if the minimum postman walk P is unique, then any postman walk W can be obtained from applying ASW and MOVE operations to P.*

Therefore, verifying the uniqueness of a walk *W* is equivalent to checking whether there is only one way of adding redundant self walks of appropriate length to *P*, and further that MOVE operations cannot be applied to this walk to obtain different walks. In the full paper, we present an algorithm using a bounded sized variant of the knapsack problem, to show

Theorem 18 *In a directed graph G, given a positive integer l, in* $O(l^3)$ *time we can determine whether there exists a unique postman walk of length l.*

References

[1] E. Arkin, J. Mitchell, H. Meijer, D. Rappaport, and S. Skiena. Decision trees for geometric objects. In *Ninth ACM Symp. Computational Geometry*, 1993.

[2] W. Bains and G. Smith. A novel method for nucleic acid sequence determination. *J. Theor. Biol.*, 135:303–307, 1988.

[3] R. Dramanac and R. Crkvenjakov. DNA sequencing by hybridization. Yugoslav Patent Application 570, 1987.

[4] J. Edmonds and E. Johnson. Matching, Euler tours, and the Chinese postman problem. *Math. Prog.*, 5:88–124, 1973.

[5] S. Fodor, J. Read, M. Pirrung, L. Stryer, A. Lu, and D. Solas. Light-directed, spatially addressable parallel chemical synthesis. *Science*, 251:767–773, 1991.

[6] I. J. Good. Normal recurring decimals. *J. London Math. Soc.*, 21:167–172, 1946.

[7] L. Guibas and A. Odlyzko. String overlaps, pattern matching, and non-transitive games. *J. Combinatorial Theory (Series A)*, 23:183–208, 1981.

[8] L. Hyafil and R. Rivest. Constructing optimal binary decision trees is NP-complete. *Information Processing Letters*, 5:15–17, 1976.

[9] M.-K. Kwan. Graphic programming using odd and even points. *Chinese Math.*, 1:273–277, 1962.

[10] Y. Lysov, V. Florent'ev, A. Khorlin, K. Khrapko, V. Shik, and A. Mirzabekov. *Dokl. Acad. Sci. USSR*, 303:1508–, 1988.

[11] B. Moret. Decision trees and diagrams. *Computing Surveys*, pages 593–623, 1982.

[12] C. Papadimitriou. The complexity of edge traversing. *J. ACM*, 23:544–554, 1976.

[13] P. Pevzner, Y. Lysov, K. Khrapko, A. Belyavski, V. Florentiev, and A. Mizabelkov. Improved chips for sequencing by hybridization. *J. Biomolecular Structure and Dynamics*, 9:399–410, 1991.

[14] P. A. Pevzner and R. J. Lipshutz. Towards DNA sequencing by hybridization. Submitted for publication, 1992.

[15] P.A. Pevzner. *l*-tuple DNA sequencing: Computer analysis. *J. Biomolecular Structure and Dynamics*, 7:63–73, 1989.

[16] H. Wilf. Strings, substrings, and the nearest integer function. *Amer. Math. Monthly*, 94:855–860, 1987.

Combinatorial Complexity of Signed Discs*

(Extended Abstract)

Diane L. Souvaine[†] Chee-Keng Yap[‡]

May 16, 1993

Abstract

Let $\mathcal{C}^+$ and $\mathcal{C}^-$ be two collections of topological discs of arbitrary radii. The collection of discs is 'topological' in the sense that their boundaries are Jordan curves and each pair of Jordan curves intersect at most twice. We prove that the region $\bigcup\mathcal{C}^+ - \bigcup\mathcal{C}^-$ has combinatorial complexity at most $10n - 30$ where $p = |\mathcal{C}^+|$, $q = |\mathcal{C}^-|$ and $n = p + q \geq 5$. Moreover, this bound is achievable. We also show bounds that are stated as functions of p and q. These are less precise.

1 Introduction

Analysis of the combinatorial complexity of geometric or topological arrangements is often a prelude to the complexity analysis of algorithms. There have been several recent papers on the combinatorial complexity of arrangements of planar curves. In particular, Kedem et al. [4] shows that if $\mathcal{C}$ is a collection of $n \geq 0$ *topological discs* then the union $\bigcup\mathcal{C}$ of these topological discs has boundary complexity at most $\max\{n, 6n - 12\}$, and that this bound is tight for all $n \geq 0$. A *geometric disc* is a standard disc which is defined by its center and radius. A *topological disc* is a bounded planar region whose boundary is a Jordan curve, called a *topological circle*. In an arrangement of topological circles, two distinct topological circles, like their geometric counterparts, may only intersect in at most two points. We introduce the notation $\beta(n)$ to denote $\max\{n, 6n - 12\}$. The construction to achieve the bound $\beta(n)$ $(n \geq 3)$ is simple, even for geometric discs.

In this paper, we extend the result of [4] to a new situation, where the topological discs are colored either *positive* or *negative*. We want to determine the combinatorial complexity of the union of the positive discs minus the union of the negative discs. The literature

*This work originated at the Parallel Algorithms and Computational Geometry Workshop (1991) at McGill University's Bellairs Research Center and was supported in part by NSF grants #DCR-84-01898, #CCR-87-03458, #CCR-88-03549 and #CCR-91-04732.

[†]Dept. of Computer Science, Rutgers University, New Brunswick, New Jersey 08903, dls@cs.rutgers.edu.

[‡]Dept. of Computer Science, New York University, New York, New York 10012, yap@cs.nyu.edu.

has some results about bichromatic arrangements of discs, under the so-called "red-blue combination lemmas" [3,1], but these results do not apply to our situation which seems to be fundamentally different. Throughout this paper, the terms "disc", "circle", and "arc" mean "topological disc", "topological circle" and "topological arc" unless otherwise stated.

Let C^+ and C^- be the collections of positive and negative discs, respectively. Let $p = |C^+|$, $q = |C^-|$ and $n = p + q \geq 1$. We study the region $R = R(C^+, C^-) := \cup C^+ - \cup C^-$ defined as the union of the positive discs minus the union of the negative discs. The boundary of a positive disc is called a *positive circle*, a connected portion of a positive circle, a *positive arc*; similarly, we speak of *negative circles* and *negative arcs*. The boundary of R can be decomposed into a collection of *maximal arcs*. The *vertices* of R are the endpoints of these maximal arcs. For simplicity, we make the following **regularity assumption:** *two Jordan curves in a collection are either disjoint or intersect transversally (at two distinct points), and three Jordan curves do not pass through a common point.* It also turns out to be more convenient to count vertices rather than maximal arcs (although in general, these two counts might not be equal.)

Let $B(C^+, C^-)$ denote the *combinatorial complexity* of the boundary of R, i.e., the number of maximal arcs of R. Let $B(p,q)$ denote the maximum value attained by $B(C^+, C^-)$ over all choices of C^+, C^- where $p = |C^+|, q = |C^-|$. We have $B(0,q) = \beta(0) = 0$, $B(p,0) = \beta(p)$. The first result is obvious and the second is a restating of the result of [4]. Let $B(n)$ be the maximum value attained by $B(p,q)$ where $n = p + q$. Henceforth, we assume $p \geq 1, q \geq 1$ and $n \geq 2$. Let $B^*(p,q), B^*(n)$ be analogously defined for geometric discs. Clearly,

$$B^*(p,q) \leq B(p,q), \qquad B^*(n) \leq B(n).$$

The first inequality is an equality when $q = 0$ (from [4]) and when $p = 0$ (obviously). It is an open question as to whether the first inequality is strict for some values of p, q.

In this extended abstract, most proofs are omitted. The full paper appears in [5].

Applications. Goodrich and Kravets [2] studied the problem of matching a pattern P to a background B where P, B are finite point sets in the plane. An obvious scenario for this is in astronomy. The problem is to translate P so as to obtain an ϵ-match with a subset of B. Let us define B_ϵ to be the union of ϵ-discs about points of B. For each p, the set of translation vectors that puts p inside B_ϵ is denoted $T(p)$. So $T(p) = B_\epsilon - p$. They were interested in constructing the intersection of all $T(p)$'s as p range over P. Suppose now we have certain circular regions that must NOT be matched to p. (This could be epsilon-discs about points of B.) Then $T(p)$ becomes a region of the form $R(C^+, C^-)$.

We can also reformulate our problem as follows. Suppose we have a collection of opaque planar discs. Each disc is colored black or white. We place each black disc on the plane $z = 1$ and each white disc on the plane $z = 0$. Now we view the configuration from an infinite distance, vertically above the discs. Then the complexity of the white region we see is $B(C_{white}, C_{black})$.

2 A $10n$-Lower Bound Construction for $B(n)$.

The reader might be tempted to conclude from the result of [4] that $B(n) \leq \beta(n)$. This proves to be false:

Lemma 1 *For all* $q \geq 1$, $B(2,q) \geq \max\{6q, 10q-10\}$. *Hence* $B(n) \geq 10n - 30$ *for all* $n \geq 3$.

Proof. This lemma can be seen directly if $q \leq 2$ or $n \leq 4$. (In particular, the reader should verify that $B(2,2) \geq 12$.) So assume $q \geq 3$ and $n \geq 5$. Our explicit construction begins with geometric discs. See Figure 1. Let A^+, A^- be two very large discs of equal radii whose centers lie on the positive y-axis and whose south-poles pass through the points $(0, 1-\varepsilon)$ and $(0,1)$, respectively, for $0 < \varepsilon \ll 1$. Moreover their radii are $\gg 1$ so that the southernmost portions of their boundaries are well-approximated by the horizontal lines $y = 1-\varepsilon$ and $y = 1$, respectively. Similarly, let B^+, B^- be two very large discs with equal radii whose centers lie on the negative y-axis and whose north-poles pass through the points $(0,\varepsilon)$ and $(0,-1)$. Again their large radii implies that the northernmost portions of their boundaries are well-approximated by the horizontal lines $y = \varepsilon$ and $y = -1$, respectively. Let $m = q - 2 \geq 1$. Let $D_1, D_2, \ldots, D_m$ be discs of radii $1+\varepsilon^2$, and whose centers lie on the x-axis, laid out in a row centered about the origin. Each D_i touches A^- and B^-; it also touches D_{i-1} (provided $i \geq 2$) and D_{i+1} (provided $i \leq m-1$). For $C^+ = \{A^+, B^+\}$ and $C^- = \{A^-, B^-, D_1, \ldots, D_m\}$,

$$B^*(C^+, C^-) = 10n - 38. \tag{1}$$

Distortion. To obtain our final complexity bound, we distort the above construction so that the discs are no longer geometric. In particular, we distort the 4 discs A^-, A^+, B^-, B^+ to form 8 new vertices as illustrated in Figure 2. This gives the stated bound. **Q.E.D.**

The following result of a simple extension of the above construction will be useful later.

Lemma 2 *For all* $q \geq 1$, $B(3,q) \geq \max\{8q+4, 10q-2\}$.

3 A Simple Upper Bound for $B(n)$

In order to prove upper bounds for $B(n)$ and $B(p,q)$ we classify vertices of the region $R = R(C^+, C^-)$ into three types (see Figure 3): a *type I* vertex is one incident on two positive arcs of R; a *Type II* vertex is one incident on two negative arcs of R; a *Type III* vertex is one incident on a negative and a positive arc.

It is clear that the number of type I vertices is at most $\beta(p)$ using the result of [4]. Similarly, the number of type II vertices is at most $\beta(q)$. The next section gives an upper bound on the number of type III vertices. But for now, we make a simple observation that leads to a non-trivial upper bound on $B(n)$:

CLAIM: the number of type I and type III vertices combined is at most $\beta(p+q)$.

To see this, let $U = (\bigcup C^{+}) \cup (\bigcup C^{-})$ be the union of the positive and negative discs. The claim follows because there is a natural injection (1-1 map) from the set of types I and III vertices to the vertices of U. Consequently,

$$
\begin{aligned}
B(p,q) &\leq \beta(q)+\beta(p+q) \\
&= \begin{cases} 6p+12q-24 & \text{if } q \geq 3 \\ 6p+7q-12 & \text{if } q=2 \\ 6p+6q-12 & \text{if } q=1, p\geq 2 \\ p+q & \text{if } q=1, p=1 \end{cases}
\end{aligned} \tag{2}
$$

Since we assume that p and q are both positive, we may further deduce that

$$B(n) \leq 12n-28, \qquad n \geq 3.$$

This upper bound leaves a substantial gap between it and the $10n$-lower bound from the last section. In order to eliminate that gap, we need to first establish some exact bounds on $B(p,q)$, especially for small values of p, q.

Lemma 3

$$
\begin{aligned}
B(1,q) &= \beta(1+q), & (3)\\
B(2,q) &= \max\{6q, 10q-10\}, & (4)\\
B(p,1) &= 6(p+q)-12, \quad (p\geq 3), & (5)\\
B(p,2) &= 6(p+q)-10, \quad (p\geq 3), & (6)\\
B(p,q) &= 6p+12q-24, \quad (p\geq 3, 3\leq q\leq p-1), & (7)
\end{aligned}
$$

The proof is included in the full paper [5]. See Figure 4 for an instance realizing Equation (7).

4 Exact bound for $B(n)$

The main result of this section will be an exact bound for $B(n)$. To achieve this result, we need a tighter bound on the number of type III vertices in an arbitrary arrangement.

Lemma 4 (Main Lemma)

(i) In any arrangement of $p \geq 1$ positive and $q \geq 1$ negative discs where $p+q \geq 3$, the number of type III vertices is at most $4(p+q)-8$. If $p=1$ or $q=1$, the upper bound is $2(p+q)-2$.

(ii) These upper bounds are achievable for all p, q.

First we prove the upper bound (part (i)). It is sufficient to consider only collections of signed discs C^+, C^- that maximize the number of type III vertices for given $p = |C^+|, q = |C^-|$. Given that a type III vertex occurs only when a positive circle intersects a negative circle at a point v which does not lie in the interior of any disc (positive or negative), no disc D of the collection can be contained in the union of the remaining discs of $C^- \cup C^-$.

The proof has two stages. In the initial stage, we prove the Main Lemma under the the following *simplifying assumption.* The *covering number* of a point v in the plane is the number of discs in $C^+ \cup C^-$ that contains v.

Simplifying assumption. *No point in the plane has covering number greater than two.*

In the second stage, we prove that any collection of signed discs can be transformed by shrinking each disc until the collection satisfies our simplifying assumption. Moreover, this transformation does not decrease the number of type III vertices.

Stage 1. Under the simplying assumption, for each disc D in $C^+ \cup C^-$, we may pick a point $v(D)$ in the interior of D such that $v(D)$ has covering number 1. We then define an *embedded* planar graph G as follows: the vertices of G are the points $v(D)$; an edge $e(D, D')$ consisting of a simple curve in $D \cup D'$ connects $v(D)$ and $v(D')$ in the graph G if and only if D and D' intersect; no two edges intersect except in sharing a common endpoint; and $e(D, D')$ avoids all other discs in $C^+ \cup C^-$. An edge $e(D, D')$ is *dichromatic* if D and D' have different signs. The number of type III vertices is exactly twice the number of dichromatic edges. Thus, part (i) of the Main Lemma amounts to the following result about graphs:

Lemma 5 *Given an embedded planar graph G on $n = p + q \geq 3$ vertices where $p \geq 1$ are colored positive and $q \geq 1$ are colored negative, the number of dichromatic edges is at most $2(p+q) - 4$. If $p = 1$ or $q = 1$, the bound can be improved to $p + q - 1$.*

Stage 2. It remains to justify the simplifying assumption. In the full paper [5], we do this by giving a terminating procedure to transform the discs in $C^+ \cup C^-$ such that the number of type III vertices is not decreased, although it may increase. At termination, the arrangement will satisfy our simplifying assumption. Briefly, we do this in two phases:

Phase one. This is illustrated in figure 5. Relative to each choice of a "reference disc" D_0, we shrink the other discs that intersect D_0 until their intersection pattern is essentially determined by their intersection with the boundary of D_0.

Phase two. It can be shown that after phase one, there are no quadruple intersections. Hence we only have to remove the triple intersections. This is illustrated in figure 6.

Lower bound on the number of Type III vertices. Included in full paper [5].

Main Result. We are ready to show the main result of this paper:

Theorem 6 *For $n \geq 5$, $B(n) = 10n - 30$.*

The lower bound was shown in section 2. For the upper bound, we argue this separately for small p or q. For $p \geq 2, q \geq 2$, we can bound the three types of vertices separately,

$$B(p,q) \leq \max\{p, 6p-12\} + \max\{q, 6q-12\} + 4(p+q) - 8$$

and some calculations lead to our stated bound.

5 Bounds for $B(p,q)$

We seek bounds on $B(p,q)$ that are exact, up to additive constants. Unlike the case of $B(n)$, the picture here is incomplete, even if we restrict p or q to be sufficiently large. Here is what we know: from Lemma 3, we have exact bounds for $B(p,q)$ when $p \leq 2$ or $q \leq 2$ and when $3 \leq q \leq p-1$. Combining Lemma 2 with some omitted details in the proof of the main theorem (see [5]), we get another the exact bound,

$$B(3,q) = 10q - 2, \qquad q \geq 3. \tag{8}$$

The difficulty in getting tight general bounds stems from the fact that we must use different techniques depending on the relative sizes of p and q. For instance, we get the following hybrid upper bound by combining Lemma 3 and the main theorem:

Lemma 7 *For $p \geq 3, q \geq 3$,*

$$\begin{aligned} B(p,q) &\leq 6p + 10q + \min\{4p-32, 2q-24\} \\ &\leq \begin{cases} 10(p+q) - 32 & \text{if } 2p-4 \leq q \\ 6p + 12q - 24 & \text{if } 2p-4 \geq q \end{cases} \end{aligned}$$

One suspects that the bound at the cross-over value, $q = 2p-4$, is not tight.

Similarly, one can derive lower bounds using hybrid constructions:

$$B(p,q) \geq 6p + 10q - 22, \qquad p \geq 5, q \geq 3.$$

This inequality comes from

$$B(p,q) \geq B(2,q) + B(p-2,0) = (10q-10) + (6p-12),$$

corresponding to doing the $10n$-construction to achieve the bound $B(2,q)$, combined with the $6n$-construction for the remaining $p-2$ positive discs. For $3 \leq p \leq q-1$, Equation (2) provided a somewhat better lower bound of $B(p,q) \geq 8p + 10q - 26$.

We can extend the bound of Equation (7) in Lemma 3 somewhat. By taking the construction in that proof, by doubling the number of positive circles, and by further deforming the negative circles, we achieve an arrangement of complexity $6p + 12q - 24$ for $p = 2q$. See Figure 7. In each of $q-2$ interstitial spaces, it is possible to insert 2 new negative discs each having complexity 12, thus maintaining the total complexity $6p+12q-24$ over a broader spectrum of values of q. See Figure 8. The last part of Lemma 3 can be extended to read:

$$B(p,q)=6p+12q-24, \quad (p\geq 3, \quad 3\leq q\leq 1.5p-4).$$

Unfortunately, we leave a gap for $B(p,q)$ when $1.5p-4\leq q\leq 2p-4$.

Another goal is to achieve a lower bound to match the upper bound

$$B(p,q)\leq 10(p+q)-32, \quad (2p-4\leq q).$$

How close can we get to that goal by expanding the previous figure? That figure achieves a complexity of $9.6(p+q)-O(1)$ for $q=1.5p-4$. Unfortunately, the highest complexity negative disc which we can insert into the current construction is of complexity 10, but there are arbitrarily many of them! Consequently, for all $1>\varepsilon>0$, there exists a $q\geq 2p-4$ such that there exists an arrangement of p positive and q negative discs with complexity $(10-\varepsilon)(p+q)-32$. It remains an open problem to find a concrete example which for arbitrary values of p and for $q\geq 2p-4$ can achieve the bound of $10(p+q)-32$.

6 Final Remarks

1. In the introduction, we view the signed disc problem as placing the discs in two parallel planes in 3-space. Suppose we have a total of $k\leq n$ distinct horizontal planes and the n discs are placed on these planes (but no plane may have both black and white discs). What is the complexity $B_k(n)$ of the white region viewed from vertically above?

2. Although we have focused on topological discs, geometric discs are interesting as well. Note that from section 2, we know that $B^*(n)=10n-c$ where c is a constant between 30 and 36. The most pressing problem is to give an explicit instance such that $B^*(p,q)<B(p,q)$. According to Lemma 3, $B(2,2)=12$, $B(2,3)=20$, $B(2,4)=30$. It is not obvious from the construction of section 2 that these bounds can be achieved geometrically. But it turns out that they are indeed geometric. For example, Figure 9 proves that $B^*(2,3)=20$. According to Equation (8), $B(3,3)=28$. An arrangement achieving this bound must simultaneously achieve $\beta(3)=6$ type I vertices, $\beta(3)=6$ type II vertices and $4(3+3)-8=16$ type III vertices. Surprisingly, this can be achieved geometrically too: $B^*(3,3)=28$. So any proof showing a difference between topology and geometry must prove an impossible configuration for 7 or more geometric discs.

References

[1] H. Edelsbrunner, L. Guibas, and M. Sharir. The complexity and construction of many faces in arrangments of lines and of segments. *Discrete and Computational Geometry*, 5:161–196, 1990.

[2] M. T. Goodrich and D. Kravets. Point set pattern matching, 1992. Preprint.

[3] L. J. Guibas, M. Sharir, and S. Sifrony. On the general motion-planning problem with 2 degrees of freedom. *Discrete and Computational Geometry*, 4:491–521, 1989.

[4] K. Kedem, R. Livne, J. Pach, and M. Sharir. On the union of Jordan regions and collision-free translational motion amidst polygonal obstacles. *Discrete and Computational Geometry*, 1:59–71, 1986.

[5] Diane Souvaine and Chee Yap. Combinatorial complexity of signed discs. DIMACS Technical Report 92-54, Center for Discrete Mathematics and Theoretical Computer Science, New Jersey, December 1992.

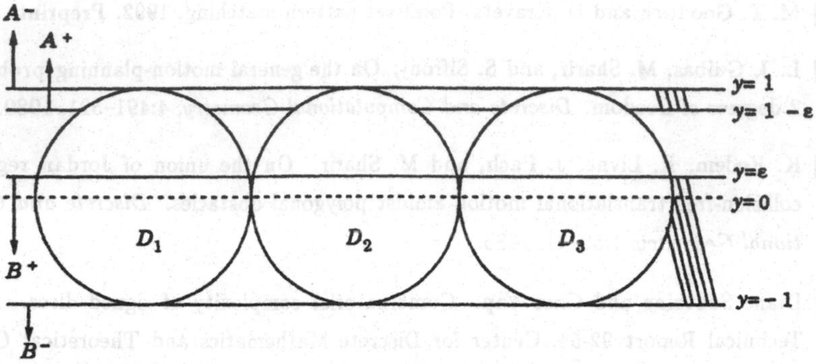

Figure 1. Arrangement with large complexity for R.

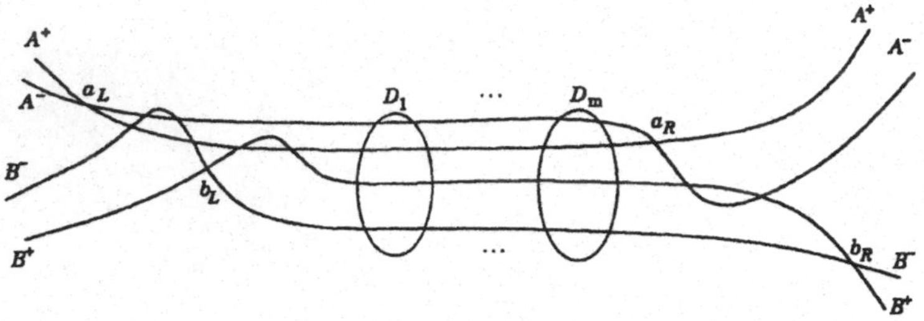

Figure 2. Distortion of construction.

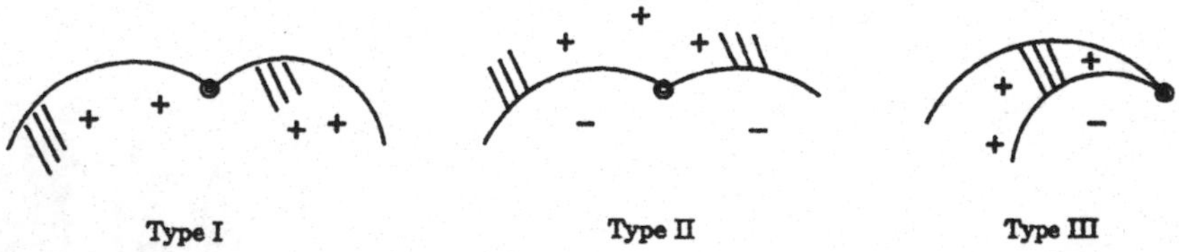

Figure 3. Three Types of Vertices.

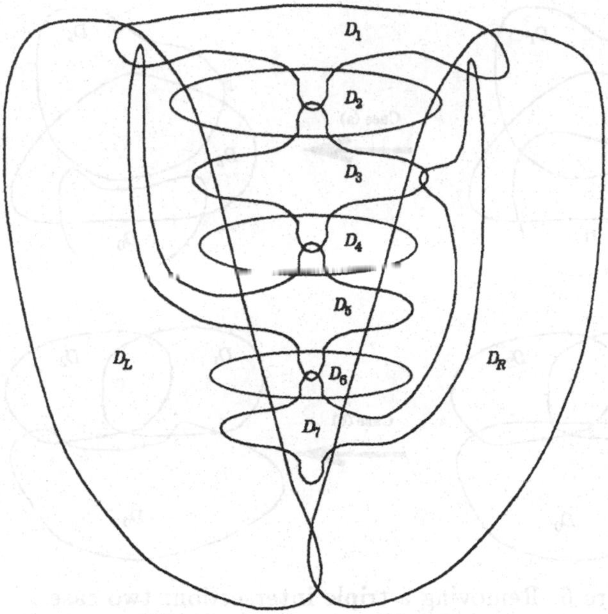

Figure 4. Construction for $6p + 12q - 24$.

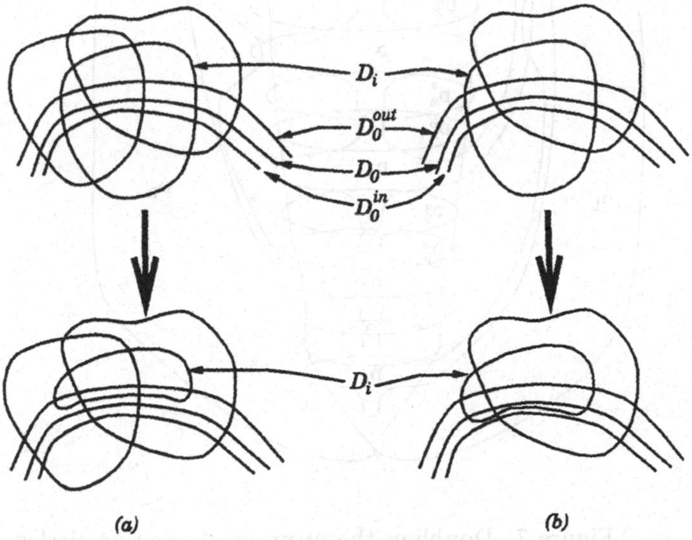

Figure 5. Shrinking disk D_i: the two cases.

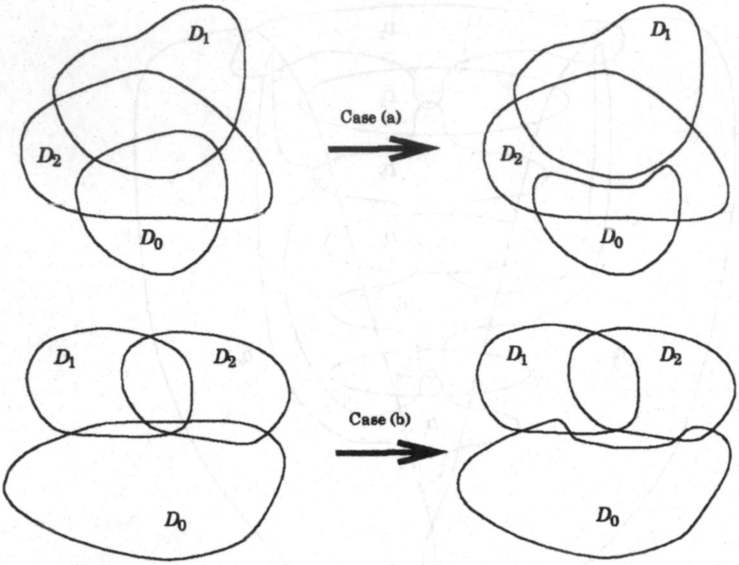

Figure 6. Removing a triple intersection: two cases.

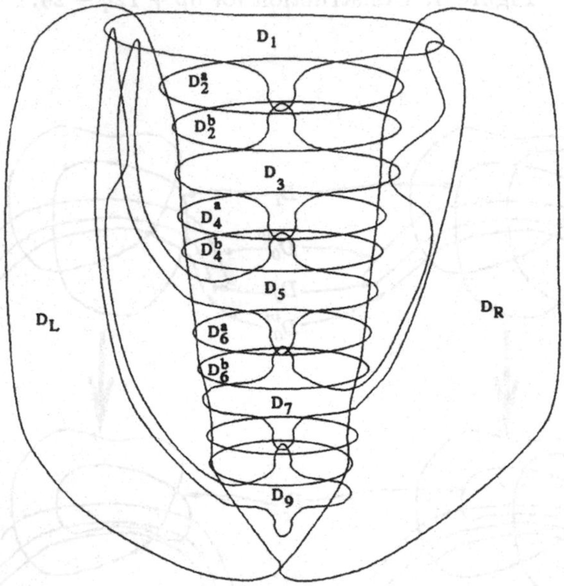

Figure 7. Doubling the number of positive circles.

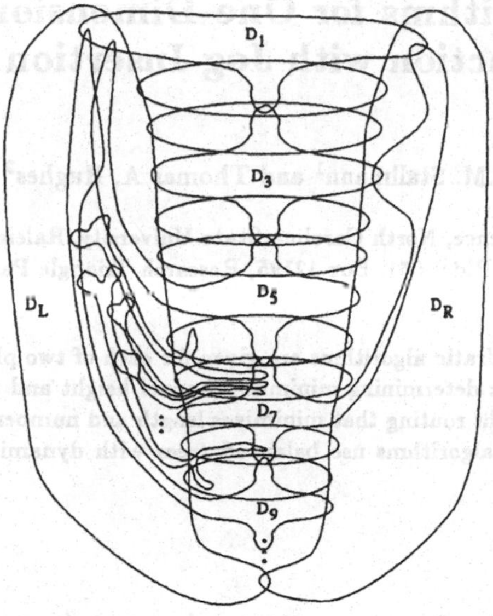

Figure 8. Tight lower bound for $B(p,q)$ for $p/2 \leq q \leq 1.5p-4$

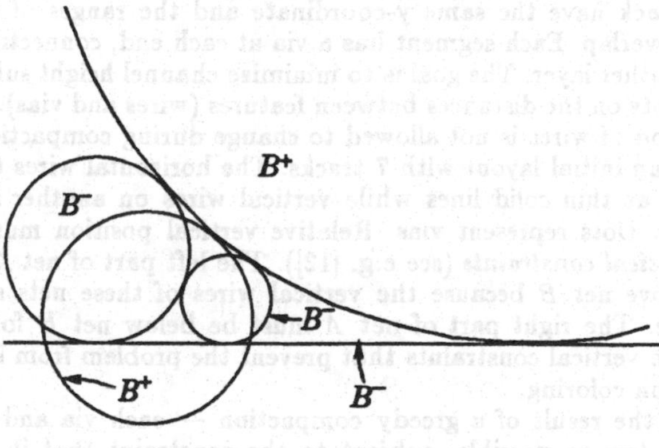

Figure 9. Construction for $B^*(2,3) \geq 20$.

Fast Algorithms for One-Dimensionsal Compaction with Jog Insertion

Matthias F.M. Stallmann[1] and Thomas A. Hughes[2]

[1] Dept. of Computer Science, North Carolina State University, Raleigh, NC 27695
[2] IBM Corp., Dept. D63/Bldg. 061, Box 12195, Research Triangle Park, NC 27709

Abstract. Subquadratic algorithms are given for each of two phases of channel compaction: determining minimum channel height and producing a minimum height routing that minimizes length and number of jogs for each wire. The algorithms use balanced trees with dynamic finger searching.

1 Introduction

Input to the one-dimensional channel compaction problem with automatic jog insertion consists of n horizontal wire segments organized into t tracks. Segments on the same track have the same y-coordinate and the ranges of their x-coordinates do not overlap. Each segment has a via at each end, connecting it to vertical wires on another layer. The goal is to minimize channel height subject to design rule constraints on the distances between features (wires and vias). The relative vertical position of wires is not allowed to change during compaction.

Figure 1(a) shows an initial layout with 7 tracks. The horizontal wires to be compacted are shown as thin solid lines while vertical wires on another layer are thick dashed lines. Dots represent vias. Relative vertical position must be maintained due to *vertical constraints* (see e.g. [12]). The left part of net A, for example, must be above net B because the vertical wires of these nets share the same x-coordinate. The right part of net A must be below net B for the same reason. It is these vertical constraints that prevent the problem from being solved by interval graph coloring.

Figure 1(b) shows the result of a greedy compaction — each via and wire segment is pushed as low as possible, subject to the constraint that it is at least unit distance away from any other via or wire segment (in the Manhattan metric). To keep our presentation simple, we assume that vias are the same width as wires. The algorithms can be modified to handle the more typical case where vias are wider than wires (see Section 5).

Greedy compaction often introduces unnecessary *jogs* or wire bends. Figure 1(c) shows the result of a wire-straightening heuristic that minimizes the number of jogs for each wire in turn, beginning with the uppermost track. While this approach does not necessarily minimize the total number of jogs, the number of jogs is reduced significantly from that of greedy compaction.

We give the fastest known algorithms for each phase of a two-phase compaction process. Phase 1 determines minimum channel height by producing (an implicit representation of) a greedy compaction. Phase 2 produces a minimum-height layout in which the number of jogs for each wire is minimized. Let n be the number of horizontal wire segments in the input routing (which has no jogs).

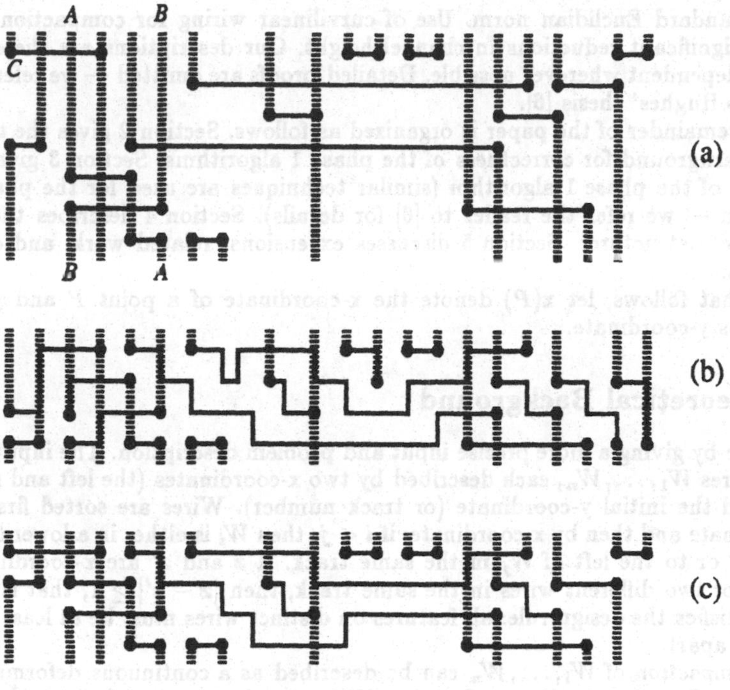

Fig. 1. An example showing two phases of compaction.

Let t be the number of tracks in the input, and let k_G and k_O be the number of wire segments in the greedy compaction of the input and the final output, respectively. Our algorithm for phase 1 runs in time $O(n \log t)$, while phase 2 requires $O(n \log t + k_O \log(k_G/k_O))$. This can be improved to $O(n \log(k_G/n))$ for phase 1 and $O(k_O \log(k_G/k_O))$ for phase 2 if the input is assumed to be *precompacted*, that is, no wire can be moved into a lower track without conflicting with other wires. In Figure 1(a), segment C is not precompacted because it can be moved to the next lower track. The latter bound is always at least as good since nt is a trivial upper bound on the number of greedy jogs. If the input is not precompacted, $\Omega(n \log t)$ is a lower bound for any algorithm that computes a routing — a special case is merging t sorted lists containing a total of n nonoverlapping wires.

The best previous bounds were $O(k_G)$ for phase 1 [17] and $O(k_G)$ for phase 2 [16] (add $n \log t$ if the input is neither sorted by x-coordinate nor precompacted). Simple examples show that k_G is in $\Omega(nt)$. Our bounds are achieved using a balanced tree with finger searching to maintain the minimum achievable height at each x-coordinate.

It appears that our phase 1 algorithms will work for distance norms other that the standard rectilinear (Manhattan) norm, allowing wire segments that are not necessarily horizontal or vertical. For example, *curvilinear* wiring allows wire segments to be straight lines or circular arcs and distances can be measured

in the standard Euclidian norm. Use of curvilinear wiring for compaction can lead to significant reductions in channel height. Our descriptions are therefore norm-independent wherever possible. Detailed proofs are omitted — we refer the reader to Hughes' thesis [6].

The remainder of the paper is organized as follows. Section 2 gives the theoretical background for correctness of the phase 1 algorithms. Section 3 gives an overview of the phase 1 algorithm (similar techniques are used for the phase 2 algorithm — we refer the reader to [6] for details). Section 4 describes the required data structures. Section 5 discusses extensions, related work, and open problems.

In what follows, let $x(P)$ denote the x-coordinate of a point P and $y(P)$ denote its y-coordinate.

2 Theoretical Background

We begin by giving a more precise input and problem description. The input is a list of wires $W_1, \ldots, W_n$, each described by two x-coordinates (the left and right vias) and the initial y-coordinate (or track number). Wires are sorted first by y-coordinate and then by x-coordinate: if $i < j$, then W_i is either in a lower track than W_j or to the left of W_j in the same track. If x and x' are x-coordinates of vias for two different wires in the same track, then $|x - x'| \geq 1$, that is, the input satisfies the design rule: all features on distinct wires must be at least unit distance apart.

A *compaction* of $W_1, \ldots, W_n$ can be described as a continuous deformation over time. Let the continuous function $W_i(z,t)$, where $z, t \in I = [0,1]$, represent the points of wire i at time t. So $W_i(z,0)$ is a point on the ith wire in its initial position and $W_i(z,1)$ is the corresponding point in the final compacted position of the wire. $W_i(0,t)$ and $W_i(1,t)$ are the left and right vias of wire i, respectively. The x-coordinates of the vias are not allowed to change, so $x(W_i(0,t)) = x(W_i(0,0))$ and $x(W_i(1,t)) = x(W_i(1,0))$ for all i, t. The objective is to minimize $\max_{1 \leq i \leq n,\, z \in I} y(W_i(z,1))$.

Not only the final position, but every intermediate position, satisfies the design rule: a compaction must satisfy $\|W_i(u,t), W_j(v,t)\| \geq 1$ for all $i \neq j$, $u, v, t \in I$, where $\|p_1, p_2\|$ denotes the distance between p_1 and p_2. Without imposing design rules on intermediate positions, the compaction problem becomes NP-hard (see [7]). Intuitively, the optimal routing under our restrictions will always have wires monotone in x. We therefore assume that $f(z) = W_i(z,t)$ is monotone in x for all $i = 1, \ldots, n$ and $t \in I$. The requirement on intermediate positions also prevents violation of vertical constraints.

Lemma 1. *If $W_j(u,1) = (x,y)$ and $W_k(v,1) = (x',y')$, where $j < k$, $|x-x'| < 1$, and u and v are arbitrary, then $y < y'$.*

The height of any via or wire segment of wire j is constrained by positions of the features of wires $1, \ldots, j-1$.

There is a *flow* of size F between x of wire j and x' of wire k if $j < k$, $|x-x'| < F+1$, and F is the maximum length of a subsequence S of $W_{j+1}, \ldots, W_{k-1}$, so that $x_i = x + i(x'-x)/(F+1)$ is between the vias of the ith wire of S. Figure 2 illustrates the definition.

Flow depends only on the x-coordinates of wires and the vertical relationships implicit in the input ordering. Repeated application of Lemma 1 yields the following.

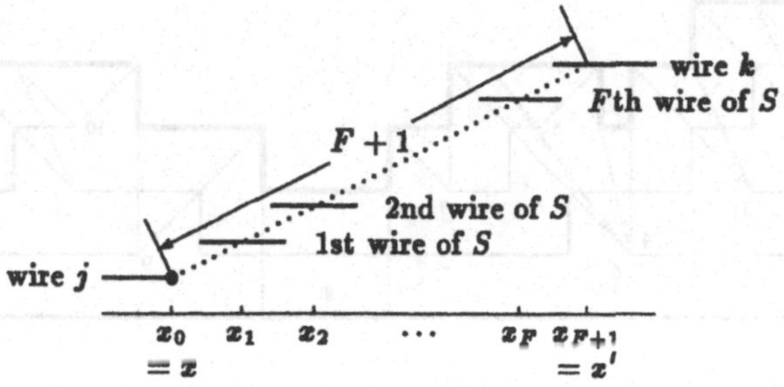

Fig. 2. Flow between two x-coordinates.

Lemma 2. *If there is a flow F between* $\mathrm{x}(W_j(u,1))$ *and* $\mathrm{x}(W_k(v,1))$, *where* $j < k$, *then* $\|W_j(u,1), W_k(v,1)\| \geq F+1$ *and* $\mathrm{y}(W_j(u,1)) < \mathrm{y}(W_k(u,1))$.

A special case occurs when the height of a wire is constrained by the bottom boundary of the channel. In all other cases, it is sufficient to consider the flow between a via of wire j and an arbitrary point on wire k. An inductive definition of the *contour* C_k, the lower bound on any part of wire $k+1$ imposed by wires $1, \ldots, k$, can now be formulated.

$C_0(x) = 1$ for all x. A *constraint* C wrt wires $1, \ldots, k$ consists of a via $V(C)$ of wire j ($1 \leq j \leq k$), a flow $F(C)$, and an interval $[\ell(C), r(C)]$ so that for any $x \in [\ell(C), r(C)]$, there is a flow of $F(C)$ between $\mathrm{x}(V(C))$ of wire j and x of wire $k+1$. Since $V(C)$'s position is bounded by C_{j-1}, $\mathrm{y}(V(C)) = C_{j-1}(\mathrm{x}(V(C)))$. In the special case of a constraint based on the bottom boundary of the channel, $\mathrm{y}(V(C)) = 0$ and $\mathrm{x}(V(C))$ is allowed to vary. The *height* of C at x, denoted $C(x)$, is 0 when $x \notin [\ell(C), r(C)]$ and is the unique $y > \mathrm{y}(V(C))$ with $\|V(C), (x, y)\| = F(C)+1$ when $x \in [\ell(C), r(C)]$ (the interval over which $C(x)$ is nonzero may be open at either end — see later discussion). Now $C_k(x) = \max_C C(x)$, where the maximum is over all constraints C wrt wires $1, \ldots, k$.

Even though more than one constraint may define the contour at any given point, we establish an ordering among constraints so that a unique constraint can be identified for each x. Let $h(C) = \mathrm{y}(V(C)) + F(C) + 1$. We say $C \succ C'$ (C *is higher than* C') if (a) $h(C) > h(C')$, or (b) $h(C) = h(C')$ and $\mathrm{y}(V(C)) > \mathrm{y}(V(C'))$, or (c) $h(C) = h(C')$, $\mathrm{y}(V(C)) = \mathrm{y}(V(C'))$, and $\mathrm{x}(V(C)) > \mathrm{x}(V(C'))$; otherwise $C \prec C'$. Constraint C is said to be *on the contour* C_k at x if $C(x) = C_k(x)$ and $C \succ C'$ for all C' having $C'(x) = C_k(x)$. Now C_k is a left-to-right list of constraints. Let $\mathrm{pred}(C)$ and $\mathrm{succ}(C)$ denote the immediate predecessor and successor of C on that list. Let $\min \ell(C) = \mathrm{x}(V(C)) - (F(C)+1)$ and $\max r(C) = \mathrm{x}(V(C)) + F(C) + 1$, so the open interval $(\min \ell(C), \max r(C))$ defines the maximum range in which C can be on the contour. Now $\ell(C) = \min \ell(C)$ if C is the leftmost constraint or $C \succ \mathrm{pred}(C)$; otherwise $\ell(C) = \max r(\mathrm{pred}(C))$. The right end point $r(C)$ is defined to be $\ell(\mathrm{succ}(C))$ ($\max r(C)$ if C is rightmost). The *interval* $I(C)$ for C, with endpoints $\ell(C)$ and $r(C)$, may be either open or closed at either end — it is open on the left (right) if $C \succ \mathrm{pred}(C)$ ($C \succ \mathrm{succ}(C)$) and closed otherwise. The interval over which $C(x)$ is nonzero is open at either end under the same conditions (in general, $C(x)$ may be nonzero over an interval larger than $I(C)$).

Figure 3(a) shows a set of wires (thin solid lines) compacted as much as

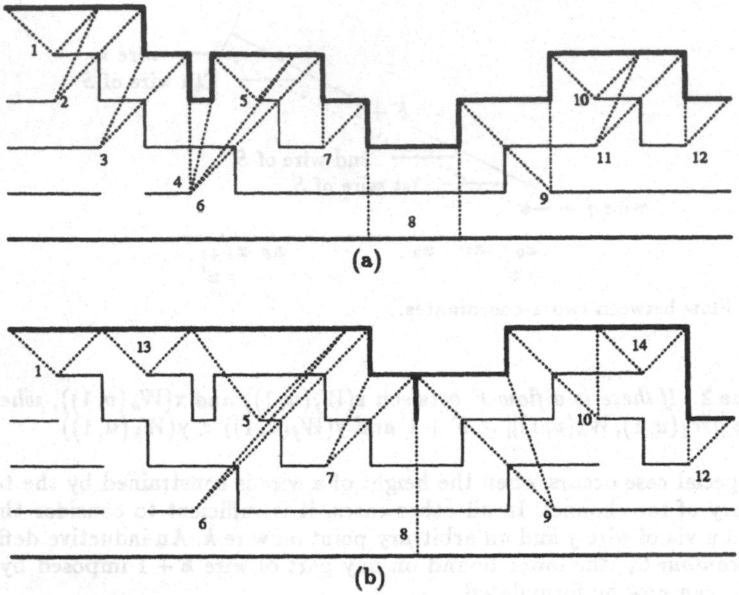

Fig. 3. **Updates to the contour after a new wire is added**

possible. Constraints numbered 1–12 from left to right are on the contour (thick solid line). The interval of each constraint is shown by dotted lines extending from the via. Figure 3(b) shows the result of adding a new wire — the new wire extends from the boundary between constraints 2 and 3 to the boundary between 10 and 11.

Lemma 2 can be used inductively to prove the following, which implies that the minimum channel height is $\max_x C_n(x)$.

Lemma 3. ***For all x, $C_k(x)$ gives the lowest possible position that can be occupied by any part of wire $k+1$.***

Our compaction algorithms differ from previous ones in two fundamental respects. First, our algorithms use an *anticipatory* contour, the lowest possible position for the next wire, instead of a contour based on the highest position of the current wires (see [6] for the advantages of using an anticipatory contour). Second, our algorithms take advantage of the similarity between C_k and C_{k+1} to maintain C_k as a balanced search tree. The differences between C_k and C_{k+1} can be summed up informally as follows: (1) Two new constraints for the vias of wire $k+1$ may be on contour C_{k+1} (constraints 13 and 14 in Figure3(b)). Each new via constraint causes the maximum constraint at the via position to be split into two parts, one part to the left and another to the right of the new via (constraints 4 and 6 resulted from such a split when the wire leading from 5 to 7 was added). Define C'_k to be C_k with the two via constraints for wire $k+1$ added (with 0 flow). (2) The flow of all constraints ranging from the new left via to the new right via is increased by 1 unit. (3) Except for the new via constraints,

no new constraint emerges on contour C_{k+1}. (4) The left to right sequence of constraints on the contour is preserved, except for deleted constraints. (5) The interval $I(C)$ may change for a constraint C based on the via positions of C, pred(C), and succ(C). (6) If C is on C'_k but not on C_{k+1} then C is part of a "narrow valley" in C'_k — there are higher constraints a small distance apart to the left and right. For example, constraint 2 is in the valley bordered by 1 and 13, 3 and 4 are in the valley bordered by 13 and 5.

The above observations can be proved using the following lemmas.

Lemma 4. *If C is on C_{k+1} at x_1 with $C_{k+1}(x_1) = y_1$, then either C is on C'_k at x_1 with $C'_k(x_1) = y_1$ or C is on C'_k at x'_1, where $C'_k(x'_1) = y'_1$, $\|(x'_1, y'_1), (x_1, y_1)\| = 1$, $\frac{y_1-y'_1}{x_1-x'_1} = \frac{y_1-y_0}{x_1-x_0}$, and $V(C) = (x_0, y_0)$.*

Lemma 5. *If $C' = succ(C)$ on C_k, $r(C) = \ell(C') = x_1$, $C(x_1) = y_1$, $C'(x_1) = y'_1$, $V(C) = (x_0, y_0)$, and $V(C') = (x'_0, y'_0)$, then $\arctan \frac{y'_1-y'_0}{x_1-x'_0} > \arctan \frac{y_1-y_0}{x_1-x_0}$.*

Lemma 6. *If wire $k+1$ is in the flow of C and C is on C_k but not C_{k+1}, then there exist $C_1 \succ C$ to the left of C (on C_k) and $C_2 \succ C$ to the right of C with $\ell(C_2) - r(C_1) < d$ (in C_k), where d is the total increase in flow for C_1 and C_2 due to wire $k+1$.*

3 Algorithm Descriptions

From the discussion of the previous section we know that the phase 1 (determining the minimum channel height) algorithm can maintain C_k as an ordered list of constraints.

Suppose for the moment that there are no deletions and let x_L and x_R be the x-coordinates of the vias of wire $k+1$, and let C_L and C_R be the constraints on C_k at these coordinates. To create C_{k+1} from C_k we first create three pieces of C'_k: Left$_k$ includes all constraints up to and including C_L; Middle$_k$ starts with D_L, the new constraint for the new left via, followed by C'_L, the part of C_L (if any) to the right of x_L, continuing to C'_R, the part of C_R to the left of x_R, and D_R, the new constraint for the right via; Right$_k$ starts with C_R and includes all constraints to the right of C_R. The flow of every constraint in Middle$_k$ is increased by 1 and then Left$_k$, Middle$_k$, and Right$_k$ are merged to form C_{k+1}.

We can combine this process for all wires on the same track. Suppose wires $j+1, \ldots, k$ occupy the same track, so the flow increases leading to C_i ($j+1 \leq i \leq k$) do not interfere with those of $C_{j+1}, \ldots, C_{i-1}$. Let Gap$_i$ denote the intersection of Right$_{i-1}$ and Left$_i$ (for $j+1 \leq i \leq k-1$; Gap$_j$ = Left$_j$ and Gap$_k$ = Right$_{k-1}$). We repeatedly split off Gap$_i$ and Middle$_i$ for $i = j, \ldots, k-1$, increasing $F(C)$ by one unit for each C in Middle$_i$. The pieces Gap$_i$ and Middle$_i$ are merged back together to form C_k.

To get the time bound, we use finger searching on a data structure that supports three primitive operations: *split*(T, x), to split a contour T into two — the constraints with intervals to the left of x and those with intervals to the right; *concatenate*(T_1, T_2), to concatenate two contours into one; and *augment*(T, δ), to add δ (positive or negative) to a numerical field for all constraints in T. Implementation details for these operations are in Section 4. Each operation takes time O(log s), where s is the number of contraints in the smaller list for *split*(T, x) and *concatenate*(T_1, T_2), and s is the size of T in *augment*(T, δ).

Since each new wire adds 2 constraints for the new vias and splits an existing constraint for each via, we can bound the number of constraints on C_k by $4k+1$. This means that the total size of all subtrees split off and merged in each track is $\leq 4n+1$ and the total time for all the *concatenate* and *split* operations is $O(\sum_{i=0}^{2n} \log s_i)$, where $\sum_{i=0}^{2n} s_i \leq (4n+1)t$ (let s_{2i} be the size of Gap_i and s_{2i+1} be the size of Middle_i). The total time is therefore $O(n \log t)$.

So far we have ignored the possibility of deletions. Deletions occurring at the boundary between Gap_i and Middle_i (Middle_i and Gap_{i+1}), such as those of constraints 2 and 11 in Figure 3, are not hard to handle. Using Lemma 6, we can argue that the deletions are contiguous, starting with either the last constraint of Gap_i (Middle_i) or the first constraint of Middle_i (Gap_{i+1}). This is the special case where C_1 is in Gap_i (Middle_i), C_2 is in Middle_i (Gap_{i+1}), and $d=1$.

We merge Gap_i with Middle_i (merging Middle_i with Gap_{i+1} is similar) by repeatedly deleting either the last constraint in Gap_i or the first constraint on Middle_i, whichever is lower, until the interval of the higher constraint no longer extends beyond (to the right if it's in Gap_i, to the left otherwise) that of the lower one. Each deletion can be implemented as a *split* involving a contour of size 1, so the total cost of all deletions is $O(n)$: a constraint can never reemerge once deleted (Lemma 4) so the total number of deletions is $\leq 4n+1$.

Deletions that occur because the increased flow in Middle_i, such as constraints 3 and 4 in Figure 3 require additional care. If C is the lowest constraint in a narrow valley, we can look at the via positions for $\text{pred}(C)$, C, and $\text{succ}(C)$ and the flow values of these three constraints, and predict how much additional flow would cause C to be deleted — call this value $dt(C)$, for ***deletion time***. In the rectilinear case, $dt(C) = (r(C) - \ell(C))/2$ if both $\text{pred}(C)$ and $\text{succ}(C)$ are higher than C, and ∞ otherwise. The *augment* operation can decrease $dt(C)$ by one unit each time $F(C)$ is increased by one unit.

To perform all the deletions, we repeatedly search Middle_i for the leftmost constraint C with $dt(C) < 1$, each time splitting off the list of constraints before C. We do this splitting before augmenting the flow. After the flow is augmented we merge (using the same algorithm as for merging Gap_i with Middle_i) all the lists that were previously split off. Constraints that need to be deleted at the ends of two neighboring lists are detected and deleted during the merge. Lemma 6 implies that any deletion within Middle_i must be in a contiguous sequence of deletions that includes a constraint C having $dt(C) < 1$. The converse of Lemma 6 also holds, so that each search and split combination yields at least one deletion. Thus an $O(n)$ bound on the number of data structure access operations is preserved.

Section 4 describes how $dt(C)$ is maintained so that search and split take time $O(\log s)$, where s is the size of the split off list.

From the preceding discussion, we have

Theorem 7. *Minimum channel height with jog insertion in a channel having n horizontal wires in t tracks can be computed in time* $O(n \log t)$.

A better time bound is possible with precompacted input. To obtain the bound $O(n \log(k_G/n))$, if suffices to get an $O(k_G)$ bound on the total size of all lists involved in *concatenate* or *split* operations (each search precedes a *split* splitting off the accessed subtree). Two important modifications to the algorithm are required.

Each time we split off and merge Middle_i, the greedy algorithm puts wire i on top of the constraints in Middle_i. If all these constraints are at different heights, each constraint adds 1 jog to the value of k_G and the number of constraints accessed in Middle$_i$ corresponds to the increase in k_G attributed to wire i. It is possible, in the rectilinear case, for neighboring constraints to have the same height. But two neighboring same-height constraints C_1 and C_2 are easily combined by inventing a new via and a new flow value for a single constraint C (see Figure 4). Same-height constraints can be detected whenever a new constraint is added or two lists are merged.

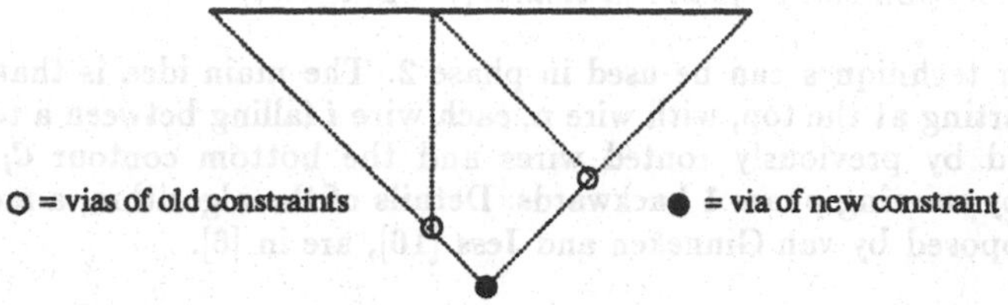

Fig. 4. Merging two consecutive constraints of the same height.

The biggest impediment to an $O(k_G)$ total size for all lists is the gaps Gap$_i$, since the constraints accessed in them do not lead to new jogs. When accessing a gap in the lowest track we create a *gap pointer*, a record that has a pointers to the first and last constraints of the gap. Initially each of these gaps will contain one constraint (the bottom of the channel). Instead of merging all the lists Gap$_i$, Middle$_i$, we create a list of all the constraints in the Middle$_i$'s with gap pointers taking the place of the gap constraints. Gap lists are updated to reflect deletions at either end. The structure of the lists with intervening gap pointers is then maintained as we add each new track.

Consider the creation of C_{i+1}, where x_L and x_R are the vias of wire $i+1$. First search for the contraint C that has $x_L \in I(C)$. Use gap pointers to treat gaps as if they were single constraints unless C is in a gap. If C is in a gap, search the gap list from the right and split off the portion affected by wire $i+1$ — the split off part yields an increase in k_G for the new wire. Search for the constraint C with $x_R \in C$, merging back into Middle$_i$ any gaps that fall completely underneath wire $i+1$. Finally, if x_R is in a gap, search the gap list from the left and split off the affected portion. This process hinges on the fact that wire $i+1$ will never fall into a single gap (if it did, the search for x_L and x_R would have to access and split off constraints not leading to new jogs).

If an endpoint, say x_L, of wire $i+1$ does not fall into a gap, the sequence of constraints including the one for x_L and extending to the left up to the nearest gap are split off and must be merged into that nearest gap (or a new gap pointer created if i is the leftmost wire in its track). The size of this list is not related directly to an increase in k_G, but can be amortized as an increase in the number of constraints in gaps.

We can use amortized analysis to show that the total size of all sublists accessed during merges and splits is $O(k_G)$. Define the potential Φ_i to be the number of jogs in the greedy compaction of wires $1, \ldots, i$ plus the total number of constraints in gap lists after the ith wire is added. The amortized sublist size for wire i (actual list size $+\Phi_{i-1} - \Phi_i$) is the number of constraints that were in a gap before wire i is added but are merged into Middle$_{i-1}$ (these decrease the number of gap constraints while increasing the number of jogs). All other accessed constraints are offset either by an increase in the number of jogs or an increase in the number of gap constraints. Thus, the total size is bounded by the total number of constraints falling under wires, which is k_G, plus Φ_n, which is at most $4n + 1 + k_G$, minus Φ_0, which is 0. Since $n \in O(k_G)$ (each wire has at least one segment), the desired bound holds.

Theorem 8. *If the input wiring is precompacted the minimum channel height with jog insertion can be found in time* $O(n \log(k_G/n))$.

Similar techniques can be used in phase 2. The main idea is that wires are routed starting at the top, with wire n, each wire i falling between a top contour determined by previously routed wires and the bottom contour C_{i-1}, reconstructed by running phase 1 backwards. Details of the algorithm, a modification of one proposed by van Ginneken and Jess [16], are in [6].

Theorem 9. *The routing of a set of n horizontal wires organized into t tracks in a channel so that the channel has minimum height and each wire has minimum length and minimum number of jogs (given routings of previous wires), takes time* $O(n \log t + k_O \log(k_G/k_O))$. *The time is* $O(k_O \log(k_G/k_O))$ *with precompacted input.*

4 Data Structure

The required operations can be supported on red-black trees (see [3] for a detailed discussion) with heterogeneous finger searching (see [14]), modified to allow numerical fields to be stored in difference form (see [1]). The leftmost and rightmost paths of each tree, called the *left path* and *right path*, respectively, have been reversed and nodes are accessed from the leftmost or rightmost leaf instead of the root. Constraints are stored at the leaves, except for the leftmost and rightmost leaf. In what follows we use the term *interior node* only for nodes that are not leaves and are not on the left or right path; data at interior nodes guides the searches. Nodes on the left or right path are used for pointers only — no data fields are stored there. The children of nodes on the left and right paths are henceforth called *path roots*. Figure 5 shows the structure of a tree storing constraints $1, \ldots, 12$.

Define $I(p)$ for an interior node p to be the union of the intervals of constraints in the subtree below p (the calculation of $I(p)$ from fields that are explicitly stored at p is discussed later). To search for the constraint C at x from the left (a search from the right is symmetric), move along the left path, accessing each path root p until we find one having $x \in I(p)$. Then continue the search in the subtree below p, following a path in which each node q has $x \in I(q)$. The number of nodes accessed in the search is proportional to the rank of the subtree containing C, which is $\log j$ if C is the jth constraint from the left. A key property of this particular implementation of a finger search tree is that any access to a node q is always preceded by accesses to all nodes on the path from the path root to q. This allows us to store numerical fields in difference form and support efficient implementation of the *augment* operation. We can implement gap pointers (for the time bound with precompacted input) as constraint nodes q for which $I(q)$ is the union of all intervals of constraints in the gap.

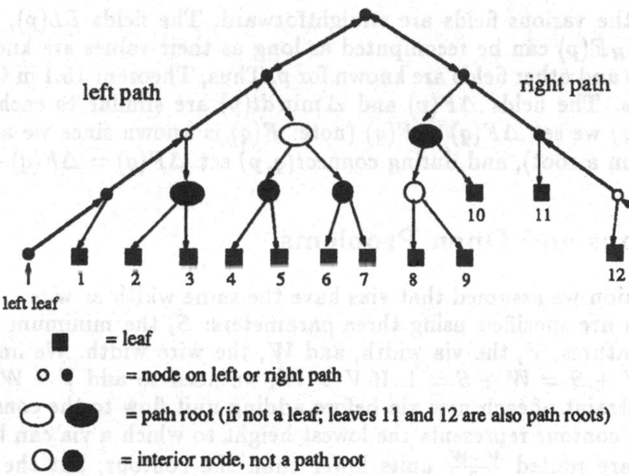

Fig. 5. **A red-black tree with heterogeneous finger searching.**

Details of the *concatenate* and *split* operations are somewhat involved. We refer the reader to Booth's thesis [1]. For what follows it suffices to observe that any tree restructuring operation is a sequence of the following types of operations:

- ***Detach(q)*: disconnect a node *q* from its parent *p***
- ***Connect(q, p)*: make *q* the left or right child of *p***

To allow efficient implementation of the *augment* operation for the flow field, we store a field $\Delta F(p)$ at each interior node and leaf *p*. The true value of $F(C)$ is the sum of all the $\Delta F(p)$ values for all nodes *z* on the path from the path root to the leaf containing *C*. Let $F(q)$ denote the sum of $\Delta F(p)$ for all *p* on the path from the path root to *q*. An *augment*(T, δ) for field *F* simply adds δ to the $\Delta F(p)$ fields of all path roots. Time is proportional to $\log s$, where *s* is the size of *T*.

Recall that $I(C)$ is computed from $F(C)$, $F(\text{pred}(C))$, $F(\text{succ}(C))$, and other values stored with *C*, pred(C), and succ(C). To compute $I(p)$ for any node *p*, we need to know the relevant values at the leftmost and rightmost leaf below *p*. We maintain pointers $LL(p)$ and $RL(p)$, pointing to the leftmost and rightmost leaves below each interior node *p*. To ensure that we can accurately compute $F(LL(p))$ and $F(RL(p))$ we also need to store $\Delta_L F(p)$ and $\Delta_R F(p)$, where $\Delta_L F(p)$ is the sum of all values $\Delta F(q)$ for *q* on the path from the left child of *p* to $LL(p)$ and $\Delta_R F(p)$ is similarly defined for the rightmost path.

The minimum value of $dt(C)$ for any constraint below *p* is also stored in difference form as $\Delta \min dt(p)$. For a path root *p*, the field $\Delta \min dt(p)$ contains the actual minimum value of $dt(C)$ below *p*. This allows a search for the leftmost *C* having $dt(C) < 1$: follow the leftmost path until a path root *p* with $\Delta \min dt(p) < 1$ is encountered, then follow a path from *p* to a leaf so that $\Delta \min dt(q) = 0$ for each node along the path.

Updates of the various fields are straightforward. The fields $LL(p)$, $RL(p)$, $\Delta_L F(p)$, and $\Delta_R F(p)$ can be recomputed as long as their values are known for the children of p and other fields are known for p. Thus, Theorem 15.1 in Cormen et al. [3] applies. The fields $\Delta F(p)$ and $\Delta \min dt(p)$ are similar to each other. During *detach(q)* we set $\Delta F(q) = F(q)$ (note: $F(q)$ is known since we accessed q via a path from a root), and during *connect(q, p)* set $\Delta F(q) = \Delta F(q) - F(p)$.

5 Extensions and Open Problems

In our presentation we assumed that vias have the same width as wires. Usually the design rules are specified using three parameters: S, the minimum separation between features, V, the via width, and W, the wire width. We implicitly assumed that $V + S = W + S = 1$. If $V > W$, we need to add $V - W$ to the flow of the constraint of each new via before adding unit flow to the constraints of Middle$_i$. The contour represents the lowest height to which a via can be compacted. Wires are routed $\frac{V-W}{2}$ units lower than the contour, but the overall channel height is still correct (see [6]).

The observations about C_k that led to our efficient phase 1 algorithms also hold for other distance norms, including octilinear (wires orthogonal or at 45 degree angles, $\|(x_1, y_1), (x_2, y_2)\| = \max(|x_1 - x_2|, |y_1 - y_2|, \frac{|x_1-x_2|+|y_1-y_2|}{\sqrt{2}}))$ and curvilinear (wires consist of line segments and circular arcs, Euclidian distance). The most difficult part of the generalization is figuring out how to calculate deletion time, $dt(C)$, for a constraint. Phase 2 becomes considerably more complicated when wires have additional degrees of freedom.

Extensions suggested by other researchers [4, 17] can easily be handled by our algorithms. Channel edges need not be straight lines. The bottom channel edge can be initialized to any shape that can be described by constraints and the shape of the top edge can be determined during phase 2. Individual wires can be constrained to have immovable vias or no jogs.

The appearance of k_G in the phase 1 time bound seems artificial, particularly when only the channel height needs to be computed. An even faster algorithm for this problem could exist. Computing the channel height without the routing is relevant in practice. Some channel routers and compactors create many alternate routings of a given channel and optimize only the one that has best *estimated* channel height (see [8]). An exact channel height computed quickly would significantly enhance such systems.

Splay trees [13] yield a simpler and cleaner implementation of the data structure operations we need. Unfortunately, our time bound would require a variant of the as yet unproven "dynamic finger conjecture" (a result for the case where there are no modifications between accesses has been proven by Cole [2]).

Our phase 2 algorithm would be more interesting if it found the global minimum number of jogs (and/or wire length). Tuan and Teo [15] give an $O(nh)$ dynamic programming algorithm, where h is the channel height, but only for the special case of river routing (the endpoints of wires are not allowed to move vertically and occur only at channel boundaries). It is not even clear whether global jog minimization is tractable in our situation.

A key open question is how far the techniques used in our algorithm can be pushed to obtain efficient algorithms for more general compaction problems. Maley's compaction algorithms [9, 10] are more general in that they allow input wires that are not monotone in x and permit modules in the layout. Run times for these algorithms have been improved to $O(n^3 \log n)$ in the worst case and $O(n^2)$ on average [11]. Our techniques appear to extend, albeit with slightly less efficient time bounds, to modules, but only if all wires on the layer to be compacted emanate from the same side of a given module. The introduction of non-horizontal wires in the input could be handled by a preprocessor that first removes all the "empty U's" (see [5] for an $O(n^2)$ algorithm).

References

1. Heather D. Booth. *Some Fast Algorithms on Graphs and Trees.* PhD thesis, Princeton University, 1991.
2. R. Cole. On the dynamic finger conjecture for splay trees. Part II: Finger searching. Technical Report 471, Courant Institute, Dept. of Computer Science, 251 Mercer St., New York, N.Y. 10012, October 1989.
3. T. H. Cormen, C. E. Leiserson, and R. L. Rivest. *Introduction to Algorithms.* MIT Press, 1990.
4. D.N. Deutsch. Compacted channel routing. In *International Conference on Computer-Aided Design, Digest of Technical Papers*, pages 223–225, 1985.
5. J.-M. Ho, M. Sarrafzadeh, and A. Suzuki. An exact algorithm for single-layer wire-length minimization. In *International Conference on Computer-Aided Design, Digest of Technical Papers*, pages 424–427, 1990.
6. Thomas A. Hughes. *Topological Routing Problems.* PhD thesis, North Carolina State University, Dept. of Electrical and Computer Engineering, Raleigh, NC 27695-7911, 1992.
7. T. Lengauer. On the solution of inequality systems relevant to IC-layout. *Journal of Algorithms*, 5:408 – 421, 1984.
8. M. Lorenzetti, M. Nifong, and J. Rose. Channel routing for compaction. In *Proc. MCNC International Workshop on Placement and Routing*, May 1988.
9. F.M. Maley. Compaction with automatic jog introduction. In *Chapel Hill Conference on VLSI*, pages 261–283, 1985.
10. F.M. Maley. *Single-Layer Wire Routing.* PhD thesis, Massachusetts Institute of Technology, 1987.
11. K. Mehlhorn and S. Näher. A faster compaction algorithm with automatic jog insertion. *IEEE Transactions on Computer-Aided Design of Integrated Circuits*, 9(2):158 – 166, 1990.
12. Bryan Preas and Michael Lorenzetti, editors. *Physical Design Automation of VLSI Systems.* Benjamin Cummings, 1988.
13. D. D. Sleator and R. E. Tarjan. Self-adjusting binary search trees. *Journal of the ACM*, 32(3):652 – 686, 1985.
14. R. E. Tarjan and C. J. Van Wyk. An $O(n \log\log n)$-time algorithm for triangulating a simple polygon. *SIAM Journal on Computing*, 17(1):143 – 178, 1988.
15. T. C. Tuan and K. H. Teo. On river routing with minimum number of jogs. *IEEE Transactions on Computer-Aided Design of Integrated Circuits*, 10(2):270 – 273, 1991.
16. L.P.P.P. van Ginneken and J.A.G. Jess. Gridless routing of general floor plans. In *IC-CAD*, pages 30–33, 1987.
17. X.-M. Xiong and E. Kuh. Nutcracker: An efficient and intelligent channel spacer. In *24th Design Automation Conference*, pages 298–304, 1987.

An Optimal Algorithm for Roundness Determination on Convex Polygons

Kurt Swanson*

Department of Computer Science, Lund University, Box 118,
S-221 00 Lund, Sweden

Abstract. In tolerancing, the Out-Of-Roundness factor determines the relative circularity of planar shapes. The measurement of concern in this work is the Minimum Radial Separation, as recommended by the American National Standards Institute (ANSI). Here presented is a further clarification of the complexity of a previously presented algorithm of Van-Ban Le and D. T. Lee to determine the Minimum Radial Separation of simple polygons, which is found to be $\Theta(n^2)$. Secondly, an optimal $O(n)$ time algorithm to compute the Minimum Radial Separation of convex polygons is presented, which represents not only a factor n improvement over the previously best known algorithm, but also a factor of $\log n$ improvement over Le and Lee's conjectured complexity for the problem.

1 Introduction

A vital control in the physical processes of production and manufacturing is the *tolerance factor*. In order to obtain a comparative measurement in qualifying adherence to design specifications, tolerance yields a relative figure for any given attribute.

Physical shape is one of many such attributes. This work is concerned with the tolerance factor known as the *Out-of-Roundness* factor, which determines the extent to which a given planar shape deviates from a circle. The Out-of-Roundness factor is determined by using one of several metrics, of which, in practice, four are commonly used: the *Least Squares Center* (LSC), the *Maximum Inscribed Center* (MIC), the *Minimum Circumscribed Center* (MCC), and the *Minimum Radial Separation* (MRS) center [7]. The LSC center designates a circle from which the sum of the squares of radial ordinates in a polar profile of the planar shape is minimized. The MIC center designates a circle of maximal radius which is inscribed, or completely contained, within the planar shape. The MCC center designates a circle of minimal radius which is circumscribed around, or completely contains, the planar shape. The MRS center designates two concentric circles, one which is circumscribed, the other inscribed, having minimal radial difference. The American National Standards Institute (ANSI Standard B89.3.1-1972), recommends the use of the MRS center to measure Out-of-Roundness. This work concerns the MRS center.

* Internet: `Kurt.Swanson@dna.lth.se`

Until recently, no exact method existed for finding the MRS center. There have been, however, several heuristical methods which would yield a general approximation to the true MRS center. Recently, the problem was solved for input in terms of a point set [2], but that result cannot be extended to simple polygons or more general planar shapes, as the MRS center, as calculated, could be found to lie outside the original boundary of a planar shape. More recently, Le and Lee [4] solved the MRS problem for input in terms of simple polygons, through methods of computational geometry. That solution relies upon intersecting the medial axis and the farthest neighbor Voronoi diagram of the polygon, and requires $O(n \log n+k)$ time, where n is the number of vertices, and k is the number of intersections between these two structures.

This work, as seen in Section 3, further clarifies the complexity, showing that the number of intersections, k, can be $\Theta(n^2)$, even when restricted to convex polygons. In Section 4.1, a new algorithm is presented that solves the MRS problem on convex polygons, in $O(n)$ time. This is an improvement of a factor of n over the previously best known algorithm, as well as a factor of $\log n$ better than a conjecture of Le and Lee [4]. The new algorithm also makes use of the medial axis and farthest neighbor Voronoi diagram, but does not require their intersection, thereby yielding the improvement in complexity.

2 Definitions

Definition 1. An *inscribed circle*, $IC_G(C, R)$, is a circle inscribed in a simple polygon G, having a center C, internal to G, with a radius R.

Definition 2. A *circumscribed circle*, $CC_G(C, R)$, is a circle that is circumscribed around a simple polygon G, having a center C, internal to G, with a radius R.

Definition 3. The *convex hull*, $CH(P)$, of a point set $P = \{p_1, p_2, \ldots, p_n\}$, is the smallest convex polygon completely containing P.

Definition 4. The *Minimal Radial Separation center*, MRS, of a simple polygon G, is defined as any point, c, internal to G, being the center of of two concentric circles, $IC_G(c, r_1)$, and $CC_G(c, r_2)$, such that the radial separation of these two circles is minimal for all points internal to G. Note that there exists at least one such point, and possibly several, in the case of non-convex polygons.

Definition 5. The *medial axis*, $MA(G)$, of a simple polygon G, is the set of all points internal to G, for which each point is closest to at least two points on the boundary of G. The medial axis, also known as the skeleton or symmetric axis, is a tree-like planar graph comprised of straight line segments and portions of parabolic curves [6]. The parabolic segments contain points which are closest to one *reflex*[2] vertex of G, and one segment of the boundary of G. Thus in the case of convex polygons, the medial axis is simply comprised of straight line segments.

[2] A reflex vertex of a polygon has an internal angle greater than π.

The medial axis divides the polygon into regions whose interiors are closer to either one boundary segment, or one reflex vertex, than any other section of the boundary.

Definition 6. The *farthest neighbor Voronoi diagram*, $FNV(P)$, of a point set P, is a division the plane into infinite bounded regions which contain all such points that are farther from one specific point in P than any other. The diagram consists of all points on the plane having equivalent maximal distances to more than one point in P. A point in P is represented in $FNV(P)$ if and only if that point lies on $CH(P)$ [9, 5].

Definition 7. An *arc bisector*, B_c, of a circular arc segment ARC_c, centered at c, is the point on the arc which partitions the segment into two halves of equal arc length.

3 An Improved Analysis of Le & Lee's Algorithm

Le and Lee have shown that while the MRS center of a simple polygon G may not be unique, that at least one MRS center must lie on the the medial axis, $MA(G)$ [4]. Thus

$$|IC_G(C_{MRS}, R) \cap G| \geq 2,$$

while

$$|CC_G(C_{MRS}, R) \cap G| \geq 1,$$

or that the inscribed circle centered at the MRS center is adjacent to at least two points on the boundary of G, but that the respective circumscribed circle is only adjacent to at least one point on the boundary of G. This leads to an $O(n \log n + k)$ algorithm of Le and Lee [4] which relies on the k intersections between the medial axis and the farthest neighbor Voronoi diagram.

A tight upper bound on the number of intersections between the medial axis and the farthest neighbor Voronoi diagram of a simple polygon has remained an open problem. Trivially, an upper bound is $O(n^2)$, since both of these structures are comprised of $O(n)$ segments, and the number of intersections between sub-segments are constant. As shown below, this bound is found to be tight, even when restricted to convex polygons.

Theorem 8. *There exist polygons for which the maximum number of intersections between the medial axis, and the farthest neighbor Voronoi diagram, is* $\Theta(n^2)$.

Proof. Trivially, we know that the number of intersections cannot be larger than $O(n^2)$, as discussed above. To show that this bound is indeed tight, consider the following example of a convex polygon, G, on n vertices, where:

- $n/2$ vertices are distributed evenly around a half circle, ARC_{C_r}, having an arc bisector B_r.

- $n/2$ vertices are distributed evenly around a second circular arc segment, ARC_{C_l}, having an arc bisector B_l, of relatively large radius and small arc length.
- The two arcs are sized and positioned such that:
  - The four points, B_l, C_l, C_r, B_r, are collinear, and appear in the stated order.
  - C_l, and C_r are separated by a very small ϵ–distance.
  - ARC_{C_l} has sufficiently large radius, such that the segments of $FNV(G)$ representing the vertices of ARC_{C_l} pass through ARC_{C_r} at most a small constant number of vertices away from B_r.

Such a polygon is shown in Figure 1, and with greater detail in Figure 2.

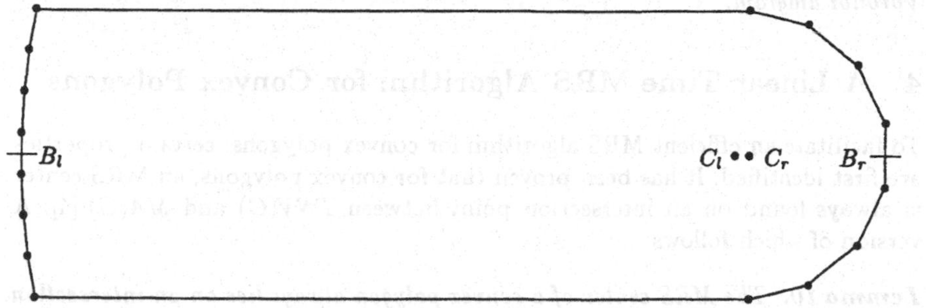

Fig. 1. A convex polygon having $\Theta(n^2)$ FNV-MA intersections.

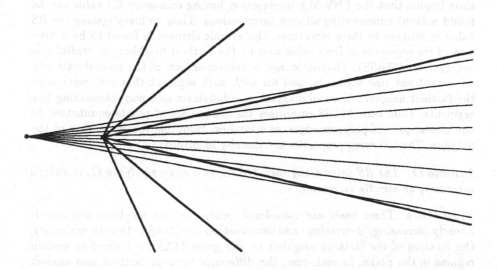

Fig. 2. Detail of Fig. 1, showing important FNV and MA intersection points.

Thus, each of the $n/2$ $FNV(G)$ segments related to the vertices of ARC_{C_l} originate at C_l and extend through the polygon within a constant number of vertices away from B_r, on ARC_{C_r}. The $n/2$ segments of $MA(G)$ corresponding to the vertices on ARC_{C_r} extend from each vertex and meet at C_r. Thus at least $n/2-c$ segments of $FNV(G)$ intersect with at least $n/4-d$ segments of $MA(G)$, for some constants c,d. Therefore, the number of intersections between the medial axis, and the farthest neighbor Voronoi diagram, of this convex polygon, is $\Theta(n^2)$. □

Through application of Theorem 8, the following corollary refines the complexity of Le and Lee's algorithm [4].

Corollary 9. *The worst case complexity of the MRS algorithm presented in [4] is $\Theta(n^2)$, dominated by the intersection of medial axis and farthest neighbor Voronoi diagram.*

4 A Linear Time MRS Algorithm for Convex Polygons

To facilitate an efficient MRS algorithm for convex polygons, certain properties are first identified. It has been proven that for convex polygons, an MRS center is always found on an intersection point between $FNV(G)$ and $MA(G)$ [4]. A version of which follows:

Lemma 10. *The MRS center of a convex polygon always lies on an intersection point between the medial axis, and the farthest neighbor Voronoi diagram of the polygon.*

To be able to find an MRS center on convex polygons in optimal, linear time implies that the FNV-MA intersection having minimum RS value can be found without enumerating all such intersections. Thus, in investigating the RS value in relation to these structures, the atomic element is found to be a division of the segments of the medial axis by the farthest neighbor, or *medial axis subsegments* (MASS). That is to say, a decomposition of the medial axis into its constituent line segments, and for each such segment that intersects with the farthest neighbor Voronoi diagram, a subdivision into non-intersecting line segments. Thus each MASS comprises the unique set of points equidistant to one closest pair of polygon edges, and farthest from one specific vertex of the polygon. The following properties are thereby identified:

Lemma 11. *The RS value along each MASS, in a convex polygon G, is strictly increasing or strictly decreasing.*

Proof Sketch. Three cases are considered: where nearest neighbor distance is linearly increasing, decreasing, and constant on the MASS. Due to convexity, the location of the farthest neighbor to any given MASS is limited to certain regions in the plane. In each case, the difference between farthest and nearest neighbor distance functions is dominated by either function, which is purely increasing or decreasing in the interval of the MASS. □

Corollary 12. ***The RS value along each MASS, is increasing if and only if the distance from the farthest neighbor to the line formed by the query point on the MASS, and its projection onto the nearest polygon edge, is increasing.***

Lemma 13. ***If any MASS is connected at the point of maximum RS value of any other MASS, that connection point defines its minimum RS value.***

Proof Sketch. Three cases are considered, depending on whether the MASS segments connect at an MA vertex, a point of the FNV, or both. Due to the limited placement of either farthest neighbor or medial axis segment in question, Corollary 12 applies to the connected MASS. □

Lemma 14. ***The RS is strictly increasing when proceeding along the medial axis, outwards from some intersection of the medial axis and the farthest neighbor Voronoi diagram.***

Proof. Since RS is minimal at the MRS, which is at an intersection of the medial axis and the farthest neighbor Voronoi diagram. Therefore the MRS is found at the endpoint of at least one MASS. Consider the division of the medial axis into MASSes as a tree structure rooted at the MRS. Since all lower level MASS segments are connected at the maximal RS point of the segment at the next higher level, RS value is strictly increasing when proceeding from the MRS to any leaf node. □

Corollary 15. ***The MRS center on convex polygons is unique.***

4.1 The Algorithm

The following algorithm is first presented in order to facilitate efficient searching within the medial axis, to find that medial axis segment containing the MRS center:

Algorithm 1 [MA-Search]

1. Perform a depth first search in the medial axis to find an edge having minimal difference between the sizes of the two subtrees it connects, amongst all edges in the medial axis tree.
2. If the ratio of sizes of the smaller subtree to the larger, at this edge, is strictly less than a predetermined constant, in the range $[0, 1/2]$, then:

   Compute the first derivative of the radial separation at both endpoints to determine which of the the subgraphs contains the MRS center. Remove the other two subgraphs from the tree. If the remaining subgraph contains more than one edge, return to step 1, otherwise stop, returning this edge.

   otherwise:

   Consider the endpoint of this edge connecting to the larger subtree. Locate this point in the farthest neighbor Voronoi diagram to determine the farthest vertex of the polygon for each incident edge. The point must lie according to one of the following situations:

No intersection: All edges are farthest from the same vertex.
Intersection with an FNV edge: Determine which half-plane each MA edge lies in.
Intersection with an FNV vertex: Merge the incident MA and FNV edges. Sweep through these edges to determine within which FNV region each MA edge lies.

Compute the first derivate of radial separation for each incident MA edge. Remove all incident MA edges and their respective subtrees, except that edge (if any) which has decreasing radial separation. If all incident edges are removed, stop, returning this point. Otherwise, return to step 1.

Analysis There are, of course, a linear number of segments in both the medial axis and the farthest neighbor Voronoi diagram of a convex polygon, since both are planar graphs (see, for example, [8]). To find an MA edge such that difference in sizes of its subtrees are minimal, a depth first search is performed on the tree, counting the number of edges found in each subtree. Next, progress through the tree, starting at any edge, comparing the separation ratio along edges. If an edge is encountered that divides the tree into two subtrees of sizes $\lfloor n/2 \rfloor - 1$, and $\lceil n/2 \rceil - 1$, the search stops there, as this is optimal. Otherwise, the search will have reached an MA vertex where all edges connect a subtree of less than $\lfloor n/2 \rfloor - 1$ edges. The edge having the largest subtree is chosen. Since every edge is visited at most twice, step 1 is $O(n)$.

Point location within the farthest neighbor Voronoi diagram can be solved in $O(\log n)$ time, using the algorithm of Edelsbrunner, et al [3]. Once found, the first derivatives of radial separation can be computed in constant time. To determine which half-plane an edge lies in, also requires constant time, or $O(n)$ to compute all incident edges. Similarly the MA and FNV edges incident on a common point can be merged and swept in $O(n)$ time. Since each step removes all but one edge, the total complexity of determining the farthest vertex of MA edges is $O(n)$. The total complexity of FNV search operations is $O(\log^2 n)$, as the algorithm performs at most a constant number of searches in each cycle, and cycles at most $O(\log n)$ times. Step 2, and the algorithm as a whole, is $O(n)$, since at least a constant fraction of medial axis edges are removed in each cycle. □

Given the above algorithm to perform searching in the medial axis in linear time, the following algorithm is derived to locate the MRS center in convex polygons.

Algorithm 2 [Find-MRS]

1. Compute the medial axis.

2. Compute the farthest neighbor Voronoi diagram.
3. Perform preprocessing necessary for search in the farthest neighbor Voronoi diagram.
4. Perform **MA-Search** to locate the medial axis segment containing the MRS center.
5. Compute the intersection of the located segment with the farthest neighbor Voronoi diagram.
6. Compute the radial separation at each intersection and return the minimum.

Analysis Linear-time algorithms exist for computing both structures on convex polygons[1]. Thus Steps 1 and 2 are both $O(n)$. In order to optimally perform point location within the farthest neighbor Voronoi diagram, O(n) preprocessing must be performed, according to the algorithm of Edelsbrunner, et al [3]. Step 3 is thus O(n). Step 4 is $O(n)$, as discussed above. The intersection of one medial axis segment and the farthest neighbor Voronoi diagram can be determined in $O(n)$ time. Step 5 is thus $O(n)$. The computation of the radial separation of all such intersections is $O(n)$. The minimum point of intersection, in terms of radial separation, is computed in $O(n)$ time. Step 6, and the algorithm as a whole, is then $O(n)$. □

Theorem 16. *The MRS center on convex polygons can be determined in* $O(n)$ *time.*

Proof. Through application of Lemma 14, Algorithm 2, as described above, locates the MRS center of a convex polygon by searching the medial axis in linear time. □

5 Summation and Discussion

It has been shown that the minimum radial separation of a convex polygon can be computed in optimal linear time, where n is the number of vertices in the polygon, through use of the medial axis, and the farthest neighbor Voronoi diagram. It has come to my attention that, independently from this work, Van-Ban Le and D. T. Lee have discovered a similarly styled solution yielding an $O(n \log n)$ algorithm.

Additionally, it has been shown that the worst case complexity of the algorithm of Le and Lee [4] to compute the minimum radial separation of a simple polygon is $\Theta(n^2)$. This is based upon the discovery that the number of intersections between the medial axis and the farthest neighbor Voronoi diagram of all simple polygons is $\Theta(n^2)$, in the worst case.

Finally, it should be noted that the lower bound of the MRS problem for star-shaped and other non-convex polygons remains an interesting problem.

6 Acknowledgements

The author wishes to thank Dr. Christos Levcopoulos for his helpful comments in optimizing the the complexity of the algorithm, and Dr. Der-Tsai Lee and Dr. Arne Andersson for many helpful comments.

References

1. A. Aggarwal, L. J. Guibas, J. Saxe, and P. W. Shor. A linear time algorithm for computing the Voronoi diagram of a convex polygon. *Discrete and Computational Geometry*, 4:591–604, 1989.
2. H. Ebara, N. Fukuyama, H. Nakano, and Y. Nakanishi. Roundness algorithms using the Voronoi diagrams. In *Abstracts of the First Canadian Conference on Computational Geometry*, page 41, 1989. Abstract only.
3. H. Edelsbrunner, L. J. Guibas, and J. Stolfi. Optimal point location in a monotone subdivision. *SIAM Journal on Computing*, 15(2):317–340, May 1986.
4. V.-B. Le and D.-T. Lee. Out-of-roundness problem revisited. *IEEE Transactions on Pattern Analysis and Machine Intelligence*, 13(3):217–223, Mar. 1991.
5. D.-T. Lee. Farthest neighbor Voronoi diagrams and applications. Technical Report 80-11-FC-04, Northwestern University, EECS department, Nov. 1980.
6. D.-T. Lee. Medial axis transformation of a planar shape. *IEEE Transactions on Pattern Analysis and Machine Intelligence*, PAMI-4(4):363–369, July 1982.
7. S. J. Levy. *Applied Geometric Tolerancing.* TAD Products Corp., 1974.
8. F. P. Preparata and M. I. Shamos. *Computational Geometry*, pages 209–211. Springer-Verlag, 1988. Second printing.
9. M. I. Shamos and D. Hoey. Closet-point problems. In *Proceedings of the 16th IEEE Symposium on Foundations of Computer Science*, pages 151–162, Oct. 1975.

Practical Algorithms on Partial k-Trees with an Application to Domination-like Problems *

Jan Arne Telle and Andrzej Proskurowski

Department of Computer and Information Science
University of Oregon, Eugene, Oregon 97403, USA
{telle,andrzej}@cs.uoregon.edu

Abstract. **Many NP-hard problems on graphs have polynomial, in fact usually linear, dynamic programming algorithms when restricted to partial k-trees (graphs of treewidth bounded by k), for fixed values of k. We investigate the practicality of such algorithms, both in terms of their complexity and their derivation, and account for the dependency on the treewidth k. We define a general procedure to derive the details of table updates in the dynamic programming solution algorithms. This procedure is based on a binary parse tree of the input graph. We give a formal description of vertex subset optimization problems in a class that includes several variants of domination, independence, efficiency and packing. We give algorithms for any problem in this class, which take a graph G, integer k and a width k tree-decomposition of G as input, and solve the problem on G in $\mathcal{O}(n2^{4k})$ steps.**

1 Introduction

A graph G is a k-tree if it is a complete graph on k vertices or if it has a vertex $v \in V(G)$ whose neighbors induce a clique of size k and $G - \{v\}$ is again a k-tree. The class of partial k-trees (the subgraphs of k-trees) is identical to the class of graphs of treewidth bounded by k. Many natural classes of graphs have bounded treewidth [15]. These classes are of algorithmic interest because many optimization problems, while inherently difficult (NP-hard) for general graphs are solvable in linear time on partial k-trees, for fixed k. These solution algorithms have two main steps, first finding a parse tree (tree-decomposition of width k [18]) of the input graph, and then computing the solution by a bottom-up traversal of the parse tree. For the first step, Bodlaender [8] gives a linear algorithm deciding if a graph is a partial k-tree and if so finding a tree-decomposition of width k, for fixed k. Unfortunately, the complexity of this algorithm as a function of the treewidth does not make it practical for larger values of k. For $k \leq 4$, however, practical algorithms based on graph rewriting do exist for the first step [4, 16, 19]. In this paper we investigate the complexity of the second step when k is not fixed. There are many approaches to finding a template for the design of algorithms on partial k-trees with time complexity polynomial, or even linear, in

* This research was supported in part by NSF grant CCR9213439

the number of vertices [17, 2]. As a rule, these approaches have tried to encompass as wide a class of problems as possible, often at the expense of increased complexity in k and also at the expense of simplicity of the resulting algorithms. Results giving explicit practical algorithms in this setting are usually confined to a few selected problems on either partial 1-trees or partial 2-trees [20, 14, 23]. In this paper we try to cover the middle ground between these extremes.

In the next section, we present a binary parse tree of the input graph which can easily be derived from a tree-decomposition. This parse tree is based on very simple graph operations which, as we show, will ease the derivation of practical solution algorithms for many problems. In section 3, we give a formal description of vertex subset optimization problems in a class that includes several variants of domination, independence, efficiency and packing. Then, in section 4, we present an algorithm template for any problem in this class. These algorithms take a graph G, integer k and a width k tree-decomposition of G as input, and solve the problem on G in $\mathcal{O}(n2^{4k})$ steps. Since these problems are NP-hard in general and a graph on n vertices is a partial n-tree, we should not expect polynomial dependence on k. In the last section we discuss and give some extensions of the algorithms.

2 Practical Algorithms on Partial k-Trees

We use standard graph terminology [9]. For a vertex $v \in V(G)$ of a graph G, let $N_G(v) = \{u : uv \in E(G)\}$, and $N_G[v] = N_G(v) \cup \{v\}$, be its open and closed neighborhoods, respectively. Let $G[S]$ denote the graph induced in G by a subset of vertices $S \subseteq V(G)$. A partial k-tree G is a partial graph of a k-tree H, meaning $V(G) = V(H)$ and $E(G) \subseteq E(H)$. A k-tree H has a perfect elimination ordering of its n vertices $peo = v_1, ..., v_n$ such that $\forall i : 1 \leq i \leq n-k$ we have $B_i = N_H[v_i] \cap \{v_i, ..., v_n\}$ inducing a $(k+1)$-clique in H. We call $B_i, 1 \leq i \leq n-k$ the $(k+1)$-*bag* of v_i in G under *peo* and each of its k-subsets is similarly called a k-bag of G under *peo*. The remaining definitions in this section are all for given G, H and $peo = v_1, ..., v_n$ as above.

We define a *peo-tree* P of G under *peo* with the node set $\{B_1, ..., B_{n-k}\}$. The node B_i has as its parent in P the node B_j such that $j > i$ is the minimum one with $|B_i \cap B_j| = k$. The $(k+1)$-ary peo-tree P is a clique tree of H and also a width k tree-decomposition of both G and H (see Figure 1 for an example.)

We sketch an algebra on simple graphs needed to define a binary parse tree T of G based on the peo-tree P. Let a graph with k distinguished vertices (also called sources, terminals, boundaries) have type G_k. For our purposes the set of sources will always be $(k+1)$-bags or k-bags of G. *Primitive* graphs are $G[B]$ for some $(k+1)$-bag B and have type G_{k+1}. The two operations are *Reduce*: $G_{k+1} \rightarrow G_k$ and *Join*: $G_{k+1} \times G_k \rightarrow G_{k+1}$. Reduce takes the vertex to be eliminated and discards it as a source. Join takes the union of its two argument graphs, where the sources of the second graph (a k-bag) are a subset of the sources of the first graph (a $(k+1)$-bag); these are the only shared vertices, and the two graphs agree

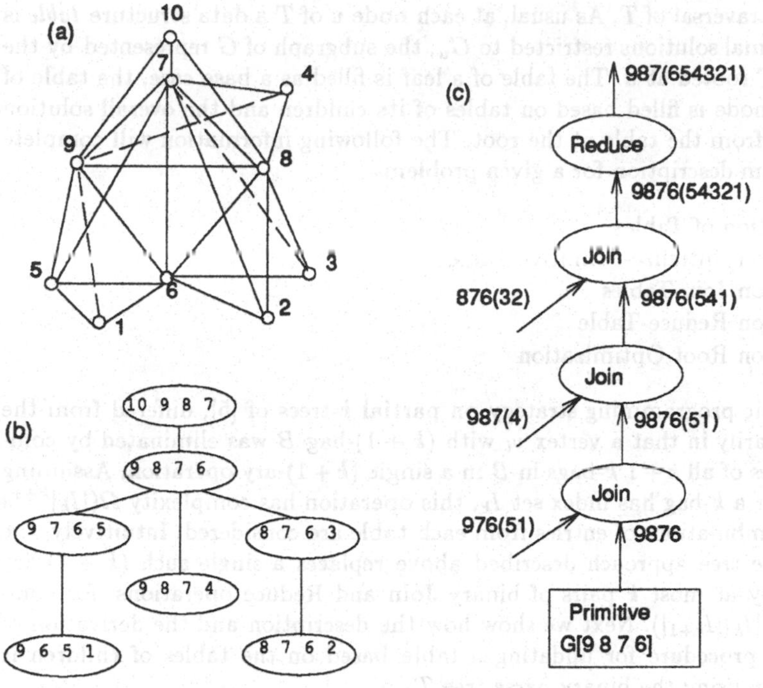

Fig. 1. (a) A partial 3-tree G, embedded in a 3-tree H, dashed edges in $E(H) - E(G)$, with peo=1,2,3,4,5,6,7,8,9,10. (b) Its peo-tree P. (c) Part of its binary parse tree T defined by node $B_6 = \{9, 8, 7, 6\}$ of P. Child-to-parent arcs of T labeled by $V(G_{child})$ with non-sources in parenthesis.

on edges between sources. The construction of the binary parse tree T is based on the peo-tree P. A node B_i of P, with c_i children and parent $p(B_i)$ defines a leaf-towards-root path of T. The path starts with the primitive graph $G[B_i]$ as the leaf endpoint, has c_i Join operations as interior nodes and terminates with a node of a Reduce operation. The Reduce operation discards the vertex v_i as a source, and its node is the second child of the node of a Join operation defined by $p(B_i)$ (see Figure 1 for an example.)

Note that T is indeed a parse tree of G since the primitive graphs in T contain all vertices and edges of G, while Join and Reduce merely identify vertices of the primitive graphs, in the order given by P, to form G. The root of T is thus a Reduce node representing G with sources $\{v_{n-k+1}, ..., v_n\}$. Since P is a tree with $n - k$ nodes, the Binary Parse Tree T of G has

- $n - k$ Primitive leaves, one for each $(k + 1)$-bag
- $n - k$ Reduce operations, one for each vertex eliminated
- $n - k - 1$ Join operations, one for each edge of T

A dynamic programming solution algorithm for a problem on G will follow a bottom-up traversal of T. As usual, at each node u of T a data structure *table* is kept of optimal solutions restricted to G_u, the subgraph of G represented by the subtree of T rooted at u. The table of a leaf is filled as a base case, the table of an interior node is filled based on tables of its children and the overall solution is obtained from the table at the root. The following information will complete the algorithm description for a given problem

- Description of Tables
- Operation Initialize-Primitive-Table
- Operation Join-Tables
- Operation Reduce-Table
- Operation Root-Optimization

The dynamic programming strategy on partial k-trees of [5], differed from the above primarily in that a vertex v_i with $(k+1)$-bag B was eliminated by combining tables of all $k+1$ k-bags in B in a single $(k+1)$-ary operation. Assuming the table for a k-bag has index set I_k, this operation has complexity $\Omega(|I_k|^{k+1})$ when all combinations of entries from each table are considered. Intuitively, the binary parse tree approach described above replaces a single such $(k+1)$-ary operation by at most k pairs of binary Join and Reduce operations, for complexity $O(k|I_k||I_{k+1}|)$. Next we show how the description and the derivation of the correct procedure for updating a table based on the tables of children is simplified by using the binary parse tree T.

Many strategies have been proposed for solving problems on graphs of bounded treewidth using a variant of the dynamic programming described above. For any such strategy, an algorithm for a given problem must describe the tables involved and also describe how tables are updated. We can classify these strategies by whether there is a procedure for automatic (mechanical) construction of a solution algorithm from a formal description of the problem, or such an algorithm has to be constructed "by hand". The EMSOL approach [3] is an automatic technique. A linear time algorithm solving a given problem can be constructed automatically from the problem description in a certain logic formalism. Although very strong for showing assymptotic complexity results, this technique is unsatisfactory for practical algorithm design since the resulting complexity in k involves towers of powers of k. Currently, there is no automatic algorithm design strategy giving algorithms which are practical for increasing values of k without hand derivation of a large part of the algorithm [6, 8, 12, 24, 1, 10].

Hand derivation of algorithms involves definition of table indices. Each table index represents an equivalence class of solutions to subproblems, equivalent in terms of forming parts of larger solutions. A solution to a subproblem at a node u of the parse tree T is restricted to G_u. This subproblem solution interacts with the solutions to larger problems only through the sources of G_u, a separator of G. Equivalent solutions affect the separator in the same manner, and hence we can define a *separator state* for each equivalence class (table index). A candidate set of separator states is verified by the correctness proof of table update procedures for all operations involved. The introduction of the operations Reduce and Join greatly simplifies this verification process.

For many problems, we can define a set L of *vertex states*, that represent the different ways that a solution to a subproblem can affect a source vertex, such that $|L|$ is independent of both n and k. A separator with k vertices has a separator state for each k-vector of vertex states so its table index set has size $|I_k| = |L|^k$. In the next section we first present a class of such problems and then derive $O(n|L|^{2k})$ dynamic programming solution algorithms for partial k-trees.

3 Formalization of Domination-like Problems

In this section, we formalize a class of problems to which we find realistic linear solution algorithms on partial k-trees given with a tree-decomposition of width k. For $S \subseteq V(G)$ let the symbols σ and ρ denote membership in S and membership in $V(G) - S$, respectively.

Definition 1. Given a graph G and a set $S \subseteq V(G)$ of *selected* vertices

- The *state* of a vertex $v \in V(G)$ is

$$state_S(v) \stackrel{df}{=} \begin{cases} \rho_i \text{ if } v \notin S \text{ and } |N_G(v) \cap S| = i \\ \sigma_i \text{ if } v \in S \text{ and } |N_G(v) \cap S| = i \end{cases}$$

- Define syntactic abbreviations $\rho_{\geq i} \equiv \rho_i, \rho_{i+1}, \ldots \quad \sigma_{\geq i} \equiv \sigma_i, \sigma_{i+1}, \ldots$

Mnemonically, σ represents a vertex selected for S and ρ a vertex rejected from S, with the subscript indicating the number of neighbors the vertex has in S. A variety of vertex subset properties can be defined by allowing only a specific set F as *final* states of vertices. For example, S is a dominating set if state ρ_0 is not allowed for any vertex, giving the final states $F = \{\rho_{\geq 1}, \sigma_{\geq 0}\}$. Optimization problems over these sets often maximize or minimize the size of the set of vertices with states in a given $M \subseteq F$. For the minimum dominating set problem, $M = \{\sigma_{\geq 0}\}$.

Definition 2. Given sets M and F of vertex states and a graph G

- $S \subseteq V(G)$ is an $[F]$-set if $\forall v \in V(G) : state_S(v) \in F$
- $minM[F]$ (or $maxM[F]$) is the problem of minimizing (or maximizing) $|\{v : state_S(v) \in M\}|$ over all $[F]$-sets S
- $minM[F](G)$ (or $maxM[F](G)$) is the corresponding parameter for G

Thus, a dominating set is a $[\rho_{\geq 1}, \sigma_{\geq 0}]$-set, with the square brackets implying the set notation. Table 1 shows some of the classical vertex subset properties [11] in our formalism. For each entry in Table 1 (except Nearly Perfect sets) there is an NP-hard problem related to minimizing or maximizing the cardinality of the set of selected vertices (i.e. with M the set of final σ-states). In related work [22], we study the general computational complexity of problems admitting the characterization given and mention one of the results here, encompassing problems not derivable from Table 1.

Our notation	$\lvert L \rvert$	Standard terminology
$[\rho_{\geq 1}, \sigma_{\geq 0}]$-set	3	Dominating Set
$[\rho_{\geq 0}, \sigma_0]$-set	2	Independent Set
$[\rho_0, \rho_1, \sigma_0]$-set	3	Strong Stable Set or 2-Packing
$[\rho_1, \sigma_0]$-set	3	Efficient Dominating Set or Perfect Code
$[\rho_{\geq 1}, \sigma_0]$-set	3	Independent Dominating Set
$[\rho_1, \sigma_{\geq 0}]$-set	3	Perfect Dominating Set
$[\rho_{\geq 1}, \sigma_{\geq 1}]$-set	4	Total Dominating Set
$[\rho_1, \sigma_1]$-set	4	Total Perfect Dominating Set
$[\rho_0, \rho_1, \sigma_{\geq 0}]$-set	3	Nearly Perfect Set
$[\rho_0, \rho_1, \sigma_0, \sigma_1]$-set	4	Total Nearly Perfect Set
$[\rho_1, \sigma_0, \sigma_1]$-set	4	Weakly Perfect Dominating Set
$max\{\rho_1\}[\rho_0, \rho_1, \rho_{\geq 2}, \sigma_{\geq 0}]$	4	Efficiency Problem

Table 1. Some classical definitions

Theorem 3. *[22] The problems* $opt\{\sigma_1\}[\rho_{\geq 1}, \sigma_1]$, $opt\{\sigma_0, \sigma_1\}[\rho_{\geq 1}, \sigma_0, \sigma_1]$, $opt\{\sigma_0\}[\rho_{\geq 2}, \sigma_0]$ *and* $opt\{\sigma_1\}[\rho_{\geq 2}, \sigma_1]$ *with opt replaced by max or min are all NP-hard.*

In the next section we give algorithms for solving any problem in the formalism just described, on a partial k-tree G.

4 Algorithm Template

Before giving an algorithm to compute a parameter $optM[F]$, as defined above, we discuss some algorithmic issues related to the concept of vertex states. For complexity reasons, we want to keep track of as few distinct vertex states as possible, and thus consider $\rho_{\geq i}$ and $\sigma_{\geq i}$ to be actual vertex states (and not merely abbreviations). To account for vertices with state in M, we require M to be a subset of F (this explains why $\rho_{\geq 0}$ is not used for the problem $max\{\rho_1\}[\rho_0, \rho_1, \rho_{\geq 2}, \sigma_{\geq 0}]$ in Table 1.) The set L of *legal* vertex states with respect to final states F is defined to be the smallest superset of F containing also, for ρ-states and σ-states respectively, any state with an integer subscript smaller than that of a final state. These states are needed when computing $optM[F]$ since a vertex may start out in a non-final state, and acquire neighbors in S as the algorithm progresses. As an example, the algorithm for $min\{\sigma_{\geq 0}\}[\rho_{\geq 1}, \sigma_{\geq 0}]$, the Minimum Dominating Set problem, has $L = \{\rho_0, \rho_{\geq 1}, \sigma_{\geq 0}\}$ and uses the following natural interpretation of $states_S$ (for $S \subseteq V(G)$)

$$states_S(v) = \begin{cases} \rho_0 & \text{if } v \notin S \text{ and } |N_G(v) \cap S| = 0 \\ \rho_{\geq 1} & \text{if } v \notin S \text{ and } |N_G(v) \cap S| \geq 1 \\ \sigma_{\geq 0} & \text{if } v \in S \end{cases}$$

In general, the algorithm computing $optM[F](G)$ for a partial k-tree G given with a tree-decomposition follows the binary parse tree T of G as outlined in section 2.

Algorithm-optM[F], where opt is either *max* or *min*.
Input: G, k, tree-decomposition of G of width k
Output: $optM[F](G)$
(1) Find a binary parse tree T of G with Primitive, Reduce and Join nodes.
(2) Initialize Primitive Tables at leaves of T.
(3) Bottom-up traversal of T using Join-Tables and Reduce-Table.
(4) Table-Optimization at root of T gives $optM[F](G)$.

For a given graph G and any fixed k, Bodlaender [8] gives a $O(n)$ (exponential in a polynomial in k) algorithm for deciding if the treewidth of G is at most k and in the affirmative case finding a tree-decomposition of G of width k. From this tree-decomposition it is straightforward to find a binary parse tree of G in time $O(nk^2)$, see e.g. [15] for how to find an embedding in a k-tree, then find a *peo* and finally follow the description given in section 2 for constructing the binary parse tree. For a problem given by the final states F, legal states L and optimizing (min or max) the set of vertices with state in M, we next describe the Tables involved and then give details of Table-Initialization, Table-Reduce, Join-Tables and Table-Root-Optimization. Each of the following subsections defines the appropriate procedure, gives the proof of its correctness and analyzes its complexity.

Table Description. Let a node u of the parse tree T represent the subgraph G_u of G with sources $B_u = \{w_1, ..., w_k\}$. The table at node u, $Table_u$, has index set $I_k = \{s = s_1, ... s_k : s_i \in L\}$, so that $|I_k| = |L|^k$. We define Ψ, with respect to G_u and s, to be the family of sets $S \subseteq V(G_u)$ such that in the graph G_u, for $w_i \in B_u$, $states_S(w_i) = s_i, 1 \leq i \leq k$ and for $v \in V(G_u) - B_u, states_S(v) \in F$. Ψ forms an equivalence class of solutions to the subproblem on G_u, and its elements are called Ψ-sets respecting G_u and s. The value of $Table_u[s]$ will be the optimum (max or min) number of vertices in $V(G_u) - B_u$ that have state in M over all Ψ-sets respecting G_u and s, and $\perp$ if no such Ψ-set exists.

$$Table_u[s] \stackrel{df}{=} \begin{cases} \perp & \text{if } \Psi = \emptyset \\ optimum_{S \in \Psi}\{|\{v \in V(G_u) - B_u : states_S(v) \in M\}|\} & \text{otherwise} \end{cases}$$

The result of an addition when one or more of the operands have the value $\perp$ is again $\perp$, and this value is considered to be smaller, respectively larger, than any integer under maximization, respectively minimization.

Table Initialization. A leaf u of T is a Primitive node and G_u is the graph $G[B_u]$, where $B_u = \{w_1, ..., w_{k+1}\}$. Following the above definition we initialize $Table_u$ in two steps

$\forall s \in I_{k+1} : Table_u[s] = \perp$
$\forall S \subseteq B_u : Table_u[s] = 0$ where $s = s_1, ..., s_{k+1}$ and $states_S(w_i) = s_i$ in $G[B_u]$

The complexity of this initialization for each leaf of T is $\mathcal{O}(|L|^{k+1} + 2^{k+2\log k})$.

Reduce Table. A Reduce node u of T has a single child a such that $B_u = \{w_1, ..., w_k\}$ and $B_a = \{w_1, ..., w_{k+1}\}$. We compute $Table_u$ based on correct $Table_a$ as follows

$$\forall \mathbf{s} \in I_k : Table_u[\mathbf{s}] = optimum\{Table_a[\mathbf{p}](+1 \text{ if } p_{k+1} \in M)\}$$

where the optimum (min or max) is taken over all $\mathbf{p} \in I_{k+1}$ such that $p_{k+1} \in F$ and $1 \leq i \leq k, p_i = s_i$. Correctness of the operation follows by noting that G_a and G_u designate the same subgraph of G, and differ only by w_{k+1} not being a source in G_u. By definition, an entry of $Table_u$ optimizes over solutions where the state of w_{k+1} is one of the final states F and w_{k+1} contributes to the entry value if it has state in M. The complexity of this operation for each Reduce node of T is $\mathcal{O}(|L|^{k+1})$.

Join Tables. A Join node u of T has children a and b such that $B_u = B_a = \{w_1, ..., w_{k+1}\}$ and B_b is a k-subset of B_a, wlog let $B_b = \{w_1, ..., w_k\}$. Moreover, G_a and G_b share exactly the subgraph $G[B_b]$. We compute $Table_u$ based on correct tables of children as follows

$$\forall \mathbf{s} \in I_{k+1} : Table_u[\mathbf{s}] = optimum\{Table_a[\mathbf{p}] + Table_b[\mathbf{q}]\}$$

where the optimum (min or max) is taken over all compatible pairs $(\mathbf{p}, \mathbf{q})$ such that $\mathbf{p} \in I_{k+1}, \mathbf{q} \in I_k, p_{k+1} = s_{k+1}$. The pair compatibility is defined by $(p_i, q_i) \in \pi(i, \mathbf{s}), 1 \leq i \leq k$, with the function $\pi(i, \mathbf{s})$ defined next. The function $\pi(i, \mathbf{s})$ gives the set of pairs of vertex states (p_i, q_i) which legally combine to form the vertex state s_i under the restriction of $\mathbf{s}$. This set of pairs depends on the number of source neighbors of w_i with σ-states. We define

$$\alpha(i, \mathbf{s}) \stackrel{df}{=} |\{w_j \in \{N_G(w_i) \cap B_a\} : s_j \text{ is a } \sigma\text{-state}\}|$$
$$\beta(i, \mathbf{s}) \stackrel{df}{=} |\{w_j \in \{N_G(w_i) \cap B_b\} : s_j \text{ is a } \sigma\text{-state}\}|$$

$$\pi(i, \mathbf{s}) \stackrel{df}{=} \begin{cases} \{(\rho_c, \rho_d) : c + d - \beta(i, \mathbf{s}) = z\} & \text{if } s_i = \rho_z \\ \{(\sigma_c, \sigma_d) : c + d - \beta(i, \mathbf{s}) = z\} & \text{if } s_i = \sigma_z \\ \{(\rho_c, \rho_d) : c + d - \beta(i, \mathbf{s}) \geq z\} \cup & \\ \quad \{(\rho_c, \rho_{\geq z}), (\rho_{\geq z}, \rho_d), (\rho_{\geq z}, \rho_{\geq z})\} & \text{if } s_i = \rho_{\geq z} \\ \{(\sigma_c, \sigma_d) : c + d - \beta(i, \mathbf{s}) \geq z\} \cup & \\ \quad \{(\sigma_c, \sigma_{\geq z}), (\sigma_{\geq z}, \sigma_d), (\sigma_{\geq z}, \sigma_{\geq z})\} & \text{if } s_i = \sigma_{\geq z} \end{cases}$$

where c and d are integer subscripts of legal states L, obeying the bounds imposed by $\alpha(i)$ and $\beta(i)$, respectively.

To show correctness of the computation, consider any $\mathbf{s} = s_1, ..., s_{k+1}$ and let $S \subseteq V(G_u)$ be a Ψ-set respecting G_u and $\mathbf{s}$ that optimizes the number of vertices among $V(G_u) - B_u$ having state in M. Let $S_a = V(G_a) \cap S$ and $S_b = V(G_b) \cap S$ and let $\mathbf{p} = p_1, ..., p_{k+1}$ (similarly $\mathbf{q} = q_1, ..., q_k$) be such that in G_a (respectively G_b) we have $states_{S_a}(w_i) = p_i$ (respectively $states_{S_b}(w_i) = q_i$). We want to show that the pair $(\mathbf{p}, \mathbf{q})$ is considered in the update of $Table_u[\mathbf{s}]$. Since $N_{G_u}(w_{k+1}) = N_{G_a}(w_{k+1})$ we have $p_{k+1} = s_{k+1}$. Next we show that $(p_i, q_i) \in \pi(i, \mathbf{s})$ for any $w_i, 1 \leq i \leq k$. Since S_a and S_b are subsets of S, if s_i is a σ-state then p_i, q_i are σ-states as well (and similarly for ρ-states). Assume $s_i = \rho_z$ so that $z = |N_G(w_i) \cap S|$. Let $|N_{G_a}(w_i) \cap S_a| = c$, $|N_{G_b}(w_i) \cap S_b| = d$, so that $(p_i, q_i) = (\rho_c, \rho_d)$. $G[B_b]$, we have Since $B_b \subseteq B_a$ we have $S_a \cap B_b = S_b \cap B_b = S \cap B_b$ and thus $|N_{G_a}(w_i) \cap S_a \cap B_b| = |N_{G_b}(w_i) \cap S_b \cap B_b| = |N_{G_u}(w_i) \cap S \cap B_b| = \beta(i)$, or

$z = |N_{G_u}(w_i) \cap S| = (c - \beta(i)) + (d - \beta(i)) + \beta(i) = c + d - \beta(i)$. We conclude $(p_i, q_i) \in \pi(i, s)$. The remaining cases where s_i equal to $\sigma_z, \sigma_{\geq z}$ or $\rho_{\geq z}$ can be argued similarly.

To complete the proof of correctness we must show that Join-Tables does not consider too many entries. Given $\mathbf{s} = s_1, ..., s_{k+1}$, let $\mathbf{p} = p_1, ..., p_{k+1}$ and $\mathbf{q} = q_1, ..., q_k$ satisfy $p_{k+1} = s_{k+1}$ and $(p_i, q_i) \in \pi(i, \mathbf{s}), 1 \leq i \leq k$. If S_a is a Ψ-set with respect to G_a and $\mathbf{p}$ while S_b is a Ψ-set with respect to G_b and $\mathbf{q}$ then $S = S_a \cup S_b$ is itself a Ψ-set with respect to G_u and $\mathbf{s}$. This since B_b forms a separator of G_u and the definition of π contains the restriction $c + d - \beta(i) = z$ (or $c + d - \beta(i) \geq z$) on the number of selected neighbors each vertex in B_b has. We conclude that Join-Tables is correct.

For each Join node of T the complexity of Join-Tables is $\mathcal{O}(|L|^{2k+1})$ since any pair of entries from tables of children is considered at most once. Many of these pairs are in fact not considered at all and a refined analysis gives the complexity $\Theta(\Sigma_{\mathbf{s} \in I_k}(\Pi_{1 \leq i \leq k}|\pi(i, \mathbf{s})|))$ with the upper bound $\mathcal{O}(max_{i,\mathbf{s}}\{|\pi(i, \mathbf{s})|\}^k |L|^{k+1})$, which depends also on the particular legal states L of the problem.

Optimize Root Table. Let r be the root of T with $B_r = \{w_1, ..., w_k\}$. We compute $optM[F](G)$ based on correct $Table_r$ as follows

$$optM[F](G) = optimum\{Table_r[\mathbf{s}] + |\{w_i \in B_r : s_i \in M\}|\}$$

where the optimum (min or max) is taken over $\mathbf{s} \in I_k$ such that $s_i \in F, 1 \leq i \leq k$. Correctness of this optimization follows from the definition of Table entries and the fact that G_r is the graph G with sources B_r. The complexity of Table optimization at the root of T is $\mathcal{O}(|L|^{k+1})$.

Overall Correctness and Complexity. Correctness of the algorithms follows from a simple induction on the parse tree T. As noted in section 2, T has $n - k$ Primitive nodes, $n - k$ Reduce nodes and $n - k - 1$ Join nodes. The algorithms find the parse tree T, execute a single respective operation at each of its nodes, and performs Table Optimization at the root. The total time complexity becomes $T(n, k, L) = \mathcal{O}(n|L|^{2k+1})$, with Join Tables being the most expensive operation.

Theorem 4. *Algorithm-optM[F], with L the set of legal states with respect to F, computes* $optM[F](G)$ *and has time complexity* $T(n, k, L) = \mathcal{O}(n|L|^{2k+1})$.

Corollary 5. *For any problem* $optM[F]$ *derived from Table 2, Algorithm-optM[F] has time complexity* $T(n, k) = \mathcal{O}(n2^{4k})$.

The corollary follows since any problem derived from Table 2 has $|L| \leq 4$. Using the refined complexity analysis of Join-Tables we can get improvements on the overall complexity, the problem Maximum Independent Set achieving complexity $\mathcal{O}(n2^{k+2\log k})$.

5 Extensions

Our technique applies to a number of more general problems, as follows.

Search Problems. To construct an $[F]$-set of G optimizing the problem parameter we add pointers from each table entry to the table entries of children achieving the optimal value.

Weighted Problems. For weighted versions of the above problems, Table entries reflect optimization over the sums of weights of vertices and we need only modify the operations Table Reduce and Table Optimization. The Reduce operation adds the weight of the reduced vertex, when its state is in M, rather than incrementing the optimum sum by one. The Root operation, with the domain of optimization unchanged, becomes

$$optM[F](G) = optimum\{Table_r[\mathbf{s}] + \Sigma weight(w_i) : w_i \in B_r \wedge s_i \in M\}$$

Digraph Problems. For the directed graph versions of these problems we define $IN_G(v) = \{u : \langle u, v\rangle \in \text{Arcs(G)}\}$ and use $IN_G(v)$, as opposed to $N_G(v)$, in the definition of $states_S(v)$, the state of vertex v with respect to a selected set $S \subseteq V(G)$. The only change in the algorithm is for the definition of $\alpha(i), \beta(i)$ in Join-Tables that will use IN_G as well.

Maximal and Minimal Sets. S is a *maximal* (minimal) $[F]$-set if no vertex can be added to (removed from) S such that the resulting set is still an $[F]$-set. Based on a hand-derived algorithm optimizing over all vertex subsets satisfying some property, Bern et al.[6] give an automatic procedure constructing an algorithm optimizing over maximal (or minimal) vertex subsets satisfying the same property. This includes an application of Myhill-Nerode finite state automata minimization techniques to minimize the resulting number of separator states. For more on the connection with finite state automata, see also [1]. Unfortunately, when the original algorithm contains $|L|^k$ separator states, this automatic technique involves simplification of a table with $|L|^k 2^{|L|^k}$ separator states, and quickly becomes infeasible for increasing values of k.

The formalization of domination-type vertex subset properties given in section 2, can be extended to encompass also maximal and minimal such sets, see [21]. This extension can be used to design algorithms on partial k-trees for problems which optimize over maximal (or minimal) $[F]$-sets. Below, we sketch a general procedure constructing a set of final vertex states $Fmin$ to identify minimal $[F]$-sets. An analogous construction will likewise give final vertex states for maximal $[F]$-sets. Define
$A_F = \{\sigma_i : \sigma_i \in F \wedge \rho_i \notin F\}$ and
$B_F = \{\rho_i : \rho_i \in F \wedge \rho_{i-1} \notin F\} \cup \{\sigma_i : \sigma_i \in F \wedge \sigma_{i-1} \notin F\}$.
It should be clear that S is a minimal $[F]$-set if and only if S is an $[F]$-set and $\forall v \in S$ either $states_S(v) \in A_F$ or $\exists u \in N_G(v) : states_S(u) \in B_F$. To account for the latter possibility, we refine the state of vertex v to carry this information about the state of its neighbors. In particular, a vertex with state in B_F is eligible to become a *mate* of neighboring vertices in S. The latter type of vertex will then have its vertex state augmented by the label *has* (a mate). The set of final states

Fmin for minimal $[F]$-sets will contain (i) any ρ-state in F, (ii) any state in A_F, (iii) any σ-state in F with the added label *has*. As an example, S is a minimal dominating set if it is a $[\rho_{\geq 1}, \sigma_0, {\sigma_{\geq 0}}^{has\rho_1}]$ -set.

To design an algorithm solving a problem optimizing over minimal $[F]$-sets we use *Fmin* as a starting point, find a corresponding set of legal vertex states and fill in details of Table Initialization, Reduce Table, Join Tables and Root Optimization. For minimal dominating sets, the legal states are $L = \{\rho_0, \rho_{\geq 1}, \sigma_0, \sigma_{\geq 1}, {\sigma_{\geq 0}}^{has\rho_1}\}$. A table index containing some state with the *has*-label is initialized to $\perp$. The Table Reduce operation for an index containing states with the *has*-label, is taken as the optimum over Table entries of the child whose indices exactly share the *has*-labels, with the added possibility of the reduced vertex having state in the above-defined set B_F and any neighbor of the reduced vertex having state without the *has*-label. In this latter case, the reduced vertex then becomes the mate of these neighbors. Moreover, the reduced vertex is not allowed to have a σ-state in $F - A_F$ without the *has*-label, as this is not a final state. The Join Tables operation has the compatibility function altered so that a resulting vertex state with a *has*-label requires the presence of a *has*-label on the corresponding vertex of at least one of the children. The Root Optimization operation considers indices having σ-states in $F - A_F$ without *has*-labels if and only if these are accompanied by a root neighbor with vertex state in B_F.

Other Problems. The extensions to our notation outlined above can also be used to design algorithms for many parameters related to *irredundant* sets, see [21]. In this notation, an irredundant set is a $[\rho_{\geq 0}, \sigma_0, {\sigma_{\geq 0}}^{has\rho_1}]$-set, and the close connection with minimal dominating sets is obvious.

We have also applied our technique to problems for which we could not find a vertex state set with size independent of k. As an example, we mention Partition into Perfect Matchings ([GT16] in [13]): For a graph G, find the minimum value of p for which there is a partition $V_1, V_2, \ldots, V_p$ of $V(G)$ such that $G[V_i], 1 \leq i \leq p$ has vertices of degree one only.

Below, we sketch a linear time algorithm for this problem on graphs of bounded treewidth (both [8] and [3] show only the existence of polynomial time algorithms). We observe that equivalent solutions to subproblems must induce identical partitions on the separator. We classify solutions by whether a given separator vertex has a mate among the other separator vertices, a mate among the reduced vertices, or no mate yet. Based on this we define a set of separator states and implementations of Initialize, Reduce, Join and Root-Optimize operations that give a linear time solution algorithm. The number of separator states as described here is $|I_k| = 3^k B(k)$, where $B(k)$ is the kth Bell number and the algorithm, although linear in n, is not very practical for increasing values of k.

References

1. K.Abrahamson and M.Fellows, Finite automata, bounded treewidth and well-quasiordering, to appear in *Contemporary Mathematics* (1992).
2. S.Arnborg, S.Hedetniemi and A.Proskurowski (editors) *Algorithms on graphs with bounded treewidth*, Special issue of *Discrete Applied Mathematics.*
3. S.Arnborg, J.Lagergren and D.Seese, Easy problems for tree-decomposable graphs, *J. of Algorithms 12(1991) 308-340.*
4. S. Arnborg and A. Proskurowski, Characterization and recognition of partial 3-trees, *SIAM J. Alg. and Discr. Methods 7 (1986) 305-314.*
5. S.Arnborg and A.Proskurowski, Linear time algorithms for NP-hard problems on graphs embedded in k-trees, *Discr. Appl. Math. 23 (1989) 11-24.*
6. M.W.Bern, E.L.Lawler and A.L.Wong, Linear-time computation of optimal sub-graphs of decomposable graphs, *J. of Algorithms 8(1987) 216-235.*
7. H.L.Bodlaender, Dynamic programming on graphs with bounded treewidth, *Proceedings ICALP 88, LNCS vol.317 (1988) 105-119.*
8. H.L. Bodlaender, A linear time algorithm for finding tree-decompositions of small treewidth, to appear in *Proceedings STOC'93.*
9. J.A.Bondy and U.S.R.Murty, *Graph theory with applications*, 1976.
10. R.B.Borie, R.G.Parker and C.A.Tovey, Automatic generation of linear algorithms from predicate calculus descriptions of problems on recursive constructed graph families, *Algorithmica*, 7:555-582, 1992.
11. E.J.Cockayne, B.L.Hartnell, S.T.Hedetniemi and R.Laskar, Perfect domination in graphs, manuscript (1992), to appear in Special issue of JCISS.
12. B.Courcelle, The monadic second-order logic of graphs I: Recognizable sets of finite graphs, *Information and Computation, 85: (1990)12-75.*
13. M.Garey and D.Johnson, *Computers and Intractability*, Freeman, San Fransisco, 1979.
14. D.Grinstead and P.Slater, A recurrence template for several parameters in series-parallel graphs, manuscript (1992).
15. J. van Leeuwen, Graph Algorithms, in *Handbook of Theoretical Computer Science vol. A*, Elsevier, Amsterdam, (1990) pg.550.
16. J.Matoušek and R.Thomas, Algorithms finding tree-decompositions of graphs, *J. of Algorithms*,12 (1991) 1-22.
17. A.Proskurowski and M.Syslo, Efficient computations in tree-like graphs, in *Computing Suppl. 7,(1990) 1-15.*
18. N. Robertson and P.D. Seymour, Graph minors II: algorithmic aspects of tree-width, *J. of Algorithms 7 (1986) 309-322.*
19. D.Sanders, On linear recognition of tree-width at most four, manuscript (1992).
20. K.Takamizawa, T.Nishizeki and N.Saito, Linear-time computability of combinatorial problems on series-parallel graphs, *J. ACM 29(1982) 623-641.*
21. J.A.Telle, Characterization of domination-type parameters in graphs, to appear in *Proceedings of 24th SouthEastern Conference on Combinatorics, Graph Theory and Computing, Congressus Numerantium.*
22. J.A.Telle, Complexity of domination-type problems in graphs, submitted BIT.
23. J.A.Telle and A.Proskurowski, Efficient sets in partial k-trees, to appear in *Domination in graphs*, Special volume of *Discrete Applied Mathematics.*
24. T.Wimer, Linear time algorithms on k-terminal graphs. Ph.D. thesis, Clemson University, South Carolina, (1988).

Greedy Algorithms for the On-Line Steiner Tree and Generalized Steiner Problems

Jeffery Westbrook * Dicky C. K. Yan †

Abstract

We study the on-line Steiner tree problem on a general metric space. We show that a class of greedy on-line algorithms are $O(\log(\frac{d}{z}s))$-competitive and no deterministic algorithm is better than $\Omega(\log(\frac{d}{z}s))$-competitive, where s is the number of regular nodes, d the maximum metric distance between any two revealed nodes and z the optimal off-line cost. Our results refine the previous known bound [8] and show that Algorithm SB of Bartal *et al.* [4] for the on-line File Allocation problem is $O(\log\log N)$-competitive on an N-node hypercube or butterfly network.

We consider the on-line generalized Steiner problem on a general metric space. We show that a class of lazy and greedy on-line algorithms are $O(\sqrt{k}\cdot\log k)$-competitive and no on-line algorithms is better than $\Omega(\log k)$-competitive, where k is the number of distinct nodes that appear in the request sequence. These are the first algorithms for this problem.

1 Introduction

In this paper, we study the on-line versions of the Steiner tree and generalized Steiner problems. The Steiner tree problem (STP) is a well known optimization problem. One is given a metric space $\mathcal{M}(M,\rho)$ with point set M and metric ρ, and a subset of $s \geq 2$ points, $\sigma \subseteq M$, called the *regular points* (or terminals). A Steiner tree is a tree of minimum weight that contains all the regular points and any number of other points, where the weight of an edge (x,y) is $\rho(x,y)$ and the weight of a tree is the sum of the weights of the edges in the tree. This problem is NP-Hard [9]. See [10, 13] for surveys on the problem on different metrics.

*Department of Computer Science, Yale University, New Haven, CT 06520-2158. Research partially supported by NSF Grant CCR-9009753.

†Department of Operations Research, Yale University, New Haven, CT 06520-0162. Research partially supported by NSF Grant DDM-8909660 and a University Fellowship from the Graduate School, Yale University.

The generalized Steiner problem (GSP) was formulated by Krarup (see [13]). One is given an undirected weighted graph with nodes which represent possible communication sites and m edges which represent links that can be made between the sites. A list σ of node pairs is given. Each pair $\{a, b\}$ has an associated positive integer r_{ab}. The problem is to find a minimum cost subgraph SS so that for each input pair $\{a, b\}$, there exists at least r_{ab} edge-disjoint paths in SS between nodes a and b. The subgraph SS need not be connected, as long as all nodes pairs have the desired connectivity. If $r_{ab} = 1$ for all the node pairs $\{a, b\} \in \sigma$ and all the node pairs contain a particular node, the problem is equivalent to STP on a graph; the generalized Steiner problem is NP-Hard. Agrawal et. al. [2] found the first approximation algorithm for this problem. The algorithm finds, in $O(m \log m \log r_{max})$ time, a network of cost at most $(2 - 2/k)\lceil \log_2(r_{max} + 1) \rceil$ times that of the optimal network, where $k \geq 2$ is the total number of distinct nodes contained in σ, and r_{max} is the maximum of the r_{ab}'s.

We study the *on-line* versions of these problems, in which the sequence of points to be connected arrives on line. In the on-line Steiner tree problem, an on-line server is presented with a metric space $\mathcal{M}$. An adversary reveals a sequence of regular nodes to the server who is required to construct on-line a connected Steiner subgraph, SS. Each time a node v is revealed, the server can expand SS, if necessary, so that the resulting SS is connected and contains v. At any time, the unrevealed part of the request sequence and its length is unknown to the server who is not allowed to remove nodes and edges once they are added to SS. The cost incurred by the server is equal to the sum of the weights of all the edges contained in SS.

Similarly, in the on-line generalized Steiner problem, the sequence of node pairs arrives on line. The on-line algorithm must connect each pair as it arrives by adding edges as necessary to SS. The pair can be connected directly or it can be connected via any number of intermediate connected components that have been previously constructed. Once a link is made, however, it cannot be changed. Again, the goal is to minimize the final cost to the server.

Given G and a request sequence σ, let $OPT(G, \sigma)$ be the cost of the optimal Steiner tree on the set of nodes in σ. In general an on-line server cannot achieve $OPT(G, \sigma)$. Let $A(G, \sigma)$ be the cost incurred by the on-line server using (deterministic) on-line algorithm A for deciding how to include a newly revealed node in the current Steiner tree, and let $R_A(G, \sigma) = A(G, \sigma)/OPT(G, \sigma)$. Then A is said to be *α-competitive* if, for any given instance G and request sequence σ, $R_A(G, \sigma) \leq \alpha$ holds. Note that α may be dependent on G and σ. Algorithm A is said to have a competitive ratio of γ if γ is the supremum of all α such that A is α-competitive. The same terms are similarly defined for on-line GSP.

Imase and Waxman [8] studied on-line STP on an undirected graph, which finds applications in communication networks. They showed that the on-line algorithm, GREEDY, that always makes the connection with the smallest additional cost is $\lceil \log s \rceil$-competitive and no deterministic on-line algorithm can be better than $(1 + \frac{1}{2}\lfloor \log_2(s-1) \rfloor)$-competitive. Alon and Azar [1] studied on-line STP on the Euclidean space, which is related to facilities planning. They showed that no on-line algorithm can be better than $\Omega(\log s/\log\log s)$-competitive.

In [4], Bartal *et al.* studied the on-line file allocation problem[3], which concerns the distribution of data files over a network of N processors. Requests arrive at the processors for reading or writing the files. The on-line server has to decide on-line, whether to discard a second copy of a file or to duplicate a file at a cost. They described a randomized algorithm, SB, that is $[(2+\sqrt{3}) \cdot c]$-competitive against an adaptive on-line adversary, where c is the competitiveness of any on-line algorithm for on-line STP on a graph against the same adversary. (See [5] for definitions of different kinds of adversaries.) Using GREEDY, SB becomes $O(\log N)$-competitive.

In this paper, we first refine the analysis of GREEDY and consider the problem on a general metric space, (Section 2). We show that GREEDY is $O(\log(\frac{d}{z}s))$-competitive on any metric space, where z is the cost of the off-line optimal Steiner tree for the sequence of vertices σ, and

$$d = \max\{\rho(v_i, v_j) | v_i, v_j \text{ are distinct nodes in } \sigma\}.$$

We show (in Section 3) that for any value of the ratio d/z, there exists a graph on which no on-line algorithm can be better than $\Omega(\log(\frac{d}{z}s))$-competitive.

Our results imply that GREEDY is $O(\log D(G))$-competitive for any undirected unweighted graph G with diameter $D(G)$, and SB[4] is $O(\log\log N)$-competitive for file allocation on a number of small diameter networks, including the N-node hypercube and butterfly networks, against an adaptive on-line adversary. The results of [5] imply there exists a deterministic $O((\log\log N)^2))$-competitive algorithm for on-line file allocation on such networks.

We consider the on-line generalized Steiner problem on a general metric space (in Section 4) under the restriction that for all node pairs $r_{ab} = 1$, i.e., we need only guarantee that each pair belong to a common tree. We show that a class of greedy algorithms are $O(\sqrt{k}\log k)$-competitive. These are the first on-line algorithms of proven performance for this problem. In Section 5, we show that when the connectivity requirements form a non-increasing sequence $\{r_{ab}\}$, the greedy on-line algorithm achieves a competitive ratio of $\Theta(k)$. The example used can be extended to show that no deterministic or randomized algorithm against an oblivious adversary can be better than $\Omega(k)$-competitive.

To simplify notation, $OPT(G, \sigma)$ and $A(G, \sigma)$ will be written as $OPT(\sigma)$ and $A(\sigma)$, respectively, when G is clear. Given ρ, a metric, we use $\rho(H)$ to represent $\sum \rho(a, b)$ where H is a graph, a tree or a path and the summation is taken over all the edges (a, b) in H.

2 On-Line STP on a General Metric Space

Given a metric space $\mathcal{M}(M, \rho)$ with point set M, metric ρ and an initial point $v_1 \in M$, the adversary is going to reveal a sequence, $\sigma = (v_1, \ldots, v_s)$, of regular points in M. Each time a point v_i ($i \geq 2$) is revealed, the on-line server is required to extend the on-line Steiner subgraph (SS), if necessary, so that that the new SS will contain v_i. The server is allowed to include Steiner points, points that are not in σ, in SS and will be charged a cost of $\rho(SS)$ after σ has been revealed. We consider the performance of the following class, $\mathcal{C}$, of on-line algorithms to which GREEDY belongs:

> When node v_i, ($2 \leq i \leq s$), is revealed, any on-line algorithm from $\mathcal{C}$ will not incur a cost greater than $\Delta_i = \min\{\rho(v_i, v_j) | 1 \leq j < i\}$.

Another on-line algorithm that belongs to $\mathcal{C}$ is one that connects v_i to the nearest previously revealed node. Let $C(\sigma) = \sum_{i=2}^{s} \Delta_i$; any algorithm from $\mathcal{C}$ will not incur a cost greater than $C(\sigma)$. Denote by T the optimal Steiner tree and $z = \rho(T) = OPT(\sigma)$ the optimal cost. It follows from Theorem 1 that any algorithm in $\mathcal{C}$ is $O(\log(\frac{d}{z}s))$-competitive. We define trees T_i ($i = 1, \ldots, s$), as follows:

- Initially, T_1 is defined to be T, with root v_1, and other trees are not defined.
- When node v_i ($i \geq 2$) is revealed, it is contained in exactly one of $T_1, \ldots, T_{i-1}$. Suppose it is contained in T_j ($1 \leq j < i$). We find the edge e that contains the mid-point of the unique path on tree T_j that runs from its root v_j to v_i. If the mid-point happens to fall on a node, we choose e to be one of the two edges incident to the node arbitrarily. We remove e from T_j and two trees are formed. The one containing v_i (v_j) will be called T_i (T_j) with v_i (v_j) as its root.

At any time, after i nodes have been revealed, we have i trees $T_1, \ldots, T_i$, with roots $v_1, \ldots, v_i$, respectively. The root is the only revealed node in a tree. These trees partition T. Consider the following on-line algorithm A that is based on the set of trees $T_1, \ldots, T_{i-1}$. Whenever a new node $v_i \in T_j$ is revealed, it is connected via the shortest path to v_j, the root of T_j. If $A(\sigma)$ denotes the cost incurred by A on sequence σ then clearly $A(\sigma) \geq C(\sigma)$. Given any tree T, with s regular nodes, $\rho(T) = z$ and only its root revealed, let $J(s, z)$ be the maximum possible value of $A(\sigma)$ over all sequences, σ, of revealing the nodes. Define $J(1, z) = 0$ for all z and $J(2, z) = z$.

Lemma 1 [8] $J(s, z) \leq z \cdot \log_2 s$

The above lemma is the basis for the proof of Imase and Waxman that GREEDY is $O(\log s)$-competitive. We use it as a building block in refining their analysis. Given T and σ, the sequence of nodes to be revealed, we can construct the binary tree $\mathcal{T}(T, \sigma) = \mathcal{T}$ to represent the recursive decomposition of T into $\{T_1, \ldots, T_s\}$. The recursion tree $\mathcal{T}$ is a full binary tree where each node has either 2 children or none. Each node in $\mathcal{T}$ corresponds to a subtree T_i, with root v_i.

A node in $\mathcal{T}$ is called ***heavy*** if the tree it represents, T_i, has $\rho(T_i) \geq d$, otherwise it is called ***light***. Let $\mathcal{T}_h$ be the subtree of $\mathcal{T}$ consisting only of heavy nodes. A leaf of $\mathcal{T}_h$ is a heavy node of $\mathcal{T}$ with two light children or no children. An internal node of $\mathcal{T}_h$ may have one heavy child and one light child.

Lemma 2 *There are no more than* ($\lfloor z/d \rfloor - 1$) *heavy nodes with two heavy children.*

Proof: Let ℓ be the number of heavy leaves and let p be the depth of $\mathcal{T}_h$. Each of the trees represented by the heavy leafs in $\mathcal{T}$ are of weight at least d and they are all disjoint. Hence $\ell \leq \lfloor z/d \rfloor$. Let n_i denote the number of heavy nodes at depth i and let l_i the number of heavy leaves at depth less than i. (The top of the tree $\mathcal{T}_h$ is at level 0 and the bottom of it is at level p). Consider the sum $m_i = n_i + l_i$. We have $m_0 = 1$, $m_p = \ell$ and $m_i \leq m_{i+1}$ for all $0 \leq i \leq p-1$. Let h_i be the number of nodes at level i with two heavy children. Then $m_i = m_{i+1} - h_i$. We have $\sum_{i=0}^{p-1} h_i = \sum_{i=0}^{p-1} m_{i+1} - m_i = m_p - m_0$. Thus the number of heavy nodes with two children is equal to $(\ell - 1)$. □

Theorem 1 $C(\sigma) \leq z \cdot (3 + \max\{\log_2 e, \log_2 \frac{s}{\lceil z/d \rceil}\})$.

Proof: The weight of the on-line tree is given by summing up the costs associated with each node in $\mathcal{T}$. By Lemma 2 there are no more than $[(z/d) - 1]$ nodes with two heavy children. Each of them has an associated cost at most d since in the tree represented by a heavy node, no two nodes are more than d apart. Hence the total cost from these nodes is no more than z.

Consider the set $\{x_1, x_2, \ldots, x_k\}$ of light nodes with heavy parents. Each x_i represents a subtree T_i of $\mathcal{T}$ of weight $\rho(T_i) < d$. The subtrees are mutually disjoint. Each tree is charged at most $[2 \cdot \rho(T_i) + 2 \cdot \rho(e_i)]$ for the split that created it, where e_i is the edge associated with the splitting up of its parent tree. Hence, the total cost associated with these splits are no more than $2 \cdot \sum_{i=1}^{k} [\rho(T_i) + \rho(e_i)] \leq 2 \cdot z$.

Let $z_i = \rho(T_i)$. By Lemma 1, the cost associated with revealing the s_i nodes in the trees T_i is bounded above by the maximum value of $\sum_{i=1}^{k} z_i \cdot \log_2 s_i$, subject to $\sum_{i=1}^{k} s_i \leq s$, $\sum_{i=1}^{k} z_i \leq z$, $s_i \geq 1$, and $0 \leq z_i \leq d$, where $1 \leq k \leq s$ and the z_i's and s_i's are variables. By considering two cases, $k \leq \lfloor z/d \rfloor$ and $k > \lfloor z/d \rfloor$, and using simple calculus, we can show that the maximum value is bounded above by $z \cdot \max\{\log_2 e, \log_2 \frac{s}{\lceil z/d \rceil}\}$ and the theorem follows. □

Corollary 1 *The algorithms in* $\mathcal{C}$ *are* $O(\log D(G))$*-competitive on any undirected unweighted graph* G *with a diameter of* $D(G)$*.*

3 Lower Bounds on Competitiveness

In [8], Imase and Waxman construct a class of graphs and corresponding request sequences, so that any deterministic on-line algorithm will incur a cost of at least $1 + \frac{1}{2}\lfloor \log_2(s-1) \rfloor$ for the on-line STP. We assume the reader is familiar with their lower bound proof. These graphs, H_n's, where n is some non-negative integer, have the following characteristics. Each edge in H_n has weight $1/(s-1)$, where $s = 2^n + 1$. The optimal Steiner tree consists of a chain of $(s-1)$ edges, running from a node v_0 to node v_1. We also have $d = z = 1$.

Theorem 2 *Given any function $r(s) : \mathcal{Z}^+ \longrightarrow \mathcal{R}^+$,*[1] *$1 \leq r(s) \leq (s-2)/2$, $\forall s$, there exists an instance G, σ of length s, such that $OPT(\sigma)/d = r(s)$, on which no deterministic on-line algorithm is better than $(\frac{1}{2}\left\lfloor \log_2 \frac{(s-1)}{r(s)+1/2} \right\rfloor)$-competitive.*

Proof: Choose any $z \geq d > 0$ such that $z/d = r(s)$. Let $q = \lfloor 2 \cdot z/d \rfloor$. Let $l = (q+1) \cdot 2^n + 1$ where n is a non-negative integer such that $(q+1) \cdot 2^n \leq (s-1) \leq (q+1) \cdot 2^{n+1}$. Let $s' = (2^n + 1)$. We construct graph G as follows: Construct q copies of H_n's and scale their edge weights so that the chain of $(s'-1)$ edges described above, running from v_0 to v_1, has weight $d/2$. Construct an extra copy of H_n and scale its edge weights so that the $(s'-1)$-chain from v_0 to v_1 has weight $(z - q \cdot d/2)$. Identify node v_0 of all the copies of H_n's to be one node.

For each copy of H_n in G, the request sequence chosen as in [8] is applied, with node v_0 revealed initially. The optimal Steiner tree consists of $(q+1)$ chains meeting at node v_0 and has weight z. There are altogether $(q+1) \cdot (s'-1) + 1 = l$ requests in the sequences mentioned above. The adversary will issue an additional $(s-l)$ requests at node v_0 so that the total request sequence is of length s. It follows from [8] that any on-line algorithm will incur a cost of at least $[1 + \frac{1}{2}\lfloor \log_2(s'-1) \rfloor]$ that of the optimal cost. It can be shown that $(s'-1) \geq \frac{1}{4} \cdot (s-1)/[r(s) + 1/2]$ from which the lemma follows. □

4 The Generalized Steiner Problem

On-line GSP on a metric space $\mathcal{M}(M, \rho)$ is similar to on-line STP except the adversary reveals a sequence of point *pairs*. Underlined letters will be used to represent point pairs. Let $\sigma = (\underline{v}_1, \ldots, \underline{v}_s)$. Each time a point pair $\underline{v} = \{a, b\}$ is revealed to the on-line server, SS has to be extended, if necessary, so that the resultant SS contains a path from a to b. Since the on-line STP is a special case of on-line GSP, the lower bound of $(1 + \frac{1}{2}\lfloor \log_2(k-1) \rfloor)$ on competitiveness holds for on-line GSP on a graph, where k is the number of distinct points that appear in σ. STP, We show that a class of greedy algorithms are $O(\sqrt{k} \log k)$-competitive for on-line GSP.

Definition: An on-line algorithm A is called *lazy* if it satisfies the following property: if when the request pair $\{x, y\}$ arrives, points x and y are already connected by some path in SS, then A will not change SS.

[1] $\mathcal{R}^+$ and $\mathcal{Z}^+$ are the sets of non-negative real numbers and positive integers respectively.

It can be shown that given any deterministic on-line algorithm A, there exists a lazy deterministic on-line algorithm A', such that $A'(\sigma) \leq A(\sigma)$ for all σ.

In general, the number of request point pairs, s, satisfies $\lceil k/2 \rceil \leq s \leq \binom{k}{2}$. However, for lazy on-line algorithms, some of the requests may be redundant. Consider the case when $s = k = 3$ and points a, b and c appear in σ. Let $\underline{v}_1 = \{a, b\}$, $\underline{v}_2 = \{b, c\}$, and $\underline{v}_3 = \{a, c\}$. When $\underline{v}_3$ is revealed, points a and c are already connected by some path in SS and the presence of the request does not make any difference to any lazy on-line algorithm. Request pairs can be classified as redundant using the following algorithm.

(1) Let $U = \{u_1, \ldots, u_k\}$ be the set of k points that appear in σ and let graph $G = (U, \phi)$.

(2) **for** $i = 1, \ldots, s$ **do**

(2.1) Reveal the point pair $\underline{v}_i = \{u_j, u_l\}$ and let edge $e = (u_j, u_l)$.

(2.2) **if** adding edge e to G creates a cycle **then**

$\underline{v}_i$ is a *redundant* request

else $\underline{v}_i$ is *non-redundant* and let $G = G + \{e\}$.

It can be easily verified by induction that at any time, points that belong to the same component in G are connected by some path in SS, independent of the on-line algorithm being used.

We observe that the number of non-redundant requests is at most $(k - 1)$, since each time we add in an edge, which corresponds to a non-redundant request, the number of components in G decreases by one. The same idea can be applied to the cases where the revealed connectivity requirements form a non-increasing sequence.

Lemma 3 *Let SS be a weighted undirected graph and a, b and c three distinct nodes in it such that there are $r \in \mathcal{Z}^+$ edge-disjoint paths between a and b and r such paths between b and c. Then there are r such paths between a and c.*

The lemma shows that the idea of non-redundant requests can be extended to cases where the connectivity requirements form a non-increasing sequence. For such request sequences, the same algorithm above can be used to classify the requests as redundant/non-redundant.

Lemma 4 *Suppose the connectivity requirements form a non-increasing sequence $\{r_{ab}\}$. Let $\alpha : \mathcal{Z}^+ \times \mathcal{Z}^+ \longrightarrow \mathcal{R}^+$ be a non-decreasing function. If a lazy on-line algorithm, A, is $\alpha(s, k)$-competitive for any σ that contains only non-redundant requests, then A is $\alpha(k-1, k)$-competitive for any general σ.*

The above lemma allows us, in considering the competitiveness of any lazy on-line algorithm, to assume that σ consists of non-redundant requests and has length $s \leq (k - 1)$.

For the rest of this section, we will investigate the case when the connectivity requirements are all equal to one. We shall show that a class of lazy on-line algorithms are $O(\sqrt{s}\log s)$-competitive for any request sequence and it follows from the above lemma that they are $O(\sqrt{k}\log k)$-competitive. We define the metric $\beta : P \times P \longrightarrow \mathcal{R}^+$, $P = \{\{a,b\} | \, a,b \in M\}$, as follows: Given point pairs $\underline{v} = \{a,b\}$ and $\underline{u} = \{x,y\}$, $\beta(\underline{u},\underline{v}) = \min\{[\rho(a,x)+\rho(b,y)], [\rho(a,y)+\rho(b,x)]\}$. One can verify that β is a metric. Let $\mathcal{C}_{GSP}$ be the class of deterministic on-line algorithms that satisfy the following properties:

1. All algorithms in $\mathcal{C}_{GSP}$ are lazy.

2. Let $\underline{v}_i = \{x,y\}$,

$$\Delta_i = \begin{cases} \rho(x,y) & i = 1 \\ \min\{\rho(x,y), \min_{1 \le j < i} \beta(\underline{v}_j, \underline{v}_i)\} & 1 < i \le s \end{cases}$$

   Algorithms in $\mathcal{C}_{GSP}$ incur a cost of no more than Δ_i when point pair $\underline{v}_i$ is revealed for $1 \le i \le s$.

The metric $\beta(\{x,y\},\{a,b\})$ measures the shortest distance in connecting point pair $\{x,y\}$ to point pair $\{a,b\}$ in the original metric ρ, with x and y connected to different points. Each time a request $\{x,y\}$ is revealed, an algorithm in $\mathcal{C}_{GSP}$ will incur a cost no greater than the smaller of the cost of directly connecting x and y and that of connecting $\{x,y\}$ to the nearest previously revealed point pair, where nearness between point pairs is as measured by β. The class $\mathcal{C}_{GSP}$ includes the greedy algorithm that connects a point pair with the smallest possible additional cost. It also includes the point pair greedy algorithm that connects a revealed point pair either directly or to the nearest previously revealed pair, whichever is less expensive.

Let $C(\sigma) = \sum_{i=1}^{s} \Delta_i$. The total cost incurred by any algorithm in $\mathcal{C}_{GSP}$ is no more than $C(\sigma)$, for any σ.

Lemma 5 *Suppose $C(\sigma) \le f(k) \cdot OPT(\sigma)$ for all σ such that the optimal off-line solution is a single Steiner tree, where $f(k)$ is some non-decreasing function of k, then $C(\sigma) \le f(k) \cdot OPT(\sigma)$ for all general request sequences.*

Proof: Suppose the optimal off-line solution for σ consists of a forest of p Steiner trees, $T_1, \ldots, T_p$. Let $\sigma_1, \ldots, \sigma_p$ be a partition of σ into subsequences such that σ_i contains all the requested points in T_i. Then we have: $C(\sigma) \le \sum_{q=1}^{p} C(\sigma_q)$, since, for each point pair $\underline{v}_i$ revealed as part of request subsequence σ_q, the value of Δ_i will be greater than if the point pair is revealed as part of σ. Let k_q be the number of points in T_q that are contained in σ, then we have
$C(\sigma) \le \sum_{q=1}^{p} f(k_q) \cdot OPT(\sigma_q) \le f(k) \cdot \sum_{q=1}^{p} OPT(\sigma_q) = f(k) \cdot OPT(\sigma)$. □

With Lemma 5 in mind, we first look at the case when the optimal off-line solution consists of a single Steiner tree T, with weight $\rho(T)$. Furthermore, we assume that it is a single chain of $(k-1)$ edges; the end points of the edges are the k revealed points in σ. We define

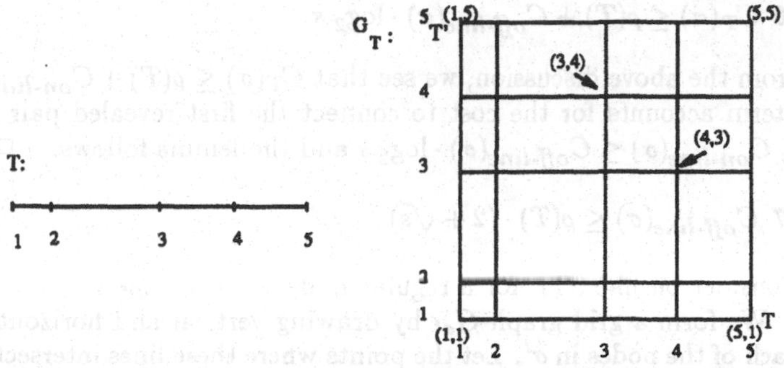

Figure 1: T and G_T

- $\rho_T(x, y)$= the length of the path in T between nodes x and y,
- $\beta_T(\{x, y\}, \{a, b\}) = \min\{[\rho_T(a, x) + \rho_T(b, y)], [\rho_T(a, y) + \rho_T(b, x)]\}$.
- For request i,

$$\Delta_{Ti} = \begin{cases} \rho_T(x, y) & i = 1 \\ \min\{\rho_T(x, y), \min_{1 \leq j < i} \beta_T(\underline{v}_j, \underline{v}_i)\} & i > 1 \end{cases}$$

- $C_T(\sigma) = \sum_{i=1}^{s} \Delta_{Ti}$

Thus we have redefined ρ and β in terms of distances on T. Clearly $\Delta_{Ti} \geq \Delta_i$ $(1 \leq i \leq s)$ and $C_T(\sigma) \geq C(\sigma)$ for all σ. To simplify notation, let $U = \{1, \ldots, k\}$ be the set of k points on T. Given a $(k-1)$-chain T, we form a grid graph G_T as follows:
Put T horizontally and let the k nodes along it be numbered from left to right, $1, \ldots, k$. Construct an identical copy of T, T', by turning T by ninety degrees anti-clockwise about node 1. Draw a vertical line at each node on T and a horizontal line at each node on T'. Denote by (i, j) the node at the intersection of the vertical line drawn at node i on T and the horizontal line drawn at node j on T'. The grid graph G_T is formed by all the segments of the lines within the square with corners $(1, 1), (1, k), (k, 1), (k, k)$. It has vertex set $V_T = U \times U$. (See Figure 1 .) We perform the following on-line STP game on G_T: The point pairs in σ are revealed one by one, with $\underline{v}_1 = \{p, q\}, p < q$, revealed in advance. Reveal node (p, q) in G_T. Each time a $\underline{v}_i = \{x, y\}$ $(i \geq 2)$ is revealed, with $x < y$, we charge the on-line server for this game a cost equal to the distance between (x, y) and its nearest revealed neighbor, (a, b), in G_T. It can be verified that this cost is equal to $\beta_T(\{a, b\}, \{x, y\}) \geq \Delta_{Ti}$. Let $C_{\textit{on-line}}(\sigma)$ be the total on-line cost charged, and $C_{\textit{off-line}}(\sigma)$ be the cost of the optimal minimum Steiner tree for the requests placed in G_T.

Lemma 6 $C_T(\sigma) \leq \rho(T) + C_{\textit{off-line}}(\sigma) \cdot \log_2 s$

Proof: From the above discussion, we see that $C_T(\sigma) \leq \rho(T) + C_{\textit{on-line}}$ where the $\rho(T)$ term accounts for the cost to connect the first revealed pair $\underline{v}_1$. By Lemma 1, $C_{\textit{on-line}}(\sigma) \leq C_{\textit{off-line}}(\sigma) \cdot \log_2 s$ and the lemma follows. □

Lemma 7 $C_{\textit{off-line}}(\sigma) \leq \rho(T) \cdot (2 + \sqrt{s})$

Proof: Consider off-line STP for a regular node set σ' on the rectilinear metric space. We form a grid graph $G_{\sigma'}$ by drawing vertical and horizontal lines through each of the nodes in σ'. Let the points where these lines intersect be the nodes of the graph. Hannan [7] showed that $G_{\sigma'}$ contains a minimum Steiner tree of the original problem. Using this result, to prove Lemma 7, we need only bound the weight of the longest minimum Steiner tree in a unit square in the rectilinear metric space. Chung and Graham [6] showed that an upper bound of $\sqrt{s} + 1 + o(1)$ holds, from which our lemma follows. □

Lemma 8 *If the optimal off-line solution is a single Steiner tree,*

$$C(\sigma) \leq 2 \cdot [1 + (2 + \sqrt{k-1}) \cdot \log_2(k-1)] \cdot OPT(\sigma)$$

Proof: If the optimal off-line solution is a single chain T, Lemmas 4, 6 and 7 and $C(\sigma) \leq C_T(\sigma)$, imply $C(\sigma) \leq [1 + (2 + \sqrt{k-1}) \cdot \log_2(k-1)] \cdot \rho(T)$. Suppose the off-line solution is a single Steiner tree T_o that is not a chain. Starting from an arbitrary leaf node, we can perform a depth-first search on T_o, marking the order in which the regular nodes are encountered on the tree. Let the nodes be encountered in the order $u_1, \ldots, u_k$. Construct the chain $\{(u_1, u_2), \ldots, (u_{k-1}, u_k)\}$, where edge (u_i, u_{i+1}) has weight $\rho_{T_o}(u_i, u_{i+1})$. The chain constructed, T, will not have weight more than twice that of T_o and $\rho(T) \leq 2 \cdot OPT(\sigma)$. Using the same lemmas and $C(\sigma) \leq C_T(\sigma)$ as before, the lemma follows. □

The next theorem follows from Lemmas 5 and 8.

Theorem 3 *The algorithms in* $\mathcal{C}_{GSP}$ *are* $2 \cdot [1 + (2 + \sqrt{k-1}) \cdot \log_2(k-1)]$*-competitive for the on-line generalized Steiner problem.*

5 Other Related Results

We consider on-line GSP with more general connectivity requirements. Let GREEDY_{GEN} be the on-line algorithm that makes the connection with the smallest additional cost. The algorithm is clearly s-competitive, and Lemma 4 implies that it is $(k-1)$-competitive for such $\{r_{ab}\}$'s. The next theorem shows that $GREEDY_{GEN}$ has a competitive ratio of $\Theta(k)$ when the connectivity requirements $\{r_{ab}\}$ form a non-increasing sequence.

Theorem 4 *GREEDY$_{GEN}$ is not better than $(k/2)$-competitive, $(k \geq 3)$, for the on-line generalized Steiner problem, even when all the connectivity requirements are 2 and the metric space is an outerplanar graph.*

Proof: Let G be an outerplanar graph on $(k+1)$ nodes $v_1, \ldots, v_{k+1}$, with edges (v_i, v_1), $(2 \leq i \leq k+1)$, each of length L, where L is some large positive integer. The other edges are (v_i, v_{i+1}), $(2 \leq i \leq k)$, each with unit length. The adversary reveals $\sigma = (\{v_1, v_2\}, \{v_1, v_4\}, \{v_1, v_5\}, \ldots, \{v_1, v_{k+1}\})$. GREEDY$_{GEN}$ will cover the whole graph, incurring a cost of $(k \cdot L + k)$. The adversary can take the cycle $\{(v_1, v_2), \ldots, (v_{k+1}, v_1)\}$, incurring a cost of $(2 \cdot L + k)$. Choosing L sufficiently large, this gives a ratio as close to $(k/2)$ as desired. □

The example used in above proof can be extended to show that no deterministic algorithm or randomized algorithm against an oblivious adversary is better than $\Omega(k)$-competitive. Another extension of the on-line generalized Steiner problem is to consider node-connectivity requirement rather than edge-connectivity requirement. We show similarly that a linear lower bound holds.

Two common network topologies for data file allocations are trees and rings [14]. Bartal *et al.*[4] stated that GREEDY is respectively 1- and 2-competitive for the on-line STP on a tree and a ring. We show that GREEDY is $[2(1 - 2^{-(s-1)})]$-competitive on a ring and no deterministic on-line algorithm achieves a smaller competitive ratio on the ring.

A variation of the Steiner tree problem is the Steiner problem on a weighted directed graph G [10, 11]. The problem is to find a directed subgraph of G of minimum weight such that there exists a directed path from a specified node to all the nodes in σ. In the on-line version of this problem, nodes in σ are revealed one by one. We show in [12] that no deterministic on-line algorithm is better than s-competitive. Since the greedy on-line algorithm is trivially s-competitive, it is also optimal for this problem. The proofs for the above results will be included in the final paper.

6 Conclusions and Further Research

We have shown that a class of algorithms, $\mathcal{C}$, are $O(\log(\frac{d}{z}s))$-competitive for the on-line STP on any metric space and no deterministic on-line algorithm can be better than $\Omega(\log(\frac{d}{z}s))$-competitive. It can be easily shown that the same bounds hold for the on-line version of the Minimum Spanning Tree problem (MSTP) [12]. For both on-line STP and on-line MSTP, it is interesting to consider a situation in which requests can arrive in blocks of $B \in \mathcal{Z}^+$. A lower bound on the competitiveness of $\Omega(\log(s/B))$ can be obtained for both problems.

We have shown that a class of lazy and greedy on-line algorithms are $O(\sqrt{k} \cdot \log k)$-competitive for the on-line generalized Steiner problem with unit connectivity requirement. There is an obvious gap between the upper bound and the lower bound of $\Omega(\log k)$. When the connectivity requirements are not all 1's, the greedy algorithm does not have a sublinear performance; we showed that it has a competitive ratio of $\Theta(k)$ when $\{r_{ab}\}$ forms a non-increasing sequence. The example we used can be extended to show that a linear lower bound holds for all deterministic and randomized algorithms against an oblivious adversary.

Other interesting open problems include closing the gap between the lower and upper bounds in the on-line Steiner tree problem on the Euclidean space, and the same problem on the rectilinear metric space.

References

[1] N. Alon and Y. Azar, On-Line Steiner Trees in the Euclidean Plane, *Proceedings of the 8th Symposium of Computational Geometry*, Berlin, 1992.

[2] A. Agrawal, P. Klein and R. Ravi, When Trees Collide: An Approximation Algorithm for the Generalized Steiner Problem in Networks, *Proceedings of the 23rd ACM Symposium on Theory of Computing*, (1991) 134-144.

[3] B. Awerbuch, Y. Bartal and A. Fiat, Competitive Distributed File Allocation, To appear, 25th ACM Symposium on Theory of Computing, 1993.

[4] Y. Bartal, A. Fiat, and Y. Rabani, Competitive Algorithms for Distributed Data Management, *Proceedings of the 24th Annual ACM Symposium on the Theory of Computing*, (1992) 39-50.

[5] S. Ben-David, A. Borodin, R. Karp, G. Tardos and A. Wigderson, On the Power of Randomization in Online Algorithms, *Proceedings of the 22nd ACM Symposium on Theory of Computing*, (1990) 379-386.

[6] F. R. K. Chung and R. L. Graham, On Steiner Trees for Bounded Point Sets, *Geometrias Dedicata*, 11 (1981) 353-361.

[7] M. Hannan, On Steiner's Problem with Rectilinear Distance, *SIAM J. App. Maths.*, 14 (1966), 255-265.

[8] M. Imaze, and B. M. Waxman, Dynamic Steiner Tree Problem, *SIAM J. Disc. Math.*, 4(3) (1991) 369-384.

[9] R. M. Karp, Reducibility among combinatorial problems, in R. E. Miller and J. W. Thatcher (eds.), *Complexity of Computer Computations*, Plenum Press, New York, 85-103.

[10] N. Maculan, The Steiner Problem in Graphs, *Ann. Dis. Maths.*, 31 (1987) 185-212.

[11] L. Nastansky, S. M. Selkow, N. F. Stewart, Cost-Minimal Trees in Directed Acyclic Graphs, *Zeitschrift für Operations Research*, 18 (1974), 59-67.

[12] J. Westbrook and D. C. K. Yan, The Performance of Greedy Algorithms for the On-Line Steiner Tree and Related Problems, Technical Report YALEU/DCS/TR-911, Yale University, November, 1992.

[13] P. Winter, Steiner Problem in Networks: A Survey, *Networks*, 17 (1987) 129-167.

[14] O. Wolfson and A. Milo, The Multicast Policy and Its Relationship to Replicated Data Placement, *ACM Trans. Database Syst.*, 16(1) (1991) 181-205.

AUTHOR INDEX

Lecture Notes in Computer Science

**For information about Vols. 1–629
please contact your bookseller or Springer-Verlag**

Vol. 630: W. R. Cleaveland (Ed.), CONCUR '92. Proceedings. X, 580 pages. 1992.

Vol. 631: M. Bruynooghe, M. Wirsing (Eds.), Programming Language Implementation and Logic Programming. Proceedings, 1992. XI, 492 pages. 1992.

Vol. 632: H. Kirchner, G. Levi (Eds.), Algebraic and Logic Programming. Proceedings, 1992. IX, 457 pages. 1992.

Vol. 633: D. Pearce, G. Wagner (Eds.), Logics in AI. Proceedings. VIII, 410 pages. 1992. (Subseries LNAI).

Vol. 634: L. Bougé, M. Cosnard, Y. Robert, D. Trystram (Eds.), Parallel Processing: CONPAR 92 – VAPP V. Proceedings. XVII, 853 pages. 1992.

Vol. 635: J. C. Derniame (Ed.), Software Process Technology. Proceedings, 1992. VIII, 253 pages. 1992.

Vol. 636: G. Comyn, N. E. Fuchs, M. J. Ratcliffe (Eds.), Logic Programming in Action. Proceedings, 1992. X, 324 pages. 1992. (Subseries LNAI).

Vol. 637: Y. Bekkers, J. Cohen (Eds.), Memory Management. Proceedings, 1992. XI, 525 pages. 1992.

Vol. 639: A. U. Frank, I. Campari, U. Formentini (Eds.), Theories and Methods of Spatio-Temporal Reasoning in Geographic Space. Proceedings, 1992. XI, 431 pages. 1992.

Vol. 640: C. Sledge (Ed.), Software Engineering Education. Proceedings, 1992. X, 451 pages. 1992.

Vol. 641: U. Kastens, P. Pfahler (Eds.), Compiler Construction. Proceedings, 1992. VIII, 320 pages. 1992.

Vol. 642: K. P. Jantke (Ed.), Analogical and Inductive Inference. Proceedings, 1992. VIII, 319 pages. 1992. (Subseries LNAI).

Vol. 643: A. Habel, Hyperedge Replacement: Grammars and Languages. X, 214 pages. 1992.

Vol. 644: A. Apostolico, M. Crochemore, Z. Galil, U. Manber (Eds.), Combinatorial Pattern Matching. Proceedings, 1992. X, 287 pages. 1992.

Vol. 645: G. Pernul, A M. Tjoa (Eds.), Entity-Relationship Approach – ER '92. Proceedings, 1992. XI, 439 pages, 1992.

Vol. 646: J. Biskup, R. Hull (Eds.), Database Theory – ICDT '92. Proceedings, 1992. IX, 449 pages. 1992.

Vol. 647: A. Segall, S. Zaks (Eds.), Distributed Algorithms. X, 380 pages. 1992.

Vol. 648: Y. Deswarte, G. Eizenberg, J.-J. Quisquater (Eds.), Computer Security – ESORICS 92. Proceedings. XI, 451 pages. 1992.

Vol. 649: A. Pettorossi (Ed.), Meta-Programming in Logic. Proceedings, 1992. XII, 535 pages. 1992.

Vol. 650: T. Ibaraki, Y. Inagaki, K. Iwama, T. Nishizeki, M. Yamashita (Eds.), Algorithms and Computation. Proceedings, 1992. XI, 510 pages. 1992.

Vol. 651: R. Koymans, Specifying Message Passing and Time-Critical Systems with Temporal Logic. IX, 164 pages. 1992.

Vol. 652: R. Shyamasundar (Ed.), Foundations of Software Technology and Theoretical Computer Science. Proceedings, 1992. XIII, 405 pages. 1992.

Vol. 653: A. Bensoussan, J.-P. Verjus (Eds.), Future Tendencies in Computer Science, Control and Applied Mathematics. Proceedings, 1992. XV, 371 pages. 1992.

Vol. 654: A. Nakamura, M. Nivat, A. Saoudi, P. S. P. Wang, K. Inoue (Eds.), Parallel Image Analysis. Proceedings, 1992. VIII, 312 pages. 1992.

Vol. 655: M. Bidoit, C. Choppy (Eds.), Recent Trends in Data Type Specification. X, 344 pages. 1993.

Vol. 656: M. Rusinowitch, J. L. Rémy (Eds.), Conditional Term Rewriting Systems. Proceedings, 1992. XI, 501 pages. 1993.

Vol. 657: E. W. Mayr (Ed.), Graph-Theoretic Concepts in Computer Science. Proceedings, 1992. VIII, 350 pages. 1993.

Vol. 658: R. A. Rueppel (Ed.), Advances in Cryptology – EUROCRYPT '92. Proceedings, 1992. X, 493 pages. 1993.

Vol. 659: G. Brewka, K. P. Jantke, P. H. Schmitt (Eds.), Nonmonotonic and Inductive Logic. Proceedings, 1991. VIII, 332 pages. 1993. (Subseries LNAI).

Vol. 660: E. Lamma, P. Mello (Eds.), Extensions of Logic Programming. Proceedings, 1992. VIII, 417 pages. 1993. (Subseries LNAI).

Vol. 661: S. J. Hanson, W. Remmele, R. L. Rivest (Eds.), Machine Learning: From Theory to Applications. VIII, 271 pages. 1993.

Vol. 662: M. Nitzberg, D. Mumford, T. Shiota, Filtering, Segmentation and Depth. VIII, 143 pages. 1993.

Vol. 663: G. v. Bochmann, D. K. Probst (Eds.), Computer Aided Verification. Proceedings, 1992. IX, 422 pages. 1993.

Vol. 664: M. Bezem, J. F. Groote (Eds.), Typed Lambda Calculi and Applications. Proceedings, 1993. VIII, 433 pages. 1993.

Vol. 665: P. Enjalbert, A. Finkel, K. W. Wagner (Eds.), STACS 93. Proceedings, 1993. XIV, 724 pages. 1993.

Vol. 666: J. W. de Bakker, W.-P. de Roever, G. Rozenberg (Eds.), Semantics: Foundations and Applications. Proceedings, 1992. VIII, 659 pages. 1993.

Vol. 667: P. B. Brazdil (Ed.), Machine Learning: ECML – 93. Proceedings, 1993. XII, 471 pages. 1993. (Subseries LNAI).

Vol. 668: M.-C. Gaudel, J.-P. Jouannaud (Eds.), TAPSOFT '93: Theory and Practice of Software Development. Proceedings, 1993. XII, 762 pages. 1993.

Vol. 669: R. S. Bird, C. C. Morgan, J. C. P. Woodcock (Eds.), Mathematics of Program Construction. Proceedings, 1992. VIII, 378 pages. 1993.

Vol. 670: J. C. P. Woodcock, P. G. Larsen (Eds.), FME '93: Industrial-Strength Formal Methods. Proceedings, 1993. XI, 689 pages. 1993.

Vol. 671: H. J. Ohlbach (Ed.), GWAI-92: Advances in Artificial Intelligence. Proceedings, 1992. XI, 397 pages. 1993. (Subseries LNAI).

Vol. 672: A. Barak, S. Guday, R. G. Wheeler, The MOSIX Distributed Operating System. X, 221 pages. 1993.

Vol. 673: G. Cohen, T. Mora, O. Moreno (Eds.), Applied Algebra, Algebraic Algorithms and Error-Correcting Codes. Proceedings, 1993. X, 355 pages 1993.

Vol. 674: G. Rozenberg (Ed.), Advances in Petri Nets 1993. VII, 457 pages. 1993.

Vol. 675: A. Mulkers, Live Data Structures in Logic Programs. VIII, 220 pages. 1993.

Vol. 676: Th. H. Reiss, Recognizing Planar Objects Using Invariant Image Features. X, 180 pages. 1993.

Vol. 677: H. Abdulrab, J.-P. Pécuchet (Eds.), Word Equations and Related Topics. Proceedings, 1991. VII, 214 pages. 1993.

Vol. 678: F. Meyer auf der Heide, B. Monien, A. L. Rosenberg (Eds.), Parallel Architectures and Their Efficient Use. Proceedings, 1992. XII, 227 pages. 1993.

Vol. 679: C. Fermüller, A. Leitsch, T. Tammet, N. Zamov, Resolution Methods for the Decision Problem. VIII, 205 pages. 1993. (Subseries LNAI).

Vol. 680: B. Hoffmann, B. Krieg-Brückner (Eds.), Program Development by Specification and Transformation. XV, 623 pages. 1993.

Vol. 681: H. Wansing, The Logic of Information Structures. IX, 163 pages. 1993. (Subseries LNAI).

Vol. 682: B. Bouchon-Meunier, L. Valverde, R. R. Yager (Eds.), IPMU '92 – Advanced Methods in Artificial Intelligence. Proceedings, 1992. IX, 367 pages. 1993.

Vol. 683: G.J. Milne, L. Pierre (Eds.), Correct Hardware Design and Verification Methods. Proceedings, 1993. VIII, 270 Pages. 1993.

Vol. 684: A. Apostolico, M. Crochemore, Z. Galil, U. Manber (Eds.), Combinatorial Pattern Matching. Proceedings, 1993. VIII, 265 pages. 1993.

Vol. 685: C. Rolland, F. Bodart, C. Cauvet (Eds.), Advanced Information Systems Engineering. Proceedings, 1993. XI, 650 pages. 1993.

Vol. 686: J. Mira, J. Cabestany, A. Prieto (Eds.), New Trends in Neural Computation. Proceedings, 1993. XVII, 746 pages. 1993.

Vol. 687: H. H. Barrett, A. F. Gmitro (Eds.), Information Processing in Medical Imaging. Proceedings, 1993. XVI, 567 pages. 1993.

Vol. 688: M. Gauthier (Ed.), Ada-Europe '93. Proceedings, 1993. VIII, 353 pages. 1993.

Vol. 689: J. Komorowski, Z. W. Ras (Eds.), Methodologies for Intelligent Systems. Proceedings, 1993. XI, 653 pages. 1993. (Subseries LNAI).

Vol. 690: C. Kirchner (Ed.), Rewriting Techniques and Applications. Proceedings, 1993. XI, 488 pages. 1993.

Vol. 691: M. Ajmone Marsan (Ed.), Application and Theory of Petri Nets 1993. Proceedings, 1993. IX, 591 pages. 1993.

Vol. 692: D. Abel, B.C. Ooi (Eds.), Advances in Spatial Databases. Proceedings, 1993. XIII, 529 pages. 1993.

Vol. 693: P. E. Lauer (Ed.), Functional Programming, Concurrency, Simulation and Automated Reasoning. Proceedings, 1991/1992. XI, 398 pages. 1993.

Vol. 694: A. Bode, M. Reeve, G. Wolf (Eds.), PARLE '93. Parallel Architectures and Languages Europe. Proceedings, 1993. XVII, 770 pages. 1993.

Vol. 695: E. P. Klement, W. Slany (Eds.), Fuzzy Logic in Artificial Intelligence. Proceedings, 1993. VIII, 192 pages. 1993. (Subseries LNAI).

Vol. 696: M. Worboys, A. F. Grundy (Eds.), Advances in Databases. Proceedings, 1993. X, 276 pages. 1993.

Vol. 697: C. Courcoubetis (Ed.), Computer Aided Verification. Proceedings, 1993. IX, 504 pages. 1993.

Vol. 698: A. Voronkov (Ed.), Logic Programming and Automated Reasoning. Proceedings, 1993. XIII, 386 pages. 1993. (Subseries LNAI).

Vol. 699: G. W. Mineau, B. Moulin, J. F. Sowa (Eds.), Conceptual Graphs for Knowledge Representation. Proceedings, 1993. IX, 451 pages. 1993. (Subseries LNAI).

Vol. 700: A. Lingas, R. Karlsson, S. Carlsson (Eds.), Automata, Languages and Programming. Proceedings, 1993. XII, 697 pages. 1993.

Vol. 701: P. Atzeni (Ed.), LOGIDATA+: Deductive Databases with Complex Objects. VIII, 273 pages. 1993.

Vol. 702: E. Börger, G. Jäger, H. Kleine Büning, S. Martini, M. M. Richter (Eds.), Computer Science Logic. Proceedings, 1992. VIII, 439 pages. 1993.

Vol. 703: M. de Berg, Ray Shooting, Depth Orders and Hidden Surface Removal. X, 201 pages. 1993.

Vol. 704: F. N. Paulisch, The Design of an Extendible Graph Editor. XV, 184 pages. 1993.

Vol. 705: H. Grünbacher, R. W. Hartenstein (Eds.), Field-Programmable Gate Arrays. Proceedings, 1992. VIII, 218 pages. 1993.

Vol. 706: H. D. Rombach, V. R. Basili, R. W. Selby (Eds.), Experimental Software Engineering Issues. Proceedings, 1992. XVIII, 261 pages. 1993.

Vol. 707: O. M. Nierstrasz (Ed.), ECOOP '93 – Object-Oriented Programming. Proceedings, 1993. XI, 531 pages. 1993.

Vol. 708: C. Laugier (Ed.), Geometric Reasoning for Perception and Action. Proceedings, 1991. VIII, 281 pages. 1993.

Vol. 709: F. Dehne, J.-R. Sack, N. Santoro, S. Whitesides (Eds.), Algorithms and Data Structures. Proceedings, 1993. XII, 634 pages. 1993.